The Geological Society of America
Special Paper 450
2009

Introduction to Paleoenvironments of Bear Lake, Utah and Idaho, and its catchment

Joseph G. Rosenbaum
U.S. Geological Survey, Box 25046, Federal Center, Denver, Colorado 80225, USA

Darrell S. Kaufman
Department of Geology, Box 4099, Northern Arizona University, Flagstaff, Arizona 86011, USA

In 1996 a group led by the late Kerry Kelts (University of Minnesota) and Robert Thompson (U.S. Geological Survey) acquired three piston cores (BL96-1, -2, and -3) from Bear Lake. The coring arose from their recognition of Bear Lake as a potential repository of long records of paleoenvironmental change. They recognized that the lake is located in an area that is sensitive to changes in regional climate patterns (Dean et al., this volume), that the lake basin is long lived (see Colman, 2006; Kaufman et al., this volume), and that, unlike many lakes in the Great Basin, Bear Lake was never dry during warm dry periods.

Bear Lake lies in the northeastern Great Basin to the northeast of Great Salt Lake, just south of the Snake River drainage, and a short distance west of the Green River drainage that makes up part of the Upper Colorado River Basin (Fig. 1). Similarity among the historic Bear Lake and Great Salt Lake hydrographs and flows on the Green River indicates that the hydrology of Bear Lake reflects regional precipitation (Fig. 2). Therefore, paleorecords from Bear Lake are important to understanding past climate for a large region, including the Upper Colorado River Basin, the source of much of the water for the southwestern United States.

Initially, paleoenvironmental studies of Bear Lake sediments focused on cores BL96-1, -2, and -3. Additional coring was conducted to elucidate the spatial distribution of sedimentary units and to extend the record back in time. The study was also expanded to include extensive study of the catchment, including the properties of catchment materials and the processes that could potentially affect the delivery of catchment materials to the lake.

Cores BL96-1, -2, and -3 were taken with a Kullenburg piston corer along an east–west profile in roughly 50, 40, and 30 m of water, respectively (Table 1, Fig. 3). These three cores, each taken as a single 4- to 5-m-long segment, provide a nearly complete composite section from ca. 26 cal ka to the late Holocene. In 1998 a number of short gravity cores were taken from the uppermost water-rich sediments that were not sampled by the 1996 cores. During 2000, cores were taken with a percussion piston corer (manufactured by UWITEC) at three locations in and around Mud Lake and at two locations in the northern end of Bear Lake (Fig. 3). Cores acquired with the percussion corer comprise as many as three overlapping segments up to 2 m in length. In 2002, additional percussion piston cores and associated gravity cores of the uppermost sediments were acquired from five sites in the northern half of the lake. In conjunction with two of the cores collected in 2000, these cores form a north–south profile along a seismic line and span water depths from less than 10 m to ~40 m. Data from this profile provide much of the evidence for lake-level variations (Smoot and Rosenbaum, this volume). Finally, during 2000, two long cores, BL00-1D and -1E (collectively referred to here simply as BL00-1), were taken at a site near the depocenter during testing of the GLAD800 coring platform (Fig. 4; Dean et al., 2002). These cores provide a record back to ca. 220 ka.

Many of the studies in this volume utilize samples from these cores. To help constrain interpretations of core data, sediment traps were deployed (Dean, this volume), modern lake and stream sediments were sampled (Rosenbaum et al., this volume; Smoot, this volume), and numerous water samples were collected (Bright, Chapter 4, this volume).

Here we briefly review the various topics presented in detail in the chapters of this volume.

Rosenbaum, J.G., and Kaufman, D.S, 2009, Introduction to *Paleoenvironments of Bear Lake, Utah and Idaho, and its catchment, in* Rosenbaum, J.G., and Kaufman, D.S., eds., Paleoenvironments of Bear Lake, Utah and Idaho, and its catchment: Geological Society of America Special Paper 450, p. v–xiii, doi: 10.1130/2009.2450(00). For permission to copy, contact editing@geosociety.org. ©2009 The Geological Society of America. All rights reserved.

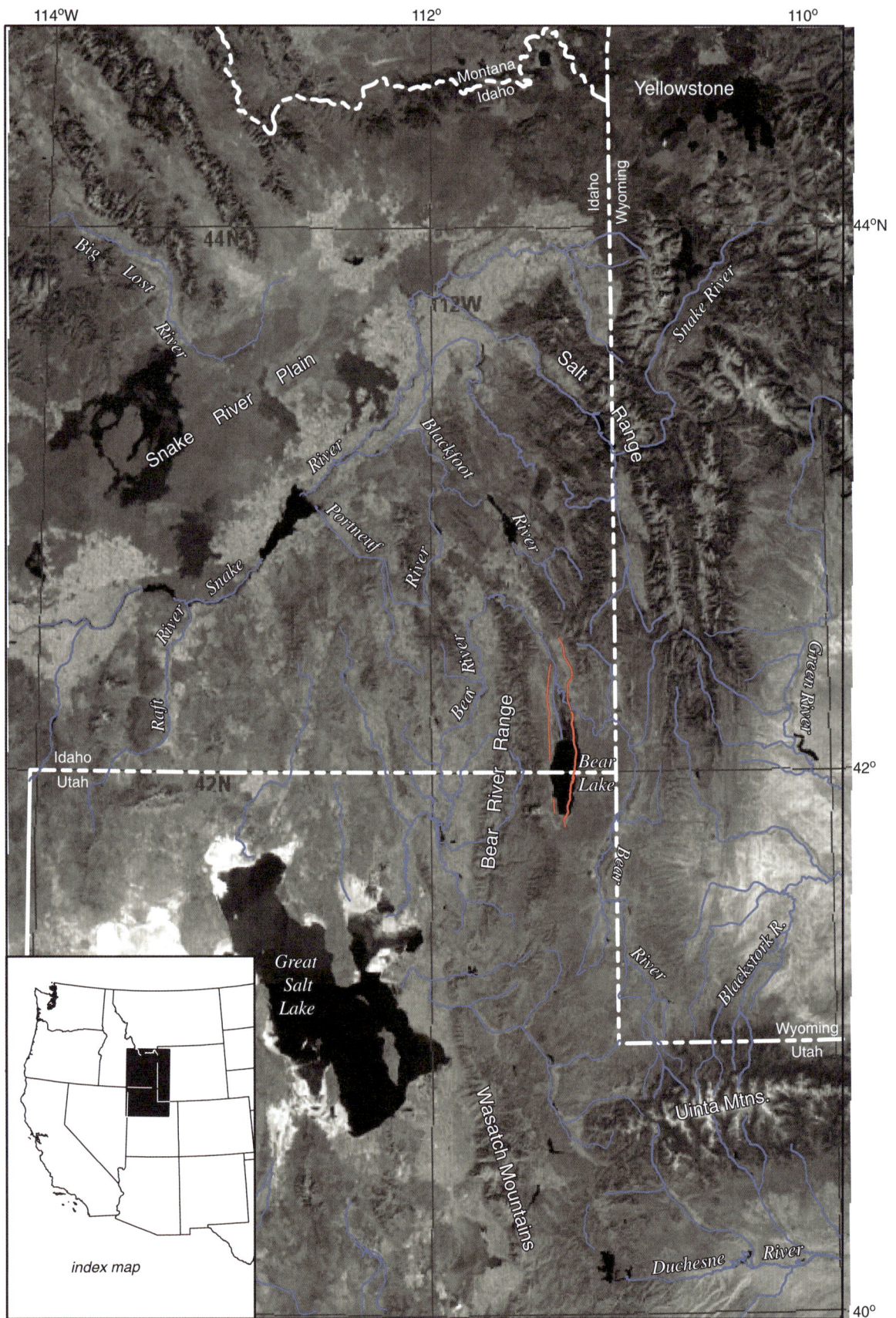

Figure 1. Location index map (inset) and regional setting, of Bear Lake, including drainages (blue) and principal faults bounding the Bear Lake Valley (red) (modified from Fig. 1 of Reheis et al., this volume).

BEAR LAKE SETTING

Geologic, Tectonic, and Geomorphic

As in other tectonically active basins in the Basin and Range Province, hundreds of meters of sediment have accumulated in Bear Lake Valley during the past several million years. Reheis et al. (this volume) describe the geology of Bear Lake Valley, including the bedrock units that underlie the adjacent uplands, and the faults that cut them. This geologic framework controls flow paths of groundwater discharged to Bear Lake (Bright, Chapter 4, this volume) and governs the types and quantities of sediment and solutes that accumulate in the lake. The modern Bear Lake Valley was shaped by faulting, particularly on the eastern valley margin. This faulting, which remains active, has southward-increasing slip rates and contributed to the migration of the Bear River both laterally, as evidenced by abandoned channels that mark the northern Bear Lake Valley, and vertically, as recorded by flights of fluvial terraces.

Climatic, Hydrologic, Limnologic, and Biologic

Dean et al. (this volume) describe the large-scale pattern of modern atmospheric and oceanic circulation that influences Bear Lake. They present the instrumental record of climate and limnology, including long-term time series of climate variables, and of physical and chemical properties of the oligotrophic lake. Dean et al. (this volume) summarize the biological components of Bear Lake, including the endemic fish population. The endemic ostracode population is described by Bright (Chapter 8, this volume), who highlights the uniqueness of the Bear Lake microcrustaceans and speculates on their coevolution with the fish.

Bright (Chapter 8, this volume) suggests factors that have enabled Bear Lake to develop its diverse endemic ecosystem. One key factor is its long-lived, relatively stable, but not invariant, hydrologic setting. The lake has survived periods of major drought without becoming highly saline like Great Salt Lake. Groundwater that discharges in Bear River Range streams (Fig. 3) is an important part of the hydrologic budget of Bear Lake. In

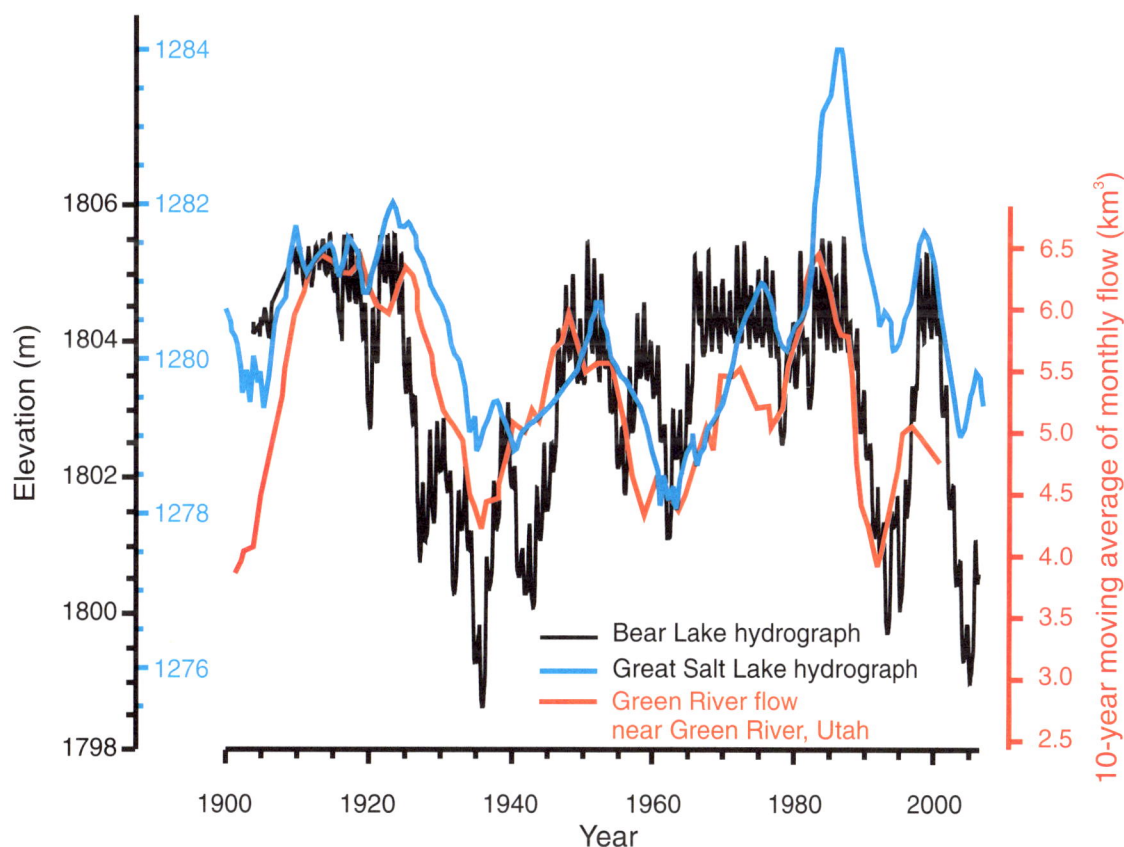

Figure 2. Elevations of Bear Lake and Great Salt Lake (in meters above sea level), and flow of the Green River near Green River, Utah. Elevation data for Bear Lake are from PacifiCorp, Salt Lake City, Utah. Note that the surface elevation of Bear Lake is limited to ~1805 m, at which point the lake spills. Elevation data for Great Salt Lake are from the U.S. Geological Survey. Green River flow is from Piechota et al. (2004).

addition to this west-side source, Bright (Chapter 4, this volume) used isotopic and major-ion compositions to identify three other major sources of groundwater in Bear Lake Valley.

During much of the Holocene, Bear Lake did not receive surface water from Bear River. Instead, inflow was dominated by streams fed by shallow groundwater in the fractured and karstic terrain of the Bear River Range west of the lake (Bright, Chapter 4, this volume). This water is charged with ions that, when concentrated by evaporation, yield alkaline lake water saturated with respect to carbonate minerals (Dean, this volume). The chemistry of Bear Lake changed after diversion of Bear River water into the lake ca. 1912. The history of this diversion and its influence on the lake and its biota are discussed by Dean et al. (this volume). The diversion produced pervasive shifts in the physical, chemical, and biological components of lake sediment, and these have been used by several authors of this volume to interpret analogous changes in the Quaternary sedimentary record.

Surface Water and Sediment Input

By analyzing the composition of surface water and sediment in and around Bear Lake, downcore changes in the physical, chemical, and biological components of the sedimentary sequence of Bear Lake can be placed into a modern context. For example, the isotopic composition of spring and stream water (Bright, Chapter 4, this volume) provides the basis for interpreting changes in the isotope values of endogenic carbonates within the lake sediment (Dean, this volume; Kaufman et al., this volume). Similarly, the mineralogy, major-element composition, and mineral-magnetic properties of sediment carried by inflowing streams are used to infer changes in the sediment provenances through time (Dean, this volume; Kaufman et al., this volume; Rosenbaum et al., this volume). These studies show that the local streams that discharge to the lake from the east and west sides carry sediment with abundant carbonate minerals (calcite and dolomite) and high magnetic susceptibility. In contrast, sediments from the headwaters of

TABLE 1. CORE LOCATIONS AND DEPTHS

Core name		Core type	Latitude (degrees N)	Longitude (degrees W)	Water depth relative to full lake level (m)	Depth to start of drive (m blf)	Depth to end of drive (m blf)
BL96-1		K	41.9527	111.3160	50	0.00	5.00
BL96-2		K	41.9527	111.3333	43	0.00	3.92
BL96-3		K	41.9532	111.3613	33	0.00	4.05
BL98-4		G	41.9640	111.3750	31.4	0.00	0.20
BL98-6		G	41.9640	111.3750	31.4	0.00	0.20
BL98-9		G	41.9654	111.3384	44.4	0.00	0.33
BL98-10		G	41.9654	111.3384	44.4	0.00	0.36
BL98-12		G	42.0581	111.3116	31.4	0.00	0.38
BLR-2K-1	(1)	U	42.1877	111.2950	0*	0.00	2.16
	(2)	U				2.16	3.55
	(3)	U				1.00	3.00
BLR-2K-2	(1)	U	42.1328	111.2974	0.8[†]	0.00	1.80
BLR-2K-3	(1)	U	42.1559	111.3038	0.6[†]	0.00	2.07
BL2K-1	(1)	U	42.1143	111.2983	6.1	0.00	1.10
BL2K-2	(1)	U	42.1063	111.2944	8.3	0.00	2.07
	(2)	U				1.50	3.57
BL2K-3	(1)	U	42.1142	111.2992	5.8	0.02	2.05
BL2002-1	(1)	U	42.1016	111.2912	9.5	0.00	1.40
	(2)	U				0.78	2.60
BL2002-2	(1)	U	42.0837	111.2975	18.0	0.00	1.37
BL2002-3		G	42.0322	111.3191	42.9	0.00	0.15
	(1)	U				0.08	1.72
	(2)	U				1.18	2.83
BL2002-4		G	42.0496	111.3106	34.5	0.00	0.39
	(1)	U				1.00	2.77
	(2)	U				2.37	4.14
	(3)	U				0.06	1.89
BL2002-5	(1)	U	42.0666	111.3046	26.8	0.10	1.91
BL00-1	(D)	GLD	41.9517	111.3083	54.8	0.00	100.47
BL00-1	(E)	GLD	41.9517	111.3083	54.8	0.00	120.65

Note: K—Kullenburg core; G—gravity core; U—UWITEC percussion core; GLD—GLAD800 drill core; blf—below lake floor.
*Core on dry land.
[†]Depth in Mud Lake.

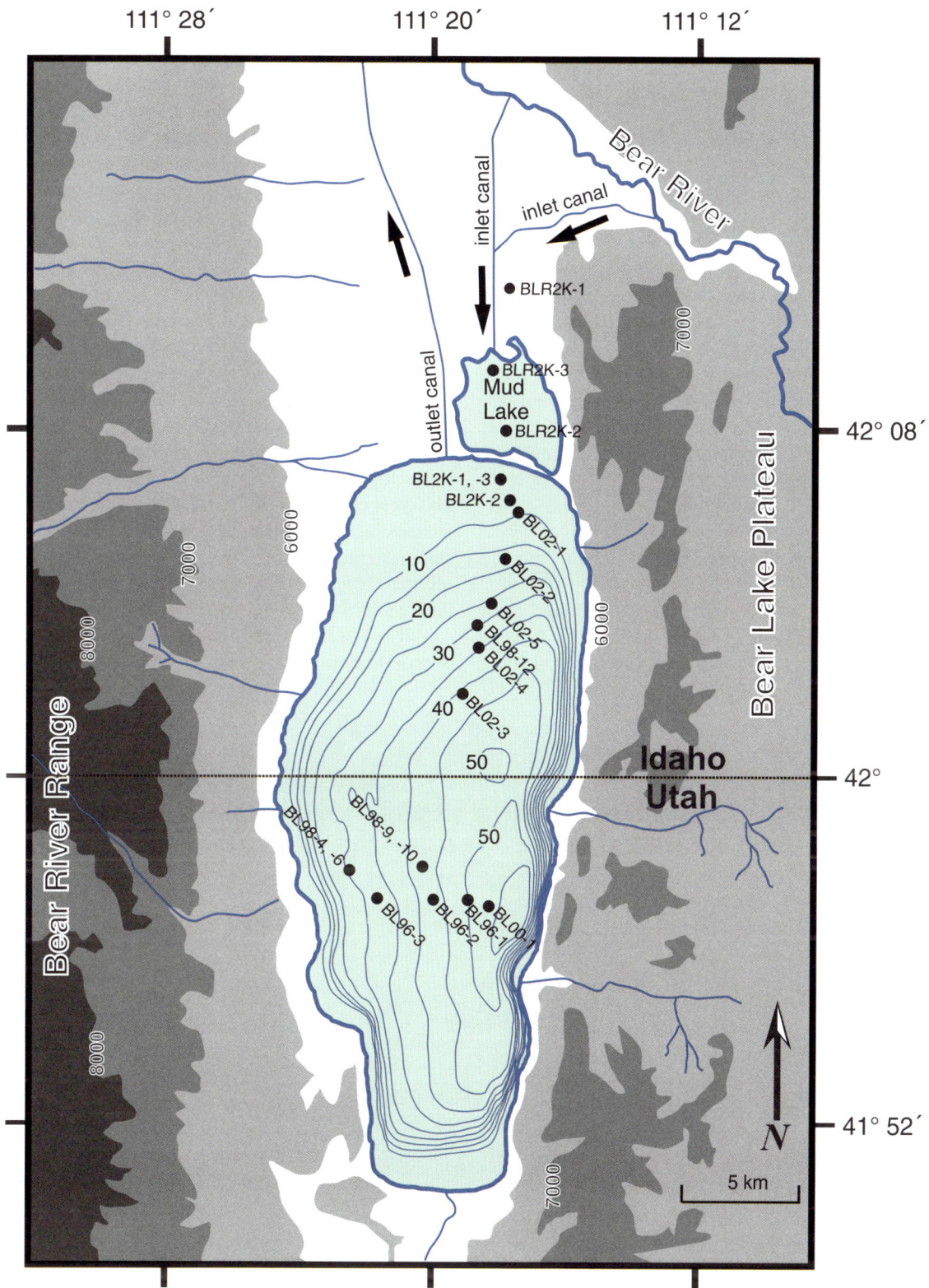

Figure 3. Core locations in and around Bear Lake. Topographic contours are in feet; bathymetric contours are in meters.

the Bear River in the Uinta Mountains lack carbonate minerals, have low susceptibility, but have high contents of hematite. Sediment traps deployed in Bear Lake record the production of high-magnesium carbonate and reveal the strong influence of sediment remobilization near the lake bottom (Dean, this volume).

SEDIMENTARY RECORD OF BEAR LAKE

Geochronology

Chronology for sediments deposited over the last 30 k.y. is based on the radiocarbon-dated stratigraphic framework developed by Colman et al. (this volume) from multiple cores. The geochronology of sediment beyond the range of radiocarbon dating is poorly constrained. A feature in the paleomagnetic data at 26.5 m depth in core BL00-1 may record the Laschamp excursion (Heil et al., this volume), and previously published data on U-series, amino acids, and tephra (Colman et al., 2006) are consistent with a relatively steady rate of sediment accumulation with an average for the last 220 k.y. of 0.54 mm yr^{-1} near the depocenter (Kaufman et al., this volume).

Allogenic and Endogenic Components

Sediment deposited in Bear Lake is strongly influenced by variable production of carbonate within the lake, by fluctuating input of fluvial and glacial-fluvial products, and by periodic retraction of the lake into a topographically closed basin. These changes are reflected in interpretations based on both allogenic lithic material (Rosenbaum et al., this volume) and endogenic carbonate (Dean, this volume) of the lake sediment. More specifically, the sediment has been studied for its paleomagnetism (Heil et al., this volume) and mineral magnetism (Heil et al., this volume; Rosenbaum and Heil, this volume; Rosenbaum et al., this volume); isotopic composition of endogenic carbonate (Dean, this volume; Kaufman et al., this volume); mineralogy and elemental geochemistry (Dean, this volume; Rosenbaum et al., this volume; Kaufman et al., this volume; Smoot, this volume); and sedimentary features (Smoot, this volume; Smoot and Rosenbaum, this volume).

Bear Lake sediment is primarily massive gray to greenish-gray calcareous silty clay, with quartz as the primary mineral in most of the sediment deposited during the last 220 k.y. (Dean, this volume; Kaufman et al., this volume). Carbonate-mineral content averaged ~30% over this entire interval and averaged 57% in six marl intervals. The endogenic component of Bear Lake sediment is the focus of the chapter by Dean, who investigated the limnological conditions that resulted in major variations in carbonate isotopic composition, geochemistry, and mineralogy on time scales ranging from seasons to millennia. These variations reflect climate changes in the region, which Dean (this volume) sets within a broader context of ocean and atmospheric circulation.

The 120 m section penetrated by core BL00-1 is only a fraction of the basin-fill sequence (Kaufman et al., this volume). The

Figure 4. The GLAD800 coring platform at the site of core BL00-1. View is to the east.

core comprises sediment that accumulated over the last two gla-cial-interglacial cycles (~220 k.y.), with essentially no hiatuses in deposition or breaks in core retrieval. We know of no other lake on the continent that has remained continuously inundated for this extended period. A large suite of analyses on the 120 m sec-tion penetrated by core BL00-1 is summarized by Kaufman et al. (this volume). Analyses have been completed at multi-centennial to millennial scales on sediment magnetic properties; oxygen, carbon, and strontium isotopes; organic- and inorganic-carbon content; palynology; mineralogy; ostracode taxonomy; and dia-tom assemblages. By combining evidence from multiple physical, chemical, and biological properties, the authors reconstructed the major paleoenvironmental changes during the last quarter-million years, and concluded that, although its influence has varied, the Bear River was connected to the lake for most of this period.

Diagenesis and Reworking

Variable post-depositional destruction of detrital Fe-oxides and formation of Fe-sulfides (e.g., greigite and pyrite) indicates changing geochemical conditions (Heil et al., this volume), with good preservation of magnetite and hematite under only the fresh-est water conditions and pyrite formation under the most saline conditions. Ostracodes decrease in abundance in the lower half of the long core, probably because they were dissolved (Kaufman et al., this volume). Sediment traps installed near the lake bottom reveal extensive resuspension of older, aragonite-bearing lake sediment (Dean, this volume). Sediment cores contain unconfor-mities, shell-rich gravel, rooting structures, and pedogenic fea-tures, indicating that lake level dropped as much as 40 m below present level during the past 18,000 years (Smoot, this volume; Smoot and Rosenbaum, this volume). Frequent migrations of the shoreline over tens of meters in elevation have reworked the sedi-ment, particularly where the lake floor has low slope.

Biological Components

Microfossils, including ostracodes, diatoms, and pollen, have been analyzed in sediment cores from Bear Lake. Changes in the abundance, preservation, and assemblages of diatoms reflect changes in hydrologic and climatic conditions over the past 19,000 years (Moser and Kimball, this volume), and the last two glacial-interglacial cycles at lower resolution (Kaufman et al., this vol-ume). The absence of diatoms in sediment deposited during glacial periods (marine oxygen isotope stage [MIS] 6 and MIS 3/2) indi-cates low-light conditions associated with increased ice cover and turbidity. Diatom assemblages of interglacial periods (MIS 5 and 1) are dominated by small, benthic/tychoplanktic fragilarioid spe-cies indicative of reduced habitat availability associated with low lake levels and more saline conditions. The MIS 3 assemblage is distinct and suggests that lake level was higher during this period than during the full interglacials.

Changes in the ostracode assemblage are described for the latest Pleistocene and Holocene (Bright, Chapter 4, this volume),

and over the length of core BL00-1 (Kaufman et al., this vol-ume). With the exception of *Cytherissa lacustris*, all of the ostra-code species in core BL00-1 and in water deeper than ~7 m are endemic. *C. lacustris* is generally absent from interglacial inter-vals, consistent with its preference for dilute lakes. It was discov-ered, however, in sediment that correlates with the interglacial MIS 7c. This is apparently the first documented occurrence of *C. lacustris* in the western contiguous United States during peak global interglacial conditions.

Core BL00-1 provides one of the most detailed and continuous records of Quaternary vegetation change in North America. Kauf-man et al. (this volume) summarize the pollen spectra using a ratio of the "warm" (juniper and oak) plus "dry" (ragweed, saltbush, and greasewood) versus "cold" (including spruce) indicators. The pol-len spectra from interglacial periods contain higher percentages of "warm" and "dry" indicators, with higher *Juniperus* percentages during the early part of each interglacial interval. Vegetation interpre-tations suggest that valley bottoms were occupied by salt-tolerant, high-desert shrubs, and that *Juniperus* woodlands expanded locally during interglaciations. Pollen spectra of glacial intervals generally have higher percentages of "cold" indicators, suggesting that for-est or forest-woodland conditions prevailed. These assemblages are similar to those described by Doner (this volume), who analyzed a shorter core from the lake extending back 19 k.y.

QUATERNARY CLIMATE CHANGE

Glaciation

Quaternary climate change has caused glaciers to advance and retreat in the alpine headwaters of the Bear Lake drainage basin. The Bear River Range on the west side of Bear Lake sup-ported glaciers that repeatedly advanced a few kilometers beyond their cirques, as mapped by Reheis et al. (this volume). The head-waters of the Bear River, in the northwestern Uinta Mountains, also supported glaciers that fluctuated with major glacial periods of the Pleistocene (Reheis et al., this volume). Meltwater charged with glacial flour from these valley glaciers left its mark on sedi-ments of the last glacial maximum in Bear Lake. A red, siliciclas-tic unit recovered in several cores and dated to between 26 and 16 ka (Colman et al., this volume) is attributed to hematite-rich material derived from glacial-fluvial outwash of Uinta Mountain glaciers (Rosenbaum et al., this volume). Rosenbaum and Heil (this volume) present a continuous record of mountain-glacier extent within the upper Bear River Basin by using rock magnetic, mineralogic, and elemental composition of sediment from Bear Lake to infer the extent of glaciers in the headwaters. They con-clude that glaciers began to form in the upper Bear River Basin ca. 26 cal ka; the glaciers reached their maximum extent around 20 ka, and receded by around 16 ka. Glaciers were probably extensive in the Uinta Mountains during MIS 6, but evidence of glacial flour has not been found in sediments of that age in Bear Lake. Magnetic property and mineralogic evidence may have been lost due to diagenetic alteration of these older sediments.

Lake-Level Fluctuations and Bear River Migrations

Quaternary climate change exerted a major control on lake-level fluctuations at Bear Lake, although other geomorphic factors that influence the geometry of the outlet and the inlet of the lake also have undoubtedly played a role. As discussed by Reheis et al. (this volume), these include changes in threshold elevation caused by aggradation, downcutting, and faulting. The highest lake deposits are early Pleistocene in age and were formed ~25 m above the present lake, prior to progressive and episodic downcutting of the lake outlet. Lake level was also linked to the changing course of the Bear River. The lake basin interacts with the river in complex ways that are modulated by climatically induced lake-level changes, the distribution of active Quaternary faults, and by the migration of the river across its outwash fan north of the present lake. The various mechanisms that caused the Bear River to shift into and out of the lake are reviewed by Reheis et al. (this volume), and the resulting paleohydrogeographic reconstructions are illustrated by Kaufman et al. (this volume).

Fluctuations in lake level have been studied in auger holes and outcrops around the lake, as described by Reheis et al. (this volume), and unexpectedly low lake levels have been inferred from sedimentological features preserved in sediment cores (Smoot, this volume). Sediment grain size decreases with increasing water depth in the modern lake. Smoot and Rosenbaum (this volume) use this relation in conjunction with other sedimentary features to reconstruct a detailed history of lake-level changes since the latest Pleistocene. This record exhibits lake-level changes of tens of meters on millennial and shorter time scales. It indicates that, during the Holocene, lake level was generally lower than present.

On a longer time scale, the closed-basin configuration of the lake is restricted to the driest interglacial periods (Kaufman et al., this volume). Analyses of core BL00-1 indicate that Bear Lake retracted into a topographically closed basin only during portions of global interglaciations (MIS 7c, 7a, 5e, 5c, and 1). During these intervals, the lake generated abundant endogenic carbonate with aragonite and high values of $\delta^{18}O$ and $^{87}Sr/^{86}Sr$.

Comparisons with Other Paleoclimate Records

The first-order fluctuations of all sediment properties analyzed in the long core from Bear Lake coincide with orbital cycles and global ice volume during the last two glacial-interglacial cycles (Dean, this volume; Kaufman et al., this volume). Millennial-scale fluctuations are pervasive throughout the last two glacial cycles and might correspond to stadial-interstadial cycles recognized in other well-known paleoclimate records, but uncertainties in age for core BL00-1 are too large to determine whether the cycles are synchronous. Millennial-scale variations in hematite content (Heil et al., this volume) and pollen assemblages (Kaufman et al., this volume) of Bear Lake sediment during the last glacial cycle resemble Dansgaard-Oeschger oscillations and Heinrich events, suggesting that the influence of millennial-scale climate oscilla-

tions influence the climate of the Great Basin. Although there are similarities in first-order trends, the timing of major lake-level changes at Bear Lake do not coincide in detail with those in the Bonneville Basin located ~100 km downstream of Bear Lake (Kaufman et al., this volume; Rosenbaum and Heil, this volume; Smoot and Rosenbaum, this volume). During the late glacial and Holocene, the lake-level history of Bear Lake is more similar to those in Pyramid Lake, Nevada, and Owens Lake, California, than to Lake Bonneville. For example, during the Younger Dryas, the level of Bear Lake seems to have lowered, whereas the level of Lake Bonneville rose (Smoot and Rosenbaum, this volume).

POTENTIAL FUTURE STUDIES

This volume presents an extensive study of Bear Lake and its catchment. Nevertheless, there remain opportunities to improve and expand on the current studies. Such opportunities fall into several categories. First, understanding of the lake's hydrology can be improved. A conspicuous gap in knowledge is the contribution of sub-lacustrine springs to the hydrologic budget. Second, different techniques (e.g., biogeochemistry, different microfossils) could be employed to derive additional paleoenvironmental proxies. For instance, it would be particularly useful to separate the effects of temperature and precipitation on changes in lake level. Third, temporal resolution of the records can be improved by denser sampling. This is especially true for the record older than ca. 26 cal ka (i.e., that part of the record sampled only in the long cores from site BL00-1). However, the resolution of the Holocene section could be improved through dense sampling of core BL00-1, which has the highest deposition rate and most continuous section. Of particular interest would be a similarly high resolution study of the last interglaciation (continuous sequences are rare in North America) and a comparison with the Holocene. Fourth, and perhaps most importantly, an improved chronology for the record beyond the range of radiocarbon dating would increase the value of existing and future proxy records immensely.

ACKNOWLEDGMENTS

Most of the funding for the studies in this volume was provided by the Earth Surface Dynamics Program of the U.S. Geological Survey. Half of the funding for drilling of the long cores at site BL00-1 during testing of the GLAD800 coring platform was provided by the National Science Foundation. Finally we would like to thank R. Reynolds for handling editorial duties for two chapters for which we are both authors.

REFERENCES CITED

Bright, J., 2009, this volume, Chapter 4, Isotope and major-ion chemistry of groundwater in Bear Lake Valley, Utah and Idaho, with emphasis on the Bear River Range, *in* Rosenbaum, J.G., and Kaufman, D.S., eds., Paleoenvironments of Bear Lake, Utah and Idaho, and its catchment: Geological Society of America Special Paper 450, doi: 10.1130/2009.2450(04).

Bright, J., 2009, this volume, Chapter 8, Ostracode endemism in Bear Lake, Utah and Idaho, *in* Rosenbaum, J.G., and Kaufman, D.S., eds., Paleoenvironments of Bear Lake, Utah and Idaho, and its catchment: Geological Society of America Special Paper 450, doi: 10.1130/2009.2450(08).

Colman, S.M., 2006, Acoustic stratigraphy of Bear Lake, Utah-Idaho—Late Quaternary sedimentation patterns in a simple half-graben: Sedimentary Geology, v. 185, p. 113–125, doi: 10.1016/j.sedgeo.2005.11.022.

Colman, S.M., Kaufman, D.S., Bright, J., Heil, C., King, J.W., Dean, W.E., Rosenbaum, J.G., Forester, R.M., Bischoff, J.L., Perkins, M., and McGeehin, J.P., 2006, Age models for a continuous 250-kyr Quaternary lacustrine record from Bear Lake, Utah-Idaho: Quaternary Science Reviews, v. 25, p. 2271–2282, doi: 10.1016/j.quascirev.2005.10.015.

Colman, S.M., Rosenbaum, J.G., Kaufman, D.S., Dean, W.E., and McGeehin, J.P., 2009, this volume, Radiocarbon ages and age models for the last 30,000 years in Bear Lake, Utah and Idaho, *in* Rosenbaum, J.G., and Kaufman, D.S., eds., Paleoenvironments of Bear Lake, Utah and Idaho, and its catchment: Geological Society of America Special Paper 450, doi: 10.1130/2009.2450(05).

Dean, W.E., 2009, this volume, Endogenic carbonate sedimentation in Bear Lake, Utah and Idaho, over the last two glacial-interglacial cycles, *in* Rosenbaum, J.G., and Kaufman, D.S., eds., Paleoenvironments of Bear Lake, Utah and Idaho, and its catchment: Geological Society of America Special Paper 450, doi: 10.1130/2009.2450(07).

Dean, W.E., Rosenbaum, J.G., Haskell, B., Kelts, K., Schnurrenberger, D., Valero-Garcés, B., Cohen, A., Davis, O., Dinter, D., and Nielson, D., 2002, Progress in Global Lake Drilling holds potential for Global Change research: Eos (Transactions, American Geophysical Union), v. 83, p. 85, 90, 91.

Dean, W.E., Wurtsbaugh, W.A., and Lamarra, V.A., 2009, this volume, Climatic and limnologic setting of Bear Lake, Utah and Idaho, *in* Rosenbaum, J.G., and Kaufman, D.S., eds., Paleoenvironments of Bear Lake, Utah and Idaho, and its catchment: Geological Society of America Special Paper 450, doi: 10.1130/2009.2450(01).

Doner, L.A., 2009, this volume, A 19,000-year vegetation and climate record for Bear Lake, Utah and Idaho, *in* Rosenbaum, J.G., and Kaufman, D.S., eds., Paleoenvironments of Bear Lake, Utah and Idaho, and its catchment: Geological Society of America Special Paper 450, doi: 10.1130/2009.2450(09).

Heil, C.W., Jr., King, J.W., Rosenbaum, J.G., Reynolds, R.L., and Colman, S.M., 2009, this volume, Paleomagnetism and environmental magnetism of GLAD800 sediment cores from Bear Lake, Utah and Idaho, *in* Rosenbaum, J.G., and Kaufman, D.S., eds., Paleoenvironments of Bear Lake, Utah and Idaho, and its catchment: Geological Society of America Special Paper 450, doi: 10.1130/2009.2450(13).

Kaufman, D.S., Bright, J., Dean, W.E., Rosenbaum, J.G., Moser, K., Anderson, R.S., Colman, S.M., Heil, C.W., Jr., Jiménez-Moreno, G., Reheis, M.C., and Simmons, K.R, 2009, this volume, A quarter-million years of paleoenvironmental change at Bear Lake, Utah and Idaho, *in* Rosenbaum, J.G., and Kaufman, D.S., eds., Paleoenvironments of Bear Lake, Utah and Idaho, and its catchment: Geological Society of America Special Paper 450, doi: 10.1130/2009.2450(14).

Moser, K.A., and Kimball, J.P., 2009, this volume, A 19,000-year record of hydrologic and climatic change inferred from diatoms from Bear Lake, Utah and Idaho, *in* Rosenbaum, J.G., and Kaufman, D.S., eds., Paleoenvironments of Bear Lake, Utah and Idaho, and its catchment: Geological Society of America Special Paper 450, doi: 10.1130/2009.2450(10).

Piechota, T., Timilsena, J., Tootle, G., and Hidalgo, H., 2004, The western U.S. drought: How bad is it?: Eos (Transactions, American Geophysical Union), v. 85, p. 301–308, doi: 10.1029/2004EO320001.

Reheis, M.C., Laabs, B.J.C., and Kaufman, D.S., 2009, this volume, Geology and geomorphology of Bear Lake Valley and upper Bear River, Utah and Idaho, *in* Rosenbaum, J.G., and Kaufman, D.S., eds., Paleoenvironments of Bear Lake, Utah and Idaho, and its catchment: Geological Society of America Special Paper 450, doi: 10.1130/2009.2450(02).

Rosenbaum, J.G., and Heil, C.W., Jr., 2009, this volume, The glacial/deglacial history of sedimentation in Bear Lake, Utah and Idaho, *in* Rosenbaum, J.G., and Kaufman, D.S., eds., Paleoenvironments of Bear Lake, Utah and Idaho, and its catchment: Geological Society of America Special Paper 450, doi: 10.1130/2009.2450(11).

Rosenbaum, J.G., Dean, W.E., Reynolds, R.L., and Reheis, M.C., 2009, this volume, Allogenic sedimentary components of Bear Lake, Utah and Idaho, *in* Rosenbaum, J.G., and Kaufman, D.S., eds., Paleoenvironments of Bear Lake, Utah and Idaho, and its catchment: Geological Society of America Special Paper 450, doi: 10.1130/2009.2450(06).

Smoot, J.P., 2009, this volume, Late Quaternary sedimentary features of Bear Lake, Utah and Idaho, *in* Rosenbaum, J.G., and Kaufman, D.S., eds., Paleoenvironments of Bear Lake, Utah and Idaho, and its catchment: Geological Society of America Special Paper 450, doi: 10.1130/2009.2450(03).

Smoot, J.P., and Rosenbaum, J.G., 2009, this volume, Sedimentary constraints on late Quaternary lake-level fluctuations at Bear Lake, Utah and Idaho, *in* Rosenbaum, J.G., and Kaufman, D.S., eds., Paleoenvironments of Bear Lake, Utah and Idaho, and its catchment: Geological Society of America Special Paper 450, doi: 10.1130/2009.2450(12).

Manuscript Accepted by the Society 15 September 2008

The Geological Society of America
Special Paper 450
2009

Climatic and limnologic setting of Bear Lake, Utah and Idaho

Walter E. Dean

U.S. Geological Survey, MS 980 Federal Center, Denver, Colorado 80225, USA

Wayne A. Wurtsbaugh

Department of Watershed Sciences and the Ecology Center, Utah State University, Logan, Utah 84322, USA

Vincent A. Lamarra

Ecosystems Research Institute, Logan, Utah 84321, USA

ABSTRACT

Bear Lake is a large alkaline lake on a high plateau on the Utah-Idaho border. The Bear River was partly diverted into the lake in the early twentieth century so that Bear Lake could serve as a reservoir to supply water for hydropower and irrigation downstream, which continues today. The northern Rocky Mountain region is within the belt of the strongest of the westerly winds that transport moisture during the winter and spring over coastal mountain ranges and into the Great Basin and Rocky Mountains. As a result of this dominant winter precipitation pattern, most of the water entering the lake is from snowmelt, but with net evaporation. The dominant solutes in the lake water are Ca^{2+}, Mg^{2+}, and HCO_3^{2-}, derived from Paleozoic carbonate rocks in the Bear River Range west of the lake. The lake is saturated with calcite, aragonite, and dolomite at all depths, and produces vast amounts of carbonate minerals. The chemistry of the lake has changed considerably over the past 100 years as a result of the diversion of Bear River. The net effect of the diversion was to dilute the lake water, especially the Mg^{2+} concentration.

Bear Lake is oligotrophic and coprecipitation of phosphate with $CaCO_3$ helps to keep productivity low. However, algal growth is colimited by nitrogen availability. Phytoplankton densities are low, with a mean summer chlorophyll *a* concentration of 0.4 mg L^{-1}. Phytoplankton are dominated by diatoms, but they have not been studied extensively (but see Moser and Kimball, this volume). Zooplankton densities usually are low (<10 L^{-1}) and highly seasonal, dominated by calanoid copepods and cladocera. Benthic invertebrate densities are extremely low; chironomid larvae are dominant at depths <30 m, and are partially replaced with ostracodes and oligochaetes in deeper water. The ostracode species in water depths >10 m are all endemic. Bear Lake has 13 species of fish, four of which are endemic.

Dean, W.E., Wurtsbaugh, W.A., and Lamarra, V.A., 2009, Climatic and limnologic setting of Bear Lake, Utah and Idaho, *in* Rosenbaum, J.G., and Kaufman, D.S., eds., Paleoenvironments of Bear Lake, Utah and Idaho, and its catchment: Geological Society of America Special Paper 450, p. 1–14, doi: 10.1130/2009.2450(01).

INTRODUCTION

Bear Lake (42°N, 111°20′ W) is an alkaline lake that occupies the southern half of the Bear Lake Valley on the Utah-Idaho border (Fig. 1). The present elevation of the lake when full is 1805 m (5922 feet) above sea level, but this level has varied considerably over the past 100 years, mainly in response to drought conditions (Fig. 2). Flow volumes of the Green River at Green River, Utah, 125 km east of the lake are low when Bear Lake level is low (Fig. 2), indicating that fluctuations of the elevation of Bear Lake are due to regional and not local conditions. The low levels of Bear Lake during the drought years of the 1990s and 2000s approached the low levels during the dust-bowl drought of the 1930s (Fig. 2).

The natural watershed of the lake is relatively small, having a basin- to lake-area ratio of ~4.8 (Wurtsbaugh and Luecke, 1997). The Bear River is the largest river in the Great Basin, and is the principal source of surface water flowing into Great Salt Lake. The Bear River originates in the Uinta Mountains 125 km southeast of Bear Lake in bedrock composed mostly of quartzite. Tributaries of the Bear River, and the streams that flow directly into Bear Lake, originate in the northern Wasatch Range, which is composed primarily of Paleozoic carbonate rocks (Hintze, 1973; Reheis et al., this volume).

Within historic times, the Bear River bypassed Bear Lake. However, part of the river's flow was diverted into Bear Lake through a series of canals, beginning in 1909 with completion in 1918 (Birdsey, 1989), making Bear Lake a reservoir to supply water for hydropower and irrigation downstream, and this continues today. Apparently the first Bear River water entered Mud Lake (Fig. 1) through a canal at the north end of Mud Lake in May 1911. Presumably some of this water overflowed into Bear Lake, but there are no records of the timing or amount of this overflow (Mitch Poulsen, Bear Lake Regional Commission, 2004, personal commun.; Connely Baldwin, PacifiCorp, 2005, personal commun.). A control structure was subsequently added to control the flow of water from Mud Lake to Bear Lake, but a minimal volume of water was diverted into Bear Lake before 1913 and possibly even later (Connely Baldwin, PacifiCorp, 2005, personal commun.).

The diversion of Bear River into Bear Lake increased the basin-to-lake-area ratio considerably, to 29.5. Since the diversion, the mean annual surface hydrologic flux (including precipitation) to the lake is estimated to be 0.48×10^9 m^3 yr^{-1} (Lamarra et al., 1986). Outflow is estimated as 0.214×10^9 m^3 yr^{-1}, which is only ~3% of the lake volume (8.0×10^9 m^3), giving an average residence time of ~37 yr. The amount of groundwater influx may be considerable but is not known. Bright (this volume, Chapter 4) concluded that the hydrologic budget of the lake is near zero.

The main purpose of this paper is to document the physical, chemical, and biological limnologic conditions in Bear Lake based mainly on unpublished data, some of them going back as far as 1975. A second objective is to examine the climatic setting of Bear Lake, both in terms of how the Bear Lake region is related to the climate of the western United States, and in terms of local meteorological conditions.

METHODS

A vertical profile of water samples at 10 m intervals in the deepest part of Bear Lake was collected using a Kemmerer water sampler between six and 13 times yearly between 1981 and the present, allowing the documentation of water quality conditions during periods of spring and fall circulation and summer and winter stratification. The samples were analyzed for total and orthophosphorus, nitrate, nitrite, ammonia, total suspended solids, and chlorophyll *a* using standard methods (American Public Health Association, 1992) at the Ecosystems Research Institute (ERI), Logan, Utah, a state and EPA certified laboratory. Prior to 1996, samples for dissolved oxygen were collected in the field and analyzed at the ERI laboratory by the standard Winkler titration method. Field temperature, dissolved oxygen, pH, conductivity, and turbidity were measured using a Hydrolab H$_2$O Water Quality Multiprobe (1996–2002), or with an In Situ MP-Troll 9000 (2003-present). In the field, each sample was split immediately into a bottle with acid preservative for the ammonia, nitrate, and total phosphorus analyses, an unpreserved bottle for the total suspended solids, nitrite, and orthophosphorus analyses, and an unpreserved bottle for chlorophyll *a* analysis. The chlorophyll *a* sample was filtered in the laboratory, frozen, and subsequently extracted in 100% buffered methanol for 24 h at room temperature. The extracts were analyzed fluorometrically with a correction for phaeophytin in a fluorometer calibrated with standard chlorophyll *a* (Holm-Hansen and Riemann, 1978).

Secchi-disk transparency was measured with a 20 cm white disk from the shaded side of a boat or under the ice. Temperature profiles were made with a Yellow Springs Instruments Model 58 thermistor. Specific conductance was made with an SBE 25 Sealogger profiler.

CLIMATIC SETTING

Regional Climatic Setting

The climate of the western United States is dominated by atmospheric circulation over the North Pacific Ocean and adjacent land areas (the North Pacific High, the Aleutian Low, and the North American Low). The seasonal strengths and positions of these pressure systems not only generate the weather and climate of the western United States (e.g., Strub et al., 1987; Thompson et al., 1993), but are part of the atmospheric teleconnections that stretch across the Northern Hemisphere (e.g., Namias et al., 1988). Extreme differences in relief in the western United States create strong elevation gradients in climate. Today, the climate of the Pacific Northwest is characterized in the spring and summer by strong, persistent, northwesterly winds generated by the juxtaposition of the North Pacific High over the eastern North Pacific and North American Low over the Great Basin, which generally

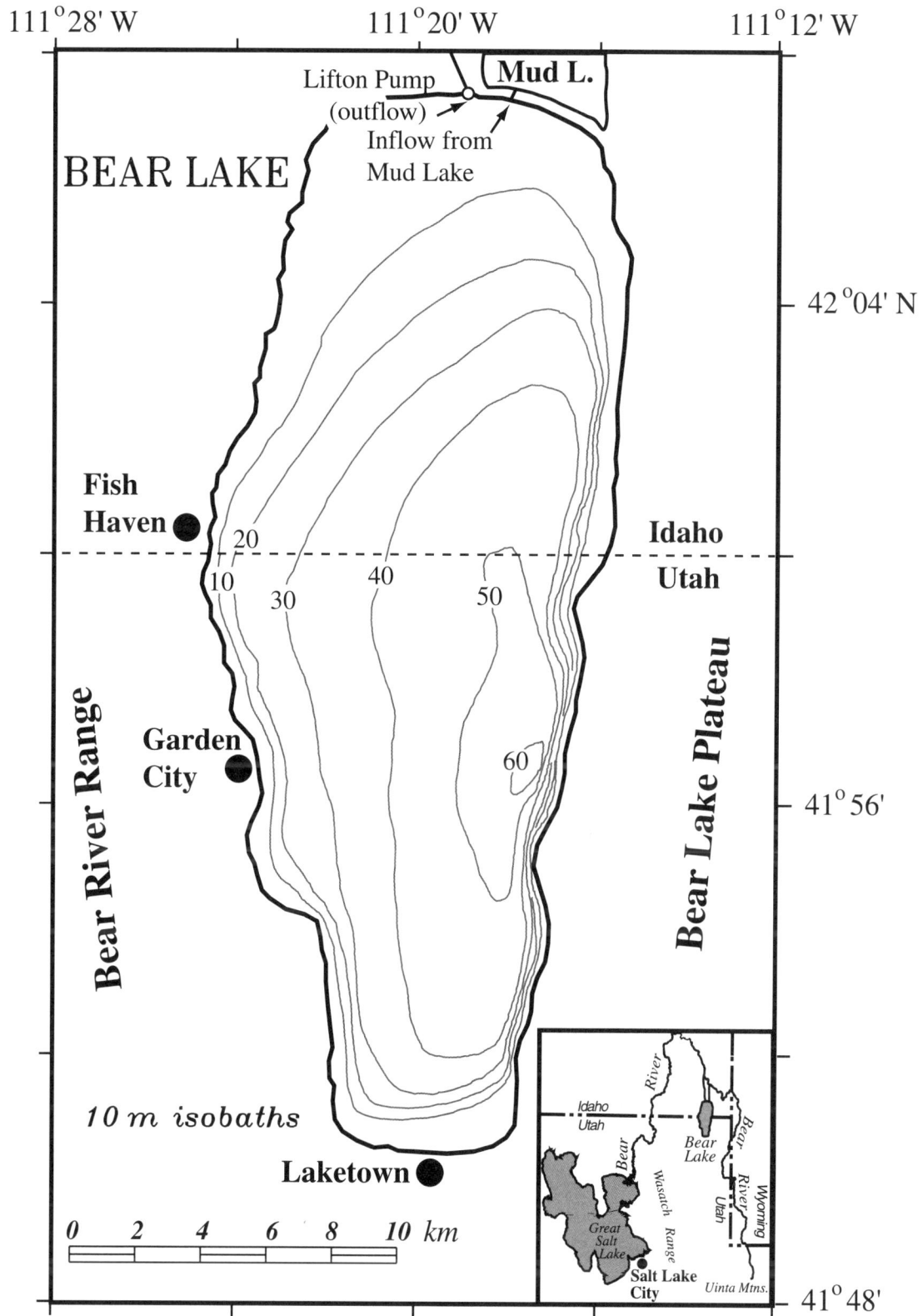

Figure 1. Bathymetric map of Bear Lake, Utah and Idaho. The inset shows the location of Bear Lake relative to Bear River and Great Salt Lake.

results in dry conditions (Thompson et al., 1993). The Aleutian Low that drives the jet stream is displaced far to the north at that time. Winters are influenced by a weakened North American Low, the migration of the North Pacific High south of 30°N, and the migration of the polar jet stream and associated Aleutian Low to an average position of ~45°N. The winter dominance of the Aleutian Low produces wet and stormy weather with zonal westerly winds (Thompson et al., 1993). These winter Pacific storms lose most of their moisture as they rise over the Sierra Nevada and Cascade Ranges so that westerly air currents reaching Utah contain little moisture.

The Bear Lake region is within the belt of the strongest of these westerly winds, which transport most of the moisture into the region. As a result of marked seasonal changes in atmospheric conditions over the eastern North Pacific, the Bear Lake region experiences hot, dry summers and cold, wet winters. The limited summer moisture arrives primarily as thunderstorms that are associated with moisture-laden monsoonal air masses from the Gulf of Mexico and Gulf of California (wrcc.dri.edu/narratives/UTAH). Summer-wet/winter-dry conditions dominate the region from the southwestern United States to the southern Rocky Mountains and Great Plains, including the basins of Wyoming due to monsoonal moisture (Whitlock et al., 1993). Summer-dry/winter-wet conditions dominate higher elevations of the northern Rocky Mountains that are able to intercept winter storms that move inland from the Pacific. However, during years with increased summer monsoonal precipitation from the Pacific and Gulf of Mexico, late summer precipitation from the southwest may reach northern Utah. Data from 500 meteorological stations in the western United States for the period 1946–1994 show that most stations in Arizona, Utah, and western Colorado exceeded the average August precipitation more than 30% of the time as the result of monsoonal moisture (Mock and Brunelle-Daines, 1999). Such departures also have occurred in the past (e.g., Thompson et al., 1993; Mock and Bartlein, 1995; Mock

and Brunelle-Daines, 1999). Therefore, Bear Lake has the potential to record changes in the strengths of monsoonal circulation and Aleutian Low circulation.

Local Temperature and Precipitation

The Bear Lake Valley is on a high plateau (1805 m) between the Bear River Range to the west and the Bear Lake Plateau to the east (Fig. 1; see Reheis et al., this volume). As described above, the valley has a continental climate with cold winters and warm to hot summers. Annual precipitation at Tony Grove Lake (elev 2415 m), in the Bear River Range west of the lake from 1979 to 2005 averages 124 cm, with the majority falling in the winter (90 cm) (wcc.nrcs.usda.gov/snow). Mean annual precipitation at Laketown is 30 cm (Fig. 3A), with the majority falling in the winter (wrcc.dri.edu/summary/Climsmut.html). The annual precipitation at Laketown over the last century has increased by ~9 cm (Fig. 3A) due mainly to an increase in winter (December–March)

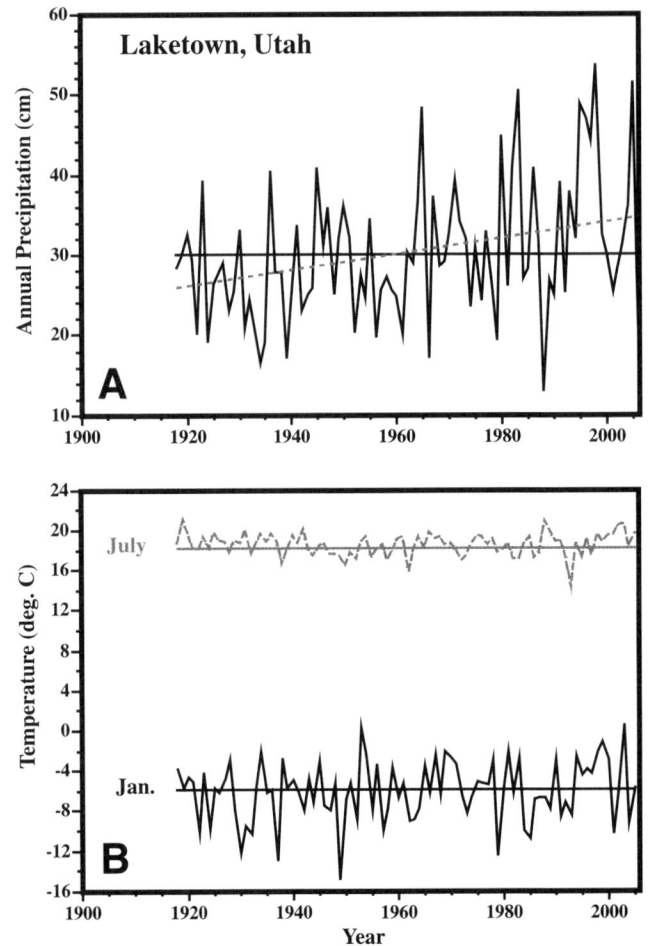

Figure 3. Annual precipitation (A) and July and January temperatures (B) since 1920 for Laketown, Utah, at the southern end of Bear Lake. Solid horizontal lines through each plot represent the mean of all data. Slanted dashed line in (A) is a linear regression through the data. Data are from wrcc.dri.edu/summary/Climsmut.html.

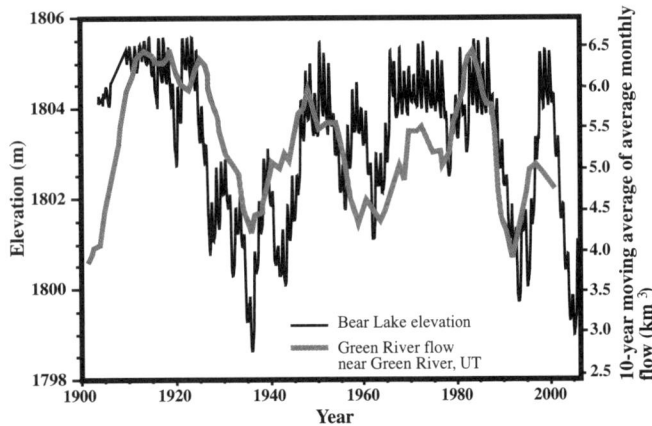

Figure 2. Elevation of Bear Lake (in meters above sea level) since 1905, and flow of the Green River near Green River, Utah. Elevation data are from PacifiCorp, Salt Lake City, Utah. Green River flow diagram is from Pieochota et al. (2004).

precipitation. Mean January temperature at Laketown, Utah, at the south end of the lake for the period 1918–2005 was –6.0 °C (Fig. 3B), and the mean July temperature for the same period was 18.2 °C (wrcc.dri.edu/summary/Climsmut.html).

Because of the dominance of winter precipitation, most of the water that enters Bear Lake, either from direct precipitation, or surface- and groundwater flow, is from snowmelt, and this is reflected in the isotopic composition of source waters to the lake (Dean et al. 2007; Bright, this volume, Chapter 4). Evaporation is poorly known, but pan estimations place it at ~100 cm yr[-1] (Kaliser, 1972), and the average of annual pan evaporation measurements at Logan, Utah (1969–2005) and Bear River Refuge (1948–1984) are 130 cm (wrcc.dri.edu/htmfiles/westevap.final.html). The average of pan measurements from May to October for the period 1935–2002 at Lifton Pump Station, Idaho, is 107 cm (http://wrcc.dri.edu), but lake evaporation would be much lower.

PHYSICAL FEATURES

Morphometry

Bear Lake is 32 km long and has a maximum width of 12 km. At full capacity (an elevation of 1805 m), the lake has a surface area of 282 km[2], a maximum depth of 63 m, and a mean depth of 28 m (Birdsey, 1989). The volume of the lake at full capacity is 8.0×10^9 m[3]. A chirp (4–24 kHz) acoustic profile (Colman, 2006) indicates that the principal structure of the basin is a half graben, with a steep, N-S oriented normal-fault margin on the east (east Bear Lake fault) and a ramp margin on the west. As a result of this structure, the lake deepens gradually from west to east, but precipitously from east to west (Fig. 1). Acoustic reflectors diverge toward the east Bear Lake fault, forming eastward-thickening sediment wedges, so that sedimentary units pinch out to the west.

Ice Cover, Thermal and Chemical Stratification

Although the Bear Lake Valley is cold in winter, the lake does not always freeze. The lake has been ice free for 25 of the past 80 years, and the frequency of freezing has decreased in the last few decades (Fig. 4). In the 11 years between 1995 and 2005, there were seven years when the lake did not freeze. The duration of ice cover is highly variable, and can range from >100 days one year to no ice cover the next (Fig. 4A). There is a general tendency for lower duration of ice cover to correspond to lower lake levels (compare Fig. 4B with Fig. 2), but there are many exceptions. For example, the lake froze over during most of the years of low lake levels in the early to mid-1990s (Fig. 4B). Ice-out usually is in April (Fig. 4A), and the timing depends on the seasonal progression of temperature and wind, generally associated with storms.

In years when Bear Lake freezes over, it behaves like a typical dimictic lake with spring and fall overturns. In years when it does not freeze over, it is monomictic with overturn in January (Wurtsbaugh and Luecke, 1997). During the annual cycle,

a thermocline forms at ~10 m in May, and gradually deepens throughout summer and fall until complete mixing occurs in late December or January (Figs. 5 and 6B). During late-summer thermal stratification, the base of the epilimnion typically is between 10 and 15 m with a broad, diffuse metalimnion (Figs. 5 and 7A). The temperature of the epilimnion ranges from 2 to 3 °C in February to 18–21 °C in August and September (Figs. 5 and 6B). The temperature of the hypolimnion is relatively stable at ~5 ± 2 °C throughout the year (range 2–8 °C; Figs. 5 and 6B).

Internal waves (seiches) are common in Bear Lake, but have not been specifically studied. SCUBA-based observers report suspended sediment in the water column where the thermocline intersects the bottom, which suggests that sediments are resuspended. Because the thermocline deepens steadily throughout the summer, and seiches are common, there is ample opportunity for the resuspension of sediments into the water column. Model and empirical analyses have shown that turbulence where the thermocline intersects the bottom can entrain nutrients from the

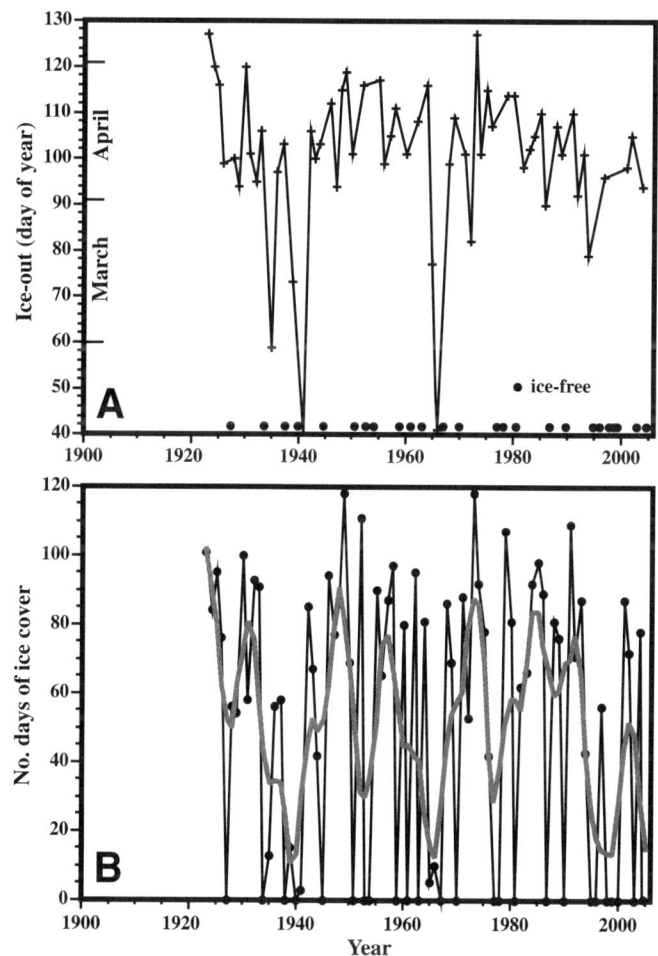

Figure 4. Day of year of ice-out (A) and number of days of ice cover (B) for Bear Lake since 1923. Heavy curved line through data in (B) is a weighted least-squares smoothing function. Data are from PacifiCorp, Salt Lake City, Utah.

sediments into the water column (Wüest and Lorke, 2003). Surface waves also entrain littoral sediments into the water column, which decreases visibility. The combination of internal and surface waves causes erosion of sediments.

Thermal stratification leads to marked chemical stratification of some parameters. Specific conductance at 25 °C in the epilimnion in late summer is ~685 µS cm^{-1}, increasing to ~700 µS cm^{-1} in the hypolimnion (Fig. 7A). The total dissolved solids (TDS in mg L^{-1}) content increases from 533 in the epilimnion to 582 in the hypolimnion (Table 1; Dean et al., 2007). Dissolved oxygen (DO) concentrations generally are high throughout the water column for most of the year, often with a maximum in the metalimnion that may be 1–2 mg L^{-1} higher than in the epilimnion due to a concentration of algal productivity. Concentrations of DO may decline below 4 mg L^{-1} in the deep hypolimnion (>50 m) by September or October (Fig. 6C) due to decomposition of produced organic matter. However, the average summer (July–September) DO at 50 m from 1981 to 2003 was 7.0 ± 1.05 mg L^{-1}, indicating that oxic conditions with high redox states predominate above the sediments.

Light Transmission

Light extinction coefficients in Bear Lake range from 0.19 to 0.28 (Neverman and Wurtsbaugh, 1992). Consequently, light intensities at the mean and maximum depths typically are 0.2% and 0.0001%, respectively, of those at the surface. Wurtsbaugh and Luecke (1997) found that Secchi depths only partly reflect changes in chlorophyll concentration (primary productivity) because of suspended carbonate particles. Secchi depth generally varies between 4 and 6 m (Fig. 6A). The average (±1 standard deviation) Secchi depth between 1975 and 2000 was 5.0 ± 1.7 m (range 1.4–12.0 m). Unusually deep Secchi depths occurred in 1996 and again in 1998 (Fig. 4), coincident with relatively high densities of the zooplankton grazer *Daphnia pulex* (Wurtsbaugh and Luecke, 1998).

There are abundant suspended CaCO$_3$ particles in the water column, which means that water transparencies are not as great as they should be for an oligotrophic lake. This is a common phenomenon in many hard-water lakes that precipitate CaCO$_3$ during the warm summer months. In these lakes, suspended and colloidal CaCO$_3$ scatters light in the blue and green wavelengths, giving these lakes a very characteristic blue color (e.g., Wetzel, 2001). Sediment-trap studies show that in Bear Lake most CaCO$_3$ precipitation as high-Mg calcite does indeed occur in the epilimnion from April through September, but below ~10 m the water column always contains particles of carbonate due to resuspension from the bottom in water <30 m (Dean et al., 2005, 2007; Dean, this volume).

LAKE CHEMISTRY

Major Ions and Carbonate Precipitation

The dominant cations in Bear Lake water today are calcium (Ca^{2+}), magnesium (Mg^{2+},) and sodium (Na$^+$) (Table 1). The dominant anion is bicarbonate (HCO$_3^-$), but there are also

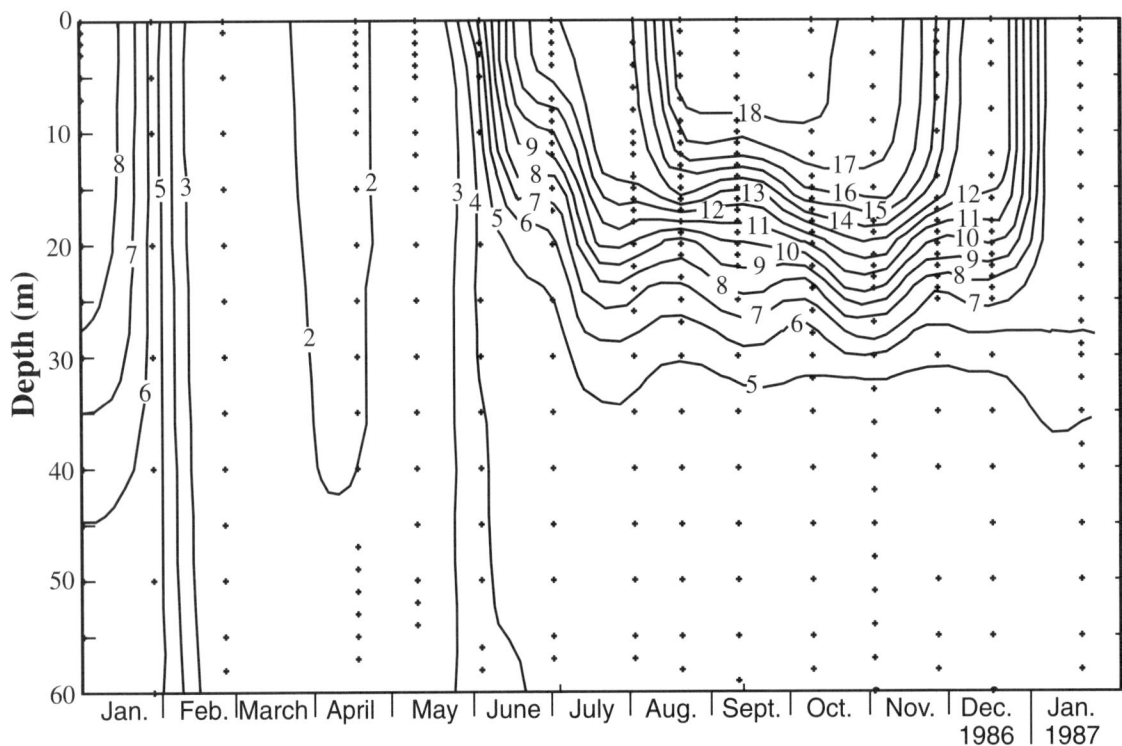

Figure 5. Temperature isopleths of Bear Lake from January 1986 through January 1987. Note that wind mixing in December and January allows the lake to cool to near 2 °C. Small crosses indicate the depths and dates of measurements.

relatively high concentrations of sulfate (SO_4^{2-}) and chloride (Cl^-). Present-day Bear Lake has two natural solute sources (west-side and east-side streams and springs) and one human-controlled solute source (Bear River). The solutes in the west-side waters (springs and streams) are Ca^{2+}-HCO_3^- dominated (Dean et al., 2007). They have high HCO_3^-:Ca^{2+} ratios (average of 4.6) due to

low Ca^{2+} concentrations. They also have low concentrations of most other ions. The west-side creeks originate as springs on the east side of the Bear River Range, fed by groundwaters flowing through cavernous Paleozoic carbonate rocks in the Bear River Range (Dean et al., 2007). The east-side waters (springs and streams) have a wide variety of compositions dominated by some

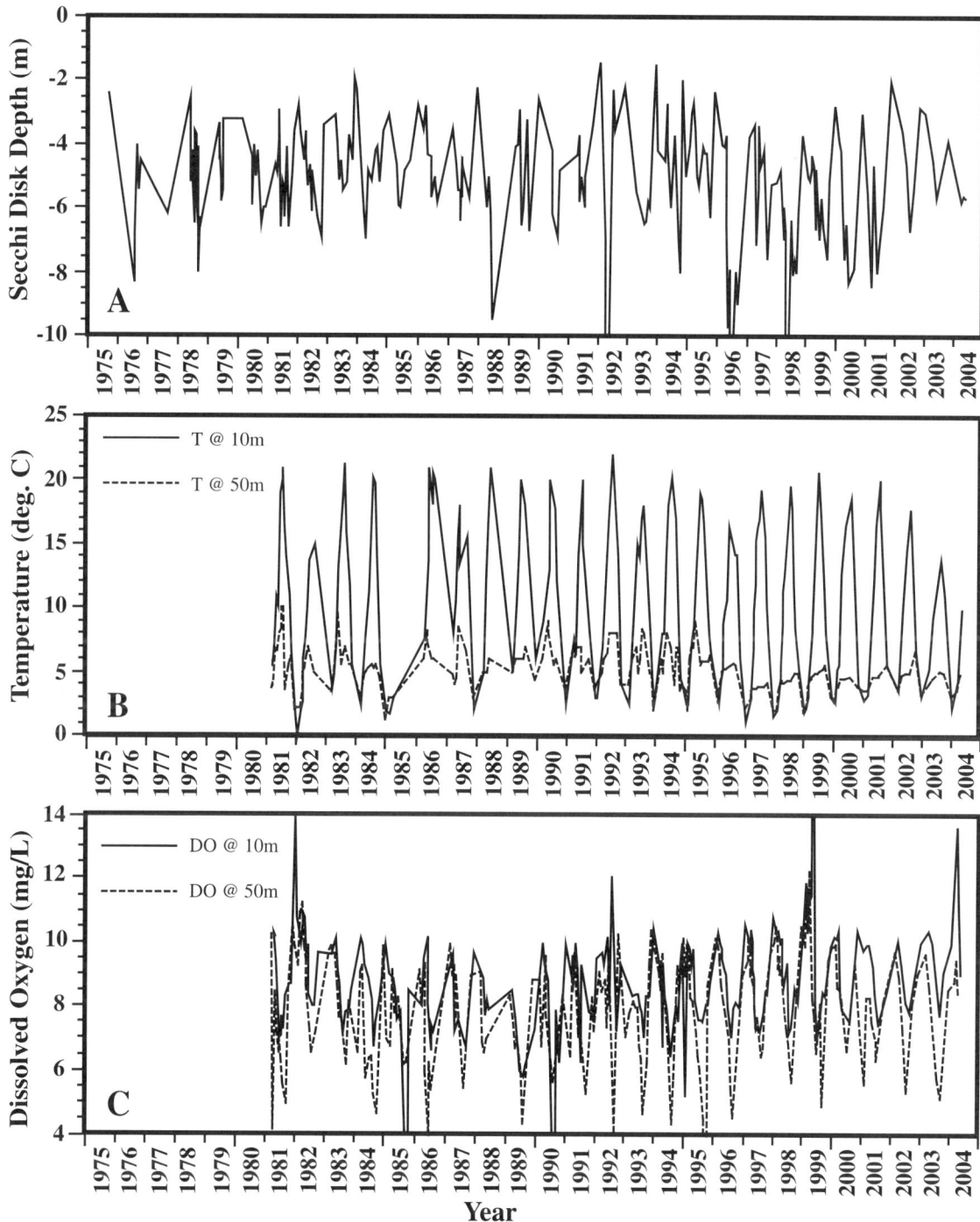

Figure 6. Secchi-disk depths (A), temperatures at 10 m and 50 m (B), and dissolved oxygen concentrations at 10 m and 50 m (C) in Bear Lake from 1975 to 2004.

Dean et al.

combination of bicarbonate, Ca^{2+}, Mg^{2+}, Na^+, SO_4^{2-}, or Cl^-. They have low HCO_3^-:Ca^{2+} ratios (average of 2.9) due to high Ca^{2+} concentrations (Dean et al., 2007). The composition of Bear River water more closely resembles that of east-side waters (Figs. 8A and 8B), which may reflect base flow from the same groundwaters that discharge to the east-side springs and streams.

The oldest chemical analysis from Bear Lake is of a sample collected in 1912 (Table 1; Kemmerer et al., 1923). We use 1912 as the nominal time of Bear River diversion, and assume that the 1912 analysis is close to the composition of the lake at the time of diversion (Dean et al., 2007). The chemistry of the lake has changed considerably over the past 100 years. Now the lake water is highly enriched in Mg^{2+} relative to Bear River, and was even more so prior to Bear River diversion (Fig. 8B, Table 1). The present lake water also is more enriched in Mg^{2+} relative to the surface streams entering the lake (Fig. 8B; Dean et al., 2007). The decrease in Ca^{2+} and increase in Mg^{2+} in the lake relative to inflowing surface streams is mostly due to precipitation of $CaCO_3$. Most likely there is, and has always been, a large Mg-rich source of groundwater entering Bear Lake. This groundwater source was even more important prior to diversion as evidenced by the extremely high Mg^{2+}:Ca^{2+} in the 1912 water (Table 1). Prior to diversion, this groundwater source prevented the lake from becoming saline or even drying up (Dean et al., 2007). This groundwater source is probably from a deep aquifer because shallow groundwaters, as sampled in springs and wells around the lake, all have compositions similar to surface waters (Dean et al., 2007). A thermal inversion was observed within 2 m of the lake bottom on one occasion in summer. A warm layer was overlain by cooler water, suggesting that a denser, possibly saltier, water was flowing into the lake from a sublacustrine spring. Although we were unable to find the source of the spring, this observation supports the hypothesis that some aspects of Bear Lake's chemistry are due to the chemical composition of sublacustrine spring inflow (Dean et al., 2007).

The net effect of the diversion of Bear River into Bear Lake was to dilute the lake water. The Mg^{2+}:Ca^{2+} and TDS in the Bear Lake water sample collected in 1912 were 38 (62.5 molar) and 1280 mg L^{-1}; today they are 1.7 (3.0 molar) and 530 mg L^{-1} (Table 1). In other words, following diversion of the Bear River into Bear Lake, the Mg^{2+}:Ca^{2+} was reduced 22-fold, whereas the TDS content decreased only 2.4-fold. There was also a significant reduction in HCO_3^- and a large increase in Ca^{2+}. The chemistry of the lake in 1912 was dominated by Mg^{2+}-HCO_3^-, which is highly unusual. Among hard-water lakes in glaciated temperate regions, Ca^{2+}-HCO_3^- lakes predominate and Mg^{2+}-HCO_3^- lakes are uncommon (Wetzel, 2001). The low Ca^{2+} concentration in the lake in 1912 can be explained by massive precipitation of $CaCO_3$ over

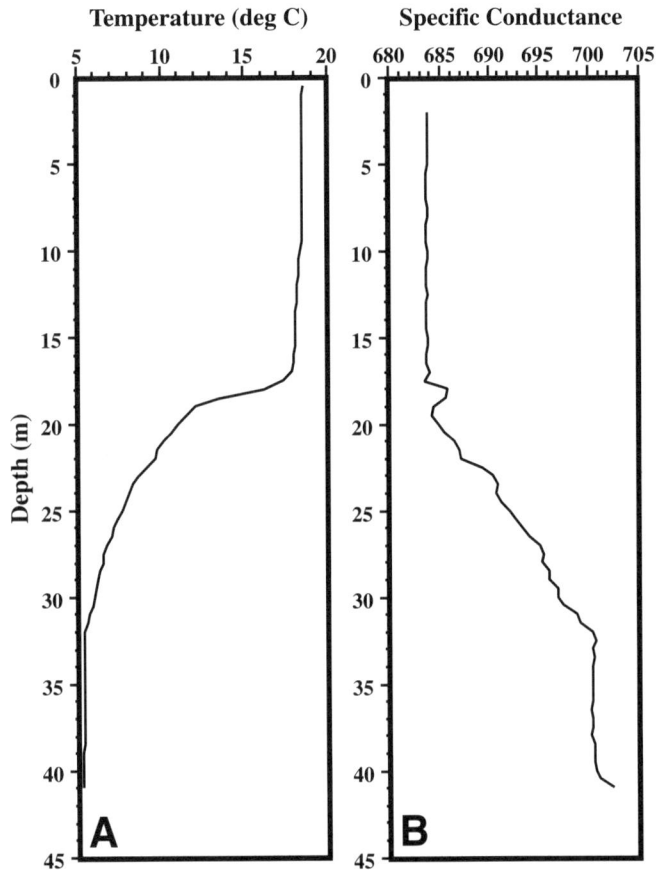

Figure 7. Temperature (A) and specific conductance (B) (in μS cm^{-1} at 25 °C) versus depth in September 2000, at a location where the water depth was 41 m.

Water	Ca	Mg	Na	K	HCO₃*	SO₄	Cl	TDS†
Lake depth 4 m	31.8	51.8	40.7	4.6	293	67.7	43.4	533
Lake depth 10 m	30.9	52.8	41.0	4.7	293	68.1	43.4	534
Lake depth 15 m	32.7	52.9	40.7	4.6	287	68.5	43.6	530
Lake depth 43 m	29.7	52.2	39.1	3.8	347	66.7	43.6	582
East Shore	26.3	47.8	35.5	4.7	298	68.0	45.1	525
Lake 1912§	4.1	152	66.3	10.5	715	96.8	78.5	1123
Lake 1952#	17.0	78	23	6	313	78	57	572
Bear River at gauging station, Idaho	69.1	23.5	31.5	2.2	256	78	40	501

TABLE 1. MAJOR DISOLVED IONS IN BEAR LAKE IN mg/L, FROM DEAN ET AL. (2007)

*HCO_3 is calculated from total alkalinity; HCO_3 for 1912 in Dean et al. (2007) is total alkalinity
†TDS—total dissolved solids (major ions)
§Kemmerer et al. (1923)
#Birdsey (1989)

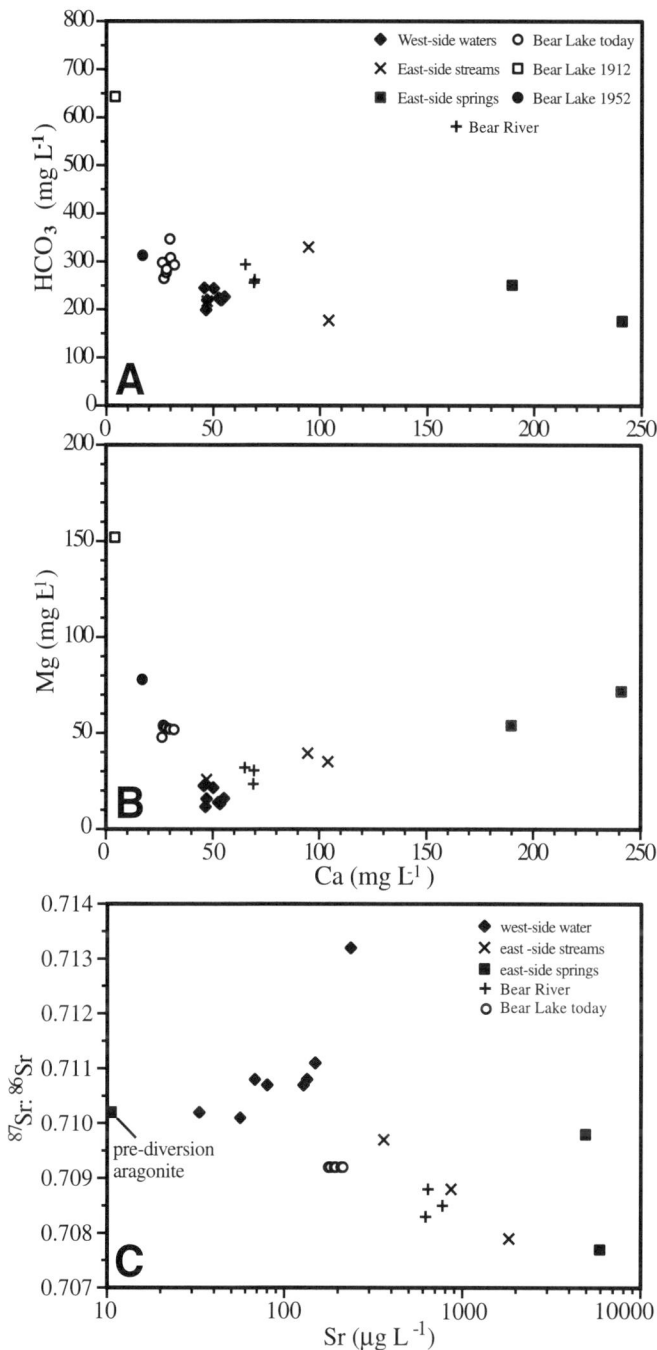

Figure 8. Crossplots of samples from Bear Lake and associated waters. (A) Total dissolved calcium (Ca) versus bicarbonate (HCO$_3$). (B) Total dissolved Ca versus total dissolved magnesium (Mg). (C) Dissolved strontium (Sr) concentration versus ^{87}Sr/^{86}Sr. Note that dissolved Sr concentration is shown on a log$_{10}$ scale. The ^{87}Sr/^{86}Sr value for the 1912 Bear Lake comes from samples of aragonite taken from cores at a stratigraphic horizon believed to just precede the diversion of Bear River into Bear Lake (Dean et al., 2007; noted as pre-diversion aragonite, Fig. 8C).

thousands of years. However, the high HCO$_3^-$ is more difficult to explain because precipitation of CaCO$_3$ should remove equal molar proportions of Ca^{2+} and HCO$_3^-$. Dean et al. (2007) concluded that excess HCO$_3^-$ probably was produced by evaporative concentration. Although the TDS of the 1912 lake was considerably higher than that of the present lake, the TDS of the lake was never very high, because if it had been, endemic fish and ostracode populations would have died out and saline minerals would have precipitated. In a region of warm summers, net evaporation, and little surface-water inflow (prior to Bear River diversion), lake level in the past varied considerably (e.g., Laabs and Kaufman, 2003; Reheis et al., this volume; Smoot and Rosenbaum, this volume) but the regional hydrology must have supplied the lake with a constant input of snowmelt-derived groundwater.

Qualitatively, the water chemistry is consistent with the abundance of Paleozoic carbonate rocks (predominantly dolomite) in the drainage basin, especially on the west side of the lake. Quantitatively, however, Bear Lake water does not reflect the chemistry of surface-water inputs (Dean et al., 2007). The present-day composition of Bear Lake more closely resembles the composition expected when old Bear Lake water (pre-diversion) is mixed with Bear River water introduced through the canals in the early twentieth century. In terms of the major ions (bicarbonate, Ca^{2+}, and Mg^{2+}), present Bear Lake is intermediate in composition between Bear River and 1912 lake water (Figs. 8A and B; Table 1).

Bicarbonate, Ca^{2+}, and Mg^{2+} are reactive ions in the lake due to precipitation of carbonate minerals, whereas the other major dissolved ions show conservative or near-conservative behavior (Dean et al., 2007). Bicarbonate and Ca^{2+} are lost to precipitation of aragonite (pre-diversion) or calcite (post-diversion; see discussion below). Magnesium is lost to solid solution in the carbonate minerals.

The relative importance of the three solute sources to Bear Lake hydrochemistry is shown with a plot of the ^{87}Sr/^{86}Sr values versus Sr concentrations (Fig. 8C). West side sources have high ^{87}Sr/^{86}Sr values, but low Sr concentrations, whereas the Bear River and east side sources have low ^{87}Sr/^{86}Sr values and high Sr concentrations. Bear Lake's ^{87}Sr/^{86}Sr values fall in between those of its sources (Fig. 8C), but indicate that there must be a significant input of west-side waters in order to compensate for their low Sr concentrations.

The lake is saturated at all depths with respect to calcite, aragonite, and dolomite (Dean et al., 2007). The high TDS and Mg^{2+}:Ca^{2+} of the lake in 1912 should have favored the precipitation of CaCO$_3$ as aragonite (Morse and Mackenzie, 1990). Sediment-core studies show that the pre-1912 sediments deposited over the past 7000 years consist of ~80% aragonite and minor low-Mg calcite, quartz, and dolomite (Dean et al., 2006; Dean, this volume). Sediment-trap studies indicate that precipitation of CaCO$_3$ occurs in the epilimnion during the late spring and summer (April through September) as high-Mg calcite. However, sediment traps placed 2 m above the bottom in 40 m water depth show that the sediment that is accumulating on the bottom of the lake today consists predominantly of aragonite, in addition to high-Mg calcite, low-Mg calcite, quartz, and minor dolomite (Dean et al., 2007; Dean, this volume). Because so little high-Mg calcite is being incorporated in sediment on the lake floor today, the dominant $CaCO_3$ mineral (aragonite) must be aragonite that was precipitated at least 50 years ago. This depositional pattern could be explained by erosion, reworking, and "focusing" of sediment into deeper water (Dean et al., 2006; Dean, this volume).

Nutrients

Bioassay experiments have shown that algal growth in Bear Lake is usually limited by nitrogen (Wurtsbaugh, 1988) as it is in many western lakes that have not suffered from excessive agricultural or atmospheric deposition of this nutrient (Stoddard, 1994). An important dissolved inorganic form of nitrogen utilized by phytoplankton is nitrate (NO_3^-). Prior to 1997, the average epilimnetic nitrate concentration in Bear Lake was 16 µg L^{-1}, but often was below the level of detection (1 µg L^{-1}). Beginning in 1997, the nitrate concentration increased by more than an order of magnitude (Fig. 9B) when high spring runoff flushed large quantities of total inorganic nitrogen into Bear Lake from the marsh and pasture lands north of the lake.

The nutrient loading in Bear Lake is mainly from Bear River. For example, Birdsey (1989) estimated that 60%–80% of phosphorus delivered to Bear Lake is from Bear River. Phosphorus may at times limit phytoplankton growth in the lake. At the high pH levels in Bear Lake (average surface pH from 1989 to 2004 was 8.43, which changed little with depth), phosphorus can precipitate as calcium phosphate (hydroxyapatite), and, more commonly, it can coprecipitate with $CaCO_3$ and/or adsorb onto $CaCO_3$ crystals (Otsuki and Wetzel, 1972; Wetzel, 2001). Coprecipitation with $CaCO_3$ can markedly decrease the amount of phosphorus available for phytoplankton, and this may limit algal productivity in Bear Lake, helping to keep it oligotrophic (Birdsey, 1985, 1989).

Prior to 1983, the total phosphorus concentration in Bear Lake was low (usually <10 µg L^{-1}; Fig. 9C) with little buildup in the hypolimnion during summer stratification, which is characteristic of oligotrophic lakes. However, the phosphorus concentration has changed considerably over the past 25 years (Fig. 9C). The total phosphorus (TP) concentration began to increase in 1983, peaked in the early 1990s at >20 µg L^{-1}, and then began to decline. However, the TP concentration increased again in the late 1990s. The most significant form of phosphorus for plant growth is soluble inorganic phosphorus (orthophosphate, PO_4^{3-}), and it is often the limiting nutrient in lakes. Prior to 1987, the concentration of orthophosphate (OP) was low (average of ~2 µg L^{-1}), but since 1987 the average OP concentration has doubled (4 µg L^{-1}), peaking in the early 1990s along with TP (Fig. 9C).

BIOLOGICAL PROPERTIES

Phytoplankton

Bear Lake is oligotrophic, with an average (±1 standard deviation) chlorophyll *a* concentration at the surface between 1980 and 1998 of 0.53 (±0.39) µg L^{-1}. Peaks in the concentration of chloro-

phyll *a* >1.0 µg L^{-1} (and as high as 5.5 µg L^{-1}) occurred throughout the water column in Bear Lake in 1999, 2000, and 2004 (Fig. 9A). This suggests that algal blooms are becoming more common in Bear Lake. In April 1999 there was an unusually large algal bloom in Bear Lake, marked by unusually high chlorophyll *a* concentrations (Fig. 9A). Sedimentary evidence for this bloom was captured in a sediment trap in a water depth of 10 m (Dean et al., 2007; Dean, his volume). Chlorophyll *a* concentrations during the summer growing season generally are higher in the metalimnion than in the epilimnion (Fig. 10). The nominal base of the photic zone (1% light intensity) ranges from 15 to 25 m, so that there is sufficient light in the metalimnion for photosynthesis, which is responsible for the metalimnetic O_2 maximum.

The phytoplankton in Bear Lake have not been studied extensively. Diatoms are the most abundant taxa, and some information about their abundance is presented by Moser and Kimball (this volume). Birdsey (1989) reported that diatoms constituted ~80% of the algal abundance.

Zooplankton

Total macrozooplankton densities are low, with seasonal peaks usually of <10 individual crustaceans L^{-1} (Wurtsbaugh and Luecke, 1997). Numerically, the community is usually dominated by the cladoceran *Bosmina longirostris*, the copepod *Epischura nevadensis,* and the colonial rotifer *Conochilus unicornis*, and occasionally by *Daphnia* spp. However, *Daphnia* often dominates the community when biomass rather than numerical densities are considered (Wurtsbaugh and Luecke, 1997). There were large blooms of *Daphnia pulex* and *Daphnia galeata* in 1995 and 1996. *Daphnia* in Bear Lake may reside on the benthic sediments during the day but move into the water column at night to feed. Therefore the dominance of *Daphnia* in the zooplankton community in those years could possibly reflect a change in the magnitude of daily vertical migration (Wurtsbaugh and Luecke, 1997).

Macrozooplankton abundance is highly seasonal, with biomass minimums in winter less than 20% of those in summer. This undoubtedly decreases grazing rates on the phytoplankton during the winter, and may explain the higher chlorophyll concentrations during that period. The seasonality and compositional changes of the zooplankton will also contribute to temporal variation in the sedimentation rate. High zooplankton grazing can effectively remove phytoplankton from the epilimnion (Lampitt et al., 1990; Pilati and Wurtsbaugh, 2003), and also increase the flux of carbonate particles because these are also ingested by some zooplankton and excreted as fecal pelets (Vanderploeg, 1981; Honjo, 1996).

Benthic Invertebrates

Little research has been done on the benthic invertebrates in Bear Lake. Erman and Helm (1971) studied community composition of the invertebrates and Wurtsbaugh and Hawkins (1990) described spatial and temporal variations of the macrobenthos that are important for fish feeding. Chironomid larvae are dominant at depths <30m and are partially replaced in deeper strata by ostracodes and oligochaetes. The ostracodes in depths >10 m are all endemic species (R. Forester, 2005, personal commun.; Bright et al., 2006; Bright, this volume, Chapter 8). Biomass is lowest in winter, and more than doubles by late summer or fall. In the one year it was studied, the mean annual biomass of macroinvertebrates decreased from 0.8 g m^{-2} in the littoral zone to less than 0.15 g m^{-2} at 50 m. Overall, the mean benthic invertebrate biomass of 0.34 g m^{-2} is among the lowest recorded for any lake (Wurtsbaugh and Hawkins, 1990). This is in part because primary production in the lake is low, but also because the soft marl

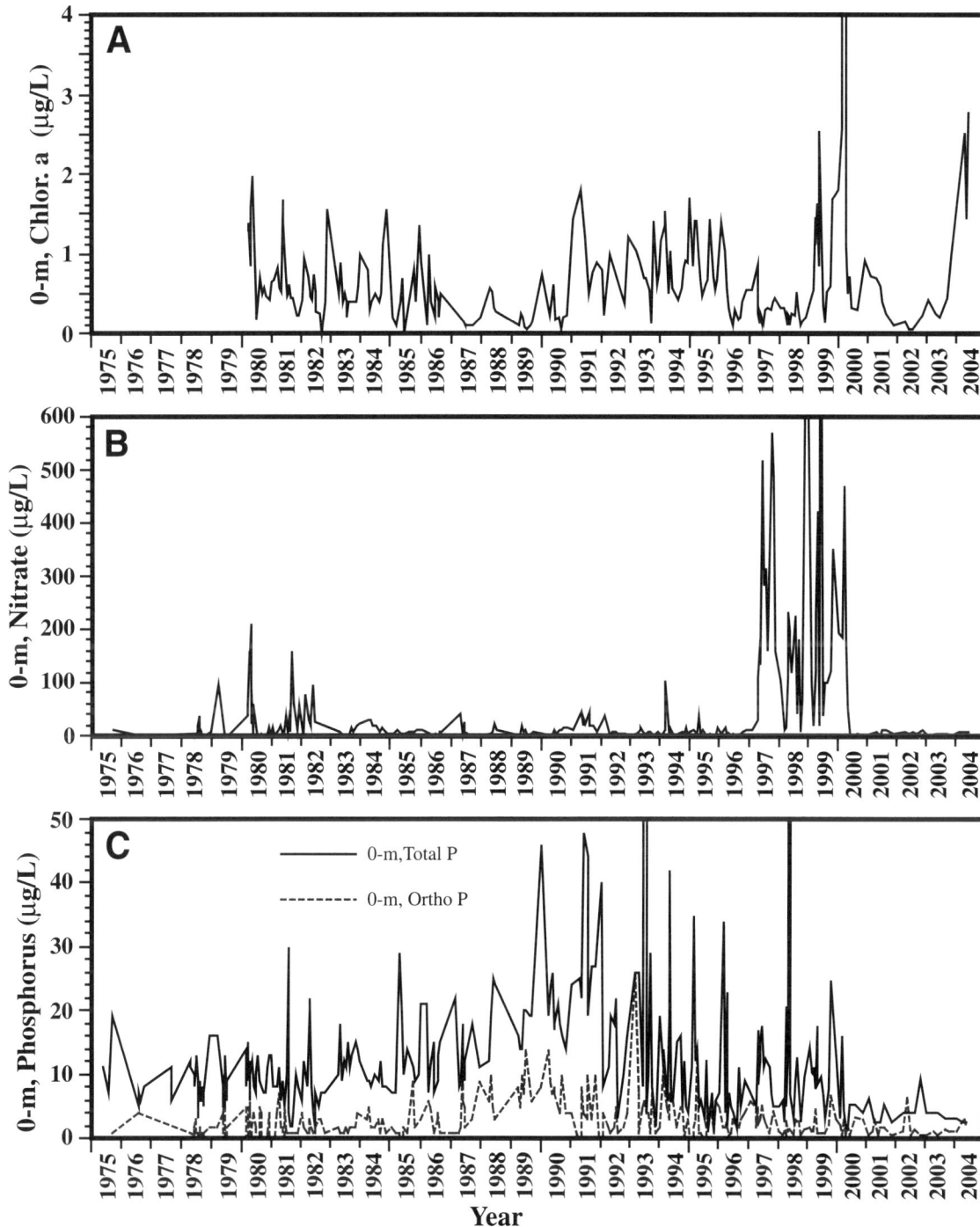

Figure 9. Concentrations of chlorophyll *a* (A), nitrate (B), and total phosphorus and orthophosphate (C) in Bear Lake at a depth of 0 m from 1975 to 2004.

sediments that dominate most of the bottom do not provide good habitats for invertebrates.

Bioturbation of the sediments by benthic invertebrates likely occurs (e.g., Martin et al., 2005), but the magnitude of disturbance may be limited because the low-organic contents of the substrate may force the invertebrates to stay close to the surface.

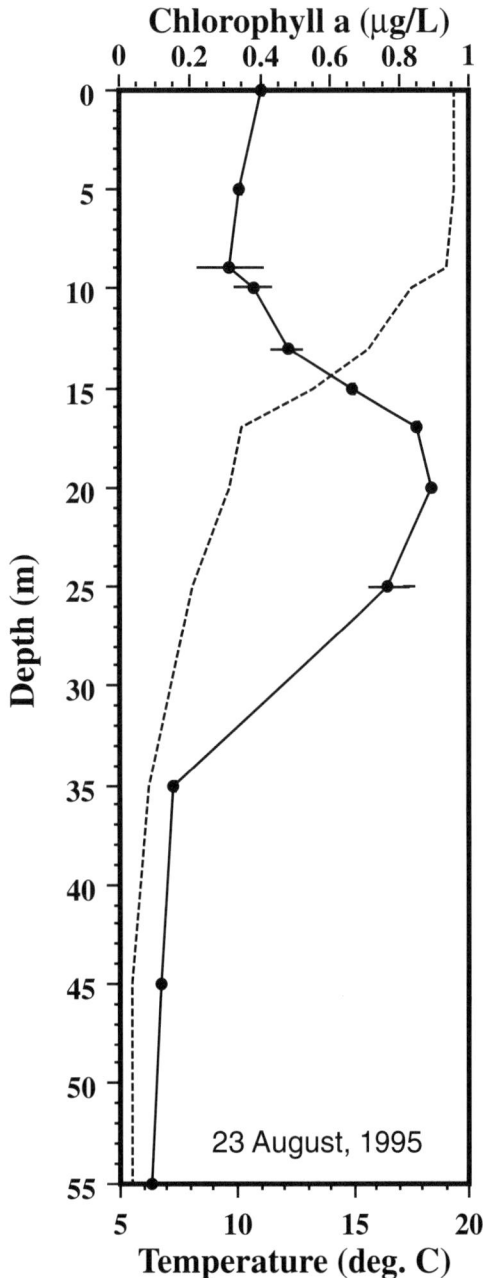

Figure 10. Depth profiles of chlorophyll *a* concentrations (solid line) and temperature (dashed line) in Bear Lake on 23 August 1995, showing the prominent deep chlorophyll layer in the metalimnion. Ranges of chlorophyll *a* concentrations are shown as bars when larger than the data points (solid circles).

Fish

Bear Lake has 13 species of fish, with four endemic species (Bonneville cisco, *Prosopium gemmifer*; Bear Lake whitefish, *Prosopium abyssicola;* Bonneville whitefish, *Prosopium spilonotus;* and Bear Lake sculpin, *Cottus extensus*). Utah suckers (*Catastromus ardens*) along with the two whitefish and sculpin are abundant and likely provide significant bioturbation of the sediments in their feeding activities at all depths in the lake. With the exception of the lake trout (*Salvelinus namaychus*), introduced species are rare. Only 1% of the fish captured in an intensive study of the lake were introduced (Wurtsbaugh and Hawkins, 1990).

The greatest biomass of fish in the lake is associated with the benthic zone, and during the summer many species are located in the littoral zone or where the metalimnion intersects the lake bottom (Wurtsbaugh and Hawkins, 1990). The metalimnion intersect provides a range of optimal temperatures for different species and often has the highest benthic invertebrate abundances. Larval sculpin living in the profundal zone of the lake undergo daily vertical migrations to the warmer metalimnion to increase digestion and growth rates (Wurtsbaugh and Neverman, 1988). The abundant sculpin also undergo an ontogenetic shift in distribution from the profundal zone when they are small, to the warmer littoral zone where growth rates are higher (Ruzycki and Wurtsbaugh, 1999).

Most fish in the lake feed primarily on benthic chironomid larvae, ostracodes, and other benthic invertebrates (Wurtsbaugh and Hawkins, 1990). Subadult cutthroat trout and the less abundant redside shiners (*Richardsonius balteatus*) rely extensively on terrestrial insects for food. Only the highly specialized zooplanktivore, the Bonneville cisco, feeds on the sparse zooplankton population (Wurtsbaugh and Hawkins, 1990). The top of the food web is dominated by the two important sport fish, adult Bonneville cutthroat trout (*Oncorhynchus clarki Utah*) and the introduced lake trout that feed initially on sculpin, but shift to preying on cisco and whitefish when they attain lengths greater than 400 mm (Ruzycki et al., 2001). The low yield of sport fish in the lake (~0.5 kg ha^{-1}y^{-1}; Nielson and Birdsey, 1989) is consistent with the low primary productivity and availability of invertebrate prey (Wurtsbaugh and Hawkins, 1990), but a lack of rock substrates may also limit habitat for fish spawning and rearing (Bouwes and Luecke, 1997; Ruzycki et al., 1998).

SUMMARY

Bear Lake is an old, alkaline, oligotrophic lake with endemic fish and ostracodes. The lake has no natural direct outlet, but for the last century the level of the lake has been artificially controlled by a series of canals connected at the north end to the Bear River. Most of the precipitation in the drainage basin falls in the winter and spring from Pacific storms so that most of the water entering the lake is from snowmelt. As a result of marked seasonal changes in atmospheric circulation over the eastern North Pacific, the Bear Lake region experiences hot, dry summers and cold, wet winters. This winter-wet, summer-dry moisture regime results in net evaporation, which keeps the lake saturated with carbonate minerals. The precipitation of large quantities of $CaCO_3$ coprecipitates phosphate, helping to keep the lake oligotrophic. Today the precipitated $CaCO_3$ is in the form of high-Mg calcite, but before the introduction of lower-salinity Bear River water, the lake precipitated aragonite. Suspended and colloidal particles of $CaCO_3$ scatter light in the blue and green wavelengths, giving the lake its characteristic blue color. Most of the supply of surface water to the lake is from west-side streams that contain relatively high concentrations of calcium, magnesium, and bicarbonate derived mainly from Paleozoic carbonate rocks in the Bear River Range west of the lake, and qualitatively the present water chemistry is consistent with this supply of solutes. However, strontium isotope studies show that present lake water more closely resembles the composition expected when old Bear Lake water (pre-diversion) is mixed with Bear River water introduced through the canals in the early twentieth century. Primary productivity is colimited by phosphate and nitrate. Secondary productivity by zooplankton also is low. Four of 13 fish species in the lake are endemic, and all ostracodes in water deeper than 10 m are endemic, which indicates that the lake is old. The greatest fish biomass is associated with the benthic zone where most fish feed on benthic invertebrates.

ACKNOWLEDGMENTS

Funding for Dean was provided by the U.S. Geological Survey–Earth Surface Dynamics Program. Funding for much of the limnological work (to Wurtsbaugh) was provided by the Utah Division of Wildlife Resources and the Ecology Center at Utah State University. We thank Mitch Poulsen, Bear Lake Regional Commission, and Connely Baldwin, PacifiCorp, for researching the early records of the Bear River diversion. Connely Baldwin also provided the data for lake elevation and ice cover. Funding for much of the water chemistry (to Lamarra) was provided by the Bear Lake Regional Commission. We thank Chris Luecke, Charles Hawkins, James Ruzycki, and Nick Bouwes for assistance in the field and insightful discussions. We thank Lesleigh Anderson, Kirsten Menking, Jim Russell, Tom Winter, and one anonymous reviewer for helpful reviews of an earlier draft of this paper.

REFERENCES CITED

American Public Health Association, 1992, Standard methods for the examination of water and wastewater (18th edition): Washington, D.C.

Birdsey, P.W., 1985, Coprecipitation of phosphorus with calcium carbonate in Bear Lake, Utah [M.S. thesis]: Logan, Utah State University, 122 p.

Birdsey, P.W., 1989, The limnology of Bear Lake, Utah-Idaho, 1912–1988: A literature review: Utah Department of Natural Resources, Division of Wildlife Resources, Publication 89-5, 113 p.

Bouwes, N., and Luecke, C., 1997, The fate of Bonneville cisco eggs in Bear Lake: Evaluating mechanisms of egg loss: Transactions of the American Fisheries Society, v. 126, p. 240–247, doi: 10.1577/1548-8659(1997)126 <0240:TFOBCE>2.3.CO;2.

Bright, J., 2009, this volume, Chapter 4, Isotope and major-ion chemistry of groundwater in Bear Lake Valley, Utah and Idaho, with emphasis on the Bear River Range, *in* Rosenbaum, J.G., and Kaufman, D.S., eds., Paleoenvironments of Bear Lake, Utah and Idaho, and its catchment: Geological Society of America Special Paper 450, doi: 10.1130/2009.2450(04).

Bright, J., 2009, this volume, Chapter 8, Ostracode endemism in Bear Lake, Utah and Idaho, *in* Rosenbaum, J.G., and Kaufman, D.S., eds., Paleoenvironments of Bear Lake, Utah and Idaho, and its catchment: Geological Society of America Special Paper 450, doi: 10.1130/2009.2450(08).

Bright, J., Kaufman, D.S., Forester, R.M., and Dean, W.E., 2006, A continuous 250,000 yr record of oxygen and carbon isotopes in ostracode and bulk-sediment carbonate from Bear Lake, Utah-Idaho: Quaternary Science Reviews, v. 25, p. 2258–2270, doi: 10.1016/j.quascirev.2005.12.011.

Colman, S.M., 2006, Acoustic stratigraphy of Bear Lake, Utah-Idaho—Late Quaternary sedimentation in a simple half-graben: Sedimentary Geology, v. 185, p. 113–125, doi: 10.1016/j.sedgeo.2005.11.022.

Dean, W.E., 2009, this volume, Endogenic carbonate sedimentation in Bear Lake, Utah and Idaho, over the last two glacial-interglacial cycles, *in* Rosenbaum, J.G., and Kaufman, D.S., eds., Paleoenvironments of Bear Lake, Utah and Idaho, and its catchment: Geological Society of America Special Paper 450, doi: 10.1130/2009.2450(07).

Dean, W., Forester, R., Colman, S., Liu, A., Skipp, G., Simmons, K, Swarzenski, P., Anderson, R., and Thornburg, D., 2005, Modern and glacial-Holocene carbonate sedimentation in Bear Lake, Utah and Idaho: U.S. Geological Survey Open-File Report 2005-1124, http://pubs.usgs.gov/of/2005/1124 (accessed 2005).

Dean, W.E., Rosenbaum, J.G., Forester, R.M., Colman, S.M., Bischoff, J.L., Liu, A., Skipp, G., and Simmons, K., 2006, Glacial to Holocene evolution of sedimentation in Bear Lake, Utah-Idaho: Sedimentary Geology, v. 185, p. 93–112, doi: 10.1016/j.sedgeo.2005.11.016.

Dean, W.E., Forester, R., Bright, J., and Anderson, R., 2007, Influence of the diversion of Bear River into Bear Lake (Utah and Idaho) on the environment of deposition of carbonate minerals: Evidence from water and sediments: Limnology and Oceanography, v. 52, p. 1094–1111.

Erman, D.C., and Helm, W.T., 1971, Comparison of some species importance values and ordination techniques used to analyze benthic invertebrate communities: Oikos, v. 22, p. 240–247, doi: 10.2307/3543733.

Hintze, L.F., 1973. Geologic history of Utah: Brigham Young University Geology Studies 20-3, 181 p.

Holm-Hansen, O., and Riemann, B., 1978, Chlorophyll-*a* determination: Improvements in methodology: Oikos, v. 30, p. 438–447, doi: 10.2307/3543338.

Honjo, S., 1996, Fluxes of particles to the interior of the open oceans, *in* Ittekkot, V., Schafer, S., Honjo, S., and Depetris, P.J., eds., Particle flux in the ocean: Chichester, UK, Wiley, p. 91–154.

Kaliser, B.N., 1972, Environmental geology of Bear Lake area, Rich County, Utah: Utah Geological and Mineralogical Survey Bulletin 96, 32 p.

Kemmerer, G., Bovard, J.F., and Boorman, W.R., 1923, Northwestern lakes of the United States; biological and chemical studies with reference to possibilities to production of fish: U.S. Bureau of Fisheries Bulletin, v. 39, p. 51–140.

Laabs, B.J.C., and Kaufman, D.S., 2003, Quaternary highstands in Bear Lake valley, Utah and Idaho: Geological Society of America Bulletin, v. 115, p. 463–478, doi: 10.1130/0016-7606(2003)115<0463:QHIBLV>2.0.CO;2.

Lamarra, V., Liff, C., and Carter, J., 1986, Hydrology of Bear Lake Basin and its impact on the trophic state of Bear Lake, Utah-Idaho: The Great Basin Naturalist, v. 46, p. 690–705.

Lampitt, R.S., Noji, T., and Von Bodungen, B., 1990, What happens to zooplankton faecal pellets? Implication for material flux: Marine Biology (Berlin), v. 104, p. 15–23, doi: 10.1007/BF01313152.

Martin, P., Boes, X., Goddeeris, B., and Fagel, N., 2005, A qualitative assessment of the influence of bioturbation in Lake Baikal sediments: Global and Planetary Change, v. 46, p. 87–99, doi: 10.1016/j.gloplacha.2004.11.012.

Mock, C.J., and Bartlein, P.J., 1995, Spatial variability of late-Quaternary paleoclimates in the western United States: Quaternary Research, v. 44, p. 425–433, doi: 10.1006/qres.1995.1087.

Mock, C.J., and Brunelle-Daines, A.R., 1999, A modern analogue of western United States summer palaeoclimate at 6000 years before present: The Holocene, v. 9, p. 541–545, doi: 10.1191/095968399668724603.

Morse, J.W., and Mackenzie, F.T., 1990, Developments in sedimentology No. 48: Geochemistry of Sedimentary Carbonates: Amsterdam, Elsevier, 707 p.

Moser, K.A., and Kimball, J.P., 2009, this volume, A 19,000-year record of hydrologic and climatic change inferred from diatoms from Bear Lake, Utah and Idaho, *in* Rosenbaum, J.G., and Kaufman, D.S., eds., Paleoenvironments of Bear Lake, Utah and Idaho, and its catchment: Geological Society of America Special Paper 450, doi: 10.1130/2009.2450(10).

Namias, J., Yuan, X., and Cayan, D.R., 1988, Persistence of North Pacific sea surface temperature and atmospheric flow patterns: Journal of Climate, v. 1, p. 682–703, doi: 10.1175/1520-0442(1988)001<0682:PONPSS>2.0.CO;2.

Neverman, D., and Wurtsbaugh, W.A., 1992, Visual feeding of juvenile Bear Lake sculpin: Transactions of the American Fisheries Society, v. 121, p. 395–398, doi: 10.1577/1548-8659(1992)121<0395:VFBJBL>2.3.CO;2.

Nielson, B.R., and Birdsey, P.W., 1989, Bear Lake cutthroat trout enhancement program 1989: Salt Lake City, Utah Division of Wildlife Resources, Annual Report, Federal Aid in Fish Restoration, F-026-R-14, 46 p.

Otsuki, A., and Wetzel, R.G., 1972, Coprecipitation of phosphate with carbonates in a marl lake: Limnology and Oceanography, v. 17, p. 763–767.

Piechota, T., Timilsena, J., Tootle, G., and Hidalgo, H., 2004, The western U.S. drought: How bad is it?: Eos (Transactions, American Geophysical Union), v. 85, p. 301–308, doi: 10.1029/2004EO320001.

Pilati, A., and Wurtsbaugh, W.A., 2003, Importance of zooplankton for the persistence of a deep chlorophyll layer: A limnocorral experiment: Limnology and Oceanography, v. 48, p. 249–260.

Reheis, M.C., Laabs, B.J.C., and Kaufman, D.S., 2009, this volume, Geology and geomorphology of Bear Lake Valley and the upper Bear River, Utah and Idaho, in Rosenbaum, J.G., and Kaufman, D.S., eds., Paleoenvironments of Bear Lake, Utah and Idaho, and its catchment: Geological Society of America Special Paper 450, doi: 10.1130/2009.2450(01).

Ruzycki, J., and Wurtsbaugh, W.A., 1999, Ontogenetic habitat shifts of juvenile Bear Lake sculpin: Transactions of the American Fisheries Society, v. 128, p. 1201–1212, doi: 10.1577/1548-8659(1999)128<1201:OHSOJB>2.0.CO;2.

Ruzycki, J., Wurtsbaugh, W.A., and Lay, C., 1998, Reproductive ecology and early life history of a lacustrine sculpin, Cottus extensus (Teleostei, Cottidae): Environmental Biology of Fishes, v. 53, p. 117–127, doi: 10.1023/A:1007436502285.

Ruzycki, J.R., Wurtsbaugh, W.A., and Luecke, C., 2001, Salmonine consumption and competition for endemic prey fishes in Bear Lake, Utah-Idaho: Transactions of the American Fisheries Society, v. 130, p. 1175–1189, doi: 10.1577/1548-8659(2001)130<1175:SCACFE>2.0.CO;2.

Smoot, J.P., and Rosenbaum, J.G., 2009, this volume, Sedimentary constraints on late Quaternary lake-level fluctuations at Bear Lake, Utah and Idaho, in Rosenbaum, J.G., and Kaufman, D.S., eds., Paleoenvironments of Bear Lake, Utah and Idaho, and its catchment: Geological Society of America Special Paper 450, doi: 10.1130/2009.2450(12).

Stoddard, J.L., 1994, Long-term changes in watershed retention of nitrogen: Its causes and aquatic consequences, in L.A. Baker, ed., Environmental chemistry of lakes and reservoirs: Washington, D.C., American Chemical Society, ACS Advances in Chemistry Series no. 237, p. 223–284.

Strub, P.T., Allen, J.S., Huyer, A., and Smith, R.L., 1987, Seasonal cycles of currents, temperatures, winds, and sea level over the northeast Pacific continental shelf: 35° N to 48° N: Journal of Geophysical Research, v. 92, p. 1507–1526, doi: 10.1029/JC092iC02p01507.

Thompson, R.S., Whitlock, C., Bartlein, P.J., Harrison, S.P., and Spaulding, W.G., 1993, Climatic changes in the western United States since 18,000 yr B.P., in Wright, H.E., Jr., Kutzbach, J.E., Webb, T., Ruddiman, W.F., Street-Perrott, F.A., and Bartlein, P.J., eds., Global climates since the last glacial maximum: Minneapolis, Minnesota, University of Minnesota Press, p. 468–513.

Vanderploeg, H.A., 1981, Seasonal particle-size selection by Diaptomus sicilis in offshore Lake Michigan: Journal of the Fisheries Research Board of Canada, v. 38, p. 504–517.

Wetzel, R.G., 2001, Limnology: Lake and River ecosystems, 3rd ed.: San Diego, California, Academic Press, 1006 p.

Whitlock, C., Bartlein, P.J., and Watts, W.A., 1993, Vegetation history of Elk Lake, in Bradbury, J.P., and Dean, W.E., eds., Elk Lake, Minnesota: Evidence for Rapid Climate Change in the North-Central United States: Geological Society of America Special Paper 276, p. 251–274.

Wüest, A., and Lorke, A., 2003, Small-scale hydrodynamics in lakes: Annual Review of Fluid Mechanics, v. 35, p. 373–412, doi: 10.1146/annurev.fluid.35.101101.161220.

Wurtsbaugh, W.A., 1988, Iron, molybdenum, and phosphorus limitation of N_2 fixation maintains nitrogen deficiency of plankton in the Great Salt Lake drainage (Utah, USA): Verhandlungen der Internationalen Vereinigung für Theoretische und Angewandt Limnologie, v. 23, p. 121–130.

Wurtsbaugh, W.A., and Hawkins, C., 1990, Trophic interactions between fish and invertebrates in Bear Lake, Utah/Idaho: Logan, Utah, Utah State University Ecology Center Special Publication.

Wurtsbaugh, W.A., and Luecke, C., 1997, Examination of the abundance and spatial distribution of forage fish in Bear Lake (Utah/Idaho): Salt Lake City, Final Report of Project F-47-R, Study 5, to the Utah Division of Wildlife Resources, 217 p.

Wurtsbaugh, W.A., and Luecke, C., 1998, Limnological relationships and population dynamics of fishes in Bear Lake (Utah/Idaho): Salt Lake City, Final Report of Project F-47-R, Study 5, to the Utah Division of Wildlife Resources, 73 p.

Wurtsbaugh, W.A., and Neverman, D., 1988, Post-feeding thermotaxis and daily vertical migration in a larval fish: Nature, v. 333, p. 846–848, doi: 10.1038/333846a0.

MANUSCRIPT ACCEPTED BY THE SOCIETY 15 SEPTEMBER 2008

The Geological Society of America
Special Paper 450
2009

Geology and geomorphology of Bear Lake Valley and upper Bear River, Utah and Idaho

Marith C. Reheis

U.S. Geological Survey, MS 980 Federal Center, Denver, Colorado 80225, USA

Benjamin J.C. Laabs

Department of Geological Sciences, ISC 216, SUNY Geneseo, New York 14454, USA

Darrell S. Kaufman

Department of Geology, Northern Arizona University, Flagstaff, Arizona 86011, USA

ABSTRACT

Bear Lake, on the Idaho-Utah border, lies in a fault-bounded valley through which the Bear River flows en route to the Great Salt Lake. Surficial deposits in the Bear Lake drainage basin provide a geologic context for interpretation of cores from Bear Lake deposits. In addition to groundwater discharge, Bear Lake received water and sediment from its own small drainage basin and sometimes from the Bear River and its glaciated headwaters. The lake basin interacts with the river in complex ways that are modulated by climatically induced lake-level changes, by the distribution of active Quaternary faults, and by the migration of the river across its fluvial fan north of the present lake.

The upper Bear River flows northward for ~150 km from its headwaters in the northwestern Uinta Mountains, generally following the strike of regional Laramide and late Cenozoic structures. These structures likely also control the flow paths of groundwater that feeds Bear Lake, and groundwater-fed streams are the largest source of water when the lake is isolated from the Bear River. The present configuration of the Bear River with respect to Bear Lake Valley may not have been established until the late Pliocene. The absence of Uinta Range–derived quartzites in fluvial gravel on the crest of the Bear Lake Plateau east of Bear Lake suggests that the present headwaters were not part of the drainage basin in the late Tertiary. Newly mapped glacial deposits in the Bear River Range west of Bear Lake indicate several advances of valley glaciers that were probably coeval with glaciations in the Uinta Mountains. Much of the meltwater from these glaciers may have reached Bear Lake via groundwater pathways through infiltration in the karst terrain of the Bear River Range.

At times during the Pleistocene, the Bear River flowed into Bear Lake and water level rose to the valley threshold at Nounan narrows. This threshold has been modified by aggradation, downcutting, and tectonics. Maximum lake levels have decreased

Reheis, M.C., Laabs, B.J.C., and Kaufman, D.S., 2009, Geology and geomorphology of Bear Lake Valley and upper Bear River, Utah and Idaho, *in* Rosenbaum, J.G., and Kaufman D.S., eds., Paleoenvironments of Bear Lake, Utah and Idaho, and its catchment: Geological Society of America Special Paper 450, p. 15–48, doi: 10.1130/2009.2450(02). For permission to copy, contact editing@geosociety.org. ©2009 The Geological Society of America. All rights reserved.

from as high as 1830 m to 1806 m above sea level since the early Pleistocene due to episodic downcutting by the Bear River. The oldest exposed lacustrine sediments in Bear Lake Valley are probably of Pliocene age. Several high-lake phases during the early and middle Pleistocene were separated by episodes of fluvial incision. Threshold incision was not constant, however, because lake highstands of as much as 8 m above bedrock threshold level resulted from aggradation and possibly landsliding at least twice during the late-middle and late Pleistocene. Abandoned stream channels within the low-lying, fault-bounded region between Bear Lake and the modern Bear River show that Bear River progressively shifted northward during the Holocene. Several factors including faulting, location of the fluvial fan, and channel migration across the fluvial fan probably interacted to produce these changes in channel position.

Late Quaternary slip rates on the east Bear Lake fault zone are estimated by using the water-level history of Bear Lake, assuming little or no displacement on dated deposits on the west side of the valley. Uplifted lacustrine deposits representing Pliocene to middle Pleistocene highstands of Bear Lake on the footwall block of the east Bear Lake fault zone provide dramatic evidence of long-term slip. Slip rates during the late Pleistocene increased from north to south along the east Bear Lake fault zone, consistent with the tectonic geomorphology. In addition, slip rates on the southern section of the fault zone have apparently decreased over the past 50 k.y.

INTRODUCTION

Sediments deposited in Bear Lake, on the Idaho-Utah border (Figs. 1 and 2), provide a record of changing inputs of water and sediment derived from voluminous calcium-bicarbonate-charged groundwater discharge and from the Bear River. In addition to these variable water and sediment sources, the river has interacted with the lake basin in complex ways that are modulated by lake-level fluctuations, by the distribution of active Quaternary faults, and by migration of the river across its fluvial fan north of the present lake. An understanding of the timing and magnitude of all these changes is required to unravel the environmental record preserved in the lake sediment.

This chapter uses published literature to review the bedrock and pre-Quaternary tectonic setting of Bear Lake Valley and the upper Bear River in the context of their influence on sediment supply and groundwater discharge to the lake. We then discuss and summarize recently published and new information on the distribution and relative age of Quaternary glacial and fluvial deposits along the Bear River upstream of Bear Lake and the implications for the timing of sediment transport by Bear River. New mapping of moraines near the crest of the Bear River Range (west of Bear Lake; Fig. 2) improves the understanding of the former extent of glaciers and their potential influence on the lake.

At intervals when the Bear River flowed directly into Bear Lake, the lake would tend to overtop its sill to the north en route to Lake Bonneville (Fig. 1). Without the input of the Bear River, the lake would fluctuate within a topographically (but not hydrologically) closed basin. This relationship is controlled by several factors, including (1) climatic changes that cause the lake to transgress and intersect the Bear River, or to retreat into a closed basin; (2) active tectonics, especially along the east Bear Lake fault zone (Fig. 2); (3) migration of Bear River across its flu-

vial fan where it debouches into the valley; and (4) aggradation and incision that affect the altitude of the basin's threshold. The relationship between the Bear River and Bear Lake during the late Quaternary is documented by preserved shoreline deposits at various altitudes (dated by radiocarbon, amino acid racemization, and tephrochronologic techniques) and by river terraces. We present new information on the progressive migration of Bear River northward away from Bear Lake in the past 8000 years based on studies of sediments preserved in abandoned channels. Finally, we discuss new information on displacements and slip rates on valley-bounding normal faults based on the dated lacustrine and marsh deposits, tectonic geomorphology, and seismic data of Colman (2006).

TECTONIC SETTING AND BEDROCK GEOLOGY

The upper Bear River flows northward for ~150 km from its headwaters in glaciated valleys on the north flank of the Uinta Mountains (Fig. 1), generally following the strike of regional structures in the Laramide overthrust belt and on the northeastern margin of the Basin and Range province. Along this margin is a northeasterly trending zone of right-stepping normal faults controlling fault-block mountains and basins that developed in response to east-west Neogene extension superimposed on Cretaceous to early Tertiary folds and thrust faults (Armstrong, 1968; McCalpin, 1993) south of the Snake River Plain. The normal faults developed along the same strikes as the older thrust faults, and in some cases coincident with them. The northeasterly extensional pattern has been related to the probable influence of the Yellowstone hotspot (Anders et al., 1989; Pierce and Morgan, 1992; Parsons et al., 1994). Pierce and Morgan (1992) subdivided a parabola-shaped region of extension around Yellowstone and the Snake River Plain into "seismic belts" by using the frequency

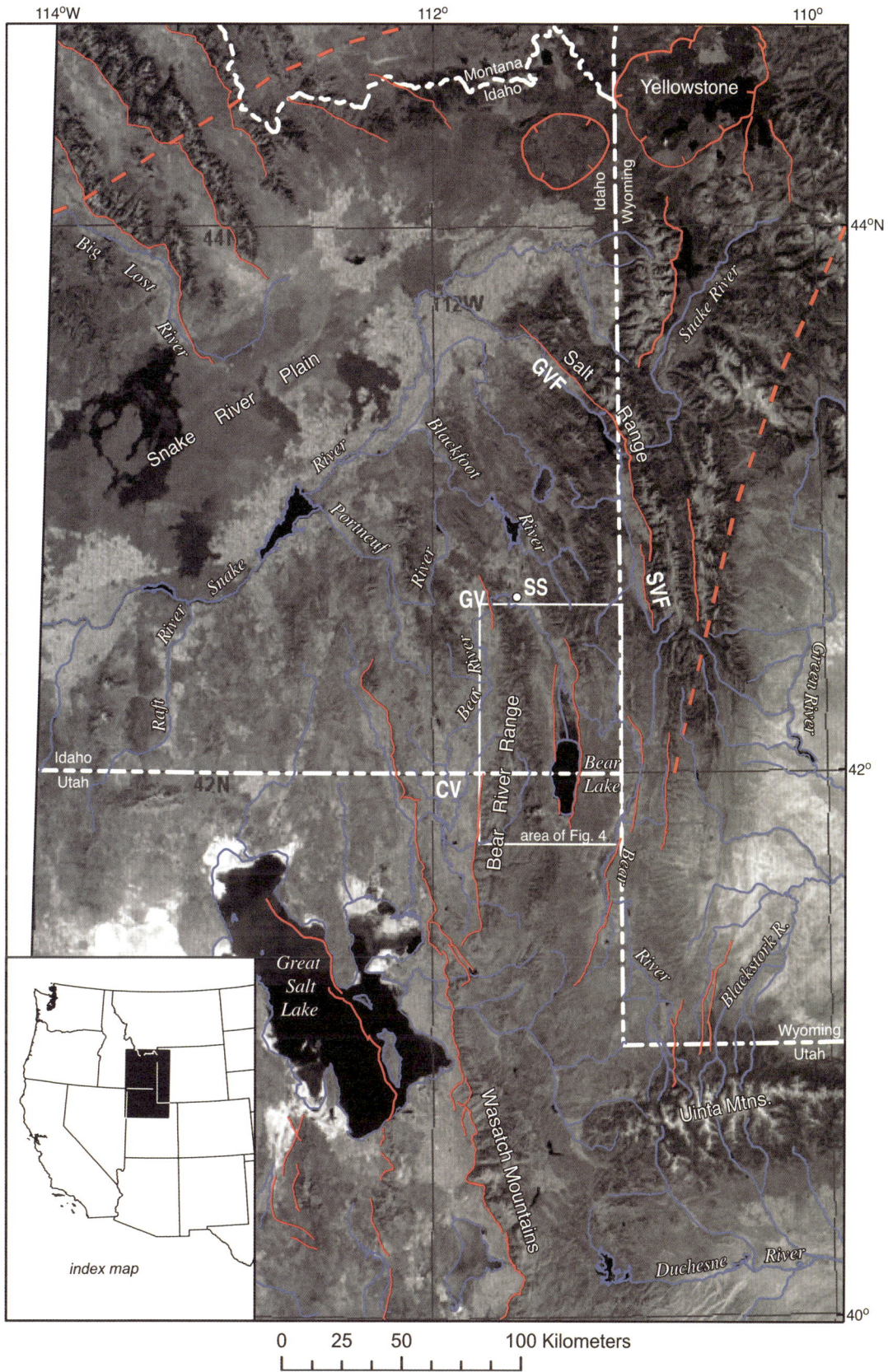

Figure 1. Location index map (inset) and regional setting, including drainages (blue) and principal Quaternary faults (red; modified from U.S. Geological Survey, 2004). Red hachured lines are boundaries of Quaternary calderas. Red dashed line is axis of high elevations within tectonic "parabola" of Pierce and Morgan (1992). CV—Cache Valley; GV—Gem Valley; GVF—Grand Valley fault; SS—Soda Springs; SVF—Star Valley fault. Box shows area of Figure 4.

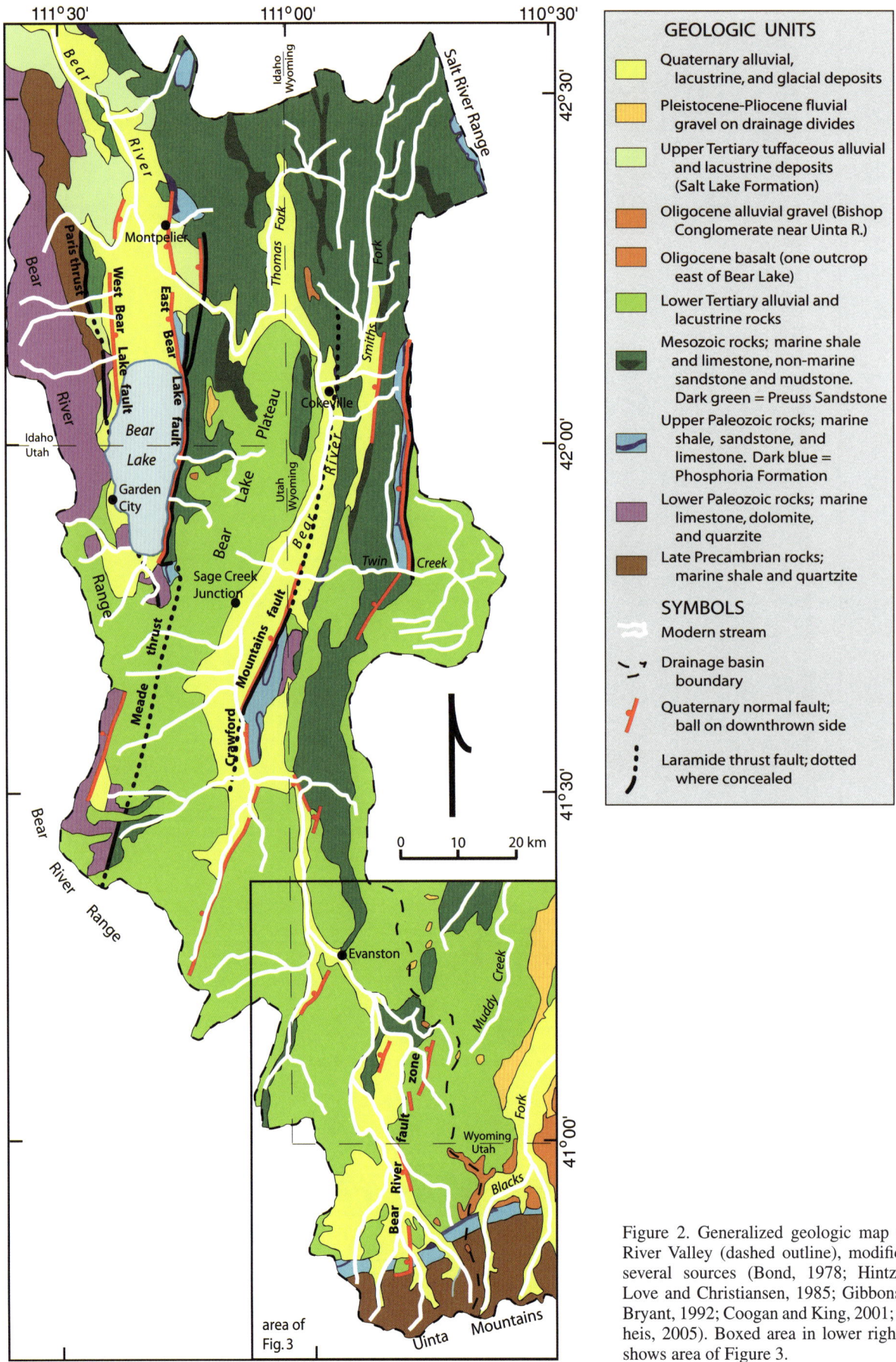

GEOLOGIC UNITS

- Quaternary alluvial, lacustrine, and glacial deposits
- Pleistocene-Pliocene fluvial gravel on drainage divides
- Upper Tertiary tuffaceous alluvial and lacustrine deposits (Salt Lake Formation)
- Oligocene alluvial gravel (Bishop Conglomerate near Uinta R.)
- Oligocene basalt (one outcrop east of Bear Lake)
- Lower Tertiary alluvial and lacustrine rocks
- Mesozoic rocks; marine shale and limestone, non-marine sandstone and mudstone. Dark green = Preuss Sandstone
- Upper Paleozoic rocks; marine shale, sandstone, and limestone. Dark blue = Phosphoria Formation
- Lower Paleozoic rocks; marine limestone, dolomite, and quarzite
- Late Precambrian rocks; marine shale and quartzite

SYMBOLS

- Modern stream
- Drainage basin boundary
- Quaternary normal fault; ball on downthrown side
- Laramide thrust fault; dotted where concealed

Figure 2. Generalized geologic map of Bear River Valley (dashed outline), modified from several sources (Bond, 1978; Hintz, 1980; Love and Christiansen, 1985; Gibbons, 1986; Bryant, 1992; Coogan and King, 2001; and Reheis, 2005). Boxed area in lower right corner shows area of Figure 3.

of recent seismic activity and the height of normal-fault scarps in each valley.

Bear Lake Valley lies within the most active seismic zone on the northeast-southwest–trending southern arm of the parabola (Fig. 1). The valley forms a complex graben between two normal faults, informally named the east and west Bear Lake fault zones (Fig. 2; McCalpin, 1993, 2003). Several reaches of the Bear River and its tributaries coincide with Pleistocene normal faults (Fig. 2). Northeast of Bear Lake, the river bends sharply west and crosses folds and faults of the Bear Lake Plateau and the Preuss Range at a nearly right angle to emerge into Bear Lake Valley, where it resumes a northward course parallel to structural trends. South of latitude 42°N, the upper Bear River lies within a less conspicuous zone of faulting that includes faults with Holocene and late Pleistocene displacement but mostly lacking high, steep range fronts (Pierce and Morgan, 1992).

The Bear River drains a wide variety of rock types and ages (Fig. 2). Its headwaters lie in Precambrian orthoquartzite, siliceous sandstone, siltstone, and shale in the core of the Uinta Range (Bryant, 1992); detritus derived from these rocks by glacial erosion and rock flour production contributed much of the sediment load of the river during glacial periods (Rosenbaum et al., this volume; Rosenbaum and Heil, this volume). Downstream, the majority of exposed bedrock consists of Paleocene–Eocene conglomerate, sandstone, and shale (Bond, 1978; Hintz, 1980; Love and Christiansen, 1985). These rocks are poorly indurated and are probably easily erodible, especially during times when vegetation cover is reduced. The Tertiary sediments overlie complexly faulted and folded Paleozoic and Mesozoic sedimentary rocks, mainly exposed on the east and north sides of the river upstream of Bear Lake. The lower Paleozoic sequence is dominated by limestone and dolomite, and the upper Paleozoic rocks include marine sandstone and phosphate-bearing rocks. In contrast, the Mesozoic rocks in the drainage basin mainly consist of continental sandstones and shales.

The local basin of Bear Lake drains a much more restricted suite of rock types and ages (Fig. 2). Although upper Paleozoic and Mesozoic rocks crop out along the east side of the lake, there is little surface runoff from the small creeks that drain the flank of the Bear Lake Plateau. The majority of streams entering the basin drain extensive areas of lower Paleozoic limestone, dolomite, and quartzite west of the lake. Bear Lake Valley also contains locally thick deposits of upper Tertiary sediment (Salt Lake Formation) deposited during the early phases of extensional faulting, suggesting persistence of the valley as a fault-controlled basin probably since the Miocene. The most recent mapping restricts the Salt Lake Formation mainly to the area north of the lake (Dover, 1995), in contrast to earlier maps that showed this unit throughout the valley (Oriel and Platt, 1980). In general, the pre-Tertiary rocks are much more resistant to erosion than the Tertiary rocks and may contribute less detritus to the Bear River. However, the older rocks also contain readily soluble carbonate and evaporite units that likely contribute to the solute load of surface streams and groundwater. Closed depressions, sinkholes, and collapse

basins in the highlands of the Bear River Range west of Bear Lake (Dover, 1995) attest to wholesale subsurface solution of Paleozoic carbonate rocks, probably facilitated by the dense network of generally north-striking thrust and normal faults.

Structural Influences on Groundwater Flow

During much of the Holocene, Bear Lake received no surface water from Bear River (Dean, this volume; Rosenbaum et al., this volume), and the input was dominated by spring-fed streams originating in the Bear River Range west of the lake (Bright, this volume). The structural setting of Bear Lake and the surrounding highlands likely controls the flow paths of groundwater to these and other springs, but not in simple or obvious ways. North-striking, west-dipping splays of the Paris and Meade thrust faults crop out on the west and east sides of Bear Lake Valley, respectively (Fig. 2), and basin-and-range-style normal faults are thought to sole into the thrust faults at depth (Coogan, 1992; Dover, 1995; Evans et al., 2003). Farther east, the west-dipping Crawford fault, a Mesozoic thrust fault reactivated as a normal fault, similarly bounds the east side of the Bear River and normal faults complementary to the Crawford fault bound the west side of the river (Bear River graben of Dover, 1995). Thus, east of Bear Lake the structural configuration should inhibit Bear River–derived groundwater from reaching the lake, except by upward infiltration along normal faults that intersect the thrust faults at depth. Discharge and solute data indicate that springs on the east side of the lake contribute only minor amounts of water (Bright, this volume; Dean et al., 2007). Flows from springs on the west side are much larger but extrabasinal sources may also contribute to the lake (Bright, this volume).

One likely source of extrabasinal water is groundwater derived from recharge from the topographically high, southern part of the Bear River Range (Fig. 2) along the western boundary of the Bear River drainage basin. Paleozoic carbonate rocks in this divide area are overlain to the north by the Wasatch Formation and are displaced by normal faults with Tertiary to Quaternary displacement that sole into the Meade thrust fault. These normal faults are part of a continuous zone of anastomosing faults that extends directly to the south end of Bear Lake Valley and merges with faults west of Bear Lake. Groundwater that originates by infiltration through the carbonate rocks could be confined by the overlying Wasatch beds, permitting the normal faults to act as conduits for groundwater that originates outside the Bear Lake drainage basin. Because the rocks in this recharge area are carbonates, isotopic compositions of groundwater from this source would probably resemble those of groundwater from the Bear River Range within the Bear Lake drainage basin.

Faults clearly play important roles in controlling groundwater discharge on a local scale. Along the west side of Bear Lake, large and small springs are located on north-striking Tertiary normal faults (Oriel and Platt, 1980; Dover, 1995); in some cases these faults also offset Quaternary deposits (Reheis, 2005). Higher in the Bear River Range, parallel north-striking faults cut gently west-dipping, lower Paleozoic carbonate rocks (Oriel and Platt, 1980; Dover, 1995), and these faults likely provide conduits for snowpack recharge into carbonate-rock aquifers, as suggested by numerous closed depressions and sinkholes (Reheis, 2005). Dye tests indicate that groundwater flow paths in the Bear River Range cross topographic divides (Spangler, 2001). Stable isotope data on spring waters (Bright, this volume) support such a groundwater source for west-side springs. Along the east side of Bear Lake, isotopic data (Bright, this volume) indicate a sharp change in the relatively small-volume groundwater sources across the east Bear Lake fault zone, which soles into the Meade thrust (Dover, 1995; Evans et al., 2003). West of the Meade thrust, near-vertically dipping upper Paleozoic and lower Mesozoic sedimentary rocks, including the phosphate-rich Phosphoria Formation, are thought to underlie the east side of Bear Lake. Solutes derived from these formations may cause the sharp change in groundwater composition. A similar abrupt change in isotopic composition of surface water (Bright, this volume) occurs along the Bear River just upstream of Cokeville (Fig. 2). This change approximately coincides with the point where the river crosses the subsurface trace of the Crawford thrust (Rubey et al., 1980; M'Gonigle and Dover, 1992).

THE UPPER BEAR RIVER DRAINAGE BASIN

Tertiary and Quaternary deposits in the upper Bear River reveal a long history of drainage-basin evolution, glaciation, and landsliding in high basin-bounding ranges, and incision and deposition of fluvial and alluvial-fan deposits. Profound changes in drainage boundaries and directions occurred in this region around the Utah-Idaho-Wyoming borders during the late Cenozoic. These changes were driven by base-level fall due to the encroachment of basin-and-range extensional faulting from the west, and by the topographic changes caused by the progressive northeastward motion of the Yellowstone hotspot and the development of the eastern Snake River Plain (Fig. 1; Pierce and Morgan, 1992).

The oldest fluvial geomorphic features preserved in the study area are the remains of coalescing piedmont gravels composed of the Oligocene Bishop Conglomerate (Bradley, 1936) that extend northward from the Uinta Mountains into the Green River Basin. These gravels presently crop out on drainage divides in the headwaters of Bear River (Fig. 2) and are much more extensively preserved to the east (Bryant, 1992). In Oligocene time, much of the study area to the north of these outcrops was probably a gentle alluvial plain that capped the sedimentary fill of the older Wasatch and Green River Formations.

With the onset of extensional faulting and the development of large calderas in the central Snake River Plain (Fig. 1) during the Miocene, deposition of the Salt Lake Formation began in valleys along the southern margin of the Snake River Plain in southeastern Idaho, including what is now northern Bear Lake Valley (Fig. 2). On the basis of fossil snail faunas, Taylor and Bright (1987) suggested that the western part of the area had affinities

to western drainages, whereas the eastern part of the area had affinities to the south; if so, the northern Bear Lake Valley was still a drainage divide in the early(?) Miocene. Previous workers (Mansfield, 1927; Williams et al., 1962; Taylor and Bright, 1987) suggested that the Bear River attained its present course, particularly its westward jog across structure upstream of Bear Lake, in Pliocene time when the valleys were presumably filled with sediments of the Salt Lake Formation. More recent geologic mapping (Dover, 1995) has reinterpreted the sediments capping the Bear Lake Plateau as the Eocene Wasatch Formation rather than the younger Salt Lake Formation as mapped by Oriel and Platt (1980), and suggested that Salt Lake Formation sediments are thin or absent within the Bear Lake Valley south of the present Bear River. If this interpretation is correct, the Bear River could not have attained its present course into Bear Lake Valley by superposition across structures buried by Pliocene sediment.

The oldest deposits of what may have been a through-flowing Bear River are preserved on the very crest of the Bear Lake Plateau (Fig. 2), as much as 360 m above the present Bear River. These remnants, discovered during the present study (Reheis, 2005), are inset below the level of an Oligocene basalt and consist of cross-bedded fluvial gravel and sand as much as 5 m thick. The dominance of well-rounded pebble- to small-cobble-sized clasts and the lateral extent of outcrop indicate that the deposits represent a river similar to Bear River in size. The deposits slope gently northward along the drainage divide for at least 8 km and also slope to the east, toward the present course of Bear River. In one locality, nested channel fills in a roadcut show that the river migrated eastward and incised successively lower channels (Reheis, 2005). The most likely candidate for this river is the ancestral Bear River; however, the distinctive pink and purple quartzites derived from the Precambrian rocks of the Uinta Range, characteristic of the Pleistocene gravels of Bear River, are absent from these gravels.

There are three possibilities to explain the absence of the Precambrian quartzite clasts. First, the present-day northward slope of the gravel remnants may not represent the gradient of the ancient river, but may be an artifact of northward tilting of the footwall block along the east Bear Lake fault zone. However, the base of the deposits descends nearly 60 m over 8 km. Tilting along the fault would have had to exceed that amount to reverse the apparent flow direction of the paleo-river. The overall topography of the Bear Lake Plateau does not support northward tilting; the highest summits along the plateau are ~2225 m in the south, 2350 m atop the Oligocene basalt remnant (Fig. 2), and 2350 m northeast of the plateau gravels. Previous workers have inferred that *southward* tilting may have occurred along the hanging-wall block (Laabs and Kaufman, 2003; McCalpin, 2003). Second, the ancestral Bear River may not have originated in its present-day headwaters in the Uinta Range. Hansen (1985) suggested that at some time in the past, these headwaters flowed into the Green River via Muddy Creek (Fig. 2), and that the Bear River subsequently captured its present headwaters. Hansen (1985) observed drainages east of Hilliard Flat (Fig. 3) that presently drain to Bear

River, but that in their upper reaches point toward upper Muddy Creek in the area of a topographic gap in the divide. In addition, later surficial geologic maps (Gibbons, 1986; Dover and M'Gonigle, 1993) show old terrace gravels (younger and lower than the Bishop Conglomerate) that lie on and near the present drainage divide (Figs. 2 and 3). The proposed capture area lies within the Bear River fault zone (West, 1993); displacement on these down-to-the-west faults and on other Pleistocene normal faults to the east along Muddy Creek could have been the proximate cause of the capture of the Uinta headwaters by an ancestral Bear River (West, 1993) that did not extend south of about Evanston prior to the capture. Third, the gravels predate the unroofing

of the distinctive Precambrian quartzites in the Uinta Range. We reject this hypothesis because the gravels are younger than the Oligocene basalt, and the Oligocene Bishop Conglomerate contains such quartzites (Dover and M'Gonigle, 1993); thus, younger gravels from the same source area should also contain them.

Addition of water from the Uinta Range would have added considerable volume to the ancestral Bear River. Further, the eastward migration and downcutting of the ancestral river atop the Bear Lake Plateau suggest that the east Bear Lake fault zone had become active, creating the depression to the west that is now Bear Lake Valley and lifting the area of the plateau along the footwall block. Thus, the Bear River may have gained a significant amount of discharge at about the same time that accommodation space was being created in Bear Lake Valley along with the potential for river diversion. We hypothesize that the subsequent diversion of the Bear River into Bear Lake Valley and the beginning of lacustrine and fine-grained basinal deposition may have been concurrent events.

The late Pliocene to early Pleistocene course of the Bear River downstream of Bear Lake Valley is speculative. Terrace deposits downstream of Bear Lake so far have been found no higher than ~40 m above river level (several roadcut exposures south of Soda Springs along U.S. Highway 30 in the northwestern part of the Fossil Valley, Idaho 7.5′quadrangle); thus, they are likely much younger than the gravels atop the Bear Lake Plateau. Previous workers have speculated that Bear River once flowed northwest to the Snake River Plain via the present-day Portneuf River Gorge (Fig. 1; Bright, 1963; Mabey, 1971) and was diverted southward by eruption of lava flows in Gem Valley west of Soda Springs, Idaho (Bouchard et al., 1998). However, given the altitude of a few Bear River terrace remnants south of Soda Springs and the estimated thickness of basalt in the Blackfoot River Canyon downstream of Blackfoot Reservoir (Mabey, 1971), it seems equally likely that the Bear River could have originally continued directly north to the Snake River via the Blackfoot River and was diverted westward by voluminous Pleistocene eruptions of the Blackfoot volcanic field north and east of Soda Springs. A few radiogenic ages on volcanic rocks in these two fields range from ca. 1.0 to 0.05 Ma (Armstrong et al., 1975; Heumann, 2004; Pickett, 2004; Scott et al., 1982). Comprehensive dating and mapping of terrace gravels and basalt flows would be required to investigate these alternative ancestral courses.

The north flank of the Uinta Mountains was extensively and repeatedly glaciated during the Pleistocene (Bryant, 1992; Munroe, 2001) and minor glacial advances may have also occurred during the Holocene (Munroe, 2000). Deposits of at least two major glaciations have been mapped in the Bear River headwaters. Outlet glaciers of a broad ice field in the western Uinta Mountains (termed the Western Uinta Ice Field by Refsnider et al., 2007) occupied valleys of the East, Hayden, and West Forks of Bear River at the maximum of the Smiths Fork glaciation (equivalent to the late Pleistocene Pinedale glaciation of the Rocky Mountains) ca. 19–18 ka. (Laabs et al., 2007). Glaciers of the Hayden and East Fork valleys coalesced on the piedmont

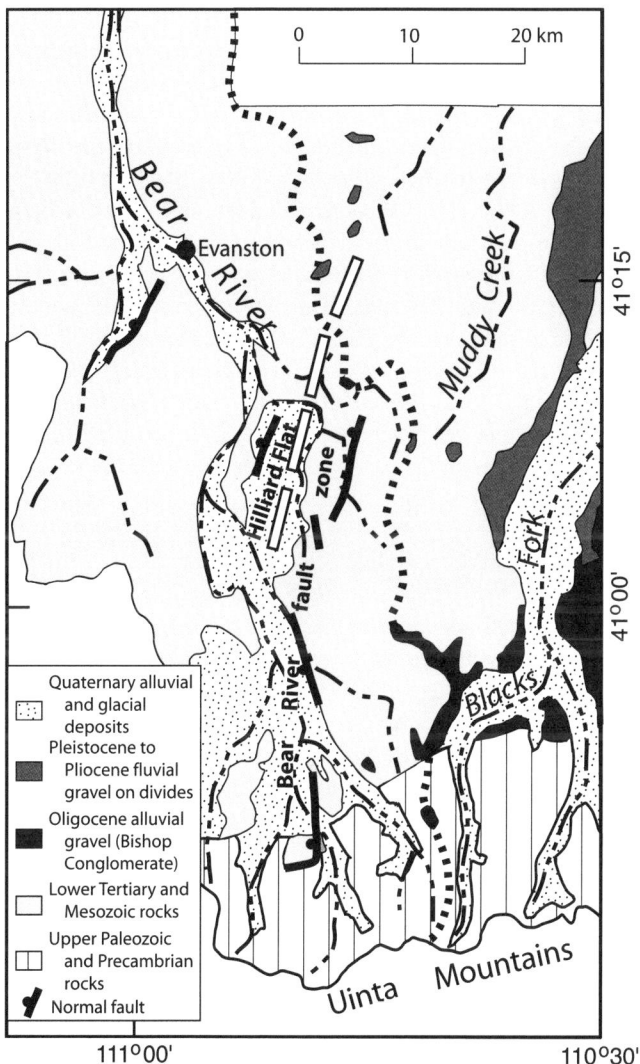

Figure 3. Generalized geologic map of capture area near Hilliard Flat showing postulated former course (white blocks) of headwaters of Bear River into Muddy Creek drainage, tributary to Green River (Hansen, 1985). Note black areas showing remnants of Bishop Conglomerate (Oligocene) and dark gray areas of younger Pliocene-Pleistocene terrace gravels capping divides, including Bear River–Muddy Creek divide (heavy dotted line) northeast of Hilliard Flat.

beyond the mouths of tributary canyons to deposit a broad area of hummocky topography. Cosmogenic [10]Be surface-exposure dating indicates that this area was abandoned at the start of ice retreat ca. 18 ka (Laabs et al., 2007), although ice retreat may have started as much as 2 k.y. later in other parts of the Uinta Range (Munroe et al., 2006). These ages compare favorably with the age for the maximum extent of glaciation, 19.7–18.9 ka as inferred from rock flour abundance in sediment cores from Bear Lake (Rosenbaum and Heil, this volume). Deposits of older Blacks Fork or pre–Blacks Fork glaciations are mainly preserved on interfluves between tributary canyons of Bear River, but one small remnant thought to be older till was mapped on the valley floor ~4 km north of the Smiths Fork till limit (Dover and M'Gonigle, 1993).

Gravelly outwash terraces are preserved downvalley of the moraine sequence in the Bear River valley (Bryant, 1992; Dover and M'Gonigle, 1993; Munroe, 2001; Reheis, 2005). The lower two gravel terraces have been traced to terminal and recessional moraines of the Smiths Fork glaciation. A suite of at least four progressively higher and older terraces is preserved in places along the river between the glacial limit and Evanston (Fig. 2; Reheis, 2005), and these terraces probably represent outwash of pre–Smiths Fork glacial advances. Downstream of Evanston, the high terraces are discontinuous and correlations among remnants have not been attempted. Relatively extensive terrace deposits only a few meters above the floor of the valley are assumed to be of late Pleistocene age (marine oxygen isotope stages 2–4) based on limited soil data and absence of meander scars; most of the valley floor is covered with Holocene alluvium (Reheis, 2005). Remnants of pre–late Pleistocene river terraces are apparently absent along the Bear River between Sage Creek Junction and Bear Lake Valley, possibly due to subsidence. The well-preserved, incisional terrace sequence along the upper Bear River and along the upper part of unglaciated Yellow Creek (Coogan and King, 2001; Reheis, 2005), which joins Bear River at Evanston, may record downcutting as the headwater adjusted to the stream capture discussed above (Hansen, 1985).

A notable feature of the surficial geology of the Bear River headwaters is the abundance and lateral extent of landslides developed in the Wasatch Formation (Bryant, 1992; Dover and M'Gonigle, 1993). These landslides have not been dated, but their appearance (undrained depressions and obvious hummocky surfaces; Reheis, 2005) suggests that most were active during the late Quaternary, and they may have contributed sediment to tributaries of Bear River.

In the northeastern Bear River drainage basin, the Smiths Fork drains the Salt River Range (Fig. 2). Cirque and small valley glaciers occupied a few north-facing positions but little outwash is preserved (Reheis, 2005). A laterally extensive, thick deposit of fluvial gravel and sand crops out near the confluence of Smiths Fork and the Bear River northeast of Cokeville (Rubey et al., 1980), with a surface elevation ~60 m above present river level. The deposits rise in altitude up Smiths Fork, indicating deposition by that stream, and also grade laterally upslope to the east and

south into alluvial fan deposits (locally faulted against bedrock along a west-down normal fault; Fig. 2) and pediment gravel. Outcrops in two gravel pits show that these deposits, locally at least 25 m thick and possibly as much as 45 m thick (Rubey et al., 1980), consist of very well rounded, well-washed, cross-bedded gravel with a few sand lenses (Reheis, 2005). This thickness and the apparent absence of buried soils suggest rapid aggradation by Smiths Fork and the adjacent Bear River. Such aggradation must have occurred in response either to local subsidence or to a significant rise in base level downstream, perhaps due to an expansion of Bear Lake in the early to middle Pleistocene (discussed below; Laabs and Kaufman, 2003).

SURFICIAL GEOLOGY OF BEAR LAKE VALLEY

Sedimentation in Bear Lake responds to a complex set of geomorphic and hydrologic influences. Previous work has focused on the shoreline record of Bear Lake, and other chapters in this volume discuss interpretations of lake fluctuations and interactions with the Bear River largely based on data from lake and marsh cores. In this section, we present new information on glacial deposits in the Bear River Range west of Bear Lake and on the fluvial terraces of the Bear River near and downstream of its entrance into Bear Lake Valley. We synthesize previously published information on the exposed shoreline deposits (Laabs, 2001; Laabs and Kaufman, 2003) with recent observations (Reheis et al., 2005; Reheis, 2005). We also present new data on Holocene marsh deposits and associated Bear River alluvium, with implications for the northward migration of Bear River since the last highstand of Bear Lake.

Glacial Deposits in the Bear River Range

Two studies documented glaciation in the Bear River Range south of the study area, west of the Bear River drainage basin (DeGraff, 1979; Williams, 1964). To our knowledge, however, no previous studies indicated glaciation in the northern Bear River Range, which is lower in altitude than the range to the south. The drainage divide of the Bear River Range west of Bear Lake (Fig. 4) lies at ~2750 m above sea level (asl) and isolated peaks east of the divide are as high as 2900 m; there is no summit plateau that would provide a source for wind-driven snow to accumulate on the lee (eastern) side. Precipitation in the Bear River Range is high, averaging 125 cm yr[-1], three-fourths as snowfall, at Tony Grove Lake (2415 m asl; http://www.wcc.nrcs.usda.gov/snow/) to the southwest of Bear Lake. Effective moisture was likely much greater during the maximum expansions of Lake Bonneville to the west due to cooler temperatures and perhaps to increased lake-effect precipitation (Munroe et al., 2006).

Three major streams head in shallow cirques and drain the east slope of the Bear River Range. Paris, Bloomington, and St. Charles Creeks contained valley glaciers as much as 10 km long (Fig. 5; Reheis, 2005). Nearly undissected, bouldery moraines that are very fresh in appearance and which have surface soils

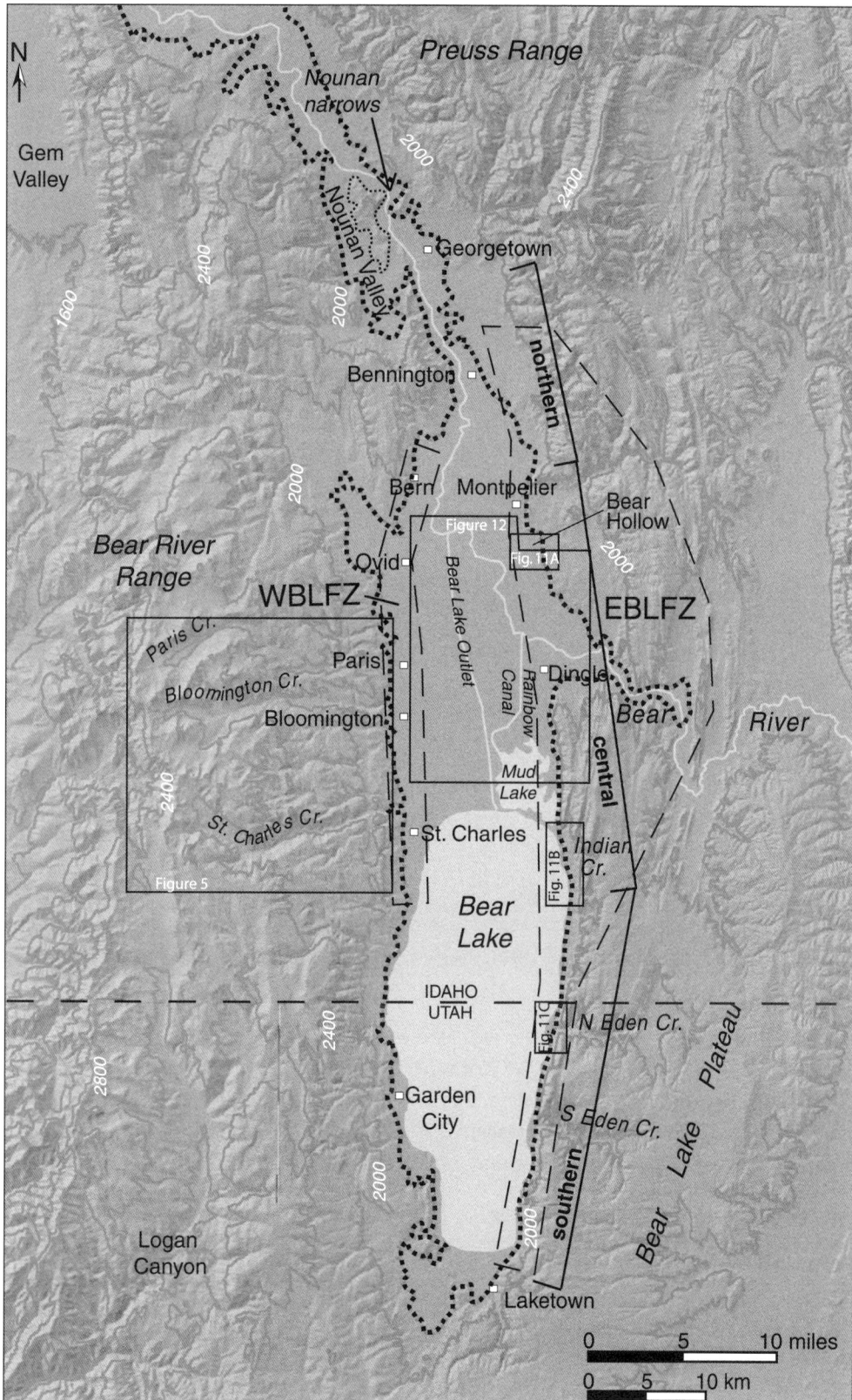

Figure 4. Shaded relief and topography of Bear Lake Valley. White rectangles show locations of towns; boxes show location of Figures 5, 11, and 12. Dashed lines enclose fault zones: west Bear Lake (WBLFZ, left) and east Bear Lake (EBLFZ, right). Brackets show north, central, and south sections of east Bear Lake fault zone. Dark gray lines are topographic contours (m asl) labeled with white text (contour interval = 400 m). Dashed heavy black contour marks elevation 1830 m asl in Bear Lake Valley, the approximate highest altitude of Bear Lake during the Pleistocene.

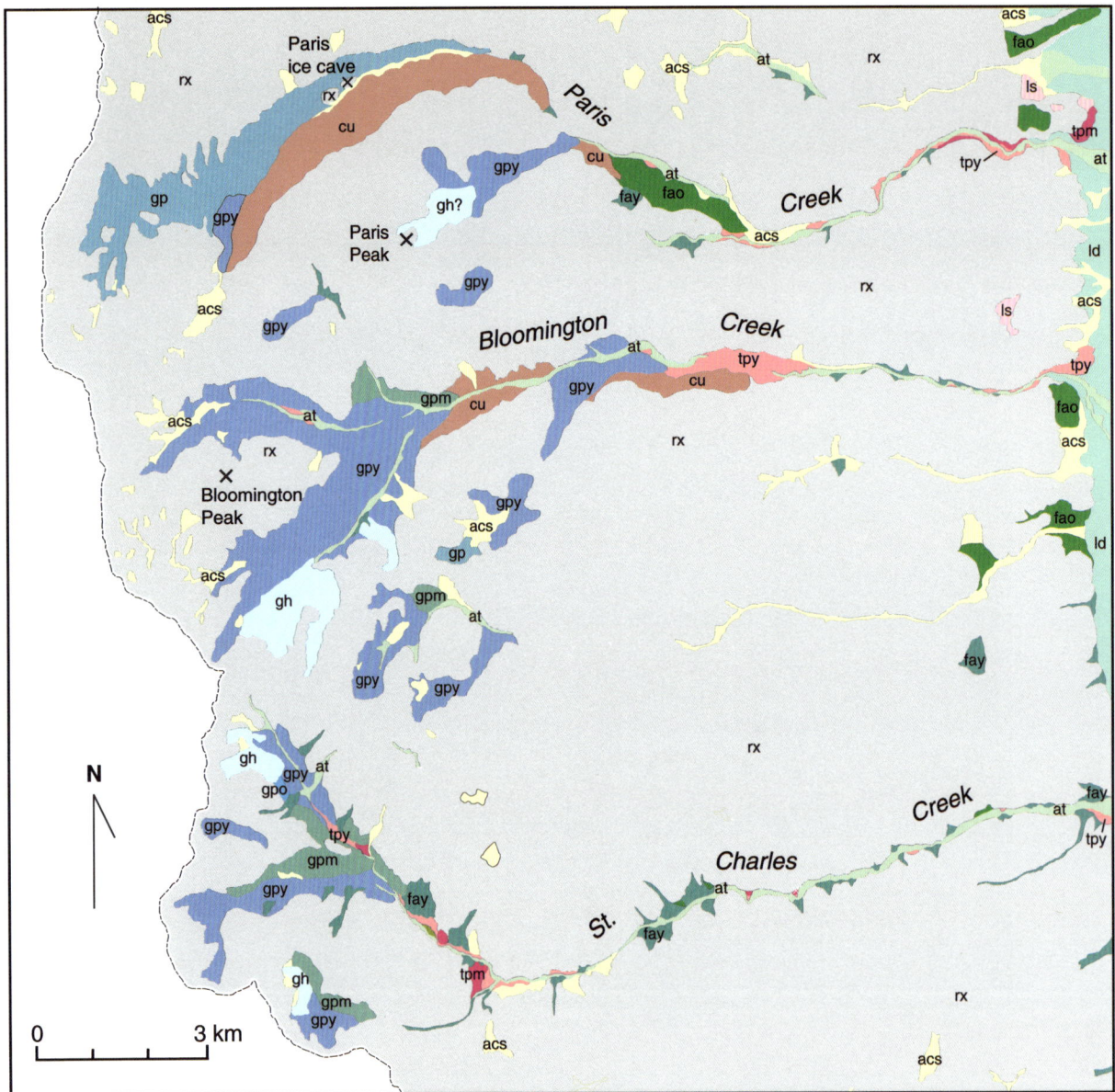

EXPLANATION

Alluvial deposits

at Fluvial channel and flood plain deposits (late Holocene)

acs Alluvium and colluvium (Holocene and late Pleistocene)--Includes fills in closed depressions (sinkholes)

fay Alluvial fans (Holocene and late Pleistocene)--Mostly undissected, smooth surfaces

tpy Fluvial terraces (late Pleistocene)--Includes outwash of Smiths Fork glaciation

tpm Fluvial terraces (middle Pleistocene)--Includes outwash of Blacks Fork glaciation

fao Alluvial fans (middle and early Pleistocene)--Incised irregular surfaces

Lake and marsh deposits

ld Lacustrine and associated marsh deposits (Holocene and Pleistocene)

Glacial deposits

gh Till of young cirque glaciers (early Holocene to latest Pleistocene)--Includes rock glacier deposits

gp Older till, undifferentiated (Pleistocene)--Subdivided into:

gpy Till of Smiths Fork glaciation (late Pleistocene)--Moraine surfaces irregular, with undrained depressions

gpm Till of Blacks Fork glaciation (middle Pleistocene)--Moraine surfaces smooth and dissected

gpo Till of pre-Blacks Fork glaciations (middle and early? Pleistocene)--Moraine forms not preserved

Mass-wasting deposits

ls Landslide deposits (Holocene and Pleistocene)

cu Undifferentiated colluvium (Pleistocene)--Mainly glacial deposits modified by mass wasting

Bedrock

rx Bedrock, undifferentiated (Tertiary through Precambrian)

Figure 5. Surficial geologic map emphasizing glacial deposits and outwash in Bear River Range (Reheis, 2005). Queried map symbol (for example, gh?) indicates uncertain identification of map unit. Note that highest points (Bloomington and Paris Peaks) do not lie on Bear Lake drainage divide, and that outwash (tpy and tpm) is of limited extent.

containing very little infiltrated fine-grained eolian sediment lie as much as 1 km from cirque headwalls. The landform properties suggest a small Holocene or latest Pleistocene advance (Figs. 6B and 6C). Well-preserved, sharp-crested, bouldery moraines with undrained depressions, and with oxidized surface soils containing eolian sediment to a depth of ~20 cm, extend 5–6 km from the cirque headwalls (Fig. 6A) and are correlated here with the late Pleistocene Smiths Fork glaciation (Munroe, 2001). Subdued, broad-crested moraines with boulders set in a finer-grained matrix than the younger moraines are locally preserved higher on valley walls and as much as 1 km downvalley from the Smiths Fork-equivalent moraines; these are correlated to the Blacks Fork glaciation, probably of middle Pleistocene age (Munroe, 2001; marine oxygen isotope stage 6). Diamictons mapped as much as 1–2 km downvalley of these older moraines may represent one or more older glacial advances. Moraine complexes of short valley glaciers are also preserved on the northeast flanks and downvalley of high peaks and ridges that lie east of the drainage divide, such as Paris Peak (Fig. 5; Reheis, 2005).

Within the glaciated terrain and especially in the cirques are numerous sinkholes and collapse features developed in the mostly carbonate bedrock (Fig. 2; Oriel and Platt, 1980). Large portions of some valleys below the cirques, such as the upper several kilometers of Paris Creek valley, have no integrated drainages or surface streams. Closed basins surround many of the sinkholes, indicating that all runoff must exit as groundwater. Sinkholes in valleys are commonly floored by fine-grained deposits (unit acs, Fig. 5) that accumulated after the glaciers withdrew; locally, small collapse features have formed in the basin floors. Other collapse features in bedrock, such as Paris Ice Cave (Fig. 6D) and a pit near the Bloomington Lake trailhead (Fig. 5), permit free drainage into open cracks. This karst topography, similar to that reported in the glaciated areas farther south in the Bear River Range (Wilson, 1979), may have permitted basal glacier meltwater and runoff to enter the groundwater directly. Such direct recharge to groundwater would efficiently deliver glacial meltwater through a carbonate aquifer into Bear Lake. Limited glacial surface runoff is consistent with the paucity of outwash terraces downvalley of moraines (Reheis, 2005). For example, terraces that probably represent late Pleistocene outwash were mapped only ~2 km downstream of terminal moraines (Fig. 5). If outwash streams were not effective at transporting sediment, this implies that much of the glacially eroded sediment is still stored within the valleys upstream of the lake. Much further work would be required to test these hypotheses and the glacial correlations.

Fluvial Terraces of Bear River

Bear River and its terraces in the vicinity of Bear Lake can be divided into three reaches based on location relative to tectonic displacements (Figs. 4 and 7): (1) the footwall block upstream of the east Bear Lake fault zone, (2) the graben area between the east and west Bear Lake fault zones, and (3) downstream of the graben area, beginning at the confluence of Ovid Creek and Bear River. The terrace deposits (Reheis, 2005) have not been previously studied, probably because they are commonly buried or obscured by thick deposits of loess on the footwall block

Figure 6. Photographs of glacial deposits in Bear River Range (see Fig. 5 for locations). (A) View downvalley of recessional moraine of Smiths Fork glaciation (gpy) in valley of Bloomington Creek. (B) Holocene moraine (gh) and part of compound cirque at head of Bloomington Creek. (C) View of outer part of Holocene moraine (gh) in Bloomington Creek. (D) View to southwest of roche moutoneé at Paris ice cave in valley of Paris Creek.

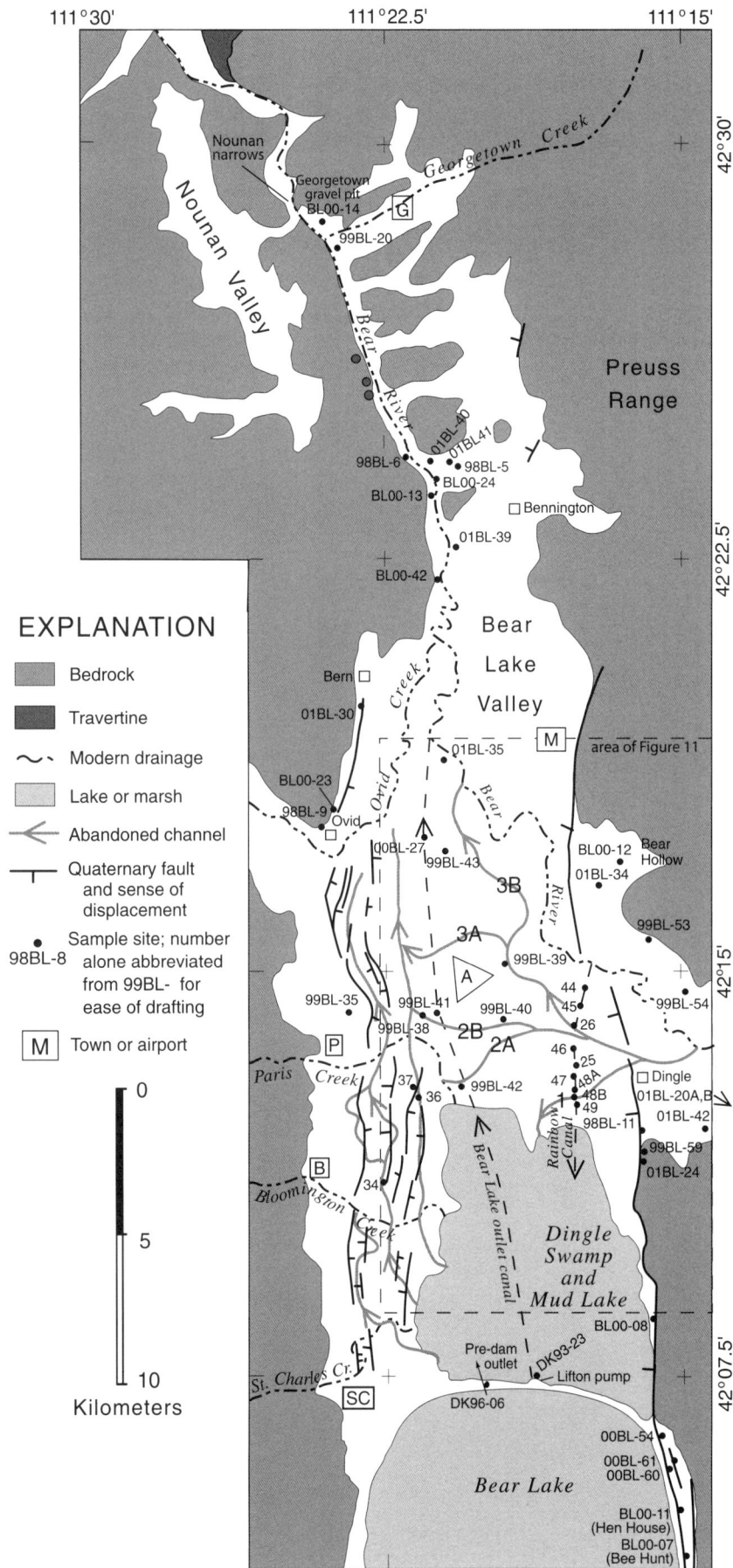

Figure 7. Sketch map showing Bear River and inferred former courses (2B, etc.), abandoned channels on the west side of the valley, faults with Holocene displacement, and selected terrace and lake-deposit study sites. Stratigraphy of some sites is shown in Figures 9 and 13. Dashed box shows area of composite aerial photograph (Fig. 12). A—airport; B—Bloomington; G—Georgetown; M—Montpelier; P—Paris; SC—St. Charles.

and by lake and marsh deposits on the hanging-wall block of the east Bear Lake fault zone. The fluvial terraces, combined with evidence of lake-level fluctuations, record a complex history of uplift, downcutting, and aggradation in response to faulting, threshold incision, lake-level fluctuations, and glacial sediment loading. We use a variety of sources to draw tentative correlations among terrace remnants and lake highstands. These include surficial mapping; stratigraphic relations among alluvium, loess, and lake sediments; and chronologic data from radiocarbon, amino-acid racemization, and tephrochronologic techniques. In this section we address the terrace mapping, dating, and correlation; in the two following sections we discuss the exposed record of lake-level fluctuations and the synthesis of these two data sets to interpret the record of downcutting and aggradation caused by tectonics, lake-level fluctuations, and outwash deposition.

Terraces along the Footwall Block

East of the east Bear Lake fault zone, terrace gravels lie on the footwall block (Fig. 7). Within this reach, a group of four terraces lie between 40 and 80 m above the modern river (Fig. 8, sites 99BL-53 and -54, 01BL-20 and -42) and locally cap fine-grained sediment interpreted as lacustrine deposits of Bear Lake (Table 1, Figs. 8 and 9). Because lacustrine deposits have not

been found at equivalent altitudes west of this fault zone, these very high terrace gravels and lake deposits probably record uplift of the footwall block. They are nested within the valley cut by the Bear River during uplift of the Bear Lake Plateau, and lie more than 200 m lower than the ancient river gravels preserved on top of the plateau (Fig. 2). A group of three younger terraces less than 15 m above the river are preserved where the Bear River enters the Bear Lake Valley.

Limited age control suggests that the older group of terraces ranges from Pliocene(?) to middle Pleistocene in age (Table 1). At site 99BL-53 north of the river (Fig. 7), a deep roadcut exposes ~5 m of terrace gravel 80 m above the modern river. The gravel caps 20 m of fine-grained lacustrine deposits that in turn overlie bedrock (Fig. 9). Reworked tephra within the fine-grained beds yielded no definitive correlations with dated rocks (A. Sarna-Wojcicki, U.S. Geological Survey, 2000, personal commun.). These deposits form a deeply dissected belt parallel to the modern river and are thickly blanketed with loess. A loess-buried terrace gravel farther upstream at site 01BL-20B may be correlative (Figs. 7 and 8). A lower terrace gravel south of the river at site 01BL-42, also buried by loess, overlies several meters of deposits interpreted as lacustrine fan-delta deposits (Figs. 8 and 9). These lake deposits conformably overlie locally

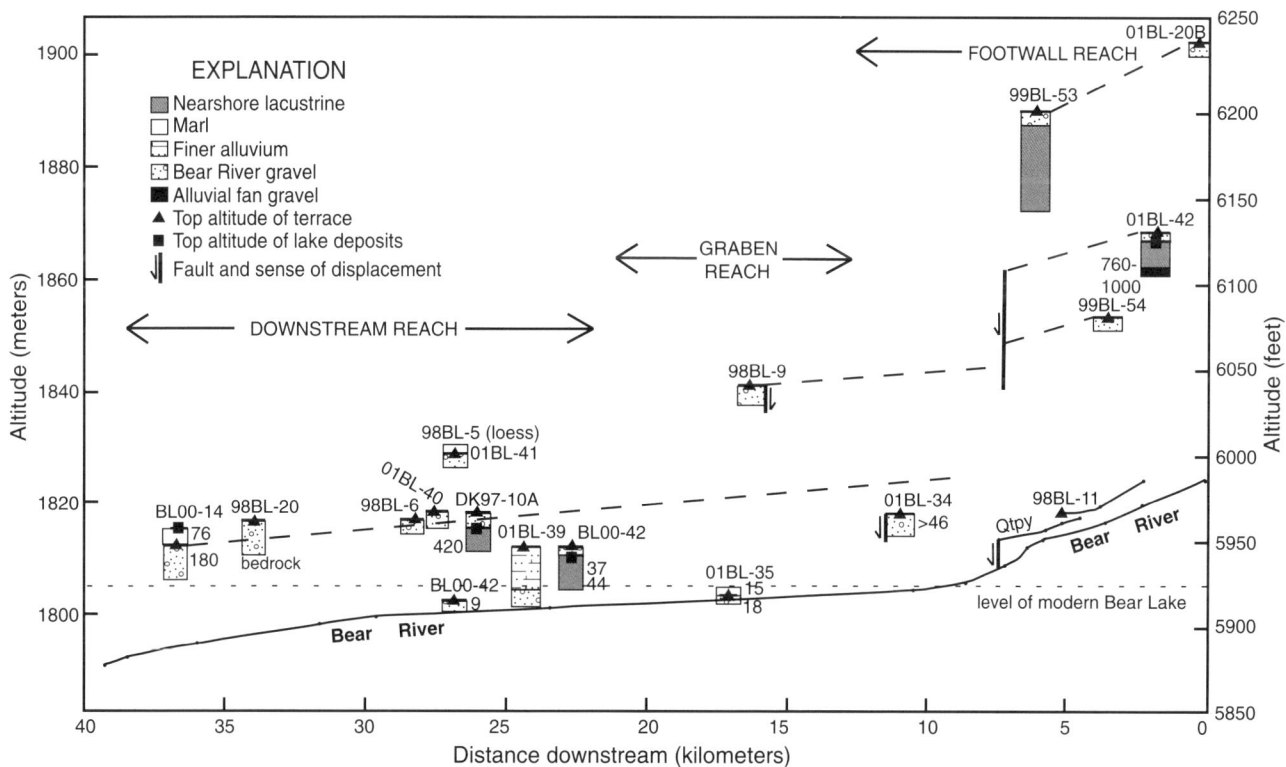

Figure 8. Bear River terrace sites and modern gradient of river showing study sites and simplified stratigraphic columns. Dashed lines show inferred terrace correlations based on height above river and limited age control; ages in ka are listed next to columns. Heavy gray arrows separate reaches of Bear River and terraces defined by major faults: footwall reach includes east Bear Lake fault zone and sites upstream, graben reach includes one site on footwall block of west Bear Lake fault zone and one site between fault zones, and downstream reach includes sites downstream of active faulting.

TABLE 1. BEAR RIVER TERRACE SITES GROUPED BY RELATIVE AGES, WITH ALTITUDES, UNDERLYING STRATIGRAPHY, AND CORRELATIONS WITH BEAR LAKE PHASES (LOCATIONS ON FIGS. 4, 7, AND 8)

Site number	Location	Position relative to fault*	Gravel altitude (m)†	Height above river (m)	Age control§	Stratigraphy and key soil horizons
BL00-24	Bennington Bridge	Downstream reach	1804	3	8.2 ± 1.6 ka (AAR)	1.5 m loess/t.g.
Willis Ranch Phase, ca. 9 cal ka, ~1814 m (many preserved barriers)						
Raspberry Square Phase, 16–15 cal ka, ~1814 m (Garden City pit)						
01BL-35	Bear River floodplain cut	Graben reach	1803	1	16.3 ka (^{14}C)	2 m marl/t.g.
BL00-42	Bear River cut (Red Pine Hollow)	Downstream reach	1812	9	<44.3 ka, <36.8 ka (^{14}C)	t.g./rippled sand
01BL-39#	Pescadero	Downstream reach	1794–1807	4	Altitude, stratigraphy	8 m sandy silt/t.g.
Jensen Spring Phase, 46–39 ka, ~1817 m (Ovid spit); 98BL-11 on FW block may be correlative fan-delta unit of Bear River						
01BL-34	Gravel pit W of US-30	Footwall block of EBLFZ	1818	14	>46 ka (^{14}C)	Bw/Bk (stage II CaCO$_3$)
99BL-56-57	Gravel pits E of Dingle cemetery	Footwall block of EBLFZ	1838	15	Soils, stratigraphy	2 m loess (Bw/Bk)/2 m loess (Bk)/t.g.
Late Bear Hollow Phase, 170–90 ka, ≥1818 m (Georgetown pit), <1830 m (FW; upper Bear Hollow pit), <<1866 m (FW; powerline)						
01BL-20A	E of old fan-delta site	Footwall block of EBLFZ	1850	26	Altitude	>4 m loess/t.g.
BL00-13	S of Bennington Bridge	Downstream reach	1818	15	>43.3 ka (^{14}C); <<419 ± 81 ka (AAR)	t.g./2.5 m marl, sand, and silt
01BL-40	N of Bennington Bridge	Downstream reach	1818	17	Soils, stratigraphy	25 cm loess/Bw/Bk
98BL-6	W side river	Downstream reach	1817	17	Altitude	Not described
99BL-20	Gravel pit S of old railroad	Downstream reach	1811–1817	21	Soils, stratigraphy	2 m loess/t.g. (Btb/Bkb, stage III CaCO$_3$)
BL00-14	Georgetown gravel pit, lower	Downstream reach	1810–1816	20	180 ± 60 ka; >76 ± 35 ka (AAR)	t.g. (Bt)/3 m marl and mud/t.g.
Middle Bear Hollow Phase, 450–380 ka, ≥1817 m (S of Bennington Bridge), ≤1824 m (FW?; culvert cut)						
99BL-54	E of Dingle siding	Footwall block of EBLFZ	1853	35	Soils, stratigraphy	Bk (stage IV CaCO$_3$)
98BL-5, 01BL-41	E of Bennington Bridge	Downstream reach	1829	27	Soils, stratigraphy	1.3 m loess (Bt/Bk)/>1 m loess (Bt/Bk)/t.g.
Early Bear Hollow Phase, 1100–820 ka, <1829 m (FW, lower Bear Hollow pit), <<1863 m (FW; old fan-delta site)						
01BL-42, 99BL-58	Old fan-delta site	Footwall block of EBLFZ	1868	46	<760–1200 ka (tephra)	>3 m loess/t.g./fan-delta/alluvial fan with tephra
98BL-9**	Ovid cemetery gravel pit	Footwall block of WBLFZ	1841	38	Soils, altitude	Eroded Bkm (stage V CaCO$_3$)
Other phases, late Pliocene–early Pleistocene (ostracodes): <1846 m (FW, 00BL-60), <<1870 m (FW, 00BL-61), <<1876 (99BL-53)						
01BL-20B	E of fan-delta site 01BL-42	Footwall block of EBLFZ	1905	81	Soils, stratigraphy	>4 m loess/alluvial fan/t.g.
99BL-53	US-30 cut, prodelta site	Footwall block of EBLFZ	1890	79	Reworked Tertiary tephra	t.g./20 m bedded sand and silt, clay at base

Note: AAR—amino acid racemization; FW—footwall; t.g.—terrace gravel.
*EBLFZ—East Bear Lake fault zone; WBLFZ—West Bear Lake fault zone; river reaches shown on Figure 8.
†Elevation in meters above sea level. Range of altitudes shows significant aggradation of fluvial deposits.
§^{14}C data given in Table 2; AAR ages from Laabs and Kaufman (2003).
#May postdate Raspberry Square phase and predate Willis Ranch phase.
**Assumed to correlate to 99BL-54; is definitely too high to correlate to sites 99BL-5 or 01BL-41 (see Fig. 8).

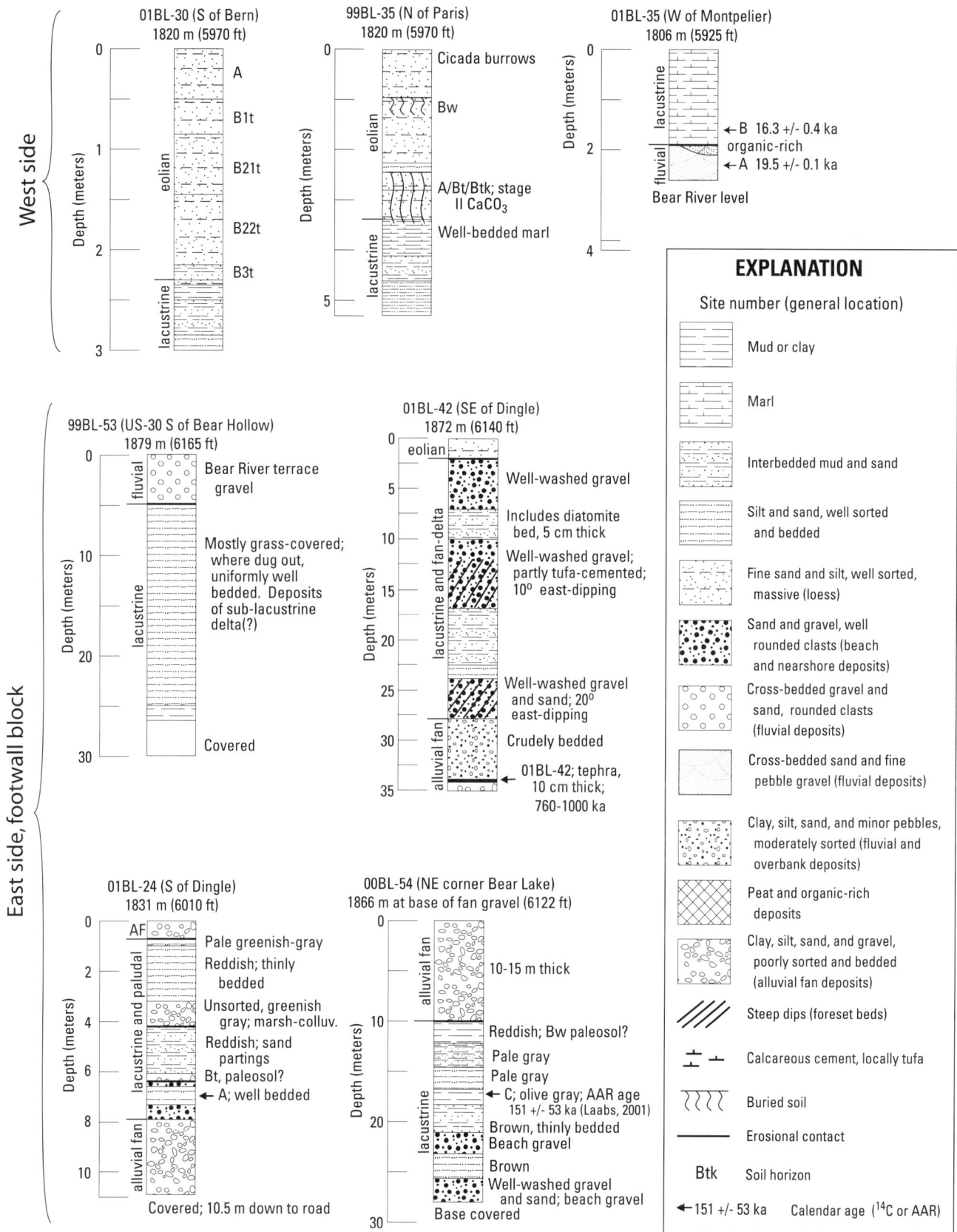

Figure 9. Selected stratigraphic sections related to Bear River terraces and lake deposits measured from outcrops and auger holes in Bear Lake Valley (data from Reheis et al., 2005). Site locations are on Figure 7. Column to right of lithology gives descriptive information such as color, sample data, and soil horizons, and column to left gives interpreted environment of deposition (AF—alluvial fan). See Table 2 for radiocarbon ages and Laabs and Kaufman (2003) for amino acid racemization (AAR) ages.

derived alluvial-fan deposits containing a 10-cm-thick, reworked, rhyolitic tephra layer that is chemically correlative with either the Bishop ash bed (760 ka) or Glass Mountain tephra (ca. 1 Ma) (A. Sarna-Wojcicki, U.S. Geological Survey, 2000, personal commun.; Sarna-Wojcicki et al., 2005). The lowest terrace in the older group (site 99BL-54) has a soil characterized by stage IV pedogenic $CaCO_3$, suggesting an early middle Pleistocene age by comparison with the carbonate morphology of soils on younger deposits (e.g., 99BL-20, Table 1). On the other side of the Bear River Valley, a terrace gravel preserved on the footwall block of the west Bear Lake fault zone (98BL-9) bears a truncated, strongly developed soil with stage V pedogenic $CaCO_3$. On the basis of this morphology and elevation above the river, it may be early Pleistocene in age.

Younger terraces in the footwall-block reach are late Pleistocene in age on the basis of soil development, one ^{14}C age, and relations to lacustrine deposits. These terraces are abruptly truncated by left-stepping strands of the east Bear Lake fault zone (Figs. 7 and 8), and fine-grained lake deposits on the hanging-wall block are in fault contact with terrace deposits. Terrace gravel ~15 m above Bear River is well exposed in 5-m-deep gravel pits at site 01BL-34, where gastropod shells yielded an age of 49,950 ± 1020 ^{14}C yr B.P. (Tables 1 and 2); we interpret this to be a minimum age for the deposit. Lower, younger terrace deposits have not been directly dated. Terrace height and soil development suggest that a terrace or fan-delta deposit at site 98BL-11 (Reheis, 2005) may correlate with or slightly postdate lacustrine deposits with ages of 46–39 ka (Laabs and Kaufman, 2003; discussed below).

Terraces and Deposits within the Graben

Within the graben reach between the east and west Bear Lake fault zones (Figs. 7 and 8), subsidence has resulted in burial of terraces by younger sediments. One cutbank of the Bear River (01BL-35, Fig. 9) exposes marl overlying cross-bedded fluvial sand; gastropod shells in these deposits yielded ages of 16,320 ± 380 (in marl) and 19,480 ± 100 (in sand) cal yr B.P. (Tables 1 and 2). No data are available for the possible reservoir effect on gastropods in fluvial and marsh deposits near Bear Lake; the reservoir effect for ostracode shells in Bear Lake cores is 370 ± 105 yr (Colman et al., this volume). Outcrops along the Rainbow Canal, transverse to the modern course of Bear River, show that river gravel and sand representing the aggradational surface of a fluvial fan extend at least 5 km south of the river (see Fig. 13A discussed below). One bivalve shell from this alluvium (site 99BL-45) yielded an age of 12,980 ± 130 cal yr B.P., and gastropod shells from sand interpreted as fluvial more than 2 m below the surface in auger hole 99BL-42 gave an age of 14,210 ± 390 cal yr B.P. These deposits descend ~3 m in altitude southward along the canal exposure and are overlain by fine-grained fluvial and marsh deposits of Holocene age.

Terraces Downstream of Active Faults

Terraces are common downstream of the northern limits of active Quaternary faults (as mapped by Reheis [2005]) in Bear Lake Valley (Table 1, Fig. 7). Because altitudes of these terraces appear to be little influenced by local fault displacement during the late Quaternary, we assume they more accurately reflect base-level and climatic changes and effects of upstream faulting than terraces to the south and east. The highest river gravel (site 01BL-41) northeast of Bennington is poorly exposed on the flank of a low plateau buried by loess. A backhoe trench at nearby site 98BL-5 (Fig. 7) exposed more than 2.2 m of loess including a surface soil and a buried soil, each with 60- to 80-cm-thick argillic horizons and stage II-III calcic horizons, suggesting an age of middle Pleistocene or older for the buried gravel.

The next lower set of terrace deposits includes five localities close to the river northward from Bennington (Table 1, Figs. 7 and 8). Their elevations fall within a range of 15–20 m above the modern river, but they are not all the same age because their stratigraphic sequences and surface soils differ. From south to north, at site BL00-13, fluvial sand and gravel unconformably overlie nearshore lacustrine deposits. A nonfinite age of >43,260 ^{14}C yr B.P. was obtained from gastropod shells in the alluvium (sample DK97-10A, Table 2) and an age of ca. 420 ka was obtained on shells in the lacustrine deposits by using amino acid racemization (AAR; Table 1; Laabs and Kaufman, 2003). At sites 01BL-40 and 98BL-6, thin loess (~25 cm) overlies terrace gravel and the surface soils are weakly developed (A/Bw/Bk, stage II $CaCO_3$), suggesting a late Pleistocene age. At site 99BL-20, 2 m of loess with a weak surface soil (A/Bw) and a stronger buried soil (Bt/Bk, stage III $CaCO_3$) overlies 6–7 m of river gravel, in turn overlying bedrock (Salt Lake Formation). At site BL00-14, in the Georgetown gravel pit, calcareous marsh or shallow lacustrine deposits 3 m thick conformably overlie nearly 6 m of fluvial sand and gravel. Gastropod shells gave AAR ages of ca. 76 ka (marsh) and 180 ka (gravel) (Table 1; Laabs and Kaufman, 2003). Such ages are compatible with the soil development observed at nearby site 99BL-20 and perhaps with the nonfinite radiocarbon age at BL00-13, but the latter two fluvial units are not buried by marsh or lacustrine deposits. Soils at sites 98BL-6 and 01BL-40 seem too weakly developed to assign a middle Pleistocene age to these sites, unless the soils were eroded.

Two adjacent terrace localities ~8 m above the modern river show evidence for incision followed by rapid aggradation. At site BL00-42 (Figs. 7 and 8; Bear River cutbank of Laabs and Kaufman, 2003), 1.4 m of alluvial sand and gravel overlies lacustrine sand containing gastropods with an age of 47,000 ± 1000 cal yr B.P. (Tables 1 and 2). At a slightly lower surface altitude 2 km downstream (site 01BL-39), terrace gravel at river level is 4 m thick and grades up into ~8 m of interbedded silt and fine sand. No datable materials were found in this outcrop, but the weak surface soil suggests an age no older than late Pleistocene. Low terraces ~3–4 m above the modern river are locally preserved. In one of these low terraces (site BL00-24), gastropod shells yielded an AAR age estimate of ca. 9 ka. These variable terrace altitudes and stratigraphic relations are discussed below in the context of shoreline fluctuations.

TABLE 2. RADIOCARBON AGES FROM OUTCROPS AND AUGER HOLES IN THE BEAR LAKE DRAINAGE BASIN, 1998–2001

Sample number*	General location	Latitude (N)	Longitude (W)	Material dated	Sample depth (cm)	^{14}C lab number†	Radiocarbon age (yr B.P.)	Calibrated age§ (yr B.P.)
99BL-26B	Rainbow Canal N of airport road bridge	42°13.992'	111°49.166'	Gastropod shells	80–100	WW-3049	1400 ± 40	1325 ± 55
99BL-34	Auger hole E of Bloomington	42°11.064'	111°22.625'	Gastropod shells	0–50	WW-2583	7760 ± 70	8530 ± 130
99BL-37	Auger hole E of Bloomington, footwall of fault cutting paleochannel	42°12.775'	111°21.814'	Shell fragments	70–80	WW-2584	1955 ± 70	1820 ± 95
				Gastropod shells	80–100	WW-2585	10,420 ± 80	12,360 ± 310
				Gastropod shells	200–230	WW-2586	9820 ± 75	11,250 ± 160
99BL-38	Auger hole, Paris-Dingle road, paleo-course BR2	42°14.140'	111°21.611'	Gastropod shells	0–20	WW-2587	2645 ± 55	2530 ± 180
				Gastropod shells	230–250	WW-2588	7985 ± 70	8830 ± 200
99BL-39	Auger hole E of airport, paleo-course BR3	42°15.119'	111°19.498'	Gastropod shells	60–70	WW-2589	2445 ± 55	2530 ± 180
99BL-42	Auger hole W of outlet canal; Bear River or Bear Lake channel?	42°12.860'	111°20.583'	Gastropod shells	0–25	WW-590	1720 ± 55	1630 ± 110
				Gastropod shells	200–220	WW-2591	12,220 ± 100	14,210 ± 390
99BL-45	Cut, Rainbow Canal; paleo-course BR3	42°14.270'	111°17.639'	Gastropod shells	60–65	WW-2592	5130 ± 65	5280 ± 150
				Gastropod shells	120–130	WW-2593	8520 ± 70	9520 ± 120
				Bivalve shells	200–220	WW-2594	11,015 ± 85	12,980 ± 130
99BL-47	Cut, Rainbow Canal	42°13.001'	111°17.762'	Gastropod shells	0–80	WW-2595	7150 ± 70	8000 ± 160
99BL-48A	Cut, Rainbow Canal	42°12.658'	111°17.048'	Gastropod shells	165–185	WW-3048	8460 ± 40	9480 ± 50
99BL-48B	Cut, Rainbow Canal	42°12.654'	111°17.706'	Gastropod shells	110–140	WW-3050	7870 ± 40	8670 ± 120
99BL-49	Auger hole, Rainbow Canal; paleo-course BR1	42°12.472'	111°17.721'	Gastropod shells	20–50	WW-2596	4000 ± 60	4460 ± 180
				Gastropod shells	270–300	WW-2597	6925 ± 70	7770 ± 120
99BL-59	Culvert cut S of Dingle	42°11.640'	111°16.082'	Shell fragments	B, 180	WW-2599	>40,800 ± 1600	N.D.
					A, 200–250	WW-2598	>39,100 ± 1100	N.D.
00BL-27	Cut, outlet canal	42°30.0'	111°21.5'	Gastropod shells	10–20	WW-3047	4880 ± 40	5600 ± 110
00BL-63A	Waterline trench S of Indian Creek	42°05.181'	111°15.309'	Gastropod and bivalve shells	Several meters	WW-3369	39,870 ± 490	N.D.
01BL-34	Borrow pit W of US-30, Bear Hollow	42°16.5'	111°17'	Gastropod shells	350	WW-3721	>45,950 ± 1020	N.D.
01BL-35	Bear River cutbank N of US 30	42°19.81'	111°23.26'	Gastropod shells	B, 150–190	WW-3723	13,675 ± 50	16,320 ± 380
					A, 190–250	WW-3722	16,350 ± 50	19,480 ± 100
DK96-01 (BL00-10)	Cisco Beach, edge of North Eden fan	41°58.45'	111°16.17'	Mollusk shell	150	WW-1557	10,420 ± 50	12,510 ± 170
DK96-06D	Fish ladder, W of Lifton pumping station	42°07.35'	111°20.15'	*Stagnicola* shell	70	NSRL-1566	7210 ± 40	8160 ± 100
DK96-06B	Fish ladder, W of Lifton pumping station	42°07.35'	111°20.15'	*Stagnicola* shell	150–160	WW-1561	5650 ± 40	6410 ± 100
DK96-06B	Fish ladder, W of Lifton pumping station	42°07.35'	111°20.15'	Charcoal	150–160	WW-1566	5530 ± 50	6410 ± 70
DK96-06A	Fish ladder, W of Lifton pumping station	42°07.35'	111°20.15'	*Valvata* shell	210–220	NSRL-10940	8520 ± 65	9500 ± 420
DK97-10A (BL00-14)	Bennington, N of Pescadero	42°23.80'	111°21.20'	*Sphaerium* shell	Several meters	WW-1559	>43260	N.D.
DK98-02A (99BL-25)	Rainbow Canal at airport road bridge	42° 14.27'	111°17.63'	*Stagnicola* shell	200–250	NSRL-10569	8350 ± 70	9290 ± 160
DK98-03A (BL00-02C)	Road cut north of North Eden canyon	41°59.53'	111°15.88'	*Stagnicola*; 2 shells	Several meters	NSRL-10570	37,900 ± 460	41,000 ± 1000
DK98-03B (BL00-02C)	Road cut north of North Eden canyon	41°59.53'	111°15.88'	*Stagnicola* shell	Several meters	NSRL-10571	38,700 ± 790	41,500 ± 1000
DK99-11 (BL00-23)	Ovid spit, N of Ovid	42°18.00'	111°23.62'	Shell fragments	180–210	NSRL-11353	41,240 ± 640	43,700 ± 1000

(Continued)

TABLE 2. RADIOCARBON AGES FROM OUTCROPS AND AUGER HOLES IN THE BEAR LAKE DRAINAGE BASIN, 1998–2001 (*Continued*)

Sample number*	General location	Latitude (N)	Longitude (W)	Material dated	Sample depth (cm)	[14]C lab number†	Radiocarbon age (yr B.P.)	Calibrated age§ (yr B.P.)
DK99-13 (BL00-42)	Bear River cutbank, W of Bear River	42°22.17'	111°21.33'	*Stagnicola* shell	170–180	NSRL-11354	36,800 ± 790	30,500 ± 1000
DK99-18A (BL00-02)	North Eden canyon, W of highway	41°59.22'	111°15.56'	Charcoal	50–150	NSRL-11355	8780 ± 90	9840 ± 130
DK99-18B (BL00-02)	North Eden canyon, W of highway	41°59.38'	111°15.95'	*Discus* shell	50–150	NSRL-11356	10,490 ± 100	12,830 ± 190
DK99-28C (BL00-14)	Georgetown gravel pit	42°28.83'	111°24.00'	*Discus* shell	100–140	NSRL-11359	>45,200	N.D.
BL00-11	Hen House, E of highway	42°05.02'	111°15.28'	*Sphaerium* shell	100–150	NSRL-12061	36,000 ± 320	39,000 ± 1000
BL00-41A	Garden City borrow pit	41°57.13'	111°23.82'	Mollusk shell	125–160	NSRL-12062	13,540 ± 70	16,310 ± 240
BL00-41B	Garden City borrow pit	41°57.13'	111°23.82'	Mollusk shell	125–160	NSRL-12063	13,280 ± 70	16,010 ± 270
BL00-42	Bear River cutbank, W of Bear River	42°22.17'	111°21.33'	*Stagnicola* shells	175–185	NSRL-12064	44,300 ± 920	47,000 ± 1000

*Site number given in parentheses if different from sample number. DK- and BL- data taken from Laabs and Kaufman (2003); other samples reported in Reheis et al. (2005).
†Letter prefixes indicate laboratory: WW—U.S. Geological Survey, Reston, Virginia; NSRL—Institute of Arctic and Alpine Research, Boulder, Colorado.
§For ages <21,000 radiocarbon years, calibrations performed using CALIB program, version 4.4 (Stuiver et al., 2005). Some ages >21,000 radiocarbon years corrected using data in Kitagawa and van der Plicht (1998). N.D.—not determined. Calibrated age values are midpoints plus or minus one-half the 2σ range.

Fluctuations of Bear Lake

Bear Lake fluctuates between a topographically closed basin when lake level is largely controlled by groundwater discharge and a topographically open basin when Bear River flows into and out of the basin (Laabs and Kaufman, 2003; Bright et al., 2006; Kaufman et al., this volume). Although Bear Lake experienced regional climate changes similar to nearby Lake Bonneville, climate change alone cannot account for Bear Lake highstands because they did not always coincide with pluvial maximum conditions (Kaufman et al., this volume). Interpreting the history of water-level changes in Bear Lake Valley is complicated by active tectonism and geomorphic processes that control basin geometry and the altitude of the basin's threshold. Because the present threshold of Bear Lake is the northern shoreline (Fig. 7), only ~2 m above modern lake level (nominally 1805.5 m asl; at times lower prior to human modifications), the altitude and location of the threshold must have been different in the past to allow higher lake levels to occur. At such times, the lake expanded to a northern, formerly higher threshold at or south of Nounan narrows (Fig. 7; Robertson, 1978; Laabs and Kaufman, 2003). Thus, maximum lake level is ultimately controlled by the altitude of this bedrock threshold, which can be changed by downcutting, fault displacement, aggradation, or landsliding (Laabs and Kaufman, 2003). An additional factor in threshold rise, which we cannot evaluate with available data, is the possible effect of travertine dams in the area of Nounan narrows (Fig. 7). Thick, extensive sheets of travertine lie on the east side of Bear River downstream of the narrows (mapped by Oriel and Platt, 1980). These sheets range from 1805 m asl to as high as 1830 m asl along side tributaries and on the west side of the river south of the Georgetown gravel pit (BL00-14, Fig. 7). We do not know the age of the travertine or whether such outcrops may once have been extensive enough to fill the valley, thus temporarily raising the threshold. If this occurred, it was likely to have been at a time of greatly reduced flow in Bear River.

In the following discussion, results from Laabs and Kaufman (2003) are augmented with stratigraphic and chronologic data from other outcrops (Reheis, 2005; Reheis et al., 2005). The paucity of exposures, uncertainty of water depth for some sediment, and post-depositional deformation hampers an accurate reconstruction of water level in Bear Lake Valley. This problem is particularly acute for the Bear Hollow phases because of tectonic subsidence and footwall uplift. We assume that sites within a given age range belong to the same lake phase, which may include more than one lake-level fluctuation. A second major assumption is that sites containing lake sediments along the west side of Bear River Valley and north of most mapped Pleistocene faults (Fig. 7) are at or near their original depositional altitudes; tectonic displacement is greatest on the east side of the valley and increases southward (discussed below).

Pliocene or Early Pleistocene Lake

Lacustrine deposits as high as 1876 m asl (99BL-53, Figs. 7 and 9) on the footwall block of the east Bear Lake fault zone represent the oldest exposed evidence of a perennial lake in Bear Lake Valley during the Pliocene or early Pleistocene (Reheis et al., 2005). Well-sorted, well-bedded, fine sand and silt with minor green clay constitute 20 m of deposits that are interpreted as lacustrine sediment deposited near the mouth of the ancestral Bear River. Although undated, these deposits are nearly 70 m above modern lake level, suggesting much more fault displacement and hence a much older age than lower deposits. Two sites near the mouth of Indian Creek, 00BL-61 and 00BL-60, are ~65 and 41 m, respectively, above modern lake level (Figs. 4 and 10).

These sites expose very similar sequences 10–12 m thick including, from base to top, green clay, gray, bedded fine sand, reddish mud, and beach gravel (sections not measured in detail). Reddish mud implies a contribution from the ancestral Bear River, if the sediment characteristics are similar to those in cores of the late Pleistocene age Bear Lake units (Rosenbaum et al., this volume). The green clay at 00BL-60 and the gray sand at 00BL-61 contain previously unknown ostracode species of the genera *Candona* and *Limnocythere* that resemble Pliocene and early Pleistocene ostracodes known from large lakes elsewhere (Table 2 *in* Reheis

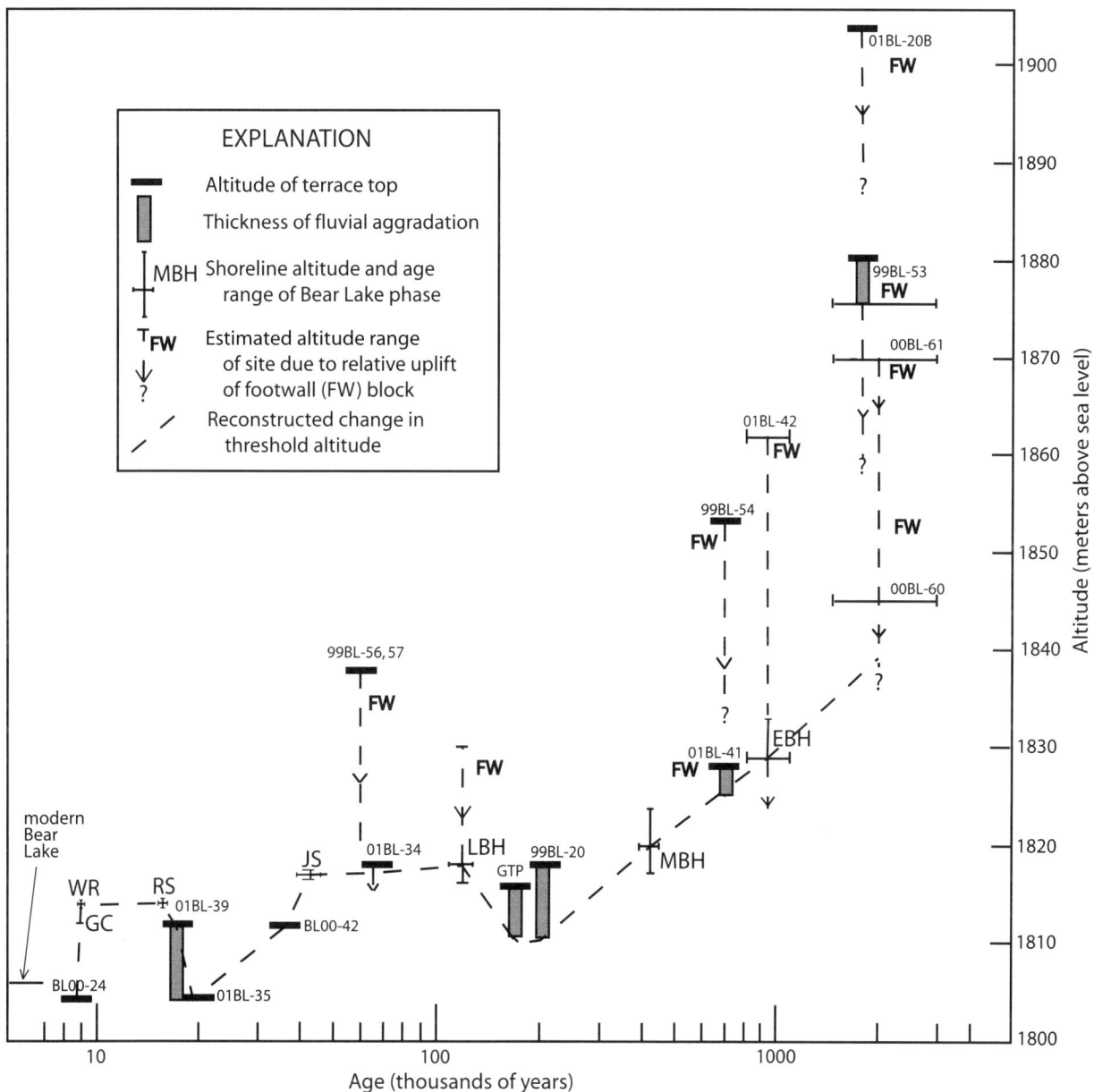

Figure 10. Synthesis of inferred Quaternary incision and aggradation history from lake levels and river terraces; key sites denoted by site name (see Figs. 4 and 7). Large capital letters indicate altitudinal phases of Bear Lake (Laabs, 2001): EBH—early Bear Hollow; GC—Garden City; GTP—Georgetown gravel pit site of Laabs and Kaufman (2003); JS—Jensen Spring; LBH—late Bear Hollow; MBH—middle Bear Hollow; RS—Raspberry Square; WR—Willis Ranch.

et al., 2005).The two sites are separated by a strand of the east Bear Lake fault zone (Reheis, 2005; see Fig. 11B), which may account for their altitude difference.

Early Bear Hollow Phase (Early Pleistocene to Early Middle Pleistocene)

Lacustrine deposits at two sites suggest a highstand of Bear Lake ca. 1200–700 ka. At Bear Hollow, deposits interpreted as fan-delta sand and gravel at 1829 m asl contain aquatic snails with AAR age estimates ranging from 1100 ± 160 ka to 830 ± 180 ka (Figs. 7 and 11A; site BL00-12 of Laabs and Kaufman, 2003). At site 01BL-42 (Figs. 7 and 9), steeply dipping gravel and sand interpreted as delta foreset beds are interbedded with horizontally stratified silt and sand as well as thin marl and diatomite beds at and below 1863 m asl (Reheis et al., 2005). These deposits conformably overlie fan gravel containing tephra correlated with either a Glass Mountain tephra or the Bishop ash (discussed earlier); thus, the lacustrine deposits must be somewhat younger than 1000 or 760 ka and hence are broadly similar in age to those at Bear Hollow. Despite being >30 m different in altitude, both sites are located on the footwall block of the east Bear Lake fault zone. A conspicuous depositional break in slope along the western (relatively unfaulted) margin of Bear Lake Valley lies at ~1830 m asl (Fig. 4); this altitude also coincides with the vertical extent of "lake" deposits mapped in this area by Mansfield (1927) and Robertson (1978). We tentatively interpret the coincidence in altitudes to indicate that the Bear Hollow deposits lie at or near their original depositional altitude; if so, they have experienced much less footwall uplift than sites to the south (discussed in tectonics section).

Middle Bear Hollow Phase (Middle Pleistocene)

Water level probably was as high as 1820 m asl (15 m above modern lake level) at least once ca. 500–300 ka, as shown by lacustrine or wetland marl and beach gravel south of Dingle exposed at Culvert cut (site 99BL-59, Fig. 7) and an adjacent site (01BL-24), and by fine-grained lacustrine deposits west of Bennington (site BL00-13; Laabs and Kaufman, 2003; Reheis et al., 2005). Gastropods at the Culvert cut and Bennington sites yielded AAR age estimates between 445 ± 105 and 385 ± 85 ka. Lacustrine deposits at the Bennington site consist mainly of marl, silt, and mud, indicating deposition below wave base. This site lies north of mapped Quaternary displacement on the east and west Bear Lake fault zones and we infer that it has been relatively unaffected by faulting. Thus the altitude of ~1817 m asl at the top of lacustrine sediments (modified slightly from the altitude given by Laabs and Kaufman, 2003) is a minimum for lake level at this time. South of Dingle, sediments exposed in the Culvert cut represent nearshore deposition. The outcrop near the fault is poor; although the deposits may lie on the hanging-wall block of the east Bear Lake fault zone, they more likely lie between two strands of the fault or are plastered onto the footwall block. Their altitude of ~1821 m asl is a few meters higher than that at the Bennington site. Nearby deposits exposed in a roadcut at site

01BL-24 (Figs. 7 and 9) extend to an altitude of ~1831 m and clearly lie on the footwall block, east of the active fault strands. These deposits include interbedded beach gravel, thinly bedded reddish sand and silt, and unsorted greenish-gray mixed mud and gravel, interpreted to represent nearshore lacustrine and marsh deposition with incursions of colluvium in this fault-marginal setting. Sand above the basal beach gravel contains an assemblage of ostracodes interpreted to represent a lacustrine setting with a nearby stream or wetland (Table 3 *in* Reheis et al., 2005). The reddish sand and silt imply that Bear River was feeding Bear Lake. The deposits in this outcrop can be traced northward to within ~200 m of those exposed in the Culvert cut (Reheis, 2005), suggesting they represent the same lacustrine phase. If so, their separation in altitude indicates at least 10 m of displacement on one fault strand.

Late Bear Hollow Phase (Late Middle to Late Pleistocene)

Lacustrine and marsh deposits of middle to late Pleistocene age crop out in three places in Bear Lake Valley: Bear Hollow, Georgetown gravel pit, and site 00BL-54 (Figs. 7, 8, and 9; latter same as Power line site of Laabs and Kaufman, 2003). AAR estimates of these deposits range from ca. 200 to 100 ka (Laabs and Kaufman, 2003). Lacustrine deposits at site 00BL-54 (Fig. 11B) extend as high as 1864 m asl. Prominent fault scarps to the west, with a combined height of nearly 35 m above the modern lake, indicate that footwall-block uplift has affected this site. Lacustrine silt and sand in this outcrop contain ostracodes endemic to Bear Lake and indicative of nearshore deposition, as well as other ostracodes indicating adjacent streams and wetlands (Reheis et al., 2005). Gastropods from this deposit yielded an AAR age of 151 ± 53 ka (Laabs and Kaufman, 2003). A second set of lake deposits at Bear Hollow (Fig. 11A), much younger than those identified as the early Bear Hollow phase, have AAR age estimates ranging from 180 ± 58 ka to 88 ± 33 ka (Laabs and Kaufman, 2003). As noted above, the Bear Hollow site is on the footwall block. At the Georgetown gravel pit (site BL00-14, Figs. 7, 8), 3 m of marl, bedded sand, and mud overlie 4 m of Bear River gravel and sand; the highest fine-grained deposits lie at ~1817 m asl. AAR age estimates for gastropods are 75 ± 35 and 90 ± 38 ka in the marl and 180 ± 58 ka in the gravel. The fine-grained deposits may be either lacustrine or spring-discharge deposits, but in either case they are interpreted to reflect a lake level at or near 1817 m (Laabs and Kaufman, 2003).

At least one very low lake level, to as low as 1781–1783 m asl (22–24 m below present lake level) may have occurred during the late Bear Hollow phase. Colman (2006) interpreted acoustic stratigraphy of Bear Lake sediments to indicate a wave-cut bench along the western side of the lake, at about the same altitude as the top of a submerged paleodelta in the northwestern part of the lake. Correlations of seismic reflectors to core data (Colman, 2006) suggest an age of somewhat younger than 97 ka for this delta and perhaps also the wave-cut bench. If this delta represents incursion of the Bear River into the lake, it implies that stream-flow must have been so small that it did not cause a rise in lake

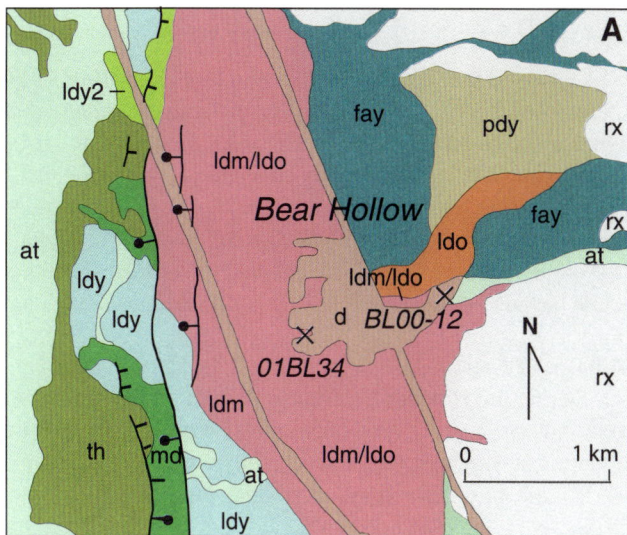

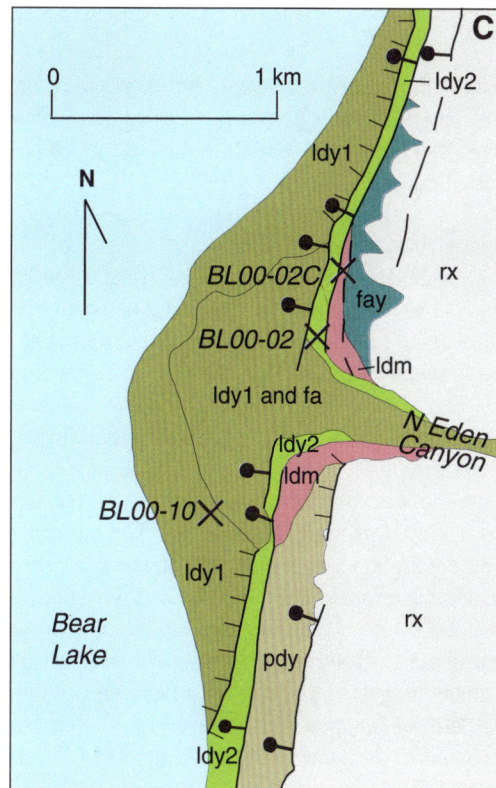

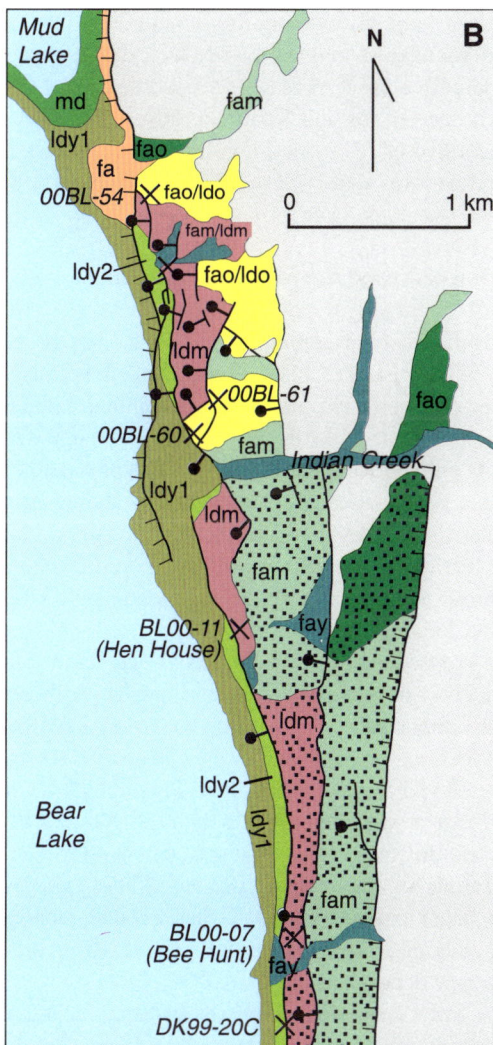

EXPLANATION

Alluvial deposits

at	Fluvial channel and flood plain deposits (late Holocene)
th	Low fluvial terraces (Holocene)
fa	Alluvial fans (Holocene)--Active surfaces, undissected
fay	Alluvial fans (Holocene and late Pleistocene)--Mostly undissected smooth surfaces
fam	Alluvial fans (late and middle Pleistocene)--Smooth surfaces incised by drainages
fao	Alluvial fans (middle and early Pleistocene)--Incised irregular surfaces
pdy	Pediment deposits (late Pleistocene)--Little dissected

Lake, fan-delta, and marsh deposits

md	Marsh deposits (Holocene and late Pleistocene)
ldy	Lacustrine and fan-delta deposits (Holocene and late Pleistocene)--Subdivided into:
ldy1	Deposits of Garden City and Lifton phases (late and middle Holocene)
ldy2	Deposits of Willis Ranch, Cisco, and Raspberry Square phases (early Holocene and late Pleistocene)
ldm	Lacustrine and fan-delta deposits (late Pleistocene)--Includes deposits of Jensen Spring phase
ldo	Lacustrine and fan-delta deposits (middle and early Pleistocene)--Includes deposits of Bear Hollow phases

Other deposits and symbols

| d | Artificial fill (historic) |
| rx | Bedrock, undifferentiated (Tertiary through Precambrian) |

-⌐ Normal fault; bar and ball on downthrown side; dashed where inferred

⅂ Fault scarp; hachures toward downthrown side

BL00-10

✕ Sample site for AAR and ¹⁴C measurements

Figure 11. Surficial geologic maps of selected sites along east Bear Lake fault zone (locations on Fig. 4). Stipple indicates alluvial fan or lacustrine deposits partly buried by loess. (A) Surficial geologic map of Bear Hollow, south of Montpelier. (B) Geologic map around Indian Creek, northeastern margin of Bear Lake. (C) Geologic map of North Eden Creek.

level such that Bear Lake expanded to the northern threshold. This period approximately coincides with major soil-forming periods noted by Oviatt et al. (1999) in the Burmester core from Lake Bonneville.

Jensen Spring Phase (Late Pleistocene)

During the Jensen Spring phase, Bear Lake extended up to ~1817 m asl (11 m above modern lake level) between ca. 47 and 39 cal ka (Table 2). Lake deposits of this age crop out at several sites on the footwall blocks of both the east and west Bear Lake fault zones, and also north of active faulting at the Bear River cutbank site (site BL00-42, Fig. 7; Laabs and Kaufman, 2003). At Ovid spit (BL00-23) near Jensen Spring, at ~1817 m asl (11 m above modern lake level), well-sorted, ripple-laminated sand forms a 2-m-high, >400-m-long spit. The eastern edge of Ovid spit is cut by a ~5-m-high scarp of the west Bear Lake fault zone. Because Quaternary slip rates on this fault are low (McCalpin, 2003), the Ovid spit probably has not experienced much footwall uplift. However, its elevation could be as much as 5 m too high, as suggested by deposits at Bear River cutbank site (BL00-42) north of mapped fault scarps (Fig. 7; Reheis, 2005), where lake sands of the same age crop out at ~1811 m asl (Laabs and Kaufman, 2003).

Lacustrine and lake-marginal deposits containing gastropods with similar ^{14}C ages crop out along the footwall block of the east Bear Lake fault zone above the east shore of Bear Lake between 1826 and 1828 m asl (Fig. 11B) (21–23 m above modern lake level; Laabs and Kaufman, 2003). Their higher elevation relative to Ovid spit indicates footwall uplift. One of the deposits (Bee Hunt, Fig. 11B) was analyzed for ostracodes and yielded a rich assemblage indicative of a marginal wetland (J. Bright, 2002, personal commun.). Gypsum-permeated spring deposits as well as beach gravel are associated with the marl at Hen House. Beach gravel also underlies the marl at North Eden Canyon (Fig. 11C).

Raspberry Square Phase (Latest Pleistocene)

No outcrop evidence indicates a lake highstand during the last glacial maximum. A terrestrial megafauna site beneath lake level at the south end of the modern lake suggests that lake level was probably lower than present ca. 22 cal ka (D. Madsen, 2001, personal commun.). Sedimentological evidence from lake cores indicates that the lake level was near the modern level during the last glacial period, ca. 26–18 cal ka (Smoot and Rosenbaum, this volume), when abundant glacial flour was deposited in the lake (Rosenbaum and Heil, this volume). The sedimentological evidence from the cores further indicates that lake level lowered to ~40 m below modern ca. 17.5–15.5 cal ka.

Nearshore sediment exposed in a gravel pit at Garden City (Fig. 4; site location in Laabs and Kaufman, 2003) indicates the lake rose to ~1814 m asl (9 m above present lake level) during the Raspberry Square phase late during the last glacial period, ca. 16–15 cal ka. This altitude is consistent with faulted fan-delta gravels of similar age at North Eden canyon (Figs. 4 and 11C); these deposits lie at ~1824 m asl on the footwall block and

1814 m asl on the hanging-wall block (McCalpin, 1993, 2003). The lake-level curve of Smoot and Rosenbaum (this volume) also shows a highstand centered ca. 15 cal ka. On the basis of their reconstruction, lake level regressed again ca. 14.8–11.8 cal ka. This regression culminated in a drop to ~30 m below modern before a rapid rise to a level above modern ca. 11.5 cal ka. At 11.0 cal ka, salinity in the lake had reached a threshold and aragonite began to precipitate (Dean, this volume).

Holocene Phases

During the Holocene, sedimentological evidence from lake cores indicates that the level of Bear Lake was typically below the modern, but with high-frequency fluctuations of 10–20 m (Smoot and Rosenbaum, this volume). On shore, the Willis Ranch phase (Williams et al., 1962) occurred ca. 9 cal ka, when Bear Lake filled the valley south of Bennington, up to ~8 m above modern lake level (1814 m asl). This timing is consistent with isotopic and other evidence in sediment cores (Liu et al., 1999; Dean et al., 2006; Dean, this volume; Smoot and Rosenbaum, this volume). Two lower and younger shorelines are not well dated, because their deposits largely consist of materials reworked from those of Willis Ranch age (Laabs and Kaufman, 2003). The Garden City shoreline formed at 1811 m asl (Williams et al., 1962) and the Lifton shoreline formed at 1808 m asl (Williams et al., 1962; Robertson, 1978) sometime later.

Synthesis of Downcutting, Aggradation, and Lake Levels

Interpreting the record of aggradation and incision by the river and highstands of Bear Lake (Table 1) is very difficult because of the effects of tectonics on terrace and shoreline altitudes and the paucity of ages on deposits older than ca. 100 ka. Although much further research could be done, a broad outline can be made. Figure 10 presents a synthesis of our current understanding of the interplay between the lake and river during the past million years. We emphasize that there probably were many more fluctuations in lake level, threshold incision, and river aggradation in addition to those we have documented.

The oldest exposed lacustrine sediments in Bear Lake Valley are probably of Pliocene or early Pleistocene age, but their depositional altitude cannot be constrained because they have been identified only on the footwall block of the east Bear Lake fault zone. Only one fluvial terrace (01BL-20B, Figs. 7 and 10) has been identified that is higher in altitude than the possible Pliocene deposits, and this may be due to displacement on presently inactive fault strands between the sites (Reheis, 2005). Lake level must then have fallen and the river established a course separated from the lake, judging from the presence of Bear River gravel overlying lacustrine deposits at site 99BL-53.

During the early Bear Hollow phase, ancestral Bear Lake reached a maximum altitude of 1829 m asl at least once between ca. 1200 and 700 ka. Cross-valley faults near Nounan narrows (Figs. 4 and 7) displace Pliocene deposits (Salt Lake Formation; Oriel and Platt, 1980). If the faults were active during the early

Pleistocene, they may have caused these high lake levels by raising the northern threshold, as suggested by Laabs and Kaufman (2003). Alternatively, if the Bear River bypassed Bear Lake prior to ca. 1.2 Ma, then southward subsidence of the valley floor along the graben-bounding faults followed by capture of the Bear River by Bear Lake due to minor water-level rise or to an avulsion of the Bear River could have caused water level to rise and expand to the valley threshold.

A prolonged period of episodic incision of the threshold to as low as ~1810 m followed the early Bear Hollow phase, with incision possibly persisting as late as 200 ka (Fig. 10). During this interval of ~600 k.y., Bear Lake filled the valley to the threshold at ~1820 m asl at least once during the middle Bear Hollow phase, at 500–300 ka. This altitude is consistent with progressive lowering of the threshold and only requires that the river rejoined the lake to contribute to lake-level rise. Several terrace remnants in the reach downstream of the graben record incision of the threshold to ~1810 m after this lake-filling episode, followed by at least 6 m of aggradation indicated by fluvial sediments at site 99BL-20 (Reheis et al., 2005) and at the Georgetown gravel pit (BL00-14; Laabs and Kaufman, 2003), just upstream of Nounan narrows (Fig. 7). This aggradation occurred ca. 200–180 ka as suggested by an AAR age estimate from shells at the Georgetown site. During this time the Bear River was probably separated from Bear Lake, because gravel containing Uinta Mountain clasts would not have aggraded in northern Bear Lake Valley if Bear River had debouched into Bear Lake.

During the late Bear Hollow phase, water level probably rose to at least 1818 m shortly after ca. 180 ka as recorded by sediments at the Georgetown pit and at Bear Hollow (Figs. 7 and 8). Such an increase in threshold altitude could be explained by one or more of the following: (1) aggradation due to a very large sediment load associated with extensive glaciation during oxygen isotope stage 6 (Blacks Fork glaciation), (2) normal faulting south of Nounan narrows, or (3) landsliding downstream of the Georgetown pit. We have not found extensive terraces of Blacks Fork-equivalent age as would be expected if aggradation were due to glacial sediment loading. Nor have we observed Quaternary deposits displaced by the cross-valley faults near Nounan narrows, or landslide deposits in the reach downstream of the Georgetown pit. The threshold appears to have remained at nearly the same altitude (at or slightly above 1817 m asl) until after the Jensen Spring phase, ca. 47–39 cal ka, suggesting that flow in Bear River was so low that it could not effectively incise a bedrock threshold. Such low flow and possibly a large reduction in ground water input is also suggested by the submerged Bear Lake shoreline at ~1782 m asl (Colman, 2006).

After the Jensen Spring phase, the Bear River began incising its threshold relatively rapidly to essentially its present gradient by ca. 18 cal ka (Fig. 10). Shortly after this time, by ca. 16 cal ka, the Bear River aggraded once again, piling up nearly 8 m of fine sand and silt at site 01BL-39 just south of Bennington (Figs. 8, 9, and 10). The rapid aggradation argues for an abrupt threshold change, possibly due to a landslide in the narrow part

of the valley west of Bennington (Laabs and Kaufman, 2003). The lake rose in concert with this aggradation to a highstand at ~1814 m asl by 16–15 cal ka during the Raspberry Square phase. In addition to the temporarily higher sill, the climate was favorable for a highstand during this time of generally cooler and moister conditions in the northern Great Basin and Snake River Plain regions (Thompson et al., 1993). Renewed threshold incision or a marked decrease in river flow caused the north shore of the lake to regress southward and Bear River to separate from Bear Lake by ca. 12 cal ka on the basis of oxygen isotope data in core BL96–2 (Liu et al., 1999; Dean et al., 2006; Dean, this volume) and ca. 12.5 cal ka on the basis of sedimentary indicators of lake level in several cores (Smoot and Rosenbaum, this volume). Then, ca. 11.5 ka, the lake rose once again (Smoot and Rosenbaum, this volume). During the early Holocene, channel migration in the delta area (discussed below) diverted all or part of the Bear River toward the lake, forming the Willis Ranch and Garden City shorelines at 1814 and 1811 m asl, respectively, by sometime after ca. 9 cal ka. Since then, the river has incised to ~1800 m asl west of Bennington (Williams et al., 1962) and to 1790 m asl at Nounan narrows (Fig. 7).

Holocene Marsh Deposits and Migration of the Bear River

Between Bear Lake and the present course of the Bear River is a low-lying region bounded by the east and west Bear Lake fault zones. This region is filled with marshes (including present-day Mud Lake), meandering channels, and scattered surfaces slightly elevated above water level (e.g., the airport area, Fig. 7). Outcrops are rare and generally restricted to artificial cuts along canals. Our observations in this region are limited to these few outcrops, sediments in hand-augered holes, and interpretation of aerial photographs (Fig. 12).

Most of the marsh is at or below the altitude of modern Bear Lake (1805.5 m) and separated from the lake only by the 2-m-high Lifton barrier beach (Figs. 4 and 7). Except for the fan-delta area where the Bear River enters Bear Lake Valley, the marsh must have been mostly submerged during the early Holocene when Bear Lake formed shorelines 6–9 m higher than present. Thus, surficial sediments in this low-lying area mostly record Holocene events, including shifting channels of the Bear River, tributary streams entering the valley from the Bear River Range on the west (St. Charles, Paris, and Ovid Creeks, Fig. 7), and probably the natural outlet of Bear Lake.

Northward Migration of the Bear River

The Bear River presently flows near the northern limit of its low valley floor, beginning to the east of the east Bear Lake fault zone (Fig. 12) and entering Bear Lake Valley where the young fault scarps display an en echelon left step. Analysis of aerial photographs shows that several abandoned channels similar in width to the modern channel lie to the south, and they become increasingly indistinct southward toward Mud Lake. These abandoned channels all appear to emanate from an abandoned course

Figure 12. Composite aerial photograph showing Bear River (solid white line), inferred paleochannels (BR1, etc., white dashed lines), faults (black lines), and locations of study sites. Although BR1 cannot be traced south of its entry into Dingle Swamp, it likely extended south to the present area of Mud Lake and thereby fed Bear Lake. See Figure 13 for stratigraphy at sites.

of the river that crosses the active fault scarp just north of Dingle. The Rainbow Canal provides a continuous, 5-km-long exposure transverse to the axis of the Bear River and to the older channels, forming a cross-section view of stratigraphy and facies changes descending from the higher part of the fluvial fan toward the marsh (Fig. 13A).

The oldest sediment exposed along the canal and encountered in auger holes elsewhere in northern Bear Lake Valley consists of fluvial sand and pebble-cobble gravel of Bear River. On the basis of topographic contours and outcrops, these deposits form a large braided fluvial fan emanating from the Bear River entrance into Bear Lake Valley. Two ^{14}C ages on shells collected

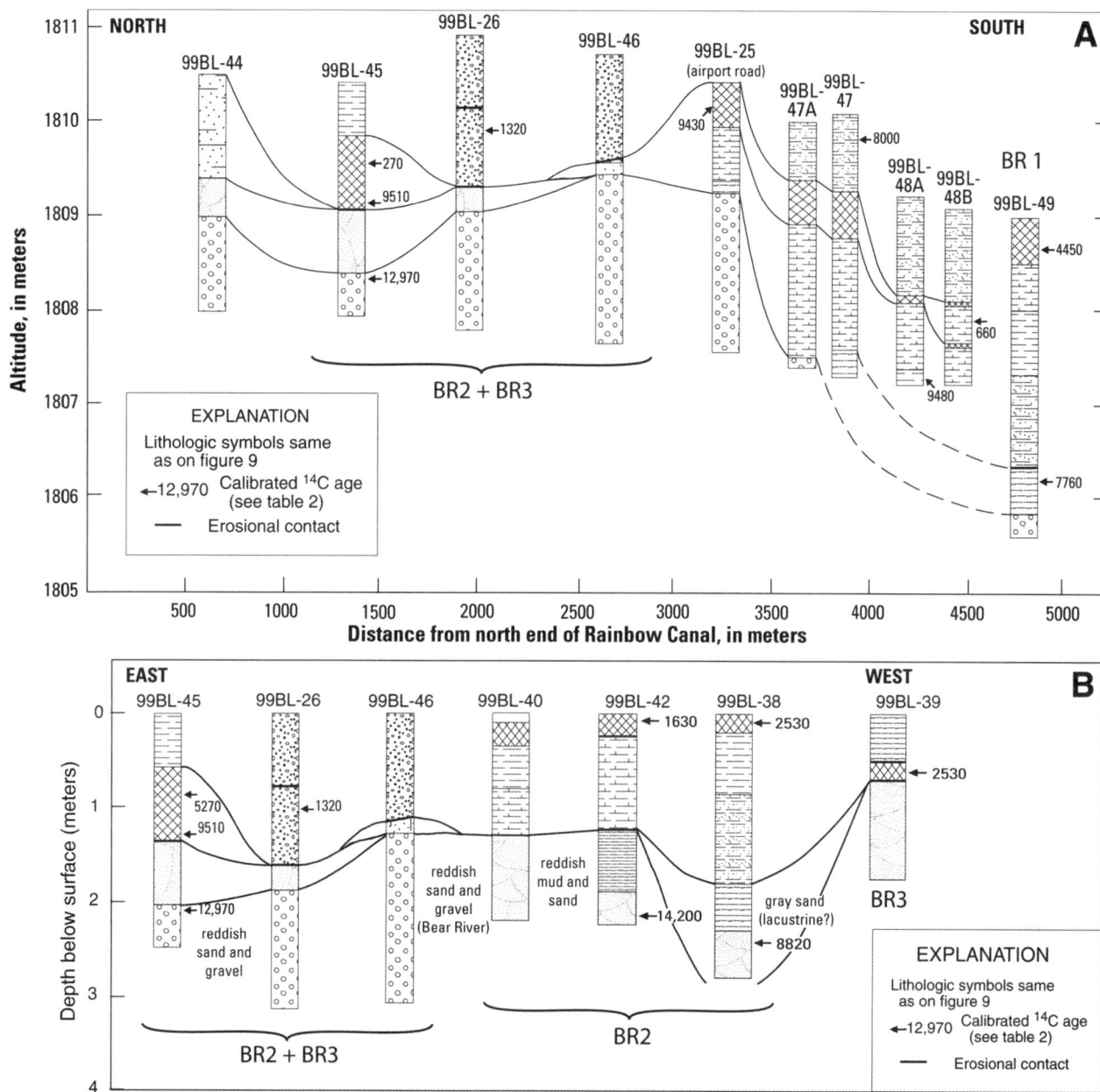

Figure 13. Stratigraphic sections of exposures and auger holes north of Bear Lake (location shown on Figs. 7 and 12). (A) Rainbow Canal. Most sections are outcrops and correlation lines were traced physically, except that 99BL-49 is an auger hole within a channel fill cut into older sediment and correlations are uncertain. Surface altitudes were plotted using differentially corrected GPS measurements. The base of the outcrop sections is the water level; thus the section bases essentially reflect the water gradient in the canal over a two-day period when flow rate in the canal remained relatively constant. Measured sections were then plotted and their altitudes slightly adjusted to yield a smoothly sloping water level at the base of the outcrop sections. BR1, 2, and 3 indicate sections relevant to reconstructing age and location of former courses of Bear River. (B) Selected stratigraphic sections from auger holes west of Rainbow Canal providing age control for BR2 and BR3.

within the gravel and sand at two different sites (99BL-45 and 99BL-42, Figs. 12 and 13 and Table 2) indicate an age of between 14 and 13 cal ka for this part of the fan, so it accumulated after the Raspberry Square highstand of Bear Lake (16–15 cal ka), presumably coincident with a lake-level decrease to below the present lake surface between ca. 15 and 12 cal ka (Smoot and Rosenbaum, this volume). Specific courses of the Bear River have not been identified during this fan-building period.

The river course farthest south, termed BR1, is the most indistinct on the aerial photographs and is crossed diagonally by the Rainbow Canal (Fig. 12). It consists of a meander belt of subdued channels partly or completely filled by younger marsh deposits. The surface trace of this channel cannot be distinguished more than ~0.5 km southwest of the canal, but at its southernmost visible extent, it appears to be directed toward modern Bear Lake. Fluvial deposits, including lenses of pebble gravel, sand, and mud, crop out at the surface along the southern part of the canal and were encountered 1.5 m below the surface in an auger hole within an infilled channel at the south end of the transect. The fluvial deposits overlie a peat bed that thins and fines southward, and the peat overlies and is intercalated with marl formed in a wetland environment (interpreted from microfossils; Reheis et al., 2005); these in turn rest on the 14 to 13 ka Bear River fan gravel. Shells from the peat and marl yielded ages of 9480–8670 cal yr B.P. (Table 2, Fig. 13A) and from the stratigraphically younger fluvial deposits, a single age of 8000 ± 160 cal yr B.P. Thus, the oldest river course is roughly 8.6–8.0 cal ka and overlaps in time with the Willis Ranch highstand of Bear Lake, and with a relatively high lake level as interpreted from sediment cores from Bear Lake (8.5–8.0 cal ka, Smoot and Rosenbaum, this volume). The correspondence in age and the orientation of the meander belt of BR1 suggest that this river course conveyed the discharge that caused Bear Lake to rise to this highstand. However, the Willis Ranch shoreline reaches 1814 m asl on the mostly unfaulted west margin of Bear Lake, an altitude well above the entire length of the Rainbow Canal (Fig. 13A). Because the fluvial deposits of BR1 either lie at the surface or are shallowly buried, these relations suggest that the deposits have been displaced downward since deposition by motion on the east Bear Lake fault zone.

Another meander belt, more sharply defined on aerial photographs, lies between the southern course and the present course of the river (Fig. 12). This middle course is directed due west across the Rainbow Canal and then splits into two main courses, termed BR2 and BR3. BR2 continues straight west, dividing into two courses, both of which appear to merge with north-trending drainages that carried the flow of the west-side creeks (St. Charles and Paris) and probably also the discharge from Bear Lake itself (Williams et al., 1962). BR3 trends northwest from the canal and also bifurcates. Fluvial deposits exposed along the northern part of the canal and encountered in auger holes to the west do not precisely limit the timing of the shifts in river courses (Figs. 12 and 13B), but do suggest that the initial jump north from BR1 occurred no later than 5.3 cal ka (99BL-45, Table 2). The shift from BR2 to BR3 occurred by ca. 2.5 cal ka (99BL-39), and the

shift to the modern course of Bear River probably occurred after ca. 1.3 cal ka (sample 99BL-26B).

North-Flowing Axial Channels

The history of the north-flowing channels carrying water from the west-side tributary streams and from Bear Lake is not well defined. Williams et al. (1962) first identified the existence of north-flowing channels or sloughs, which they interpreted as former outlets of Bear Lake, and inferred that discharge had shifted eastward through time. However, Williams et al. (1962) did not recognize the fault-bounded nature of these channels. Several ages on shells from the auger holes of this study and from McCalpin's (1993, 2003) trenching studies on the west Bear Lake fault zone near Bloomington provide rough constraints on the establishment of these drainages following the retreat of Bear Lake in latest Pleistocene time. Due east of Bloomington, ^{14}C ages indicate that a low-energy fluvial channel occupied a north-flowing course sometime after 13 cal ka and was abandoned by ca. 7.4–6.7 cal ka (McCalpin, 2003), a time that McCalpin interpreted as closely dating the most recent earthquake on this part of the west Bear Lake fault zone. On aerial photographs (WAC on Fig. 12, and shown as western channel on Fig. 7), this channel can be traced southward through a complex set of grabens at least as far as St. Charles Creek, and one indistinct remnant south of this creek also suggests that this drainage conducted discharge from the area of present-day Mud Lake and Bear Lake itself (Williams et al., 1962; Reheis, 2005). North of Bloomington Creek at site 99BL-34, this channel is incised into lacustrine deposits thinly overlain by ~50 cm of marsh or possibly overbank fluvial deposits that yielded an age of ca. 8.5 cal ka (Table 2), consistent with McCalpin's (2003) age constraints for this channel. To the north, the same channel (Figs. 7 and 12) continues northward transverse to the modern course of Paris Creek and can be traced nearly to Ovid Creek.

Another north-trending channel parallel to and just east of the channel described above is less confined to the graben structures (EAC on Fig. 12). Auger holes 99BL-37 and -36 within this channel yielded stratigraphic relations and ^{14}C ages that indicated this channel was cut after ca. 12 cal ka and was filled and abandoned by ca. 2 cal ka (Table 2). Because the BR2 and BR3 river courses merge with this eastern north-trending channel, we infer that the channel formed after the western north-trending channel identified near the Bloomington scarp and is coeval with BR2 and BR3, that is, older than ca. 5.3 cal ka to as young as 1.3 cal ka.

Processes Affecting Channel Shifts

Several factors, including climate change, faulting, location of the fluvial gravel fan, and fluvial geomorphic processes may have interacted to produce the Holocene changes in channel position and lake level. BR1 may simply represent an avulsion of Bear River to the south across its post-glacial fluvial fan, and this diversion resulted in the river's merging with the lake, causing the lake to rise to a higher level. Displacement to the south along the east and west Bear Lake fault zones may have encouraged the diversion. On the west Bear Lake fault, two or more events resulting in ~6 m of displacement probably occurred between ca. 12 and 7 ka (McCalpin, 1993, 2003).

The location of the north-flowing, west-side axial channel north of Bear Lake was probably controlled by two factors. First, the Bear River fluvial fan formed a natural dam that was lowest on the west side of the valley (Williams et al., 1962), and the fan was probably higher in elevation before Holocene displacement on the east Bear Lake fault zone. We interpret the stratigraphic and chronologic data and map relations to indicate that northward drainage from the tributary creeks and Bear Lake was established after deposition of the fluvial fan of Bear River and probably just after the land was exposed following the ca. 9 ka Willis Ranch highstand of Bear Lake (as suggested by Williams et al., 1962). This northward drainage followed active graben structures on the west side of the valley.

The former river courses show that the Bear River has migrated northward in the past 8000 years and that it eventually abandoned the west-side axial drainage system. The reasons for this northward shift include (1) gradual channel migration, (2) earthquake events on the east Bear Lake fault zone, and (3) inactivity of the west Bear Lake fault zone since the early Holocene such that graben formation ceased to aid north-flowing drainage. The second possibility is supported by the coincidence of the shift from BR1 to BR2–3 near Dingle with a left step in the fault (Fig. 11). The third possibility is supported by McCalpin's (2003) work showing that the last earthquake event on the Bloomington scarp predates ca. 7 ka.

NEOTECTONICS

Bear Lake Valley resides in the zone of active extension affected by the Yellowstone hotspot (Fig. 1; Pierce and Morgan, 1992), and forms a relatively simple half-graben between the lystric eastern Bear Lake fault zone and the steeply dipping, antithetic western Bear Lake fault zone (Fig. 2; Evans et al., 2003; Colman, 2006). We examine the tectonic geomorphology of range fronts bounding the valley to characterize the pattern of normal faulting in Bear Lake Valley, and use ages of faulted lacustrine and wetland deposits to estimate minimum average slip rates during the past ca. 200,000 years on the east Bear Lake fault zone.

Studies of surficial deposits surrounding Bear Lake have yielded evidence for late Quaternary activity on both the east and west Bear Lake fault zones. McCalpin (1993, 2003) trenched and dated faulted colluvial and lacustrine sediment at the mouth of North Eden canyon where it intersects the east Bear Lake fault zone (Fig. 14A) and calculated a late Quaternary slip rate of 1.1 m k.y.$^{-1}$. McCalpin (1993, 2003) also studied faulted lacustrine, fluvial, and marsh deposits on the west Bear Lake fault zone (Fig. 14D) and estimated a slip rate of ca. 0.5 m k.y.$^{-1}$ since ca. 13 ka; he found that the last earthquake event on this fault zone occurred at or just before ca. 7 ka. Surficial mapping has expanded knowledge of the distribution of fault scarps and youngest displaced deposits along

both faults (Reheis, 2005). Seismic profiling along the lake bottom has also revealed evidence of recent faulting (Colman, 2001, 2006; Denny and Colman, 2003).

Tectonic Geomorphology

Several morphological properties of East Bear Lake fault scarps (Figs. 4, 7, and 14) were studied to assess the pattern and relative activity of Quaternary normal faulting in Bear Lake Valley. Fault scarps were examined in the field and on aerial photographs, 1:24,000 U.S. Geological Survey topographic maps, and digital elevation models (DEMs). Mountain front–piedmont slope angle, length, and range-front sinuosity were measured using geographic information software (Laabs, 2001).

Along-strike changes in mountain-front morphology can be used both to identify fault-zone discontinuities (dePolo et al., 1990) and to indicate different ages of faulting (Crone and Haller, 1991). Distinct, along-strike changes in scarp height, slope angle, and orientation were used to divide the east Bear Lake fault zone into fault sections (Fig. 4; modified from segments as defined in Laabs, 2001). The range-front length and strike and the length of the mountain-piedmont junction of each segment were measured where applicable. The sinuosity was calculated as the ratio of the length of the mountain-piedmont junction to the length of the range front (Burbank and Anderson, 2001). We avoided river-eroded reaches while measuring sinuosity.

The morphology of fault scarps in Bear Lake Valley suggests that the east Bear Lake fault zone consists of three sections (Fig. 4) similar to those defined by McCalpin (1993), bounded by changes in fault geometry and recency of motion. The northern section of the east Bear Lake fault zone extends north of Montpelier, Idaho, where scarp strike changes from north-northwest to north-northeast (Fig. 7) and late Quaternary displacement has been minimal or zero. Only two possible short faults cutting late Pleistocene or Holocene deposits have been observed north of Bennington (Fig. 7), and unlike the fault zone to the south, these strike northeast. The central section lies between Montpelier and Indian Creek (Fig. 4). Along this section, several strands of the fault splay off to the north-northeast into bedrock where they show little or no evidence of late Quaternary displacement (Reheis, 2005), whereas the active strands strike north-northwest and bound steep, straight bedrock range fronts or exhibit youthful scarps in late Pleistocene and Holocene deposits (Figs. 14B and 14C). South of Indian Creek, the southern section of the east Bear Lake fault zone strikes north and is essentially confined to a single active trace or to two parallel, closely spaced faults between the range front and Bear Lake (Fig. 14A). Seismic data show additional faults offshore, some forming (paired) grabens (Colman, 2006).

The southern section of the east Bear Lake fault zone and that part of the central section south of Dingle (Figs. 4 and 7) reveal morphological evidence for "maximal" activity based on parameters outlined in McCalpin (1996) for mountain range fronts in semiarid or arid regions. Piedmont landforms include undissected alluvial and debris fans, talus cones, and triangular-faceted ridges (Figs. 14B and 14C). Alluvial fans at the mouths of North and South Eden Canyons are dissected by modern streams and cut by individual fault scarps as much as 14 m high (Fig. 14A; McCalpin, 1993, 2003). Drainage development on the mountain front is incipient, forming shallow V-shaped valleys in bedrock (Fig. 4). The low sinuosity (1.23) and steepness of the Bear Lake Plateau front (30–45°) and fresh scarps that mark surface ruptures suggest that faulting is active (McCalpin, 1996). Finally, a bathymetric low in Bear Lake west of South Eden Canyon suggests that normal faulting on the southern east Bear Lake fault zone has been recently active (Denny and Colman, 2003; Colman, 2006).

The northern section of the east Bear Lake fault zone (Fig. 4) reveals morphological evidence for "minimal" activity according to classifications of McCalpin (1996). Piedmont landforms include deeply dissected alluvial fans near the range fronts and deeply dissected pediments on Tertiary bedrock in the distal zones of alluvial fans (Oriel and Platt, 1980). Exposures of unfaulted alluvial fan gravel between Bennington and Georgetown reveal soils with Bk horizons as much as 2.5 m thick (K and Km horizons). Normal-fault scarps in alluvial fan deposits and pediments are scarce to absent. Alluvium-filled valleys are present in drainages of the Preuss Range, and the mountain front is generally draped by coalesced alluvial fans, suggesting that erosion and deposition have obliterated evidence of fault displacement. The

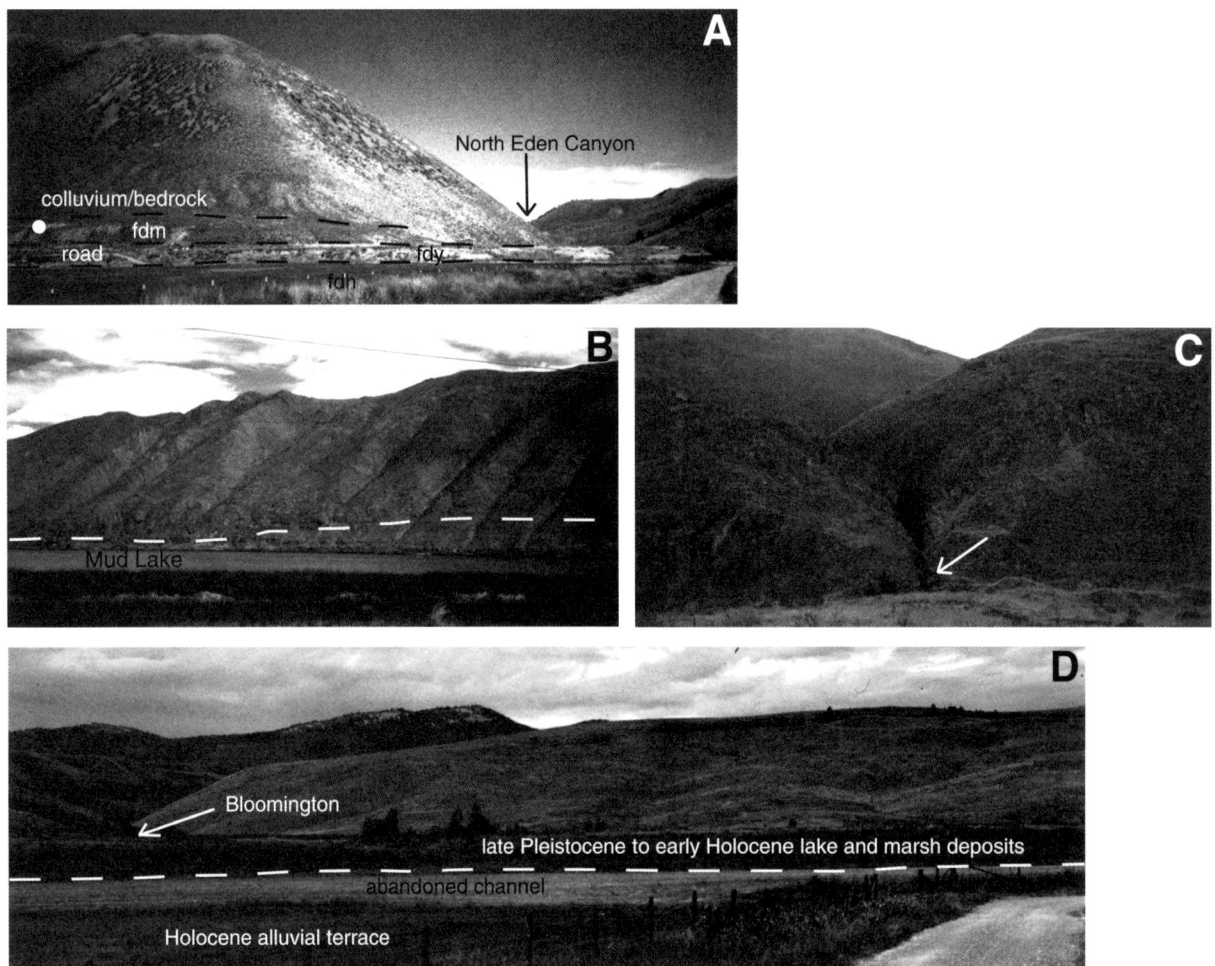

Figure 14. Photographs of fault scarps (see Figs. 4 and 7 for locations). (A) Faults separate fan-delta units at North Eden Creek on southern section of east Bear Lake fault zone; compare with map in Figure 11C. Black dashed lines show approximate locations of fault splays separating units fdm (fan-delta deposit of Jensen Spring phase, 49–37 cal ka), fdy (fan-delta deposit of Raspberry Square phase, 16–15 cal ka), and fdh (fan-delta deposit of Willis Ranch and younger phases, <8.5 cal ka). White dot is location of gastropod samples dated ca. 41 cal ka (Table 2). Fault trenches of McCalpin (2003) located above and below gravel road in lower left. (B) Central section of east Bear Lake fault zone looking east across Mud Lake; dashed white line shows active fault trace. (C) Wineglass canyon and fault scarp on east Bear Lake fault zone adjacent to Mud Lake. Arrow indicates outcrop exposing Holocene alluvial-fan and shoreface deposits faulted against bedrock at site 00BL-37 (Reheis, 2005). (D) View west of fault scarp near Bloomington on west Bear Lake fault zone. Scarp is 6–8 m high; white dashed line marks approximate fault trace. Fault trench of McCalpin (2003) located just off photo to left.

relatively high sinuosity (2.12) and lower slope angles (15–30°) of the Preuss Range front indicate that it has undergone slow or minimal tectonic activity in recent times (McCalpin, 1996).

On the west Bear Lake fault zone, fault scarps in Quaternary deposits are abundant between Ovid and St. Charles (Figs. 4 and 14D). Such scarps have not been identified north of Bern and are rare along the west side of Bear Lake, although fault-line scarps along steep bedrock fronts are present and seismic profiles within the lake (Colman, 2001, 2006; Denny and Colman, 2003) indicate that minor faults displace lake-bottom sediments along the western margin. In the absence of more definite evidence for changes in fault behavior along strike, we do not divide the west Bear Lake fault zone into sections. Previous studies have suggested that the west Bear Lake fault zone extends north of Bern either as several Neogene fault strands that displace Tertiary deposits (Oriel and Platt, 1980) or as a fault with morphologic evidence for recent activity (Laabs, 2001). However, the mountain-piedmont junction in this area is difficult to define and most of it has been eroded by the Bear River. In addition, the gentle eastward dip of deposits of the Miocene-Pliocene Salt Lake Formation beneath pediments east of the river (Oriel and Platt, 1980; Laabs, 2001) implies that displacement along this possible northward extension of the west Bear Lake fault zone must have been very limited (as it is within the lake basin; Colman, 2006); otherwise, the dips would be horizontal or even westward toward the footwall block of the west Bear Lake fault zone.

Slip Rates on the East Bear Lake Fault Zone

We use the water-level history developed above, founded on AAR and ^{14}C geochronology of Laabs and Kaufman (2003) and Reheis et al. (2005), to estimate late Quaternary slip rates on the east Bear Lake fault zone (Table 3). Uplifted fluvial and lacustrine deposits representing Pliocene(?) to middle Pleisto-

cene highstands of Bear Lake on the footwall block of the east Bear Lake fault zone, including terrace gravel on the crest of the Bear Lake Plateau (Fig. 2), provide dramatic evidence of long-term slip, but their counterparts have not been identified in undeformed locations or on the hanging-wall block and thus cannot be used to calculate slip rates. Estimates of depth beneath the lake to certain prominent seismic reflectors, along with the estimated ages of the reflectors and their overlying sediment thicknesses taken from Colman (2006), allow constraints to be placed on the total slip rate and component of footwall uplift between North Eden Creek and Indian Creek (Figs. 4 and 14) during different time intervals since ca. 200 ka.

During the late Bear Hollow phase, water level rose to at least 1818 m asl sometime between ca. 200 and 100 ka. Marl of this age at the Georgetown pit (BL00-14, Fig. 7) lies in a little-deformed area of the valley and is assumed to represent a reference datum. Two other sites of the same lake phase lie on the footwall block of the east Bear Lake fault zone and thus constrain footwall uplift. Fan-delta deposits at Bear Hollow (BL00-12, Figs. 7 and 11A) have a similar age range and lie on the footwall block of the north-central section of the east Bear Lake fault zone at ~1830 m asl. Shorezone deposits at site 00BL-54 (Figs. 7 and 11B) in the south-central section of the fault zone extend as high as 1864 m asl. Fault scarps west of the Bear Hollow and 00BL-54 sites show that displacement is a minimum of 6 m and 35 m, respectively (Table 3, Figs. 11A and 11B). We estimate the component of footwall uplift as 7–17 m to the north at Bear Hollow and 41–51 m at site 00BL-54 in the past 100–200 k.y., yielding uplift rates of 0.04–0.15 and 0.23–0.51 m k.y.$^{-1}$, respectively. Total slip rates can be estimated assuming that footwall uplift has been either 20% of total fault slip (Stein and Barrientos, 1985; Demsey, 1987; Morley, 1995), yielding minimum slip rates, or 10% of total slip, yielding maximum slip rates (Jackson and McKenzie, 1983). These calculations yield minimum slip

TABLE 3. COMPARISON OF SLIP RATES ALONG THE EAST BEAR LAKE FAULT ZONE (SEE FIGS. 4, 7, AND 11 FOR LOCATIONS)

Site number or location	Age range (ka)	Footwall uplift (m)	Footwall uplift rate (m/k.y.)	Max. total slip (m)	Max. slip rate (m/k.y.)	Min. total slip (m)	Min. slip rate (m/k.y.)
Central section of fault zone							
BL00-12 (Bear Hollow)	200–100	7–17	0.04–0.15	70–170*	0.4–1.5*	35–85[†]	0.2–0.8[†]
98BL-11 (Dingle)	47–39	N.D.	N.D.	≥6	≥0.12–0.15	≥6	≥0.12–0.15
North of 98BL-11	16–15	N.D.	N.D.	≥4	≥0.25	≥4	≥0.25
Central-southern border area of fault zone							
00BL-54 (North of Indian Creek)	200–100	41–51	0.2–0.5	410–510*	2.0–5.1*	205–255[†]	1.0–2.6[†]
00BL-54 and seismic reflector R7	200–100	41–51	0.2–0.5	110[‡]	1.1[‡]	60[‡]	0.3[‡]
Southern section of fault zone							
BL00-07, DK99-20C, BL00-02C (Hen House, Bee Hunt, North Eden)	47–39	9–12	0.2–0.3	90–120*	1.9–3.1*	45–60[†]	1.0–1.5[†]
BL00-07 (Hen House) and seismic reflector R3	47–35	9	0.2–0.3	65[‡]	1.9[‡]	45[‡]	1.0[‡]
BL00-02C (North Eden) and seismic reflector R3	47–35	12	0.3	91[‡]	2.6[‡]	31[‡]	0.8[‡]
North Eden (McCalpin, 2003)	47–39	N.D.	N.D.	>22.9	>1.4–1.5	>22.9	>1.4–1.5
North Eden (McCalpin, 2003)	16–15	N.D.	N.D.	10.5	0.7	10.5	0.7

*Assumes footwall uplift is ~10% of magnitude of hanging-wall subsidence (Jackson and McKenzie, 1983).

[†]Assumes footwall uplift is ~20% of magnitude of hanging-wall subsidence (Stein and Barrientos, 1985; Demsey, 1987; Morley, 1995).

[‡]Maximum slip and slip rates calculated using elevation of site above lake level plus estimated modern water depth offshore plus estimated thickness of sediment above marker seismic horizon of similar age (R3, about 35 ka; R7, about 97 ka; seismic data and horizon age from Colman et al., this volume). Minimum slip and slip rates exclude modern water depth. See discussion in text. N.D.—no data.

rates of 0.2–0.8 m k.y.$^{-1}$ and maximum rates of 0.4–1.5 m k.y.$^{-1}$ for the north-central section of the east Bear Lake fault zone, and minimum rates of 1.0–2.6 m k.y.$^{-1}$ and maximum rates of 2.0–5.1 m k.y.$^{-1}$ for the south-central section of the fault zone.

During the Jensen Spring phase, ca. 47–39 ka based on ^{14}C ages, Bear Lake attained an altitude of ~1817 m (Table 1). At site 98BL-11 south of Dingle (Fig. 7), fan-delta deposits correlated with the Jensen Spring phase on the basis of soil morphology lie at ~1817 m asl and are cut by a 6-m-high fault scarp; assuming a similar age range for these deposits yields a minimum slip rate of 0.12–0.15 m k.y.$^{-1}$ for one strand of the central section of the fault zone. Deposits of the Jensen Spring phase at three localities on the footwall block of the southern section of the east Bear Lake fault zone—Bee Hunt, Hen House, and North Eden Canyon (Figs. 11B and 11C)—lie above the east shore of Bear Lake between 1826 and 1829 m asl (Laabs and Kaufman, 2003; Reheis, 2005). Their higher elevations represent footwall uplift of ~9–12 m (increasing to the south and highest at North Eden Canyon) since ca. 47–39 ka, and yield uplift rates of 0.19–0.31 m k.y.$^{-1}$ (Table 3). Again assuming this is either 20% (Stein and Barrientos, 1985; Demsey, 1987; Morley, 1995) or 10% (Jackson and McKenzie, 1983) of the total displacement, estimated total slip rates are 1.9–3.1 m k.y.$^{-1}$ (maximum) and 1.0–1.5 m k.y.$^{-1}$ (minimum) along the southern section of the fault zone, increasing to the south. These slip rates are somewhat higher than, but entirely consistent with, a minimum slip rate of ~1.4–1.5 m k.y.$^{-1}$ calculated from the minimum displacement of beach gravel of the Jensen Spring phase exposed in fault trenches at North Eden Canyon (McCalpin, 1993, 2003). Slip rates estimated from fault trenches at North Eden Canyon are minimum values because additional faults lie offshore, and because the Jensen Spring-equivalent beach deposits were not exposed in the fault trench west of the western fault trace.

During the Raspberry Square phase, ca. 16–15 cal ka, interpreted nearshore deposits at Garden City on the west shore (Fig. 4) indicate that Bear Lake rose to ~1814 m asl (Table 1). Faulted fan-delta gravels of similar age at North Eden Canyon on the east shore (Fig. 11C) at an altitude of ~1815 m are inset below a fault scarp below the older Jensen Spring-equivalent deposits; these inset deposits have been displaced a cumulative 10.5 m across two fault strands (McCalpin, 1993, 2003). Fan-delta deposits to the north at the mouth of Bear River between Dingle and site 98BL-11 are mapped as correlative with the Raspberry Square phase on the basis of soil morphology (Reheis, 2005) and are cut by a 4-m-high fault scarp. These displacements yield slip rates of 0.7 m k.y.$^{-1}$ and 0.25 m k.y.$^{-1}$, respectively, for the southern and central sections of the east Bear Lake fault zone (Table 3).

To estimate total slip rates and footwall-uplift rates for the east Bear Lake fault zone near Indian Creek and at North Eden Creek, we can compare slip-rate estimates based on an assumed ratio of footwall uplift to total displacement to those made by combining altitudes of footwall shoreline sites with seismic data of Colman (2006) (Table 3, Fig. 15). Colman (2006) identified and mapped prominent seismic reflectors and their overlying sediment thicknesses beneath Bear Lake, and correlated these markers with sediments in core BL00-1E. The estimated ages of two of these reflectors are 35 ka for R3 and 97 ka for R7. Assuming that R3 represents a time near the end of the Jensen Spring phase (47–39 ka) and that R7 represents a time near the end of the late Bear Hollow phase (200–100 ka) allows us to calculate maximum total slip by summing (1) the altitude of similar-aged deposits above modern lake level, (2) modern water depth offshore, and (3) depth to the seismic reflector. The calculation yields maximum total slip because the seismic-reflector strata must have been deposited at some water depth; hence the use of modern water depth is an overestimate for that part of the hanging-wall slip not accounted for by sediment thickness. Minimum slip estimates exclude modern water depth. For site BL00-7 (Hen House) just south of Indian Creek (Fig. 11B), these calculations yield minimum and maximum slip rates of 1.0 and 1.9 m k.y.$^{-1}$, and for site BL00-02C at North Eden Canyon, 0.8 and 2.6 m k.y.$^{-1}$. Comparing maximum total displacement with footwall uplift at these sites (Table 3) indicates that footwall uplift has been a minimum of 13% of total displacement since ca. 50 ka, a result that supports the assumed 10%–20% range for the footwall-uplift component.

Applying the same method to reflector R7 and site 00BL-54 (200–100 ka) north of Indian Creek yields minimum and maximum slip rates of 0.3 and 1.1 m k.y.$^{-1}$ and a value of 35%–45% footwall uplift relative to maximum total displacement (Table 3). Although the slip rates seem reasonable, the proportion of footwall uplift is much larger than the 13% estimated for younger deposits. The discrepancy could be caused by three factors: (1) the age of deposits at 00BL-54 is not the same as the age estimated for R7; (2) the depth to R7, here estimated by extrapolating incomplete isopachs of sediment thickness northward to Indian Creek from Figure 6 of Colman (2006), is incorrect; or (3) the estimated proportion of footwall uplift to total slip is incorrect over the longer time scale. For example, in a study of the Borah Peak, Idaho, earthquake, Stein and Barrientos (1985) pointed out that measured footwall-to-hanging-wall displacement (1:5) caused by the seismic event was similar to that indicated by geologic data during the past ca. 15,000 years but the cumulative Pliocene displacement across the fault suggested that interseismic deformation increased uplift to approximately equal subsidence.

Discussion of Neotectonic Effects

The presence of steeper and straighter mountain fronts to the south along the east Bear Lake fault zone (Fig. 4), along with greater slip rates on the central and southern sections of the fault relative to the northern section (Table 3), means either that the focus of normal faulting has shifted southward with time, or that slip rates have always been higher to the south. The northern section of the east Bear Lake fault zone was probably more active during the Pliocene than later. In this area, Quaternary pediment gravel lying on Tertiary mudstone and tuff of the Salt Lake Formation has not been faulted and dips gently westward toward the valley floor (Oriel and Platt, 1980; Laabs, 2001). In addition, the subdued, sinuous morphology of the Preuss Range front suggests that the north segment of the east Bear Lake fault zone has had minimal tectonic activity during the Quaternary. In contrast, the east-dipping Salt Lake Formation strata suggest that this part of the fault was active in the Tertiary. The distribution of fault scarps (Reheis, 2005) and the morphology of range fronts suggest that the southward increase in slip rate coincides with confinement of the fault zone to fewer, more closely spaced faults. The morphology of the southern section of the east Bear Lake fault zone, along with seismic data of Colman (2006) and results of McCalpin (1993, 2003), indicates that it is the focus of most recent normal faulting in Bear Lake Valley. Our interpretation of a southward shift in fault activity since Pliocene time is similar to observations of Anders et al. (1989) on the Grand Valley and Star Valley faults to the northeast of Bear Lake (Fig. 1) and to general conclusions about patterns of fault activity associated with the track of the Yellowstone hotspot (Pierce and Morgan, 1992).

Slip rates on the southern section of the east Bear Lake fault zone have apparently decreased over the past 50 k.y. (Table 3), from at least 1.0 m k.y.$^{-1}$ since 50 ka to ~0.7 m k.y.$^{-1}$ since 15 ka (data of McCalpin, 2003). McCalpin (1993) suggested several hypotheses to account for apparently lower slip rates during the past 15 k.y. than the long-term, late Cenozoic rates on the east

Bear Lake fault zone. These include (1) a seismic cycle greater than 15 k.y., (2) an earlier time of fault initiation, and (3) a possible "slip deficit" caused by some factor that has restrained slip since 15 ka. We cannot assess these hypotheses independently, but suggest that slip could have been restrained during the past 15 k.y. by relatively low lake levels that reduced water loading in the valley.

CONCLUSIONS

Our studies of the surficial geology, geomorphology, and neotectonics shed light on the complex interactions that drive the relationship between the Bear River and Bear Lake. The chemistry of water and the lithology of sediments in Bear Lake largely reflect whether Bear River was merged with Bear Lake to form a flow-through, river-dominated lake, or was separated from Bear Lake, resulting in a topographically closed lake fed largely by groundwater flow. Whether the river and lake are merged depends on faulting, fluvial geomorphic processes, changes in threshold altitude, and changes in paleoclimate.

An ancestral Bear River course, probably of Pliocene age, that lacks Uinta Range–derived clasts is preserved atop the Bear River Plateau east of Bear Lake. This ancestral river likely began

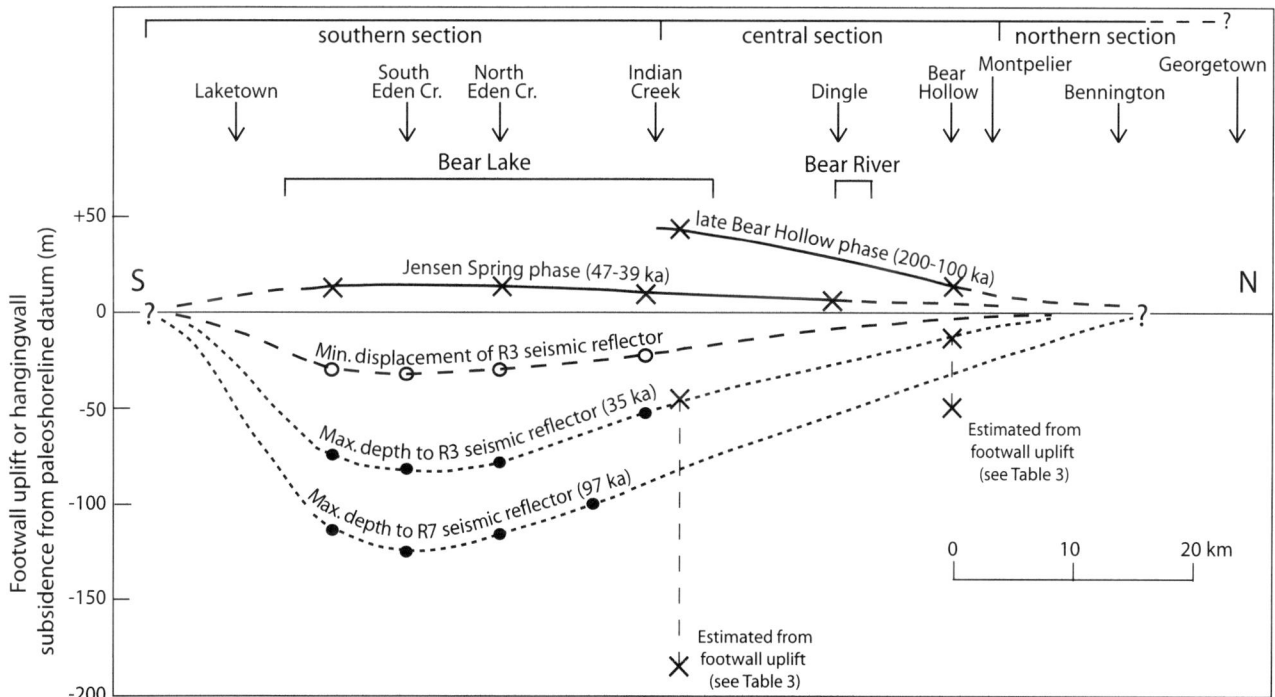

Figure 15. Diagram showing relative displacement of footwall (positive) and hanging-wall (negative) blocks along east Bear Lake fault zone. Footwall sites (X's) based on sites discussed in text (Figs. 3 and 7) and in Reheis (2005). Hanging-wall sites of late Bear Hollow phase (X's; 200–100 ka) north of Indian Creek and at Bear Hollow estimated from footwall uplift:subsidence ratios of 1:5 (maximum) and 1:1 (minimum) (Stein and Barrientos, 1985). Other hanging-wall sites derived from seismic data of Colman (2006). Maximum estimates of hanging-wall subsidence (filled circles) obtained by summing (1) water depth, (2) thickness of sediment overlying seismic reflector, and (3) difference between modern lake level (1805 m asl) and reconstructed lake highstand during Jensen Spring and late Bear Hollow phases (Fig. 10). Minimum estimates (open circles) exclude water depth. Minimum estimates not shown for late Bear Hollow phase (R7 reflector) for clarity, because those points generally fall on the same line as that of maximum depth to the R3 reflector.

to incise and migrate eastward in response to uplift along the east Bear Lake fault zone at the same time that subsidence of the hanging-wall block began to create an accommodation space that became Bear Lake. Capture of the modern headwaters in the Uinta Range, which were formerly tributary to the Green River, added considerable volume to this ancestral Bear River. The subsequent diversion of the Bear River into Bear Lake Valley and the beginning of lacustrine deposition may have been concurrent.

Our studies have established the following broad framework for the record of aggradation and incision by the river and highstands of Bear Lake. Reconstruction of this record is complicated by the differential effects of tectonics on terrace and shoreline altitudes and the paucity of ages on deposits older than ca. 100 ka. The oldest exposed lacustrine sediments in Bear Lake Valley are probably of Pliocene or early Pleistocene age, but their depositional altitude cannot be constrained. Lake level must then have fallen and the river established a course separated from the lake, as indicated by Bear River gravel overlying lacustrine deposits.

During the early Bear Hollow phase, ancestral Bear Lake reached a maximum altitude of 1829 m asl at least once between ca. 1200 and 700 ka, constrained by a threshold at that altitude near Nounan narrows. A prolonged, ~600 k.y. period of episodic incision of the threshold to as low as ~1810 m asl followed this phase, with incision possibly persisting as late as 200 ka. Bear Lake filled the valley to the threshold at ~1820 m asl at least once during the middle Bear Hollow phase, 500–300 ka. This altitude is consistent with progressive lowering of the threshold and only requires that the river merged with the lake to cause the lake to rise. Several terrace remnants in the reach downstream of the graben record incision of the threshold to ~1810 m asl after this lake-filling episode, followed by at least 6 m of aggradation just upstream of Nounan narrows by ca. 180 ka. During this time the Bear River was probably separated from Bear Lake.

During the late Bear Hollow phase, water level rose to at least 1818 m asl shortly after ca. 180 ka. Such an increase in threshold altitude could be explained by one or more of the following: (1) aggradation due to a very large sediment load associated with extensive glaciation during oxygen isotope stage 6, (2) normal faulting to the south, or (3) landsliding upstream of Nounan narrows. This lake phase probably coincided with the Blacks Fork glaciation in the Uinta Range; correlative moraines in the Bear River Range west of the lake indicate that valley glaciers there were as much as 6–7 km long. The lake threshold appears to have remained at nearly the same altitude (at or slightly above 1817 m asl) until the Jensen Spring phase, ca. 47–39 cal ka, suggesting that flow in Bear River was so low that it could not effectively incise a bedrock threshold.

After the Jensen Spring phase, the Bear River incised its threshold relatively rapidly to essentially its present gradient by ca. 18 cal ka, but by ca. 16 cal ka had aggraded 8 m of sediment again. The rapid aggradation may have been caused by a landslide in a narrow part of the valley west of Bennington (Laabs and Kaufman, 2003). The lake simultaneously rose to a highstand at ~1814 m asl by 16–15 cal ka during the Raspberry Square phase.

This period coincided with the start of the last deglaciation in the Uinta Range and likely the Bear River Range. Renewed threshold incision or a marked decrease in river flow caused the north shore of the lake to regress southward and Bear River to separate from Bear Lake by ca. 12 cal ka (Dean et al., 2006; Dean, this volume), but the lake once again rose during the early Holocene when channel migration in the delta area diverted all or part of the Bear River toward the lake. Since then, the river has incised to ~1790 m asl at Nounan narrows.

Dating of sediments within the former river courses visible in aerial photography of the extensive modern marsh between Bear Lake and the Bear River shows that the Bear River has migrated northward during the past 8000 years. In addition, two west-side axial drainages, partly controlled by grabens, were eventually abandoned; these drainages had previously carried streamflow northward from tributaries and from Bear Lake. The causes of the northward shift and abandonment of the west-side drainages include gradual channel migration, earthquake events on the east Bear Lake fault zone, and inactivity of the west Bear Lake fault zone since the early Holocene such that graben formation ceased to aid north-flowing drainage.

We interpret the presence of steeper and straighter mountain fronts to the south along the east Bear Lake fault zone, along with greater slip rates on the central and southern sections of the fault relative to the northern section, to indicate either that the focus of normal faulting has shifted southward with time, or less likely, that slip rates have always been higher to the south. The distribution of fault scarps and the morphology of range fronts suggest that the southward increase in slip rate coincides with confinement of the fault zone to fewer, more closely spaced faults. Slip rates on the southern section of the east Bear Lake fault zone have apparently decreased over the past 50 k.y., from at least 1.0 m k.y.$^{-1}$ since 50 ka to ~0.7 m k.y.$^{-1}$ since 15 ka (McCalpin, 2003).

ACKNOWLEDGMENTS

We acknowledge the contributions of many people to this study. Thanks go to Rick Forester for his paleoenvironmental interpretations of ostracode fauna; Jack McGeehin for radiocarbon ages; Lanya Ross, Dhiren Khona (U.S. Geological Survey), Jordon Bright, and Laura Levy (Northern Arizona University) as field assistants; and Mark Manone for technical assistance with GIS analysis. We appreciate the insightful reviews by Steve Colman, Ken Pierce, Jeff Munroe, and Joel Pederson; their comments have greatly improved the paper.

REFERENCES CITED

Anders, M.H., Geissman, J.W., Piety, L.A., and Sullivan, J.T., 1989, Parabolic distribution of circumeastern Snake River plain seismicity and latest Quaternary faulting; Migration pattern and association with the Yellowstone hot spot: Journal of Geophysical Research, v. 94, p. 1589–1621, doi: 10.1029/JB094iB02p01589.

Armstrong, R.L., 1968, The Sevier orogenic belt in Nevada and Utah: Geological Society of America Bulletin, v. 79, p. 429–458, doi: 10.1130/0016-7606(1968)79[429:SOBINA]2.0.CO;2.

Armstrong, R.L., Leeman, W.P., and Malde, H.E., 1975, K-Ar dating, Quaternary and Neogene volcanic rocks of the Snake River Plain, Idaho: American Journal of Science, v. 275, p. 225–251.

Bond, J.G., 1978, Geologic map of Idaho: Idaho Bureau of Mines and Geology Map GM-1, scale 1:500,000.

Bouchard, D.P., Kaufman, D.S., Hochberg, A., and Quade, J., 1998, Quaternary history of the Thatcher Basin, Idaho, reconstructed from the $^{87}Sr/^{86}Sr$ and amino acid composition of lacustrine fossils: Implications for the diversion of the Bear River into the Bonneville Basin: Palaeogeography, Palaeoclimatology, Palaeoecology, v. 141, p. 95–114, doi: 10.1016/S0031-0182(98)00005-4.

Bradley, W.H., 1936, Geomorphology of the north flank of the Uinta Mountains: U.S. Geological Survey Professional Paper 185-I, p. 163–204.

Bright, R.C., 1963, Pleistocene Lakes Thatcher and Bonneville, southeastern Idaho [Ph.D. thesis]: Minneapolis, University of Minnesota, 171 p.

Bright, J., Kaufman, D.S., Forester, R.M., and Dean, W.E., 2006, A continuous 250,000 yr record of oxygen and carbon isotopes in ostracode and bulk-sediment carbonate from Bear Lake, Idaho-Utah: Quaternary Science Reviews, v. 25, p. 2258–2270, doi: 10.1016/j.quascirev.2005.12.011.

Bryant, B., 1992, Geologic and structure maps of the Salt Lake City 1° × 2° quadrangle, Utah and Wyoming: U.S. Geological Survey Miscellaneous Investigations Series Map I-1997, scale 1:125,000.

Burbank, D.W., and Anderson, R.S., 2001, Tectonic geomorphology: Malden, Mass., Blackwell Science, 274 p.

Colman, S.M., 2001, Seismic-stratigraphic framework for drill cores and paleoclimate records in Bear Lake, Utah-Idaho: Eos (Transactions, American Geophysical Union), v. 82, no. 47, p. F755.

Colman, S.M., 2006, Acoustic stratigraphy of Bear Lake, Utah-Idaho—Late Quaternary sedimentation patterns in a simple half-graben: Sedimentary Geology, v. 185, p. 113–125, doi: 10.1016/j.sedgeo.2005.11.022.

Colman, S.M., Rosenbaum, J.G., Kaufman, D.S., Dean, W.E., and McGeehin, J.P., 2009, this volume, Radiocarbon ages and age models for the last 30,000 years in Bear Lake, Utah and Idaho, *in* Rosenbaum, J.G., and Kaufman, D.S., eds., Paleoenvironments of Bear Lake, Utah and Idaho, and its catchment: Geological Society of America Special Paper 450, doi: 10.1130/2009.2450(05).

Coogan, J.C., 1992, Structural evolution of piggyback basins in the Wyoming-Idaho-Utah thrust belt, *in* Link, P.K., Kuntz, M.A., and Platt, L.B., eds., Regional geology of eastern Idaho and western Wyoming: Geological Society of America Memoir 179, p. 55–82.

Coogan, J.C., and King, J.K., 2001, Progress report, geologic map of the Ogden 30′× 60′ quadrangle: Utah Geological Survey Open-File Report 380, scale 1:100,000.

Crone, A.J., and Haller, K.M., 1991, Segmentation and the coseismic behavior of Basin and Range normal faults: Examples from east-central Idaho and southwestern Montana: Journal of Structural Geology, v. 13, p. 151–164, doi: 10.1016/0191-8141(91)90063-O.

Dean, W.E., 2009, this volume, Endogenic carbonate sedimentation in Bear Lake, Utah and Idaho over the last two glacial-interglacial cycles, *in* Rosenbaum, J.G., and Kaufman, D.S., eds., Paleoenvironments of Bear Lake, Utah and Idaho, and its catchment: Geological Society of America Special Paper 450, doi: 10.1130/2009.2450(07).

Dean, W.E., Rosenbaum, J., Skipp, G., Colman, S., Forester, R., Liu, A., Simmons, K., and Bischoff, J., 2006, Unusual Holocene and late Pleistocene carbonate sedimentation in Bear Lake, Utah and Idaho, USA: Sedimentary Geology, v. 185, p. 93–112, doi: 10.1016/j.sedgeo.2005.11.016.

Dean, W.E., Forester, R.M., Bright, J., and Anderson, R.Y., 2007, Influence of the diversion of the Bear River into Bear Lake (Utah and Idaho) on the environment of deposition of carbonate minerals: Evidence from water and sediments: Limnology and Oceanography, v. 53, p. 1094–1111.

DeGraff, J.V., 1979, Numerical analysis of cirques in the Bear River Range, north-central Utah: Modern Geology, v. 7, p. 43–51.

Demsey, K., 1987, Holocene faulting and tectonic geomorphology along the Wassuk Range, west-central Nevada [M.S. thesis]: Tucson, University of Arizona, 64 p.

Denny, J.F., and Colman, S.M., 2003, Geophysical survey of Bear Lake, Utah-Idaho, September 2002: U.S. Geological Survey Open-File Report 03-150.

dePolo, C.M., Clark, D.G., Slemmons, D.B., and Ramelli, A.R., 1990, Historical surface faulting in the Basin and Range province, western North America: Implications for fault segmentation: Journal of Structural Geology, v. 13, p. 123–136.

Dover, J.H., 1995, Geologic map of the Logan 30′ × 60′ quadrangle, Cache and Rich counties, Utah, and Lincoln and Uinta counties, Wyoming: U.S.

Geological Survey Miscellaneous Investigations Series Map I-2210, scale 1:100,000.

Dover, J.H., and M'Gonigle, J.W., 1993, Geologic map of the Evanston 30′× 60′ quadrangle, Uinta and Sweetwater counties, Wyoming: U.S. Geological Survey Miscellaneous Investigations Series Map I-2168, scale 1:100,000.

Evans, J.P., Martindale, D., and Kendrick, R.D.J., 2003, Geologic setting of the 1884 Bear Lake, Idaho, earthquake: Rupture in the hanging wall of a Basin and Range normal fault revealed by historical and geological analyses: Bulletin of the Seismological Society of America, v. 93, p. 1621–1632, doi: 10.1785/0120020159.

Gibbons, A.B., 1986, Surficial materials map of the Evanston 30′ × 60′ quadrangle, Uinta and Sweetwater Counties, Wyoming: U.S. Geological Survey Coal Investigations Series Map C-103, scale 1:100,000.

Hansen, W.R., 1985, Drainage development of the Green River Basin in southwestern Wyoming and its bearing on fish biogeography, neotectonics, and paleoclimates: The Mountain Geologist, v. 22, no. 4, p. 192–204.

Heumann, A., 2004, Timescales of evolved magma generation at Blackfoot Lava Field, SE Idaho, USA: IAV-CEI General Assembly, Abstracts.

Hintz, L.F., 1980, Geologic map of Utah: Utah Geological Survey, scale 1:500,000.

Jackson, J., and McKenzie, D., 1983, The geometrical evolution of normal fault systems: Journal of Structural Geology, v. 5, p. 471–482, doi: 10.1016/0191-8141(83)90053-6.

Kitagawa, H., and van der Plicht, J., 1998, Atmospheric radiocarbon calibration to 45,000 yr B.P.: Late glacial fluctuations and cosmogenic isotope production: Science, v. 279, p. 1187–1190, doi: 10.1126/science.279.5354.1187.

Laabs, B.J.C., 2001, Quaternary lake-level history and tectonic geomorphology in Bear Lake Valley, Utah and Idaho [M.S. thesis]: Flagstaff, Northern Arizona University, 132 p.

Laabs, B.J.C., and Kaufman, D.S., 2003, Quaternary highstands in Bear Lake Valley, Utah and Idaho: Geological Society of America Bulletin, v. 115, p. 463–478, doi: 10.1130/0016-7606(2003)115<0463:QHIBLV>2.0.CO;2.

Laabs, B.J.C., Munroe, J.S., Rosenbaum, J.G., Refsnider, K.A., Mickelson, D.M., Singer, B.S., and Caffee, M.W., 2007, Chronology of the Last Glacial Maximum in the upper Bear River basin, Utah: Arctic, Antarctic, and Alpine Research, v. 39, p. 537–548, doi: 10.1657/1523-0430(06-089)[LAABS]2.0.CO;2.

Liu, A., Kelts, K., Rosenbaum, J.G., Dean, W.E., Kaufman, D.S., and Bright, J., 1999, Stable-isotope stratigraphy of carbonate-rich deposits from Bear Lake, Utah-Idaho: Eos (Transactions, American Geophysical Union), v. 80, p. F1168.

Love, J.D., and Christiansen, A.C., 1985, Geologic map of Wyoming: U.S. Geological Survey, scale 1:500,000.

Mabey, D.R., 1971, Geophysical data relating to a possible Pleistocene overflow of Lake Bonneville at Gem Valley, southeastern Idaho: U.S. Geological Survey Professional Paper 750-B, p. B122–B127.

Mansfield, G.R., 1927, Geography, geology, and mineral resources of part of southeastern Idaho: U.S. Geological Survey Professional Paper 152, p. 453.

McCalpin, J.P., 1993, Neotectonics of the northeastern Basin and Range margin, western USA: Zeitschrift für Geomorphologie N.F., Suppl. Bd. 94, p. 137–157.

McCalpin, J.P., 1996, Paleoseismology: San Diego, Academic Press, 588 p.

McCalpin, J.P., 2003, Neotectonics of Bear Lake Valley, Utah and Idaho; a preliminary assessment: Utah Geological Survey Miscellaneous Publication 03-4, 43 p.

M'Gonigle, J.W., and Dover, J.H., 1992, Geologic map of the Kemmerer 30′ × 60′ quadrangle, Lincoln, Uinta, and Sweetwater Counties, Wyoming: U.S. Geological Survey Miscellaneous Investigations Series Map I-2079, scale 1:100,000.

Morley, C.K., 1995, Developments in the structural geology of rifts over the last decade and their impact on hydrocarbon exploration, *in* Lambiase, J.J., ed., Hydrocarbon habitat in rift basins: Geological Society [London], Special Publication 80, p. 1–32.

Munroe, J.S., 2000, Radiocarbon dates provide constraints on the extent of neoglaciation in the Uinta Mountains, northeastern Utah: Geological Society of America Abstracts with Programs, v. 32, no. 5, p. 34.

Munroe, J.S., 2001, Late Quaternary history of the northern Uinta Mountains, northeastern Utah [Ph.D. thesis]: Madison, University of Wisconsin, 398 p.

Munroe, J.S., Laabs, B.J.C., Shakun, J.D., Singer, B.S., Michelson, D.M., Refsnider, K.A., and Caffee, M.W., 2006, Latest Pleistocene advance of alpine glaciers in the southwestern Uinta Mountains, Utah, USA: Evidence for the influence of local moisture sources: Geology, v. 34, p. 841–844, doi: 10.1130/G22681.1.

Oriel, S.S., and Platt, L.B., 1980, Geologic map of the Preston 1° × 2° quadrangle, southeastern Idaho and western Wyoming: U.S. Geological Survey Miscellaneous Investigations Series Map I-112, scale 1:250,000.

Oviatt, C.G., Thompson, R.S., Kaufman, D.S., Bright, J., and Forester, R.M., 1999, Reinterpretation of the Burmester Core, Bonneville Basin, Utah: Quaternary Research, v. 52, p. 180–184, doi: 10.1006/qres.1999.2058.

Parsons, T., Thompson, G.A., and Sleep, N.H., 1994, Mantle plume influence on the Neogene uplift and extension of the U.S. western Cordillera: Geology, v. 22, p. 83–86, doi: 10.1130/0091-7613(1994)022<0083:MPIOTN>2.3.CO;2.

Pickett, K.E., 2004, Physical volcanology, petrography, and geochemistry of basalts in the bimodal Blackfoot Volcanic Field, southeastern Idaho [M.S. thesis]: Pocatello, Idaho State University, 92 p.

Pierce, K.L., and Morgan, L.A., 1992, The track of the Yellowstone hot spot: Volcanism, faulting, and uplift, *in* Link, P.K., Kuntz, M.A., and Platt, L.B., eds., Regional geology of eastern Idaho and western Wyoming: Geological Society of America Memoir 179, p. 1–53.

Refsnider, K.A., Laabs, B.J.C., and Mickelson, D.M., 2007, Glacial geology and equilibrium line altitude reconstructions for the Provo River drainage, Uinta Mountains, Utah, U.S.A.: Arctic, Antarctic, and Alpine Research, v. 39, p. 529–536, doi: 10.1657/1523-0430(06-060) [REFSNIDER]2.0.CO;2.

Reheis, M.C., 2005, Surficial geologic map of the upper Bear River and Bear Lake drainage basins, Idaho, Utah, and Wyoming: U.S. Geological Survey Scientific Investigations Series Map 2890, scales 1:150,000 and 1:50,000 (http://pubs.usgs.gov/sim/2005/2890/).

Reheis, M.C., Laabs, B.J.C., Forester, R.M., McGeehin, J.P., Kaufman, D.S., and Bright, J., 2005, Surficial deposits in the Bear Lake basin: U.S. Geological Survey Open-File Report 2005-1088.

Robertson, G.C., 1978, Surficial deposits and geologic history, northern Bear Lake Valley, Idaho [M.S. thesis]: Logan, Utah State University, 162 p.

Rosenbaum, J.G., and Heil, C.W., Jr., 2009, this volume, The glacial/deglacial history of sedimentation in Bear Lake, Utah and Idaho, *in* Rosenbaum, J.G., and Kaufman, D.S., eds., Paleoenvironments of Bear Lake, Utah and Idaho, and its catchment: Geological Society of America Special Paper 450, doi: 10.1130/2009.2450(11).

Rosenbaum, J.G., Dean, W.E., Reynolds, R.L., and Reheis, M.C., 2009, this volume, Allogenic sedimentary components of Bear Lake, Utah and Idaho, *in* Rosenbaum, J.G., and Kaufman, D.S., eds., Paleoenvironments of Bear Lake, Utah and Idaho, and its catchment: Geological Society of America Special Paper 450, doi: 10.1130/2009.2450(06).

Rubey, W.W., Oriel, S.S., and Tracey, J.I., Jr., 1980, Geologic map and structure sections of the Cokeville 30-minute quadrangle, Lincoln and Sublette counties, Wyoming: U.S. Geological Survey Miscellaneous Investigations Series Map I-1129, scale 1:62,500.

Sarna-Wojcicki, A.M., Reheis, M.C., Pringle, M.S., Fleck, R.J., Burbank, D.M., Meyer, C.E., Slate, J.L., Wan, E., Budahn, J.R., Troxel, B.W., and Walker, J.P., 2005, Tephra layers of Blind Spring Valley and related upper Pliocene and Pleistocene tephra layers, California, Nevada, and Utah: Isotopic ages, correlation, and magnetostratigraphy: U.S. Geological Survey Professional Paper 1701, p. 63.

Scott, W.E., Pierce, K.L., Bradbury, J.P., and Forester, R.M., 1982, Revised Quaternary stratigraphy and chronology in the American Falls area, southeastern Idaho, *in* Bonnichsen, B., and Breckenridge, R.M., eds., Cenozoic geology of Idaho: Idaho Bureau of Mines and Geology, p. 581–595.

Smoot, J.P., and Rosenbaum, J.G., 2009, this volume, Sedimentary constraints on late Quaternary lake-level fluctuations at Bear Lake, Utah and Idaho, *in* Rosenbaum, J.G., and Kaufman, D.S., eds., Paleoenvironments of Bear Lake, Utah and Idaho, and its catchment: Geological Society of America Special Paper 450, doi: 10.1130/2009.2450(12).

Spangler, L.E., 2001, Delineation of recharge areas for karst springs in Logan Canyon, Bear River Range, northern Utah, *in* Kuniasky, E.L., ed., U.S. Geological Survey Karst Interest Group Proceedings: U.S. Geological Survey Water-Resources Investigations Report 01-4011, p. 186–193.

Stein, R.S., and Barrientos, S.E., 1985, Planar high-angle faulting in the Basin and Range: Geodetic analysis of the 1983 Borah Peak, Idaho, earthquake: Journal of Geophysical Research, v. 90, no. B13, p. 11,355–11,366, doi: 10.1029/JB090iB13p11355.

Stuiver, M., Reimer, P.J., and Reimer, R., 2004, CALIB Radiocarbon Calibration, version 4.4: http://depts.washington.edu/qil/calib/ (accessed 2005).

Taylor, D.W., and Bright, R.C., 1987, Drainage history of the Bonneville Basin, Cenozoic geology of Western Utah: Salt Lake City, Utah Geological Association, p. 239–256.

Thompson, R.S., Whitlock, C., Bartlein, P.J., Harrison, S.P., and Spaulding, W.G., 1993, Climatic changes in the western United States since 18,000 yr B.P, *in* Wright, H.E., Jr., Kutzbach, J.E., Webb, T.I.I.I., Ruddiman, W.F., Street-Perrott, F.A., and Bartlein, P.J., eds., Global climates since the last glacial maximum: Minneapolis, University of Minnesota Press, p. 468–513.

U.S. Geological Survey, 2004, Quaternary fault and fold database for the United States: U.S. Geological Survey, http://qfaults.cr.usgs.gov/ (accessed 2005).

West, M.W., 1993, Extensional reactivation of thrust faults accompanied by coseismic surface rupture, southwestern Wyoming and north-central Utah: Geological Society of America Bulletin, v. 105, p. 1137–1150, doi: 10.1130/0016-7606(1993)105<1137:EROTFA>2.3.CO;2.

Williams, E.J., 1964, Geomorphic features and history of the lower part of Logan Canyon, Utah [M.S. thesis]: Logan, Utah State University, 64 p.

Williams, J.S., Willard, A.D., and Parker, V., 1962, Recent history of Bear Lake Valley, Utah-Idaho: American Journal of Science, v. 260, p. 24–36.

Wilson, J.R., 1979, Glaciokarst in the Bear River Range, Utah: National Speleological Society Bulletin, v. 41, p. 89–94.

MANUSCRIPT ACCEPTED BY THE SOCIETY 15 SEPTEMBER 2008

The Geological Society of America
Special Paper 450
2009

Late Quaternary sedimentary features of Bear Lake, Utah and Idaho

Joseph P. Smoot

U.S. Geological Survey, M.S. 926A National Center, Reston, Virginia 20192, USA

ABSTRACT

Bear Lake sediments were predominantly aragonite for most of the Holocene, reflecting a hydrologically closed lake fed by groundwater and small streams. During the late Pleistocene, the Bear River flowed into Bear Lake and the lake waters spilled back into the Bear River drainage. At that time, sediment deposition was dominated by siliciclastic sediment and calcite. Lake-level fluctuation during the Holocene and late Pleistocene produced three types of aragonite deposits in the central lake area that are differentiated primarily by grain size, sorting, and diatom assemblage. Lake-margin deposits during this period consisted of sandy deposits including well-developed shoreface deposits on margins adjacent to relatively steep gradient lake floors and thin, graded shell gravel on margins adjacent to very low gradient lake-floor areas. Throughout the period of aragonite deposition, episodic drops in lake level resulted in erosion of shallow-water deposits, which were redeposited into the deeper lake. These sediment-focusing episodes are recognized by mixing of different mineralogies and crystal habits and mixing of a range of diatom fauna into poorly sorted mud layers. Lake-level drops are also indicated by erosional gaps in the shallow-water records and the occurrence of shoreline deposits in areas now covered by as much as 30 m of water. Calcite precipitation occurred for a short interval of time during the Holocene in response to an influx of Bear River water ca. 8 ka. The Pleistocene sedimentary record of Bear Lake until ca. 18 ka is dominated by siliciclastic glacial flour derived from glaciers in the Uinta Mountains. The Bear Lake deep-water siliciclastic deposits are thoroughly bioturbated, whereas shallow-water deposits transitional to deltas in the northern part of the basin are upward-coarsening sequences of laminated mud, silt, and sand. A major drop in lake level occurred ca. 18 ka, resulting in subaerial exposure of the lake floor in areas now covered by over 40 m of water. The subaerial surfaces are indicated by root casts and gypsum-rich soil features. Bear Lake remained at this low state with a minor transgression until ca. 15 ka. A new influx of Bear River water produced a major lake transgression and deposited a thin calcite deposit. Bear Lake quickly dropped to a shallow-water state, accumulating a mixture of calcite and siliciclastic sediment that contains at least two intervals of root-disrupted horizons indicating lake-level drops to more than 40 m below the modern highstand. About 11,500 yr B.P., the lake level rose again through an influx of Bear River water producing another thin calcite layer. The Bear River ceased to flow into the basin and the lake salinity increased, resulting in the aragonite deposition that

Smoot, J.P., 2009, Late Quaternary sedimentary features of Bear Lake, Utah and Idaho, *in* Rosenbaum, J.G., and Kaufman, D.S., eds., Paleoenvironments of Bear Lake, Utah and Idaho, and its catchment: Geological Society of America Special Paper 450, p. 49–104, doi: 10.1130/2009.2450(03). For permission to copy, contact editing@geosociety.org. ©2009 The Geological Society of America. All rights reserved.

persisted until modern human activity. The climatic record of Bear Lake sediment is difficult to ascertain by using standard chemical and biological techniques because of variations in the inflow hydrology and the significant amount of erosion and redeposition of chemical and biological sediment components.

INTRODUCTION

The history of Bear Lake hydrology and climate resides within its sedimentary record. All mineralogical, chemical, biological, and magnetic measurements are derived from these deposits as recovered from cores and grab samples. The sedimentological framework is essential for interpreting the significance of the various measurements and for providing additional information on past physical conditions of Bear Lake.

This chapter provides sedimentary descriptions of all cores collected from Bear Lake except the GLAD 800 cores (Dean, this volume; Kaufman et al., this volume) and short gravity cores that were sampled in the field. Grab samples and trenches in and around Bear Lake are also included in this study. Data derived from these sources provide an overview of shallow-water to deep-water sedimentation for about the last 26,000 yr.

Physiographic Setting

Bear Lake is nestled in a valley between Paleozoic carbonate rocks of the steep mountains bordering the west side and the Mesozoic and Tertiary sandstones, conglomerates, and shales of the Bear Lake Plateau to the east (see Reheis et al., this volume). Faults that bound the valley offset drainages and other geomorphic features (McCalpin, 1993; Reheis et al., this volume), including the lake floor (Colman, 2005, 2006), indicating active Holocene tectonism. In its natural state, Bear Lake is primarily fed by spring-fed streams draining the adjacent highlands, and there has been little if any surface outflow (Fig. 1). A series of canals, completed in 1912, divert water from the Bear River into the lake and allow lake waters to drain northward back into the Bear River. The lake waters are alkaline, but not very saline, supporting a population of fish and invertebrates with many endemic species (Dean et al., this volume).

Sedimentation

Most of the information on modern sedimentation in Bear Lake is from studies presented in this volume. The bulk of the sediment currently being deposited in Bear Lake is either chemically precipitated or biologically formed (Fig. 2). Prior to completion of the canals, aragonite was the principal carbonate mineral precipitated for most of the Holocene, but low-magnesium calcite is now the primary mineral (Dean et al., 2006; Dean, this volume). The most important biological components of the sediments include diatom tests (Moser and Kimball, this volume) and ostracode shells (Bright, this volume). Very little clastic sediment is being introduced into the lake from the surface drain-

ages (Rosenbaum et al., this volume). Geomorphic features that resemble deltas at the mouths of some creeks (particularly North and South Eden Creek) appear to have formed during the Pleistocene (see Colman, 2006). Wave transport disperses clastic sediment around the lake, mostly reworking older deposits. Shorelines are composed of siliciclastic sand, and boulders and cobbles reflecting the lithology of adjacent mountains. In shoreline areas more distant from the mountains, the coarse material includes abundant shells of snails, clams, and ostracodes.

The primary mechanical sedimentation processes occurring in Bear Lake today are wave transport and sediment gravity flows, mostly induced by wave activity (Fig. 2). Direct evidence of wave sorting in the modern lake is mostly limited to depths of less than 10 m. Grain-size distributions in surface sediments (see Smoot and Rosenbaum, this volume) suggest that wave winnowing occurs to depths of 30 m or more. Coarser grain sizes persist at greater depths where the lake floor is steeper, suggesting that underwater gravity flows are more important in those areas. A model for maximum wave depth indicates that silt-sized material is deposited well beyond the depth for wave movement even in the areas with low bottom slopes (Smoot and Rosenbaum, this volume). Therefore, it is believed that underwater gravity currents are an important mechanism for redistributing sediment throughout the lake. The resedimentation of shallow-water deposits lakeward is called sediment focusing (Davis and Ford, 1982; Hilton, 1985). Dean et al. (2006) and Dean (this volume) observed that near-surface sediment traps collected calcite precipitated at the surface, but sediment traps closer to the lake floor collected mostly aragonite, presumably older sediment that was mechanically resuspended.

Analytical Techniques

Three cores (BL96-1, -2, -3) were obtained with a Kullenburg piston apparatus, and the other ten cores (BL2K-2, -3; BLR2K-1, -2, -3; BL02-1, -2, -3, -4, -5) were collected with a UWITEC piston corer that is hammered into the sediment (Rosenbaum and Kaufman, this volume). Shoreline sample localities were trenched to just below the water table by hand, and 1 gal rectangular cans (25 cm deep, 18 cm wide) were pushed into two overlapping samples from the surface (total depth ~45 cm). All cores were split lengthwise with a steel wire and the surfaces were cleaned of smear marks and irregularities with a straight razor. Initial descriptions were made with a 10× hand lens and additional information was obtained by using a Nikon zoom stereoscopic microscope (up to 6×) with a 10-power ocular lens. The cores were photographed with 35 mm film in 14–20 cm overlapping segments, and the film was scanned at 3000 dpi to

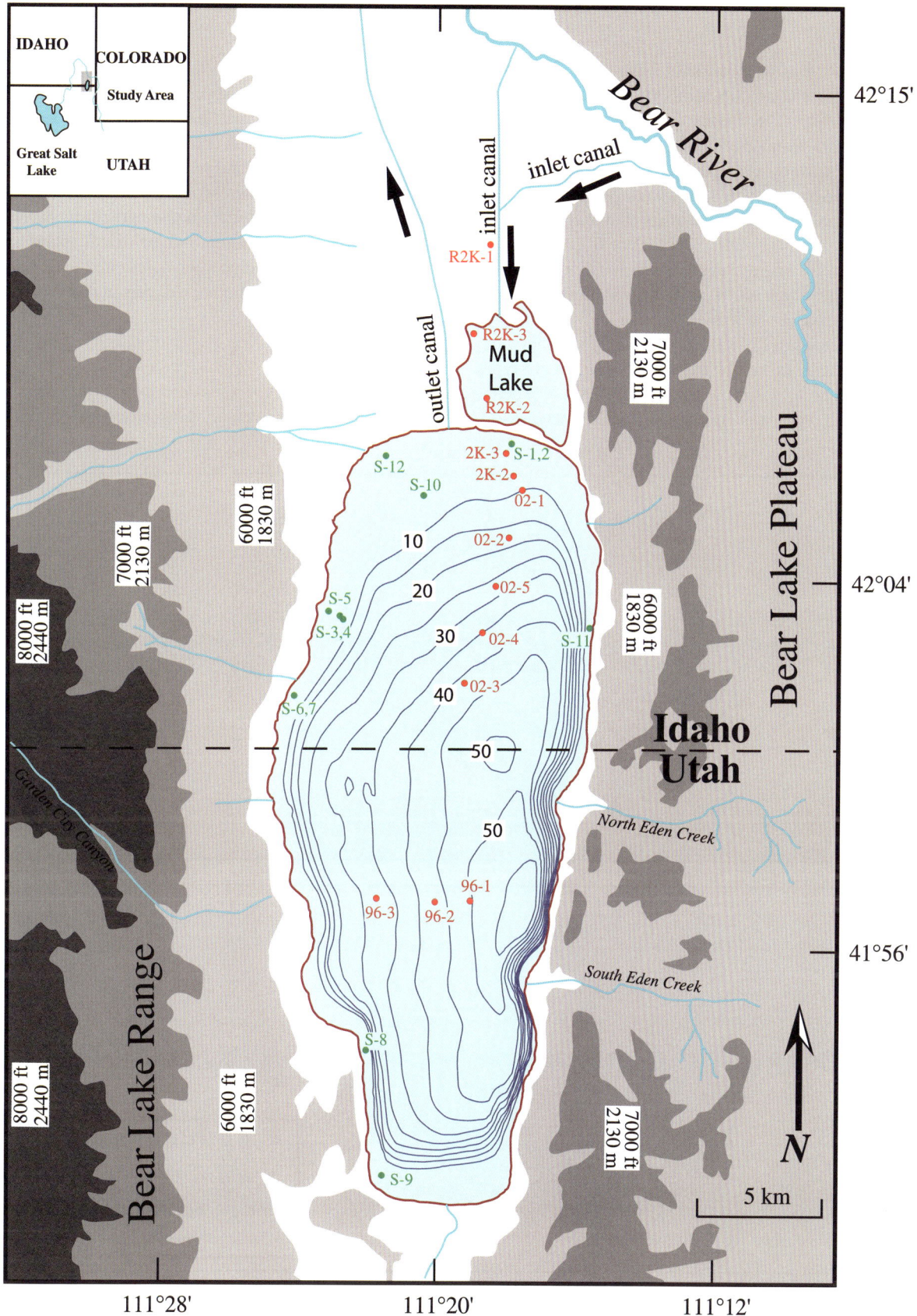

Figure 1. Map of Bear Lake and surrounding areas. Elevation zones are divided into 1000 foot intervals. Isobaths are at 5 m intervals except for depths less than 10 m. Red dots show locations of cores described in this paper and green dots show locations of trenches used in this paper.

4000 dpi. X-radiographs were made of 1-cm-thick slabs of the three Kullenburg cores in a Picker Minishot 1 using 17-inch Agfa Structurix D4DW film. The images were digitized at 4000 dpi with a UMAX 2100 scanner using the UTA-2100XL transparency adaptor. The images were examined with a hand lens and stereoscopic microscope. Smear slides were collected at 10 cm intervals and at visible changes in sediment character in all cores. A subsequent set of smear slides was collected in the Kullenburg cores at intervals where changes in fabric are visible in the X-radiographs. The smear slides were examined with a Zeiss binocular petrographic scope with 10× ocular and 40× objective lens. Selected smear slides (~100 samples) were counted along a 10 μm grid to establish variability in diatoms and grain types. Three hundred diatoms were counted on each slide, whereas abundance of grain types was estimated by using a shadow diagram (Terry and Chilingar, 1955). Polished thin sections were made at select intervals from chips imbedded with Spurr resin as described by Kemp et al. (2001). These were examined both by petrographic microscope and SEM. About 250 samples were dried and sieved into silt and sand fractions that were examined under the stereoscopic microscope. Diatoms species observed in smear slides were identified from photographs by Katrina Moser. Other biological identification, mineralogical analyses, biological identification, and radiocarbon dates cited in this paper were provided by the authors of other chapters in this volume.

Core Descriptions

The locations for the cores described in this paper are shown in Figure 1. The sedimentary characteristics of the cores are

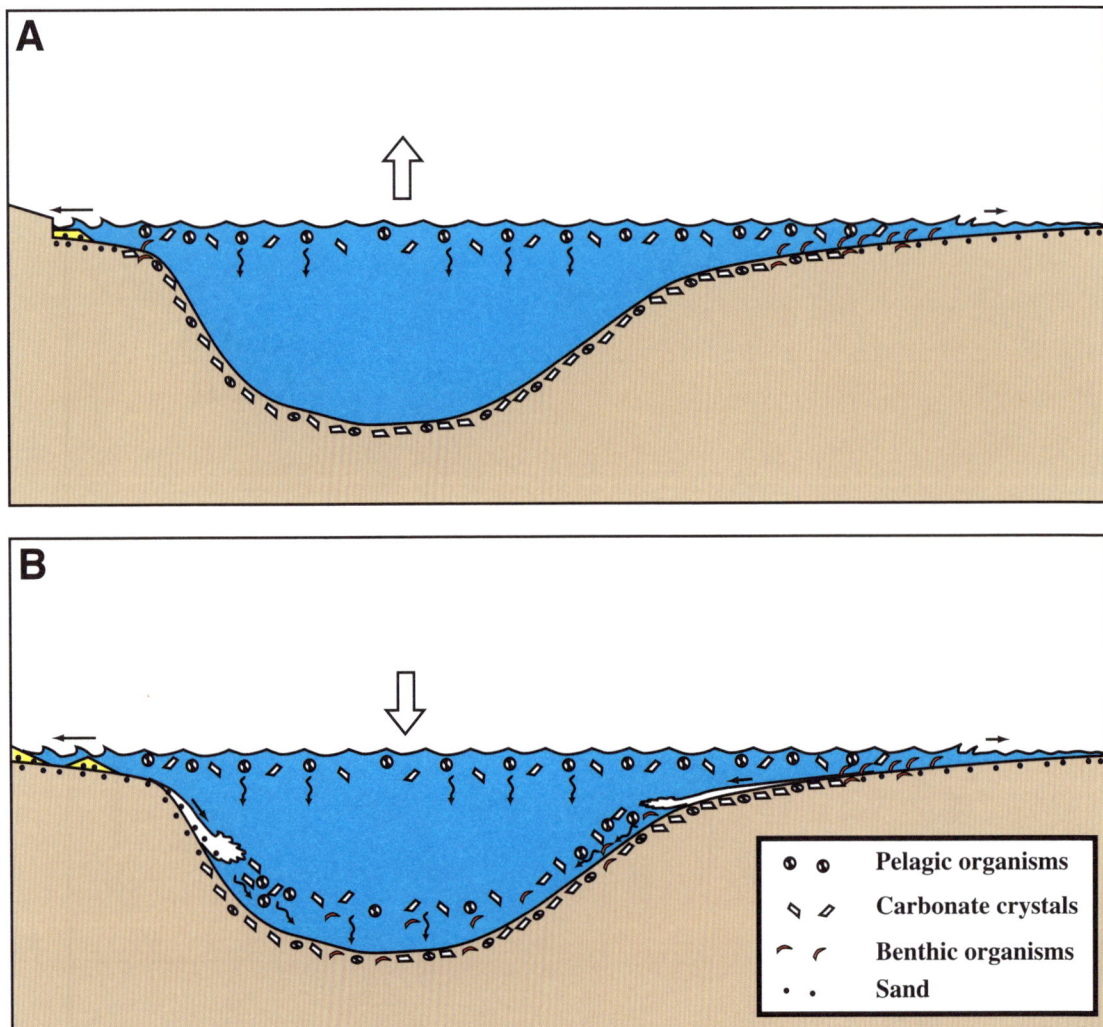

Figure 2. Schematic illustration of modern sedimentation in Bear Lake. (A) Rising lake level. Aragonite precipitation and pelagic organisms dominate deeper water and benthic organisms are mostly restricted to shallow water. Shoreline cuts into older deposits, particularly on margins with steep bottom slopes. (B) Falling lake level. Sediment focusing through wave transport and gravity flows is mixed with aragonite precipitation and biogenic sedimentation in deep water. Shoreline deposits build out into lake.

presented schematically in Figures 3–6. The details about sedimentary designations and sedimentary features are provided in the following sections. The text is organized to describe first the Holocene record from the cores and the trenches, and then the more poorly constrained late Pleistocene core record from 32 ka. The conditions that produced the assemblage of sedimentary features in the different settings and time periods are discussed following their descriptions.

MODERN AND HOLOCENE SEDIMENTS

The Holocene sediments deposited prior to diversion of Bear River into Bear Lake are dominated by aragonite. Calcite is the dominant mineral in sediment formed after diversion of the river, and in sediments deposited during a short interval in the early Holocene (Dean et al., 2006). Holocene and modern marsh deposits north of the lake largely consist of a mixture of calcite and siliciclastic clay and silt.

Deep-Water Aragonite Deposits

The variability of Holocene deep-water aragonite sediments is best illustrated in core BL96-1 (Fig. 3). Although the modern calcite-rich sediments are missing, the core contains a sedimentary record from ca. 7500 yr B.P. to less than 1000 yr B.P. (Colman et al., this volume). The sediments are uniformly fine grained and obvious layering is absent. Subtle changes in the sediment texture are discernable, as are variations in the bioturbation patterns, particularly in X-radiographs. The aragonitic sediments in core BL96-1 are divided into three varieties that are gradational in character.

Aragonite Type I

This type of sediment is characterized by centimeter-scale alternation of smooth tan mud and more poorly sorted tan mud with silt-sized black grains (Fig. 7). Random burrows with diameters ranging from 0.3 cm to 0.5 cm disrupt the poorly sorted intervals. These intervals are also less dense than the smooth intervals in X-radiographs. The smooth tan intervals typically have smaller (less than 0.2 cm) burrows that are mostly oriented parallel to the bedding surface (Fig. 8). Smear slides show a subtle contrast between the smooth and poorly sorted intervals (Fig. 9). Aragonite needles with a small range of crystal sizes (2–5 μm) dominate the smooth intervals, whereas the poorly sorted intervals include a variety of aragonite crystal sizes and shapes, as well as a significant component of calcite crystals and grain aggregates. Smear slides of smooth intervals also show that diatoms are dominated by one or two species, particularly *Navicula oblonga* and *Cymbella messiana*, whereas the poorly sorted layers also contain notable increases of a wider variety of diatom genera. A variation in the bioturbation is limited to the upper 40 cm of BL96-1 and coeval intervals in other cores. Random burrows have diameters as much as 1 cm and show dense outer rims in X-radiographs. Another variation within mud assigned

to Aragonite I includes scattered coarse-silt- to sand-sized ostracodes in the poorly sorted intervals.

Aragonite Type II

This type of sediment is generally finer grained than Type I and characteristically has alternations of tan mud with no readily visible structure and tan mud with dark tubes (Fig. 10). The dark tubes are burrows 1–3 mm in diameter. The dark color is due to the presence of sulfide minerals that either line the burrows (Fig. 11) or are distributed in the sedimentary fill of the burrows. The burrows are typically oriented parallel to the bedding surface in the lower part of a dark tube interval and become larger and more randomly oriented toward the top of an interval (Fig. 11). The tan mud intervals that lack dark tubes have small (~1 mm in diameter) bedding-parallel burrows and the intervals are slightly denser in X-radiographs than the aragonite mud with dark tubes. Smear slides show mixtures of variously sized aragonite needles and some calcite aggregates in both sediment types, and a mixed assemblage of diatoms (Fig. 12). The intervals without dark tubes have a larger percentage of small pelagic diatoms, such as *Stephanodiscus* and *Cyclotella*, than the intervals with dark tubes. In contrast, the dark-tube intervals have a greater variety of diatom types, particularly the larger diatoms such as *Navicula oblonga* and *Cymbella mesiana*. The intervals with dark tubes also have a greater variety of aragonite crystal sizes and more calcite than the intervals without dark tubes. The dark-tube intervals are transitional in character to the poorly sorted intervals in Aragonite I.

Aragonite Type III

This type of sediment is characterized by yellowish silty to sandy mud alternating with tan mud (Fig. 13). The tan mud intervals are, in places, similar to the smooth tan mud of Aragonite I, and in other places resemble the poorly sorted mud of Type I aragonite or dark-tube intervals of Type II aragonites. In all occurrences, silt- to sand-sized ostracodes are a common component. In places, the contact of the yellowish silty interval with an underlying tan mud is very sharp with a concentration of sand grains. Random burrows (0.3–0.5 cm in diameter) occur throughout the yellowish intervals and sand-filled burrows extend from the sharp contacts into the underlying tan mud layers (Fig. 14). Small horizontal burrows are common in the denser tan mud layers. The yellowish silty intervals contain coarse silt- to sand-sized siliciclastic grains, rock fragments, and ostracodes in addition to a mixture of different aragonite crystals and calcite (Fig. 15). Diatoms in the silty intervals are characteristically fragmented and generally poorly preserved. The poorly sorted intervals of Type I aragonite that contain ostracodes could be reassigned to Type III aragonite, but the matrix is less silty and the diatoms are less fragmentary than is typical for Type III aragonite.

Interpretations

The relative grain sizes of each aragonite type and their distribution in the cores (Figs. 3, 4, and 5) suggest that Type II aragonite

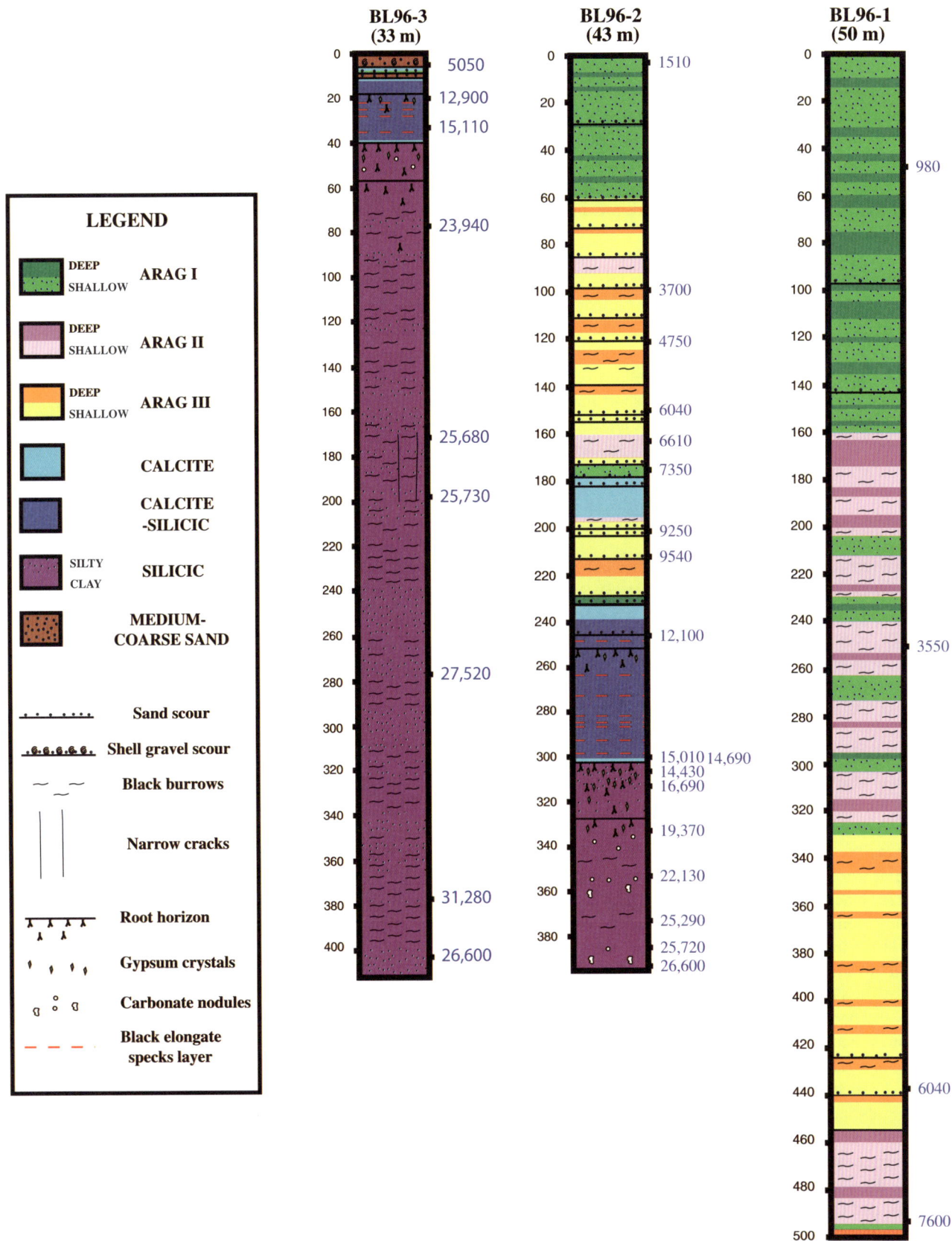

LEGEND

	DEEP	**ARAG I**
	SHALLOW	
	DEEP	**ARAG II**
	SHALLOW	
	DEEP	**ARAG III**
	SHALLOW	
		CALCITE
		CALCITE -SILICIC
	SILTY CLAY	**SILICIC**
		MEDIUM- COARSE SAND

- ·· ·· ·· ·· **Sand scour**
- ·o·o·o·o· **Shell gravel scour**
- ～ ～ **Black burrows**
- | | **Narrow cracks**
- ⊥⊤⊥⊤⊥ **Root horizon**
- ♦ ♦ ♦ **Gypsum crystals**
- ◌ ◦ ◌ **Carbonate nodules**
- – – – **Black elongate specks layer**

BL96-3 (33 m)

5050
12,900
15,110
23,940
25,680
25,730
27,520
31,280
26,600

BL96-2 (43 m)

1510
3700
4750
6040
6610
7350
9250
9540
12,100
15,010 14,690
14,430
16,690
19,370
22,130
25,290
25,720
26,600

BL96-1 (50 m)

980
3550
6040
7600

Figure 3. Schematic illustration of sediment types and sedimentary structures in the three Kullenburg cores whose locations are shown in Figure 1. Blue numbers are ages in calendar years derived from radiocarbon ages of pollen separates (Colman et al., this volume). Scales are in centimeters below the surface. Water depths of the coring sites (in parentheses below the core locations) are corrected to the modern highstand level (1805.5 m elevation).

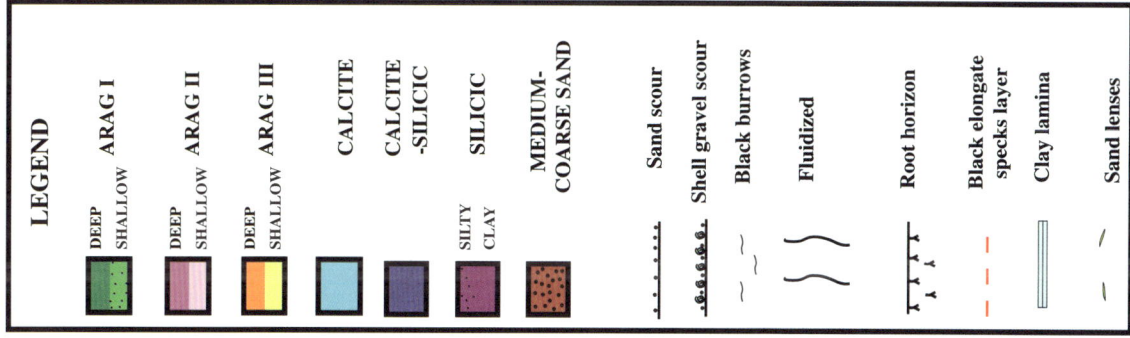

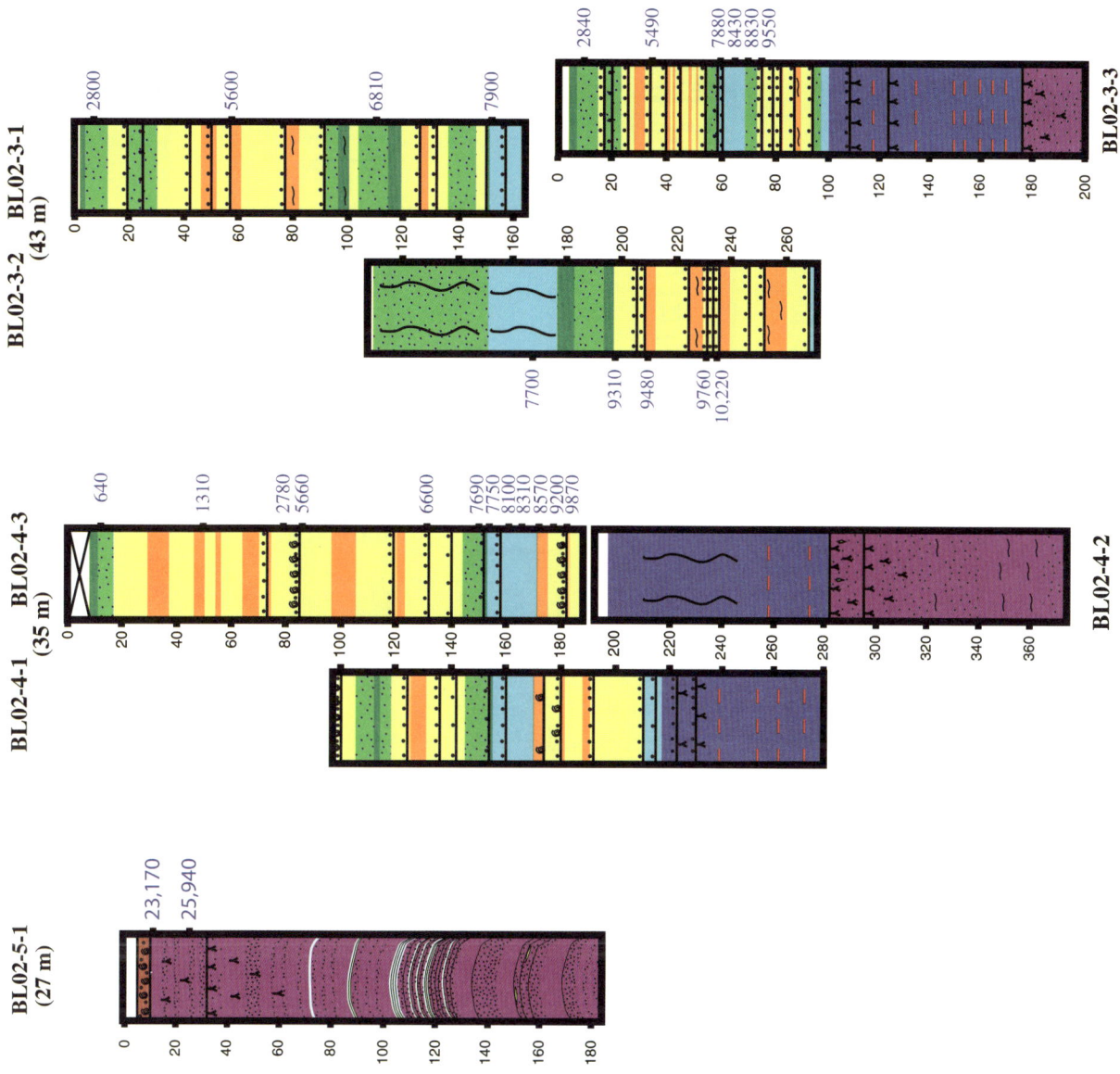

Figure 4. Schematic illustration of sediment types and sedimentary structures in UWITEC cores at deep-water sites (see Fig. 1). Scales are in centimeters and include composite depths of multiple cores. Blue numbers are ages as described in Figure 3. Water depths of the coring sites (in parentheses below the core locations) are corrected as described in Figure 3.

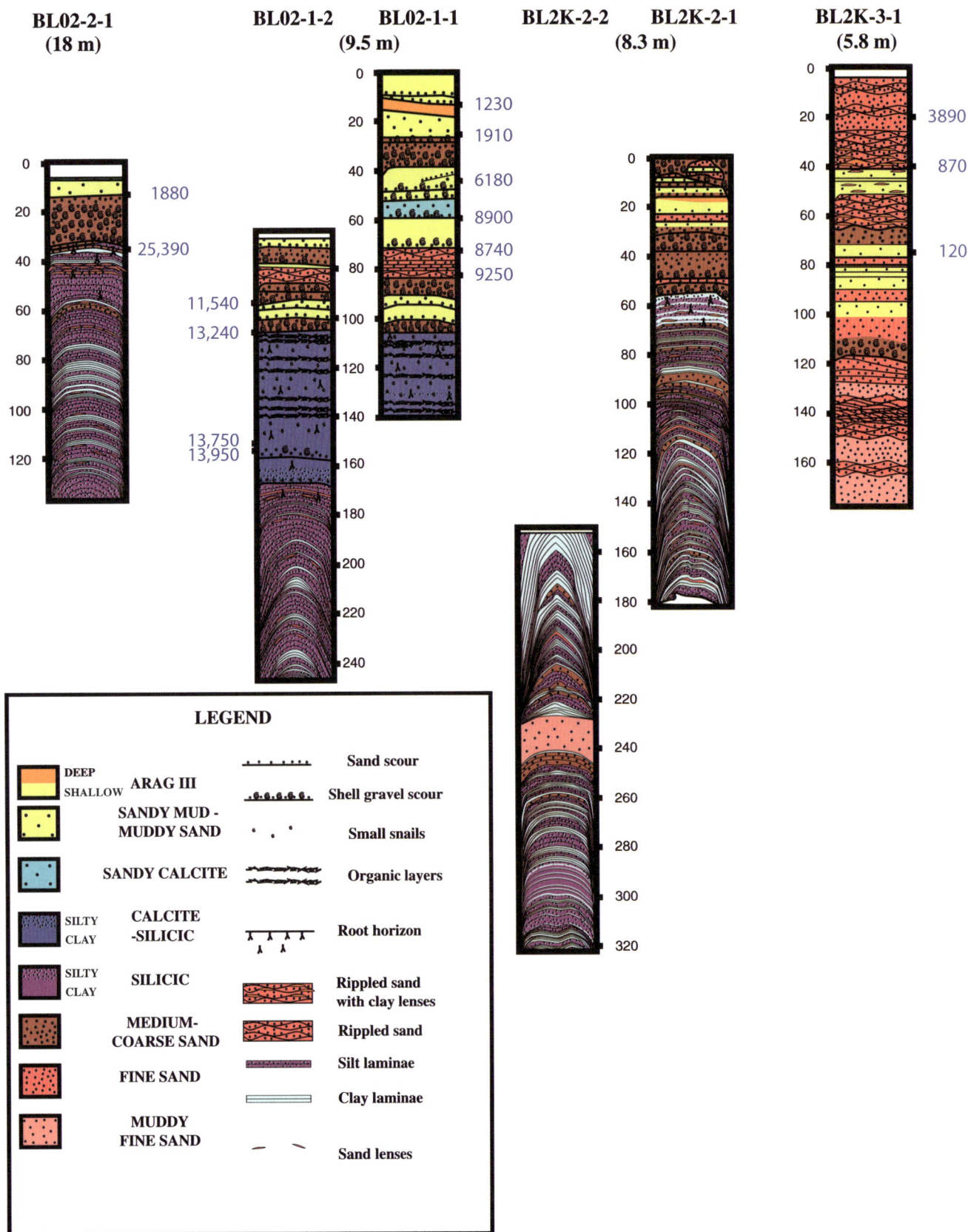

BL02-2-1 (18 m) **BL02-1-2** **BL02-1-1** (9.5 m) **BL2K-2-2** **BL2K-2-1** (8.3 m) **BL2K-3-1** (5.8 m)

BL02-2-1: 1880, 25,390

BL02-1-2: 11,540, 13,240, 13,750, 13,950

BL02-1-1: 1230, 1910, 6180, 8900, 8740, 9250

BL2K-3-1: 3890, 870, 120

LEGEND

DEEP / SHALLOW	**ARAG III**	Sand scour
	SANDY MUD - MUDDY SAND	Shell gravel scour
	SANDY CALCITE	Small snails
SILTY CLAY	**CALCITE -SILICIC**	Organic layers
SILTY CLAY	**SILICIC**	Root horizon
	MEDIUM- COARSE SAND	Rippled sand with clay lenses
	FINE SAND	Rippled sand
	MUDDY FINE SAND	Silt laminae
		Clay laminae
		Sand lenses

Figure 5. Schematic illustration of sediment types and sedimentary structures in UWITEC cores at shallow-water sites (see Fig. 1). Scales are in centimeters and include composite depths of multiple cores. Blue numbers are ages as described in Figure 3. Water depths of the coring sites (in parentheses below the core locations) are corrected as described in Figure 3.

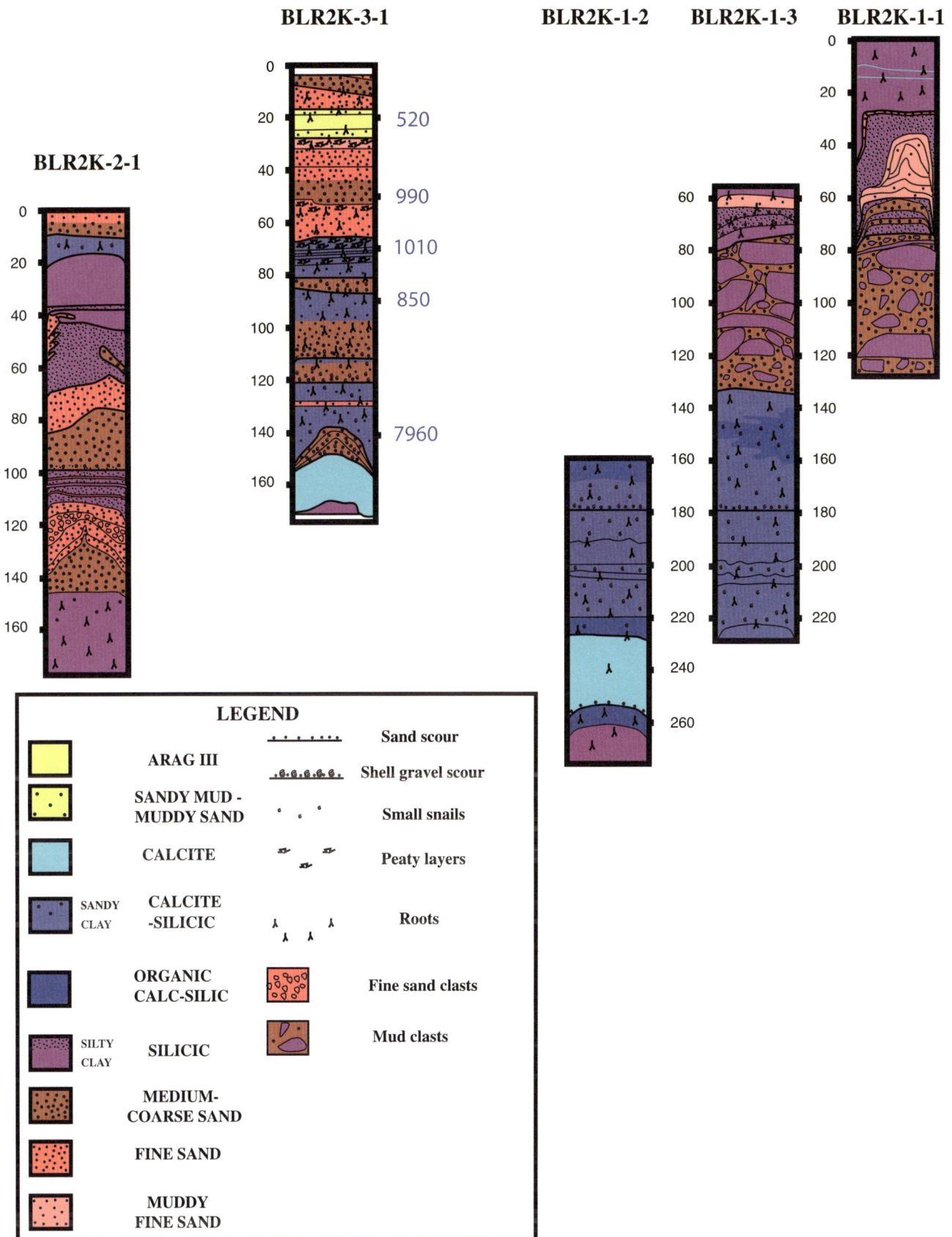

BLR2K-3-1

BLR2K-1-2   BLR2K-1-3   BLR2K-1-1

520

990

1010

850

7960

BLR2K-2-1

LEGEND

	ARAG III	····· Sand scour
	SANDY MUD - MUDDY SAND	·0·0·0·0·0· Shell gravel scour
	CALCITE	· · · Small snails
SANDY CLAY	CALCITE -SILICIC	Peaty layers
	ORGANIC CALC-SILIC	Roots
SILTY CLAY	SILICIC	Fine sand clasts
	MEDIUM- COARSE SAND	Mud clasts
	FINE SAND	
	MUDDY FINE SAND	

Figure 6. Schematic illustration of sediment types and sedimentary structures in UWITEC cores north of Bear Lake (see Fig. 1). Scales are in centimeters and include composite depths of multiple cores. Blue numbers are ages as described in Figure 3. Water depths of the coring sites (in parentheses below the core locations) are corrected as described in Figure 3.

represents the deepest water conditions and Type III aragonite represents the shallowest. Type I aragonite is found in the upper part of BL02-4 (35 m water depth), indicating that it formed at that depth in the pre-canal historical lake. Type III aragonite at the top of BL02-1 (9.5 m water depth) is less than 1000 yr old, and was probably deposited in the historical, pre-canal lake at similar depths. The abundance of broken diatoms in Type III aragonite is probably due to reworking by waves. Type II aragonite was not observed in any cores with historical-age deposits. The small pelagic diatoms in the Type II aragonite without dark tubes include *Stephanodiscus medius*, which is an indicator of cool fresh water (Moser and Kimball, this volume). Part of the overall

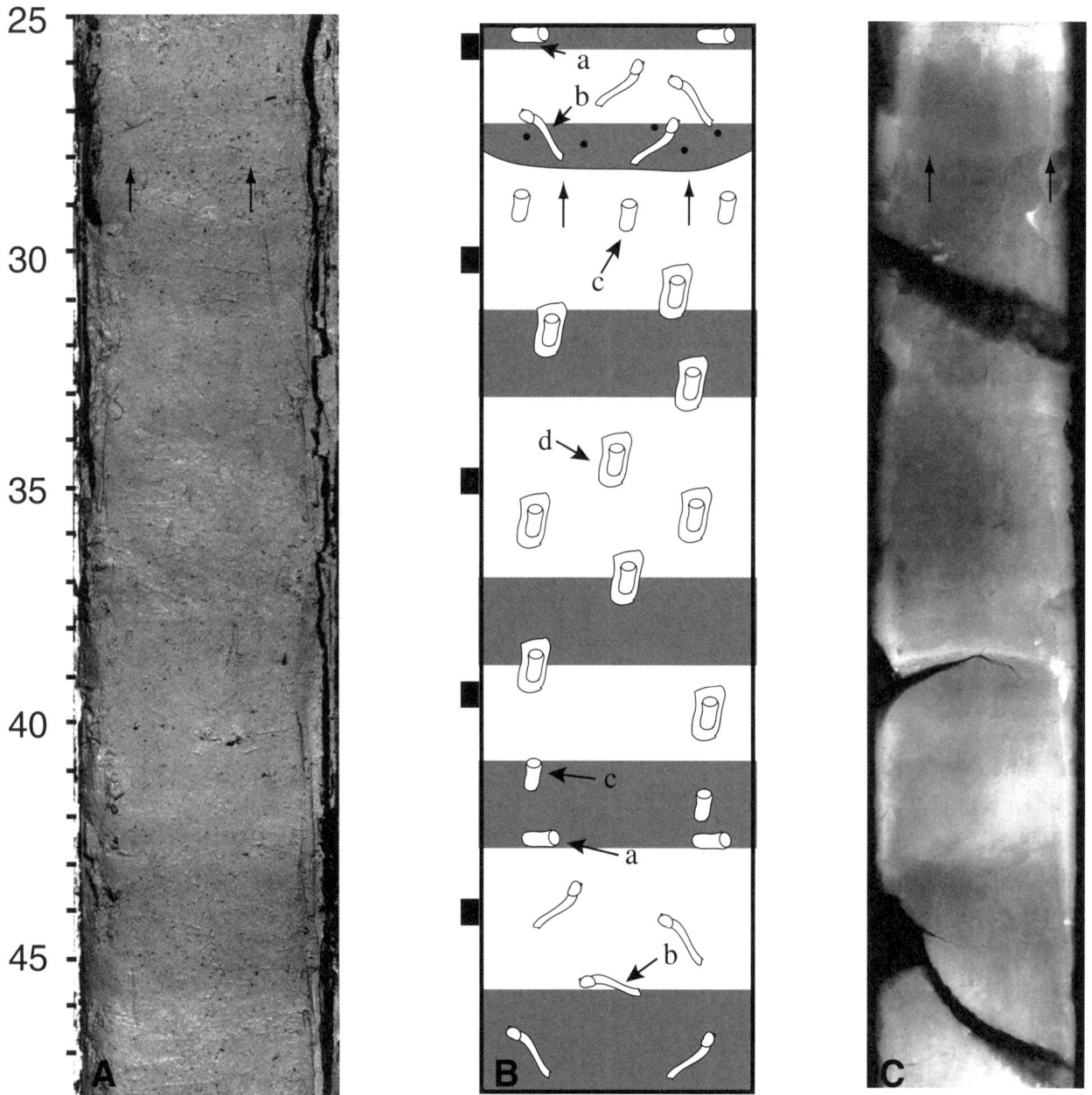

Figure 7. Aragonite I showing core segment on cut surface (A), schematic drawing of X-radiograph image (B), and X-radiograph (C) of BL96-2. The smooth tan intervals appear as denser (lighter) intervals in the X-radiograph (C) and poorly sorted intervals are less dense (darker). High-density intervals are shaded in B. Arrows point to scour surface filled with sand composed mostly of ostracodes. The features shown in B (not to scale) are small horizontal burrows (a), large random burrows (b), small random burrows (c), and random burrows with dense rims (d). Some of the density differences in C are due to irregularities in the sample thickness. The cut-surface photo was digitally enhanced to contrast the layers. Scales are in centimeters. The X-radiograph scale (B and C) differs slightly from the core (A) due to differential shrinkage from drying.

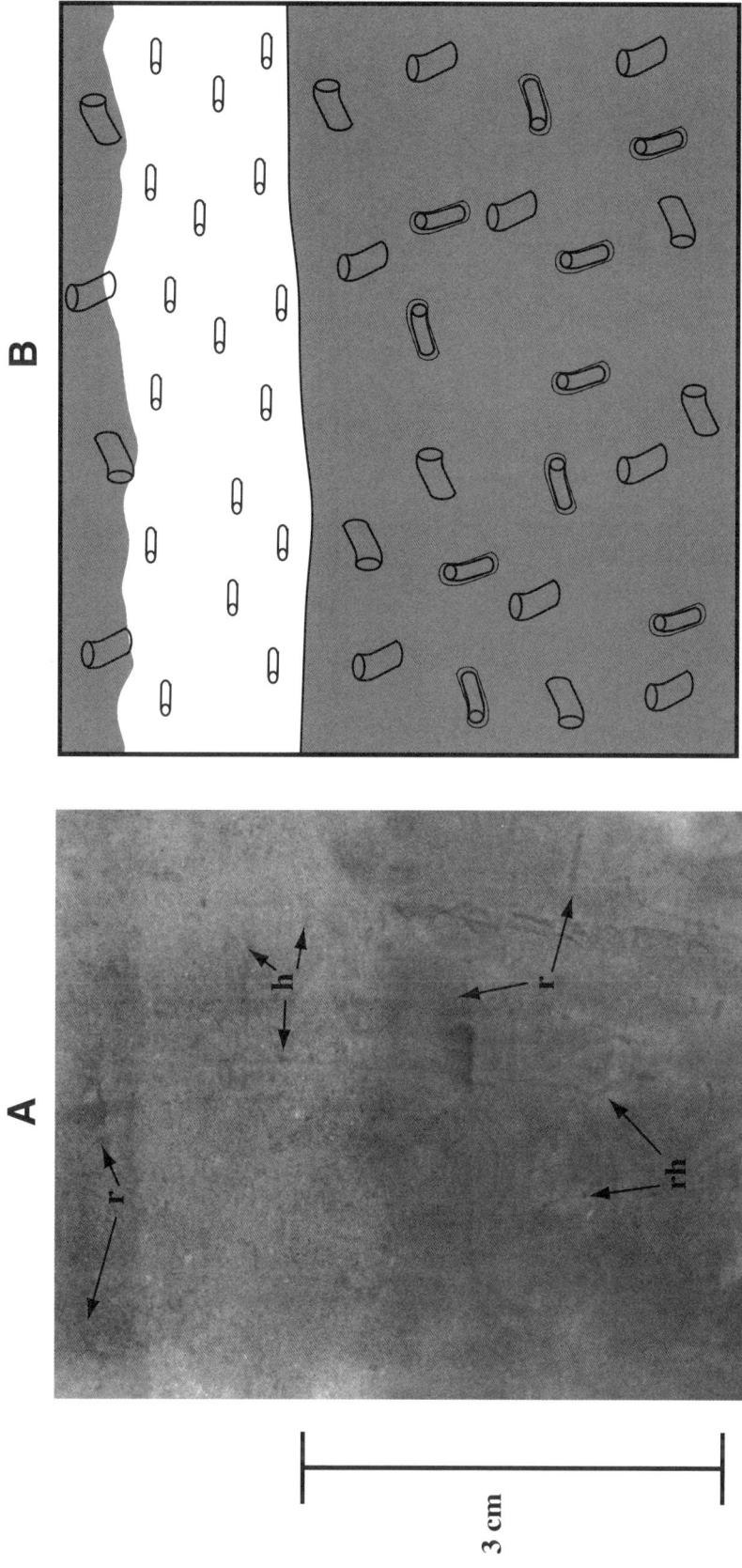

Figure 8. Aragonite I in BL96-1 showing detail of bioturbation patterns in X-radiograph (A) and schematic drawing (B). The denser (lighter) fine-grained interval contains small horizontal burrows (h). The less dense (darker), poorly sorted interval contains random burrows (r). Some random burrows have a denser outer rim (rh). The schematic depiction of sedimentary features does not show them to scale.

finer grain size in comparison to Type I aragonite is the relative dearth of the larger diatoms rather than a finer aragonite crystal size. It is possible that Aragonite II may represent a shift in lake temperature and/or salinity rather than a greater depth than Aragonite I. The argument against this interpretation is that Type II aragonite is apparently age correlative to Type I aragonite in cores from shallower water (Smoot and Rosenbaum, this volume).

The alternations of layers with relatively more and less mixing of diatoms and different mineral types and sizes is interpreted as indicating different degrees of sediment focusing. At Pyramid Lake (Smoot, 2003; Smoot and Benson, 2004), sediment focusing increases during falling lake levels. By analogy, rising lake levels would show less evidence of sediment focusing, particularly in deeper water, accounting for the

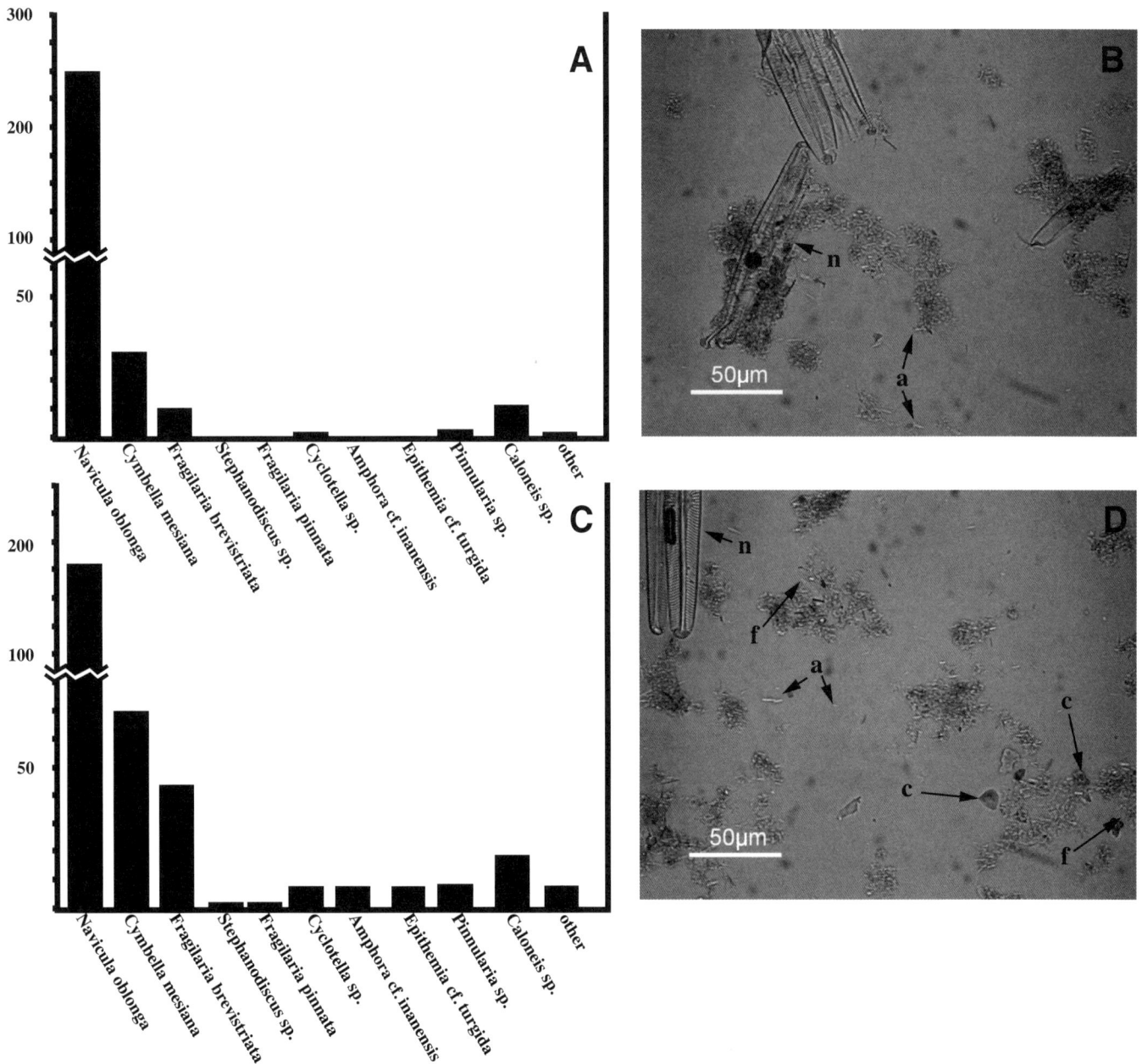

Figure 9. Aragonite I showing comparison of diatom count for a smear slide in a smooth tan interval (A and B) to that for a poorly sorted interval (C and D) in BL96-1. The smooth interval smear slides are dominated by one or two diatom species (A). The smear slide (B) shows *Navicula oblonga* (n) and 2–5 μm aragonite needles (a). The grainy intervals have more diverse diatom assemblages (C). The smear slide (D) shows *Navicula oblonga* (n) plus *Fragilaria brevistriata* (f) and calcite crystals (c) mixed with 2–20 μm aragonite needles (a). Note the compression of graphs and change of scale for counts over 100.

simpler mineralogies and diatom assemblages. The pronounced grading of some layers in Type III aragonite is consistent with deposition by turbidity currents. The variation in burrow styles in Type I and Type II aragonites is coincident with this shift in sediment focusing. The limitation of burrows to bedding planes may reflect an oxygen deficit (as in Smoot and Benson, 1998), whereas random burrowing represents a more oxygen-rich lake floor. Another possibility is that the substrate controlled the burrow style, with more random burrowing occurring in coarser sediment. The dark burrows in Type II aragonite may indicate more oxygen depletion than in Type I or Type III aragonite, or they may reflect a different organism that provided an organic

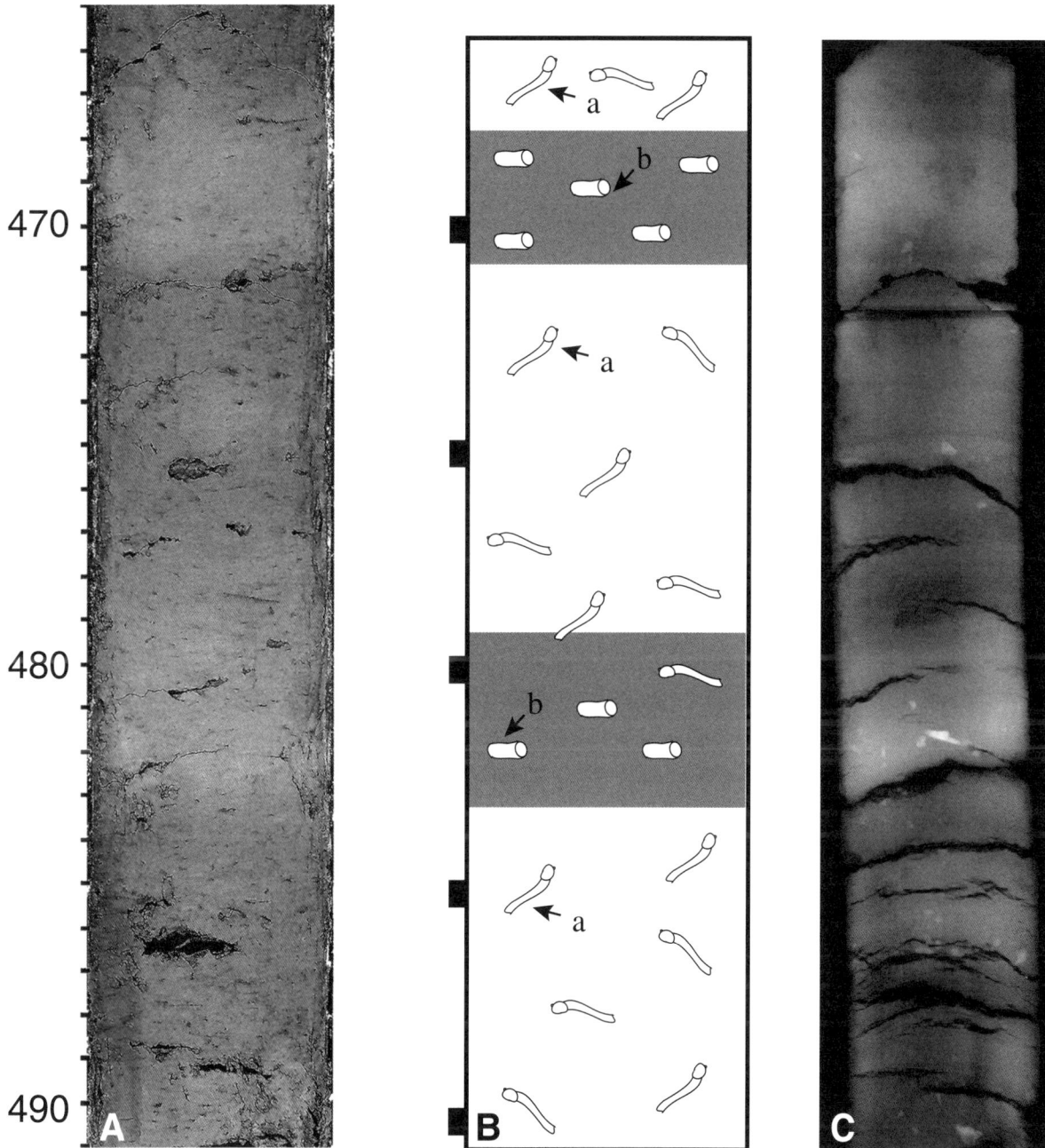

Figure 10. Aragonite II showing core segment cut surface (A), schematic drawing of X-radiograph image (B), and X-radiograph (C) of BL96-1. Random burrows (a) dominate intervals with dark tubes. These intervals are slightly less dense (darker) in C. Intervals without dark tubes are dominated by small, horizontal burrows (b). These intervals are denser (lighter) in C and shaded in B. Sedimentary features schematically shown in B are not to scale. Some of the density differences in C are due to variations in sample thickness. The cut-surface photo was digitally enhanced to contrast the layers. Scales are in centimeters. The X-radiograph scale (B and C) differs slightly from the core (A) due to differential shrinkage from drying.

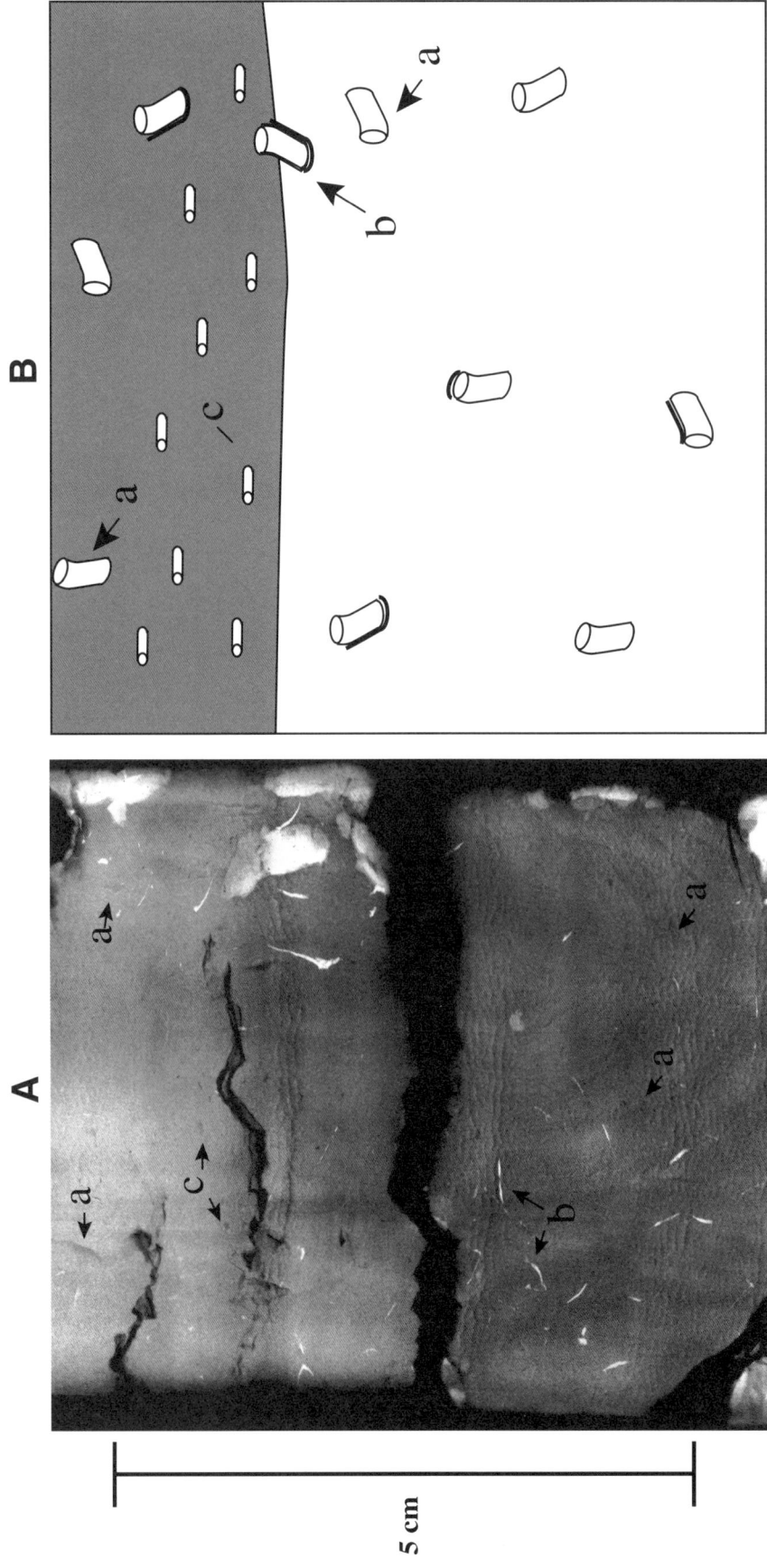

Figure 11. Aragonite II in BL96-1 showing detail of the bioturbation pattern in an X-radiograph (A) and schematic drawing (B). Sedimentary features in B are not drawn to scale. The dark-tube intervals are slightly less dense (darker in A) than the interval without dark tubes (light in A and shaded in B). The less dense intervals contain random burrows (a), some which have partial linings of sulfide minerals (b) that are very dense (white). The denser interval is dominated by small horizontal burrows (c), but some larger random burrows also occur.

substrate for sulfide precipitation (like a slime coating). The actual organisms that produced the burrows are not known. The limited literature on burrowing organisms in lakes does not provide sufficient criteria to recognize different types or even to differentiate between different burrowing strategies of the same type of organism.

Deep-Water Calcite Deposits

Gray mud dominated by euhedral to subhedral calcite crystals (2-10 µm) is found in the Holocene portion of cores BL96-2, BL96-3, BL02-3, BL02-4, and BL02-1. This layer is underlain and overlain by aragonite mud (Fig. 16). Radiocarbon ages are

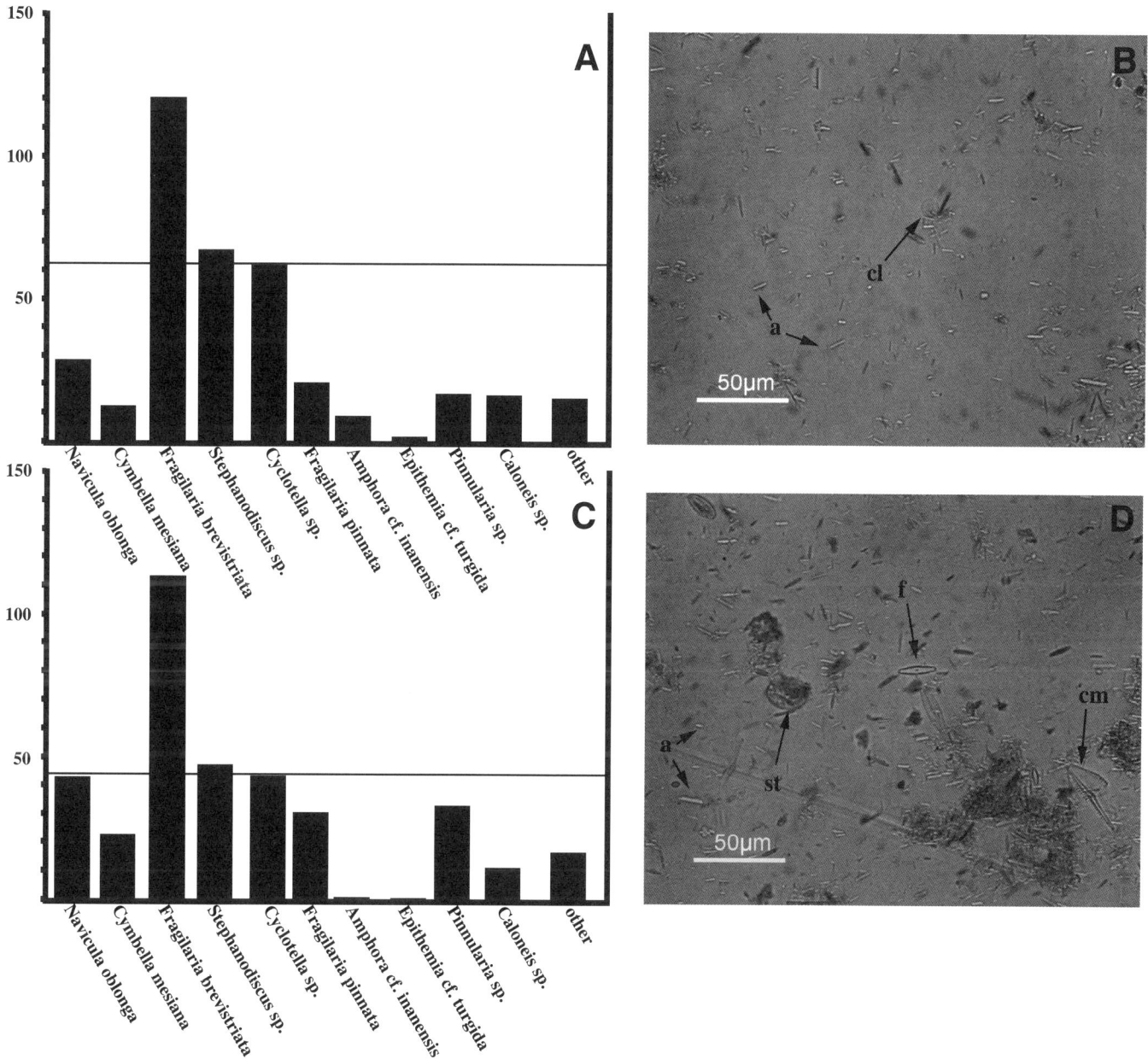

Figure 12. Aragonite II in BL96-1 showing comparison of diatom count for a dense interval without dark tubes (A and B) with that for an interval with dark tubes (C and D) of BL96-1. Both intervals show a variety of diatoms, although A has a higher percentage of small pelagic diatoms such as *Cyclotella* and *Stephanodiscus*, whereas in C, the small pelagic diatoms are mixed with nearly equal amounts of other diatoms such as *Navicula oblonga* and *Fragilaria pinnata*. Smear slide of the dense interval (B) shows variable aragonite crystal sizes (a) and the pelagic diatom *Cyclotella* (cl); compare this with the smear slide for a random burrow interval (D), which shows a wider range of aragonite crystal sizes (a) and a mixture of diatoms including *Stephanodiscus* (st), *Fragilaria brevistriata* (f), and *Cymbella mesiana* (cm).

consistent with the calcitic mud's representing the same lakewide sedimentation event. Cores BLR2K-1 and BLR2K-3 have similar beds that may also be correlative. In the cores from the deepest water (BL96-2, BL02-3, and BL02-4), the calcite-rich interval has a similar stratigraphy (Fig. 17). At the base, there is a 1- to 2-cm-thick bed that appears very dense in X-radiographs and

contains small burrows (1–3 mm in diameter) that are oriented parallel to the bedding surface. Smear slides show this layer to be composed entirely of uniform euhedral calcite crystals and small planktonic diatoms (Fig. 18). This thin bed is overlain by 5–10 cm of more poorly sorted calcitic mud that has more randomly distributed burrows. Smear slides of this poorly sorted

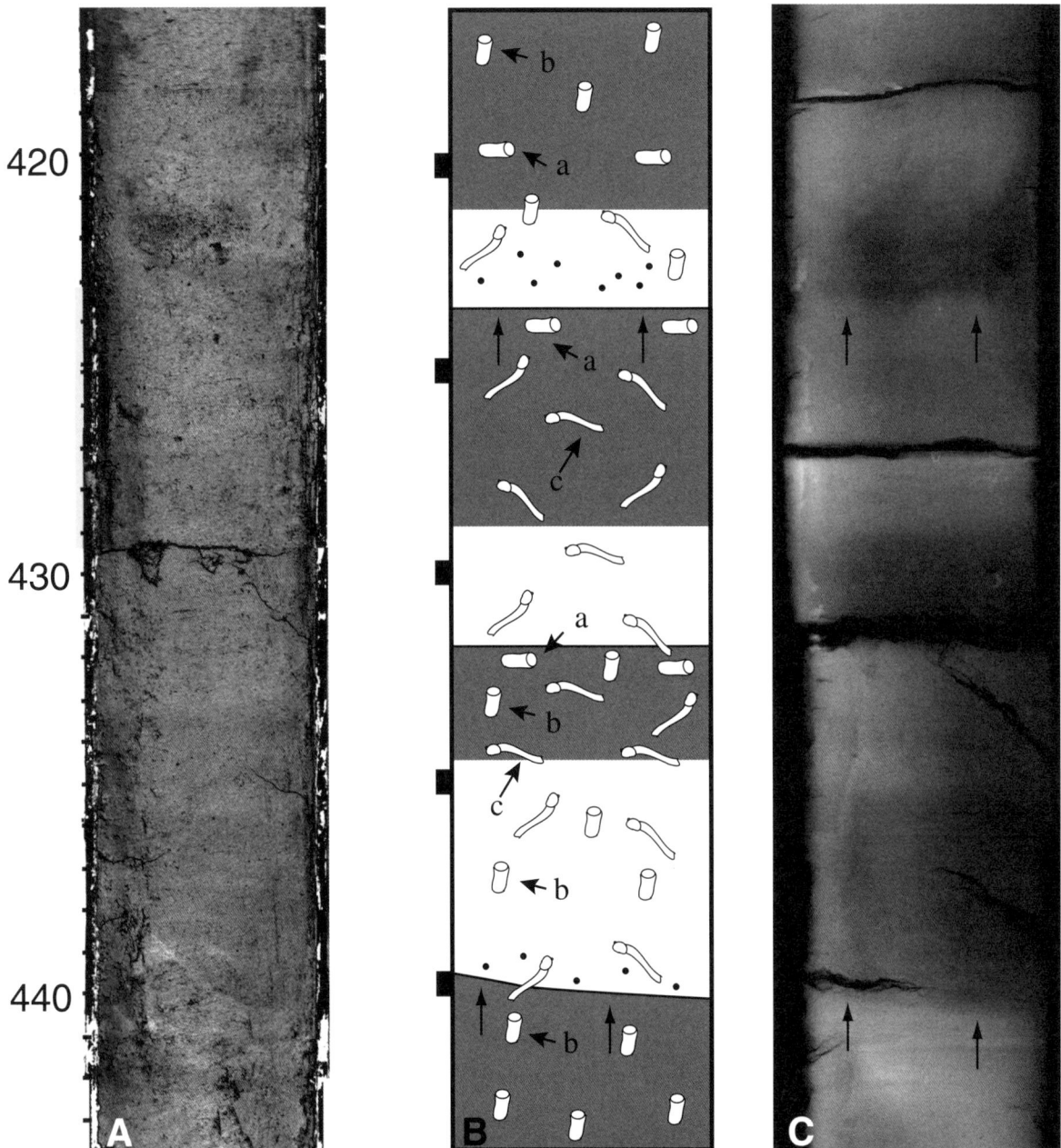

Figure 13. Aragonite III showing core segment in cut surface (A), schematic drawing of X-radiograph (B), and X-radiograph (C) of BL96-1. Sedimentary features in B are not drawn to scale. Coarse-grained intervals are less dense (dark) and have sharp, often irregular bases (vertical arrows). Burrows are less variable than in the other aragonite types but there are more small horizontal burrows (a) in the denser, finer-grained intervals (shaded in B). The coarser-grained interval contains random burrows including small, simple burrows (b) and larger more sinuous black burrows with high-density sulfide rims (c). Some of the density differences in C are due to irregularities in the sample thickness. The cut-surface photo was digitally enhanced to contrast the layers. Scales are in centimeters. The X-radiograph scale (B and C) differs slightly from the core (A) due to differential shrinkage from drying.

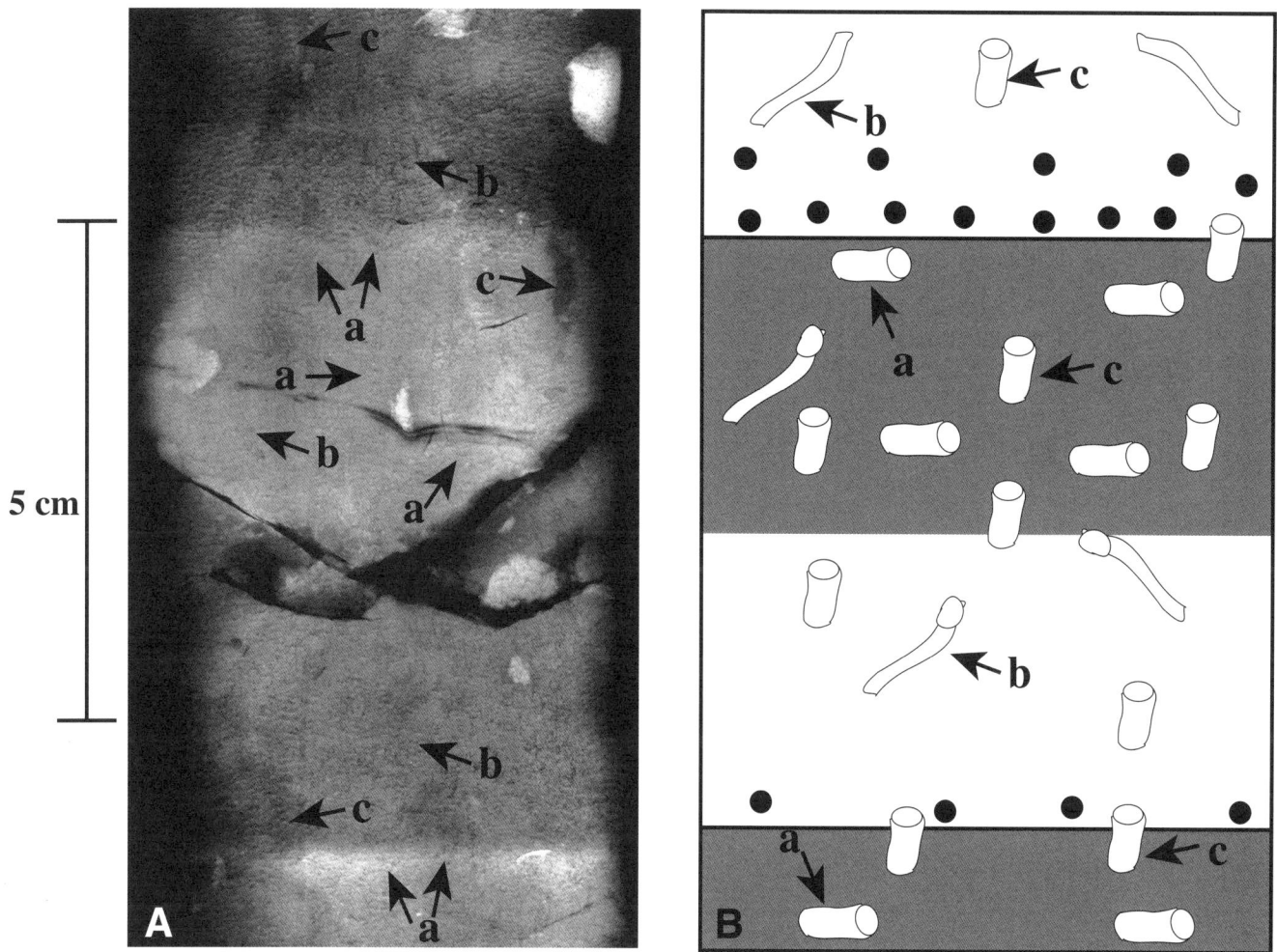

Figure 14. Aragonite III in BL96-2 showing detail of the bioturbation pattern in an X-radiograph (A) and a schematic drawing (B). Coarse-grained intervals (darker in A) have a sandy base overlying a sharp contact. Fine-grained intervals are denser (light in A and shaded in B) and have more abundant small, horizontal burrows (a). Random burrows cross all units and range from small, sinuous burrows (b) to large burrows (c).

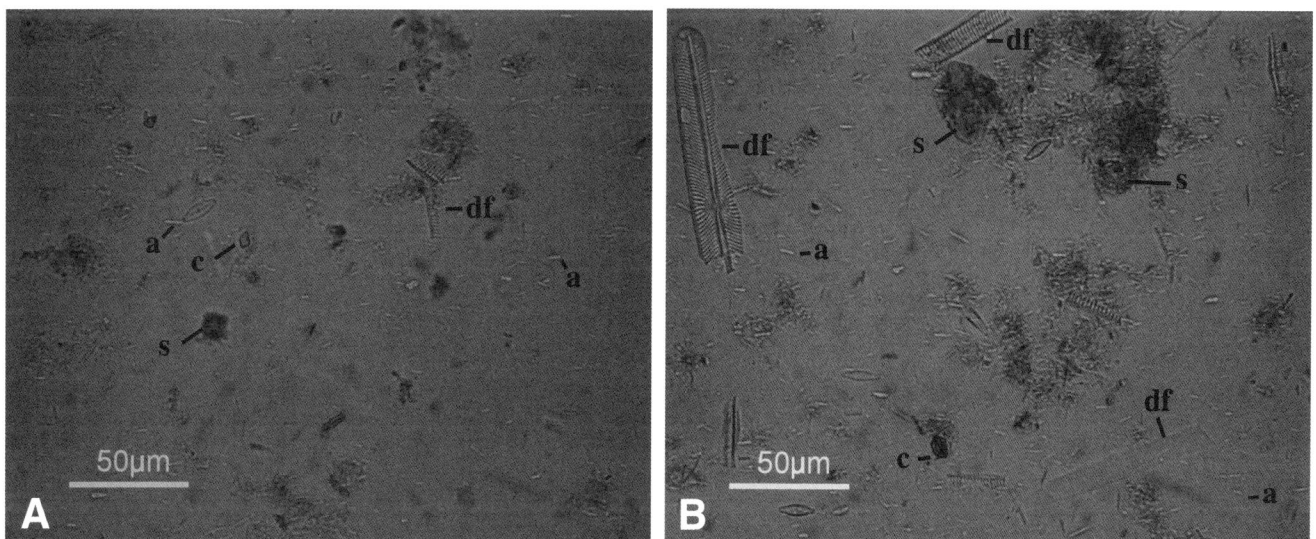

Figure 15. Aragonite III in BL96-1 showing comparison of smear slides from a fine-grained interval (A) and a coarse-grained interval (B). Both slides have a mixture of diatoms and a mixture of aragonite (a) and calcite (c). Diatom fragments (df) and silt-sized rock fragments (s) are present in both slides, but the abundance and size range in the coarse-grained interval (B) is much greater.

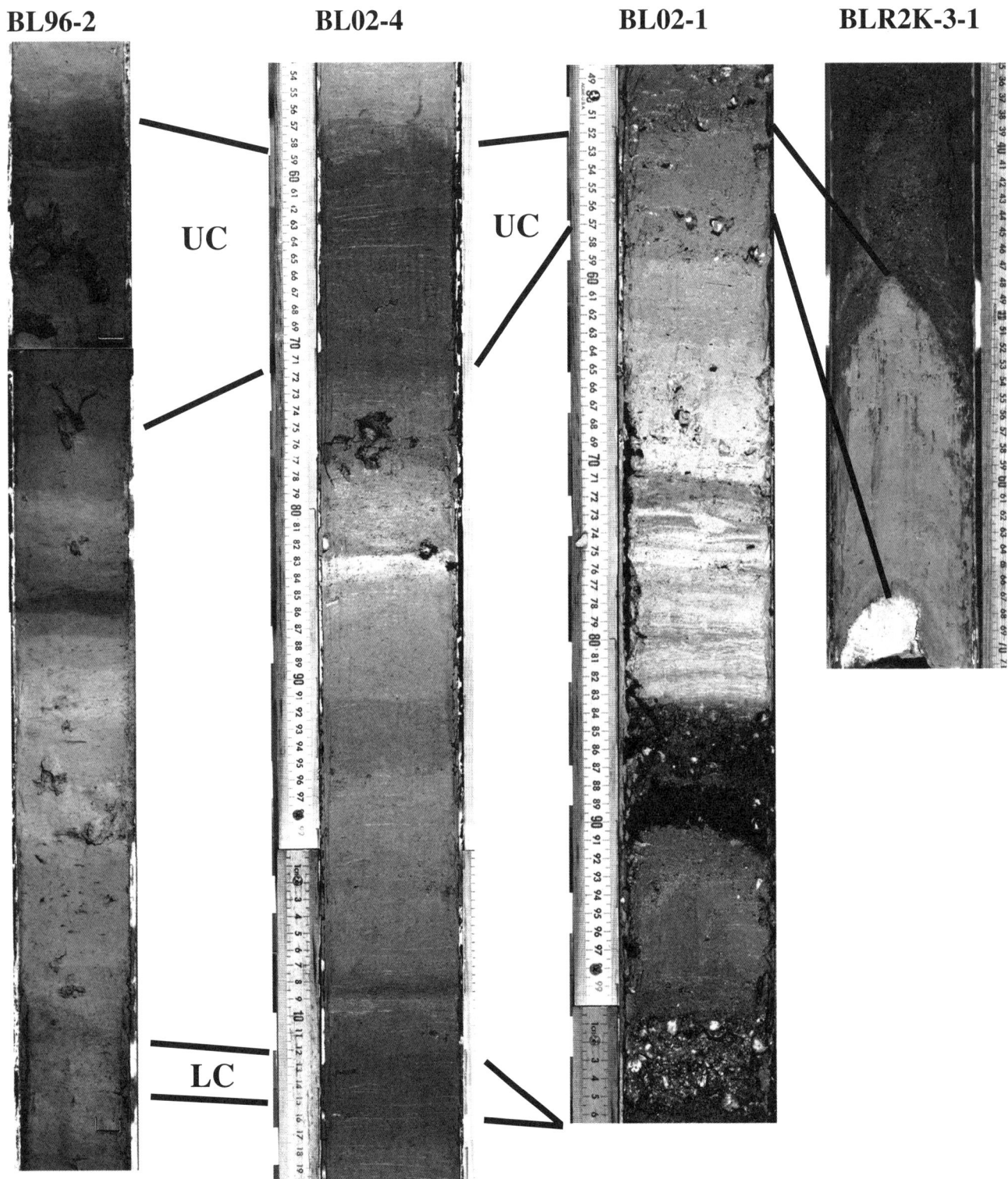

Figure 16. Correlation of four core segments containing Holocene deep-water calcite intervals (UC) from 43 m water depth (left), 34.9 m, 9.5 m, and Mud Lake (right). The calcite portions show only slight changes, whereas the underlying aragonite changes radically between the cores. A Pleistocene calcite layer (LC) underlies an aragonite interval in the deep-water cores. The calcite layer in the Mud Lake core overlies Pleistocene siliciclastic silt and is overlain by Holocene calcitic marsh deposits. The photos were enhanced digitally to contrast the layers. The scales are in centimeters.

unit show that the calcite also includes a mixture of rock fragments, scattered aragonite needles, and siliciclastic silt and that the diatoms include a number of shallow-water forms (Fig. 18). The mixed calcitic layer is sharply overlain by an ostracode-rich sand that is 1–2 cm thick. This sand is gradationally overlain by a mixed calcite and aragonite layer similar to the underlying one, but with more aragonite. Aragonite in this layer gradually increases upward to a poorly sorted Type I aragonite. In BL96-3 and BL02-1, the calcitic interval is thinner (1–5 cm thick) and is composed of calcite mixed with aragonite. These occurrences are rich in silt- to sand-sized ostracodes. The calcite-rich mud layers in BLR2K-1 and BLR2K-3 are 10–15 cm thick and dominated by euhedral calcite crystals and small planktonic diatoms. Silt- to sand-sized ostracodes are common at these locations and siliciclastic silt is an important component.

Applying the logic of the aragonite mud to calcite, the basal layer of calcite crystals represents a lake transgression. The rapid transition from aragonite to calcite is consistent with a major freshening of the lake. The remaining thickness of mixed calcite and aragonite represents sediment focusing in a falling lake before it changed to a strictly aragonite-precipitating lake. Geochemical data (Dean et al., 2006; Dean, this volume) indicate that the calcite layer precipitated in response to an influx of Bear River water. The occurrence of the calcitic layer north of Bear Lake in Mud Lake cores (Fig. 1) is consistent with a lake that extended well beyond the modern boundaries. The mixed calcite and aragonite probably does not represent the co-precipitation of these minerals, but rather the mixing of old aragonite with precipitated and reworked calcite as the lake level dropped. The ostracode-rich sand layer near the top of the calcitic layer is interpreted as indicating a drop in lake level. It could represent a lag deposit of storm wave activity, but was probably deposited as a wave-formed turbidite (as in Smoot and Benson, 1998, p. 140). The calcitic interval in the shallow-water localities is probably mostly reworked material representing the falling lake. The thicker calcitic interval in the cores within the marsh area

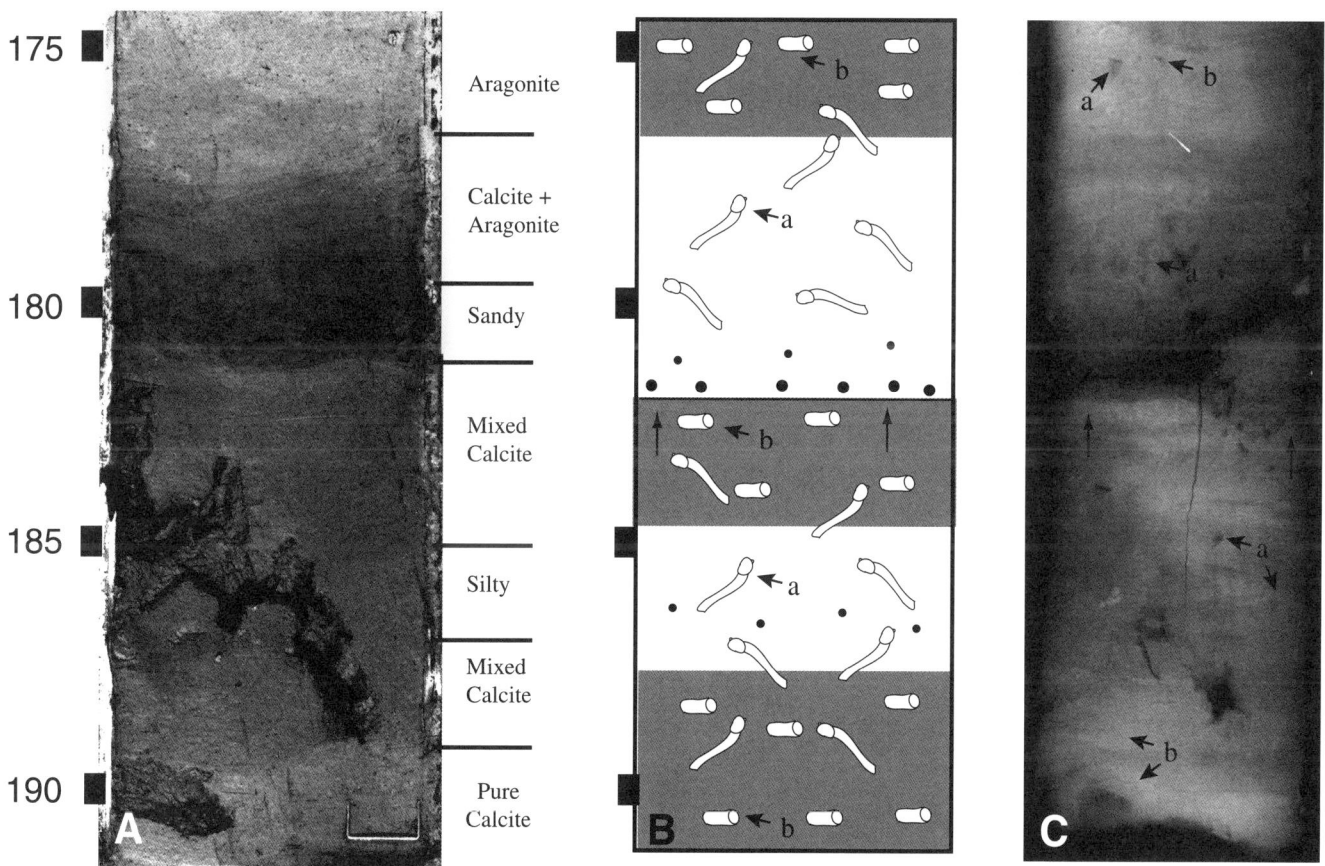

Figure 17. Deep-water calcite in core segment cut surface (A), schematic drawing of X-radiograph image (B), and X-radiograph (C) of BL96-2. Pure calcite is at base overlain by calcite mixed with minor aragonite. Silt-sized ostracode shells and rock fragments define a diffuse graded layer within the mixed interval. A sandy layer overlying a sharp surface (vertical arrows) separates sediment that is mostly calcite from sediment containing ~50% aragonite. The X-radiograph shows variations in sediment density (more dense is lighter) and pervasive bioturbation by random burrows (a). Small horizontal burrows (b) are abundant in the densest layers. Some of the density differences in C are due to variations in the sample thickness. The cut-surface photo was digitally enhanced to contrast the layers. Scales are in centimeters. The X-radiograph scale (B and C) differs slightly from the core (A) due to differential shrinkage from drying.

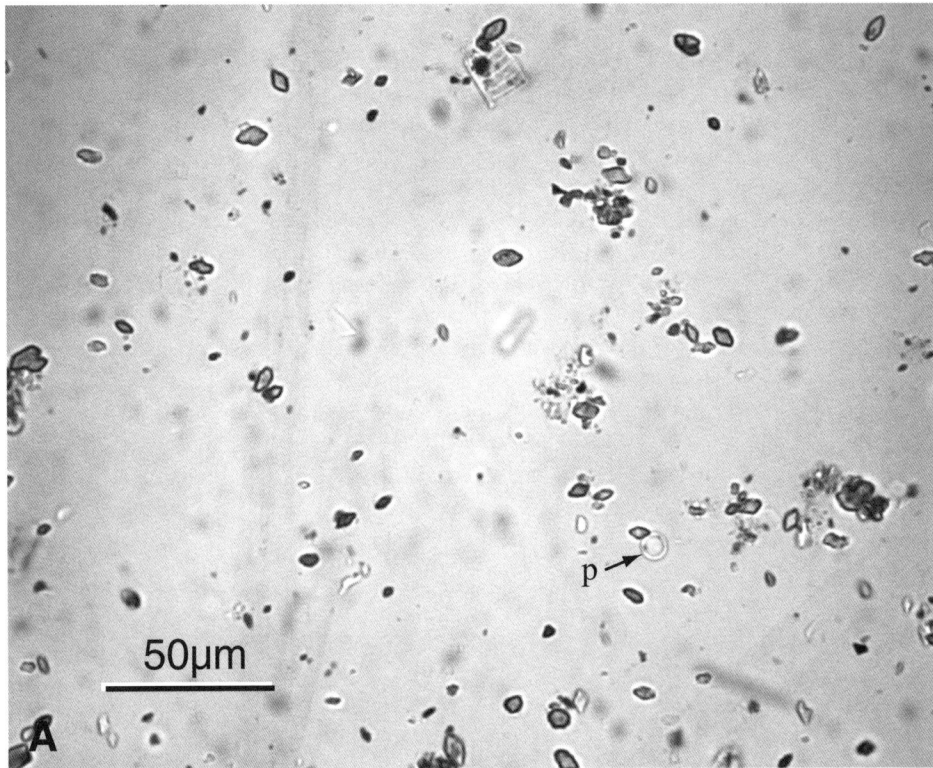

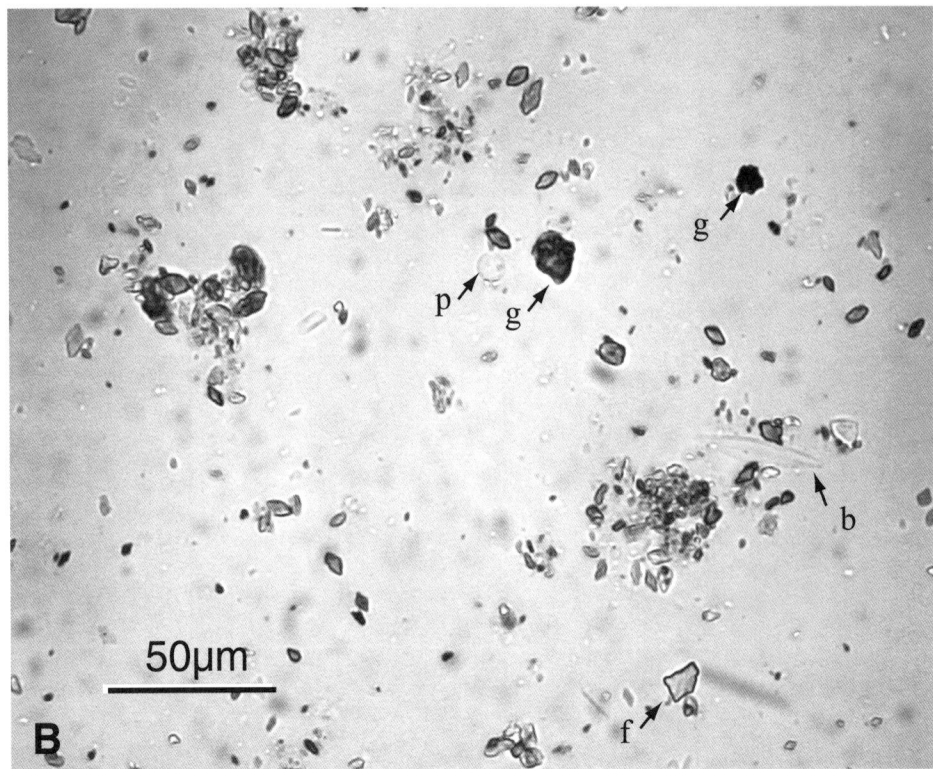

Figure 18. Deep-water calcite from BL96-2 showing comparison of pure calcite smear slide (A) with that of the silty mixed calcite (B). The pure calcite is composed of euhedral to subhedral calcite crystals and pelagic diatoms (p). The mixed calcite also contains benthic diatoms (b) and more irregular calcite composite grains (g) and calcite crystal fragments (f).

may indicate less erosion because the lake rapidly fell below that level, or the layer may have started thicker in that area due to mixing of detrital material closer to the mouth of the Bear River.

Shallow-Water and Shoreline Deposits

Cores taken in lake depths less than 35 m have sandy Holocene deposits that are generally less than 1 m thick (Figs. 3 and 5). These sediment thicknesses are consistent with the reflection profiles of Colman (2005, 2006), which illustrate a profound thinning of Holocene deposits at depths of less than 30 m. Modern shoreline deposits were examined from shallow trenches and short cores at different sites along the lake edge (Fig. 1), mostly during a prolonged drought in the summer of 2004. Three cores were also taken in the area north of Bear Lake (Fig. 1). Reheis et al. (2005, this volume) described 2–4 m thick sequences exposed along the Rainbow Canal north of the lake. They noted that Pleistocene deposits were mostly sand and gravel, but that Holocene deposits included peat, peaty mud, and the calcitic marl unit described earlier.

Bioturbated Sandy Mud and Muddy Sand

These are the most common deposits observed in cores and are transitional in character to Type III aragonite deposits. These deposits typically have vague layers or patches of different grain sizes with no distinct boundaries (Fig. 19). The sand component is commonly dominated by ostracode shells, particularly in cores taken in deeper water. In cores taken near the modern shoreline, the sand component may be dominated by siliciclastic sand. The patchy character includes well-defined burrow tubes as much as 1 cm in diameter. Bioturbated muddy sand with patches of plant debris and roots were observed in shallow cores along the northern shore of Bear Lake (see Fig. 23B in section on shoreline deposits) in less than 1 m of water. Such muddy sand is a common component of deposits in BL2K-1, -2, and -3, and BL02-1. Bioturbated sandy mud occurs in those cores, as well as in BL02-2 and BL96-3.

Bioturbated muddy sand and sandy mud represent variations on the same theme. Sediment transport was relatively infrequent, allowing organisms to homogenize the sediment, which therefore shows only vague bedding contacts. Differences in the sand-to-mud

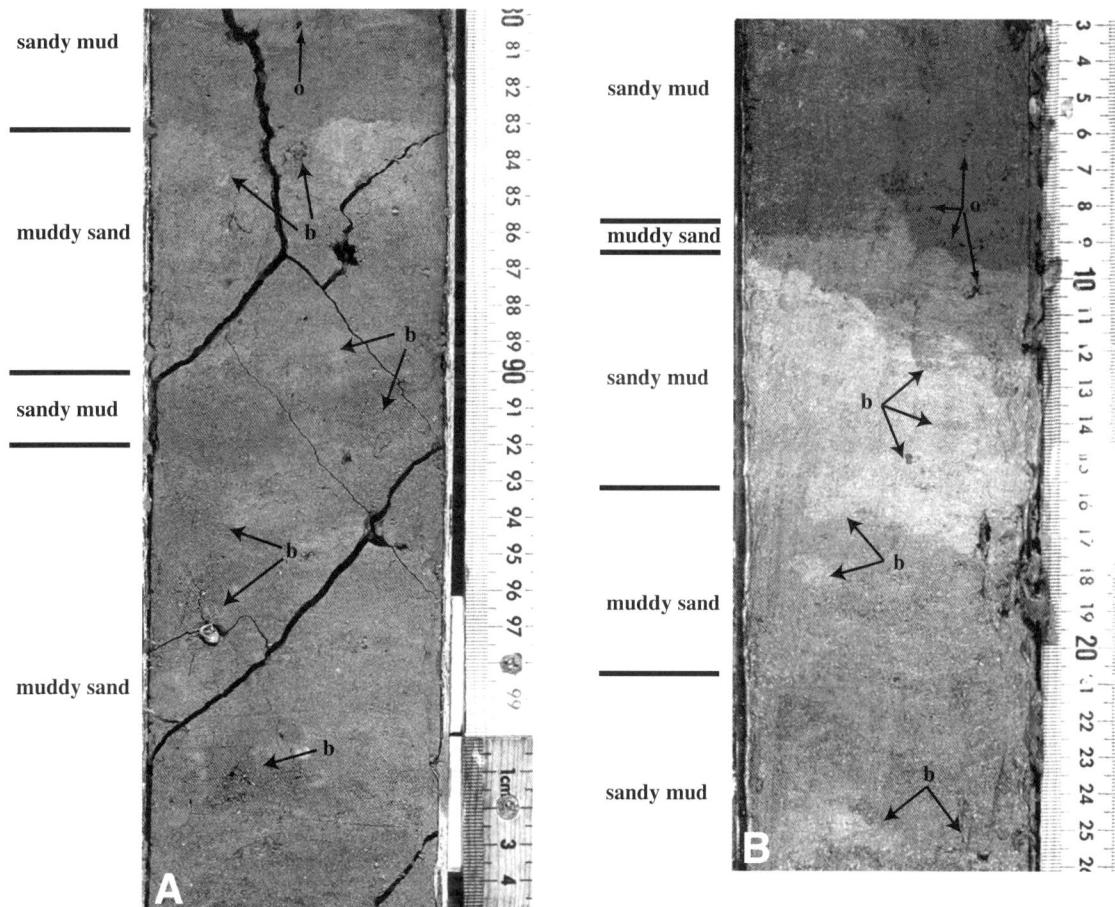

Figure 19. Shallow-water sandy mud and muddy sand in cores at BL2K-3-1 (A) and BL02-1 (B). Numerous burrows (b) homogenize the bedding including sand-mud contacts. Organic flecks (o) probably represent root hairs and fragments of aquatic plants. Scales are in centimeters.

ratio are interpreted as indicating the relative impact of winnowing of fine-grained material by wave action. Trenches and can cores from the north shore of the lake show abrupt transitions with muddy sediment beneath a surface veneer of sand or shell gravel. An abundance of sediment-filled burrows and open burrows within the upper 10 cm attests to the activity of worms and insects. Aquatic plants densely colonize the surface with small, hairlike holdfasts (e.g., Hutchinson, 1975), and their leaves provide dark organic material to the sediment. Terrestrial plants also occur less abundantly into water depths of several tens of centimeters, but are more common in areas that are subaerially exposed during lake fluctuations.

Rippled Sand Deposits

Where the lake-floor slope is relatively steep, oscillatory ripples are common in a few meters of water. These appear as well-defined sand lenses or sand layers with interspersed mud lenses. Sand lenses are characteristically well sorted and may be composed almost entirely of shells or shell fragments. Thin elongate lenses have flat bases and broad, rounded tops (Figs. 20 A and 20B), whereas thicker shorter lenses have more trough-shaped bases and steeper sides (Fig. 20C). Sand layers with mud lenses are actually superimposed sand lenses, each of which may comprise very different grain sizes, with a mud parting filling the troughs (Fig. 20B). These were observed in shallow can cores taken in less than 1 m of water on the west-central shore of the lake. The surface was composed of long-crested oscillatory ripples whose crests shifted with passing waves, but whose troughs were still and muddy. Ripple troughs may also be filled by coarser material such as shells that also form lenticular bodies (Fig. 21).

Sand lenses are formed by migration of wave-formed ripples on the lake floor. Thin, flat-bottomed, elongated lenses represent rolling-grain ripples which are the lowest-velocity bedforms produced by waves (see Harms et al., 1982, Chapter 2, p. 25–41). Thicker lenses with trough-shaped bases are produced by fully turbulent ripples under higher wave-bottom shear stress. The wave energy that produces the ripple deposits is intermittent and variable in strength. In areas where wave energy is felt only during the largest storms, isolated sand lenses in mud will form. These lenses will be preserved if the bottom bioturbation is insufficient to mix the radically different grain sizes. In areas where wind velocities are sufficient to cause wave sand movement frequently, beds composed of superimposed ripple lenses will form. Obviously, shallower water conditions favor this development as do steeper lake floors and long fetch distances. In transition to shorelines, wave-formed ripples may be erased by planar lamination during large storms or they may be intercalated with wave-formed bars (see below).

Shoreline Deposits

The modern shorelines around Bear Lake are mostly sandy deposits that are composed of a relatively narrow beach face and low-relief nearshore bars. Locally, these deposits are inset within wave-cut terraces eroded into older deposits mostly during lake-level rises. Where the lake floor has a very low slope in the northwest corner of the lake, the shoreline is a broad, flat expanse of shell gravel. Older shoreline deposits above the historical lake level include strandlines of boulders and cobbles.

On the northwestern corner of Bear Lake, the lake floor has a very low slope (Fig. 1), so shallow depths occur for kilometers away from the shoreline. Storm waves entering this part of the lake rapidly lose energy as they drag on the lake floor. Under these conditions, the shoreline deposits form very thin, shell-rich sheets (Fig. 22), reflecting the effect of bottom drag on all waves entering the area. The deposits observed in trenches were 10–20 cm thick and consisted mostly of snail shells at the base with coarse siliciclastic sand matrix grading upward to fine sand with scattered snail and clam shells (Fig. 23). These graded layers were traceable shoreward as a continuous sheet for kilometers. The sheet had a sharp base overlying bioturbated sandy mud and muddy sand. In the occurrences closer to the historical high-stand shoreline, the underlying mud had polygonal desiccation cracks filled with shelly sand. Graded shell beds that resemble these shoreline deposits were observed in several cores (Fig. 24) including BL2K-2, BL2K-3, BL02-1, BL02-2, and BL02-5. Two thin (3 cm) graded shell layers in BL02-4 may also be shoreline deposits, but they appear to be thinner and muddier. These deposits could represent storm wave deposits or wave-formed turbidite deposits farther offshore.

Sheet-like graded beds are formed on a shoreline by the rise and fall of lake level exposing large areas to wave action (Fig. 23A). On the very low slopes of the northwestern part of the lake, storm waves break kilometers from the edge of the lake. where they erode muddy sediment and concentrate coarse-grained material (mostly shells). Smaller waves can move closer to the shore, moving smaller grain sizes. Only mud and silt are moved at the actual water edge. The progressive landward loss of energy produces the characteristic graded bed. There is a hint of very low relief changes in gravel thickness over distances of tens of meters that could reflect a very low relief bar form. During falling lake levels, the wave sheets build lakeward, overlying bioturbated sandy mud (Fig. 23B). During rising lake levels, the gravel sheet may overlap previously mud-cracked mud (Fig. 23C).

The eastern side of Bear Lake has a steep transition from 40 to 50 m depth to the lake margin (Fig. 1), and the western side has a steep transition from 20 to 30 m deep to the lake margin. Under these conditions, storm waves impinge on the shoreline with much more energy than for the northwestern low-slope lake floor. The northeastern and southern ends of the lakes are transitional into low-slope shorelines as described above. The shoreline deposits adjacent to steep lake floors are thicker and more pronounced than the deposits formed adjacent to low slopes. These deposits consist of thinner sediment veneers over wave-cut terraces or thicker accumulations of bars and shoreface sand (Fig. 25).

Small wave-cut terraces are commonly overlain either by imbricated gravel or by lakeward-dipping tabular foresets. The imbricate gravel consists of coarser ridges (cobbles or boulders), with both lakeward and shoreward imbrication overlying finer

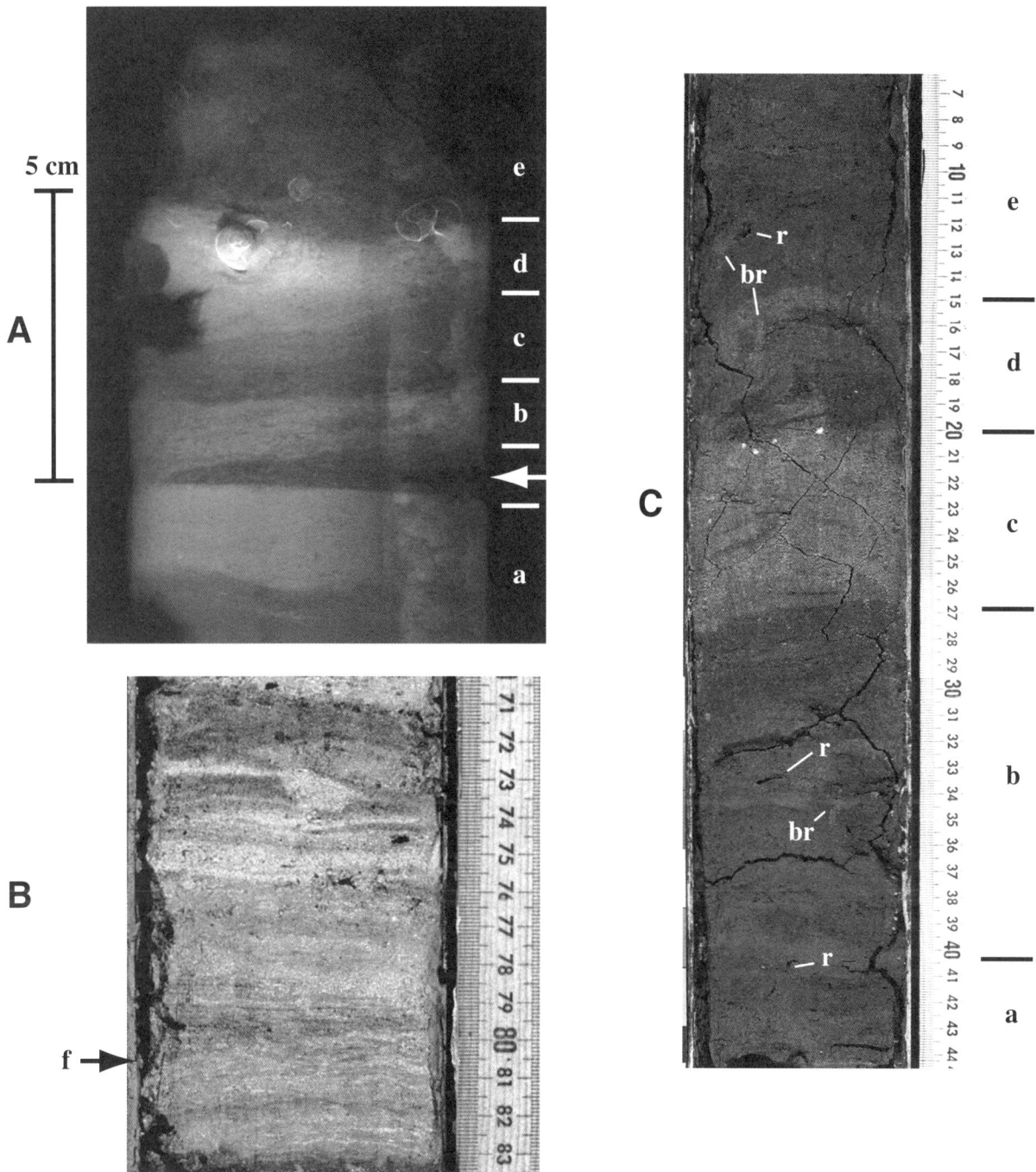

Figure 20. Wave-rippled deposits. (A) X-radiograph of top of BL96-3 showing convex-upward ripple lens of ostracode shells (arrow). The lens overlies bioturbated aragonite mud (a) and is draped with calcite mud (b). Graded aragonite sand (c) is overlain by sandy aragonite mud (d). Graded aragonite sand with snail shells at the base (e) is at the surface. (B) Quartz-rich fine sand (light) interbedded with aragonitic sandy mud (dark) in BL02-1. Wavy character is due to sand lenticularity reflecting rolling-grain ripples. Note mud lenses at ~80 cm (f) that were probably deposited in ripple troughs (flasers). (C) Ripple sequence in BL2K-3-1 consists of isolated very fine sand lenses in sandy mud (a), very fine sand with fine sand and muddy sand lenses (b), fine to medium sand with cross-lamination (c), fine sand with medium to coarse sand lenses (d), and muddy sand with fine sand lenses (e). Note burrows (br) and carbonaceous roots (r). Scales are in centimeters.

gravel with a shoreward imbrication. These produce steplike lenses perpendicular to the shore. The lakeward-dipping tabular sets may be sand, shelly sand, or gravel built over wave-rippled sand (Fig. 26). The foresets thicken lakeward from the erosional surface to as much as 30–40 cm.

Lakeward-dipping steep foresets are produced by waves piling up on shore and then sweeping sediment lakeward (Smoot and Lowenstein, 1997, p. 241–242). These are only produced on a small scale at Bear Lake, reflecting the generally shallow shoreline slopes and low sediment availability.

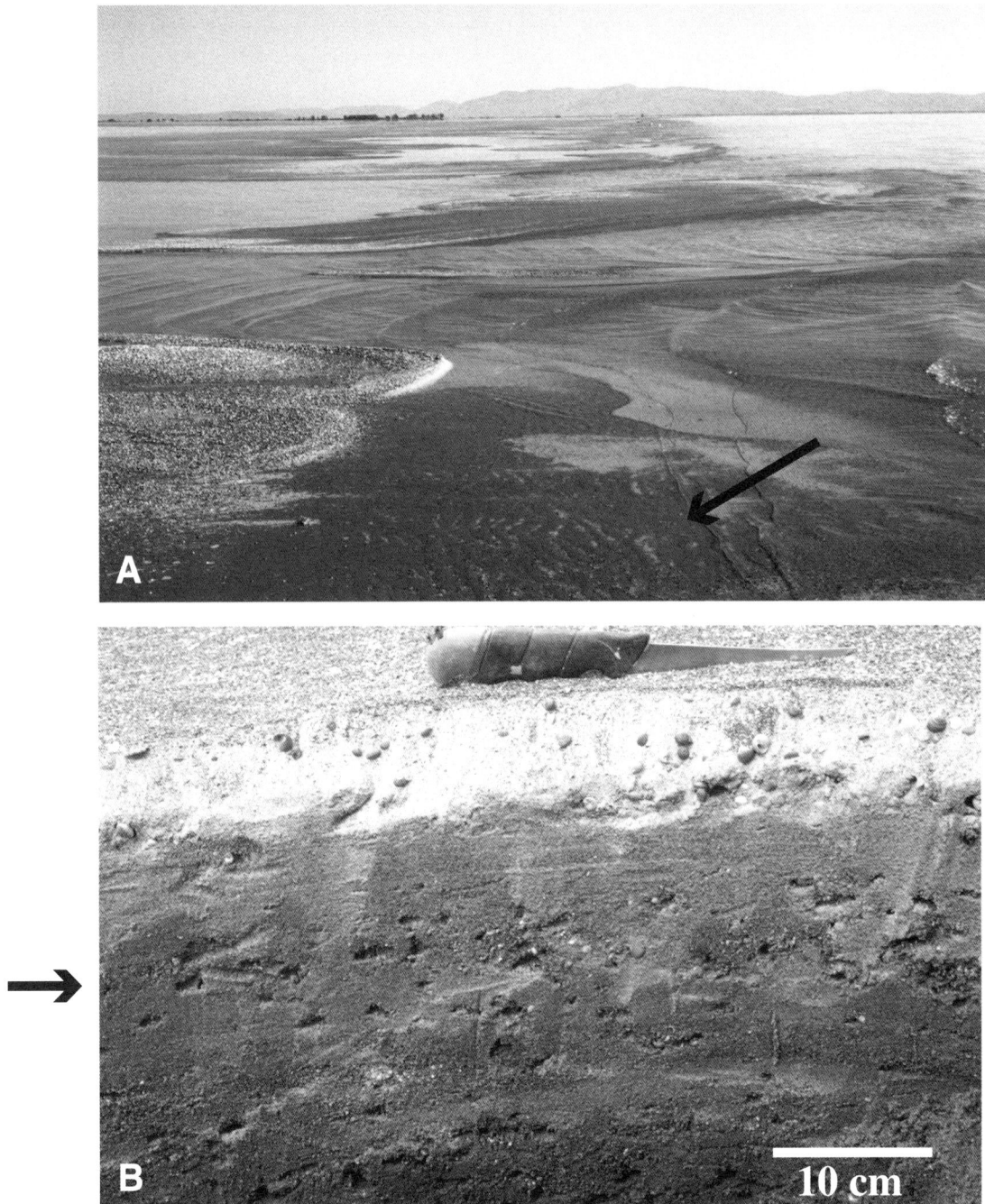

Figure 21. (A) View of shoreline at locality S7 showing wave-formed bars of coarse, shelly sand with ripples forming in shallow water in the foreground (arrow). Note the coarse material (light bands) trapped in the ripple troughs. The foreground of picture is ~2 m wide. (B) Cross section of rippled sand with coarse sand in the troughs (arrow) from a trench just shoreward of A.

Sandy beachfront deposits consist of sandy bars that have steep shoreward-dipping fronts and lower-angle lakeward dips that overlap at the lake edge into a ridge-and-furrow strandline. This type of deposit is generally thicker on the eastern side of the lake than on the western side. In cross section, the bars produce landward-dipping tabular foresets ranging in thickness from ~15 cm to over 50 cm. Each foreset bed is graded, with coarsest grain sizes at the base and progressively finer grain sizes toward the top. Adjacent beds vary in average grain size and composition. The foresets overlie wave-rippled sand deposits. The lakeward surface acts as a shoreface, accreting low-angle planar lamination that thickens gently lakeward where it grades into rippled sand (Fig. 27). The shoreface sands are generally finer grained than the bar foresets, but may include coarser layers or erosional insets of shoreward-dipping tabular sets 5–20 cm thick. The bar and shoreface deposits in broad beach areas are commonly modified by wind, which erodes sand from the crests and redeposits it into the troughs.

As surface waves impinge on shallower water, they become progressively more asymmetric with a stronger shoreward component. The offshore bars are composite bedforms produced by these asymmetric waves. The size and thickness depends upon the wave strength, water depth, and availability of sediment. Most of the coarse sediment in the modern Bear Lake is derived from reworking of older deposits. There is not much evidence for well-developed long-shore drift or influx of river-deposited sediment. The barforms are pushed shoreward during storms and then modified by lower-energy waves. Once a bar is pushed into shallow enough water, its back side becomes the surface for wave run-up and that side builds lakeward as a shoreface deposit.

Marsh Deposits

The cores taken in Mud Lake (Figs. 1 and 6) include sediments from the present-day shallow lake, and deposits formed when the area was a heavily vegetated swamp before completion

Figure 22. Gravel sheet at shoreline of locality S10. The bottom slope is very flat, with water depth remaining shallow for several hundred meters from shore. Snail shells and sand form a continuous layer for more than a kilometer inland. The picture foreground is ~2 m wide.

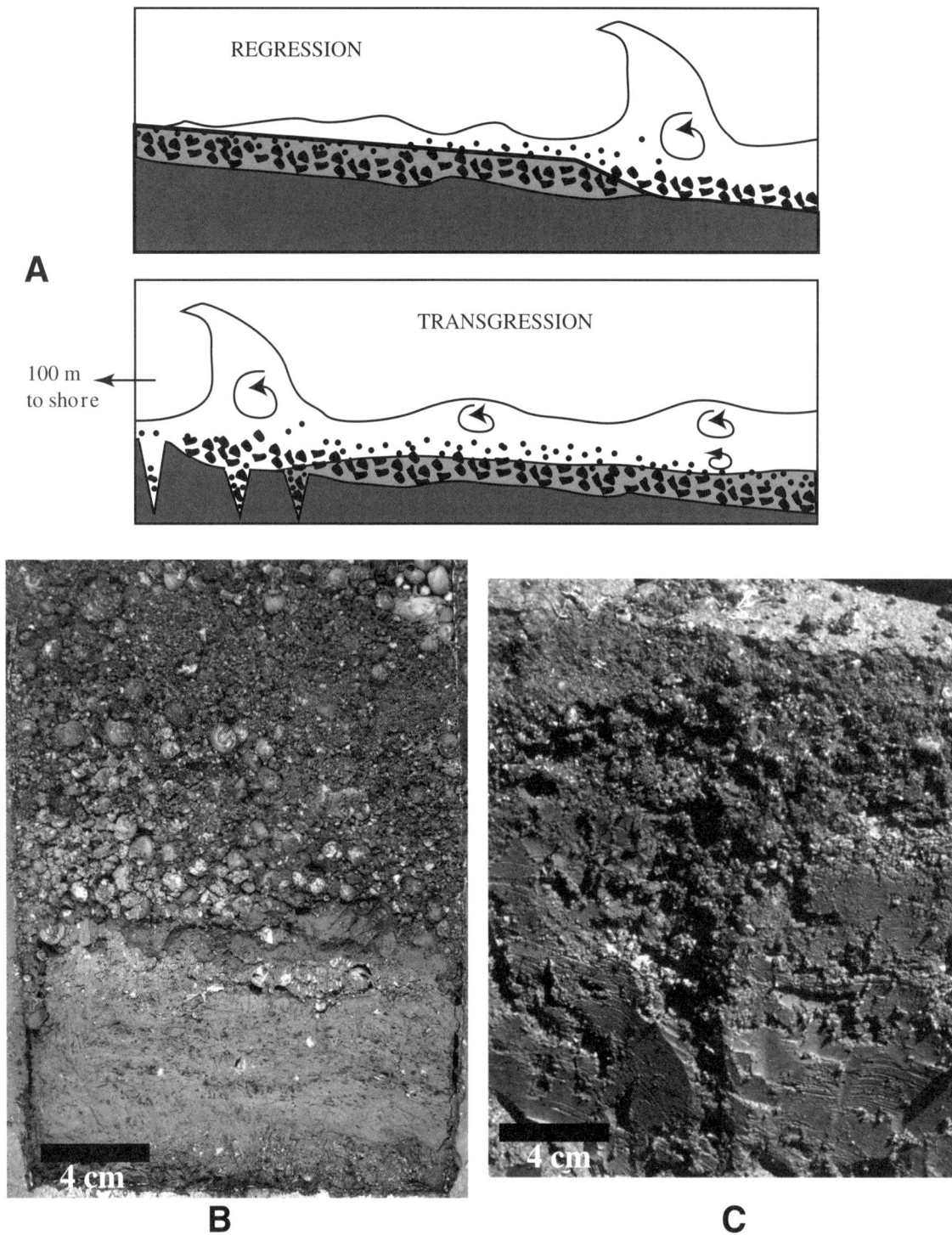

Figure 23. Cross sections through shoreline gravel sheet at locality S10. (A) Schematic drawing of how gravel sheets are formed during a dropping and rising lake level. Storm waves break hundreds of meters from shore, eroding the surface and depositing coarse sand and shells. Waves lose energy shoreward, depositing sand and mud. Transgression over previous subaerial deposits is initially fine grained then coarser. (B) Can core taken at modern shoreline showing graded shell gravel overlying burrowed sandy mud with small aquatic plant rootlets. (C) Trench in gravel sheet ~600 m shoreward showing graded shell gravel overlying sandy mud with desiccation cracks.

of the canals. A breccia composed of irregular mud clasts occurs 74 cm below the surface in a core taken near the canal (Fig. 28). The clasts reflect sedimentary units below the breccia in the same core. The sediment above the breccia is mostly a reddish sandy mud with abundant burrows and carbonaceous root casts. Thin sand layers are mostly siliciclastic sand. The mud includes a mixture of subhedral calcite crystals, with very few diatoms visible in the smear slides. Below the breccia, the mud is more organic rich and less siliciclastic. Diatoms are abundant, dominated by a mixture of attached diatoms (Moser and Kimball, this volume). Carbonized roots and plant fragments are locally abundant, in places making a muddy peat deposit. Shells and shell fragments, mostly small snails, are randomly scattered in the mud and concentrated into thin layers. Thin gray to tan layers are mostly calcite with a mixture of diatoms, charophyte fragments and oogonia, and silicate grains (Fig. 29). Sand beds with sharp bases contain mostly siliciclastic grains (Fig. 28), but also contain abundant shell fragments and snails. These beds are commonly graded, but are fairly well sorted into layers. There appears to be a loss of Holocene section in the areas closer to the barrier bar separating Mud Lake from Bear Lake. The sand beds also appear to be thicker and more common in that direction.

The pre-canal marsh environment was dominated by biological production and carbonate precipitation. Siliciclastic material was probably introduced by floods from the mountain streams, windblown dust, and storm wash over the barrier ridge of northern Bear Lake. Storm wash would account for the apparent increase of sand layers in the southern cores and for the high sorting of those sands. The mud breccia in BLR2K-1 is interpreted as material dredged from the canal and dumped into the adjacent swamp. After construction of the canal, there was a source for muddy sediment from the Bear River. The muddy sequence in BLR2K-1 was above water level at the time of coring, but may be a local area of high sedimentation. The dearth of diatoms within the sediments supports the idea of high sedimentation rates. The aragonite layer in BLR2K-3 suggests that

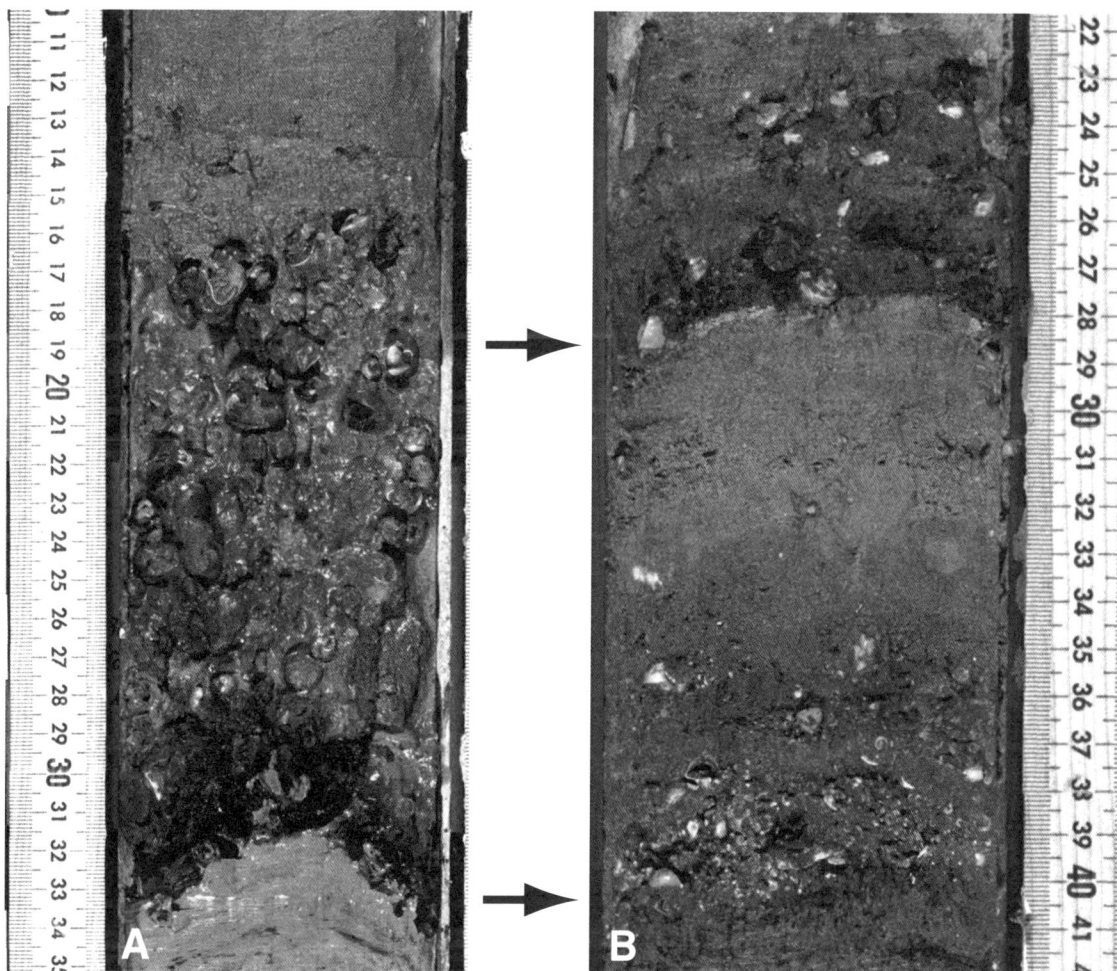

Figure 24. Graded shell gravel beds in BL02-2-1 (A) and BL02-1 (B). The graded gravel in B (arrows) is very similar to the shoreline gravel illustrated in Figure 23. The gravel in A is muddier due to infiltration of material during deposition of deeper-water sediment following a transgression. Scales are in centimeters.

Figure 25. Shoreline deposits at locality S8 (A) and locality S11 (B). (A) Recently notched shoreline from a minor transgression has small wave-built platforms building lakeward (arrows) over rippled sand. (B) Beach face developed on a stranded bar deposit. Note the depression behind the beach front (with tire tracks) and the erosional terrace in the background (with cars parked on top).

the lake rose above the barrier bar and flooded the swamp. The 5.2 ka age of that layer (Fig. 6) indicates that it is a fairly recent deposit. The calcite-precipitating lake stage of the early Holocene also flooded the swamp (Fig. 28). Sandy deposits overlying the gray calcitic mud probably represent lake lowering and re-establishment of the vegetation.

Modern Dam Breach

In BL2K-3, a 35-cm-thick sequence of fining-upward sand with ripple cross-lamination is overlain by an upward-coarsening sand sequence with climbing-ripple cross-lamination (Fig. 30). This sand overlies burrowed muddy sand and is overlain by bioturbated mud and graded shell gravel. Radiocarbon ages through

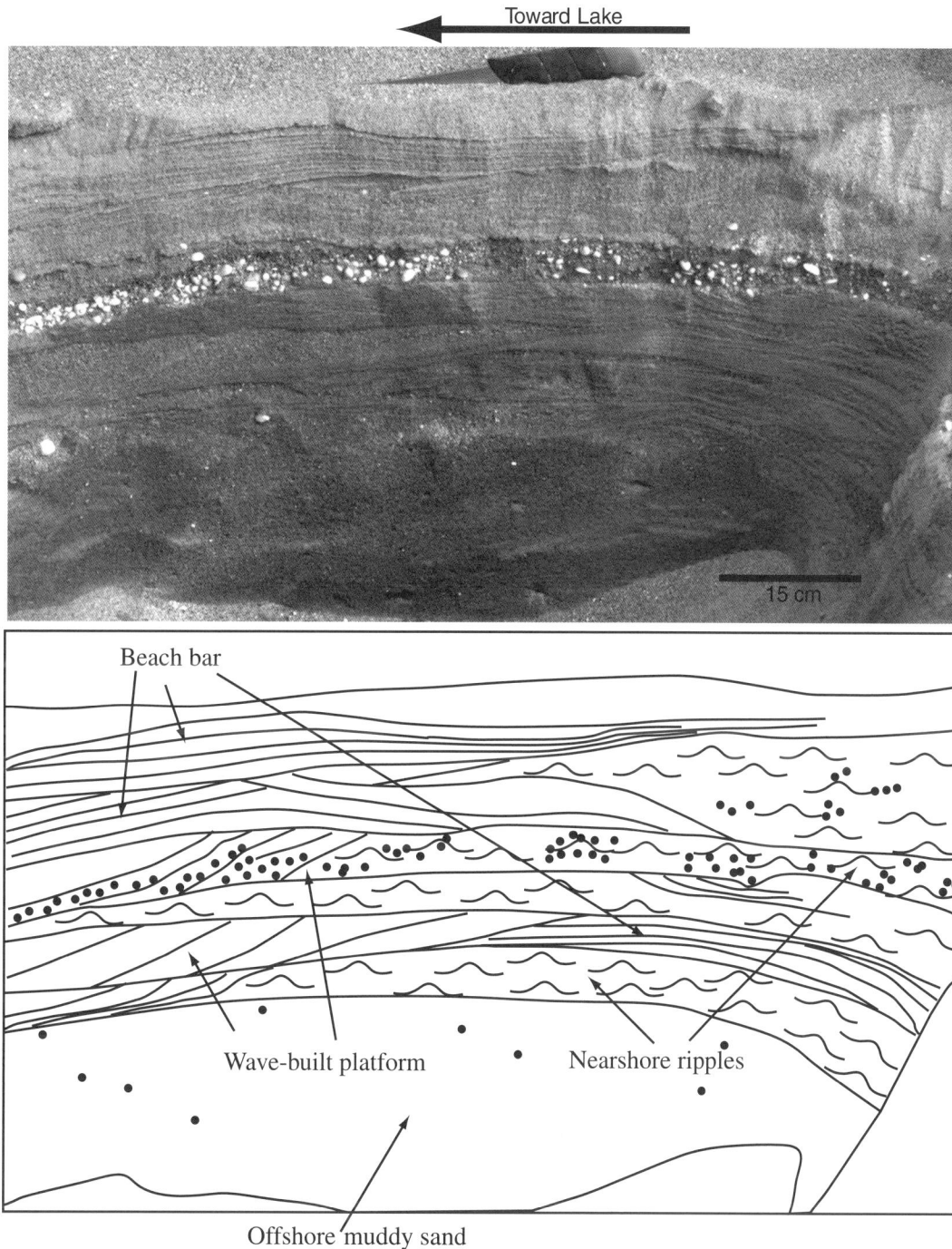

Figure 26. Trench cross section of shoreline deposits at locality S8. Base of trench is burrowed muddy sand. Series of lakeward-dipping tabular foresets are small wave-built platforms of quartz sand and shells. Beach-bar deposits include shoreward dipping sets with wave-rippled sand.

this sequence (Fig. 5) are inverted. The upward-fining sand layer is interpreted as the initial pulse of sediment when a dam used to regulate Mud Lake water levels was breached during a spring flood, causing a rush of Mud Lake water and sediment to enter Bear Lake in 1993. The upward-coarsening sequence is due to the continued erosion of older deposits from the dam breach and their deposition as a clastic wedge on the lake floor. The stratigraphically reversed ages reflect the erosion of older deposits and their redeposition into the lake. Prior to the breached-dam flood, burrowed muddy sand formed at this locality. The shallower water conditions produced by the addition of sediment caused this locality subsequently to be more nearshore in character.

Holocene Depositional Model

The reconstruction of Holocene sedimentary facies at Bear Lake relies heavily on the calibrated ages derived from radiocarbon ages (Colman et al., this volume). The ages based on pollen separates (pollen+ of Colman et al., this volume) are considered more reliable than those for shells (370 yr reservoir effect correction from Colman et al., this volume). Comparison of coeval core records along north-south or east-west transects clearly show radical changes in the thickness of sedimentary packages, with much thicker records in deeper water (Figs. 31 and 32). This observation, coupled with the abundance of reworked sediment and the presence of sharp bedding contacts overlain by sand, suggests that erosion of the shallower deposits and redeposition in deeper water have been important components of the sedimentary record. The cores available for sedimentary analysis typically lack the modern surface sediments and often have sediment thousands of years old at the top. This, combined with the modern canal-fed lake chemistry, complicates direct comparisons to modern depositional conditions. Geochemical data for BL02-4 (Dean, this volume) suggest that the uppermost sediments were deposited after the opening of the canals, that BL2K-3 contains a thick sand sequence that is correlative to a recent event, and that the upper part of BLR2K-1 has a breccia attributed to the canal-digging process. A collection of surface samples along several transects from shallower to deeper water provided grain-size distributions of insoluble clastic sediment (Smoot and Rosenbaum, this volume). These data suggest that the modern surface sediments are consistently finer grained than most of the Holocene

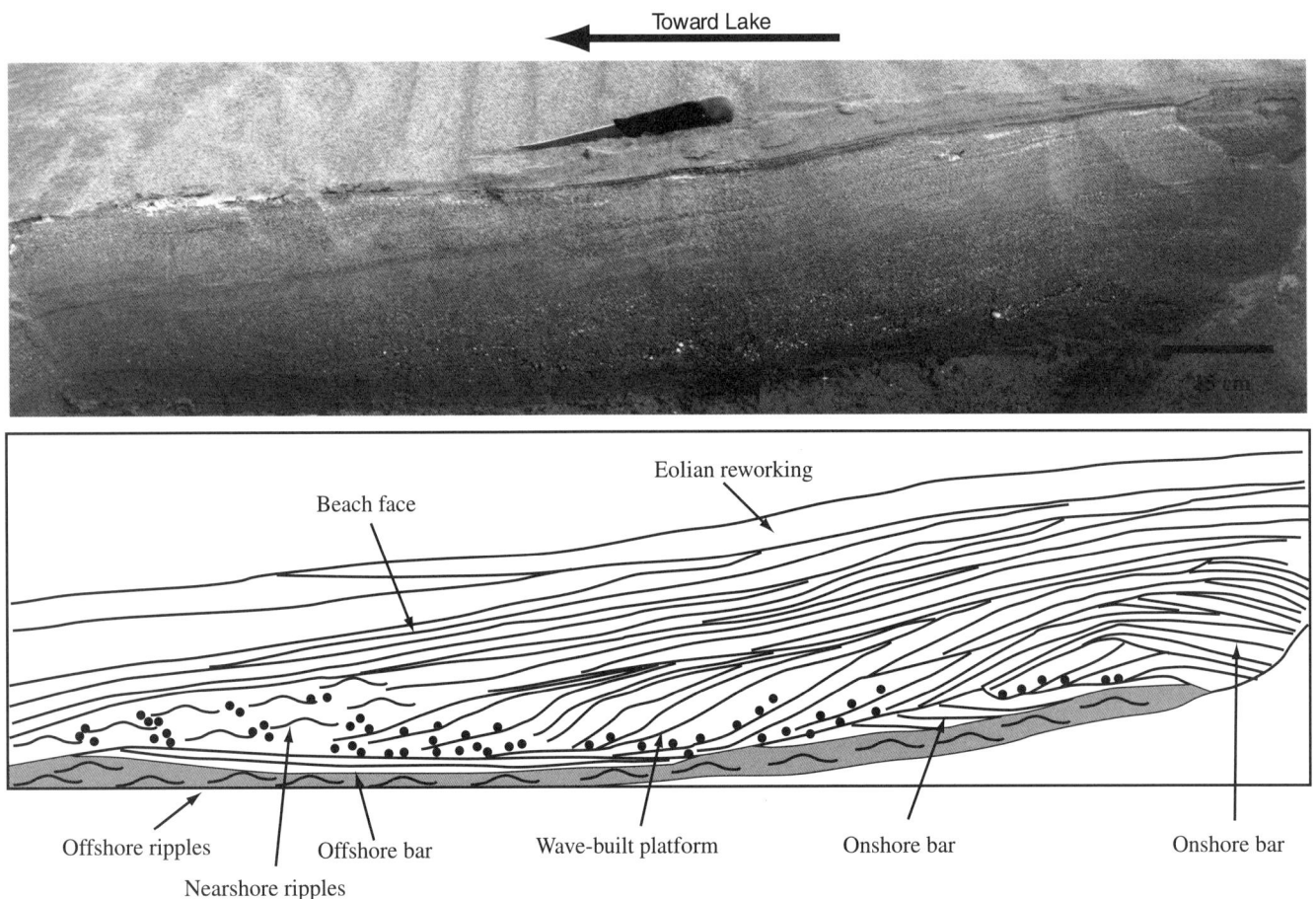

Figure 27. Trench cross section of shoreline deposits at locality S6. Beach-face deposit of predominantly quartz sand is superimposed on a shoreline bar that was modified by wave-built platforms. Offshore rippled fine sand forms lenses in muddy sand.

sediments and that the grain-size distribution is not uniform to depth. Therefore, the depositional model for the Bear Lake Holocene record is forced to use relative depth relationships with only broad constraints on absolute depth.

According to the distribution of the sediments in cores (Figs. 31 and 32) and trenches, the ideal Bear Lake depositional model has Aragonite II forming in the deepest water (greater than 50 m) and Aragonite I (35–50 m) followed by Aragonite III (20–35 m) forming in progressively shallower water (Fig. 33). Sandy mud and muddy sand are deposited at depths less than 20 m, with rippled sand beds grading to shoreline sands on the steeper lake margins and muddy sand directly overlain by sheet gravels on the low-angle lake margins. Although this type of arrangement is suggested by the core records, the reality of the facies distribution is much more complicated. The correlation used in Figures 31 and 32 assumes that most of the deviation is due to the degree of erosion, with progressively more erosion at shallower depths. Erosion is more pronounced along the east-west transect than the north-south transect, perhaps reflecting the different slopes. Shoreline deposits are notably absent in east-west seismic profiles in the middle of the lake, suggesting they are systematically eroded during transgressions. Aragonite II is the least well behaved of the

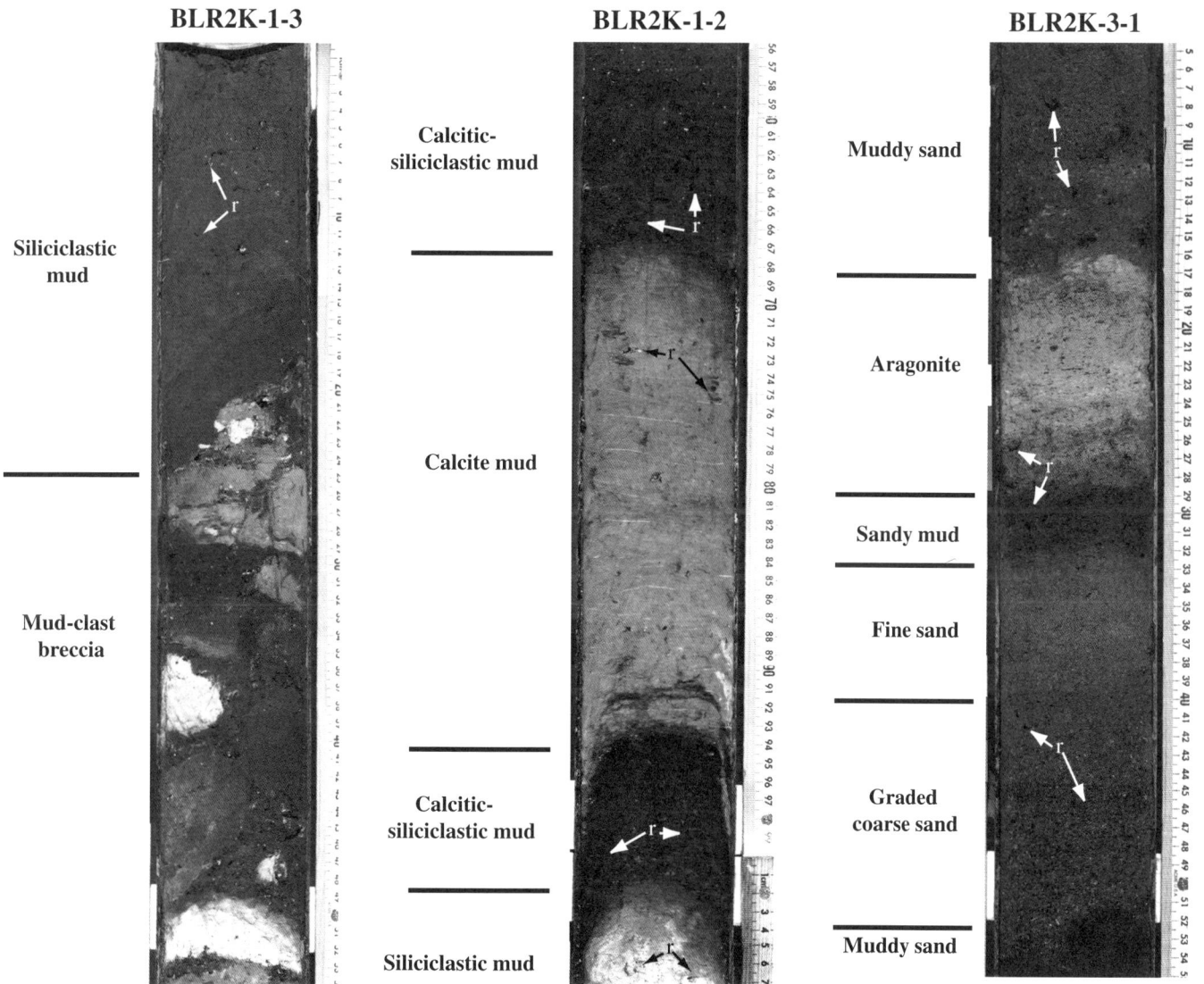

Figure 28. Core segments from the area north of Bear Lake at localities shown in Figure 1. In BLR2K-1-3, lower breccia includes fragments of mud and muddy sand from Holocene and Pleistocene deposits in the area. Upper brown silty mud with root casts (r) shows very little diatom or calcite content. In BLR2K-1-2, calcitic siliciclastic mud is organic rich and contains abundant small snails (white dots). Calcite mud contains pelagic diatoms similar to those in the Holocene calcite layer in Bear Lake. Siliciclastic mud at the base appears to be mostly glacial flour. BLR2K-3-1 reflects proximity to narrow bar separating the marsh from the lake. The graded sand and fine sand are composed mostly of quartz and shell fragments, similar to the bar deposit. Sandy mud and muddy sand contain calcite crystals, small snails, and organic matter, similar to typical marsh deposits. The aragonite layer is burrowed sandy mud, similar to shallow-water lake deposits. All scales are in centimeters.

Figure 29. Core segments showing variations in typical marsh deposits. (A) Calcitic siliciclastic mud in BLR2K-1-3 with abundant carbonized roots (r) and small snails (white dots). Faint banding in upper part of sample is due to organic-poor, calcite rich layers (lighter). Light layer at base is mud containing abundant glacial flour. (B) Calcitic siliciclastic mud in BLR2K-3-1 with abundant carbonized roots (r) and small snails (white dots). Snails are concentrated into a layer near base. Banding represents organic-poor, calcite-rich layers (light) and peat-like, organic-rich patches, and irregular bands (very dark). (C) Smear slide of calcitic siliciclastic mud showing calcite crystals (cc), rock-fragment grains (rf), and a charophyte oogonium (ch). Tiny white dots are silicate grains (glacial flour?). Scales in A and B are in centimeters.

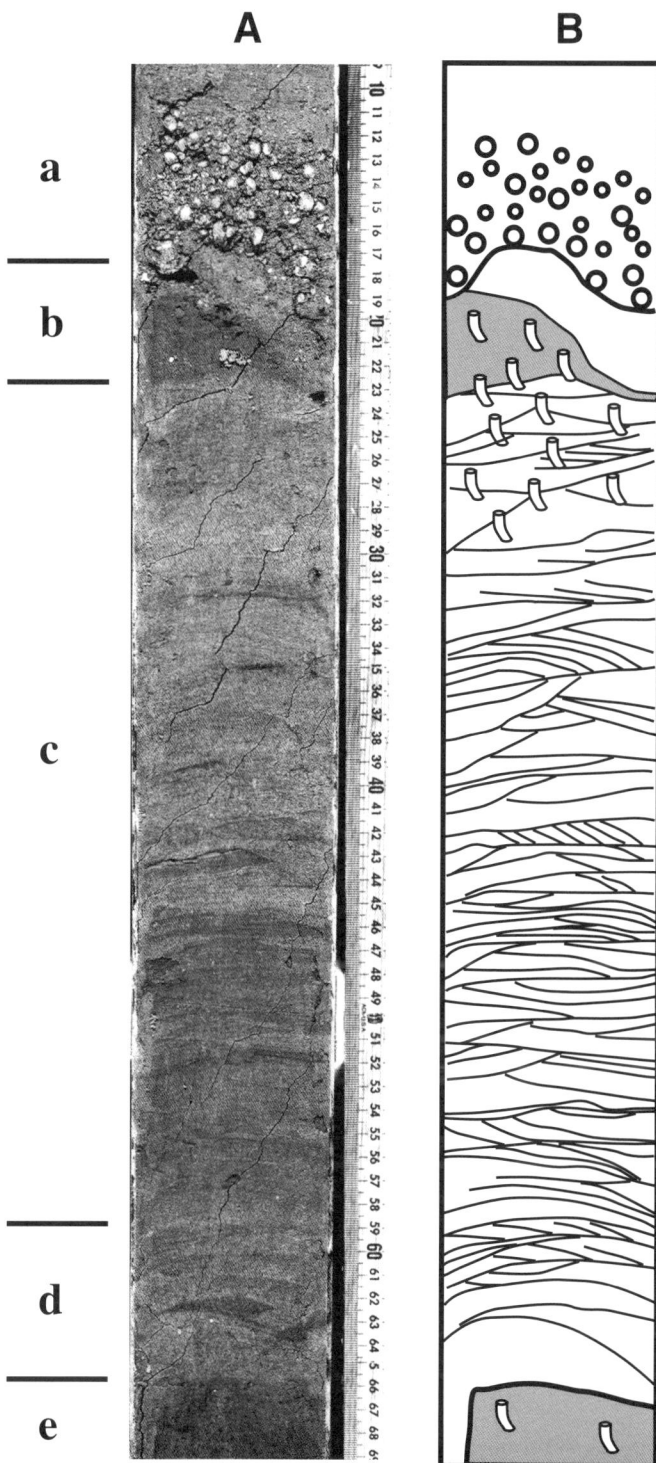

Figure 30. Core segment (A) and schematic sketch (B) from BL2K-3-1. (a) Graded shell gravel at the top is a shoreline deposit similar to gravel sheet in Figure 22. (b) Bioturbated sandy mud (shaded in B) with wave-sorted sand cap. (c) Upward-coarsening sequence of ripple cross-lamination. The upper part was modified by burrowing. (d) Upward-fining sequence of ripple cross-laminated sand. (e) Bioturbated sandy mud. Scale is in centimeters.

carbonate facies, requiring nearly systematic removal of equivalent strata in the shallower cores. This may indicate that Aragonite II represents a range of depths and may be more indicative of a change in water chemistry or temperature. It is interesting to note that the two Aragonite II horizons in BL96-2 and their correlative counterparts in BL96-1 are the only intervals that registered as distinctly different diatom populations in the study of Moser and Kimball (this volume). This suggests that the other Aragonite II intervals in BL96-1 may not be as distinct from Aragonite I. Another possible explanation is that Aragonite II represents very short lived lake-level rises during periods of rapid shifts in lake depth. This explanation would also explain the relatively poor mixing of Aragonite II in contrast to Aragonite I and the association of interbedded Aragonite I with the Aragonite II intervals that are equivalent to Aragonite III in BL96-2 and BL02-3.

The deepest-water core record (BL96-1) indicates there were four types of deep-lake conditions at Bear Lake during the Holocene. Aragonite I deposition appears to overlap with the modern lake conditions to ca. 2.5 ka. Aragonite II deposition occurred during two intervals, 2.5–4.5 ka and 6.5–7.5 ka, the younger one intermittently switching to Aragonite I. Aragonite III deposition dominated over a prolonged period around 5–6.5 ka and it also is found in 9–10 ka deposits in BL 96-2 and BL02-3, which were not penetrated in BL96-1. Calcite deposition (not penetrated in BL96-1) occurred ca. 8–8.5 ka. Aragonite III occurrences in BL96-1 indicate lake conditions much shallower than the modern lake. This interpretation is supported by the presence of sheet gravel shoreline deposits in cores at 35 m below the modern highstand in those core records equivalent to the Aragonite III in BL96-1. In contrast, the calcite depositional event is indicative of a major lake-level rise to depths greater than the modern highstand. The age range of the calcite is roughly coincident with that of the Willis Ranch shoreline, which is 9 m above the modern highstand (Laabs and Kaufman, 2003; Smoot and Rosenbaum, this volume). This rise in lake level coincides with an influx of Bear River water (Dean et al., 2006; Dean, this volume). Aragonite I deposits in BL96-1 appear to be coeval with conditions similar to the modern Bear Lake and conditions that were shallower. If Aragonite II represents aragonite deposition in a deeper lake, it suggests two time intervals with intermittent rises above the modern highstand without Bear River influence. The Aragonite II at 3.5 and 7.0 ka are the most likely associated with a deeper lake, whereas the other occurrences are more debatable.

The area now covered by Mud Lake was a shallow marsh for most of the Holocene. It was inundated with lake water during the calcite interval and possibly briefly during a lake-level rise ca. 500 yr B.P. The latter record is a thin aragonite bed in BLR2K-3 that may also represent a single depositional event during a storm rather than a lake-level rise. The area near the northern shoreline of Bear Lake experienced frequent episodes of storm washover depositing the graded sand beds found there. Following the opening of the canals, the marsh was replaced by Mud Lake and deposition of siliciclastic silt and mud was more rapid than before.

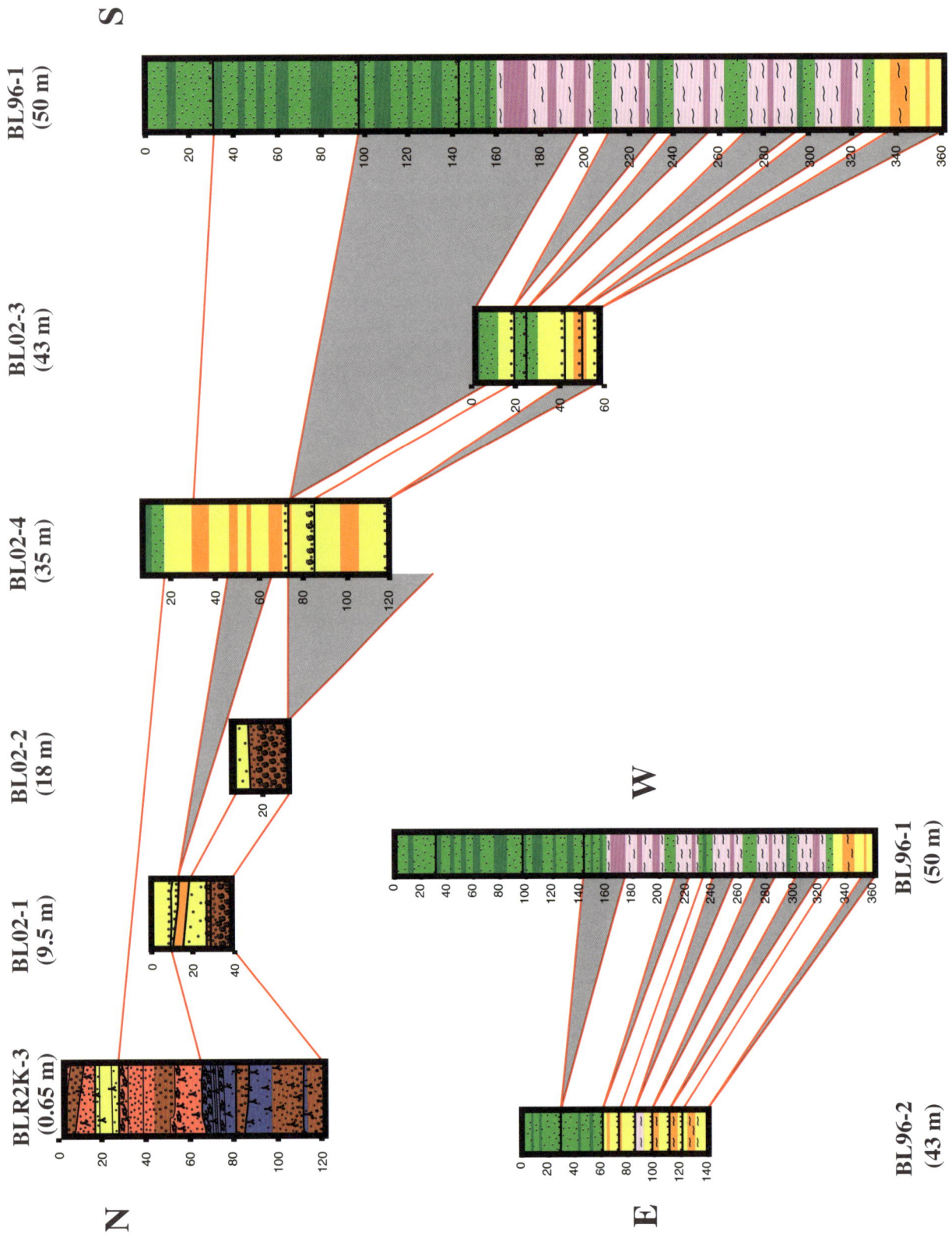

Figure 31. Correlation of core segments younger than 5 ka along a north-south and an east-west transect. Sediment types and sedimentary structures are provided in the legends for Figures 3–6 (p. 54–57). Note the change of scale for the east-west correlation. Red lines represent correlated horizons and gray shading indicates portions of sections missing, presumably due to erosion at the shallower core site. Section thicknesses may be composites of two or more cores. Correlations are from Smoot and Rosenbaum (this volume). Water depths of the coring sites (in parentheses below the core locations) are corrected to the modern highstand level (1805.5 m elevation). Vertical scales are depths below the surface in cm.

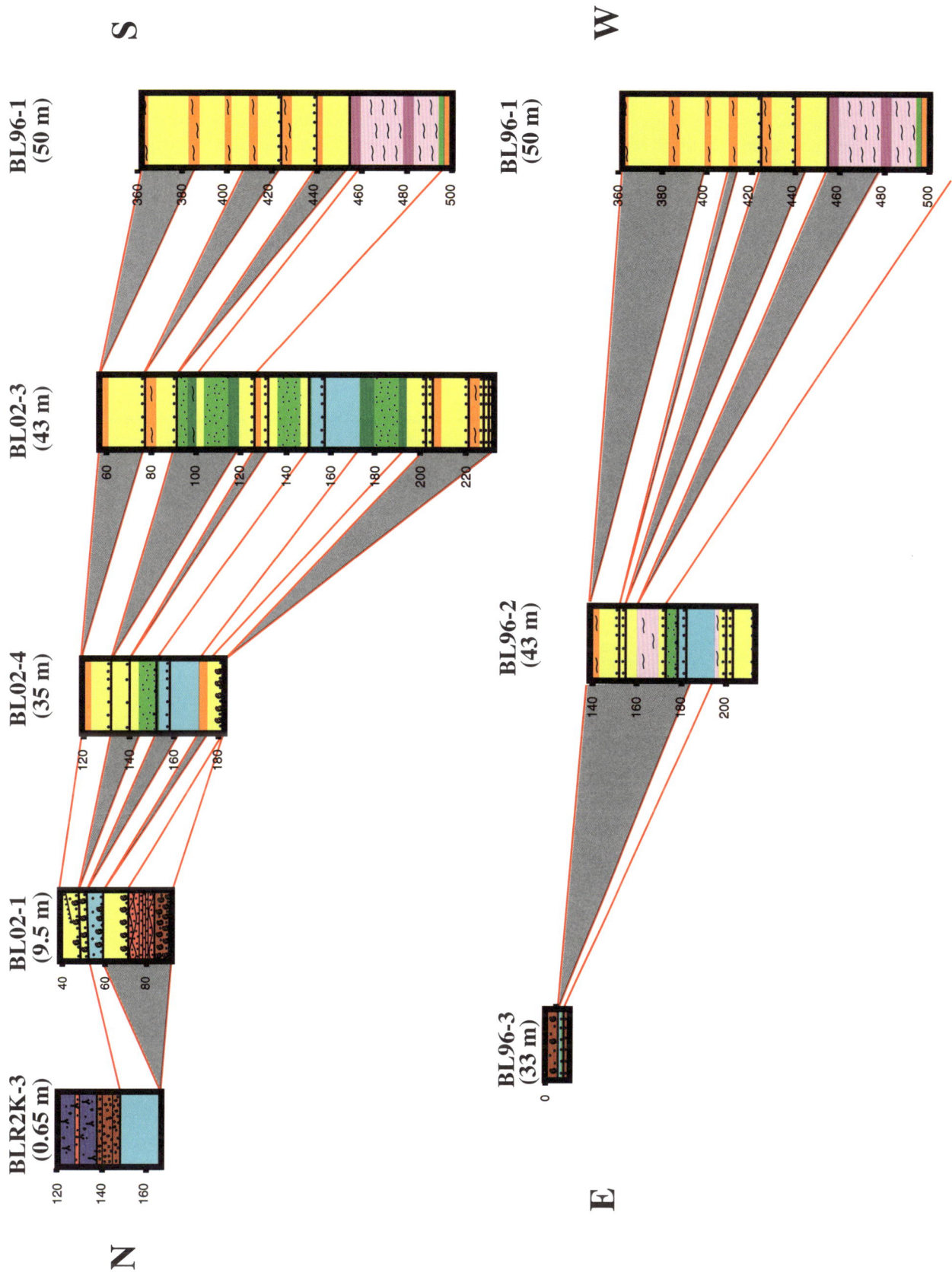

Figure 32. Correlation of core segments representing the time interval 10–5 ka along a north-south and an east-west transect. Formatting, scales, and data sources are the same as for Figure 31.

PLEISTOCENE SEDIMENTS

Pleistocene age sediments are found in a few exposures of shoreline deposits around the lake margin (Laabs and Kaufman, 2003; Reheis et al., this volume) and in cores (Figs. 3–6). The longest sequence of Pleistocene sediments collected was in the GLAD800 cores, which penetrated 120 m of sediment spanning ~220 k.y. (Dean, this volume; Kaufman et al., this volume). These cores, however, have not been thoroughly examined. The preliminary data suggest that the character of these deposits is similar to the range of Pleistocene sediment types observed in shallower cores, including some aragonite intervals (Dean, this volume; Kaufman et al., this volume). The longest core through Pleistocene sediments that has been thoroughly examined (BL96-3) is ~4 m long and dates to ca. 26 ka (Colman et al., this volume; Smoot and Rosenbaum, this volume). The shoreline exposures

were not examined for this study but have been described by Robertson (1978) and Laabs and Kaufman (2003).

Deep-Water Sediments

The majority of the deep-water sediments in the cores examined consist of siliciclastic mud and mud consisting of a mixture of siliciclastic sediment and calcite. The youngest Pleistocene deep-water deposits are aragonite, mostly Aragonite III. The Pleistocene aragonite deposits are underlain by a thin calcitic mud. Another thin calcitic mud occurs lower in the Pleistocene section below the mixed calcitic-siliciclastic mud interval. The aragonite deposits are effectively identical to those of the Holocene and will not be redescribed. The calcitic mud intervals are also like the Holocene calcitic interval previously described. Both are characterized by a basal nearly pure calcite that becomes more

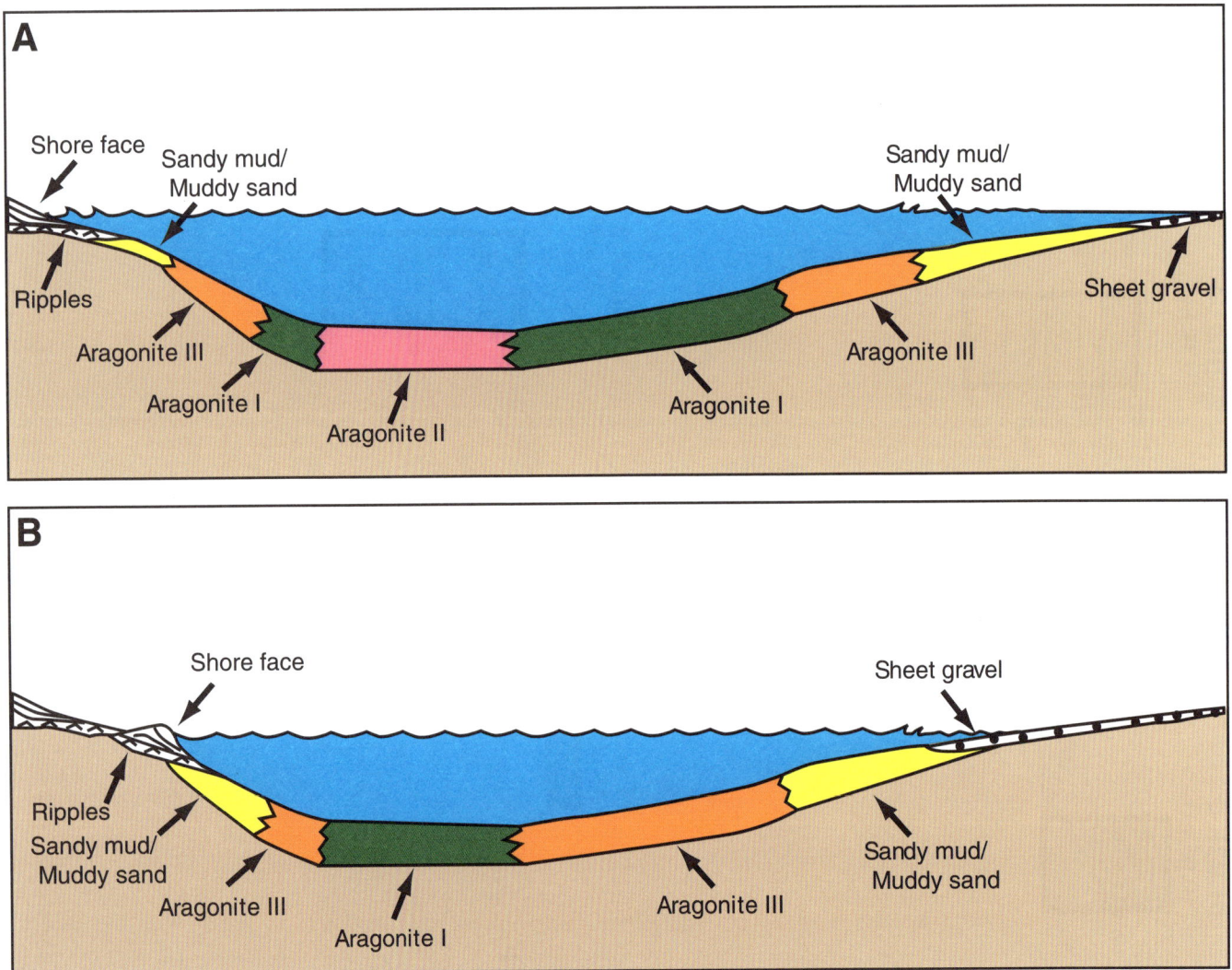

Figure 33. Schematic illustration of aragonite depositional facies in Bear Lake during the Holocene for conditions slightly deeper than the modern lake (A) and conditions slightly shallower than the modern lake (B). The left side represents a lake margin with a relatively steep slope (as in the eastern side of the lake), whereas the right side represents a shallow slope (as in the northern side).

siliciclastic rich with rock fragments upward. These will not have any further description.

Mixed Calcitic-Siliciclastic Mud

Mixed calcitic-siliciclastic mud is mostly gray colored with vague decimeter-scale lighter and darker bands and horizons of dark blebs (Fig. 34). This sediment type occurs in BL96-2 from ~235–330 cm depth, and is also found in BL96-3, BL2002-4, and BL2002-3. It contains 30%–40% calcite mixed with silty clay (Dean, this volume). The calcite is a mixture of subhedral to anhedral crystals (2-10 µm) and a variety of polycrystalline or irregular calcite grains (2-100 µm) (Fig. 35). The silt and clay are a mixture of silicate minerals and rock fragments with only minor clay minerals. Diatoms are varied and include both attached and pelagic forms. The relative amount of calcite is variable, with the darker, more calcite-poor layers containing more rock fragments and fewer visible diatoms. There is an ostracode-rich sandy layer with a sharp contact at 242 cm in BL96-2, but the typical sediment is mostly silt and clay. Burrows are typically small (1–3 mm diameter) and random, and some zones of burrows appear more horizontal. The most conspicuous layering in mixed calcitic-siliciclastic mud consists of bands containing abundant head capsules and sulfide-coated egg casings of the aquatic crustacean *Daphnia*. The latter resemble concentrations of dark blebs on the split core surfaces (Fig. 34). In the X-radiographs, the *Daphnia* concentration zones appear to display an upward size increase of very dense bleb-shaped features (Fig. 36). The largest blebs (as much as 1 mm) are horizontally elongated and may be tubes, whereas the smaller blebs (~0.5 mm) are clearly *Daphnia* egg casings.

The depositional conditions for the mixed calcitic-siliciclastic mud are difficult to determine. The mixture of carbonate crystals,

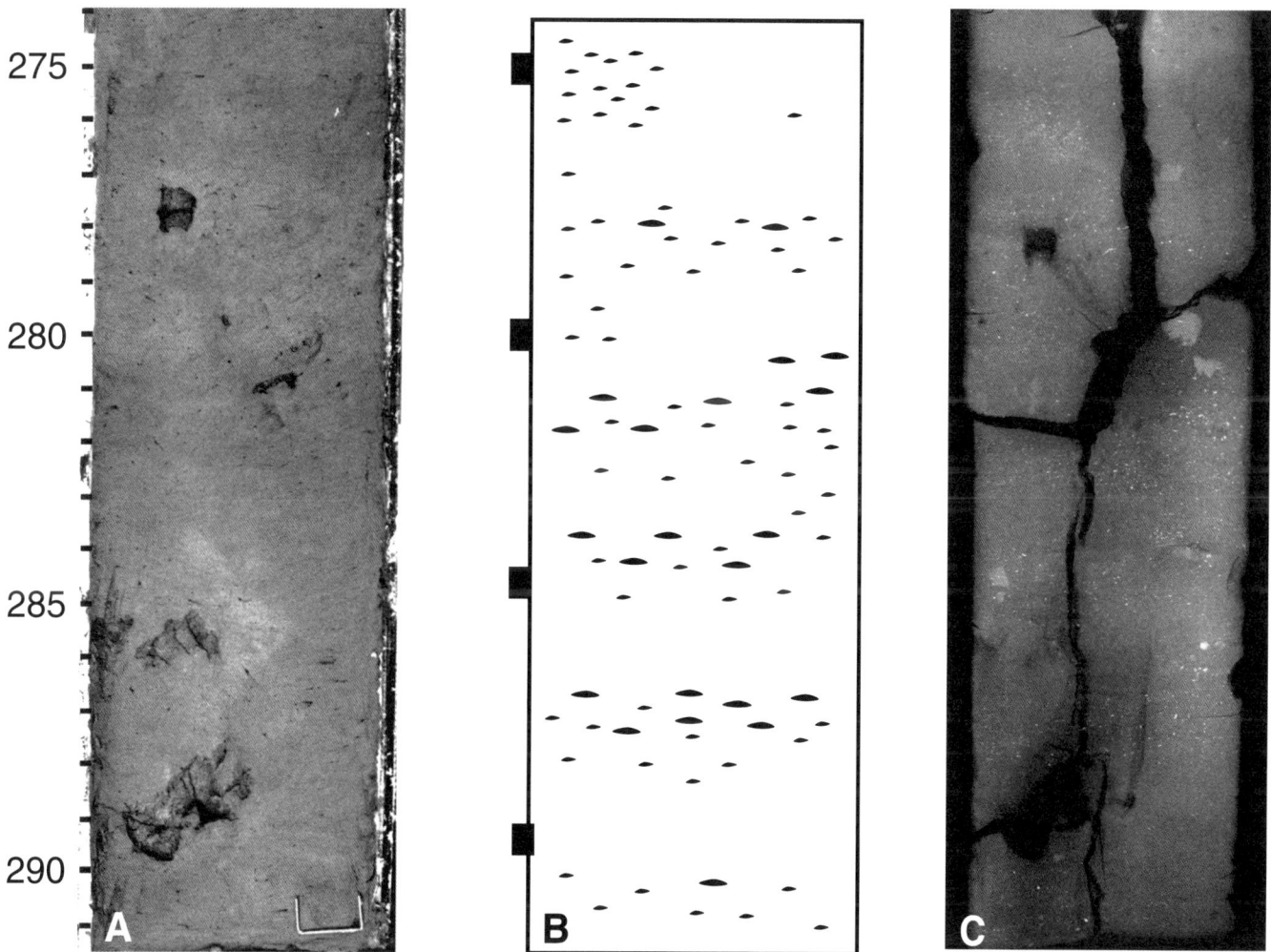

Figure 34. Mixed calcitic and siliciclastic mud in core segment cut surface (A), schematic drawing of X-radiograph image (B), and X-radiograph (C) from BL96-2. An irregular banding is defined by concentrations of sulfide-coated blebs (*Daphnia* egg casings and possible horizontal tubes) that appear black on the cut face and bright white in the X-radiograph. Scales are in centimeters. Some of the density differences in C are due to irregularities in the sample thickness. The large light patches in C are fragments under the sample. The cut-surface photo was digitally enhanced to contrast the layers. The irregular light patches in A are due to uneven oxidation around sample plugs (holes).

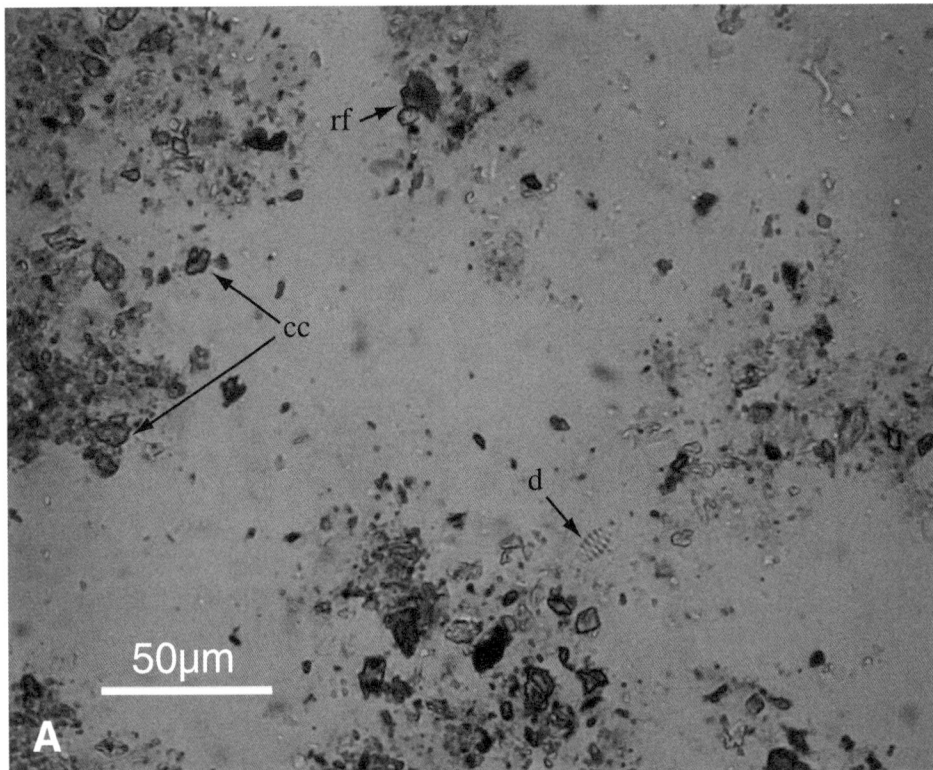

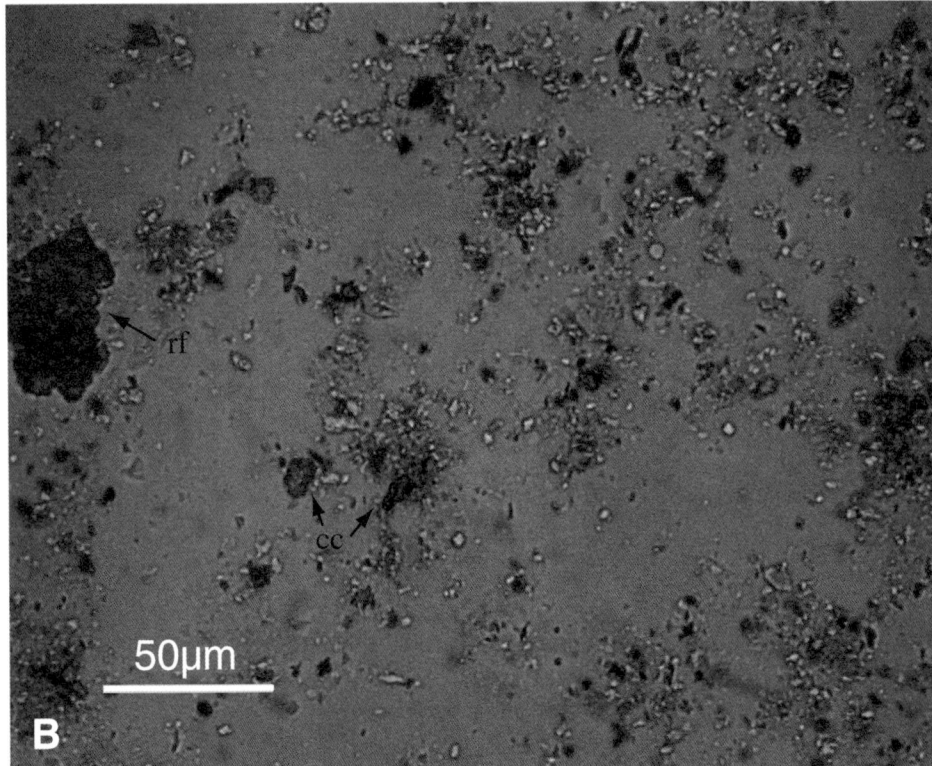

Figure 35. Smear slides of mixed calcitic and siliciclastic mud. (A) More calcite rich interval with calcite crystals (cc), small rock fragments (rf), and diatoms (d). (B) Calcite-poor interval with calcite crystals (cc) and larger, more plentiful rock fragments (rf). Diatoms are less abundant.

siliciclastic grains, and carbonate grains suggests that sediment reworking was important. This is supported by the mixed diatom assemblage. The relative abundance of ostracodes is similar to that of the shallower-water deposits in the Holocene section. Because *Daphnia* live and produce eggs at the water surface, the concentration of their remains in the sediment provides no relative indication of water depth. The concentration does suggest that, at least intermittently, sedimentation rates were very low. Subaqueous plants produce tiny root hairs that are concentrated at the sediment surface. It is possible that the tiny, tube-shaped features associated with the *Daphnia* concentrations are actually sulfide-coated root casts of subaqueous plants. If so, the mixed calcitic-siliciclastic mud intervals represent a clear and probably shallow lake and the dearth of sandy mud and sand layers indicates it had a small surface area.

Siliciclastic Mud

The thickest section of deep-water siliciclastic mud is found in BL96-3, where it constitutes the lower 320 cm of the core. These deposits consist of alternations of reddish fine mud and slightly greenish to brownish silty mud (Fig. 37). The silt includes some carbonate clasts, but is mostly quartz and rock fragments. The clay-sized particles are predominantly small grains (rock flour?). Diatoms are uncommon in smear slides, and probably constitute a very small fraction of the sediment. Mud and silty mud alternations occur in 2–20 cm intervals. The coarser layer usually has an abrupt contact with the underlying finer layer, whereas the transition to finer material may be more gradational. Bioturbation patterns in the sediment are intense. Randomly oriented burrows ranging from 1 to 5 mm in diameter are filled with sediment and may have partial lining of sulfide minerals. The lat-

ter features appear as dense streaks in X-radiographs. The finer-grained intervals appear to have more horizontal burrows that are generally smaller than the random burrows. Many of these horizontal burrows appear as dark flattened ovals in the cut face of cores and as denser features in the X-radiographs. These features suggest sulfide concentrations in the burrows. At a larger scale, the distribution of coarser and finer sediment appears to define rhythmic upward-coarsening successions (Fig. 38). The thickest fine-grained intervals alternate with the thinnest coarse-grained intervals. Over a 20–60 cm thickness, the fine-grained portions of alternations become progressively thinner as the coarse-grained portions become progressively thicker. A thick fine-grained interval abruptly overlies this succession, indicating the next upward-coarsening sequence. These upward-coarsening successions are also recognizable in the grain-size data (Rosenbaum and Heil, this volume).

Siliciclastic mud in Bear Lake was largely derived from the Bear River, including a large amount of glacial flour from the Uinta Mountains (Rosenbaum and Heil, this volume). The smear-slide observations confirm that most of the clay-sized material in siliciclastic mud units is not clay minerals from weathering, but small silicate grains and rock fragments. The lack of bedding in deep-water siliciclastic mud is due to the intense bioturbation and suggests low sedimentation rates. The variability in bioturbation style may indicate changes in the accumulation rate, oxidation state of the lake floor, or substrate preference. The coincidence of changes in bioturbation style with changes in grain size does little to distinguish between the different possibilities. The 2- to 20-cm-thick alternations of mud and silty mud characteristic of the deep-water siliciclastic mud and the larger-scale upward-coarsening sequences have several possible

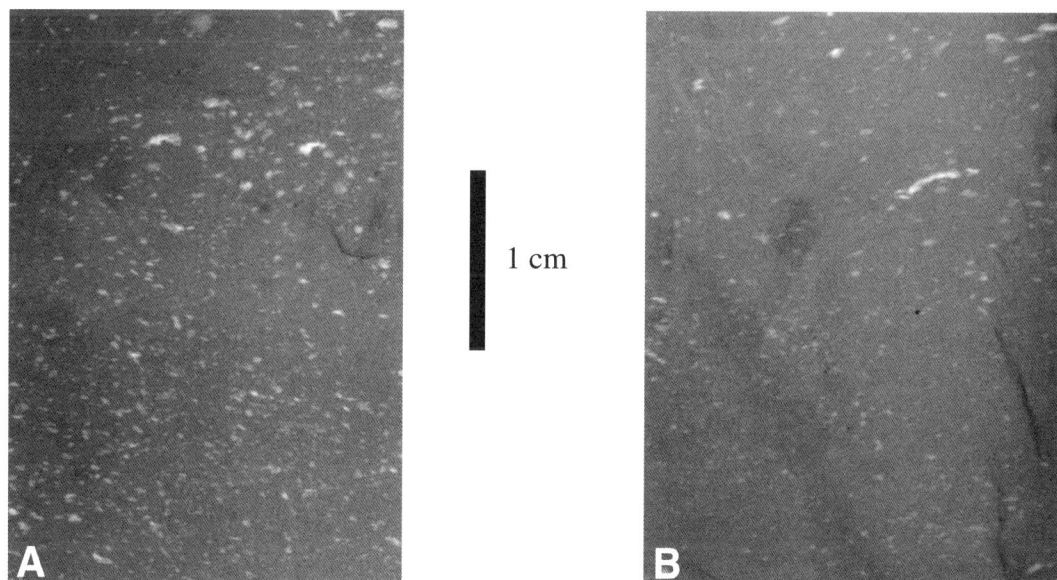

Figure 36. X-radiographs showing two examples of upward increase of size in sulfide-coated blebs in mixed calcitic and siliciclastic mud. In A, the larger blebs are only slightly more elongate, whereas in B, some larger blebs are very tubelike.

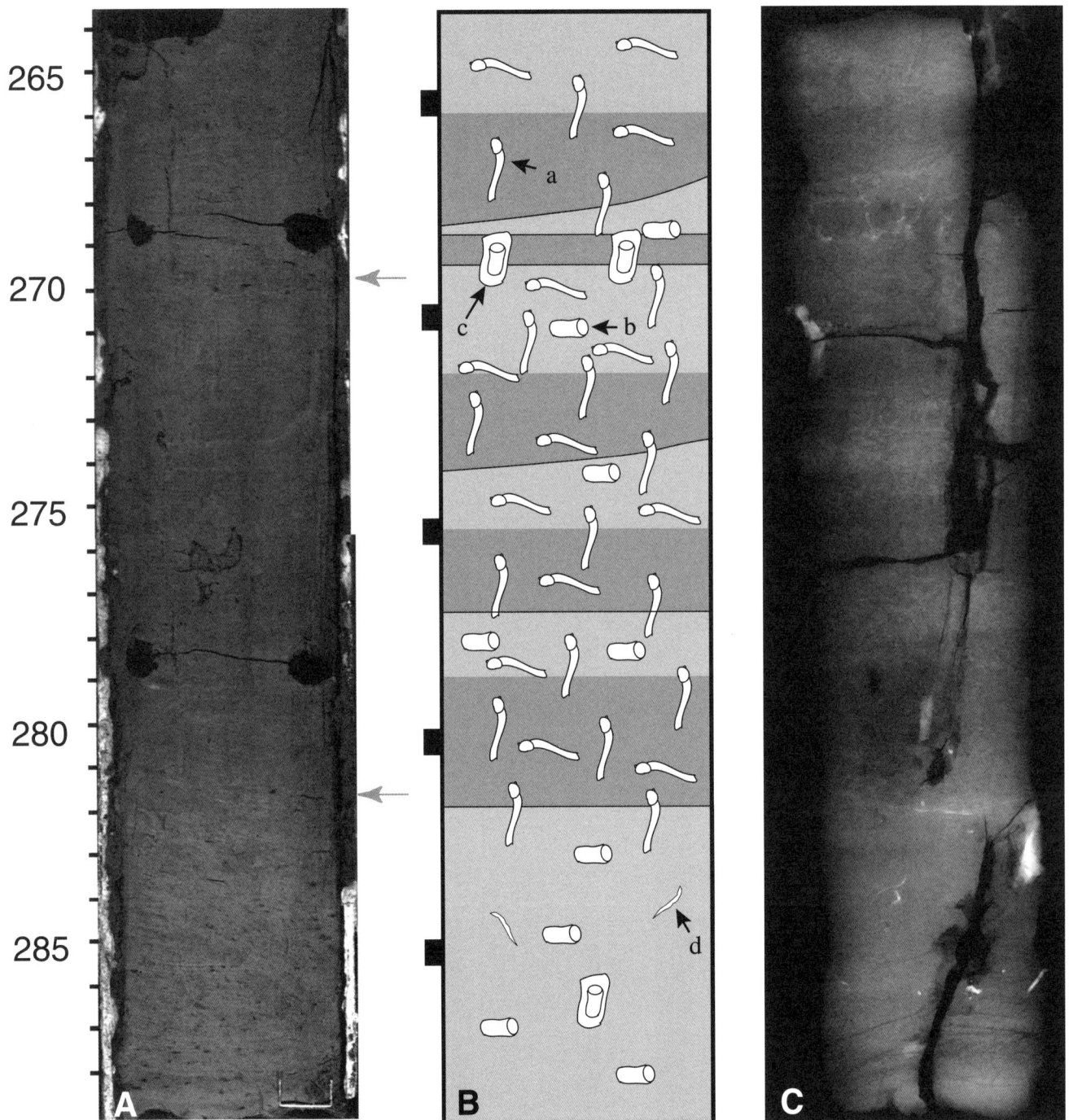

Figure 37. Siliciclastic mud in core segment cut surface (A), schematic drawing of X-radiograph image (B), and X-radiograph (C) of BL96-3. In the cut surface (A), the section appears to consist of a 10 cm greenish silty band (darker between gray arrows) overlain and underlain by reddish clay bands with black streaks. The X-radiograph (C) shows that the reddish intervals are denser (lighter), reflecting finer grain size, but each interval comprises smaller alternations of coarser and finer material. The schematic drawing (B) illustrates the sharp basal contacts of coarser layers (darker shading) and distribution of burrow types including large sinuous burrows (a), small horizontal burrows (b), and random burrows with denser rims (c). Bright gashes (d) are probably partial sulfide mineral coatings on large burrows. The sedimentary features shown in B are not to scale. Scales are in centimeters. The X-radiograph scale (B and C) differs slightly from the core (A) due to differential shrinkage from drying. Some of the density differences in C are due to irregularities in the sample thickness. The cut-surface photo was digitally enhanced to contrast the layers.

causes. Rosenbaum and Heil (this volume) note that the indicators of glacial flour are less pronounced in coarser layers, and suggest that the variations in grain size may reflect advance and retreat of glaciers in the Uinta Mountains. Although this is a very reasonable interpretation of the data, it neglects the observation that the clay-sized material in the presumed glacial-advance layers is identical to that of the presumed glacial-retreat layers. The shifts in indicators of glacial flour noted by Rosenbaum and Heil (this volume) may actually be indicating dilution by locally derived coarse sediment rather than changes in the influx of glacial flour. Changes in grain size in the deep-water siliciclastic mud might reflect changes in the discharge of the Bear River that may have caused turbiditic underflows to travel deeper into the lake (as in Sturm, 1979). The change in grain size may also represent slight changes in water depth, with coarser sediment deposited when lake level lowered, allowing coarser material to

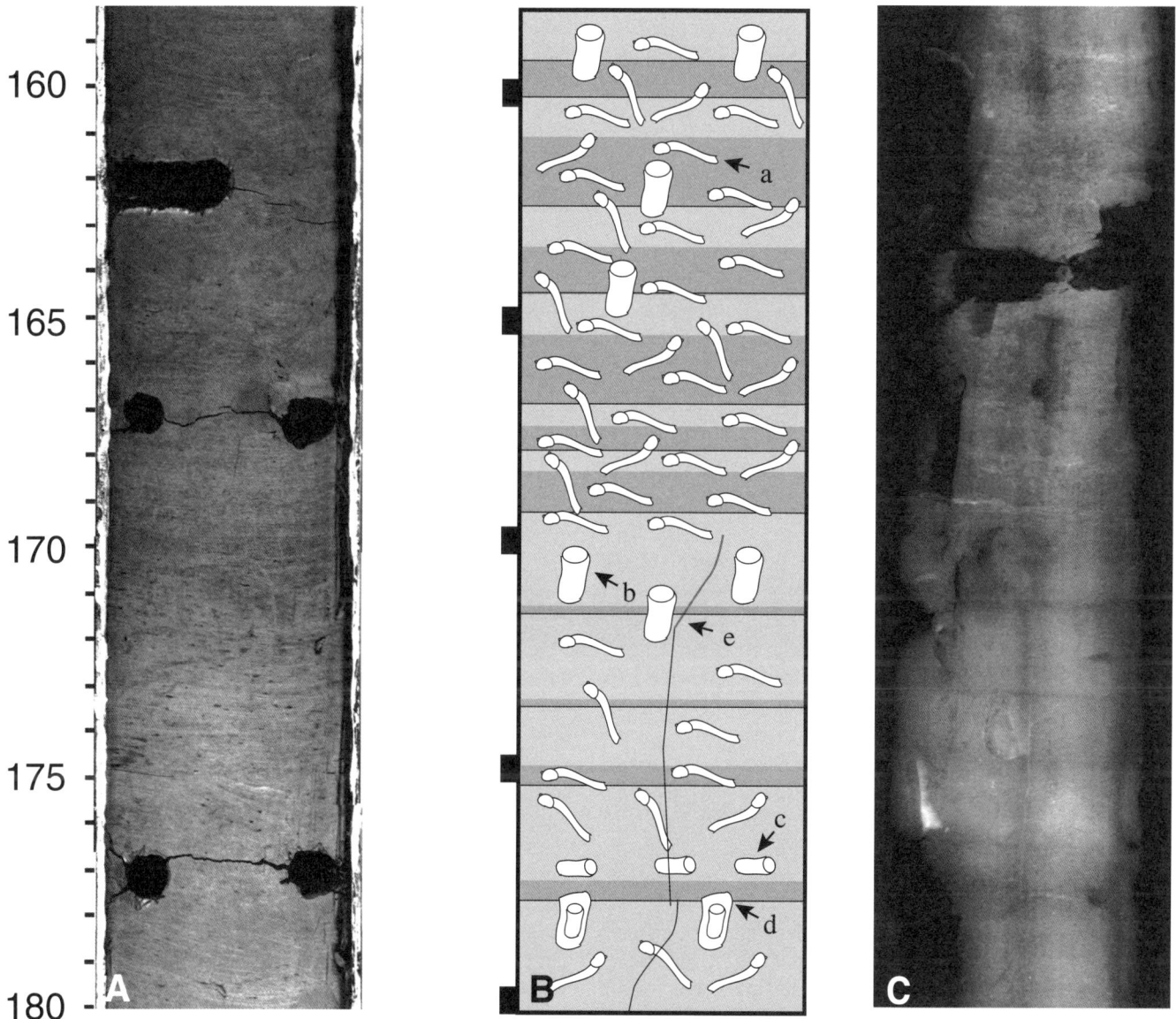

Figure 38. Siliciclastic mud in core segment cut surface (A), schematic drawing of X-radiograph image (B), and X-radiograph (C) of BL96-3. In the cut surface (A), there is an upward-coarsening sequence of reddish clay (lighter) to greenish silty mud (darker). The top is the base of another upward-coarsening sequence. The X-radiograph (C) shows that the sequence is composed of finer banding with a gradual increase in the thickness and number of less dense, coarser-grained beds (darker). The schematic drawing (B) illustrates the sharp basal contacts of coarser layers (darker shading) and distribution of burrow types, including large sinuous horizontal burrows (a) most abundant in coarser layers, random burrows (b), small horizontal burrows (c), and random burrows with denser rims (d). Hairline vertical cracks (e) cut bedding. The sedimentary features shown in B are not to scale. Scales are in centimeters. The X-radiograph scale (B and C) differs slightly from the core (A) due to differential shrinkage from drying. Some of the density differences in C are due to irregularities in the sample thickness. The cut-surface photo was digitally enhanced to contrast the layers.

move into the core localities. On the other hand, geochemical data suggest that the lake was spilling during deposition of the siliciclastic mud (Dean et al., 2006; Dean, this volume), which would limit the degree of lake-level fluctuation. A spilling lake at this time of sedimentation is also suggested by geomorphic data (Reheis et al., this volume) and shoreline data (Laabs and Kaufman, 2003; Smoot and Rosenbaum, this volume). A similar change of grain size could also have been caused by the rapid growth of delta fronts, combining components of increased discharge and increased proximity of inflow. The processes for making shifts in grain size could all have been precipitated by the advance and retreat of glaciers in the Uinta Mountains, which controlled the discharge and sediment supply of the Bear River. However, they were not necessarily a direct response to the glaciers themselves.

Shallow-Water Sediments

Carbonaceous Calcitic-Siliciclastic Mud

In core BL02-1, there is a 50-cm-thick interval of gray mixed calcitic-siliciclastic mud with bands of carbonaceous material (Fig. 39). This mud contains abundant small snail shells and irregular clumps of calcite and charophyte oogonia. Black carbonaceous tubes form both horizontal patterns and vertical disruptions. The horizontal tubes appear to be concentrated in horizons. Random burrows give the mud an irregular mottled appearance.

The organic-rich calcitic-siliciclastic mud deposit in BL02-1 is similar to the Holocene marsh deposits described earlier. The snails, charophytes, and carbonate crystals look to be the same and the concentrations of organic materials are similar. Carbonaceous tubes that are probably roots are common as was also observed in the Holocene marsh deposits. The concentrations of horizontal tubes are interpreted as roots of submerged plants.

Laminated Siliciclastic Deposits

Siliciclastic deposits that are laminated alternations of silt, sand, and clay were found in BL02-1, BL02-2, BL02-5, and BL2K-2 (Figs. 4 and 5). Most of the cores through this type of sediment were deformed by the coring process, in some cases causing extensive bowing and flowage of the originally horizontal laminae (Fig. 40). The lamina range from 0.5 to 5 mm in thickness. There is a range of lamina types. Graded laminae, which are continuous across the width of each core, are the ubiquitous sedimentary style. These laminae vary from mostly silt with a thin clay parting to mostly clay with a thin silty base (Fig. 41). Sand laminae are internally well sorted with sharp upper and lower contacts (Fig. 42). They commonly pinch out over the width of the core (10 cm) defining convex-upward lenses with flat bases. Although core deformation may account for some of the discontinuity of sand beds, most of them appear to be independent of their position along the deformation axis and independent of the thickness of the layer. Bioturbation is pervasive throughout the laminated beds, but burrows tend to

mottle within the layering. Some thick sand beds appear structureless, which may be due to bioturbation. Within the laminated siliciclastic sequences the distribution of graded clay-rich laminae, graded silty laminae, and sand laminae appear to define upward-coarsening successions that are 2–15 cm thick (Fig. 40). Graded laminae become progressively thicker and coarser in upward-coarsening sequences, with sandy laminae at the top. These upward coarsening sequences are also stacked into successions 20–40 cm thick where each successive sequence has less clay at the base and more sand at the top. An example of laminated siliciclastic sediment that is transitional to the deep-water siliciclastic mud is in core BL02-5 from 27 m below the modern highstand. Intervals with graded clay-rich laminae and silty laminae are overlain and underlain by siliciclastic mud with vague, wispy bands of light (clay) and dark (silty) sediment (Fig. 43). The transition between these bedding styles is characterized by an increased density of bioturbation features disrupting the layering both above and below the layered intervals. There are no sand laminae in BL02-5.

The predominance of repeated graded silt layers with less common clay laminae is similar to prodeltaic deposits in glacial lakes (Smith and Ashley, 1985). In this setting, each graded layer represents an influx of sediment related to a melting event, such as summer warming. If each graded lamina were a varve, it would suggest a much higher sedimentation rate for the laminated deposits than is indicated by radiocarbon ages in the deeper-water siliciclastic sediments. The lenticular sand laminae are interpreted as wave deposits. The sharp bases and broad convex cross sections with no internal lamination are characteristic of cross sections of rolling-grain ripples (i.e., Harms et al., 1982, Chapter 2, p. 25–41). Rolling-grain ripples are low-relief, long-crested ripples that develop at the lowest velocities for wave movement. Upward-coarsening sequences in the laminated siliciclastic sediment may represent changes in the inflow or shifts in lake depth changing the loci of sediment input. Modern lakes fed by seasonal melting commonly show sequential variations in grain size of similar scale to those in Bear Lake. These changes may be due to increases in discharge to the lake during longer melting periods (i.e., Lamoureux and Bradley, 1996) or they could be due to more random processes such as delta lobe switching (i.e., Lamoureux, 1999). The occurrence of wave deposits in BL2K-2 and BL02-3 indicates that the lake depth was no more than 3 m above the historical highstand (Smoot and Rosenbaum, this volume). The interbedding of sand lenses and clay suggests that the occurrence of rippled sand is related more to decrease of silt input than to decrease of depth. The deposits in BL02-5 were deposited below wave base and in an environment transitional to the deep-water siliciclastic mud. The increase of bioturbation with depth is probably due to the decrease of sedimentation rates away from the delta.

Subaerial Features

Black carbonaceous tubes that are coated with pyrite cut vertically across lamination in siliciclastic sediment in BL2K-2 (Fig. 44).

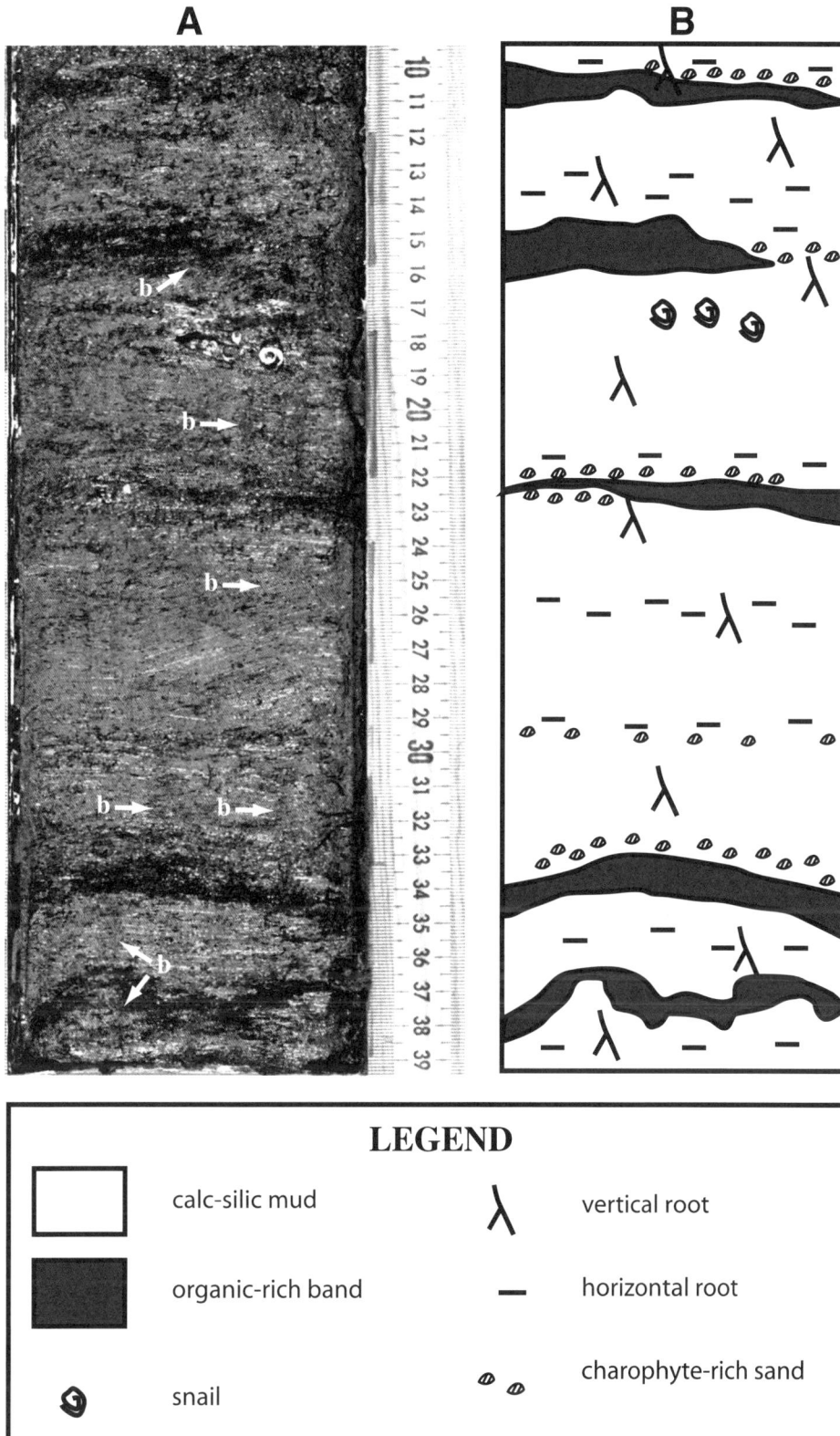

Figure 39. Mixed calcitic and siliciclastic mud with dark carbonaceous bands in core BL02-1. Mottled appearance of sediment is due to abundant random burrows. Some of these are indicated with "b." Carbonaceous tubes are oriented vertically or are concentrated in layers as horizontal features. Charophytes are mixed with unidentifiable calcite clumps, producing sandy clumps and layers. Scale is in centimeters.

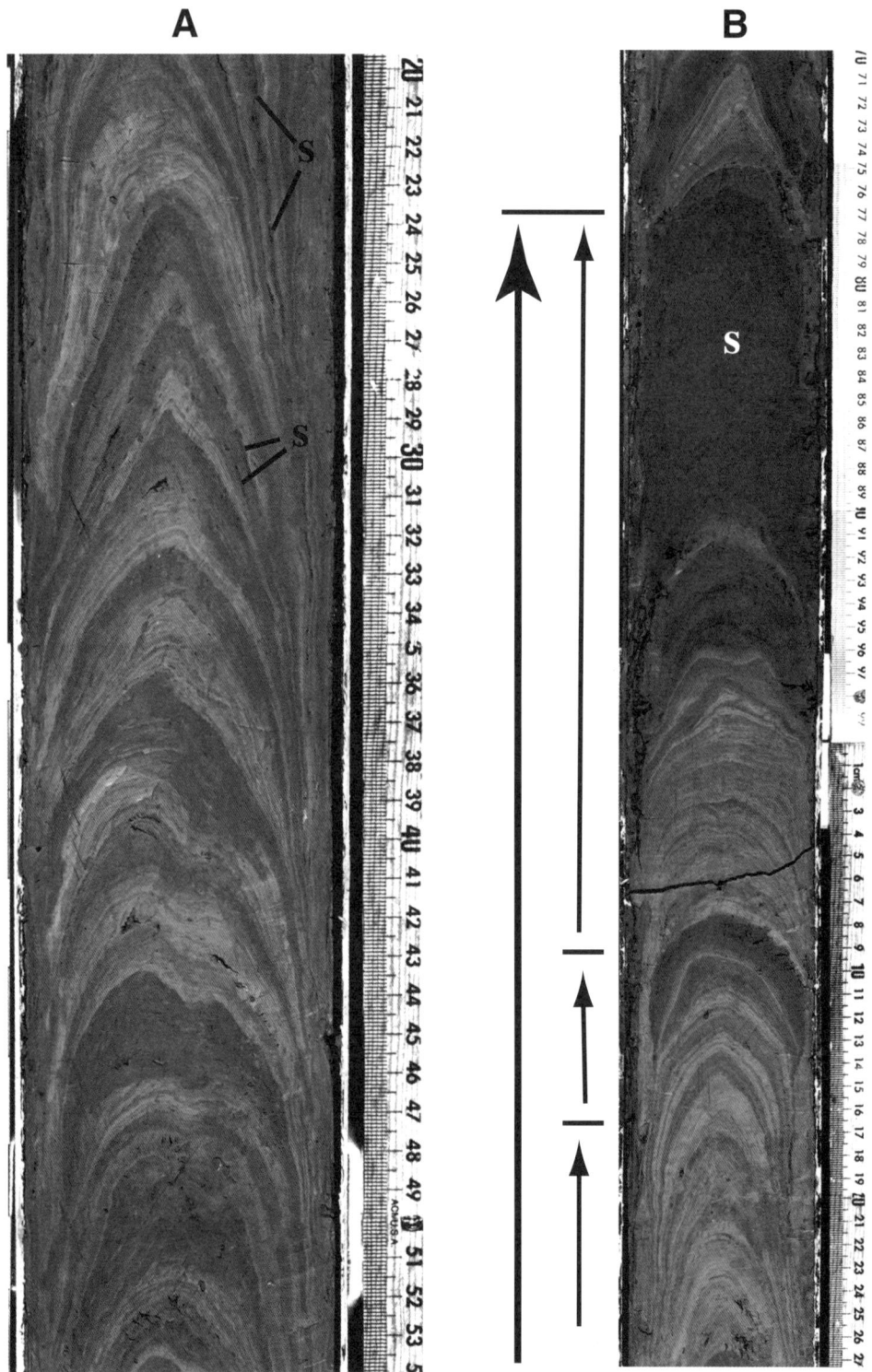

Figure 40. Core segments of laminated siliclastic interval in BL2K-2-1 (A) and BL2K-2-2 (B). The segment in A shows alternations of clay laminae (light), silt laminae (darker), and sand lenses (darkest, some are labeled "s"). The segment in B shows three small upward-coarsening sequences (arrows) that are part of a larger upward-coarsening sequence (heavy arrow). Sediments range from clay laminae (lightest) to unbedded medium sand (darkest). Layering was deformed by the coring process. Scale is in centimeters.

The tubes are thickest (~1 mm in diameter) and most abundant immediately below the contact of a graded shell gravel that marks the base of aragonitic sediment. The tubes become progressively smaller and less abundant downward over ~20 cm. Each tube commonly displays abrupt kinks, and some branch downward. Similar carbonaceous tube sequences ("root horizons" in Fig. 5) are in BL02-1 also below a graded shell gravel marking the transition to aragonitic sediment and in BL02-2 below the carbonaceous calcitic-siliciclastic mud deposit. In cores BL96-2 and BL96-3, the contact between siliciclastic mud

and calcitic-siliciclastic mud contains numerous millimeter to sub-millimeter tubes coated with framboidal pyrite, which have a morphology similar to that of the carbonaceous tubes in BL2K-2, but they are hollow (Fig. 45). The tubes in BL96-2 and BL96-3 are abruptly concentrated at the contact then gradually become smaller and less abundant downward over 5–20 cm (Fig. 44). The sulfide-lined tubes in BL96-2 and BL96-3 are easily identified in X-radiograph images as bright dense objects. They are also recognizable in split cores, including BL02-3, BL02-4, and BL02-5, as dark tube-shaped lines. Each of these cores shows

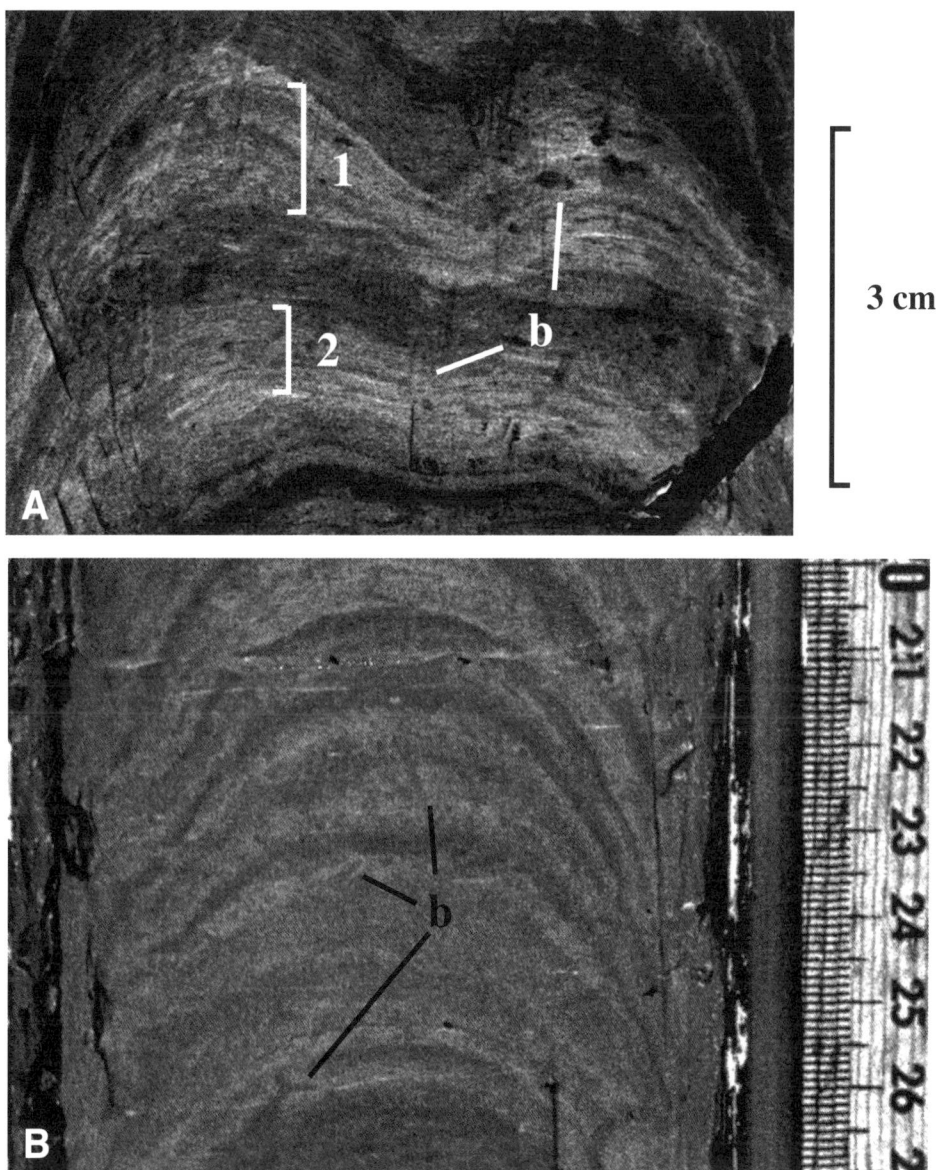

Figure 41. Clay- and silt-dominated graded layers in laminated siliciclastic sediments of BL02-2. (A) Laminated clay (light) with thin silt bases (dark lines). There are four graded laminae in the thickness marked "1" and seven graded laminae in the thickness marked "2." Note small burrows (b). (B) Silt (dark) grading to clay (light). There are eight graded laminae in the center of the picture. Note small burrows (b) and mottling of light clay layers. Scale is in centimeters. Layering was deformed by the coring process.

multiple horizons of these tubes including two within the upper part of the siliciclastic mud intervals and one or two within the calcitic-siliciclastic mud interval (Figs. 4 and 5). In BL96-2, a Scanning Electron Microscope (SEM) image of the mud containing the pyrite-lined tubes also contained numerous tiny hollow tubes (5–10 μm diameters) that are lined with thin sheets of clay (Fig. 45). In BL96-3, the lower tube horizon in the siliciclastic mud interval includes sediment-filled tubes as much as 1 cm in diameter that taper downward and branch (Fig. 46). The siliciclastic mud intervals associated with the pyrite-lined tubes in BL96-2, BL96-3, BL02-3, and BL02-4 show an abundance of clay minerals in smear slides in contrast to the rest of the siliciclastic mud deposits.

The morphology of the carbonaceous tubes in BL2K-2 indicates that they were plant roots. Plant stems as thin as these features would not cut across lamination, including sand layers, nor would they show the systematic decrease in size and abundance downward from the contact with the shell gravel. Given the morphology and distribution of the sulfide-coated tubes in BL96-2 and -3, the tubes in these samples are also interpreted as roots. The small diameters of most of the tubes suggest that the vegetation was mostly grasses or small bushes, although the larger features in BL96-3 could represent small trees. The clay-lined tubes are consistent with eluviated clay linings around root hairs (e.g., Kubiena, 1970, Plate 10). This suggests, along with the vertical orientation of the tubes, that the plants grew in a subaerial

environment (Retallack, 2001, p. 13–19). The occurrence of clay minerals, rather than the glacial flour, in the horizons containing the structures interpreted as roots is also suggestive of soil formation. The downward decrease of root diameters and abundances is interpreted as representing protrusion of the roots down from the surface, but it could also reflect proximity to a larger taproot.

Gypsum crystals ranging from 0.05 to 0.3 mm long are associated with the black sulfide-coated tubes in two intervals in BL96-2 (Fig. 45). The thickest interval of gypsum (305–330 cm) has the largest crystals near the top and progressively smaller crystals downward. The crystals are organized into vertically elongate concentrations and horizontal bands (Fig. 47). The lowest occurrence consists of small, irregular concentrations of crystals. BL96-3 also has gypsum crystals associated with the black tubes, but the best developed occurrence (19–23 cm) consists of small, irregular patches of small crystals (around 0.5 mm), and the other interval (40–50 cm) is known only from silt-sized crystals in smear slides. Ovate carbonate concretions (1–5 mm diameter) are associated with the sulfide-coated tubes at the interval starting at 40 cm in BL96-3 (Fig. 48), and in the lower part of the interval starting at 330 cm in BL96-2.

The gypsum crystal concentrations, which coarsen upward, are similar to modern gypsum soil profiles (Dan and Yaalon, 1982). In gypsum soils, the gypsum is introduced as windblown dust and is then worked downward into the soil profile by rainfall (Drever and Smith, 1978). The crystals reprecipitate during

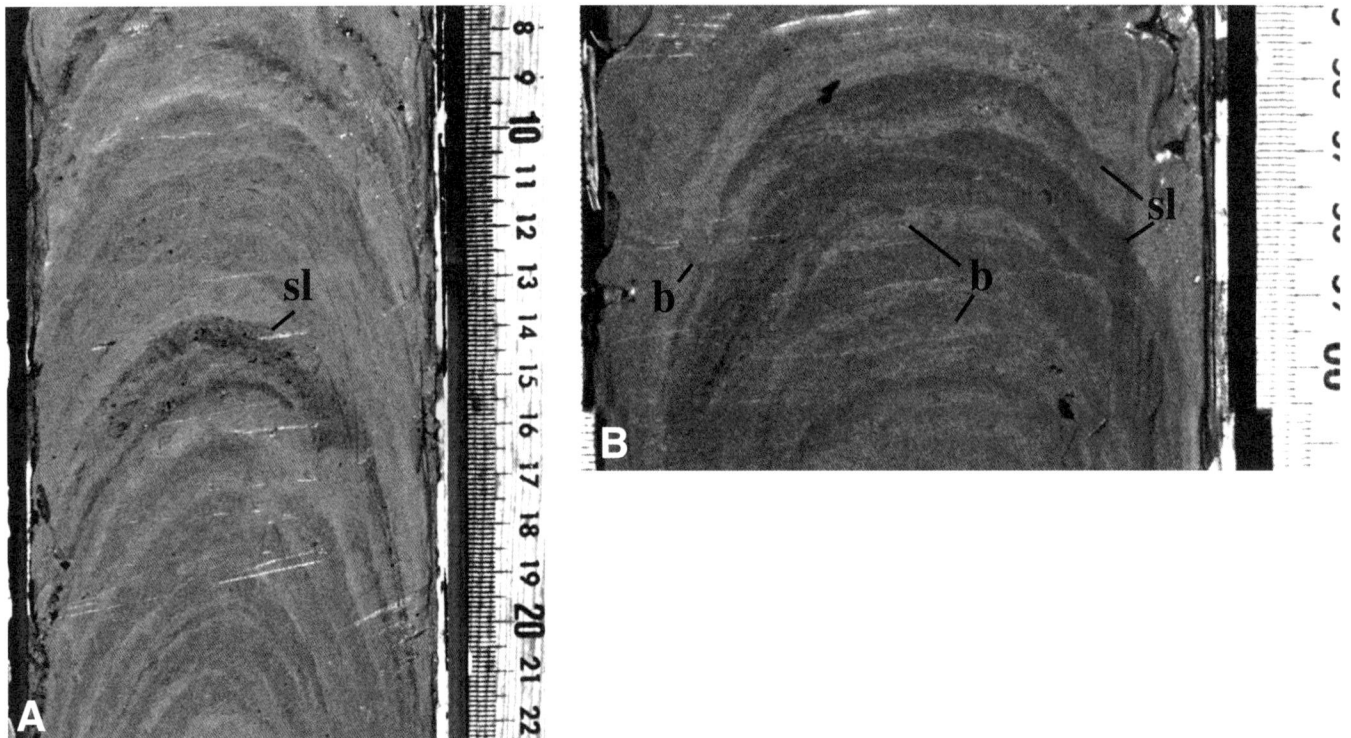

Figure 42. Core segments of BL02-2 with laminated siliciclastic sediment showing sand lenses. (A) Medium sand lenses (sl) at top of upward-coarsening sequence overlain by a thicker clay unit. (B) Fine sand lenses (sl) with thin clay partings. Note burrows (b) of various sizes. Scales are in centimeters. Layering was deformed by the coring process.

dry intervals, with larger crystals forming closer to the surface. Similarly, the carbonate nodules may represent caliche development in a soil horizon by reprecipitation of carbonate dust or silt-sized detrital limestone. The large calcitic concretions near the base of BL96-2, however, may not be related to the rooted zone, and could represent a later diagenetic feature.

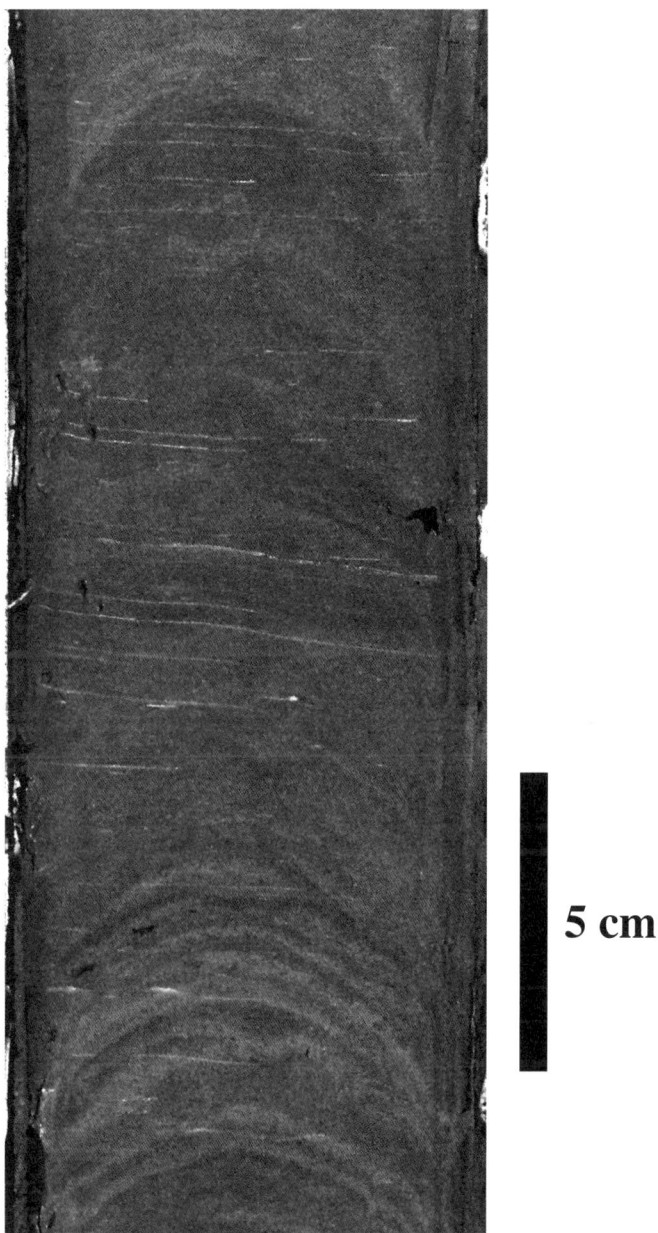

Figure 43. Core segment of BL02-5 showing transition between laminated siliciclastic sediment and unbedded siliciclastic sediment. Laminated interval is comprised of silt and clay (light). Dark upper interval is mottled silty mud with vague bands of coarser-grained (darker) and finer-grained (lighter) sediment.

Pleistocene Depositional Model

Age constraints on the Pleistocene sedimentation are poorer than those for the Holocene. Several dated intervals are in apparent conflict with other ages including two samples from the same depth in BL96-2 (Fig. 3). The change from siliciclastic to calcitic sedimentation and calcitic to aragonitic sedimentation are the most important correlative features that are consistent with the radiocarbon age. The reconstruction presented here assumes that the horizons marking the initiation of the structures interpreted as roots are correlative, and that they represent significant time breaks in the record. The correlation of core segments representing the age interval 15–10 ka (Fig. 49) and the age interval 26–15 ka (Fig. 50) rejects the two 14 ka ages in BL96-2 in favor of the older ages, and it also rejects both of the radiocarbon ages in BL02-5 (Fig. 4).

The youngest Pleistocene deposits in Bear Lake are aragonite deposits similar to those of the Holocene. Aragonite III is the most common sediment in deep-water cores (BL96-2, BL02-3) indicating conditions much shallower than the modern lake. There was at least one transgression that brought lake levels near the modern depth, very close to 10 ka as indicated by Aragonite III deposits in BL02-1 at depths similar to modern occurrences (~10 m). The graded shell gravel at the tops of BL2K-2, BL02-1, BL02-2, and BL02-5 may actually represent much younger shorelines that have eroded to the Pleistocene deposits. The oldest aragonite deposits, which include Aragonite II in the deepest core sites, are ca. 11.5 ka. These gradationally overlie calcite deposits and are interpreted as indicators of increasing lake salinity after the Bear River ceased to flow into the basin (Dean et al., 2006; Dean, this volume).

Calcitic sediment deposition occurred in Bear Lake over the period of 15–11.5 ka. Deep-water calcite similar to the Holocene deposit was precipitated at the very beginning and very end of this depositional interval, and it chemically indicates that the lake was receiving Bear River water. It is not known if the lake became as deep as inferred for the Holocene calcite during either of these intervals. Both calcites are found only in cores whose locations are deeper than 30 m below the modern highstand. Shoreline shell gravel overlying the correlative strata in the cores from shallower-water localities suggests that the younger calcite could have been eroded. Deposits of the Raspberry Square phase (Laabs and Kaufman, 2003) at ~9 m above the modern highstand are 15.9 k.y. old (Smoot and Rosenbaum, this volume). This could be equivalent to the older of Pleistocene calcite deposits, or the younger one if the shells that were dated had been reworked. The bulk of the calcitic interval consists of a mixture of calcite and siliciclastic mud. The sedimentation rate was probably low despite the increase of siliciclastic sediment, as indicated by concentrations of *Daphnia* egg casings and body parts. If the elongated blebs represent aquatic vegetation, the water must have been relatively clear to allow growth. The siliciclastic component includes glacial flour, but this may have been derived from older deposits exposed on the lake margin and reworked by local drainages. This model explains the conflicting evidence of clastic deposition with low sedimentation rates and low turbidity. The rooted horizons indicate that at two times during the depositional interval, the lake fell to at least 43 m below the modern highstand. Because the mixed calcitic-siliciclastic mud deposits immediately above and below the root-disrupted horizons are nearly the same, the lake-level changes to produce the exposure surfaces must have been relatively small. The coeval occurrence of marsh-like deposits in BL02-1 indicates that the maximum lake level was lower than 9 m below the modern highstand (Fig. 51). If the marsh deposits in BL02-1 formed in a depression above the lake margin, it could have always been much shallower.

Siliciclastic mud contains abundant glacial flour brought in from the Bear River. The oldest deposits examined in this report date to 26 ka and the youngest siliciclastic sediments to 16 ka (Smoot and Rosenbaum, this volume). The lake during this period

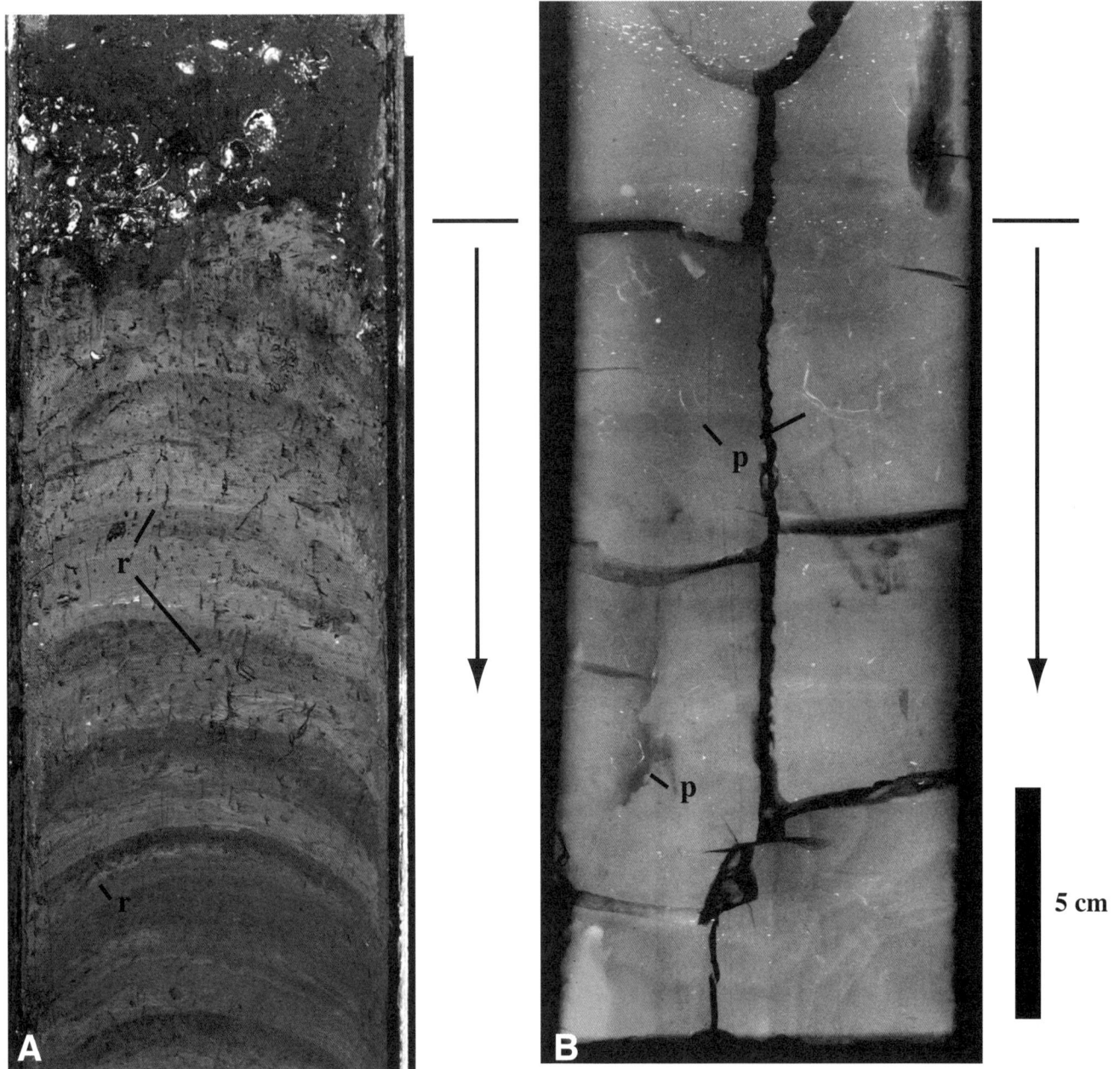

Figure 44. Core segments showing root structures. (A) Core BL2K-2 showing carbonaceous black roots (r) decreasing in abundance away from an overlying graded shell-gravel contact (arrow). The carbonaceous cylinders are coated with framboidal pyrite. Note how vertical features crosscut lamination. (B) X-radiograph of BL96-2 showing hollow cylinders of framboidal pyrite (p) decreasing in abundance away from a contact with calcitic siliciclastic mud with blebs (arrow). The cylinders are interpreted as former root casts similar to those in A.

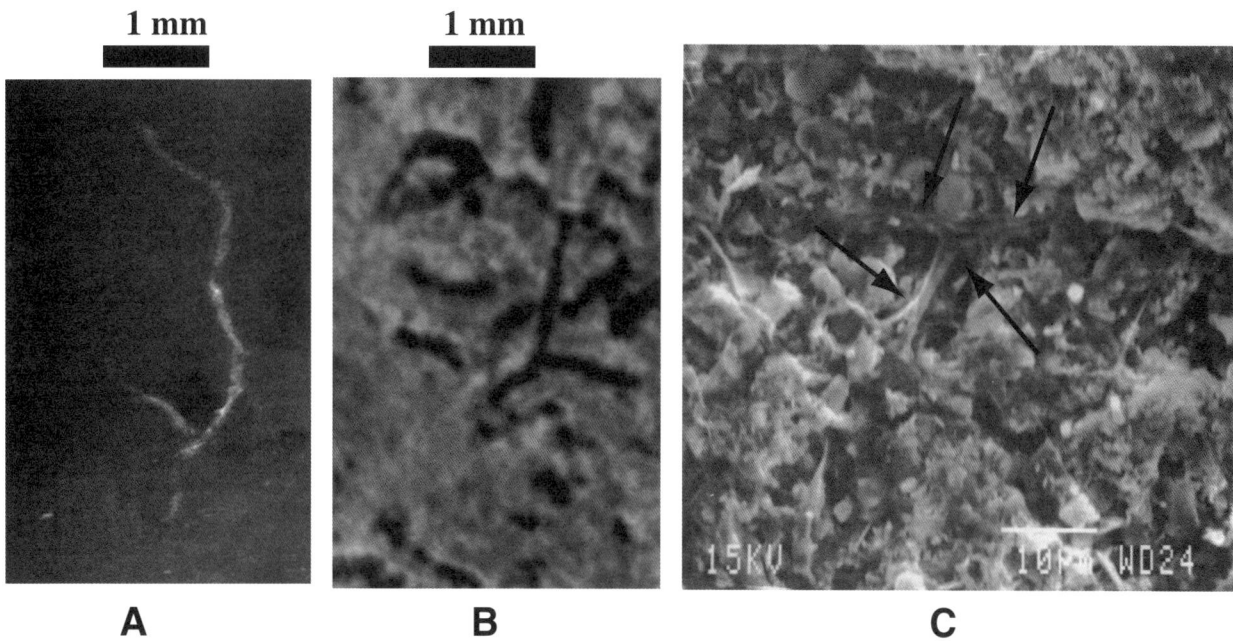

Figure 45. Details of root structures in cores. Comparison of X-radiograph of hollow pyrite tube from BL96-2 (A) and cut surface of carbonaceous tube from BL2K-2 (B), both showing characteristic branching. (C) SEM (Scanning Electron Microscope) image from BL96-2 showing a sheet of clay partially defining a branching set of tubes (arrows). The sheet of clay is interpreted as a cutan lining a root hair cast. (D) SEM backscatter image from BL96-2 showing framboidal pyrite coating a hollow cylinder. Surrounding matrix is full of euhedral gypsum crystals (g).

was probably spilling to the north and re-entering the Bear River (Reheis et al., this volume). Over most of this time span, laminated siliciclastic mud formed in the area adjacent to the northern margin of Bear Lake as distal equivalents to deltaic deposits of the Bear River (Fig. 51). To the south, the deposits were dominated by mud and silt over most of the basin floor. Fluctuations in the Bear River discharge and distributary channel avulsion caused shifts in the distance of silt sedimentation from the delta mouth. The lake margins probably had sandy shorelines with sandy mud and muddy sand in transition to the siliciclastic mud. At ca. 18 ka, the lake level dropped abruptly, leaving an exposed muddy bottom to at least 43 m below the modern highstand. Vegetation became established on the surface and gypsum-rich soils

formed. At ca. 17 ka, the lake level rose to at least 33 m below the modern highstand, and then rapidly dropped to the former low, allowing a second vegetated surface with gypsum soils to form. The gypsum horizons suggest at least semiarid conditions during the lowstand intervals. The relatively rapid transition from spilling, glacier-fed lake to a nearly completely desiccated flat was probably caused by diversion of Bear River flow northward from the basin during an extended drought. The low-flow Bear River incised into delta deposits making avulsion away from the lake basin very possible. The minor transgression between the two gypsum-bearing soils may not have been caused by a return of Bear River water to the basin. Increased discharge of local drainages may have reworked the older subaerial deposits. When

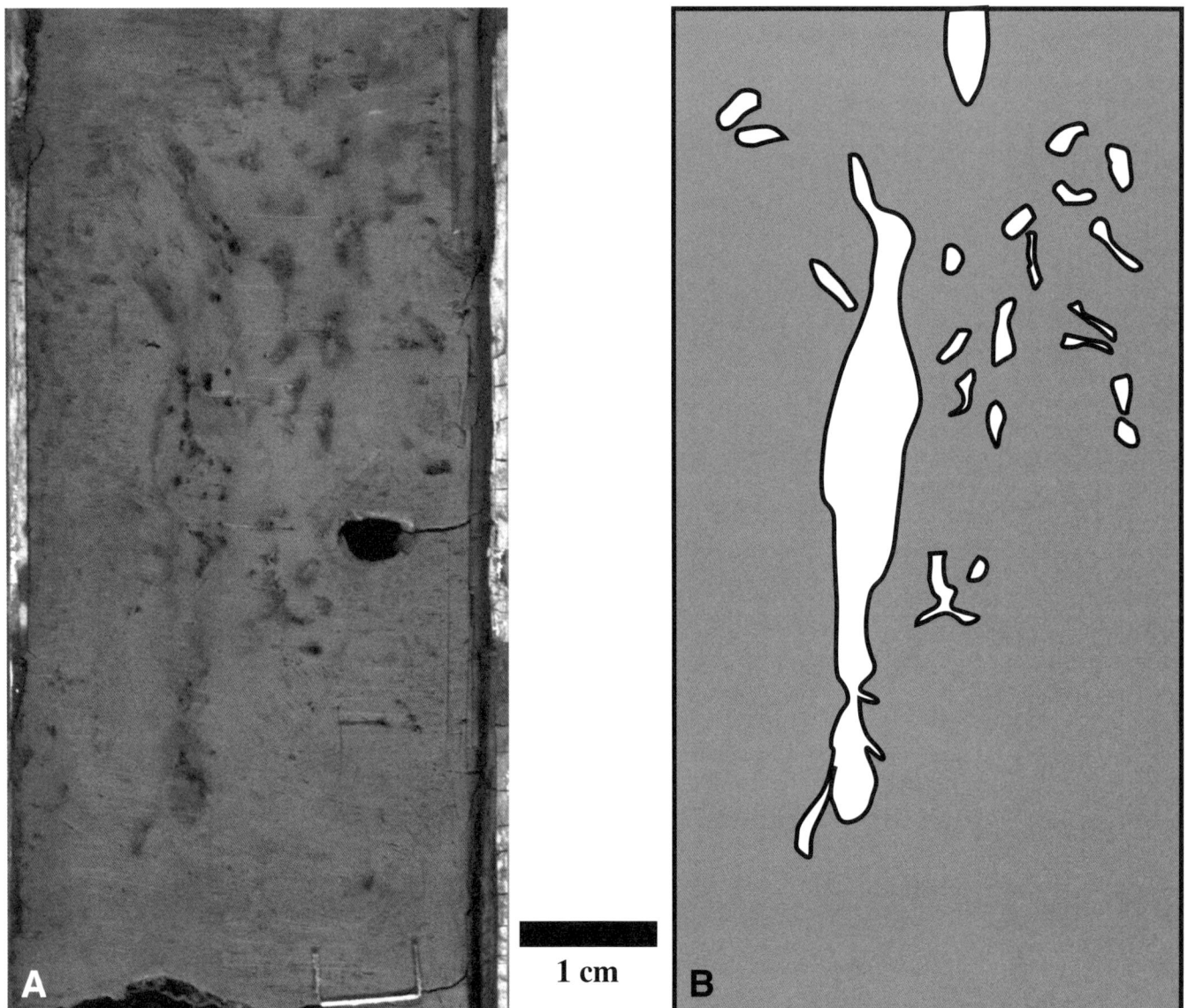

Figure 46. Cut surface (A) and schematic drawing (B) of sediment-filled root casts in BL96-3. Root outlines (darker gray in A and white in B) are defined by changes in grain texture and rims of iron oxide. Small blebs of pyrite (black) occur within the largest cast. Only largest root casts are shown in the schematic sketch.

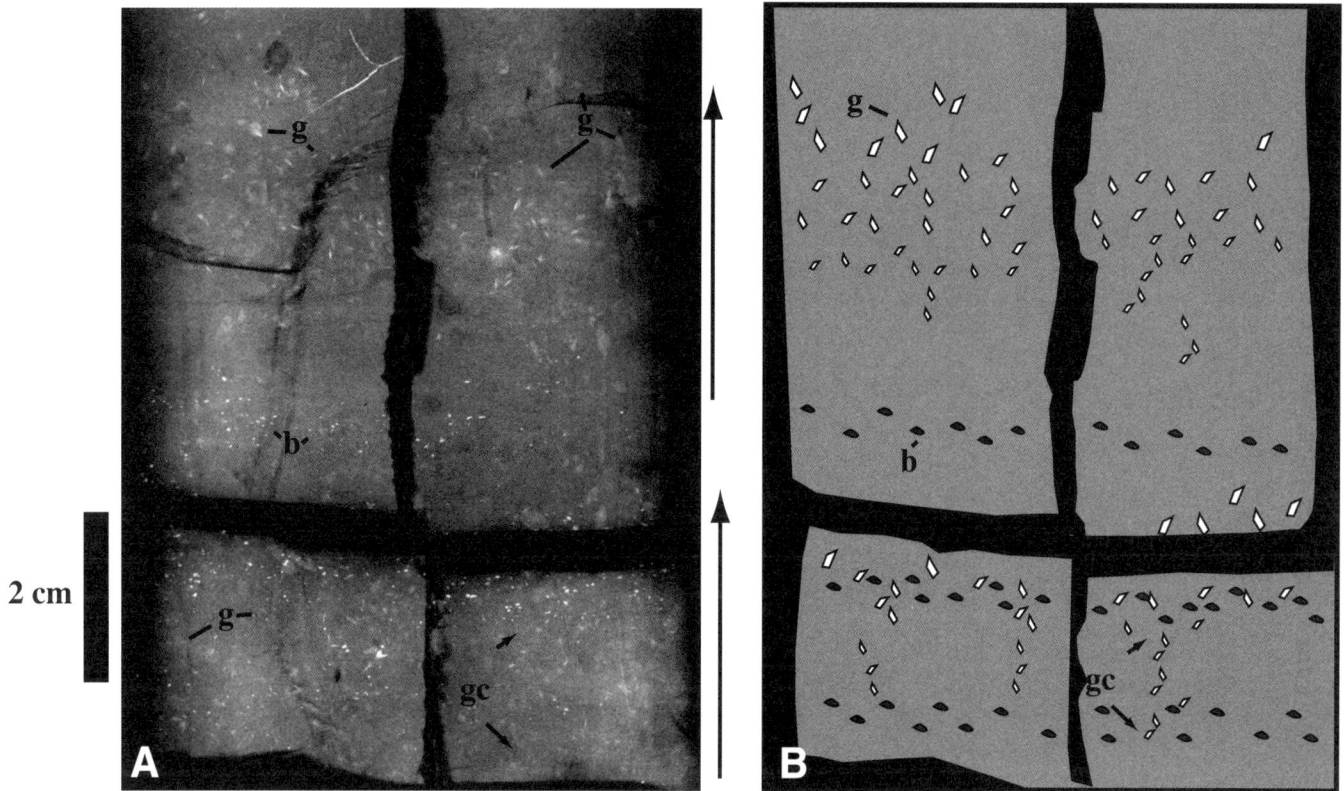

Figure 47. Gypsum crystals in core segment at BL96-2 in X-radiograph (A) and schematic sketch (B). Gypsum crystals (g) occur as two upward-coarsening sequences. Smaller crystals are arranged in vertical columns (gc) which probably follow roots or soil structures. Background sediment includes numerous pyrite-coated blebs (b), which are mostly *Daphnia* egg casings.

BL96-3

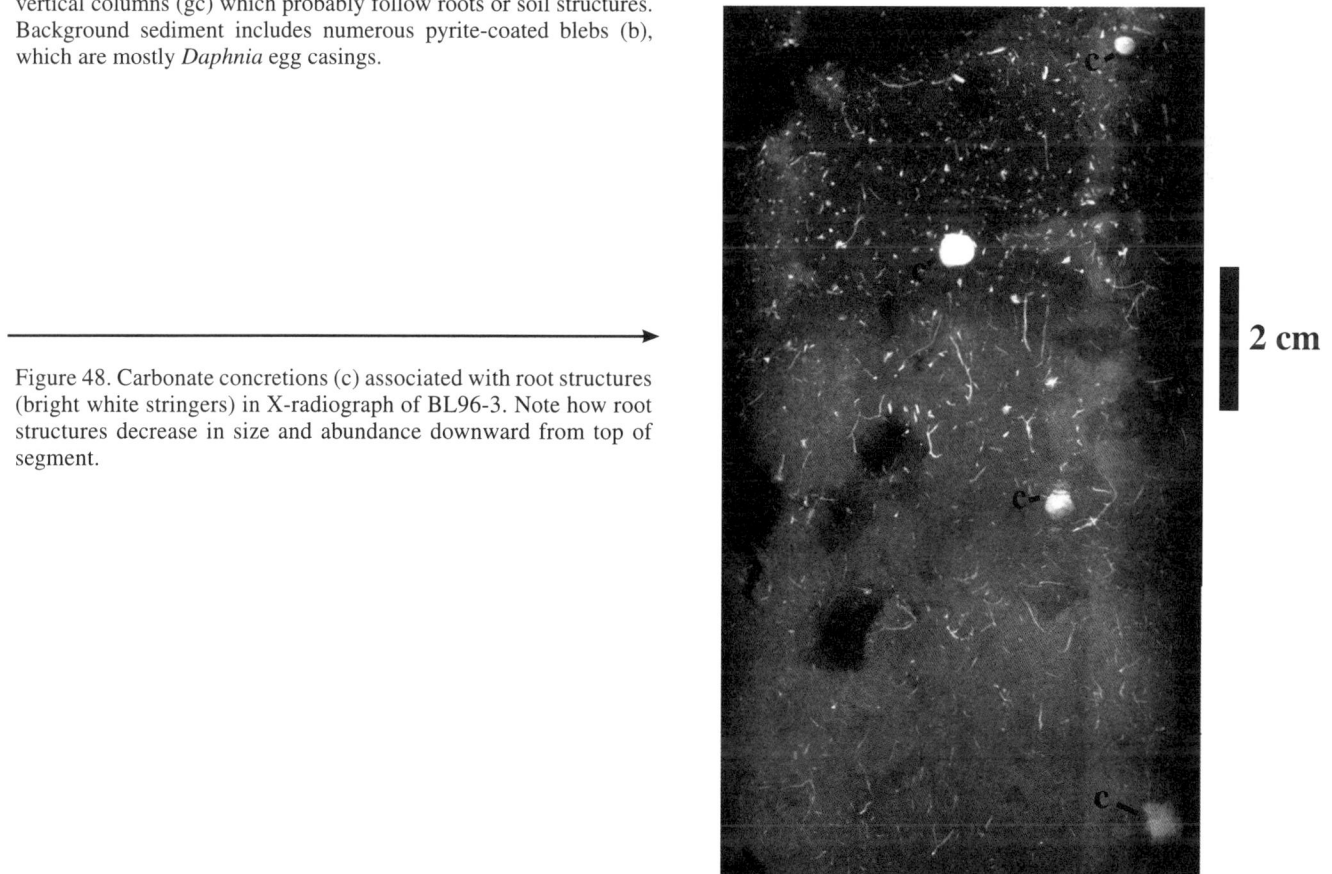

Figure 48. Carbonate concretions (c) associated with root structures (bright white stringers) in X-radiograph of BL96-3. Note how root structures decrease in size and abundance downward from top of segment.

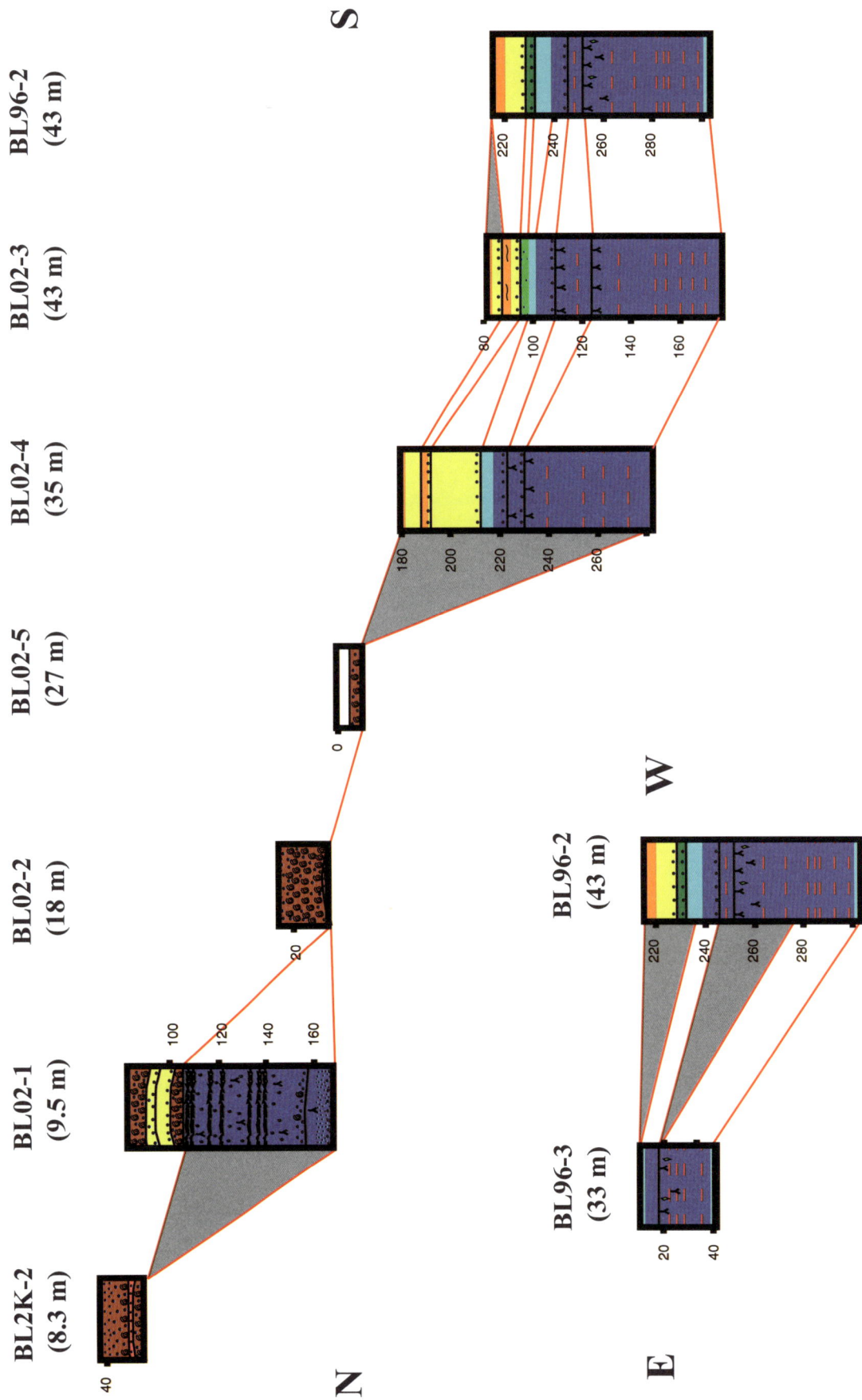

Figure 49. Correlation of core segments representing the time interval 15–10 ka along a north-south and an east-west transect. Formatting, scales, and data sources are the same as for Figure 31.

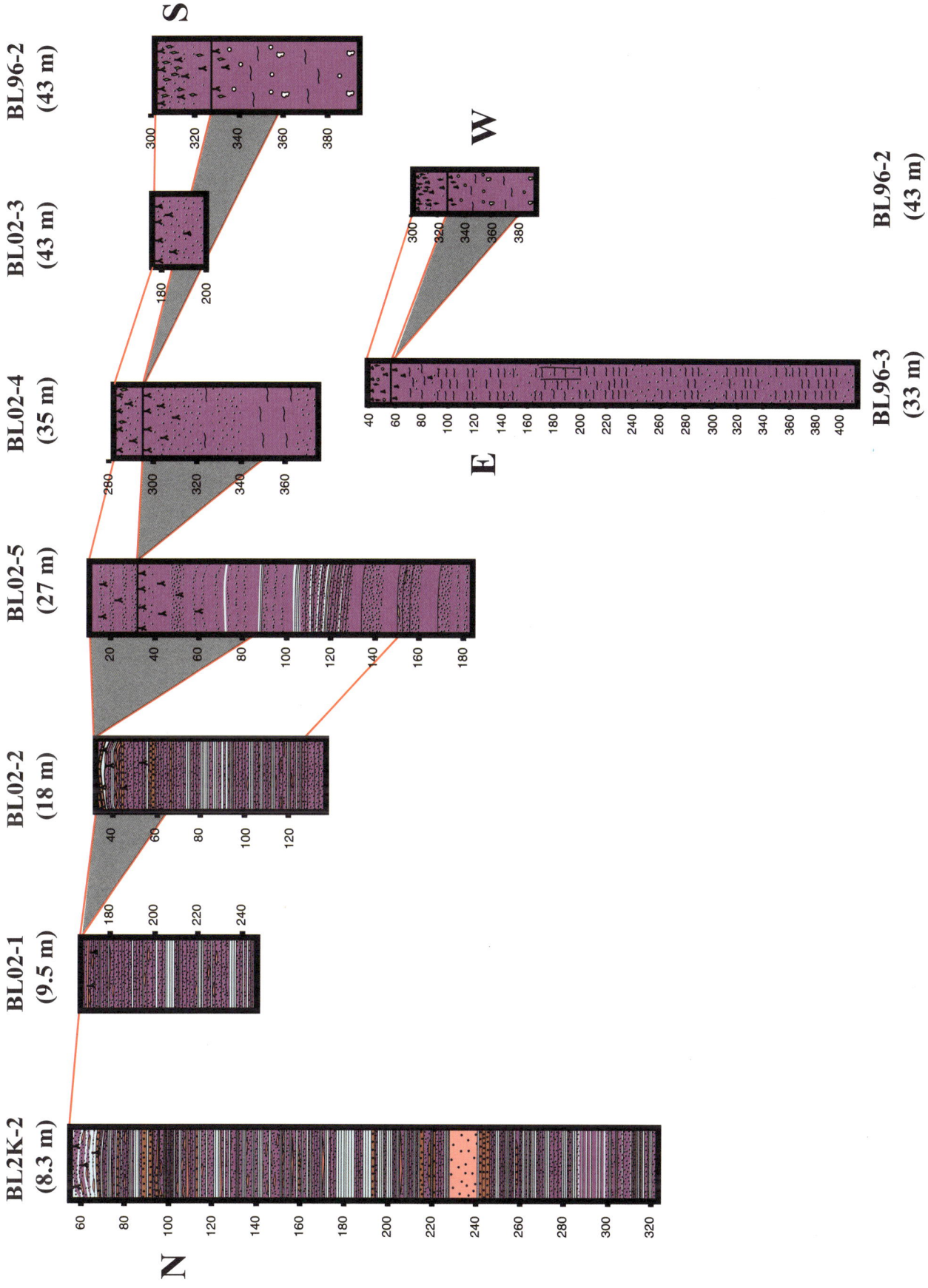

Figure 50. Correlation of core segments older than 15 ka along a north-south and an east-west transect. Formatting, scales, and data sources are the same as for Figure 31. Note the change of scale for the east-west correlation.

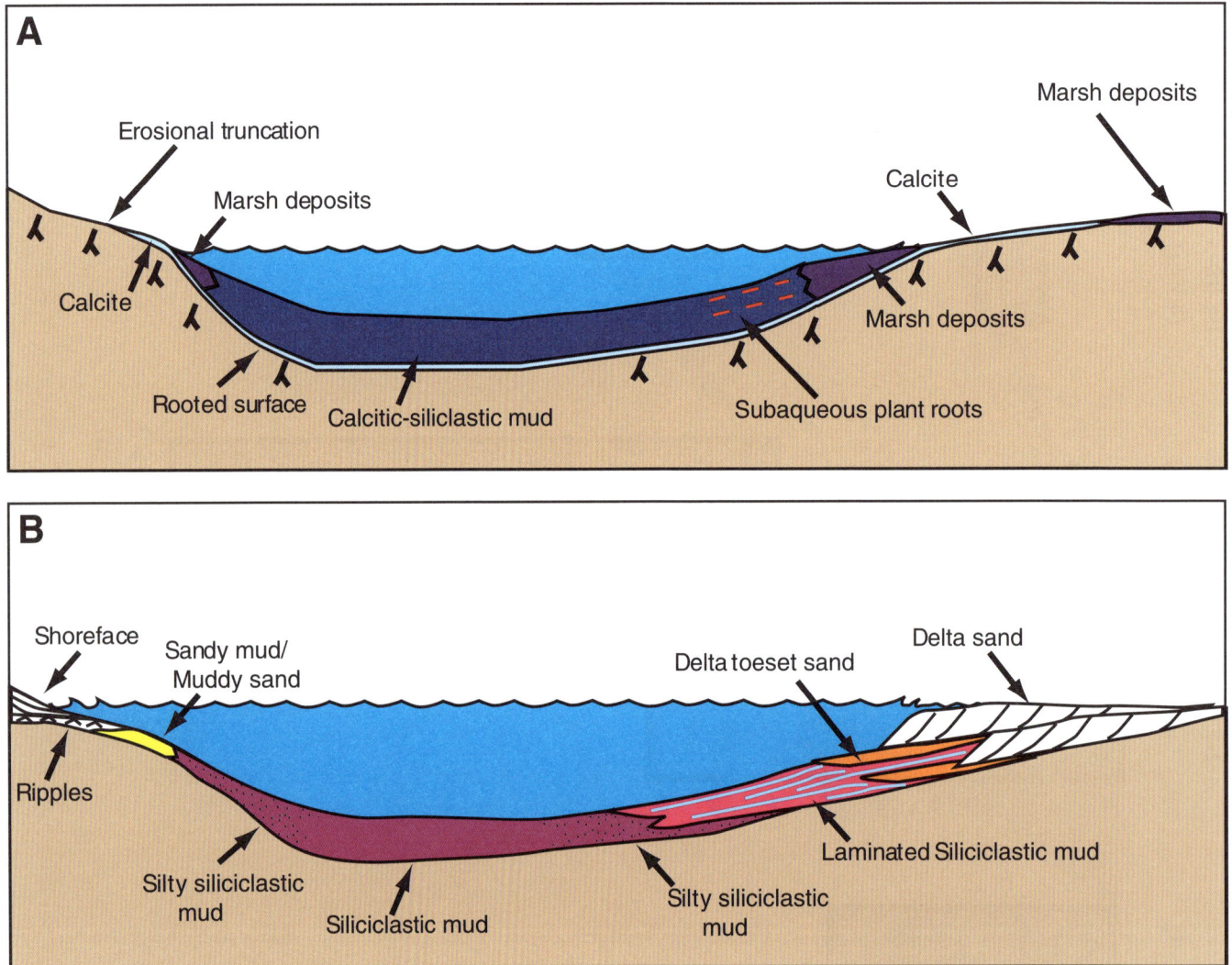

Figure 51. Schematic illustration of Pleistocene depositional facies in Bear Lake during calcitic-siliciclastic mud deposition (A) and siliciclastic mud deposition (B). The right side represents the northern margin of the basin. The left side is shown steeper to illustrate changes that also occur along the east-west transect. (A) The time of Bear Lake deposition following the highstand, which produced a widespread calcite layer overlying the lowstand rooted surface, itself superimposed on siliciclastic mud. Calcite and siliciclastic mud are eroded from the subaerial deposits and redeposited into the shallower lake. Subaqueous plants grow in the broad shallow areas to the north. (B) The time of siliciclastic deposition in Bear Lake before the first subaerial surface. Deltas fed by the Bear River form on the north margin of the lake in the area that is now Mud Lake. Siliciclastic mud becomes sandy near shorelines south of the delta area.

the lake level next rose ca. 15 ka, it signaled a return of Bear River inflow, but with virtually no siliciclastic sediment. This may indicate that only a minor distributary of the river entered the basin and that the lake-level rise was rapid enough to make sediment reworking unimportant.

DISCUSSION

Although the sedimentary record of Bear Lake records changes in climate over the past 26,000 years, the signals of change were modified by changes in the basin hydrology that were not directly linked to climate. Up to ca. 18 ka, the Bear

River flowed into the basin, draining glaciers in the Uinta Mountains. Climate changes while the lake was spilling to the north were probably not well recorded in the sediment. The dry episode was probably an exaggerated representation of the actual climate change, because the loss of Bear River inflow contributed to the regression. In the 4000 years before the next major lake-level rise, the lack of Bear River inflow may have prevented the recording of wetter conditions in the drainage area. The return of Bear River water was probably related to a wetter climate, but the contrast with the exposed deposits was more a measure of having fluvial input into the basin. The transition to an aragonite precipitating lake was not exclusively due to a transition from spilling lake to closed lake, because there is no record of aragonite precipitation associated with any of the rooted horizons. One possible explanation is that the root structures and gypsum-bearing soils are misinterpreted and that the lake did not experience the extreme desiccation (e.g., Dean et al., 2006). This is unlikely given the range of evidence for subaerial exposure, so the cause must be related to a change in the hydrology. It is possible that the groundwater systems that feed the modern lake were not established until the end of the Pleistocene. The well-documented active faulting around the lake margin could have provided the needed changes in groundwater flow paths.

Another important implication of this sedimentological study is that since ca. 18 ka, a large portion of the preserved sediment record in the basin is redeposited material, including both chemical precipitates and organisms. Furthermore, there are numerous gaps in the lacustrine record where the sediment was eroded and redeposited in the deeper parts of the basin. The fact that chemical and biological proxy measurements recognize variations in the lake level indicates that the actual lake changes are not completely masked by this mixing. Nonetheless, the sensitivity of these records has likely been reduced, because the chemical and biological proxy records presented in this volume show very little of the lake-level fluctuations inferred from the sedimentology.

SUMMARY

The sedimentary record of Bear Lake over the past 26,000 years can be divided into six episodes. The first episode is siliciclastic sedimentation as the Bear River deposited glacial flour derived from the Uinta Mountains. This episode ended ca. 18 ka, with a radical drop in lake depth probably due to a drought and diversion of the Bear River away from the basin. The second episode was a dry interval, with subaerial surfaces forming in areas now covered with over 40 m of water for most of the time between 18 and 15 ka. The third episode also included Bear River inflow, but clastic sediment was only a minor component. Calcite precipitation marked the initial transgression ca. 15 ka and the final transgression of this interval ca 11.5 ka. During the remainder of this time interval, the lake appeared to be shallow with intermittent periods of lowstands exposing the surfaces again to depths more than 40 m below the modern highstand. The fourth episode, ca 11.5–8 ka, is marked by aragonite precipitation from waters with no Bear River input. The lake was mostly shallower than the modern lake, with shorelines as much as 30 m below the modern highstand. The fifth episode marks a return of Bear River inflow and a major lake transgression ca. 8.5–8 ka. Once again, however, there was no siliciclastic input from the river and only calcite precipitation. Finally, the sixth episode is a shift to the modern Bear Lake hydrology, with no Bear River input and aragonite precipitation until the opening of the canal system in 1912. During this depositional episode lake level fluctuated between a few meters above the modern highstand to lowstands as much as 35 m below the modern highstand.

ACKNOWLEDGMENTS

All research for this paper was conducted with funding from the U.S. Geological Survey Western Lakes Catchment Project with Joe Rosenbaum as project chief. X-radiograph analysis of suspected gypsum samples was provided by Daniel Webster. Harvey Belkin provided access to an SEM and instructions for using it. Katrina Moser identified diatoms from photographs of smear slides, and Rick Forester identified the fragments of *Daphnia* and their egg casings. I thank Gail Ashley, Harland Goldstein, Tim Lowenstein, John Rayburn, and Joe Rosenbaum for their detailed reviews of this manuscript which greatly improved its clarity.

REFERENCES CITED

Bright, J., 2009, this volume, Ostracode endemism in Bear lake, Utah and Idaho, *in* Rosenbaum, J.G., and Kaufman, D.S., eds., Paleoenvironments of Bear Lake, Utah and Idaho, and its catchment: Geological Society of America Special Paper 450, doi: 10.1130/2009.2450(08).

Colman, S.M., 2005, Stratigraphy of lacustrine sediments cored in 1996, Bear Lake, Utah and Idaho, U.S. Geological Survey Open-File Report 2005-1288, 15 p.

Colman, S.M., 2006, Acoustic stratigraphy of Bear Lake, Utah-Idaho—Late Quaternary sedimentation patterns in a simple half-graben: Sedimentary Geology, v. 185, p. 113–125, doi: 10.1016/j.sedgeo.2005.11.022.

Colman, S.M., Rosenbaum, J.G., Kaufman, D.S., Dean, W.E., and McGeehin, J.P., 2009, this volume, Radiocarbon ages and age models for the past 30,000 years in Bear Lake, Utah and Idaho, *in* Rosenbaum, J.G., and Kaufman, D.S., eds., Paleoenvironments of Bear Lake, Utah and Idaho, and its catchment: Geological Society of America Special Paper 450, doi: 10.1130/2009.2450(05).

Dan, J., and Yaalon, D.H., 1982, Automorphic saline soils in Israel, *in* Yaalon, D.H., ed., Aridic soils and geomorphic processes: Catena Supplement, v. 1, p. 103–115.

Davis, M.B., and Ford, M.S., 1982, Sediment focusing in Mirror Lake, New Hampshire: Limnology and Oceanography, v. 27, p. 137–150.

Dean, W.E., 2009, this volume, Endogenic carbonate sedimentation in Bear Lake, Utah and Idaho, over the last two glacial-interglacial cycles, *in* Rosenbaum, J.G., and Kaufman, D.S., eds., Paleoenvironments of Bear Lake, Utah and Idaho, and its catchment: Geological Society of America Special Paper 450, doi: 10.1130/2009.2450(07).

Dean, W., Rosenbaum, J., Skipp, G., Colman, S., Forester, R., Liu, A., Simmons, K., and Bischoff, J., 2006, Unusual Holocene and late Pleistocene carbonate sedimentation in Bear Lake, Utah and Idaho, USA: Sedimentary Geology, v. 185, p. 93–112, doi: 10.1016/j.sedgeo.2005.11.016.

Drever, J.I., and Smith, C.L., 1978, Repeated wetting and drying of the soil zone as an influence on the chemistry of groundwater in arid terrains: American Journal of Science, v. 278, p. 1448–1454.

Harms, J.C., Southard, J.B., and Walker, R.G., 1982, Structures and sequences in clastic rocks: Tulsa, Oklahoma, SEPM Short Course no. 9, 249 p.

Hilton, J., 1985, A conceptual framework for predicting the occurrence of sediment focusing and sediment redistribution in small lakes: Limnology and Oceanography, v. 30, p. 1131–1143.

Hutchinson, G.E., 1975, Limnological botany, v. 3 of A treatise on limnology: New York, John Wiley and Sons, 660 p.

Kaufman, D.S., Bright, J., Dean, W.E., Rosenbaum, J.G., Moser, K., Anderson, R.S., Colman, S.M., Heil, C.W., Jr., Jiménez-Moreno, G., Reheis, M.C., and Simmons, K.R., 2009, this volume, A quarter-million years of paleoenvironmental change at Bear Lake, Utah and Idaho, *in* Rosenbaum, J.G., and Kaufman, D.S., eds., Paleoenvironments of Bear Lake, Utah and Idaho, and its catchment: Geological Society of America Special Paper 450, doi: 10.1130/2009.2450(14).

Kemp, A.S., Dean, J., Pearce, R.B., and Pike, J., 2001, Recognition and analysis of bedding and sediment fabric features, *in* Last, W.M., and Smol, J.P., eds., Physical and geochemical methods, v. 2 of Tracking environmental change using lake sediments: Dordrecht, Kluwer Academic Publishers, p. 7–22.

Kubiena, W.L., 1970, Micromorphological features of soil morphology: New Brunswick, New Jersey, Rutgers University Press, 254 p.

Laabs, B.J.C., and Kaufman, D.S., 2003, Quaternary highstands in Bear Lake Valley, Utah and Idaho: Geological Society of America Bulletin, v. 115, p. 463–478, doi: 10.1130/0016-7606(2003)115<0463:QHIBLV>2.0.CO;2.

Lamoureux, S.F., 1999, Spatial and interannual variations in sedimentation patterns recorded in nonglacial varved sediments from the Canadian High Arctic: Journal of Paleolimnology, v. 21, p. 73–84, doi: 10.1023/A:1008064315316.

Lamoureux, S.F., and Bradley, R.S., 1996, A late Holocene varved sediment record of environmental change from northern Ellesmere Island, Canada: Journal of Paleolimnology, v. 16, p. 239–255, doi: 10.1007/BF00176939.

McCalpin, J.P., 1993, Neotectonics of the northeastern Basin and Range margin, western U.S.A.: Zeitschrift für Geomorphologie N.F., Suppl. Bd., v. 94, p. 137–157.

Moser, K.A., and Kimball, J.P., 2009, this volume, A 19,000-year record of hydrologic and climatic change inferred from diatoms from Bear Lake, Utah and Idaho, *in* Rosenbaum, J.G., and Kaufman, D.S., eds., Paleoenvironments of Bear Lake, Utah and Idaho, and its catchment: Geological Society of America Special Paper 450, doi: 10.1130/2009.2450(10).

Reheis, M.C., Laabs, B.J.C., Forester, R.M., McGeehin, J.P., Kaufman, D.S., and Bright, J., 2005, Surficial deposits in the Bear Lake basin: U.S. Geological Survey Open-File Report 2005-1088, 30 p.

Reheis, M.C., Laabs, B.J.C., and Kaufman, D.S., 2009, this volume, Geology and geomorphology of Bear Lake Valley and upper Bear River, Utah and Idaho, *in* Rosenbaum, J.G., and Kaufman, D.S., eds., Paleoenvironments of Bear Lake, Utah and Idaho, and its catchment: Geological Society of America Special Paper 450, doi: 10.1130/2009.2450(02).

Retallack, G.J., 2001, Soils of the past: Oxford, Blackwell Science, 404 p.

Robertson, G.C., 1978, Surficial deposits and geologic history, northern Bear Lake Valley, Idaho [M.S. thesis]: Logan, Utah State University, 162 p.

Rosenbaum, J.G., and Heil, C.W., Jr., 2009, this volume, The glacial/deglacial history of sedimentation in Bear Lake, Utah and Idaho, *in* Rosenbaum, J.G., and Kaufman, D.S., eds., Paleoenvironments of Bear Lake, Utah and Idaho, and its catchment: Geological Society of America Special Paper 450, doi: 10.1130/2009.2450(11).

Rosenbaum, J.G., and Kaufman, D.S., 2009, this volume, Introduction to *Paleoenvironments of Bear Lake, Utah and Idaho, and its catchment, in* Rosenbaum, J.G., and Kaufman, D.S., eds., Paleoenvironments of Bear Lake, Utah and Idaho, and its catchment: Geological Society of America Special Paper 450, doi: 10.1130/2009.2450(00).

Rosenbaum, J.G., Dean, W.E., Reynolds, R.L., and Reheis, M.C., 2009, this volume, Allogenic sedimentary components of Bear Lake, Utah and Idaho, *in* Rosenbaum, J.G., and Kaufman, D.S., eds., Paleoenvironments of Bear Lake, Utah and Idaho, and its catchment: Geological Society of America Special Paper 450, doi: 10.1130/2009.2450(06).

Smith, N.D., and Ashley, G., 1985, Proglacial lacustrine environment, *in* Ashley, G., Shaw, J., and Smith, N.D., eds., Glacial sedimentary environments: Tulsa, Oklahoma, SEPM Short Course no. 16, p. 135–216.

Smoot, J.P., 2003, Impact of sedimentation styles on paleoclimate proxies in late Pleistocene through Holocene lakes in the western U.S.: Tucson, International Limnogeology Congress, 3rd, Abstracts, p. 273.

Smoot, J.P., and Benson, L.V., 1998, Sedimentary structures as indicators of paleoclimatic fluctuations: Pyramid Lake, Nevada, *in* Pitman, J.K., and Carroll, A.R., eds., Modern and ancient lake systems: Utah Geological Association, Guidebook 26, Salt Lake City, Utah, p. 131–161.

Smoot, J.P., and Benson, L.V., 2004, Mechanical mixing of climate proxies by sediment focusing in Pyramid Lake, Nevada: A cautionary tale: Geological Society of America Abstracts with Programs, v. 36, p. 473.

Smoot, J.P., and Lowenstein, T.K., 1997, Depositional environments of nonmarine evaporites, *in* Melvin, J.L., ed., Evaporites, petroleum and mineral resources: Amsterdam, Elsevier, Developments in Sedimentology, v. 50, p. 189–347.

Smoot, J.P., and Rosenbaum, J.G., 2009, this volume, Sedimentary constraints on late Quaternary lake-level fluctuations at Bear Lake, Utah and Idaho, *in* Rosenbaum, J.G., and Kaufman, D.S., eds., Paleoenvironments of Bear Lake, Utah and Idaho, and its catchment: Geological Society of America Special Paper 450, doi: 10.1130/2009.2450(12).

Sturm, M., 1979, Origin and composition of clastic varves, *in* Schluchter, C., ed., Moraines and varves: Rotterdam, A.A. Balkema, p. 281–285.

Terry, R.D., and Chilingar, G.V., 1955, Summary of "Concerning some additional aids in studying sedimentary formations," by M.S. Shvetsov: Journal of Sedimentary Petrology, v. 25, p. 229–234.

MANUSCRIPT ACCEPTED BY THE SOCIETY 15 SEPTEMBER 2008

The Geological Society of America
Special Paper 450
2009

Isotope and major-ion chemistry of groundwater in Bear Lake Valley, Utah and Idaho, with emphasis on the Bear River Range

Jordon Bright

Department of Geology, Box 4099, Northern Arizona University, Flagstaff, Arizona 86011, USA

ABSTRACT

Major-ion chemistry, strontium isotope ratios ($^{87}Sr/^{86}Sr$), stable isotope ratios ($\delta^{18}O$, δ^2H), and tritium were analyzed for water samples from the southern Bear Lake Valley, Utah and Idaho, to characterize the types and distribution of groundwater sources and their relation to Bear Lake's pre-diversion chemistry. Four groundwater types were identified: (1) Ca-Mg-HCO_3 water with $^{87}Sr/^{86}Sr$ values of ~0.71050 and modern tritium concentrations was found in the mountainous carbonate terrain of the Bear River Range. Magnesium (Mg) and bicarbonate (HCO_3) concentrations at Swan Creek Spring are discharge dependent and result from differential carbonate bedrock dissolution within the Bear River Range. (2) Cl-rich groundwater with elevated barium and strontium concentrations and $^{87}Sr/^{86}Sr$ values between 0.71021 and 0.71322 was found in the southwestern part of the valley. This groundwater discharges at several small, fault-controlled springs along the margin of the lake and contains solutes derived from the Wasatch Formation. (3) SO_4-rich groundwater with $^{87}Sr/^{86}Sr$ values of ~0.70865, and lacking detectable tritium, discharges from two springs in the northeast quadrant of the study area and along the East Bear Lake fault. (4) Ca-Mg-HCO_3-SO_4-Cl water with $^{87}Sr/^{86}Sr$ values of ~0.71060 and submodern tritium concentrations discharges from several small springs emanating from the Wasatch Formation on the Bear Lake Plateau.

The $\delta^{18}O$ and δ^2H values from springs and streams discharging in the Bear River Range fall along the Global Meteoric Water Line (GMWL), but are more negative at the southern end of the valley and at lower elevations. The $\delta^{18}O$ and δ^2H values from springs discharging on the Bear Lake Plateau plot on an evaporation line slightly below the GMWL. Stable isotope data suggest that precipitation falling in Bear Lake Valley is affected by orographic effects as storms pass over the Bear River Range, and by evaporation prior to recharging the Bear Lake Plateau aquifers.

Approximately 99% of the solutes constituting Bear Lake's pre-diversion chemistry were derived from stream discharge and shallow groundwater sources located within the Bear River Range. Lake-marginal springs exposed during the recent low lake levels and springs and streams draining the Bear Lake Plateau did not contribute significantly to the pre-diversion chemistry of Bear Lake.

Bright, J., 2009, Isotope and major-ion chemistry of groundwater in Bear Lake Valley, Utah and Idaho, with emphasis on the Bear River Range, *in* Rosenbaum, J.G., and Kaufman, D.S., eds., Paleoenvironments of Bear Lake, Utah and Idaho, and its catchment: Geological Society of America Special Paper 450, p. 105–132, doi: 10.1130/2009.2450(04). For permission to copy, contact editing@geosociety.org. ©2009 The Geological Society of America. All rights reserved.

INTRODUCTION

Bear Lake is a large (>280 km²), deep (>60 m), turquoise-blue lake straddling the border of north-central Utah and south-eastern Idaho (Fig. 1). Bear Lake is situated in the rain shadow of the Bear River Range and had a small local watershed prior to the 1912 diversion of the Bear River (pre-diversion watershed:lake area = 4.5:1; Lamarra et al., 1986). Only a handful of small streams drain the surrounding highlands, yet the hydrologic budget of the lake is balanced, or nearly so (Lamarra et al., 1986; Bright et al., 2006). Long sediment cores (100 and 120 m) from the lake extend back over 250,000 years (Bright et al., 2006; Kaufman et al., this volume) and seismic evidence reveals that the valley, and likely the lake, has been in existence much longer (Colman,

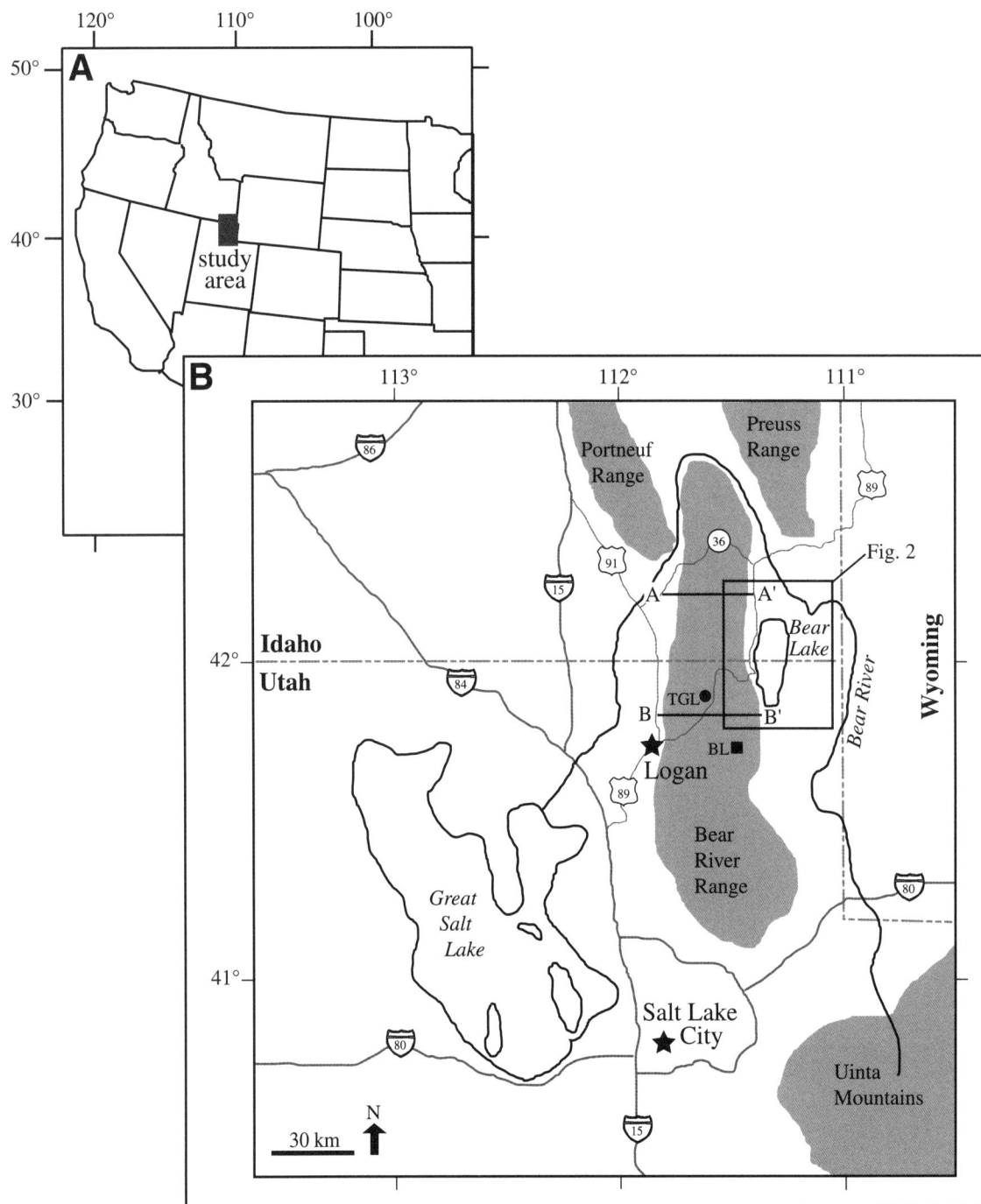

Figure 1. Bear Lake study area relative to (A) the western United States, and (B) the Great Salt Lake. TGL—Tony Grove Lake; BL—Bug Lake; cross sections A–A′ and B–B′ are shown in Figure 9.

2006). There are no evaporite minerals (gypsum or halite) in the long cores, indicating that the lake has survived major changes in climate without becoming saline, like the Great Salt Lake, or drying out. The lake's existence and survival are thought to be strongly dependent on groundwater, but only two small studies of groundwater in the area have been published (Kaliser, 1972; Wylie et al., 2005).

Previously published papers and other papers in this volume discuss Bear Lake's carbonate-rich sedimentary sequence and the unusual geochemistry of Bear Lake, both before and after the 1912 diversion that connected Bear River to the lake via a series of canals (e.g., Dean et al., 2006, 2007; Bright et al., 2006; Fig. 2). These previously published papers focused on Bear Lake itself, and on a subset of the available geochemical and isotopic data from the surrounding watershed as they pertained to the lake. This paper describes groundwater chemistry and distribution surrounding Bear Lake in a more spatially comprehensive manner using the entire chemical and isotopic data set, and incorporates data from studies by Kaliser (1972) and Wylie et al. (2005). The impact of groundwater on Bear Lake's pre-diversion chemistry is revisited in context of this more comprehensive assessment.

The Bear River currently bypasses the Bear Lake watershed (Figs. 1 and 2) but it has played a major role in the history of Bear Lake (e.g., Bright et al., 2006; Dean et al., 2006, 2007; Dean, this volume; Kaufman et al., this volume). Although not discussed in this paper, pertinent Bear River major-ion and strontium data are included in this paper for completeness, and for comparison to other water sources within the Bear Lake watershed.

Throughout this paper, locations with formal names are referenced by their name followed by a site number in parentheses. Many sites do not have formal names and are referenced by a general description followed by a site number in parentheses. Site numbers correspond to the numbers on Figure 2, for example, "Swan Creek Spring (22)," or "lake-marginal spring (26)."

Geologic Setting

Bear Lake Valley (Figs. 1 and 2) is situated within the Laramide overthrust belt, along the eastern margin of the Basin and Range geologic province. Traces of the Paris, Willard, Meade and Laketown thrust faults crop out sequentially in a west-to-east fashion across the valley (Fig. 2; Oriel and Platt, 1980; Dover, 1995; Willis, 1999; Liu et al., 2005). Bear Lake Valley is a north-south–trending, southeast-dipping, half-graben bounded on the west and east by the West Bear Lake Fault and East Bear Lake Fault, respectively (Fig. 2; McCalpin, 1993; Reheis et al., this volume). The majority of offset is on the East Bear Lake Fault, as illustrated by eastward-thickening lacustrine deposits within Bear Lake (Colman, 2006) and truncated spurs and prominent fault scarps along the eastern margin of the lake (Kaliser, 1972; McCalpin, 1993; Reheis et al., this volume). The East Bear Lake Fault and possibly the West Bear Lake Fault are thought to sole into the deeper thrust faults and may provide conduits for groundwater movement (Reheis et al., this volume).

The western margin of Bear Lake Valley is bounded by the Bear River Range. The Bear River Range has a maximum elevation 3042 m and in the study area is composed primarily of westward-dipping Paleozoic (Cambrian–Permian) marine limestone and dolomite, with lesser amounts of quartzite and shale (Wilson, 1979; Oriel and Platt, 1980; Dover, 1995; Spangler, 2001). Both the Eocene Wasatch Formation and Mio-Pliocene Salt Lake Formation crop out at lower elevations along the eastern flank of the Bear River Range, north of Bear Lake. The Wasatch Formation is prevalent at the southern end the valley, where it overlies the local Paleozoic sequence (Oriel and Platt, 1980; Dover, 1995; Reheis et al., this volume).

The eastern margin of Bear Lake Valley incorporates the western portion of the Bear Lake Plateau. The Bear Lake Plateau has a maximum elevation of 2349 m and is composed of early Mesozoic (Triassic and Jurassic) limestone and sandstone that is exposed in the three major drainages (Indian, North Eden, and South Eden Creeks; Fig. 2) and along the main ridgeline immediately east of Bear Lake. Several of the Paleozoic marine carbonate units in the Bear River Range crop out at the northern and southern ends of the Bear Lake Plateau (Oriel and Platt, 1980; Dover, 1995). The Bear Lake Plateau is mantled by the Eocene Wasatch Formation (Dover, 1995), although Oriel and Platt (1980) map the exposures on the Idaho portion of the Bear Lake Plateau as the Salt Lake Formation (see Coogan, 1992; Reheis et al., this volume). Coogan (1992) extensively mapped the Bear Lake Plateau and assigned four informal names to the Wasatch Formation sediments. The majority of the sediments were classified as the Diamictite (gravel and massive mudstone) and Main Body Members (sandstone and mudstone fluvial sequence). Less extensive exposures of the Quartzite Conglomerate (gravel and sandstone) and Limestone (lacustrine limestone with abundant coarse clasts) members are also present on the eastern and western margins, respectively, of the plateau. For additional discussion of the local geology, see Reheis et al. (this volume).

Precipitation, and Spring and Stream Discharge

Regional precipitation is dominated by winter storms that originate in the central and northern Pacific Ocean and move west to east across the study area. The average maximum accumulated precipitation (1979–2005) in the central Bear River Range (Tony Grove Lake, 2583 m) is ~125 cm yr⁻¹, and decreases southward to ~77 cm yr⁻¹ at Bug Lake (2423 m; rcc.nrcs.usda.gov/snotel; Fig. 1). No precipitation data exist for the Bear Lake Plateau, but three stations (15, 24, Laketown) bordering Bear Lake report mean annual precipitation values of ~30 cm yr⁻¹ (wrcc.sage.dri.edu/summary/climsmut and /climsmid). The majority (~60%) of precipitation at Bear Lake falls as snow during the months of October through April.

Large areas of sinkholes and solution basins are common in the Bear River Range (Fig. 2; Wilson, 1979), facilitating infiltration to the aquifer systems within the mountain range. Groundwater moves across topographic drainage divides and has

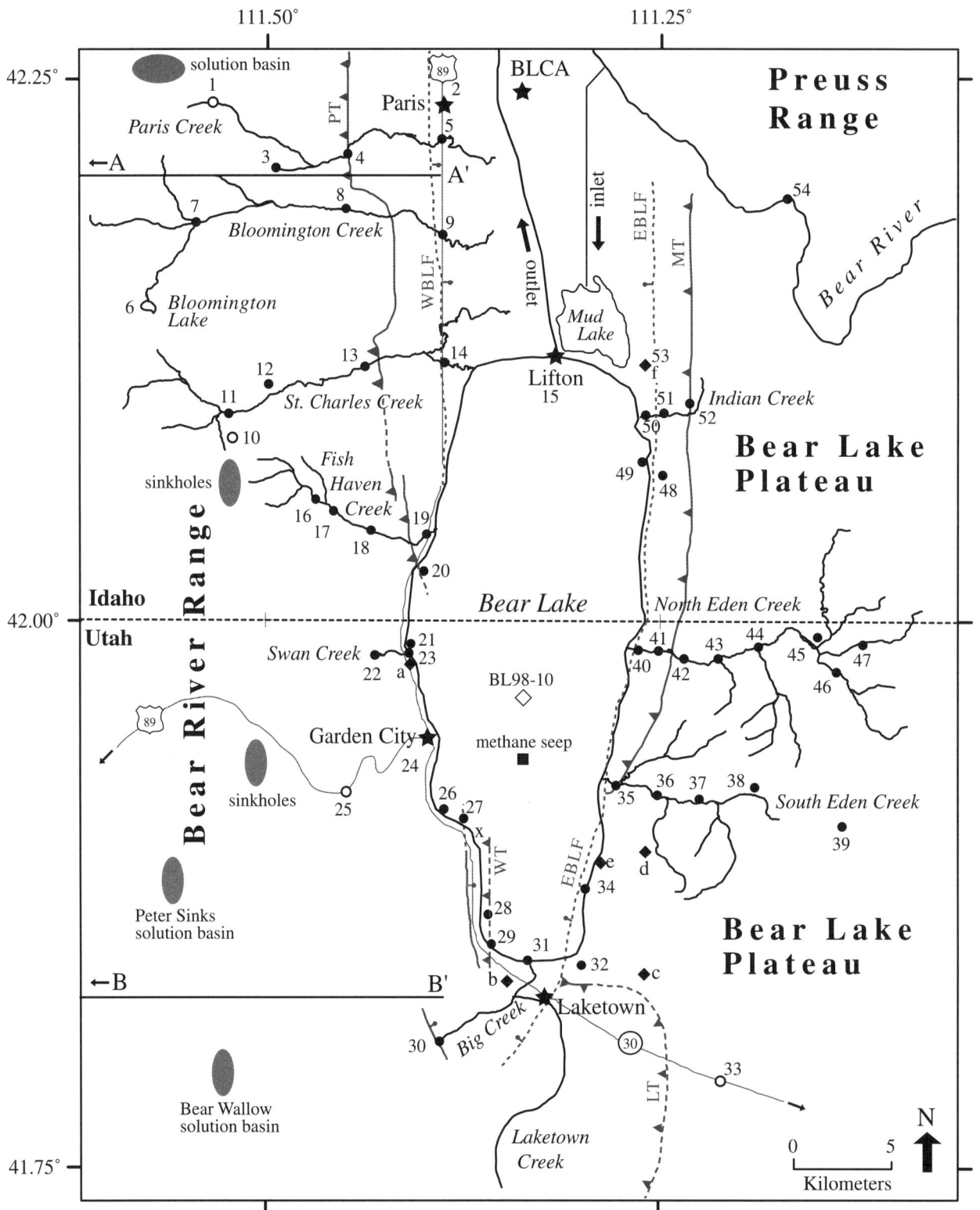

Figure 2. Locations of water samples from Bear Lake Valley. Numbers correspond to chemical and isotopic data in Tables 2–6. Solid circles are isotope and chemistry sampling sites. Open circles are snow-pit sampling sites. Solid diamonds labeled a–f represent approximate locations of sites from Kaliser (1972). Solid square in Bear Lake is location of a large methane seep (see Fig. 14 and text for discussion). Open diamond in Bear Lake is location of sediment core BL98-10. "x" is location of a dry tufa mound, along southwest shore of Bear Lake. PT—Paris Thrust; WT—Willard Thrust; MT—Meade Thrust; LT—Laketown Thrust; WBLF—West Bear Lake Fault; EBLF—East Bear Lake Fault. Approximate locations of faults shown in gray; solid where known, dashed where inferred. Teeth are on upper thrust plate, ball and pillar are on downthrown blocks. BLCA—Bear Lake County Airport; cross sections A–A′ and B–B′ are shown in Figure 9.

transmission times of less than a month in the central Bear River Range (Spangler, 2001), immediately west of the study area. Groundwater transmission times in the eastern Bear River Range area are currently unknown, but are likely similar to those reported by Spangler (2001).

The Paleozoic carbonate units that make up most of the lithologies in the Bear River Range are fractured, faulted, and karsted, which facilitates groundwater movement. For example, Logan Cave Spring and Ricks Spring, both located west of our study area, discharge from a prominent bedrock joint and along a fault, respectively (Spangler, 2001). Solution caverns in the Bear River Range may be well developed, such as Minnetonka Cave (St. Charles Canyon; not shown on Fig. 2), which is more than 600 m long with individual rooms up to 100 m long and 30 m high. Hydrologic studies by Rice and Spangler (1999) in the northern Wasatch Range, an area with a geologic and lithologic setting similar to the study area, suggest a duality in spring discharge. Their study showed that rapidly moving snowmelt pulses passed through that groundwater system within days and were superimposed on older (3–13 yr) base flow discharge. A similar situation likely occurs within the eastern Bear River Range study area. Locally, the Paleozoic Brigham (Geertzen Canyon) Quartzite is fractured and produces water (Kaliser, 1972; Wylie et al., 2005), but much less than the carbonate bedrock. Quartzites and shales within the Bear River Range are likely barriers to local groundwater movement (Wylie et al., 2005).

Spring Discharge

Springs emanating from carbonate terrain in the eastern Bear River Range are numerous and their discharges differ by an order of magnitude (Table 1). Swan Creek Spring (22) has an average discharge of ~1.69 m^3 s^{-1} (Mundorff, 1971) and a maximum discharge of ~9.1 m^3 s^{-1} (epa.gov/storet/dw_home.html; station 4907200). In comparison, Bloomington Spring (not shown on Fig. 2) has an average discharge of ~0.03 m^3 s^{-1} (Idaho Department of Environmental Quality, 2002), and four other large springs in the Bear River Range west of the study area have discharges on the order of 0.1–2.1 m^3 s^{-1} (Table 1; Spangler, 2001).

Springs discharging from the Wasatch Formation on the Bear Lake Plateau are small, generally 1–2 orders of magnitude less than spring discharges in the Bear River Range. All of the Bear Lake Plateau Wasatch Formation springs sampled in this study (39, 45–47) discharge from the Main Body Member. The sandstone lenses of this member may be potential aquifers but they are discontinuous and confined by surrounding mudstone. None of the local Bear Lake Plateau springs are gauged, but several springs emanating from the Wasatch Formation east and south of the study area have discharges of <3 × 10^{-3} m^3 s^{-1}, but most estimates are in the 3 × 10^{-4} to 5 × 10^{-4} m^3 s^{-1} range (waterdata.usgs. gov/nwis/gwsi). Big Spring (30), situated on a fault in the southwest corner of the study area, is the only significant spring discharging from Wasatch Formation terrain. Coogan (1992) does not describe the Wasatch Formation in the Big Spring area other than to generalize it as being fine-grained strata (and therefore probably the Main Body or possibly the Limestone Member). Big Spring is not gauged, but its discharge is larger (see stream discussion below) than that of most other springs in the Bear River Range (Table 1).

The Mesozoic Twin Creek Limestone, which is exposed primarily in the North Eden Creek drainage, is considered a confining unit within the local geologic sequence (capp.water.usgs.gov/ gwa/gwa.html). Only one small spring (37) in the South Eden Creek drainage emanates from the Twin Creek Limestone within the study area. Springs emanating from the Twin Creek Limestone east of the study area are small and have discharges of <1.5 × 10^{-3} m^3 s^{-1} (waterdata.usgs.gov/nwis/gwsi).

The Mesozoic Nugget Sandstone crops out conspicuously along the ridgeline immediately east of Bear Lake (Dover, 1995) and is the only unit in the Bear Lake Plateau classified as an aquifer (capp.water.usgs.gov/gwa/gwa.html). No springs

TABLE 1. SPRING AND STREAM DISCHARGE IN THE BEAR RIVER RANGE AND BEAR LAKE PLATEAU

Name	Discharge (m^3s^{-1})	Source	Maximum discharge (m^3s^{-1})	Source
Swan Creek Spring	1.69[†]	epa.gov/storet/dw_home.html; 4907200	9.08	epa.gov/storet/dw_home.html; 4907200
Bloomington Spring	0.03[†]	Idaho Dept. Env. Quality (2002)	N.D.	N.D.
Bloomington Creek	0.89[†]	waterdata.usgs.gov; station 10058600	6.43	waterdata.usgs.gov; station 10058600
Paris Creek	0.30[†]	waterdata.usgs.gov; station 10060500	5.01	waterdata.usgs.gov; station 10060500
St. Charles Creek	1.64[†]	waterdata.usgs.gov; station 10054600	11.13	waterdata.usgs.gov; station 10054600
North Eden Creek	0.11[†]	epa.gov/storet/dw_home.html: station 4907120	0.34	epa.gov/storet/dw_home.html: 4907120
Big Creek	0.71[†]	epa.gov/storet/dw_home.html: station 4907100	1.90	epa.gov/storet/dw_home.html: station 4907100
Ricks Spring*	0.06–2.10	Mundorff (1971)	4.25	Wilson (1979)
Dewitt Spring*	0.28–0.99	Spangler (2001)	0.99	Spangler (2001)
Logan Cave Spring*	0.03–0.28	Spangler (2001)	0.71	Wilson (1979)
Wood Camp Hollow Spring*	0.08–>1.13	Spangler (2001)	1.84	Wilson (1979)

Note: N.D.—no data.

*Springs located in central Bear River Range, outside of study area. "Discharge" values for these springs are ranges as provided in references.

[†]Mean discharge values for the time of record. Bloomington Spring interval not reported. Swan Creek Spring, periodically from 1979 to 2004; Bloomington Creek, 1960–1986; Paris Creek, 1942–1946; St. Charles Creek, 1961–1966; Big Creek, periodically from 1979 to 2004.

discharge directly from the Nugget Sandstone within the study area, although Falula Spring (32) may discharge at the contact between the Nugget Sandstone and valley alluvium (Kaliser, 1972). Springs discharging from the Nugget Sandstone to the east of the study area are typically small (<1 × 10^{-3} m^3 s^{-1}; waterdata. usgs.gov/nwis/gwsi).

During 2000–2004, the Bear Lake watershed experienced its most intense drought since the 1930s. The negative moisture balance and the release of stored water caused lake level to fall nearly 6 m. As a result, several lake-marginal springs emerged, especially around the southwest margin of the lake (26–28; Fig. 2). These springs appear to discharge along several faults that define the southern margin of the lake (Fig. 2). Discharge from these springs was localized at well-defined orifices and water flowed into the lake. Wet areas several meters above lake level also developed (29, 34, 49), but these areas lacked distinct orifices and water did not reach the lake. The persistence of these beach seeps over several years suggests that they are areas of diffuse groundwater discharge. Stable isotope values from sites 29 and 34 are more positive than those from the beach springs with distinct orifices, indicating a higher degree of evaporative enrichment. These seeps are not actively precipitating tufa, but a crystalline precipitate with a salty taste was present at site 34. The chemistry of these diffuse beach seeps is reported but not discussed further because evaporation and mineral precipitation potentially alter their composition such that they may no longer be representative of the local geohydrology. A small tufa mound on an otherwise sandy expanse of beach was exposed at site 34 during the recent low lake levels as well. The tufa suggests that groundwater discharge does (or did) occur at this site. A similar tufa mound that would be flooded at full lake level is located at site X (Fig. 2) on the southwest margin of the lake. No water was associated with this mound when visited in April 2004.

Stream Discharge

Streams in the study area receive their water from two principal sources, and potentially a third source: (1) Essentially instantaneous overland flow and spring discharge from snowmelt during the spring, and to a lesser extent from infrequent summer rainstorms. Consequently, runoff to Bear Lake is strongly correlated with precipitation in the Bear River Range (Fig. 3). (2) Delayed discharge of infiltrated snowmelt (and rain) that sustains local springs and streams throughout most of the year. Stream sediment from Bloomington Creek, Swan Creek, and North Eden Creek contains grains of "popcorn tufa" (Bright et al., 2006; Dean et al., 2006), which indicate that calcite-saturated groundwater is discharged and degasses along the streambeds. (3) Extrabasinal groundwater sourced outside of the study area may discharge along faults, fractures, or bedding planes within the study area.

Stream discharge in the Bear River Range is not currently gauged, but available data for Paris Creek, Bloomington Creek, and St. Charles Creek indicate average discharges are on the order of 0.3–1.6 m^3 s^{-1}, with peak discharges in excess of 6.3 m^3 s^{-1} (Table 1; waterdata.usgs.gov). Big Creek is sourced at Big Spring and flows several kilometers before reaching the lake. Laketown Creek (Fig. 2) is often ephemeral in its lower reaches and does not contribute significantly to Big Creek discharge. The discharge values at the mouth of Big Creek (31) range from <0.03 m^3 s^{-1} to 2.1 m^3 s^{-1}, and average 0.7 m^3 s^{-1} (Table 1). These values must represent minimum values for the actual discharge at Big Spring because of agricultural diversions upstream of the Big Creek gauging station.

Stream discharge on the Bear Lake Plateau is not gauged, but recent (1999, 2004) monthly estimates of instantaneous discharge for North Eden Creek were 0.15 and 0.06 m^3 s^{-1}, respectively, and peak estimated discharges were 0.33 and 0.17 m^3 s^{-1}, respectively

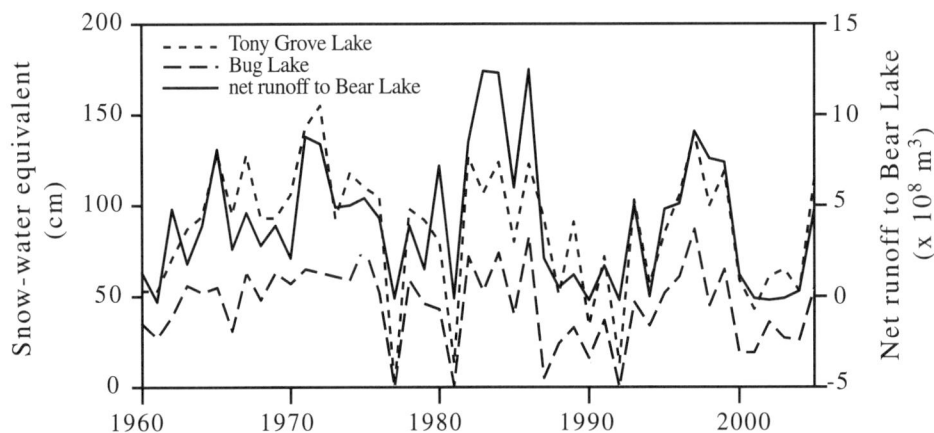

Figure 3. Snow-water-equivalent data for May at Tony Grove Lake (short-dashed line) and Bug Lake (long-dashed line) plotted against net runoff to Bear Lake (solid black line) for the years 1960–2005. Net runoff to Bear Lake = ΔS – (I + O + E); where ΔS = annual change in Bear Lake storage from elevation capacity curves (m^3 yr^{-1}); I = annual inlet canal inflow (m^3 yr^{-1}); O = annual outlet canal release (m^3 yr^{-1}); E = evaporation (m^3 yr^{-1}). Net runoff data provided by Connely Baldwin (PacifiCorp).

(Table 1; epa.gov/storet/dw_home.html; station 4907120). The 50% reduction in discharge from 1998–99 to 2003–04 illustrates the severity of the recent drought. In 2004, South Eden Creek was dry and Laketown Creek (Fig. 2) stopped flowing several kilometers upstream of the lake.

METHODS

Oxygen and Hydrogen Isotopes

Water samples for oxygen and hydrogen isotope analyses were collected between May 2003 and August 2004 ($n = 110$). Samples were collected in 25 ml heavy-gauge PVC bottles or 125 ml amber glass bottles and chilled until analysis. Snow samples (May 2003 and April 2004) were collected from the Bear River Range ($n = 4$) and Bear Lake Plateau ($n = 1$). Pits were dug into large drifts down to the ground surface and snow from the entire pit face was scraped into heavy-duty Ziplock® freezer bags and melted. Upon melting, samples were transferred to 25 ml heavy-gauge PVC bottles. Rain samples ($n = 5$) were collected during two frontal storms during 8–9 September 2003. Rain water was collected in 25 ml heavy-gauge PVC bottles as runoff from roofs at the Limber Pine trailhead (25; summit of Hwy 89 in the Bear River Range) and in the towns of Garden City (24), Utah, and Paris (2) and Lifton (15), Idaho (Fig. 2). Oxygen and hydrogen isotopes were measured on a gas-source isotope ratio mass spectrometer (Finnigan Delta S) at the University of Arizona. For oxygen, 3 ml water samples were equilibrated for 9 h with CO_2 gas at ~15 °C in an automated equilibration device coupled to the mass spectrometer. For hydrogen, samples were reacted at 750 °C with chromium metal using a Finnigan H/Device coupled to the mass spectrometer. Standardization was based on international reference materials Vienna Standard Mean Ocean Water (VSMOW) and Standard Light Arctic Precipitation (SLAP). Isotopic ratios are reported using standard del (δ) notation as per mil (‰) differences between the sample and a standard, where $\delta‰ = [(R_{sample}/R_{std})-1] \times 10^3$ and R = ratio of $^{18}O{:}^{16}O$ or $^2H{:}^1H$ and R_{std} refers to the standards VSMOW or SLAP. Precision is 0.08‰ or better for $\delta^{18}O$ and 0.9‰ or better for δ^2H on the basis of repeated internal standards.

Major-Ion Chemistry

Twenty-six water samples for major-ion chemistry were collected from springs and streams in the Bear River Range and Bear Lake Plateau during 1999, 2000, and 2004. Water samples were collected as follows: 0.2 μm filtered, HNO_3 acidified water samples for cation analyses; 0.2 μm filtered, unacidified samples for anion analyses; and raw, unfiltered samples for total carbonate alkalinity. Samples were kept chilled on ice or refrigerated until analyzed. The 1999 samples were analyzed at the University of Minnesota. The 2000 and 2004 samples were analyzed at the U.S. Geological Survey (USGS) in Denver, Colorado; the methods of this analysis are described in Fishman and Friedman

(1985). Previously published major-ion data included in this study from Kaliser (1972) and Wylie et al. (2005), and data for the Bear River published in Dean et al. (2007), are reported in Appendix 1[1] for completeness.

Strontium Isotopes

Water samples for strontium isotope ($^{87}Sr/^{86}Sr$) analysis were collected in 1996, 1999, 2000, and 2004 ($n = 46$). All analyses were done on filtered, unacidified water samples that were collected in acid-washed 125 ml PVC bottles. Twenty-eight outcrop samples from 13 bedrock units were collected from the Bear River Range and Bear Lake Plateau. Rock samples were leached in 5 M acetic acid. The water samples and the rock sample leachates were centrifuged, loaded onto a cation-exchange column, and extracted with hydrochloric acid. Samples were loaded on a single tantalum filament with phosphoric acid. All isotope ratios were measured with an automated VG54 sector multi-collector, thermal-ionization mass spectrometer in dynamic mode in the USGS isotope geology laboratory, Denver, Colorado. Mass-dependent fractionation was corrected assuming a $^{87}Sr/^{86}Sr$ value of 0.1194. Strontium isotope ratios are reported relative to the SRM-987 standard value of 0.71025. Precision is usually ±0.00001. Although not discussed here, numerous $^{87}Sr/^{86}Sr$ values and Sr concentrations for samples from the Bear River are reported in Appendix DR2 (see footnote 1) for completeness.

Tritium

Five unfiltered, unacidified tritium samples were collected in 1 L heavy-gauge PVC bottles with no headspace. Tritium concentration was measured by liquid scintillation spectrophotometry on samples that were first distilled to remove nonvolatile solutes, and then enriched by electrolysis by a factor of about nine. Enriched samples were mixed 1:1 with Ultimagold Low Level Tritium (R) cocktail, and counted for 1500 min in a Quantulus 1220 Spectrophotometer in an underground counting laboratory at the University of Arizona. The detection limit under these conditions is 0.5 tritium units (TU).

RESULTS AND DISCUSSION

Oxygen and Hydrogen Isotopes

The stable isotopic compositions of precipitation and groundwater collected for this study and other previously published data from the surrounding area (Friedman et al., 2002) are plotted together against the Global Meteoric Water Line (GMWL; Craig,

---

[1] GSA Data Repository item 2009047, Appendix DR1, including previously published major-ion data for springs in the Bear River Range and Bear Lake Plateau, and from the Bear River, and Appendix 2, including $^{87}Sr/^{86}Sr$ values and strontium concentrations for water samples taken from the Bear River, is available at http://www.geosociety.org/pubs/ft2009.htm or by request to editing@geosociety.org.

1961) in Figure 4A. The majority of the data plot very near the GMWL. The high isotope ratios measured in snow collected from the Bear River Range in May 2003 (Table 2) were probably the result of melting and refreezing of the snow, which preferentially removed the lighter isotopes (Cooper, 1998). Although collected in April 2004, and also subjected to melting and refreezing, the snow sample (33) from the Bear Lake Plateau has the most negative isotope values in the precipitation data (Table 2).

The isotopic composition of rain collected within the study area during a two-day frontal storm event (8–9 September 2003) shows no consistent differences between sample elevations or days of collection (Table 2). The data for rain plot to the right of the GMWL, indicating evaporation prior to reaching the ground surface (Fig. 4A). The isotopic composition of the rain is considerably heavier (more enriched) than the isotopic composition of the local groundwater discharge. Whereas rain may make up nearly 30% of the annual precipitation it does not contribute significantly to the isotopic composition of the local groundwater (e.g., Winograd et al., 1998).

The results from the spring and stream samples collected during this study fall into two distinctive groups (Fig. 4B). The Bear River Range data plot very near the GMWL ($\delta^2H = 8.2(\delta^{18}O) + 13.2$; $r^2 = 0.80$, $n = 66$) and the Bear Lake Plateau data plot slightly below the GMWL ($\delta^2H = 6.2(\delta^{18}O) - 27.6$; $r^2 = 0.89$, $n = 40$). The average $\delta^{18}O$ values overlap for the Bear River Range and Bear Lake Plateau data ($-17.4 \pm 0.5‰$ and $-17.6 \pm 0.7‰$, respectively), but water from the Bear Lake Plateau has a lower (more negative) average δ^2H value ($-136.2 \pm 4.1‰$; Table 2) than

water from the Bear River Range ($-129.7 \pm 4.2‰$; Table 2). The lower average δ^2H value from the Bear Lake Plateau spring and stream samples may result from the more isotopically negative precipitation (site 33; Table 2) that falls on the Bear Lake Plateau. Additional precipitation isotope samples from the Bear Lake Plateau are needed to verify this relationship, however.

Hydrogen isotope values of water from springs and streams in the Bear River Range become more negative from north to south (Fig. 5A) and, with the exception of Paris Creek, δ^2H values are more negative at lower elevations within individual drainage basins (Fig. 5B). In contrast to the Bear River Range, there are no apparent elevational or latitudinal trends in $\delta^{18}O$ or δ^2H values on the Bear Lake Plateau. North Eden Creek shows a slight decrease in $\delta^{18}O$ and δ^2H values along its path, but the difference between samples is small (Table 2). Indian Creek (51) and springs 48 and 53 show decreased stable isotope values in the vicinity of the East Bear Lake Fault (Table 2, Fig. 2), suggesting that the fault may be a conduit for an isotopically depleted groundwater source.

Major-Ion Chemistry

Most reported major-ion analyses for water samples collected in Bear Lake Valley have charge balances <5%, but in an effort to report a comprehensive data set, data from five stations with charge balances >5% are included in Table 3. Four of the five stations represent unique locations that were sampled only once, and for that reason they are included. The higher charge balances were most likely due to errors in the alkalinity measurements.

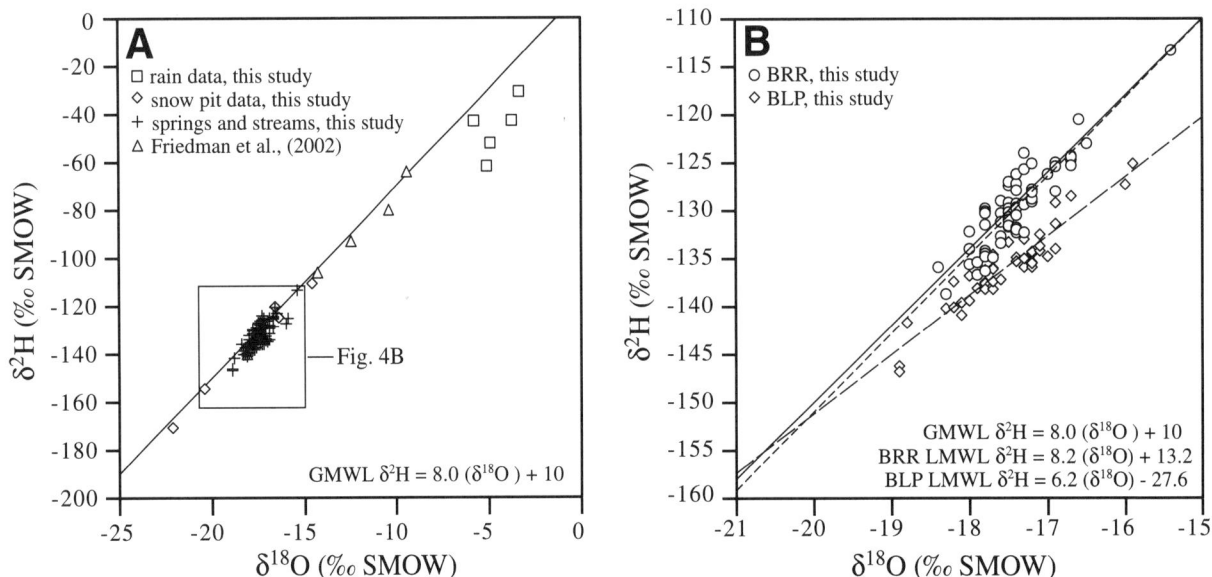

Figure 4. Stable isotope composition of water samples within and around Bear Lake Valley. (A) $\delta^{18}O$ and δ^2H values in precipitation, springs, and streams in Bear Lake Valley. (B) Increased detail of the $\delta^{18}O$ and δ^2H values from spring and stream samples in Bear Lake Valley (this study). Solid line represents Global Meteoric Water Line (GMWL); short-dashed line represents Local Meteoric Water Line (LMWL) for Bear River Range (BRR) samples; long-dashed line represents LMWL for Bear Lake Plateau (BLP) water samples. Data from this study are presented in Table 2. SMOW—Standard Mean Ocean Water.

TABLE 2. OXYGEN (δ^{18}O) AND HYDROGEN (δ^2H) ISOTOPE VALUES FROM SPRINGS AND STREAMS IN BEAR LAKE VALLEY

Site (Fig. 2)	Lat. (°N)	Long. (°W)	Elev. (m)	Site name	May 2003 δ^{18}O (‰)	May 2003 δ^2H (‰)	Sep. 2003 δ^{18}O (‰)	Sep. 2003 δ^2H (‰)	Apr. 2004 δ^{18}O (‰)	Apr. 2004 δ^2H (‰)	Aug. 2004 δ^{18}O (‰)	Aug. 2004 δ^2H (‰)	Average δ^{18}O (‰)	Average δ^2H (‰)
Precipitation														
Snow samples														
1	42.236	111.529	2315	Paris Canyon	-14.6	-110.5	N.D.	N.D.	N.D.	N.D.	N.D.	N.D.	N.D.	N.D.
10	42.088	111.531	2214	St. Charles Canyon	-16.4	-124.7	N.D.	N.D.	N.D.	N.D.	N.D.	N.D.	N.D.	N.D.
25	41.925	111.471	2288	Limber Pine Trailhead	-16.6	-120.2	N.D.	N.D.	-20.4	-154.4	N.D.	N.D.	N.D.	N.D.
33	41.805	111.248	2200	Hwy. 30	N.D.	N.D.	N.D.	N.D.	-22.1	-170.6	N.D.	N.D.	N.D.	N.D.
Rain samples														
2	42.236	111.407	1820	Paris, Idaho	N.D.	N.D.	-5.1	-61.7	N.D.	N.D.	N.D.	N.D.	N.D.	N.D.
15	42.123	111.313	1806	Lifton, Idaho	N.D.	N.D.	-3.7	-42.7	N.D.	N.D.	N.D.	N.D.	N.D.	N.D.
24	41.948	111.395	1819	Garden City, Utah	N.D.	N.D.	-4.9	-52.0	N.D.	N.D.	N.D.	N.D.	N.D.	N.D.
25	41.925	111.471	2288	Limber Pine Trailhead	N.D.	N.D.	-3.3	-30.7	N.D.	N.D.	N.D.	N.D.	N.D.	N.D.
25	41.925	111.471	2288	Limber Pine Trailhead	N.D.	N.D.	-5.2	-44.5	N.D.	N.D.	N.D.	N.D.	N.D.	N.D.
Bear River Range														
Paris Creek														
3	42.206	111.498	2001	Paris Spring	-16.9	-125.0	-17.4	-129.5	N.D.	N.D.	-17.4	-127.2	-17.2	-127.2
4	42.212	111.449	1946	Paris Creek	-16.9	-125.4	-16.9	-128.0	N.D.	N.D.	-15.4	-113.3	-16.4	-122.2
5	42.219	111.400	1813	Paris Creek	-16.7	-124.7	-17.2	-129.1	N.D.	N.D.	-17.2	-125.1	-17.0	-126.3
Bloomington Creek														
6	42.146	111.575	2499	Bloomington Lake*	N.D.	N.D.	-13.7	-110.0	N.D.	N.D.	-14.3	-110.5	-14.0	-110.3
7	42.182	111.544	2202	Bloomington Creek	-16.5	-123.0	-16.7	-125.3	N.D.	N.D.	-16.6	-120.5	-16.6	-122.9
8	42.188	111.447	1879	Bloomington Creek	-16.7	-124.3	-17.4	-129.3	N.D.	N.D.	N.D.	N.D.	-17.1	-126.8
9	42.183	111.401	1815	Bloomington Creek	-16.7	-124.4	-17.2	-128.8	-17.5	-129.2	-17.3	-124.0	-17.2	-126.6
St. Charles Creek														
11	42.096	111.530	2092	St. Charles Creek	-17.0	-126.2	-17.5	-129.8	N.D.	N.D.	-17.3	-125.7	-17.3	-127.2
12	42.105	111.495	1975	Blue Pond Spring	-17.3	-129.4	-17.5	-130.2	-17.6	-132.7	-17.4	-126.2	-17.5	-129.6
13	42.113	111.446	1940	St. Charles Creek	-17.2	-127.8	-17.5	-131.1	N.D.	N.D.	-17.5	-127.4	-17.4	-128.8
14	42.124	111.391	1817	St. Charles Creek	-17.2	-128.1	-17.4	-131.8	N.D.	N.D.	-17.5	-127.0	-17.4	-129.0
Fish Haven Creek														
16	42.054	111.467	2123	Fish Haven Creek	-17.5	-130.7	-17.3	-132.3	N.D.	N.D.	-17.4	-127.9	-17.4	-130.3
17	42.052	111.459	2086	Sadduccee Spring	-17.4	-130.5	-17.6	-133.4	N.D.	N.D.	-17.6	-129.0	-17.5	-131.0
18	42.043	111.437	1960	Fish Haven Creek	-17.5	-131.0	-17.8	-134.7	N.D.	N.D.	-17.8	-129.8	-17.7	-131.8
19	42.037	111.410	1815	Fish Haven Creek	-17.5	-131.0	-17.7	-134.9	N.D.	N.D.	-17.8	-130.1	-17.7	-132.0
20	42.026	111.402	1806	spring	N.D.	N.D.	N.D.	N.D.	-17.8	-134.8	-17.8	-131.5	-17.8	-133.2
Swan Creek														
21	41.985	111.406	1807	spring	N.D.	N.D.	N.D.	N.D.	-17.8	-136.3	-18.0	-134.0	-17.9	-135.2
22	41.985	111.427	1891	Swan Creek Spring	-17.4	-129.7	-17.8	-134.2	-17.5	-131.7	-17.6	-130.3	-17.6	-131.5
23	41.985	111.410	1813	Swan Creek	-17.4	-129.8	-17.8	-134.5	N.D.	N.D.	-17.8	-130.3	-17.7	-131.5
Big Creek														
30	41.809	111.389	1824	Big Spring	-18.0	-135.6	-17.9	-136.7	-17.9	-135.4	-18.0	-132.2	-18.0	-135.0
31	41.846	111.337	1806	Big Creek*	N.D.	N.D.	-17.9	N.D.	-16.9	-131.4	-14.5	-114.2	-15.7	-122.8

(Continued)

TABLE 2. OXYGEN (δ^{18}O) AND HYDROGEN (δ^2H) ISOTOPE VALUES FROM SPRING AND STREAMS IN BEAR LAKE VALLEY (Continued)

Site (Fig. 2)	Lat. (°N)	Long. (°W)	Elev. (m)	Site name	May 2003 δ^{18}O (‰)	May 2003 δ^2H (‰)	Sep. 2003 δ^{18}O (‰)	Sep. 2003 δ^2H (‰)	Apr. 2004 δ^{18}O (‰)	Apr. 2004 δ^2H (‰)	Aug. 2004 δ^{18}O (‰)	Aug. 2004 δ^2H (‰)	Average δ^{18}O (‰)	Average δ^2H (‰)
Miscellaneous Bear River Range springs														
26	41.915	111.389	1806	spring	-17.5	-131.6	-17.4	-132.3	-17.4	-130.5	-17.4	-127.9	-17.4	-130.6
27	41.909	111.372	1806	spring	N.D.	N.D.	N.D.	N.D.	-17.4	-132.0	-17.6	-131.2	-17.5	-131.6
28	41.865	111.360	1801	spring	N.D.	N.D.	N.D.	N.D.	-18.3	-138.7	-18.4	-135.9	-18.4	-137.3
29	41.851	111.356	1805	beach seep†	N.D.	N.D.	N.D.	N.D.	N.D.	N.D.	-6.8	-74.0	N.D.	N.D.
Average (n = 66)													-17.4 ± 0.5	-129.7 ± 4.2
Bear Lake Plateau														
South Eden Creek														
36	41.920	111.254	1891	South Eden Creek	-15.9	-125.1	-16.0	-127.3	N.D.	N.D.	N.D.	N.D.	-15.9	-126.2
37	41.918	111.228	1937	spring (Twin Creek Ls)	-17.7	-136.0	-17.6	-137.2	N.D.	N.D.	N.D.	N.D.	-17.6	-136.6
38	41.921	111.189	2001	spring (Wasatch Fm.)	-17.2	-134.5	-17.3	-135.9	N.D.	N.D.	-17.3	-133.0	-17.3	-134.5
North Eden Creek														
41	41.986	111.255	1842	North Eden Creek	-17.1	-133.7	-17.2	-135.5	N.D.	N.D.	-16.7	-128.5	-17.0	-132.6
42	41.983	111.233	1873	North Eden Creek	-17.0	-134.8	-17.2	-135.4	N.D.	N.D.	N.D.	N.D.	-17.1	-135.1
43	41.984	111.212	1885	North Eden Creek	-17.1	-134.2	-17.4	-135.3	N.D.	N.D.	N.D.	N.D.	-17.3	-134.8
44	41.988	111.189	1896	North Eden Creek	-17.4	-134.9	-17.2	-135.9	N.D.	N.D.	N.D.	N.D.	-17.3	-135.4
45	41.993	111.148	2001	spring (Wasatch Fm.)	-17.8	-137.6	-17.7	-138.2	N.D.	N.D.	N.D.	N.D.	-17.8	-137.9
46	41.997	111.140	2001	spring (Wasatch Fm.)	N.D.	N.D.	N.D.	N.D.	-17.7	-136.1	-17.7	-134.6	-17.7	-135.4
47	41.991	111.119	1983	spring (Wasatch Fm.)	-17.9	-138.1	-18.0	-139.4	-17.8	-137.5	-17.8	-136.2	-17.9	-137.8
Indian Creek														
50	42.094	111.256	1830	Indian Creek	-17.7	-137.4	-17.8	-138.2	N.D.	N.D.	-18.0	-136.8	-17.8	-137.5
51	42.095	111.247	1879	Indian Creek	-18.1	-139.6	-18.2	-140.1	N.D.	N.D.	N.D.	N.D.	-18.2	-139.8
52	42.096	111.232	1964	spring	-17.2	-134.3	-17.3	-135.0	N.D.	N.D.	N.D.	N.D.	-17.3	-134.7
Miscellaneous Bear Lake Plateau springs														
32	41.842	111.302	1812	Falula Spring	-16.9	-134.0	-17.1	-132.5	N.D.	N.D.	N.D.	N.D.	-17.0	-133.3
34	41.878	111.294	1803	beach seep†	N.D.	N.D.	-12.4	-111.4	N.D.	N.D.	N.D.	N.D.	N.D.	N.D.
39	41.910	111.139	2054	Rabbit Spring	N.D.	N.D.	N.D.	N.D.	N.D.	N.D.	-17.5	-133.3	-17.5	-133.3
48	42.075	111.250	1835	spring	-18.3	-140.2	-18.1	-140.9	N.D.	N.D.	-18.2	-137.4	-18.2	-139.5
49	N.D.	N.D.	1804	beach seep†	N.D.	N.D.	-17.6	-135.5	N.D.	N.D.	N.D.	N.D.	N.D.	N.D.
53	42.115	111.264	1824	Mud Lake Hot Spring	-18.9	-146.2	-18.9	-146.8	N.D.	N.D.	-18.8	-141.7	-18.9	-144.9
Average (n = 40)													-17.6 ± 0.7	-136.2 ± 4.1

Note: Lat.—latitude; Long.—longitude; N.D.—no data.
*Stable isotope values not used in Bear River Range average due to lacustrine setting and evaporative enrichment.
†Stable isotope values not used in averages due to diffuse character and full exposure on beach.

The remaining ions in the five questionable samples have concentrations similar to those of neighboring samples that have good charge balances (<5%), further suggesting that the alkalinity measurements are the cause of the imbalances. The data from samples with high charge balances do not appreciably affect the conclusions of this study, however.

Bear River Range

Water temperatures measured at spring orifices and streams discharging from the Bear River Range ranged from ~5.5 to ~16.0 °C. Mountain springs (3, 12, 17, 22) are consistently colder than lake-marginal and low-elevation springs (20, 21, 26–28, 30; Table 4, Fig. 6). A warm lake-marginal spring (a; Figs. 2 and 6) was also encountered by Kaliser (1972). Groundwater sourced directly from snowmelt infiltration and passing quickly through the Bear River Range along shallow flow paths should have temperatures near or slightly below the local mean annual temperature (MAT). Groundwater following deeper flow paths may be heated at depth and have temperatures above the MAT. The MAT in Minnetonka Cave (St. Charles Canyon; not shown on Fig. 2) is 4 °C, and the MAT at Bear Lake is ~6 °C. These temperatures are similar to the high elevation Bear River Range spring and stream temperatures, suggesting shallow flow paths (see section on tritium results below). Geothermal data for the Bear River Range are not available, but geothermal gradients measured in wells on the Bear Lake Plateau range from 19 to 37 °C km^{-1} (Blackett, 2004; www.smu.edu/geothermal), and the majority of wells in Cache Valley, west of Bear Lake Valley, have geothermal gradients of ~30 °C km^{-1} (Blackett, 2004). Assuming a geothermal gradient of ~30 °C km^{-1} for the Bear Lake region, the lake-marginal spring discharge temperatures that are ~10 °C above MAT suggest flow depths on the order of a few hundred meters.

Total dissolved-solids concentrations (TDS) of Bear River Range springs and streams typically range from ~250 to 350 mg L^{-1}, with the higher TDS values typically occurring at lake-marginal springs (Table 3). Bear River Range water samples are dominated by calcium (Ca), magnesium (Mg), and bicarbonate (HCO$_3$; Fig. 7). These three ions constitute >90% of the TDS (Table 5), as expected from the dissolution of carbonate bedrock that dominates the Bear River Range. Sodium (Na), chloride (Cl), and sulfate (SO$_4$) typically constitute <2% of the TDS in water samples north of Fish Haven Creek, but their proportions increase in Swan Creek Spring (22), Big Spring (30), and the lake-marginal springs (20, 21, 26–28) where they constitute anywhere from 3% to 15% of the TDS (Table 5). Consequently, water samples collected in the carbonate terrain north of Fish Haven Creek have high Ca:(SO$_4$+Cl) and Mg:(SO$_4$+Cl) values, but water from Swan Creek Spring (22), Big Spring (30), and the lake-marginal springs (20, 21, 26–28, a) cluster with noticeably lower Ca:(SO$_4$+Cl) and Mg:(SO$_4$+Cl) values (Fig. 8A). In addition, the lower Ca:(SO$_4$+Cl) values occur with higher ^{87}Sr/^{86}Sr values and lower average δ^{18}O values (Figs. 8B and 8C).

Barium (Ba) and strontium (Sr) concentrations are also low in water samples taken north of Garden City (Table 6), indicating that these ions are minor constituents in the Bear River Range carbonate bedrock. Water samples north of Garden City (3, 9, 12–14, 19, 22) cluster with low Ba and Sr concentrations and the lake-marginal springs (20, 21, 26–28) and Big Spring (30) cluster with significantly higher values (Fig. 8D). The increased Cl, SO$_4$, Ba, and Sr in the low-elevation springs indicate different water-rock interactions (e.g., different source area). The concentrations of these ions increase toward the south, where the Wasatch Formation is more prevalent. Water discharging from springs emanating from the Wasatch Formation on the Bear Lake

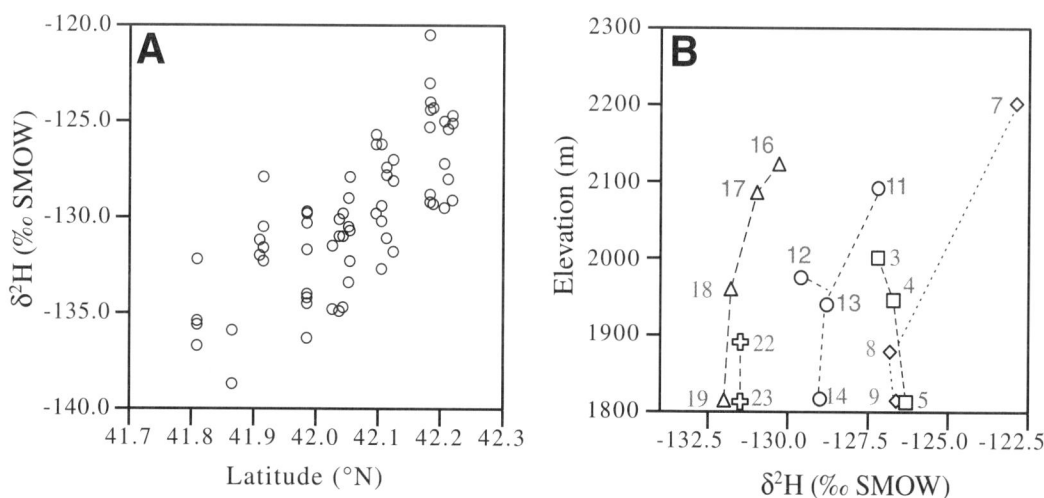

Figure 5. Hydrogen isotope (δ^{2}H) values in spring and stream discharge for the Bear River Range. (A) Individual δ^{2}H values plotted by latitude. (B) Averaged δ^{2}H values for sample sites north of and including Swan Creek Spring plotted by elevation within individual drainages. Numbers refer to sampling locations on Figure 2 and in Table 2. Dashed lines connect samples within a common drainage. Note the progressive decrease in average δ^{2}H values southward along the Bear River Range. SMOW—Standard Mean Ocean Water.

TABLE 3. LOCATION, ELEVATION, AND MAJOR-ELEMENT CHEMISTRY FOR SPRINGS AND STREAMS IN BEAR LAKE VALLEY

Site (Fig. 2)	Lat. (°N)	Long. (°W)	Elev. (m)	Site name	Date (m/d/yr)	TDS (mg L⁻¹)	Ca (mg L⁻¹)	Mg (mg L⁻¹)	Na (mg L⁻¹)	K (mg L⁻¹)	HCO₃ (mg L⁻¹)	SO₄ (mg L⁻¹)	Cl (mg L⁻¹)	SiO₂ (mg L⁻¹)	NO₃ (mg L⁻¹)	Balance* (% error)
Bear River Range																
3	42.206	111.498	2001	Paris Spring	8/1/00	292.2	46.6	11.7	2.1	0.3	228.2	2.0	1.5	6.1	N.D.	-6.0
9	42.183	111.401	1815	Bloomington Creek	9/22/99	283.9	49.7	19.3	2.5	0.4	207.4	2.9	2.1	3.9	N.D.	8.7
12	42.105	111.495	1975	Blue Pond Spring	4/5/04	256.6	41.1	17.5	1.5	0.4	192.1	2.4	1.7	2.7	1.3	4.7
13	42.113	111.446	1940	St. Charles Creek	8/1/00	329.7	45.8	22.6	1.5	0.4	256.2	2.0	1.4	4.4	N.D.	-0.6
14	42.124	111.391	1817	St. Charles Creek	9/22/99	317.3	53.2	24.9	2.1	0.3	231.8	3.0	1.8	4.6	N.D.	10.2
19	42.037	111.410	1815	Fish Haven Creek	9/22/99	272.5	35.5	19.5	2.7	0.5	207.4	5.0	2.3	4.4	N.D.	-0.8
20	42.026	111.402	1806	spring	4/5/04	368.6	54.4	21.2	7.6	1.3	268.1	9.1	7.0	4.8	1.6	0.3
21	41.985	111.406	1806	spring	4/5/04	359.7	52.4	23.2	6.9	1.4	258.1	10.0	7.8	5.5	0.5	2.1
22	41.985	111.427	1891	Swan Creek Spring	9/22/99	275.6	50.5	18.0	3.1	0.4	195.2	4.8	3.8	4.6	N.D.	9.7
22	41.985	111.427	1891	Swan Creek Spring	8/1/00	297.7	47.1	15.8	2.8	0.5	223.8	3.6	3.9	6.1	N.D.	-0.9
22	41.985	111.427	1891	Swan Creek Spring	4/5/04	296.6	53.4	13.2	4.3	0.5	214.1	3.0	8.2	2.8	1.0	1.9
26	41.915	111.389	1806	spring	8/1/04	443.1	74.5	20.4	9.8	0.8	324.6	5.0	7.6	18.7	N.D.	1.9
27	41.909	111.372	1805	spring	4/5/04	524.2	60.4	27.0	38.1	2.7	304.1	46.0	46.0	6.8	2.2	-2.0
28	41.864	111.360	1801	spring	4/5/04	369.1	50.2	21.6	19.3	2.0	234.1	19.0	23.0	4.9	1.6	2.8
29	41.851	111.356	1805	beach seep	8/23/04	1931.2	137.0	154.0	255.0	8.2	262.6	778.0	336.0	33.4	N.D.	1.3
30	41.809	111.389	1824	Big Spring	8/1/00	307.7	52.6	13.9	4.9	0.6	224.2	5.1	6.6	7.9	N.D.	0.3
30	41.809	111.389	1824	Big Spring	4/5/04	298.6	53.9	14.6	5.1	0.7	212.1	5.7	6.6	4.0	1.2	4.3
31	41.846	111.337	1806	Big Creek	9/22/99	320.4	55.2	16.0	6.4	1.0	225.7	6.7	9.1	4.3	N.D.	3.3
Bear Lake Plateau																
34	41.878	111.294	1803	beach seep	9/12/03	1694.0	120.0	54.0	340.0	20.0	190.0	660.0	310.0	2.0	N.D.	0.2
35	N.D.	N.D.	N.D.	South Eden Creek	9/22/99	659.5	94.6	39.5	32.5	2.0	329.4	105.5	55.4	9.0	N.D.	1.5
39	42.910	111.139	2054	Rabbit Spring	8/23/04	439.5	66.1	25.9	28.2	1.3	214.9	36.0	67.0	15.8	26.0	4.1
40	N.D.	111.255	N.D.	North Eden Creek	9/22/99	462.3	66.9	32.2	32.6	2.1	207.4	69.6	51.9	7.7	N.D.	8.4
41	41.986	111.140	1842	North Eden Creek	4/5/04	404.3	47.1	25.8	27.0	2.2	217.2	40.8	44.4	8.6	2.6	0.4
46	42.977	111.140	2001	spring	4/5/04	455.9	63.8	26.1	31.9	1.1	228.1	31.0	74.0	7.1	21.0	2.0
47	42.991	111.119	1983	spring	4/5/04	402.0	31.7	19.8	17.1	1.4	248.1	21.0	33.0	4.7	4.5	0.5
48	42.075	111.250	1835	spring	8/1/00	1371.6	240.8	71.9	26.0	2.4	191.7	827.0	11.5	17.6	N.D.	-3.9
49	N.D.	N.D.	1804	beach seep	4/5/04	3172.3	555.0	198.0	80.5	1.8	396.2	1900.0	41.0	15.4	N.D.	0.3
50	42.094	111.256	1830	Indian Creek	8/23/04	601.4	104.0	35.1	21.1	1.8	176.7	240.0	22.4	15.3	N.D.	3.0
53	42.115	111.264	1824	Mud Lake Hot Spring	9/22/99	1558.9	173.2	56.5	163.7	40.8	268.4	779.4	77.3	17.6	N.D.	-3.1
53	42.115	111.264	1824	Mud Lake Hot Spring	8/1/00	1539.6	189.7	54.0	150.3	41.8	261.4	772.0	70.8	32.1	N.D.	-1.9

Note: Lat.—latitude; Long.—longitude; N.D.—no data.
*Charge balance calculated as [(cation sum − anion sum)/(cation sum + anion sum)] × 100, where units of measurement are in milliequivalents.

TABLE 4. WATER TEMPERATURES FROM SPRINGS IN THE BEAR RIVER RANGE AND BEAR LAKE PLATEAU

Site (Fig. 2)	Lat. (°N)	Long. (°W)	Elev. (m)	Site name	Sep. 2003 (°C)	Apr. 2004 (°C)	Aug. 2004 (°C)
Bear River Range							
3	42.206	111.498	2001	Paris Spring	5.2	N.D.	5.5
12	42.105	111.495	1975	Blue Pond Spring	6.3	6.2	8.5
17	42.052	111.459	2086	Sadduccee Spring	5.9	N.D.	6.0
20	42.026	111.402	1806	spring	N.D.	9.8	11.0
21	41.985	111.406	1807	spring	N.D.	15.0	16.0
22	41.985	111.427	1891	Swan Creek Spring	7.1	7.2	7.0
26	41.915	111.389	1806	spring	N.D.	9.4	10.0
27	41.909	111.372	1806	spring	N.D.	11.0	11.0
28	41.865	111.360	1801	spring	N.D.	13.0	12.0
30	41.809	111.389	1824	Big Spring	12.0	10.2	10.0
Bear Lake Plateau							
39	41.910	111.139	2054	Rabbit Spring	N.D.	N.D.	9.5
45	41.993	111.148	2001	spring	10.1	N.D.	N.D.
46	41.997	111.140	2001	spring	7.9	7.4	8.0
47	41.991	111.119	1983	spring	7.0	7.8	8.0
48	42.075	111.250	1835	spring	12.3	N.D.	13.0
53	42.115	111.264	1824	Mud Lake Hot Spring	42.1	N.D.	44.0

Note: Lat.—latitude; Long.—longitude; N.D.—no data.

Plateau (b, c, d, 39, 46, 47) is characterized by high Cl, SO_4, Ba, and Sr concentrations (Tables 3 and 6, Fig. 7, Appendix 1 [see footnote 1]). Consequently, within the limits of this study, the Wasatch Formation at the south end of Bear Lake Valley is the best candidate for the source of the elevated SO_4, Cl, Ba, and Sr concentrations found in Big Spring and the lake-marginal springs located along the western margin of Bear Lake.

Bear Lake Plateau

Spring discharge on the Bear Lake Plateau is typically cold, ~8–13 °C, except for Mud Lake Hot Spring where water temperatures are ~42 °C (Table 4). These temperatures are higher

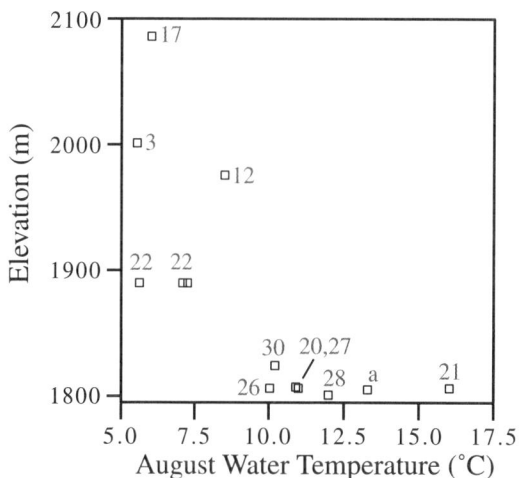

Figure 6. Elevation versus August 2004 water temperatures for spring sites along the eastern Bear River Range. Numbers and letter "a" refer to sample sites on Figure 2 and in Table 3.

than the local MAT by only a few degrees (except Mud Lake Hot Spring) and suggest that flow paths are not especially deep. Total dissolved-solids concentrations on the Bear Lake Plateau typically range between 400 and 600 mg L^{-1}, except for Mud Lake Hot Spring (53) and spring 48 where TDS concentrations approach 1600 mg L^{-1} and 3200 mg L^{-1}, respectively (Table 3).

Water samples from springs emanating from the Wasatch Formation (b, c, d, 39, 46, 47) are Ca-Mg-HCO_3 waters but have moderate Na, Cl, and SO_4 concentrations (Tables 3 and 5; Appendix 1: Fig. 7 [see footnote 1]). Strontium concentrations are slightly higher than barium concentrations in three Wasatch Formation springs (39, 46, 47; Table 6). Springs that discharge near the East Bear Lake Fault (48, 53) contain very high concentrations of strontium and very little barium.

One of the more unusual springs on the Bear Lake Plateau is Mud Lake Hot Spring (f, 53), with a high Li concentration (Table 6). The high Li concentration probably indicates a hydrothermal source (White, 1957). Applying Na/Li, Li (Fouillac and Michard, 1981), and Na-K-Ca (Fournier and Truesdell, 1973) chemical geothermometry calculations to the most complete hot spring data (August 2000) suggests relatively similar water-rock interaction temperatures of 110°, 112°, and 107 °C, respectively. On the basis of these temperatures a fourth, chalcedony-based, silica geothermometry calculation was performed (Fournier, 1981). The silica thermometry result indicates a water-rock reaction temperature of only 52 °C. The lower silica result may suggest dilution or mixing of the thermal waters with another cooler water prior to reaching the surface (Fournier, 1981). Mixing with dilute surface runoff should have a negligible effect on the Na/Li, Li, and Na-K-Ca geothermometers (e.g., Fournier, 1981), but the proximity of Mud Lake Hot Spring to the East Bear Lake Fault makes the mixing of thermal water with a groundwater chemistry similar to site 48, or possibly site 50, a distinct possibility.

Calculating what the unaltered thermal water composition may be by assuming that the Mud Lake Hot Spring chemistry is a mixture of 95% initial thermal water chemistry and 5% water chemistry from site 48 or 50 produces Na/Li, Li, and Na-K-Ca geothermometry values of ~110°, 114°, and 110 °C, respectively. These values are similar to the original values, and still internally consistent. Subtracting progressively larger proportions of site 48 water chemistry from the Mud Lake Hot Spring chemistry produces progressively higher Na/Li, Li, and Na-K-Ca water-rock interaction temperatures, but the results lose their internal

consistency and the silica values always remain lower than the other geothermometry values for all correction calculations. The loss of consistency among the Na/Li, Li, and Na-K-Ca results at higher mixture ratios suggests that the Mud Lake Hot Spring water is probably not composed of a high percentage of another water chemistry, East Bear Lake Fault-related or otherwise. The consistently lower silica-derived temperature is best explained by a loss of silica as the thermal waters rise toward the surface. Fournier (1981) states that solutions below ~100 °C can remain supersaturated with respect to silica for an unspecified number

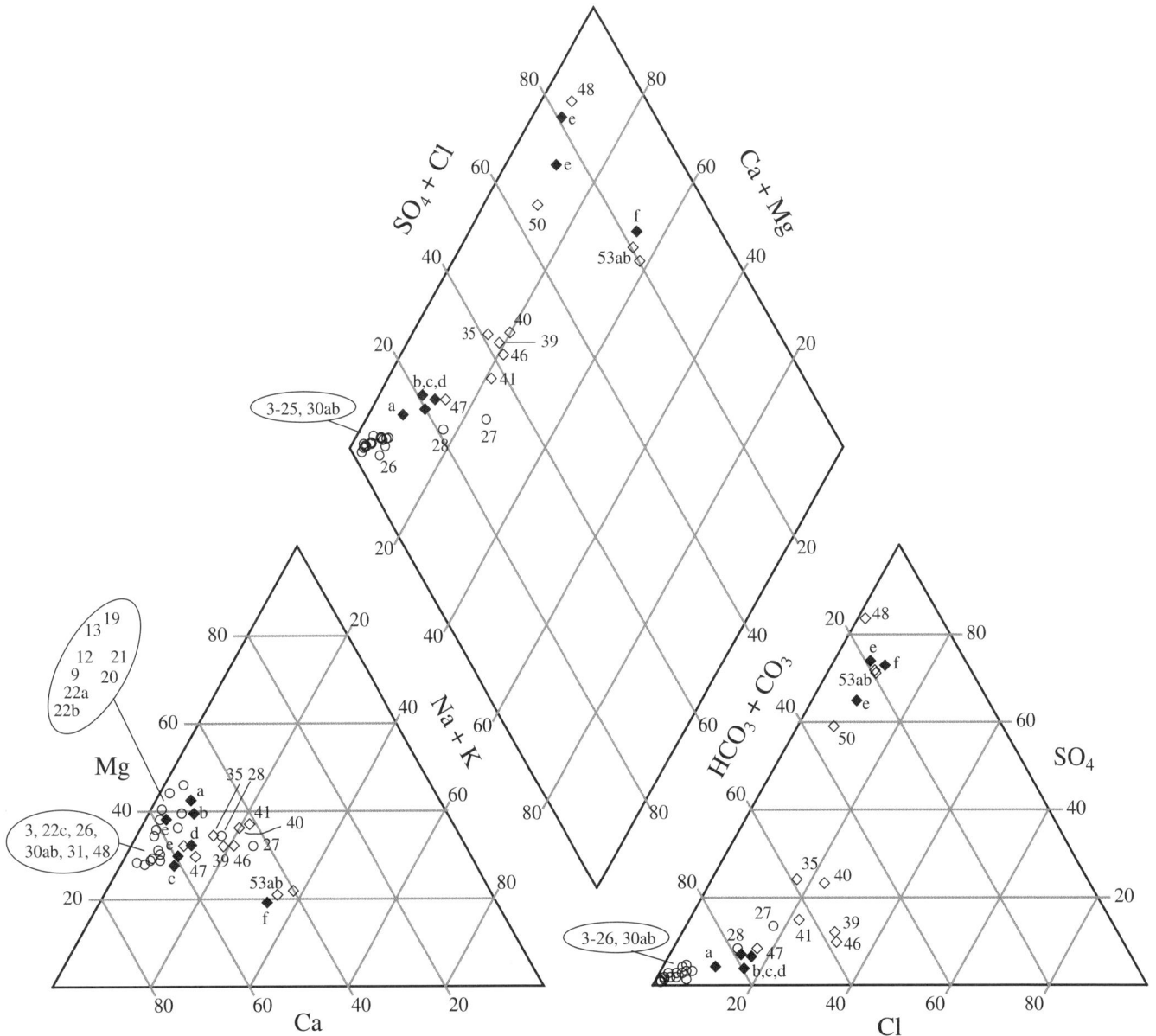

Figure 7. Piper trilinear plot of Bear Lake Valley water chemistry samples. Numbers and letters refer to sampling sites on Figure 2 and Table 3. Circles represent Bear River Range samples; open diamonds represent Bear Lake Plateau samples; solid diamonds with letters represent data from Kaliser (1972); a—lake-marginal spring; b–d—Wasatch Formation springs; e—Ca-SO$_4$-rich well water along East Bear Lake Fault; f—Mud Lake Hot Spring.

of years. The lack of detectable tritium in Mud Lake Hot Spring water (see below) suggests that the spring discharge is at least several decades old, and perhaps much older, and silica may have precipitated out of solution prior to reaching the surface.

Another unique spring on the Bear Lake Plateau is site 48, with high Ca and SO_4 concentrations (Table 3). This spring is situated near the East Bear Lake Fault. A well situated south of South Eden Creek and near the East Bear Lake Fault (site e in Fig. 2) was sampled by Kaliser (1972) and has a similar major-ion chemistry to spring 48, suggesting that Ca-SO_4-rich water extends southward along the eastern margin of the lake, possibly in association with the East Bear Lake Fault.

Of the three streams that discharge from the Bear Lake Plateau, North and South Eden Creeks (35, 40, 41) have relatively similar chemistry. The solutes in these streams are predominantly Ca-Mg-HCO_3-SO_4 with moderate Cl concentrations (Tables 3 and 5, Fig. 7). Strontium concentrations in both streams are higher in the vicinity of the East Bear Lake Fault (35, 40; Table 6, Fig. 2) than they are at the headwater springs east of the East Bear Lake Fault (39, 46, 47). The chemistry of Indian Creek (50) is distinct from that of North and South Eden Creeks by having high SO_4

and low Cl concentrations, and by having Ba and Sr concentrations more similar to Mud Lake Hot Spring (53) and the small sulfate-rich spring at site 48 than to either North or South Eden Creeks (Table 6, Fig. 7). These data indicate that an SO_4- and Sr-rich water discharges along the East Bear Lake Fault.

Strontium Isotopes

Strontium isotope ratios ($^{87}Sr/^{86}Sr$) are useful for determining water-rock interactions and serve as a groundwater tracer. Strontium isotopes do not fractionate between the solid and aqueous phase during weathering. Consequently, water-rock interactions result in a water with the same $^{87}Sr/^{86}Sr$ value as the rock (Bullen and Kendall, 1998).

Bear River Range

With the exception of one sample, the Paleozoic carbonate bedrock units of the Bear River Range have $^{87}Sr/^{86}Sr$ values ranging from 0.70811 to 0.71038, and average ~0.70928 ($n = 12$; Table 7). Most Bear River Range bedrock samples have $^{87}Sr/^{86}Sr$ values that are consistent with established Paleozoic seawater

TABLE 5. IONIC COMPOSITION AND TRITIUM CONCENTRATION FOR SPRINGS AND STREAMS IN BEAR LAKE VALLEY

Site (Fig. 2)	Site name	$(Ca+Mg+HCO_3)$* (%)	Na* (%)	Cl* (%)	SO_4* (%)	Tritium[†] (TU)
Bear River Range						
3	Paris Spring	97.1	1.6	0.7	0.4	N.D.
9	Bloomington Creek	96.3	2.0	1.1	0.5	N.D.
12	Blue Pond Spring	97.0	1.4	1.0	0.4	N.D.
13	St. Charles Creek	97.8	1.1	0.6	0.3	N.D.
14	St. Charles Creek	97.2	1.4	0.8	0.5	N.D.
19	Fish Haven Creek	95.5	2.3	1.1	0.9	N.D.
20	spring	91.1	4.5	2.8	1.2	N.D.
21	spring	90.8	4.2	3.1	1.4	N.D.
22	Swan Creek Spring (1999)	94.5	2.4	2.0	0.9	N.D.
22	Swan Creek Spring (2000)	95.2	2.1	1.9	0.7	N.D.
22	Swan Creek Spring (2004)	92.1	3.3	3.9	0.5	11.5 ± 0.43
26	spring	91.9	4.9	2.4	0.6	N.D.
27	spring	68.4	14.9	11.7	4.3	N.D.
28	spring	77.5	10.9	8.4	2.6	N.D.
29	beach seep	32.7	25.8	22.1	18.9	N.D.
30	Big Spring (2000)	92.2	3.5	3.2	0.8	N.D.
30	Big Spring (2004)	91.7	3.7	3.2	1.0	5.0 ± 0.32
31	Big Creek–Laketown Creek	90.0	4.4	4.1	1.1	N.D.
Bear Lake Plateau						
34	beach seep	21.2	37.7	22.3	17.5	N.D.
35	South Eden Creek	69.5	10.4	11.5	8.1	N.D.
39	Rabbit Spring	63.9	12.6	19.4	3.8	N.D.
40	North Eden Creek (1999)	63.6	14.1	14.5	7.2	N.D.
41	North Eden Creek (2004)	66.7	13.4	14.4	4.8	N.D.
46	spring (Wasatch Fm.)	62.6	13.6	20.4	3.1	N.D.
47	spring (Wasatch Fm.)	76.9	8.9	11.1	2.6	2.6 ± 0.30
48	spring	54.5	5.1	1.4	38.7	<0.6
49	beach seep	53.8	6.6	2.2	37.3	N.D.
50	Indian Creek	62.8	8.3	5.7	22.7	N.D.
53	Mud Lake Hot Spring (1999)	37.4	24.1	7.4	27.5	N.D.
53	Mud Lake Hot Spring (2004)	38.9	22.6	6.9	27.8	<0.6

Note: TU—tritium units; N.D.—no data.
*"%X" calculated from Table 4, where units of measurement are in millimoles.
[†]Tritium samples collected in September 2003.

[87]Sr/[86]Sr reconstructions, although five samples yielded [87]Sr/[86]Sr values that were slightly higher than the highest Paleozoic seawater values (Burke et al., 1982; Denison et al., 1998; Veizer et al., 1999). These higher [87]Sr/[86]Sr values are likely due to post-depositional alteration (Burke et al., 1982; Clauer et al., 1989; Denison et al., 1994). In comparison, the [87]Sr/[86]Sr values from the Bear River Range shale and quartzite bedrock samples were more radiogenic, averaging 0.71543 ($n = 9$; Table 7), which is reflective of their continental sources (Palmer and Edmond, 1992).

Water samples from the Bear River Range typically have low Sr concentrations (Fig. 8D) and [87]Sr/[86]Sr ≥ 0.71000 (range from 0.71005 to 0.71322; Table 6, Fig. 8E). Although carbonate dissolution is nearly the sole source of solutes for the Bear River Range springs and streams, spring and stream [87]Sr/[86]Sr values are higher than for the local carbonate bedrock. This suggests hydrologic interaction with the shale and quartzite units within the Bear River Range. Leaching of more radiogenic strontium from these units as they force groundwater to the surface leads to spring discharge with [87]Sr/[86]Sr values higher than for the carbonate bedrock average. The [87]Sr/[86]Sr value of the Wasatch Formation (see below) is higher than the local Paleozoic carbonates and may be responsible for the general southward increase in the Sr concentrations and the higher [87]Sr/[86]Sr values of the small, lake-marginal springs located along the southwestern margin of the lake (20, 21, 27, 28; Table 6; Figs. 8D and 8E).

Bear Lake Plateau

One sample of the Wasatch Formation (reddish sandstone) from the South Eden Creek drainage has a [87]Sr/[86]Sr value of 0.71367 (Table 7). Several samples from the Twin Creek Limestone from both North and South Eden Creek drainages produced relatively similar [87]Sr/[86]Sr values from 0.70712 to 0.70790 (Table 7), which match well with reconstructed Jurassic seawater [87]Sr/[86]Sr values (Burke et al., 1982; Denison et al., 1998; Veizer et al., 1999). Two samples of the Nugget Sandstone from the North Eden Creek drainage yielded slightly higher [87]Sr/[86]Sr values of 0.70986 and 0.71066 (Table 7).

Strontium isotope ratios in aqueous samples from the Bear Lake Plateau fall into two distinct groups separated by a [87]Sr/[86]Sr value of 0.71000. Spatially, the 0.71000 boundary appears to coincide with the East Bear Lake Fault. All water samples east of the fault have [87]Sr/[86]Sr values >0.71000, and water samples near, or west of the fault have [87]Sr/[86]Sr values <0.71000. For example, the [87]Sr/[86]Sr values in water from North Eden Creek is uniformly ~0.71000 from its headwaters (45, 46, 47) to near its mouth, then deceases to <0.71000 at sample sites near the East Bear Lake Fault (41, 40; Table 6). In addition, the water at the mouth of North Eden Creek (40) contains nearly three times more Sr than the springs feeding the stream (46, 47; Table 6). A similar trend in [87]Sr/[86]Sr values and Sr concentrations occurs in the South Eden Creek data (35, 39; Table 6).

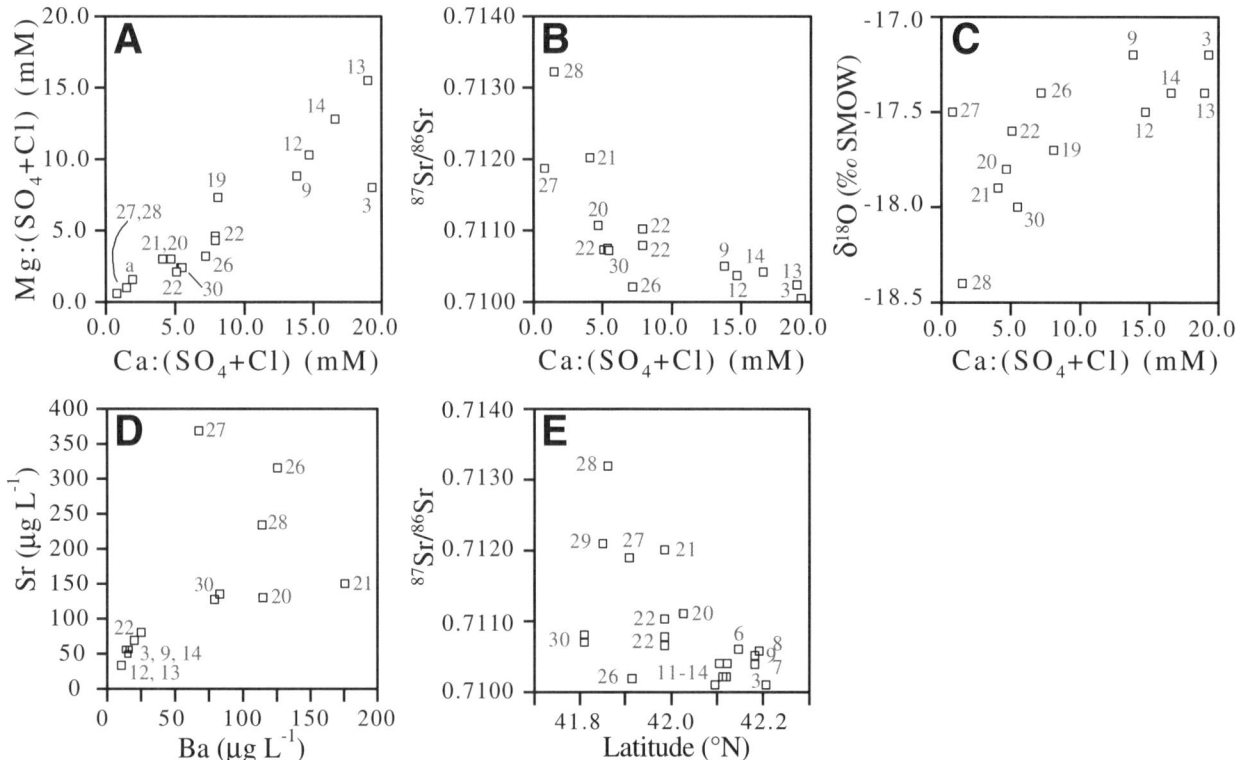

Figure 8. Solute and isotopic data for springs and streams along the eastern Bear River Range. (A) Mg:(SO$_4$+Cl) versus Ca:(SO$_4$+Cl). (B) [87]Sr/[86]Sr versus Ca:(SO$_4$+Cl). (C) δ^{18}O versus Ca:(SO$_4$+Cl). (D) Sr versus Ba. (E) [87]Sr/[86]Sr versus latitude. Numbers refer to sample sites on Figure 2. Data are presented in Tables 2, 3, and 6. SMOW—Standard Mean Ocean Water.

Streamflow at the mouth of Indian Creek (50) and the sulfate-rich spring at site 48 have low $^{87}Sr/^{86}Sr$ values of ~0.70770 and very high Sr concentrations (Table 6). A spring at the head of Indian Creek (52) has an equally low $^{87}Sr/^{86}Sr$ value (0.70783), but Sr concentration was not measured. These low $^{87}Sr/^{86}Sr$ values are similar to those in the Twin Creek Limestone that is prevalent in the area. In contrast, just to the north of Indian Creek, Mud Lake Hot Spring (53) has a $^{87}Sr/^{86}Sr$ value of ~0.70977 and a Sr concentration in excess of 4500 µg L^{-1} (Table 6). Falula Spring (32), located at the southeast corner of Bear Lake, has a $^{87}Sr/^{86}Sr$ value of 0.71072 (Table 6), comparable to one of the Nugget Sandstone bedrock analyses. This value supports the proposal by Kaliser (1972) that Falula Spring may be fed by the Nugget Sandstone aquifer.

Tritium

Tritium concentrations in the atmosphere peaked in 1963–1964, at the end of atomic bomb testing, and have since decreased

TABLE 6. LITHIUM (Li), BARIUM (Ba), AND STRONTIUM (Sr) CONCENTRATIONS AND $^{87}Sr/^{86}Sr$ VALUES FOR SPRINGS AND STREAMS IN BEAR LAKE VALLEY

Site (Fig. 2)	Lat. (°N)	Long. (°W)	Elev. (m)	Site name	Date (m/d/yr)	Li (µg L^{-1})	Ba (µg L^{-1})	Sr (µg L^{-1})	$^{87}Sr/^{86}Sr$
Bear River Range									
3	42.206	111.498	2001	Paris Spring	8/1/00	4	14	56	0.71005
6	42.146	111.575	2501	Bloomington Lake	8/23/04	2	5	13	0.71056
7	42.182	111.544	2202	Bloomington Creek	8/23/04	N.D.	N.D.	N.D.	0.71037
8	42.188	111.447	1879	Bloomington Creek	4/5/04	N.D.	N.D.	N.D.	0.71056
9	42.183	111.401	1815	Bloomington Creek	9/22/99	N.D.	16	57	0.71046
9	42.183	111.401	1815	Bloomington Creek	8/23/04	N.D.	N.D.	N.D.	0.71050
11	42.096	111.530	2092	St. Charles Creek	8/23/04	N.D.	N.D.	N.D.	0.71009
12	42.105	111.495	1975	Blue Pond Spring	4/5/04	0	11	33	0.71037
13	42.113	111.446	1940	St. Charles Creek	8/23/04	N.D.	N.D.	N.D.	0.71024
14	42.124	111.391	1817	St. Charles Creek	1996	N.D.	N.D.	N.D.	0.71042
14	42.124	111.391	1817	St. Charles Creek	9/22/99	N.D.	15	50	0.71036
14	42.124	111.391	1817	St. Charles Creek	8/1/00	0	11	33	0.71015
19	42.037	111.410	1815	Fish Haven Creek	9/22/99	N.D.	16	59	N.D.
20	42.026	111.402	1806	spring	4/5/04	6	115	129	0.71107
21	41.985	111.406	1806	spring	4/5/04	6	176	150	0.71202
22	41.985	111.427	1891	Swan Creek Spring	1996	N.D.	N.D.	N.D.	0.71104
22	41.985	111.427	1891	Swan Creek Spring	9/22/99	N.D.	25	81	0.71102
22	41.985	111.427	1891	Swan Creek Spring	8/1/00	0	20	68	0.71079
22	41.985	111.427	1891	Swan Creek Spring	4/5/04	2	25	80	0.71073
26	41.915	111.389	1806	spring	8/1/00	10	126	315	0.71021
27	41.909	111.372	1806	spring	4/5/04	30	68	369	0.71187
28	41.865	111.360	1801	spring	4/5/04	15	114	235	0.71322
29	41.851	111.356	1805	beach seep	8/23/04	138	72	1060	0.71210
30	41.809	111.389	1824	Big Spring	8/1/00	6	79	128	0.71072
30	41.809	111.389	1824	Big Spring	4/5/04	4	83	134	0.71075
31	41.846	111.337	1806	Big Creek	9/22/99	N.D.	88	149	0.71106
31	41.846	111.337	1806	Big Creek	4/5/04	N.D.	N.D.	N.D.	0.71104
Bear Lake Plateau									
32	41.842	111.302	1812	Falula Spring	5/28/03	N.D.	N.D.	N.D.	0.71072
34	41.878	111.294	1803	beach seep	5/28/03	130	90	3400	0.70950
35	N.D.	N.D	N.D.	South Eden Creek	9/22/99	N.D.	117	861	0.70880
37	41.918	111.228	1937	spring (Twin Creek Ls)	4/5/04	N.D.	N.D.	N.D.	0.71114
38	41.921	111.189	2001	spring (Wasatch Fm.)	4/5/04	N.D.	N.D.	N.D.	0.71132
39	42.910	111.139	2054	Rabbit Spring	8/23/04	18	268	318	0.71060
40	N.D.	N.D.	N.D.	North Eden Creek	9/22/99	N.D.	156	645	0.70901
41	41.986	111.255	1842	North Eden Creek	8/23/04	22	119	369	0.70974
42	41.983	111.233	1873	North Eden Creek	5/28/03	N.D.	N.D.	N.D.	0.71013
43	41.984	111.212	1885	North Eden Creek	5/28/03	N.D.	N.D.	N.D.	0.71044
44	41.988	111.189	1896	North Eden Creek	5/28/03	N.D.	N.D.	N.D.	0.71044
45	41.993	111.148	2001	spring (Wasatch Fm.)	5/28/03	N.D.	N.D.	N.D.	0.71074
46	42.977	111.140	2001	spring (Wasatch Fm.)	4/5/04	16	239	292	0.71057
47	42.991	111.119	1983	spring (Wasatch Fm.)	4/5/04	16	186	250	0.71057
48	42.075	111.250	1835	spring	8/1/00	19	12	5933	0.70767
49	N.D.	N.D.	1804	beach seep	4/5/04	82	18	1850	0.70844
50	42.094	111.256	1830	Indian Creek	8/23/04	16	41	1830	0.70785
52	42.096	111.232	1964	spring	5/28/03	N.D.	N.D.	N.D.	0.70783
53	42.115	111.264	1824	Mud Lake Hot Spring	9/22/99	N.D.	25	4652	0.70976
53	42.115	111.264	1824	Mud Lake Hot Spring	8/1/00	265	28	4930	0.70978

Note: Lat.—latitude; Long.—longitude; N.D.—no data.

to pre-bomb era levels (Clark and Fritz, 1997). Clark and Fritz (1997) defined continental tritium values as follows: tritium values <0.8 tritium units (TU) are considered pre-bomb recharge, values of 5–15 TU are considered modern (<5–10 yr) recharge, and values >30 TU are considered to be recharge from the 1960s and 1970s. No long-term tritium data are available for the immediate study area, although the averaged annual tritium concentration in precipitation falling in Albuquerque, New Mexico, during 1990–2002 ranged from ~5–20 TU (IAEA, 2004) with the majority of the concentrations (82%) ranging between 6 and 13 TU. Rice and Spangler (1999) reported a single value of ~9 TU from winter precipitation collected in 1986 in Mantua Valley, ~100 km southwest of Bear Lake Valley. Sixteen precipitation samples (rain and snow) collected between September 2002 and June 2005 in Utah County, ~200 km southwest of Bear Lake, ranged from 2.1 to 11.7 TU, and averaged 7.2 TU (A. Mayo, 2008, personal commun.). A reasonable estimate for tritium concentration in modern precipitation falling in Bear Lake Valley is ~6–13 TU.

Of the five samples analyzed for tritium from Bear Lake Valley, two were from the largest springs in the Bear River Range: Swan Creek Spring (22) and Big Spring (30). Swan Creek Spring yielded a modern value of 11.5 TU (Table 5). Big Spring, which discharges along a fault in the Wasatch Formation at the south-western end of the valley, had a lower value of 5.0 TU (Table 5). Three tritium samples were collected from the Bear Lake Plateau: one from a spring emanating from the Wasatch Formation at the head of North Eden Creek (47), and two from low-elevation springs near or along the East Bear Lake Fault (Mud Lake Hot Spring (53) and site 48). The spring at the head of North Eden Creek had a value of 2.6 TU (Table 5). Neither Mud Lake Hot Spring (53) nor the spring at site 48 contained detectable amounts of tritium (<0.6 TU; Table 5), suggesting that no modern recharge is present at these springs.

DISCUSSION

Stable Isotope Distribution in Springs and Streams of Bear Lake Valley

The $\delta^{18}O$ and δ^2H values of spring discharge, especially base flow, in the Bear River Range are likely homogenized values approaching the weighted average $\delta^{18}O$ and δ^2H values of winter precipitation over some interval of time (e.g., Winograd et al., 1998). Linking stable isotope values from spring discharge to recent spot-collections of winter precipitation is difficult because aquifers can store water for months to years (e.g., Rice and Spangler, 1999), resulting in stable isotope values in spring discharge

TABLE 7. STRONTIUM ISOTOPE RATIOS ($^{86}Sr/^{86}Sr$) OF BEDROCK UNITS IN THE BEAR RIVER RANGE AND BEAR LAKE PLATEAU

Unit	Location	Age	$^{87}Sr/^{86}Sr$	St. dev. (±)
Carbonate				
Twin Creek Limestone	South Eden Canyon, Bear Lake Plateau	Jurassic	0.70790	0.00001
Twin Creek Limestone	North Eden Canyon, Bear Lake Plateau	Jurassic	0.70712	0.00001
Twin Creek Limestone	North Eden Canyon, Bear Lake Plateau	Jurassic	0.70743	0.00001
Twin Creek Limestone	North Eden Canyon, Bear Lake Plateau	Jurassic	0.70720	0.00001
Lodgepole Limestone	Logan Canyon, Bear River Range	Mississippian	0.70811	0.00001
Hyrum Dolomite	Logan Canyon, Bear River Range	Devonian	0.70896	0.00001
Laketown Dolomite	Logan Canyon (float), Bear River Range	Silurian	0.70879	0.00001
Laketown Dolomite	Logan Canyon (float), Bear River Range	Silurian	0.70854	0.00001
Fish Haven Dolomite	Logan Canyon (float), Bear River Range	Ordovician	0.71011	0.00001
Garden City Limestone	Logan Canyon, Bear River Range	Ordovician	0.70965	0.00001
Garden City Limestone	Logan Canyon, Bear River Range	Ordovician	0.70901	0.00001
Garden City Limestone	Logan Canyon, Bear River Range	Ordovician	0.70918	0.00001
Bloomington Formation (oolitic limestone)	Logan Canyon, Bear River Range	Cambrian	0.71038	0.00001
Blacksmith Dolomite	Logan Canyon, Bear River Range	Cambrian	0.70975	0.00002
limestone facies of Langston Dolomite?	Bloomington Canyon, Bear River Range	Cambrian	0.70978	0.00001
Blacksmith Dolomite	Bloomington Canyon, Bear River Range	Cambrian	0.70906	0.00001
	average, Logan Canyon, Bear River Range (*n* = 12)		**0.70928**	**0.00067**
Non-carbonate				
Wasatch Formation	South Eden Canyon, Bear Lake Plateau	Tertiary (Eocene)	0.71367	0.00001
Nugget Sandstone	North Eden Canyon, Bear Lake Plateau	Triassic/Jurassic	0.70986	0.00001
Nugget Sandstone	North Eden Canyon, Bear Lake Plateau	Triassic/Jurassic	0.71066	0.00001
Swan Peak Quartzite	Bloomington Lake, Bear River Range	Ordovician	0.71262	0.00001
Swan Peak Quartzite	Logan Canyon, Bear River Range	Ordovician	0.71211	0.00001
Swan Peak Quartzite	St. Charles Canyon, Bear River Range	Ordovician	0.71153	0.00003
Swan Peak Quartzite, above shale	Logan Canyon, Bear River Range	Ordovician	0.71280	0.00001
Swan Peak Quartzite, interbedded with shale	Logan Canyon, Bear River Range	Ordovician	0.72206	0.00004
shale, base of Swan Peak Quartzite	Logan Canyon, Bear River Range	Ordovician	0.71544	0.00001
Bloomington Formation (shale)	Logan Canyon, Bear River Range	Cambrian	0.71743	0.00001
Geertzen Canyon Quartzite	Garden City, Bear River Range	Cambrian	0.71654	0.00003
Geertzen Canyon Quartzite	St. Charles Canyon, Bear River Range	Cambrian	0.71831	0.00013
	average, Logan Canyon, Bear River Range (*n* = 9)		**0.71543**	**0.00351**

Note: Unit names and ages from Dover (1995).

that are not representative of recent precipitation. The lack of continuous, long-term precipitation monitoring in the eastern Bear River Range hinders the discussion of the Bear River Range groundwater and stream stable isotope values, but two systematic trends are apparent. One is the progressive southward decrease in $\delta^{18}O$ and δ^2H from Bear River Range springs and streams given the geographically limited study area (Fig. 5A). And the other is the more negative stable isotope values in lower-elevation spring and stream samples within a particular watershed (Fig. 5B).

Maximum elevations in the eastern Bear River Range decrease southward by ~170 m, from Paris Peak (2918 m) at the head of Paris Canyon to Temple Peak (2751 m) located roughly 11 km due west of Big Spring. Intuitively, higher elevations should accumulate more isotopically negative precipitation (primarily snow) due to altitude-dependent fractionation effects (e.g., Gat, 1980; Rózanski et al., 1993; Poage and Chamberlain, 2001). Therefore, the higher elevations in the northern portion of the study area and within any individual watershed should gener-

ate more isotopically negative runoff than the lower elevations or more southerly locations.

The most likely explanations for the spatial distribution of groundwater $\delta^{18}O$ and δ^2H values in the eastern Bear River Range are (1) the location of the study area, which is on the leeward side of the Bear River Range; and (2) the topography of the Bear River Range on a regional, rather than local, scale. The negative correlation between stable isotopes in precipitation and altitude occurs as air masses rise, cool, condense, and rain out while passing over a topographic barrier—the Bear River Range in particular. Once an air mass impacts the western, windward slope of the Bear River Range it must traverse an additional 30–35 km before reaching the study area (Fig. 9). Additional rainout while crossing the remaining topography of the Bear River Range would produce increasingly negative precipitation at increasing distance from the windward range front (e.g., Moran et al., 2007, and references therein). The topography of the Bear River Range is such that the windward range front due west of the southern study area

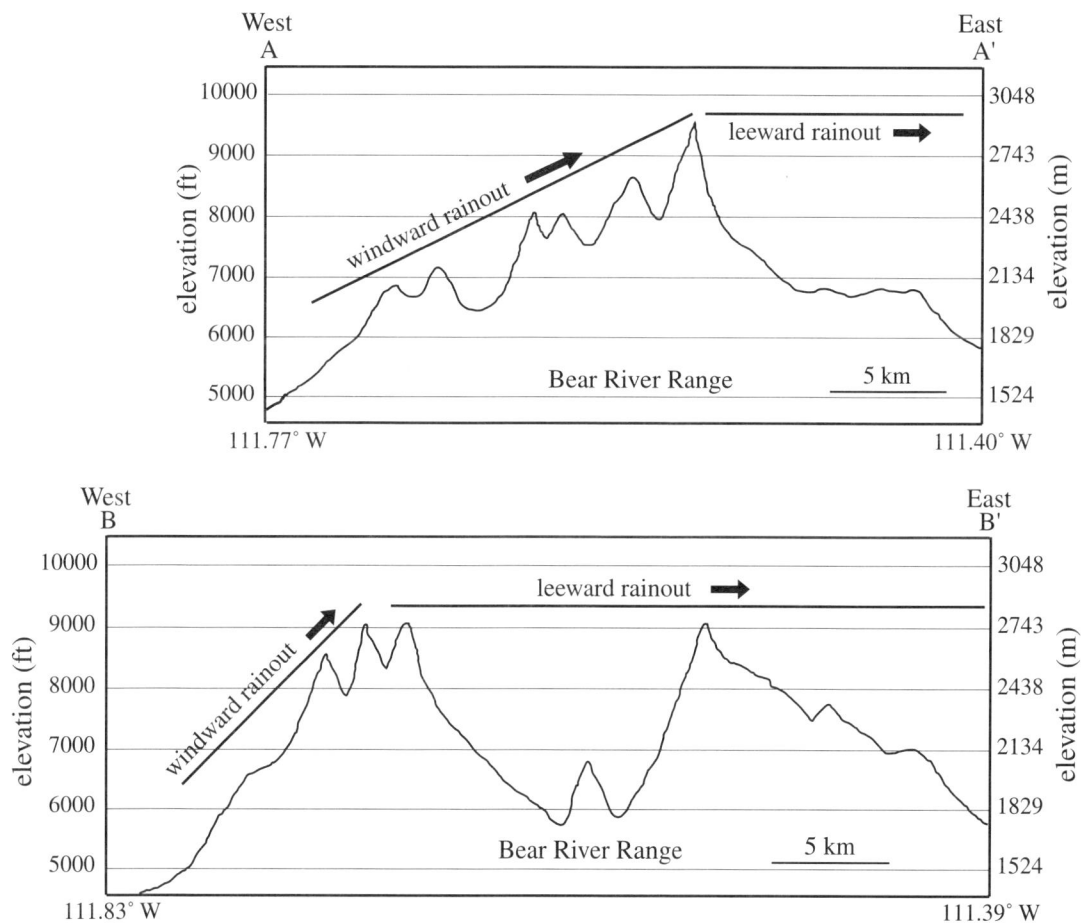

Figure 9. Elevational cross sections A–A′ and B–B′ through the Bear River Range. See Figures 1 and 2 for locations of sections. Vertical exaggeration 8×. Moisture moving west to east along cross section B–B′ reaches maximum elevation farther west and experiences greater rainout before reaching Bear Lake than does precipitation passing over cross section A–A′. As a result, moisture condensing in the southern study area has experienced greater isotopic distillation (is more negative) than moisture condensing farther to the north. See text for discussion.

is very steep and maximum elevations are reached ~25 km west of the study area (Figs. 2 and 9; cross section B–B'). Farther to the north the elevational gradient of the windward range front is much more gentle and maximum elevations are not reached until ~12 km west of the study area (Figs. 2 and 9; cross section A–A'). Consequently, precipitation falling in the southern portion of the study area will have been subjected to greater leeward rainout and will be more isotopically negative (more depleted) than precipitation falling in the northern portion of the study area.

The decrease in $\delta^{18}O$ and δ^2H observed in springs and streams within an individual drainage may be explained, at least partially, by the same leeward rainout process. There is ~11 km of lateral distance between the maximum and minimum elevations in the northern portion of the study area (Fig. 9; cross section A–A'). If leeward rainout does occur, then the stable isotope values in precipitation would progressively decrease west of the topographic high. High-elevation springs and stream catchments in any given drainage would be recharged by slightly isotopically heavier precipitation than the more distant, lower-elevation springs and stream catchments. This relationship is supported by the limited data on hand, with the exception of Paris Creek drainage. A more comprehensive study of the stable isotope variability in local precipitation and spring discharge in the eastern Bear River Range is needed to further test this hypothesis, however.

The tritium data suggest that spring discharge from the Wasatch Formation on the Bear Lake Plateau is predominantly pre-bomb era water mixed with a small amount of modern recharge (Site 47; 2.6 TU). The age of the groundwater discharging at springs 48 and 53 along the East Bear Lake Fault is not known but the lack of detectable tritium indicates that no modern precipitation is present. Currently, there are no radiometric or other chronologic data to refine the "pre-bomb era" age for these springs. Consequently, a portion of the spring discharge on the Bear Lake Plateau is recharge that could be significantly older and no longer representative of modern climate dynamics.

The $\delta^{18}O$ and δ^2H data from the Bear Lake Plateau springs fall below the GMWL on an evaporation line with a slope of 6.2 that crosses the GMWL at approximately $\delta^{18}O = -20‰$ and $\delta^2H = -150‰$ (Fig. 4B). These values indicate that the original isotopic composition of Bear Lake Plateau precipitation is significantly more depleted than the average precipitation collected at the Bear Lake County Airport (Fig. 2; Friedman et al., 2002). Precipitation that falls on the Bear Lake Plateau is sourced by storms passing over the Bear River Range, and the lower isotopic values are likely the result of continued rainout as storms pass over the range (e.g., Mayo and Loucks, 1995; Moran et al., 2007). Once storms pass over the Bear River Range the distance between cloud-base and the ground increases and the humidity is likely lower. Precipitation falling on the Bear Lake Plateau probably experiences evaporation (or sublimation) during air-fall before reaching the ground. Additional evaporation of snowmelt could occur during the spring if infiltration rates on the Bear Lake Plateau are slow. On the basis of the data presented here, it is reasonable to conclude that the stable isotope values on the Bear

Lake Plateau (excluding sites 48 and 53) result from local orographic and evaporative effects, although long-term Bear Lake Plateau precipitation isotope data and additional groundwater-age determinations are needed to test this hypothesis.

Solute Behavior of Swan Creek Spring

Swan Creek Spring (22) is one of the largest springs in Utah (Mundorff, 1971). The impressive discharge of Swan Creek Spring is indicative of a large and well-developed karst conduit system within the Bear River Range. The sensitivity of Swan Creek Spring to rainfall events (Kaliser, 1972) indicates a strong linkage to the surface. Infiltration into the Swan Creek Spring aquifer is likely quick, and given spring's cold temperature, passes quickly through the mountain range along shallow, and possibly short, flow paths. The spring is fed by a large solution channel (Kaliser, 1972) and is located along one of a series of north-south–trending faults in the Bear River Range, west of Bear Lake (Dover, 1995). The location of this spring along a fault provides an opportunity to study the effect of faulting on spring chemistry. The EPA has monitored Swan Creek Spring for nearly 30 years (epa.gov/storet/dw_home.html; station 4907200) and has generated a large chemical data set. Many of the EPA analyses do not include K, SO_4 or Cl, so several charge balance errors are greater than 5%. The majority of the more complete analyses have charge balance errors less than 5%, however. The most complete EPA analyses for Swan Creek Spring are presented in Table 8.

The molar ratio of calcium (Ca) to magnesium (Mg) in ideal dolomite is 1.0, with calcian dolomites having Ca:Mg values slightly above 1.0 (Goldsmith and Graf, 1958; Sperber et al., 1984). All but one of the dolomites (characterized by slight to very slight effervescence) in the Bear Lake drainage have Ca:Mg values slightly above 1.0 (Table 9). Aqueous dissolution of the local limestone and the average local dolomite follows the equations:

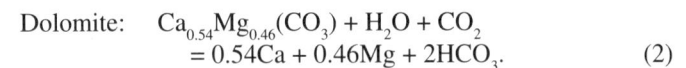

Calcite: $CaCO_3 + H_2O + CO_2 = Ca + 2HCO_3,$ (1)

Dolomite: $Ca_{0.54}Mg_{0.46}(CO_3) + H_2O + CO_2$
             $= 0.54Ca + 0.46Mg + 2HCO_3.$ (2)

Water dissolving equal amounts of the local limestone and dolomite would acquire 1 mol of Ca from limestone and ~0.54 mol of Ca and 0.46 mol of Mg from dolomite, resulting in water with a Ca:Mg value of 3.35 (e.g., Szramek et al., 2007). The Ca:Mg value decreases as the mixture becomes more enriched in dolomite.

Dolomite dissolution kinetics are not well understood (Morse and Arvidson, 2002) but two variables likely explain most of the Swan Creek Spring solute behavior. First, dolomite is more soluble than calcite at temperatures below 15 °C (Langmuir, 1997), and second, dissolution of dolomite does not appear to be congruent, especially early in the dissolution process. The $CaCO_3$ phase of dolomite is apparently more soluble than the $MgCO_3$ phase, with the dissolution of $MgCO_3$ being a slower, rate-limiting step (Busenberg and Plummer, 1982; Morse and Arvidson, 2002). The higher Mg concentrations in Swan Creek Spring base flow (Fig. 10) suggests that this groundwater has been in contact with the dolomitic bedrock for an extended period of time, likely on the order of several years (e.g., Herman and White, 1985). The relatively constant Ca concentrations and decreased Mg concentrations at higher discharge (Fig. 10) likely reflect cold snowmelt passing through the karst conduit network within the Bear River Range (e.g., White, 2002; Ozyurt and Bayari, 2007), and

increased dissolution of limestone and the $CaCO_3$ phase of dolomite during the peak snowmelt months.

Sodium (Na) and chloride (Cl) concentrations in Swan Creek Spring exhibit unexpected behavior in that the highest Na and Cl concentrations occur in a relatively narrow discharge window of ~1–2.5 m^3s^{-1} (Fig. 11). Chloride concentrations at Swan Creek Spring generally track the snow-water-equivalent data in the Bear River Range for the same years (Fig. 12). Assuming that analytical errors are not the cause, then there are three likely causes for the discharge-dependent increase in Na and Cl at Swan Spring. These potential sources include an atmospheric

TABLE 8. DISCHARGE AND MAJOR-ION CHEMISTRY DATA FOR SWAN CREEK SPRING, UTAH

Collection (mo-yr)	Discharge ($m^3 s^{-1}$)	Ca (mg L^{-1})	Mg (mg L^{-1})	Na (mg L^{-1})	K (mg L^{-1})	HCO$_3$ (mg L^{-1})	SO$_4$ (mg L^{-1})	Cl (mg L^{-1})	balance* (% error)	Ca (mM)	Mg (mM)	HCO$_3$ (mM)	Na (mM)	Cl (mM)
2-75	N.D.	48	16	2.0	1	228	7.0	4.0	-2.0	1.20	0.66	3.74	0.09	0.11
5-75	N.D.	51	10	1.0	1	202	5.0	3.0	-0.9	1.27	0.41	3.31	0.04	0.08
6-75	N.D.	42	9	1.0	1	156	4.0	3.0	3.4	1.05	0.37	2.56	0.04	0.08
10-75	N.D.	46	17	2.0	1	222	8.0	2.0	-0.7	1.15	0.70	3.64	0.09	0.06
11-76	N.D.	45	16	3.0	1	212	8.0	3.0	0.1	1.12	0.66	3.47	0.13	0.08
1-77	N.D.	45	18	2.0	1	226	9.0	2.0	-1.3	1.12	0.74	3.70	0.09	0.06
3-77	N.D.	48	17	2.0	2	228	16.0	3.0	-2.6	1.20	0.70	3.74	0.09	0.08
5-77	N.D.	49	19	3.0	1	236	10.0	4.0	-0.2	1.22	0.78	3.87	0.13	0.11
7-77	N.D.	51	19	4.0	1	244	7.0	4.0	0.5	1.27	0.78	4.00	0.17	0.11
11-77	N.D.	46	17	4.0	N.D.	218	8.0	3.0	0.7	1.15	0.70	3.57	0.17	0.08
3-78	N.D.	42	19	5.0	1	222	16.0	3.0	-1.8	1.05	0.78	3.64	0.22	0.08
7-78	N.D.	40	14	2.0	N.D.	184	10.0	2.0	-0.8	1.00	0.58	3.02	0.09	0.06
8-78	N.D.	50	14	3.0	N.D.	222	5.0	3.0	-0.5	1.25	0.58	3.64	0.13	0.08
3-79	N.D.	46	18	3.0	1	216	8.0	3.0	1.9	1.15	0.74	3.54	0.13	0.08
8-79	0.42	50	16	3.0	N.D.	208	10.0	4.0	2.9	1.25	0.66	3.41	0.13	0.11
11-79	N.D.	42	18	4.0	1	N.D.	13.0	5.0	-1.3	1.05	0.74	N.D.	0.17	0.14
2-80	N.D.	46	20	3.0	N.D.	216	9.0	5.0	2.6	1.15	0.82	3.54	0.13	0.14
5-80	N.D.	50	7	1.0	2	169	4.0	3.0	3.9	1.25	0.29	2.77	0.04	0.08
8-80	N.D.	46	14	2.0	1	200	10.0	4.0	-0.4	1.15	0.58	3.28	0.09	0.11
10-80	N.D.	42	17	3.0	N.D.	202	10.0	4.0	0.0	1.05	0.70	3.31	0.13	0.11
12-80	N.D.	32	18	3.0	N.D.	172	11.0	3.0	1.3	0.80	0.74	2.82	0.13	0.08
2-81	N.D.	49	17	3.0	N.D.	224	11.0	2.0	0.3	1.22	0.70	3.67	0.13	0.06
12-81	0.37	32	20	4.0	N.D.	182	15.0	3.0	0.7	0.80	0.82	2.98	0.17	0.08
7-82	N.D.	38	15	2.0	N.D.	182	10.0	3.0	-0.8	0.95	0.62	2.98	0.09	0.08
9-82	N.D.	45	15	3.0	N.D.	218	13.0	3.0	-4.1	1.12	0.62	3.57	0.13	0.08
10-82	N.D.	48	17	3.0	N.D.	216	13.0	4.0	0.1	1.20	0.70	3.54	0.13	0.11
12-82	N.D.	47	17	7.0	N.D.	214	10.0	5.0	2.4	1.17	0.70	3.51	0.30	0.14
2-83	N.D.	46	17	2.0	N.D.	212	13.0	4.0	-0.8	1.15	0.70	3.47	0.09	0.11
6-83	N.D.	47	8	3.0	N.D.	182	5.0	3.0	-0.3	1.17	0.33	2.98	0.13	0.08
8-98	N.D.	43.8	16.7	3.0	N.D.	226	N.D.	3.5	-1.5	1.09	0.69	3.70	0.13	0.10
9-98	0.75	49.8	17.4	2.8	N.D.	222	N.D.	4.0	3.7	1.24	0.72	3.64	0.12	0.11
10-98	1.31	48.3	18	3.3	N.D.	222	N.D.	4.0	3.6	1.21	0.74	3.64	0.14	0.11
12-98	1.04	49	18	4.1	N.D.	222	N.D.	6.0	3.8	1.22	0.74	3.64	0.18	0.17
1-99	0.82	48	19.3	3.2	N.D.	232	15.2	5.0	-1.5	1.20	0.79	3.80	0.14	0.14
2-99	0.79	49	19.3	4.1	N.D.	228	N.D.	6.5	3.7	1.22	0.79	3.74	0.18	0.18
3-99	1.16	51.1	19	8.6	N.D.	234	N.D.	13.5	3.0	1.28	0.78	3.84	0.37	0.38
4-99	2.06	57	16.7	9.1	N.D.	226	N.D.	15.5	5.4	1.42	0.69	3.70	0.40	0.44
5-99	2.41	58.1	13.7	5.5	N.D.	212	N.D.	10.0	6.5	1.45	0.56	3.47	0.24	0.28
5-99	7.08	55.2	8.4	2.7	N.D.	198	N.D.	4.5	2.6	1.38	0.34	3.25	0.12	0.13
6-99	9.08	48.9	8.2	2.7	N.D.	181	N.D.	4.5	2.1	1.22	0.34	2.97	0.12	0.13
6-99	3.11	46.7	9.8	2.4	N.D.	176	N.D.	N.D.	5.7	1.17	0.40	2.88	0.10	N.D.
7-03	0.42	46.8	16.3	3.4	N.D.	220	N.D.	N.D.	3.0	1.17	0.67	3.61	0.15	N.D.
8-03	0.10	50.5	17.6	3.9	N.D.	228	N.D.	N.D.	5.1	1.26	0.72	3.74	0.17	N.D.
9-03	0.48	50.2	18	3.6	N.D.	241	N.D.	N.D.	2.2	1.25	0.74	3.95	0.16	N.D.
10-03	0.24	47	17.6	3.2	N.D.	236	N.D.	N.D.	0.9	1.17	0.72	3.87	0.14	N.D.
12-03	0.79	45.6	18.8	3.4	N.D.	224	N.D.	N.D.	4.1	1.14	0.77	3.67	0.15	N.D.
1-04	0.63	43.1	10.5	3.2	N.D.	214	N.D.	N.D.	-5.4	1.08	0.43	3.51	0.14	N.D.
2-04	0.30	47.4	18.8	3.4	N.D.	218	N.D.	N.D.	6.5	1.18	0.77	3.57	0.15	N.D.
3-04	1.03	50.2	17.6	7.4	N.D.	216	23.5	12.6	-1.4	1.25	0.72	3.54	0.32	0.36
4-04	2.09	55.2	13.5	4.7	N.D.	208	N.D.	N.D.	8.7	1.38	0.56	3.41	0.20	N.D.
4-04	1.59	46.2	13.6	4.0	N.D.	204	N.D.	N.D.	3.7	1.15	0.56	3.34	0.17	N.D.
5-04	3.70	49.8	10.5	3.0	N.D.	188	N.D.	N.D.	6.1	1.24	0.43	3.08	0.13	N.D.
5-04	1.16	47.1	11.4	3.3	N.D.	181	N.D.	N.D.	7.2	1.18	0.47	2.97	0.14	N.D.
6-04	1.97	48.3	11	3.2	N.D.	164	N.D.	N.D.	12.5	1.21	0.45	2.69	0.14	N.D.
6-04	0.78	50.9	13.6	3.7	N.D.	185	N.D.	N.D.	11.5	1.27	0.56	3.03	0.16	N.D.

Note: Data from Environmental Protection Agency (www.epa.gov/storet/dw_home.html; station 4907200); N.D.—no data.
*Charge balance percent error calculated as [(cation sum − anion sum)/(cation sum + anion sum)] x 100, where units of measurement are in milliequivalents.

influx (e.g., dust from the Great Salt Lake basin), an influx of road salt from Highway 89, or another aquifer-solute source. Atmospheric Cl concentrations in precipitation at Logan, Utah (Fig. 1) tend to track the snow-water-equivalent data for the Bear River Range (Fig. 13). The Cl concentration in Logan precipitation is roughly an order of magnitude lower than the Cl concentrations in Swan Creek Spring, however. Some of the Cl at Swan Creek Spring is undoubtedly derived from atmospheric sources, but the atmospheric influx is not large enough to cause the Cl fluctuations in the spring discharge. Road maintenance along Highway 89 is another likely source of Na and Cl. Highway 89 is maintained throughout the year and salted during the winter. Salt-laden snowmelt could easily infiltrate into the Swan Creek Spring groundwater basin through the solution caverns that are in close proximity to Highway 89 (Fig. 2). Finally, solutes derived from the southern end of the valley may also be responsible. The Wasatch Formation is prevalent at the southern end of the valley, and spring water emanating from it has high Na and Cl concentrations. Wasatch Formation groundwater from the southern valley may be able to move northward along the faults that bound the western margin of the Bear River Range and discharge at Swan Creek Spring.

Differentiating between the road salt and Wasatch Formation groundwater hypotheses for the Na-Cl behavior at Swan Creek Spring should be possible using SO_4 data. Wasatch Formation groundwater is also high in SO_4 whereas typical road salt would only be a source of Cl. The road salt hypothesis would be supported if there were a no correlation between SO_4 and spring discharge, whereas the Wasatch Formation groundwater hypothesis would be favored if these variables did covary. Unfortunately this approach is not currently possible due to a lack of SO_4 data for Swan Creek Spring (Table 8).

Paris Spring and Blue Pond Spring

Groundwater discharge at Paris Spring (3) is probably controlled by the Lead Bell Shale (Wylie et al., 2005) and discharge at Blue Pond Spring (12) is likely fault controlled (Oriel and Platt, 1980). Time-series data for the solute composition and discharge rates of these springs are not available. Several observations are possible with the limited data that are available, however. The immediate bedrock lithology surrounding Paris Spring is limestone and dolomite (Oriel and Platt, 1980). Paris Spring's major-ion chemistry has been reported twice, once from a September collection (this study) and once from an August 2002 collection (Wylie et al., 2005). Both collections were taken in the late summer and should reflect base flow conditions. The Ca and Mg concentrations from both collections suggest that dolomite

TABLE 9. CALCIUM (Ca) AND MAGNESIUM (Mg) ASSAYS
FOR CARBONATE BEDROCK IN BEAR LAKE DRAINAGE

Formation	Lithology	Effervescence	CaO (wt%)	MgO (wt%)	Ca (mol)	Mg (mol)	Ca:Mg
Laketown Dolomite	dolomite	very slightly	22.68	15.95	0.40	0.40	1.00
Nounan Limestone	dolomite	very slightly	31.74	20.36	0.57	0.51	1.12
Nounan Limestone	dolomite	very slightly	33.71	21.52	0.60	0.53	1.13
Fish Haven Dolomite	dolomite	very slightly	31.51	19.77	0.56	0.49	1.14
Jefferson Dolomite	dolomite	very slightly	28.87	17.75	0.51	0.44	1.16
Wells Formation	dolomite	very slightly	31.31	19.37	0.56	0.48	1.17
Nounan Limestone*	dolomite	slightly	32.01	19.44	0.57	0.48	1.19
Langston Limestone[†]	limestone	very slightly	31.58	18.89	0.56	0.47	1.19
Bloomington Formation	limestone	very slightly	31.45	18.14	0.56	0.45	1.24
Fish Haven Dolomite/Laketown Dolomite	dolomite	very slightly	32.70	18.35	0.58	0.46	1.26
St. Charles Limestone[§]	dolomite	slightly	31.35	17.57	0.58	0.44	1.27
Brazer Limestone[#]	dolomite	very slightly	26.00	13.20	0.56	0.33	1.39
Bloomington Formation	limestone	strongly	43.26	9.07	0.46	0.23	3.35
Ute Limestone**/Langston Limestone[†]	dolomite	strongly	44.42	2.26	0.77	0.06	13.17
Bloomington Formation	limestone	strongly	47.84	2.43	0.85	0.06	14.17
Blacksmith Limestone[††]	dolomite	very strongly	49.41	2.25	0.88	0.06	14.67
Madison Limestone	limestone	strongly	59.62	1.19	1.06	0.03	35.33
Garden City Limestone[§§]	limestone	strongly	49.51	0.62	0.88	0.02	44.00
Garden City Limestone[§§]	limestone	strongly	50.12	0.52	0.89	0.01	89.00
St. Charles Limestone[§]	limestone	strongly	50.64	0.47	0.90	0.01	90.00
Wells Formation	limestone	strongly	52.60	0.40	0.94	0.01	94.00
Average Ca, Mg, and Ca:Mg of slightly and very slightly reactive units (n = 12)					**0.56**	**0.44**	**1.19 ± 0.10**

Note: Formation, effervescence, lithology, %CaO, %MgO data from Kaliser (1972).
*Nounan Limestone is currently named Nounan Dolomite (Dover, 1995).
[†]Langston Limestone is currently named Langston Dolomite (Dover, 1995).
[§]St. Charles Limestone is currently named St. Charles Formation (Dover, 1995).
[#]Brazer Limestone is currently named Brazer Dolomite (Dover, 1995).
**Ute Limestone is currently named Ute Formation (Dover, 1995).
[††]Blacksmith Limestone is currently named Blacksmith Dolomite (Dover, 1995).
[§§]Garden City Limestone is currently named Garden City Formation (Dover, 1995).

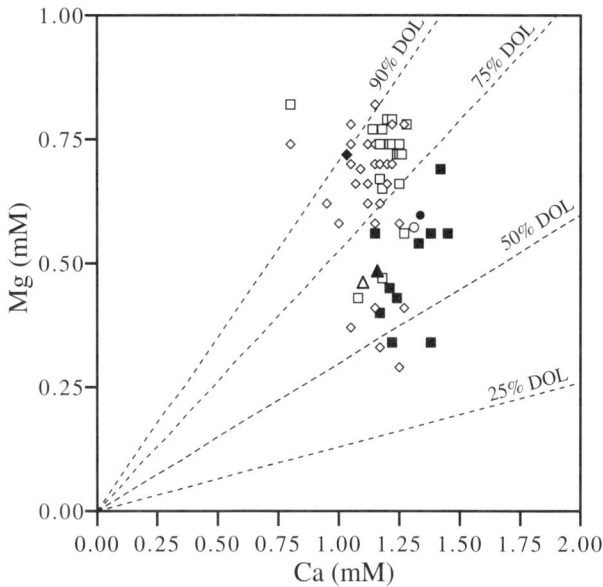

Figure 10. Magnesium (Mg) versus calcium (Ca) concentrations for four large springs in the Bear River Range. Open and solid squares represent data from Swan Creek Spring where spring discharge is known. Open diamonds represent data from Swan Creek Spring where spring discharge is not known. Solid diamond represents data from Blue Pond Spring, spring discharge not known. Open and solid circles represent data from Big Spring, spring discharge not known. Open and solid triangles represent data from Paris Spring, spring discharge not known. Dashed lines represent various Ca:Mg ratios created by the dissolution of different amounts of dolomite (DOL) and limestone. See text for discussion. Data are presented in Tables 3 and 8 and in Appendix DR1 (see footnote 1).

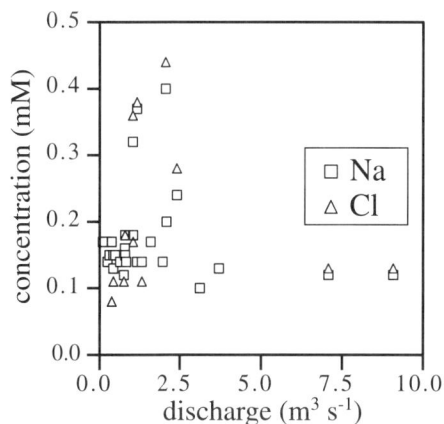

Figure 11. Sodium (Na) and chloride (Cl) concentrations versus spring discharge at Swan Creek Spring. Note the peak in concentrations for both ions during discharges of ~2.0 m³ s⁻¹.

dissolution provides the majority of the base flow solutes (Fig. 10). In contrast, the bedrock lithology in the area of Blue Pond Spring is exclusively dolomite (Oriel and Platt, 1980). The major-solute chemistry of Blue Pond Spring (12) has been analyzed only once, during what should have been peak discharge conditions during April, 2004. The Ca and Mg concentrations from that collection indicate that Blue Pond Spring's solute chemistry is primarily derived from the dissolution of dolomite (Fig. 10), even during peak discharge. This implies that the groundwater basin that feeds Blue Pond Spring may be relatively small and local, or alternatively, if Blue Pond Spring's groundwater basin is large then the conduit system that feeds the spring is developed within dolomite. Repeated sampling and gauging of Paris Spring and Blue Pond Spring should be conducted during different seasons to test for discharge-dependent changes in major-ion chemistry, like that observed at Swan Creek Spring. Such information would be crucial for further understanding the karst development within the Bear River Range.

Big Spring

Streams in the eastern Bear River Range are conspicuously located in the carbonate terrain north of Garden City (Fig. 2). Perennial streams and large springs are absent in the Wasatch Formation terrain between Big Spring (30) and Garden City. Solution basins in the Bear River Range west and north of Garden City are thought to be primary recharge areas and conduits for snowmelt for the more northern springs and streams (Reheis et al., this volume). Similar solution basins (Bear Wallow and Peter Sinks) and one sinkhole region are mapped along the Bear River Range ridge crest west and southwest of Garden City (Dover, 1995; Fig. 2), yet with the exception of Big Spring, there are no substantial springs or streams in the area. The groundwater divide between Big Spring and Swan Creek Spring may lie relatively close to Big Spring such that a large portion of the infiltration from the southern portion of the valley flows northward and discharges at Swan Creek Spring. Additionally, the paucity of large springs and streams south of Garden City might be explained by the presence of the Wasatch Formation in this area: with its relatively low permeability it acts as a confining bed where it overlies the local Paleozoic carbonates so groundwater cannot reach the surface. Big Spring (30), the only major groundwater discharge point within the Wasatch Formation, emanates from a fault that may penetrate to the Paleozoic carbonate aquifer.

In solute chemistry, Big Spring is more similar to other large springs discharging from carbonate terrain, such as Paris Spring (3), Blue Pond Spring (12), and Swan Creek Spring (22) than to springs emanating from the Wasatch Formation (e.g., 39, 46, 47; Figs. 7 and 8). The $^{87}Sr/^{86}Sr$ value of Big Spring is indistinguishable from values of springs sourced in either Paleozoic carbonate or Wasatch Formation rocks, however, and is not a useful indicator of Big Spring's source. The tritium value from Big Spring (~5 TU) is half of the modern value from Swan

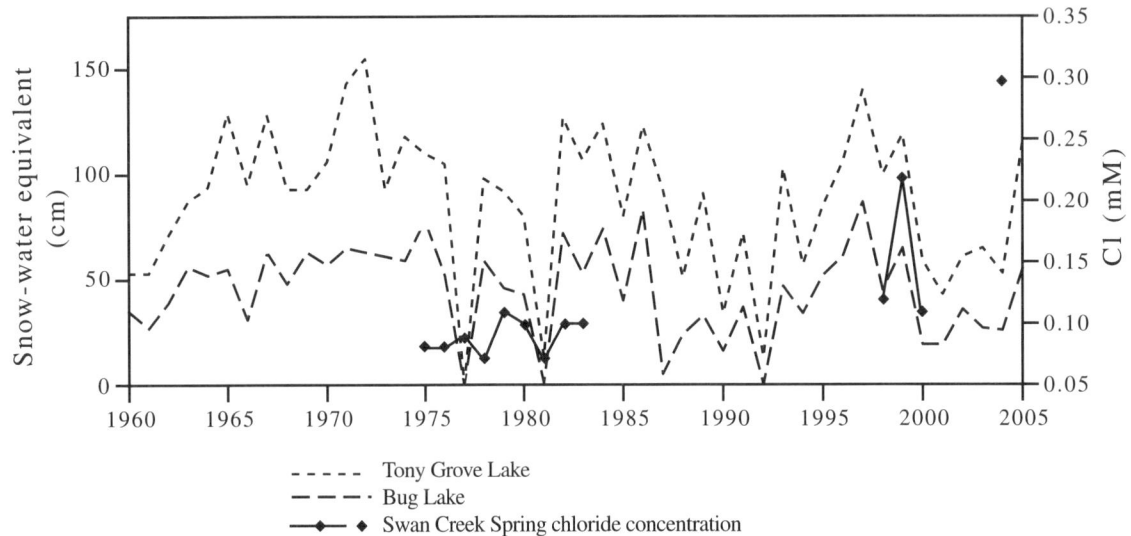

Figure 12. Snow-water-equivalent data for Tony Grove Lake (short-dashed line), Bug Lake (long-dashed line), and chloride concentration data from Swan Creek Spring (solid line with solid symbols).

Creek Spring, however, but nearly twice that of another Wasatch Formation spring (47).

The large volume of water issuing from Big Spring also suggests a strong connection to the Paleozoic carbonate aquifer, but note that the Ca and Mg concentrations at Big Spring during assumed peak discharge (April) and base flow (September) conditions are not discharge dependent (Fig. 10). The groundwater basin and conduit-fracture network that feeds Big Spring may be significantly different from the system that feeds Swan Creek Spring. Additional data are needed to test this hypothesis, however.

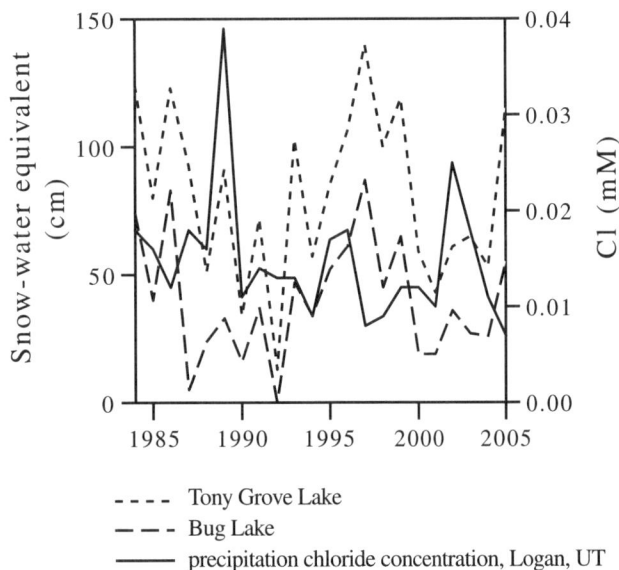

Figure 13. Snow-water-equivalent data for Tony Grove Lake (short-dashed line), Bug Lake (long-dashed line), and chloride concentration in precipitation at Logan, Utah (solid line).

Impacts of Groundwater on Bear Lake's Hydrologic and Solute Balance

The modern hydrologic balance of Bear Lake (excluding the Bear River) is probably balanced, or nearly so (Lamarra et al., 1986; Bright, 2003; Bright et al., 2006) even though the value of Swan Creek discharge used in previous estimates was underestimated by a factor of nearly three. Mundorff (1971) reported an average Swan Creek Spring discharge of ~0.9 m^3 s^{-1}, whereas Bright (2003) estimated ~0.3 m^3 s^{-1}. Given the higher Swan Creek Spring discharge, the hydrologic balance of Bear Lake is more easily explained by surface runoff and shallow subsurface sources, without the need for substantial amounts of groundwater influxes from lake-marginal or sublacustrine springs. Lake-marginal influxes to the lake do occur, but given the isotopic and solute chemistry of the eastern and southern lake-marginal springs reported here, they are evidently of minor importance.

Dean et al. (2007) used an analysis of pre-diversion lake chemistry and the $^{87}Sr/^{86}Sr$ values of modern water to conclude that ~99% of the solutes in Bear Lake prior to the diversion of the Bear River were derived from the streams sourced in the Bear River Range. This estimate is based on $^{87}Sr/^{86}Sr$ and Cl balances using mainly stream-solute data. The Dean et al. (2007) estimate did not include the lake-marginal springs that although small, contain concentrated solutes. The solute and isotope data from the lake-marginal springs presented in this study reinforce their interpretation. Using the average $^{87}Sr/^{86}Sr$ value of pre-diversion aragonite (0.71031; Table 10A) in sediment core BL96-10 as a reference, the influx of solutes (Na, K, and Cl) from the local streams and lake-marginal springs was mixed in various proportions to generate a hypothetical water body with a $^{87}Sr/^{86}Sr$ value of 0.71031 and with similar Na, K, and Cl concentration factors in relation to the lake's 1912 chemistry (Table 10B and 10C).

TABLE 10A. STRONTIUM ISOTOPE ($^{87}Sr/^{86}Sr$) VALUES
IN BEAR LAKE SEDIMENT CORE BL96-10

Depth (cm)	Status	$^{87}Sr/^{86}Sr$ ratio	St. dev. (±)
0	post-diversion	0.70942	0.00001
5	post-diversion	0.70942	0.00001
7	post-diversion	0.70943	0.00001
9	post-diversion	0.70951	0.00001
11	post-diversion	0.70975	0.00001
13	pre-diversion	0.71023	0.00001
16	pre-diversion	0.71031	0.00001
20	pre-diversion	0.71034	0.00001
25	pre-diversion	0.71035	0.00001
29	pre-diversion	0.71032	0.00001
	pre-diversion avg.	**0.71031**	

TABLE 10B. MODELING DATA FOR BEAR LAKE SOLUTE SOURCES

Site (Fig. 2)	Name	Sr (µg L^{-1})	f Sr total*	$^{87}Sr/^{86}Sr$	Contribution[†]	Cl (mg L^{-1})	Na (mg L^{-1})	K (mg L^{-1})
West streams								
14	St. Charles Creek	50	0.18727	0.71036	0.13303	1.8	2.1	0.3
22	Swan Creek Spring/Creek	68	0.25468	0.71079	0.18103	5.2	4.3	0.5
31	Big Creek	149	0.55805	0.71106	0.39681	9.1	6.4	1.0
	sum	**267**			**0.71087**			
	average	**89**				**5.3**	**4.2**	**0.6**
West springs								
20	spring	129	0.10768	0.71107	0.07657	7.0	7.6	1.3
21	spring	150	0.12521	0.71202	0.08915	7.8	6.9	1.4
26	spring	315	0.26294	0.71021	0.18674	7.6	9.8	0.8
27	spring	369	0.30801	0.71187	0.21926	46.0	38.1	2.7
28	spring	235	0.19616	0.71322	0.13991	23.0	19.3	2.0
	sum	**1198**			**0.71163**			
	average	**240**				**18.3**	**16.3**	**1.6**
East streams								
35	South Eden Creek	861	0.25809	0.70880	0.18293	55.4	32.5	2.0
40	North Eden Creek	645	0.19335	0.70901	0.13709	51.9	32.6	2.1
50	Indian Creek	1830	0.54856	0.70785	0.38830	22.4	21.1	1.8
	sum	**3336**			**0.70832**			
	average	**1112**				**43.2**	**28.7**	**2.0**
East springs								
48	spring	5933		0.70767		11.5	26.0	2.4
53	spring	4930		0.70977		74.1	157.0	41.3

*f Sr total = fraction of Sr total, calculated by (sample Sr concentration/Sr sum) for each geographic grouping.
[†]Contribution = (f Sr total * $^{87}Sr/^{86}Sr$) for each site.

TABLE 10C. TYPICAL MIXING MODEL RESULTS FOR PRE-DIVERSION BEAR LAKE

Source	Sr (µg L^{-1})	$^{87}Sr/^{86}Sr$ (avg.)	Cl (mg L^{-1})	Na (mg L^{-1})	K (mg L^{-1})	Proportion	Cl:Na	Cl:K	Na:K
West streams	89	0.71087	5.3	4.2	0.6	0.9906			
East streams	1112	0.70832	43.2	28.7	2.0	0.0018			
West springs	240	0.71163	18.3	16.3	1.6	0.0014			
spring 48 (Fig. 2)	5933	0.70767	11.5	26.0	2.4	0.0023			
spring 53 (Fig. 2)	4930	0.70977	74.1	157.0	41.3	0.0039			
Mixture		**0.71031**	**5.8**	**5.0**	**0.8**	**1.0000**	1.2	7.3	6.3
Pre-diversion lake*		0.71031	78.5	66.3	10.5		1.2	7.5	6.3
Concentration factor[†]			13.5	13.3	13.1				

*Pre-diversion lake major-ion concentrations from Birdsey (1989).
[†]Concentration factor = Pre-diversion lake concentration for given ion/mixture concentration for same ion.

These results indicate that the western lake-marginal springs (20, 21, 26–28) and eastern streams (35, 40, 50) and springs (48, 53) collectively contributed only ~1% of the $^{87}Sr/^{86}Sr$ in the pre-diversion lake (Table 10C). The mixing models also reveal that Mud Lake Hot Spring–type water (53) is necessary to generate the K concentrations and Na, K, and Cl solute ratios reported for the pre-diversion lake (Table 10C). No other water sources reported in this study have the K concentrations needed to balance the model. This conclusion is based on the assumption that Na, K, and Cl behave conservatively in the lake (see Dean et al., 2007). Mud Lake Hot Spring (53) is, and has been, separated from the lake by a sandbar but prior to the construction of the water control structures at Lifton (15) ca. 1912 (Fig. 2) there was a small outlet on the west side of the lake that connected the lake to the Bear River via the marshes surrounding Mud Lake (McConnell et al., 1957). Solutes from Mud Lake Hot Spring must have entered the lake though this outlet, percolated through the sandbar, or, alternatively, there is hot spring-type water entering the lake from unlocated springs within the lake basin or possibly from groundwater leakage along the East Bear Lake Fault.

Sublacustrine springs were thought to discharge in the lake because isolated portions of the lake surface typically do not freeze during the winter. During calm lake conditions the surface of the lake at these ice-free areas visibly roils, preventing the lake surface from freezing in the winter. Sonar images at one of these ice-free areas revealed a strong reflector emanating from the lake floor (Figs. 2 and 14). Subsequent investigations by divers, however, indicated no detectable discharge of water at these sites. Bubble trails associated with the sonar reflections are composed of isotopically depleted methane gas (Dean, this volume). A density contrast between ambient lake water and methane-charged water is most likely responsible for the sonar reflection and surface-water disturbances. Another peculiar location in Bear Lake, termed "the rock pile" (Dean, this volume), may be an example of diffuse, sublacustrine spring discharge. Divers detected no noticeable groundwater influx at this site, however, and the site may no longer be active. To date, no large-volume springs have been identified on the floor of the lake. Sublacustrine spring discharge, if occurring, is probably of minor importance to the hydrologic balance of Bear Lake.

SUMMARY

1. The two primary rock types in southern Bear Lake Valley—Paleozoic marine carbonates, which are exposed primarily north of Garden City, and the Wasatch Formation, which is exposed at the southern end of Bear Lake Valley and on the Bear Lake Plateau—contain groundwater with two distinct solute compositions. The Ca-Mg-HCO$_3$ carbonate-terrain water contrasts sharply with the Ca-Mg-HCO$_3$-SO$_4$-Cl and Ba- and Sr-enriched water associated with Wasatch Formation springs. Water discharging at several fault-related, lake-marginal springs along the western margin of Bear Lake appears to be a mix of carbonate bedrock- and Wasatch Formation-sourced solutes. The

Ca-SO$_4$-rich groundwater in the northeast quadrant of Bear Lake is a third distinct water type in the watershed. Its distribution appears to extend southward along the eastern flank of Bear Lake along the East Bear Lake Fault. A hot spring with a Ca-Na-SO$_4$-Cl chemistry is located near the northeast quadrant of Bear Lake, also along the East Bear Lake Fault. Extrabasinal solute sources may be important to Bear Lake Valley, but additional data are needed to adequately address that issue. Faulting exerts a strong control on spring locations, including Swan Creek Spring, Big Spring, and Mud Lake Hot Spring. Faults in Bear Lake Valley are important conduits for groundwater flow.

2. The groundwater north of Garden City is derived from modern recharge with shallow flow paths. The solute chemistry of Swan Creek Spring varies in response to its discharge. Solutes derived primarily from dolomite dissolution dominate base flow and solutes derived from increased limestone dissolution dominate peak discharge. Paris Spring and Blue Pond Spring have solute chemistries that reflect the dissolution of the dominant carbonate bedrock in their source areas. Discharge-dependent increases in Na and Cl at Swan Creek Spring may be anthropogenic, or related to a northward migration of Wasatch Formation solutes

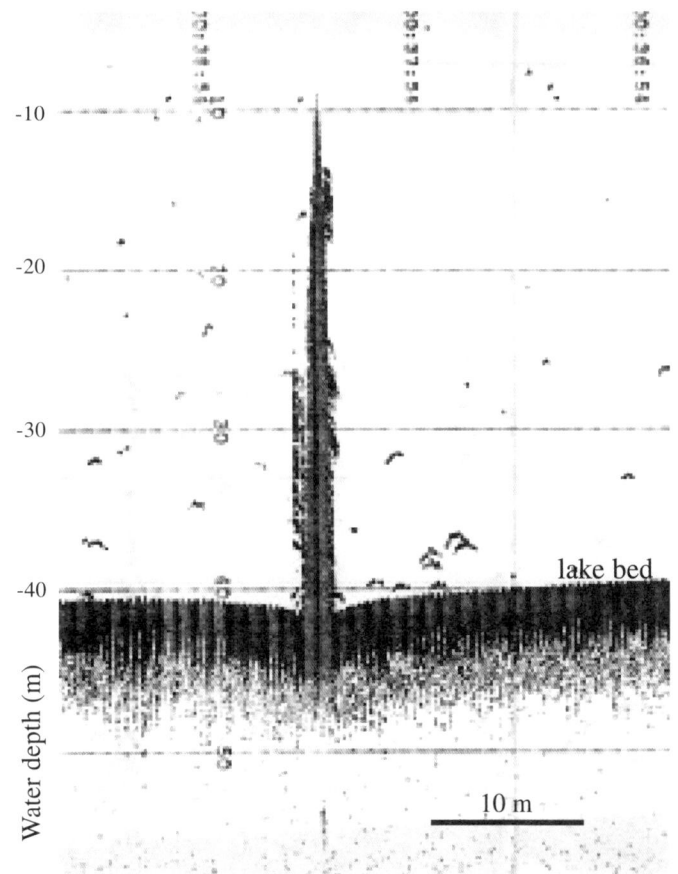

Figure 14. Sonar image of a sublacustrine methane seep initially thought to be a large spring. See Figure 2 for location.

along range-bounding faults. Spring discharge at Big Spring is fault controlled, and its solute composition is a mixture of both Paleozoic carbonate and Wasatch Formation derived ions.

3. The karsted Bear River Range aquifer north of Garden City is the primary recharge and discharge area for water and solutes entering Bear Lake. The solution features west of Bear Lake are important recharge areas, but recharge outside of the immediate watershed is also likely. Faulting exerts a strong control on the local and regional hydrology, serving in some cases as conduits and other cases as barriers for groundwater flow. A portion of the infiltration from the solution basins located in the southern part of the Bear River Range is probably discharged at Big Spring, but the remainder may be diverted to the north, where it discharges at Swan Creek Spring, or it may be diverted away from Bear Lake.

4. Groundwater in the Bear River Range is modern, but only a small portion of the groundwater in the Bear Lake Plateau is modern. Stable isotope ($\delta^{18}O$, $\delta^{2}H$) data indicate that the topography of the Bear River Range exerts a major control on the distribution of stable isotope values of groundwater in southern Bear Lake Valley.

5. The hydrologic balance of Bear Lake is apparently maintained by surface runoff and by shallow groundwater sourced within the Bear River Range. Groundwater leakage around the margins of the lake or through the lake floor is (was) probably not a major source of solutes into Bear Lake, although a unique K-rich water source is needed to generate the pre-diversion lake chemistry. A significant influx of solutes from the eastern and southern parts of Bear Lake Valley is incompatible with the solute and $^{87}Sr/^{86}Sr$ balance for the pre-diversion lake. If the inflow to Bear Lake has been overestimated and subsurface groundwater influx is a sizable component of the lake's hydrologic balance, then the northwest quadrant of the valley is the only source area with a compatible solute composition.

ACKNOWLEDGMENTS

The Bear Lake project was funded by the U.S. Geological Survey (USGS) Earth Surface Dynamics Program. Scott Tolentino and Bryce Nelson of the Utah Department of Natural Resources (Bear Lake Station, Garden City, Idaho) helped immensely to locate and sample the lake-marginal springs. Scott Tolentino provided the methane seep sonar reflection image. Peter Kolesar (Utah State University) and Alan Riggs (USGS, Denver) dove on many of the sublacustrine sonar reflectors and bubble trails. Kathleen Simmons (USGS, Denver) provided the numerous strontium isotope measurements. The Johnson family (Laketown, Utah) graciously allowed access to Big Spring and North Eden Canyon. Rick Forester and Kelly Conrad (USGS) and Scott Tolentino helped collect the spring samples. This study has benefited greatly from discussions with Rick Forester, Darrell Kaufman, Walter Dean, Marith Reheis, and Scott Tolentino. Insightful reviews by Alan Mayo (Brigham Young University) and Larry Spangler (USGS, Salt Lake City) greatly improved the manuscript.

REFERENCES CITED

Blackett, R.E., 2004, Geothermal gradient data for Utah: Salt Lake City, Utah, Utah Geological Survey, 49 p.

Bright, J., 2003, A 240,000-yr record of oxygen-isotope and ostracode-faunal fluctuations, Bear Lake, Utah-Idaho [M.S. thesis]: Flagstaff, Arizona, Northern Arizona University, 130 p.

Bright, J., Kaufman, D.S., Forester, R.M., and Dean, W.E., 2006, A continuous 250,000 yr record of oxygen and carbon isotopes in ostracode and bulk-sediment carbonate from Bear Lake, Utah-Idaho: Quaternary Science Reviews, v. 25, p. 2258–2270, doi: 10.1016/j.quascirev.2005.12.011.

Bullen, T.D., and Kendall, C., 1998, Tracing of weathering reactions and water flow-paths: A multi-isotope approach, *in* Kendall, C., and McDonnell, J.J., eds., Isotope tracers in catchment hydrology: Amsterdam, Elsevier, p. 611–646.

Burke, W.H., Denison, R.E., Hetherington, E.A., Koepnick, R.B., Nelson, H.F., and Otto, J.B., 1982, Variation in seawater $^{87}Sr/^{86}Sr$ throughout Phanerozoic time: Geology, v. 10, p. 516–519, doi: 10.1130/0091-7613(1982)10<516:VOSSTP>2.0.CO;2.

Busenberg, E., and Plummer, L.N., 1982, The kinetics of dissolution of dolomite in CO_2-H_2O systems at 1.5 to 65 °C and 0 to 1 ATM PCO_2: American Journal of Science, v. 282, p. 45–78.

Clark, I.D., and Fritz, P., 1997, Environmental isotopes in hydrogeology: New York, Lewis Publishers, 328 p.

Clauer, N., Chaudhuri, S., and Subramanium, R., 1989, Strontium isotopes as indicators of diagenetic recrystallization scales within carbonate rocks: Chemical Geology, v. 80, p. 27–34.

Colman, S.M., 2006, Acoustic stratigraphy of Bear Lake, Utah-Idaho—Late Quaternary sedimentation patterns in a simple half-graben: Sedimentary Geology, v. 185, p. 113–125, doi: 10.1016/j.sedgeo.2005.11.022.

Coogan, J.C., 1992, Structural evolution of piggyback basins in the Wyoming-Idaho-Utah thrust belt, *in* Link, P.K., Kuntz, M.A., and Platt, L.B., eds., Regional geology of eastern Idaho and western Wyoming: Geological Society of America Memoir 179, p. 55–81.

Cooper, L.W., 1998, Isotopic fractionation in snow cover, *in* Kendall, C., and McDonnell, J.J., eds., Isotope tracers in catchment hydrology: Amsterdam, Elsevier, p. 119–136.

Craig, H., 1961, Isotope variations in meteoric water: Science, v. 133, p. 1702–1703, doi: 10.1126/science.133.3465.1702.

Dean, W.E., 2009, this volume, Endogenic carbonate sedimentation in Bear Lake, Utah and Idaho, over the last two glacial-interglacial cycles, *in* Rosenbaum, J.G., and Kaufman, D.S., eds., Paleoenvironments of Bear Lake, Utah and Idaho, and its catchment: Geological Society of America Special Paper 450, doi: 10.1130/2009.2450(07).

Dean, W.E., Rosenbaum, J., Skipp, G., Colman, S., Forester, R., Liu, A., Simmons, K., and Bischoff, J., 2006, Unusual Holocene and late Pleistocene carbonate sedimentation in Bear Lake, Utah and Idaho, USA: Sedimentary Geology, v. 185, p. 93–112, doi: 10.1016/j.sedgeo.2005.11.016.

Dean, W.E., Forester, R., Bright, J., Anderson, R., and Simmons, K., 2007, Influence of the diversion of the Bear River into Bear Lake (Utah and Idaho) on the environment of deposition of carbonate minerals: Evidence from water and sediments: Limnology and Oceanography, v. 52, p. 1094–1111.

Denison, R.E., Koepnick, R.B., Fletcher, A., Howell, M.W., and Callaway, W.S., 1994, Criteria for the retention of original $^{87}Sr/^{86}Sr$ in ancient shelf limestones: Chemical Geology, v. 112, p. 131–143, doi: 10.1016/0009-2541(94)90110-4.

Denison, R.E., Koepnick, R.B., Burke, W.H., and Hetherington, E.A., 1998, Construction of the Cambrian and Ordovician seawater $^{87}Sr/^{86}Sr$ curve: Chemical Geology, v. 152, p. 325–340, doi: 10.1016/S0009-2541(98)00119-3.

Dover, J.H., 1995, Geologic map of the Logan 30′ × 60′ quadrangle, Cache and Rich counties, Utah, and Lincoln and Uinta counties, Wyoming: Denver, Colorado, U.S. Geological Survey Miscellaneous Investigations Series Map I-2210, scale, 1:100,000.

Fishman, M.J., and Friedman, L.C., 1985, Methods for the determination of inorganic substances in water and fluvial sediments: U.S. Geological Survey Techniques of Technical Water-Resources Investigations Book 5, Ch. A1.

Fouillac, C., and Michard, G., 1981, Sodium/lithium ratio in water applied to geothermometry of geothermal reservoirs: Geothermics, v. 10, p. 55–70, doi: 10.1016/0375-6505(81)90025-0.

Fournier, R.O., 1981, Application of water geochemistry to geothermal exploration and reservoir engineering, *in* Rybach, L., and Muffler, L.J.P., eds.,

Geothermal systems: Principles and case histories: New York, John Wiley and Sons, p. 109–143.

Fournier, R.O., and Truesdell, A.H., 1973, An empirical Na-K-Ca geothermometer for natural waters: Geochimica et Cosmochimica Acta, v. 37, p. 1255–1275, doi: 10.1016/0016-7037(73)90060-4.

Friedman, I., Smith, G.I., Johnson, C.A., and Moscati, R.J., 2002, Stable isotope composition of waters in the Great Basin, United States, 2. Modern precipitation: Journal of Geophysical Research, v. 107, D19, p. 4401, doi: 10.1029/2001JD000566.

Gat, J.R., 1980, Isotopes of hydrogen and oxygen in precipitation, *in* Fritz, P., and Fontes, J.-Ch., eds., The terrestrial environment, v. 1 of Handbook of environmental isotope geochemistry: Amsterdam, Elsevier, p. 21–48.

Goldsmith, J.R., and Graf, D.L., 1958, Structural and compositional variations in some natural dolomites: Journal of Geology, v. 66, p. 678–693.

Herman, J.S., and White, W.B., 1985, Dissolution kinetics of dolomite: Effects of lithology and fluid flow velocity: Geochimica et Cosmochimica Acta, v. 49, p. 2017–2026, doi: 10.1016/0016-7037(85)90060-2.

Idaho Department of Environmental Quality, 2002, City of Bloomington (PWS 6040007) Source Water Assessment Final Report, 20 p.

Isotope Hydrology Information System, 2004, The ISOHIS database, http://isohis.iaea.org (accessed January 2007).

Kaliser, B.N., 1972, Environmental geology of Bear Lake area, Rich County, Utah: Utah Geological and Mineralogical Survey Bulletin, v. 96, 32 p.

Kaufman, D.S., Bright, J., Dean, W.E., Rosenbaum, J.G., Moser, K., Anderson, R.S., Colman, S.M., Heil, C.W., Jr., Jiménez-Moreno, G., Reheis, M.C., and Simmons, K.R., 2009, this volume, A quarter-million years of paleoenvironmental change at Bear Lake, Utah and Idaho, *in* Rosenbaum, J.G., and Kaufman, D.S., eds., Paleoenvironments of Bear Lake, Utah and Idaho, and its catchment: Geological Society of America Special Paper 450, doi: 10.1130/2009.2450(14).

Lamarra, V., Liff, C., and Carter, J., 1986, Hydrology of Bear Lake basin and its impact on the trophic state of Bear Lake, Utah-Idaho: The Great Basin Naturalist, v. 46, p. 690–705.

Langmuir, D., 1997, Aqueous environmental geochemistry: Upper Saddle River, New Jersey, Prentice-Hall Inc., 600 p.

Liu, S.-F., Nummedal, D., Yin, P.-G., and Luo, H.-J., 2005, Linkage of Sevier thrusting episodes and Late Cretaceous foreland basin megasequences across southern Wyoming (USA): Basin Research, v. 17, p. 487–506.

Mayo, A.L., and Loucks, M.D., 1995, Solute and isotopic geochemistry and ground water flow in the central Wasatch Range, Utah: Journal of Hydrology (Amsterdam), v. 172, p. 31–59, doi: 10.1016/0022-1694(95)02748-E.

McCalpin, J.P., 1993, Neotectonics of the northeastern Basin and Range margin, western USA: Zeitschrift für Geomorphologie, N.F., Suppl. Bd., v. 94, p. 137–157.

McConnell, W.J., Clark, W.J., and Sigler, W.F., 1957, Bear Lake: Its Fish and Fishing: Utah State Department of Fish and Game, Idaho Department of Fish and Game, Wildlife Management Department of Utah State Agricultural College, 76 p.

Moran, T.A., Marshall, S.J., Evans, E.C., and Sinclair, K.E., 2007, Altitudinal gradients of stable isotopes in lee-slope precipitation in the Canadian Rocky Mountains: Arctic, Antarctic, and Alpine Research, v. 39, p. 455–467, doi: 10.1657/1523-0430(06-022)[MORAN]2.0.CO;2.

Morse, J.W., and Arvidson, R.S., 2002, The dissolution kinetics of major sedimentary carbonate minerals: Earth-Science Reviews, v. 58, p. 51–84, doi: 10.1016/S0012-8252(01)00083-6.

Mundorff, J.C., 1971, Nonthermal springs of Utah: Utah Geological and Mineralogical Survey Bulletin 16, 70 p.

Oriel, S.S., and Platt, L.B., 1980, Geologic map of the Preston 1° × 2° quadrangle, southeastern Idaho and western Wyoming: U.S. Geological Survey Miscellaneous Investigations Series, Map I-1127, scale 1:250,000.

Ozyurt, N.N., and Bayari, C.S., 2007, Temporal variation of chemical and isotopic signals in major discharges of an alpine karst aquifer in Turkey: Implications with respect to response of karst aquifers to recharge: Hydrogeology Journal, doi: 10.1007/s10040-007-0217-6.

Palmer, M.R., and Edmond, J.M., 1992, Controls over the strontium isotope composition of river water: Geochimica et Cosmochimica Acta, v. 56, p. 2099–2111, doi: 10.1016/0016-7037(92)90332-D.

Poage, M.A., and Chamberlain, C.P., 2001, Empirical relationships between elevation and the stable isotope composition of precipitation and surface waters: Considerations for studies of paleoelevational change: American Journal of Science, v. 301, p. 1–15, doi: 10.2475/ajs.301.1.1.

Reheis, M.C., Laabs, B.J.C., and Kaufman, D.S., 2009, this volume, Geology and geomorphology of Bear Lake Valley and upper Bear River, Utah and Idaho, *in* Rosenbaum, J.G., and Kaufman, D.S., eds., Paleoenvironments of Bear Lake, Utah and Idaho, and its catchment: Geological Society of America Special Paper 450, doi: 10.1130/2009.2450(02).

Rice, K.C., and Spangler, L.E., 1999, Hydrology and geochemistry of carbonate springs in Mantua Valley, northern Utah, *in* Spangler, L.E., and Allen C.J., eds., Geology of northern Utah and vicinity: Salt Lake City, Utah Geological Association Publication 27, p. 337–352.

Rózanski, K., Araguás-Araguás, L., and Gonfiantini, R., 1993, Isotopic patterns in modern global precipitation, *in* Stewart, P.K., Lohmann, K.C., McKenzie, J., and Savin, S., eds., Climate change in continental isotopic records: Washington D.C., American Geophysical Union, Geophysical Monograph 78, p. 1–36.

Spangler, L.E., 2001, Delineation of recharge areas for karst springs in Logan Canyon, Bear River Range, northern Utah, *in* Kuniansky, E.L., ed., U.S. Geological Survey Karst Interest Group Proceedings: Atlanta, Water-Resources Investigations Report 01-4011, p. 186–193.

Sperber, C.M., Wilkinson, B.H., and Peacor, D.R., 1984, Rock composition, dolomite stoichiometry, and rock/water reactions in dolomitic carbonate rocks: Journal of Geology, v. 92, p. 609–622.

Szramek, K., McIntosh, J.C., Williams, E.L., Kanduc, T., Ogrinc, N., and Walter, L., 2007, Relative weathering intensity of calcite versus dolomite in carbonate-bearing temperate watersheds: Carbonate geochemistry and fluxes from catchments within the St. Lawrence and Danube river basins: Carbonate Geochemistry Geophysics Geosystems, v. 8, p. Q04002, doi: 10.1029/2006GC001337.

Veizer, J., Ala, D., Azmy, K., Bruckschen, P., Buhl, D., Bruhn, F., Carden, G.A.F., Diener, A., Ebneth, S., Godderis, Y., Jasper, T., Korte, C., Pawellek, F., Podlaha, O.G., and Strauss, H., 1999, $^{87}Sr/^{86}Sr$, $\delta^{13}C$ and $\delta^{18}O$ evolution of Phanerozoic seawater: Chemical Geology, v. 161, p. 59–88, doi: 10.1016/S0009-2541(99)00081-9.

White, D.E., 1957, Thermal waters of volcanic origin: Geological Society of America Bulletin, v. 68, p. 1637–1658, doi: 10.1130/0016-7606(1957)68 [1637:TWOVO]2.0.CO;2.

White, W.B., 2002, Karst hydrology: Recent developments and open questions: Engineering Geology, v. 65, p. 85–105, doi: 10.1016/S0013-7952(01)00116-8.

Willis, G.C., 1999, The Utah thrust system—An overview, *in* Spangler, L.E., and Allen, C.J., eds., Geology of northern Utah and vicinity: Salt Lake City, Utah Geological Association Publication 27, p. 1–9.

Wilson, J.R., 1979, Glaciokarst in the Bear River Range, Utah: National Speleological Society Bulletin, v. 41, p. 89–94.

Winograd, I.J., Riggs, A.C., and Coplen, T.B., 1998, The relative contribution of summer and cool-season precipitation to groundwater recharge, Spring Mountains, Nevada: Hydrogeology Journal, v. 6, p. 77–93, doi: 10.1007/s100400050135.

Wylie, A.H., Otto, B.R., and Martin, M.J., 2005, Hydrogeologic analysis of the water supply for Bloomington and Paris, Bear Lake County, Idaho: Moscow, Idaho, Idaho Geological Survey Information Circular 58, 10 p.

MANUSCRIPT ACCEPTED BY THE SOCIETY 15 SEPTEMBER 2008

The Geological Society of America
Special Paper 450
2009

Radiocarbon ages and age models for the past 30,000 years in Bear Lake, Utah and Idaho

Steven M. Colman

Large Lakes Observatory and Department of Geological Sciences, University of Minnesota Duluth, Minnesota 55812, USA

Joseph G. Rosenbaum

U.S. Geological Survey, MS 980, Box 25046, Denver Federal Center, Denver, Colorado 80225, USA

Darrell S. Kaufman

Department of Geology, Northern Arizona University, Flagstaff, Arizona 86011, USA

Walter E. Dean

U.S. Geological Survey, MS 980, Box 25046, Denver Federal Center, Denver, Colorado 80225, USA

John P. McGeehin

U.S. Geological Survey, National Center MS 955, Reston, Virginia 20192, USA

ABSTRACT

Radiocarbon analyses of pollen, ostracodes, and total organic carbon (TOC) provide a reliable chronology for the sediments deposited in Bear Lake over the past 30,000 years. The differences in apparent age between TOC, pollen, and carbonate fractions are consistent and in accord with the origins of these fractions. Comparisons among different fractions indicate that pollen sample ages are the most reliable, at least for the past 15,000 years. The post-glacial radiocarbon data also agree with ages independently estimated from aspartic acid racemization in ostracodes. Ages in the red, siliclastic unit, inferred to be of last glacial age, appear to be several thousand years too old, probably because of a high proportion of reworked, refractory organic carbon in the pollen samples.

Age-depth models for five piston cores and the Bear Lake drill core (BL00-1) were constructed by using two methods: quadratic equations and smooth cubic-spline fits. The two types of age models differ only in detail for individual cores, and each approach has its own advantages. Specific lithological horizons were dated in several cores and correlated among them, producing robust average ages for these horizons. The age of the correlated horizons in the red, siliclastic unit can be estimated from the age model for BL00-1, which is controlled by ages above and below the red, siliclastic unit. These ages were then transferred to the correlative horizons in the shorter piston cores, providing control for the sections of the age models in those cores in the red, siliclastic unit.

Colman, S.M., Rosenbaum, J.G., Kaufman, D.S., Dean, W.E., and McGeehin, J.P., 2009, Radiocarbon ages and age models for the past 30,000 years in Bear Lake, Utah and Idaho, *in* Rosenbaum, J.G., and Kaufman, D.S., eds., Paleoenvironments of Bear Lake, Utah and Idaho, and its catchment: Geological Society of America Special Paper 450, p. 133–144, doi: 10.1130/2009.2450(05). For permission to copy, contact editing@geosociety.org. ©2009 The Geological Society of America. All rights reserved.

These age models are the backbone for reconstructions of past environmental conditions in Bear Lake. In general, sedimentation rates in Bear Lake have been quite uniform, mostly between 0.3 and 0.8 mm yr^{-1} in the Holocene, and close to 0.5 mm yr^{-1} for the longer sedimentary record in the drill core from the deepest part of the lake.

INTRODUCTION

A major goal of our research at Bear Lake was to reconstruct a history of environmental change in the basin. To this end, a wide variety of paleoenvironmental proxies were measured (Rosenbaum and Kaufman, this volume). Changes in these different proxies with time form the basis for the paleoenvironmental reconstruction for Bear Lake. Of course, the other necessary component of any environmental history is an accurate chronology. An accurate chronology depends on many variables, including the sample material, analytical accuracy and precision, and the way in which continuous age models are constructed.

Here we report the results of radiocarbon analyses of sediments in several cores from Bear Lake. In the absence of macrofossils, we analyzed several different fractions of the bulk sediments and compared the resulting ages with each other and with independently estimated ages derived from amino acid analyses in ostracodes. We next rejected certain ages as outliers for several different reasons, and developed depth scales that account for multiple, overlapping cores and loss of surface materials. Finally, we developed continuous age models for each core, using multiple fit methods, to form the chronological framework for other paleoenvironmental studies.

A preliminary version of the present study, confined to the 1996 piston cores, was published as a U.S. Geological Survey Open-File Report (Colman et al., 2005). Some of the radiocarbon ages discussed here were used in the age model for the long (120 m) Bear Lake drill core (Colman et al., 2006), as discussed later in this paper. No other studies of the chronology of Bear Lake cores exist, except for a study of recent sedimentation by ^{210}Pb methods (Smoak and Swarzenski, 2004).

METHODS

Coring

A variety of cores were obtained at several different times in Bear Lake, as described by Rosenbaum and Kaufman (this volume). Detailed radiocarbon dating and age modeling were conducted for five of these core sites: BL96-1, BL96-2, BL96-3, BL02-3, and BL02-4 (Fig. 1). Radiocarbon ages for other cores and materials from Bear Lake are reported in Table 1, but are not discussed further here because these cores were taken in shallow water and many of them contain discontinuities or unconformities. The five cores from relatively deep water were analyzed in

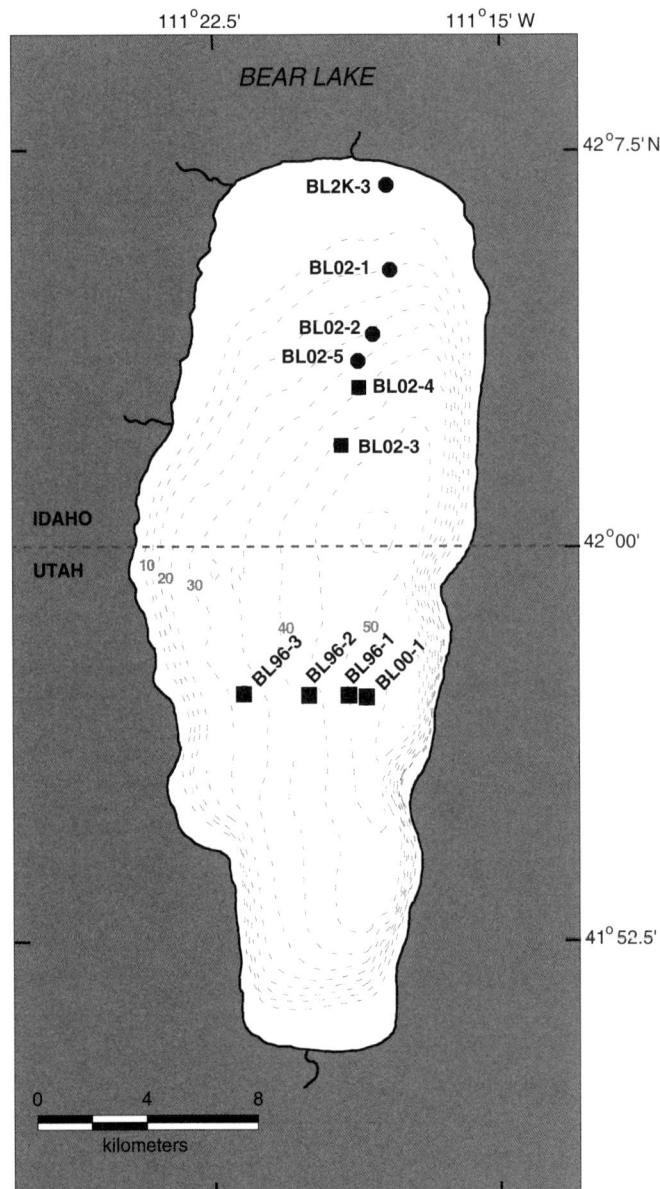

Figure 1. Map of Bear Lake, showing the bathymetry of the lake and location of the cores collected in this study. Cores from sites discussed here in detail are indicated by squares. Core BLR2K-3 is located in a shallow lake (Mud Lake) north of Bear Lake. Bathymetric contour interval 5 m, beginning at 10 m.

TABLE 1. RADIOCARBON MEASUREMENTS MADE IN THIS STUDY

Core (bold) and sample designation	Corrected total depth (cm)[a]	Material	δ^{13}C (per mil)[b]	Age (^{14}C yr)	Error (1σ yr)	Lab number[c]	Calibrated age (cal yr B.P.)[d]	Calibrated 1σ range (cal yr B.P.)
BL96-1								
1-A 11/12	7	ostracodes	−1.09	905	130	OS-19513	830	700–930
1-A 16/17	12	ostracodes	0	890	130	OS-19507	820	700–920
1-A 27	23	ostracodes	0	900	75	OS-19496	820	740–910
1-A 37/38	33	ostracodes	0	14,500	1700	OS-19509	17,200[e]	15,190–19,330
1-A 52/53	48	ostracodes	−0.31	1450	45	OS-18663	1350	1310–1370
1-A-52/53	48	pollen	−25	1070	40	WW-2774	980	930–1050
1-C-55	252	pollen	−25	3320	40	WW-1754	3550	3480–3610
1-E-39	437	pollen	−25	5260	40	WW-1752	6040	5940–6170
1-E-95	493	TOC	−25	6740	50	WW-1384	7600	7570–7660
BL96-2								
2-A-13	3	pollen	−25	3020	50	WW-1755	3230[e]	3160–3330
2-A-13	3	TOC	−25	1620	50	WW-1758	1510	1420–1560
2-B-9	99	pollen	−25	3435	60	WW-2600	3700	3620–3830
2-B-31	121	ostracodes	0	4620	60	WW-2803	5370	5290–5470
2-B-31	121	pollen	−25	4230	40	WW-2775	4750	4660–4850
2-B-61	151	ostracodes	0	5460	50	WW-2799	6260	6210–6300
2-B-61	151	pollen	−25	5260	40	WW-2776	6040	5940–6170
2-B-73	163	pollen	−25	5810	50	WW-1756	6610	6540–6670
2-B-85	175	pollen	−25	6420	50	WW-1757	7350	7320–7420
2-B-85	175	TOC	−25	6970	50	WW-1759	7800	7730–7920
2-C-10	201	pollen	−25	8265	70	WW-2601	9250	9130–9400
2-C-21	212	ostracodes	0	9070	60	WW-2800	10,230	10,190–10,270
2-C-21	212	pollen	−25	8580	40	WW-2777	9540	9520–9560
2-C-55	246	pollen	−25	10,300	60	WW-1773	12,100	11,980–12,340
2-D-7	299	pollen	−25	12,710	50	WW-1774	15,010	14,900–15,130
2-D-7	299	TOC	−25	13,110	60	WW-1760	15,500	15,320–15,660
2-D-8	300	pollen	−25	12,545	90	WW-2602	14,690	14,470–14,940
2-D-15	307	rotifer	−29.2	12,400	80	OS-18559	14,430	14,190–14,610
2-D-21	313	pollen	−28.14	14,000	100	OS-35973	16,690[f]	16,460–16,920
2-D-41	333	pollen	−27.92	16,200	75	OS-35624	19,370[f]	19,300–19,470
2-D-61	353	pollen	−25.98	18,550	140	OS-35974	22,130[f]	22,000–22,310
2-D-73	365	pollen	−25	8380	40	WW-2783	9420[e]	9320–9470
2-D-81	373	pollen	−25.5	21,000	110	OS-35625	25,290[f]	25,000–25,520
2-D-93	385	pollen	−25.34	21,300	150	OS-36022	25,720[f]	25,560–25,930
2-D-101	393	pollen	−25	22,600	60	WW-2778	26,600[f]	26,540–26,660
BL96-3								
3-A-18	5	pollen	−25.31	4440	40	OS-36023	5050[e]	4970–5270
3-A-33	20	pollen	−25	10,940	75	WW-2607	12,900[e]	12,850–12,940
3-A-46	33	pollen	−25.82	12,800	100	OS-35976	15,110[e]	14,950–15,260
3-A-90	77	pollen	−25	19,980	60	WW-2779	23,940[f]	23,820–24,060
3-B-85	172	pollen	−24.53	21,800	140	OS-36267	25,680[f]	25,540–25,820
3-C-13	201	pollen	−25	21,850	230	WW-2605	25,730[f]	25,500–25,960
3-C-89	277	pollen	−24.54	23,400	130	OS-35626	27,520[f]	27,390–27,650
3-D-89	377	pollen	−25.04	26,700	170	OS-36024	31,280[f]	31,110–31,450
3-E-15	403	pollen	−25	22,150	210	WW-2606	26,080	25,870–26,290
BL98-09								
09-Top	1	pollen	−25	1510	40	WW-2771	1390	1340–1490
09-Top	1	ostracodes	0	600	70	WW-2801	600	540–650
09-30+	30	pollen	−25	980	40	WW-2772	870	800–930
09-30+	30	ostracodes	0	1350	50	WW-2802	1280	1190–1310
BLR2K-3								
R3-1 20-21	20.5	wood	−25	472	92	WW-3885	520	500–540
R3-1 50-51	50.5	wood	−25	1076	86	WW-3886	990	940–1050
R3-1 70-71	70.5	wood	−25	1105	88	WW-3887	1010	970–1060
R3-1 90-91	90.5	wood	−25	950	114	WW-3888	850	800–930
R3-1 140-142	141.0	gastropods	0	7125	80	WW-3859	7960	7880–8000
BL2K-3-1								
3-1 20-21	20.5	wood	−25	3588	116	WW-3883	3890	3780–3980
3-1 40-41	40.5	wood	−25	969	116	WW-3884	870	800–930
3-1 75-76	75.5	wood	−25	100	80	WW-3826	120	0–260
3-1 110-111	110.5	shells	0	3307	140	WW-3982	3540	3460–3620
BL02-1								
1-1 13	13.0	pollen	−25	1295	80	WW-4571	1230	1180–1280
1-1 25	25.0	pollen	−25	1955	80	WW-4572	1910	1870–1970
1-1 31-32	31.5	gastropods	0	6975	90	WW-4655	7810	7740–7920
1-1-44	44.0	pollen	−25	5390	140	WW-4573	6180	6030–6280
1-1-59	59.0	pollen	−25	8040	140	WW-4574	8900	8780–9020
1-1-72	72.0	pollen	−25	7880	260	WW-4575	8740	8560–8980
1-1-82	82.0	pollen	−25	8260	140	WW-4576	9250	9130–9400
1-2 29	29.0	pollen	−25	10,040	80	WW-4577	11,540	11,400–11,690
1-2 37-38	37.5	gastropods	0	10,290	120	WW-4656	12,080	11,840–12,340
1-2 41	41.0	pollen	−25	11,375	90	WW-4578	13,240	13,200–13,290
1-2 86	86.0	pollen	−25	11,880	90	WW-4579	13,750	13,700–13,800
1-2 89	89.0	pollen	−25	12,100	80	WW-4580	13,950	13,890–14,010
BL02-2								
2-1 13	12.0	pollen	−25	1930	80	WW-4581	1880	1830–1920
2-1 21-22	21.5	gastropods	0	7530	80	WW-4657	8360	8330–8390
2-1 35	35.0	pollen	−25	21,550	190	WW-4582	25,390	25,200–25,580

(Continued)

TABLE 1. RADIOCARBON MEASUREMENTS MADE IN THIS STUDY (*Continued*)

Core (bold) and sample designation	Corrected total depth (cm)[a]	Material	δ¹³C (per mil)[b]	Age (¹⁴C yr)	Error (1σ yr)	Lab number[c]	Calibrated age (cal yr B.P.)[d]	Calibrated 1σ range (cal yr B.P.)
BL02-3								
3-1 7	21.0	pollen	−25	2700	70	WW-4596	2800	2760–2840
3-1 58	72.0	pollen	−25	4850	70	WW-4595	5600	5490–5640
3-1-110	124.0	pollen	−25	5975	90	WW-4593	6810	6750–6880
3-1-152	166.0	pollen	−25	7075	70	WW-4594	7900	7860–7950
3-2 60-61	186.5	pollen	−25	6865	80	WW-4262	7700	7660–7740
3-2 90-91	216.5	pollen	−25	8290	80	WW-4263	9310	9150–9400
3-2 102-103	228.5	pollen	−25	8445	60	WW-4264	9480	9450–9500
3-2 123-124	249.5	pollen	−25	8760	80	WW-4265	9760	9680–9890
3-2 127-128	253.5	pollen	−25	9040	80	WW-4266	10,220	10,200–10,230
3-3 10	disturbed	pollen	−25	2750	70	WW-4597	2840	2790–2870
3-3 35	disturbed	pollen	−25	4770	240	WW-4614	5490	5320–5600
3-3 60	disturbed	pollen	−25	7040	70	WW-4598	7880	7850–7930
3-3 65-66	disturbed	pollen	−25	7635	100	WW-4267	8430	8380–8510
3-3 70-71	disturbed	pollen	−25	7950	60	WW-4268	8830	8670–8980
3-3 75-76	disturbed	pollen	−25	8600	80	WW-4269	9550	9530–9600
BL02-4								
4M 8-9	9.0	pollen	−25	1575	80	WW-4914	1470[e]	1420–1520
4M 10-11	11.0	pollen	−25	1040	70	WW-4910	950[e]	930–970
4M 20-21	21.0	pollen	−25	920	140	WW-4911	840	780–920
4M 30-31	31.0	pollen	−25	930	80	WW-4912	850	800–910
4M 38-39.5	38.8	pollen	−25	1270	80	WW-4913	1210	1180–1270
4-3 12	12.0	pollen	−25	680	160	WW-4615	640[e]	560–690
4-3 50	50.0	pollen	−25	1390	70	WW-4600	1310	1290–1330
4-3 79-80	79.5	pollen	−25	2670	80	WW-4270	2780	2750–2840
4-3 83-84	83.5	gastropods	0	2760	90	WW-4658	2860[e]	2790–2920
4-3 83-84	83.5	gastropods	0	3295	60	WW-4659	3520[e]	3480–3560
4-3 86-87	86.5	pollen	−25	4935	80	WW-4271	5660	5610–5710
4-3-131	131.0	pollen	−25	5800	70	WW-4599	6600	6560–6660
4-3 150-151	150.5	pollen	−25	6865	60	WW-4272	7690	7670–7720
4-3 154-155	155.0	pollen	−25	6910	110	WW-5048	7750	7680–7790
4-3 160-161	161.0	pollen	−25	7289	88	WW-5049	8100	8050–8160
4-3 166-167	167.0	pollen	−25	7490	90	WW-5050	8310	8210–8380
4-3 172-173	173.0	pollen	−25	7792	88	WW-5051	8570	8490–8630
4-3 177-178	177.5	pollen	−25	8230	80	WW-4273	9200	9130–9270
4-3 180-181	180.5	gastropods	0	8865	100	WW-4660	10,000[e]	9900–10160
4-3 182-183	182.5	pollen	−25	8820	80	WW-4274	9870	9740–10120
BL02-5								
5-1 6-8	7.0	gastropods	0	6115	80	WW-4661	7000	6910–7150
5-1 11	11.0	pollen	−25	19,460	440	WW-4601	23,170	22,760–23,500
5-1 25	25.0	pollen	−25	22,030	290	WW-4602	25,940	25,650–26,230
BL00-1								
D 6H2 109-110	1800	pollen	−25	24,280	110	WW-6452	28,530	28,420–28,640
D 7H1 19-20	1859	pollen	−25	23,340	100	WW-6453	27,450	27,350–27,550
04EJ-105 Sealy Spring (surface)	–	spring carbonate	−25	8785	70	WW-4915	9800	9710–9890

[a]Top of 1-cm interval. From Rosenbaum et al. (this volume).

[b]Whole values indicated as "0" and "−25" were assumed for ostracodes and organic carbon, respectively; other values were measured.

[c]See text (Methods) for explanation.

[d]Calibrations from program CALIB 5.01 (Stuiver et al., 1998), using 1σ errors and the median probability age. Ages older than 21,381 ¹⁴C yr B.P. were calibrated with the relation given in Bard et al. (1998), and their 1σ ¹⁴C errors were retained (see text).

[e]Rejected for various reasons (see text for explanation) and not used in age models.

[f]Ages from red, siliclastic zone (Rosenbaum and Heil, this volume): not initially rejected, but eventually eliminated from age models (see text).

detail for paleolimnological proxy analyses, so these cores are the focus here. With one exception (BL02-4) these cores contain no major unconformities, although they may contain minor hiatuses (Smoot, this volume). In light of the results presented here, the age model for the upper part of the 2000 Bear Lake drill core, BL00-1 (Colman et al., 2006), was re-examined.

The three 1996 cores were collected with the University of Minnesota Kullenberg-type piston coring system (Kelts et al., 1986). As is commonly the case for this type of core, the uppermost sediments were not recovered. The two 2002 cores consist of multiple overlapping segments obtained with an Austrian UWITEC piston coring system. A small gravity core designed to sample the sediment-water interface was also used at all 2002 sites. Composite depth scales for the surface core and multiple sections of the piston cores were constructed by using key marker horizons (Dean, this volume; Rosenbaum et al., this volume).

Radiocarbon Dating

Ideally, radiocarbon chronologies for temperate lakes are based on dating of small terrestrial macrofossils, but, as is the case in many large lakes, macrofossils in Bear Lake are rare. We were unable to find any macrofossils suitable for dating in our cores, except for mollusks and detrital wood in some shallow-water cores. Consequently, for the deep-water cores that contain a nearly continuous sedimentary record, we focused on three types of material that were separated from the sediments and analyzed by accelerator-mass spectrometer (AMS) methods:

(1) total organic carbon (TOC), (2) biogenic carbonate (ostra-codes and mollusks) hand-picked from the sediments, and (3) material remaining after minerogenic sediment was removed by standard pollen-preparation procedures (Faegri and Iverson, 1975), here called "pollen+."

For TOC, samples of bulk sediment were acidified with organic-free HCl and filtered through a nominal 1 μm diameter, precleaned quartz-fiber filter. Pollen+ samples were prepared using standard palynological methods (Faegri and Iverson, 1975). The processed material contains charcoal and other refractory organic material in addition to pollen, but visual inspection indicated that non-pollen materials were minor components of most of the samples. Ostracodes were separated from the sediment by hand pick-ing, following the procedures described in Colman et al. (1990).

The samples were then converted to CO_2 by combustion (TOC and pollen+) or dissolution in phosphoric acid using stan-dard methods (Jones et al., 1989; Slota et al., 1987). Carbon diox-ide from the samples was reduced to elemental graphite over hot iron in the presence of hydrogen (Vogel et al., 1984). The graph-ite targets were prepared and analyzed at the NOSAMS facility in Woods Hole (OS- numbers in Table 1) or they were prepared at the U.S. Geological Survey (WW- numbers in Table 1) and run at the Lawrence Livermore's Center for Accelerator Mass Spec-trometry (CAMS). Ages were calculated according to the meth-ods of Stuiver and Pollach (1977), using either measured $\delta^{13}C$ values or, in some cases, assumed $\delta^{13}C$ values (–25 for pollen or TOC, 0 for biogenic carbonate; Table 1).

Calibrated ages were calculated with the CALIB 5.01 pro-gram (Stuiver et al., 1998), using the terrestrial calibration data set. 1σ errors were used in the calibration procedure, and the median probability was used as the age estimate. Ages greater than 21,381 ^{14}C yr B.P. were converted to calibrated years using the relationship developed by Bard et al. (1998). The equation used is:

$$A = -3.0126*10^{-6}*C^2 + 1.2896*C - 1005,$$

where A is calibrated age and C is the age in radiocarbon years. Their 1σ ^{14}C errors were retained. Although other calibration schemes are available for ages older than the CALIB 5.01 data set, the Bard et al. (1998) equation was used for consistency with pre-vious analyses of Bear Lake data (Colman et al., 2005, 2006).

Reservoir corrections for TOC and carbonate samples were used in the calibration exercise, as described in the next section.

Age-Depth Modeling

In order to produce continuous records of various paleoen-vironmental proxies, age models are required for each core. We generated age models for the five cores discussed in detail, as well as for the upper part of the Bear Lake drill core (BL00-1) using two methods: (1) polynomial regression, and (2) a generalized additive model (GAM) regression using smooth cubic splines (Heegaard et al., 2005). Core BL02-4 was divided into three sec-tions, separated by two unconformities, for the age models.

Polynomials of various orders were fit to the data by regres-sion. In the case of each of the five cores, a second-order polyno-mial (quadratic) produced the best fit, as judged from R^2 values. These quadratic equations were all calculated with a zero-order coefficient; i.e., the core-top age was not specified. As shown later, they all have quite high R^2 values and do not have serious problems at the ends of the depth range that are common with polynomial fits. Ages at any depth are easily calculated from the equations. A disadvantage of this method is that the uncertainties associated with the curve fit are difficult to define.

Our second method of constructing age models uses newly developed statistical methods, which weigh data by their uncer-tainty and include both the uncertainty in the measurements and the uncertainty introduced by the regression procedure (Heegaard et al., 2005). These methods use weighted, nonparametric regres-sion within generalized additive models (GAM). Functions are fitted to the data using multiple smooth cubic splines; the degree of smoothing is determined by the number of spline functions (k). The methods (here called "spline fits") also produce confi-dence limits for the age model, based on uncertainties in both the control points and the regression procedure. We found that using a value for k equal to about half the number of control points yielded a good balance between smoothness and precision of fit. This balance follows the recommendation to use "the simplest parsimonious model," i.e., the "simplest statistically significant solution that uses the fewest terms in the model and the fewest degrees of freedom in the fitted smoother" (Birks and Heegaard, 2003). Because the sediment-water interface was recovered in the BL02-4 short core, an age of –52 yr B.P. (1950–2002) was used as a control point at zero depth. The long drill core (BL00-1) also appeared to recover nearly the entire sedimentary section, so, considering the scale of the ~250,000 yr record in the drill core, a simple (0,0) control point was used in the age-depth model. For all other cores, no control point was used for the core top.

One disadvantage of the spline-fit procedure is that it does not produce a single age equation with an associated R^2 value, and depths must be converted to ages by using tabulated data produced by the method. On the other hand, an advantage of this procedure is that it generates confidence intervals that can be used to infer the reliability of the age model with depth.

RESULTS

Different fractions of the same samples allow comparisons among TOC, pollen+, and biogenic carbonate (ostracodes). On general principles, the pollen+ samples are thought to be the most reliable, even though they may contain fragments of refractory organic matter, which may be significantly older than the enclosing sediment, in addition to pollen. In the glacial part of the section, however, this may not be true, as discussed in the next section. TOC samples contain all grain sizes and molecular forms of carbon and are likely to include detrital organic carbon that has been washed into the lake. Both biogenic and authigenic carbonate samples are sub-ject to reservoir effects, the size of which were not known a priori.

Biogenic and authigenic carbonate samples share this limitation equally; authigenic carbonate was not analyzed because of the additional potential problem of contamination with detrital carbonate.

The ^{14}C ages are subject to various sources of error, including contamination and mixed ^{14}C sources. TOC and biogenic carbonate are subject to inputs of carbon from two different reservoirs that may be depleted in ^{14}C compared to the atmosphere. First, reworked terrestrial organic matter and diagenetic activity (e.g., carbon bound to clay minerals) may contribute to TOC. Second, ^{14}C-depleted water may be used by aquatic organisms that produce both biogenic carbonate and organic carbon, thus affecting radiocarbon ages of both materials. Depletion of ^{14}C in lake water is a complex function of local bedrock, groundwater flow rates, mixing, and other aspects of the lake's hydrological budget.

Two pairs of pollen+ and TOC samples (Table 2) indicate a consistent difference between the two kinds of samples: pollen+ samples average 480 ± 105 yr younger than corresponding TOC samples. This result is consistent with the assumption that TOC samples contain more detrital organic carbon than the pollen+ samples. Although it unlikely that the difference between TOC and the pollen+ samples was constant through time, we corrected the relatively few TOC analyses by 480 ± 105 yr, treating it as a "reservoir effect" in the calibration process.

Five pairs of pollen+ and ostracode samples (Table 2) show a remarkably consistent relationship: pollen+ samples average 370 ± 105 yr younger than the ostracode samples. The consistency of this result suggests that difference is due to the radiocarbon content of the water being slightly out of equilibrium with that of the atmosphere; i.e., there is a reservoir effect of ~370 yr. Although the magnitude of the reservoir effect likely varies with time, no consistent trend with time was seen in our data set, and this reservoir correction was used in the calibration procedure for carbonate samples.

A number of samples produced anomalous results and were treated as outliers for various reasons ("Rejected" in Fig. 2). Most commonly, these were anomalously old ages near the tops of several cores (one in BL96-1, one in BL96-2, three in BL96-3, and three in BL02-4). In the case of BL02-4, the uppermost three ages (on pollen+), already mentioned, are at or above the geochemically defined horizon that marks the diversion of the Bear River into Bear Lake ca. 1912 (Dean, this volume). These samples apparently contain older materials transported during that

diversion and were thus rejected as a group. Diversion-related reworking may explain anomalously old ages near the tops of other cores, although, of the cores examined here, the diversion horizon has been identified only in BL02-3 and BL02-4. Core BL02-4 also contains two shallow-water, graded shell layers (Smoot, this volume) that are clearly unconformities. We therefore rejected three ages on gastropod shells that were probably reworked within these layers. In some cases (e.g., BL96-3), the Holocene section of the core is thin and clearly reworked (Colman, 2006), so ages in these reworked sediments were rejected (Fig. 2). Thin, reworked Holocene sediments tend to occur in most cores taken in present water depths of less than ~30 m.

In addition to the radiocarbon ages in Figure 2, a few other age constraints were used in the age models. The multiple cores taken at the sites of BL02-3 and BL02-4 ensured that the uppermost sediments and the sediment-water interface were recovered (see Methods). For these cores, various geochemical data are available that reveal the core depth at which the diversion of the Bear River into Bear Lake is recorded (Dean, this volume). We used the depth and approximate age (1912 common era; 38 cal yr B.P.) of this event ("Diversion" in Fig. 2). The depth of the diversion horizon in BL02-3 is only slightly above a much older radiocarbon age, suggesting some erosion between the two, but in BL02-4, the diversion horizon is compatible with the core top and the ages below. In addition, the sediments in the lower part of BL02-4 contain distinctive profiles of magnetic properties that allow close correlations with similar profiles in core BL96-2 (Rosenbaum et al., this volume). Three horizons from BL96-2, with interpolated ages from that core, were correlated to BL02-4 in this way, assuming the horizons were not time transgressive. These correlations were used as control points ("Mag correlation" in Fig. 2) in the lower part of core BL02-4.

Numerous ages were obtained from the red, siliclastic zone that is interpreted as containing rock flour deposited near the time of the last glacial maximum (LGM; Rosenbaum et al., 2002, this volume; Rosenbaum and Heil, this volume), especially in cores BL96-2 and BL96-3. These ages (indicated in Table 1) are generally in stratigraphic order (Fig. 2) and there was no a priori reason to suspect them. However, as discussed in the next section, we concluded that these ages are significantly too old and we developed an alternative strategy for age models in the red, siliclastic zone.

TABLE 2. AGES FOR POLLEN+ SAMPLES COMPARED
TO THOSE FOR TOTAL ORGANIC CARBON AND OSTRACODE SAMPLES

Pollen+	TOC	Ostracodes	TOC-Pollen	Ostracodes-Pollen
1070		1450		380
4230		4620		390
5260		5460		200
8580		9070		490
980		1350		370
6420	6970		550	
12,710	13,110		400	
		Mean[a]	480	370
		St. Dev.[a]	105	105

Note: Data from Table 1; all values in ^{14}C yr BP. TOC—Total Organic Carbon.
[a]Rounded to nearest 10 years (Mean) or 5 years (Standard Deviation).

EVALUATION OF RADIOCARBON AGES

Although some ages were rejected for the reasons discussed in the previous section, we made several other attempts to evaluate the overall accuracy of the remaining ages before developing age models for the cores. The radiocarbon ages that are less than 15 cal ka compare well with age estimates based on amino acid racemization. We used an independently derived equation for racemization of aspartic acid in ostracodes (Kaufman, 2000), with an effective temperature of 4.6 °C for the bottom of Bear Lake, to derive amino acid age estimates (Fig. 3). The two types of age estimates are entirely consistent (Fig. 3), lending strong support for the validity of ages that are <15 cal ka.

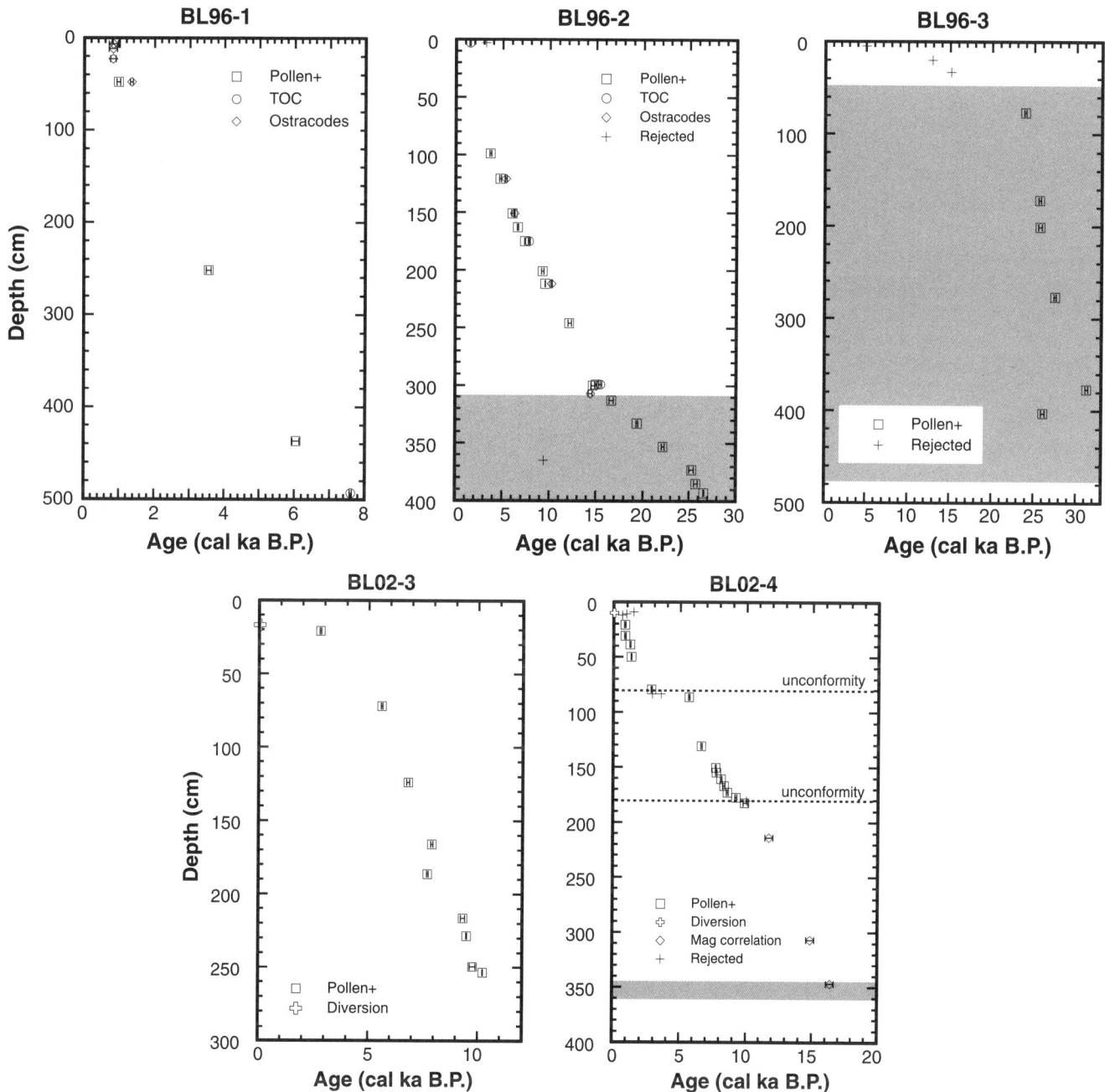

Figure 2. Radiocarbon ages for five Bear Lake cores plotted against depth in the sediment sequence. Depth is the independent variable in these plots. Data from Table 1. Ostracode and total organic carbon (TOC) ages were corrected for reservoir effects (see text and Table 2). One rejected age for core BL96-1 is off scale of plot. 1σ errors are shown. See text for discussion of rejected, diversion, and "Mag correlation" ages. Shaded areas indicate the red, siliclastic unit of Rosenbaum et al. (this volume).

A second method of evaluating the radiocarbon ages involves comparisons of ages for several distinct horizons that occur in two or more cores and that can be confidently correlated on the basis of detailed analytical data. Age estimates for each of these horizons can be estimated from the age models for each core. Where the horizon is dated in multiple cores, a more robust age for each horizon can then be generated by calculating a mean of the ages derived from each core. These horizons and their depths in various cores are given in Table 3. In our initial attempts (not shown) at age modeling to produce age estimates for the horizons, two results emerged. First, ages for these horizons in sediments from different cores above the red, siliclastic unit were remarkably consistent, in most cases essentially identical within the errors. Second, model ages for horizons within the red, siliclastic unit were commonly incompatible between cores BL96-2 and BL96-3. This discrepancy was our first indication that a problem existed with the ages from the red, siliclastic unit.

Another problem with the ages in the red, siliclastic zone involves the age of the local LGM. Taken at face value, the ages in the red, siliclastic zone (BL96-2 and BL96-3, Fig. 2), combined with geochemical and magnetic indicators of rock flour (Rosenbaum and Heil, this volume; Rosenbaum et al., 2002), suggest that the local LGM occurred as much as 25 k.y. ago. In contrast, cosmogenic radionuclide exposure ages for terminal moraines in the headwaters of the Bear River (Uinta Mountains) are much younger. Cosmogenic-exposure analyses from two LGM moraines in the Uinta Mountains yield age estimates of 17.1 ± 0.7 ka ($n = 5$) and 18.5 ± 0.7 ka ($n = 6$) (Laabs et al., 2007). On the basis of cosmogenic-exposure dating through-out the western United States, Licciardi et al. (2004) suggested that there were two pulses at the LGM, at 17 and at 21 cal ka, although Pierce (2004) points out that the cosmogenic-exposure ages are consistently younger than radiocarbon ages for comparable events. In any case, the radiocarbon ages from the red, siliclastic zone yield an apparent age of the LGM at Bear Lake that appears to be significantly too old compared to nearby estimates for the age of the LGM.

Even though the ages for the red, siliclastic unit in BL96-2 and BL96-3 progressively increase in age with depth, the organic carbon content in these samples is very low. Furthermore, the pollen+ samples from this interval that have been examined under the microscope contain little pollen, and that pollen is badly degraded (R. Thompson, 2004, personal commun.). This suggests that the material is mainly refractory organic matter, and as such, may be significantly older than the sediment in which it was deposited. For this and the other reasons discussed above (the ages appear to be too old compared with other LGM records), we decided to reject the ages from the red, siliclastic unit in the age models for the cores.

AGE MODELS AND DISCUSSION

Colman et al. (2006) developed an age model for the entire 120 m of Bear Lake drill core BL00-1. This model used the radiocarbon ages from the red, siliclastic unit in BL96-2 and BL96-3, correlated to BL00-1 on the basis of magnetic susceptibility profiles. These ages produced a kink in the age model indicating relatively slow rates of sedimentation for the glacial times represented by the red, siliclastic unit. Because we now believe that the radiocarbon ages from the red, siliclastic unit are several thousand years too old, we have recalculated the spline-fit age model for the upper 20 m of the drill core, excluding those ages (Fig. 4). In addition to providing a more accurate age model for the upper part of the drill core, the lithologic horizons (Table 3, Fig. 4) are well correlated among the cores. Thus we can use their ages—derived from the age model for the upper part of BL00-1—for age control in the red, siliclastic unit. These ages are given in Table 3.

Above the red, siliclastic unit, control for the age model of the upper 20 m of the drill core comes from radiocarbon ages from BL96-1 and BL96-2, correlated to BL00-1 on the basis of magnetic susceptibility profiles (Rosenbaum et al., this volume). The lowest radiocarbon age in BL96-3 is below the red, siliclastic unit, and since the Colman et al. (2006) study, we have obtained two new pollen+ radiocarbon ages from below the red, siliclastic unit in BL00-1 (Table 1). Microscopic examination of the three associated samples showed that they contain much more well-preserved pollen than samples from the red, siliclastic unit, and significantly, all three resulting ages are notably younger than ages in the red, siliclastic unit (Fig. 4). These ages are used to constrain the age model through the red, siliclastic unit to 20 m depth in the drill core.

For the five piston cores in Bear Lake, age models were constructed as described in the Methods section. Two types of control points were used for these age models (Fig. 5). Above the red, siliclastic zone, the ages shown in Figure 2 were used

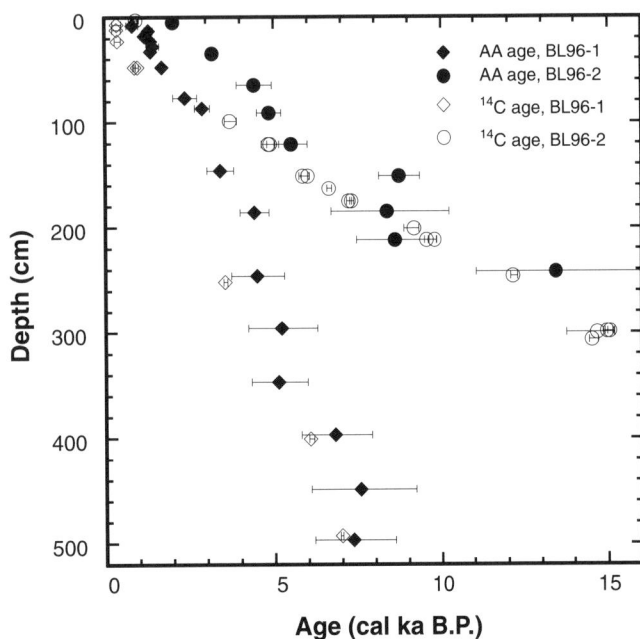

Figure 3. Comparison between ages estimated from radiocarbon (^{14}C) and amino acid racemization (AA) methods for two Bear Lake cores. See text for explanation.

TABLE 3. AGES OF PROMINENT STRATIGRAPHIC HORIZONS PRESENT IN MORE THAN ONE CORE, DERIVED FROM SPLINE-FIT AGE MODELS

Horizon	Data type	Comment	Depth (m) in BL96-3	Depth (m) in BL02-4	Depth (m) in BL96-2	Depth (m) in BL96-1	Depth (m) in BL00-1	Age or mean age[a]	Error[a]
AA1	AA-14C	AA-^{14}C match			0.10	0.55		1.43	0.16
D1	Diatoms	*N. oblongata* spike			0.25	0.97		1.86	0.06
D2	Diatoms	Top, zone 3b			0.63	1.67		2.75	0.20
Mag1	Mag. susc.					2.42	3.24	3.41	
Mag2	Mag. susc.					3.46	4.48	4.59	
Mag3	Mag. susc.					3.97	5.22	5.38	
D3	Diatoms	Bottom, zone 3b			1.50	4.27		6.09	0.17
AA2	AA-14C	AA-^{14}C match			1.40	4.50		6.11	0.53
Mag4	Mag. susc.					4.45		6.37	
Min1	Mineralogy	Midpoint, calcite decrease	<0.4	1.55	1.80		5.79	7.81	0.03
R1	Seismic	Base of upper marl	<0.4		1.90		7.30	8.42	
Min2	Mineralogy	Midpoint, calcite increase	<0.4	1.70	1.95		7.60	8.63	0.13
Mag5	Mag. susc.			2.14	2.35		8.65	11.12	0.11
Min3	Mineralogy	Begin decrease in calcite	<0.4	2.14	2.37		8.39–9.89	11.18	0.20
Mag6	Mag. susc.			3.07	2.92		9.82	14.60	0.26
R2	Seismic	Top of red unit	<0.40		3.10		10.40	16.31	0.39
Min4	Mineralogy	Begin calcite, qtz decline	<0.40	3.47	3.47		10.03–10.24	15.60	0.38
Mag7	Mag. susc.		<0.40	3.47	3.47		10.42	16.36	0.39
Mag8	HIRM		0.76		3.62		11.02	17.74	0.42
Mag9	HIRM		0.86		3.70		11.24	18.22	0.43
Mag10	HIRM		1.05		3.88		11.38	18.52	0.44
Mag11	Mag. susc.		1.47				12.10	19.98	0.47
Mag12	Mag. susc.		1.65				12.46	20.67	0.49
Mag13	Mag. susc.		1.85				12.85	21.37	0.50
Mag14	Mag. susc.		2.01				13.11	21.81	0.52
Mag15	Mag. susc.		2.78				14.44	23.86	0.58
Mag16	Mag. susc.		3.55				15.78	25.54	0.64
Mag17	HIRM		4.01				16.68	26.47	0.68

Note: Gray shading indicates horizons in or bounding the red, siliclastic unit. HIRM—hard isothermal remanent magnetization; Mag. susc.—magnetic susceptibility.

[a]Age and error are estimated differently for horizons above and below the top of the red, siliclastic unit. For those above, the age is the average of ages derived from spline-fit age-depth model (Fig. 5) for the different cores in which the depth is given (blank error column indicates age from only one core). For horizons below top of the red, siliclastic unit, the age and error are derived from the age model for BL00-1 (Fig. 4).

directly as control points (BL96-1 and BL02-3 do not penetrate to the unit). Within the red, siliclastic unit and in the lowest section of BL02-4, ages of the lithological horizons derived from the drill core (Fig. 4) were transferred to other cores according to the depth correlations in Table 3.

The two different types of age models (quadratic and spline fit) for each core are very similar (Fig. 5), and in many cases (e.g., BL96-3 and BL02-4) are nearly identical. No regression is shown for BL00-1 (see Fig. 4). The spline-fit model is better able to handle multiple or irregular changes in apparent sedimentation rate, although such changes are minor in the deep-water Bear Lake cores. The spline-fit method has the advantage of providing confidence intervals for the age model, although the confidence intervals are relatively large because they account for uncertainties in both the control points and the regression procedure. The average 95% confidence interval for 8000 cal yr B.P. (five cores, not BL96-3), is about ± 280 yr (Fig. 5). This age has relatively high data density; sparser data lead to larger confidence intervals.

Because the primary purpose of this study is to provide chronologies for paleoenvironmental proxies, mass accumulation rates (e.g., g cm^{-2} yr^{-1}) are not calculated here. Density data are available (Rosenbaum et al., this volume), and these data have been used to calculate carbon and carbonate mass accumulation rates (Dean et al., 2006). However, sedimentation rates (in mm yr^{-1}) are still of interest. Sedimentation rates in the cores generally decrease with depth, for most cores creating a convex-upward shape in the age-depth plots (Fig. 3). Presumably, this is due at least in part to progressive compaction and diagenesis of the sediments. In the uppermost, least compacted sediments, ^{210}Pb data indicate sedimentation rates in two cores of 0.76 and 0.91 mm yr^{-1} (Smoak and Swarzenski, 2004). For the Holocene section in the deep part of the lake, sedimentation rates are as high as 0.64 mm yr^{-1} (BL96-1) to 0.80 mm yr^{-1} (BL00-1). At shallower sites, Holocene sedimentation rates are 0.30 mm yr^{-1} (BL96-2) to 0.37 mm yr^{-1} (BL02-4). Slow apparent sedimentation rates occur in the upper part of relatively shallow water cores (BL96-3 and BL02-3) because of reworking and (or) erosion of the surface sediments. From the drill core (BL00-1), sedimentation rates for the past 220,000 years or so average ~0.5 ± 0.03 mm yr^{-1} (Colman et al., 2006; Kaufman et al., this volume) and show remarkably little variation with time.

CONCLUSIONS

Radiocarbon analyses of total organic carbon, pollen, and ostracodes provide a reliable chronology of post-glacial sediments deposited in Bear Lake. The differences in apparent age between TOC, pollen, and carbonate fractions are consistent and in accord with the origins of these fractions. The data are also in accord with ages independently estimated from aspartic acid racemization in ostracodes. Ages in the red, siliclastic unit, inferred to be of last glacial age, are several thousand years too old, seemingly because of a high proportion of refractory organic carbon in the pollen samples.

Age-depth models for five piston cores and the Bear Lake drill core (BL00-1) were constructed using two methods: quadratic equations and smooth cubic-spline fits. The two types of age models, each of which has its own advantages, are compatible for the Bear Lake cores, differing only in detail. Specific horizons defined by paleontologic, mineralogic, or magnetic properties were dated in several cores and correlated among them. The average of the interpolated ages for these horizons provides more robust age estimates. The age of the correlated horizons in the red, siliclastic unit can be estimated from the age model for BL00-1, which is controlled by ages above and below the red, siliclastic unit. These ages can then be transferred to the correlative horizons in the shorter piston cores, providing control for the sections of the age models in those cores in the red, siliclastic unit. These age models are the backbone for reconstructions of past environmental conditions in Bear Lake. In general, sedimentation rates in the deeper parts of Bear Lake have been quite

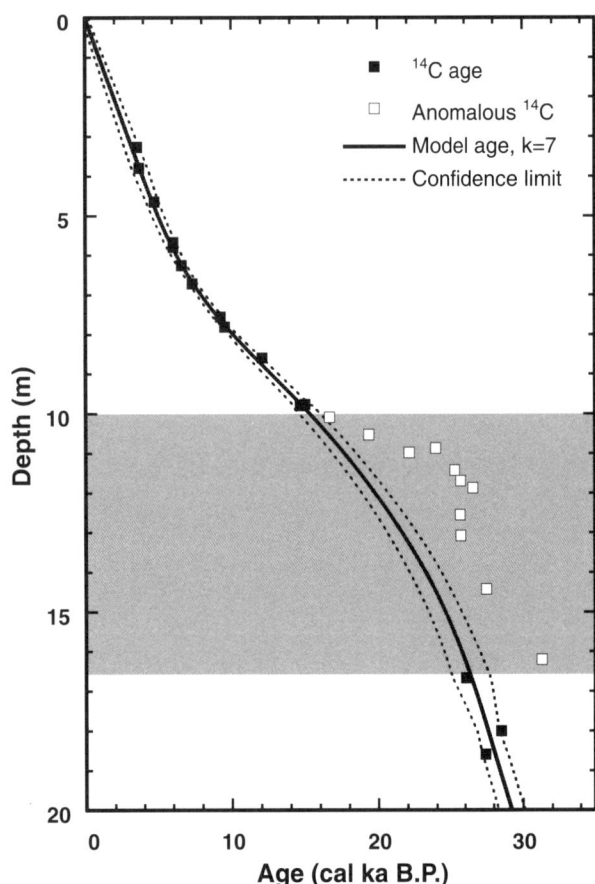

Figure 4. Age model for the upper part of core BL00-1. Shaded area indicates the red, siliclastic unit of Rosenbaum et al. (this volume). Depth is the independent variable in this plot. Control for the age model comes from radiocarbon ages above and below the red, siliclastic unit, as described in the text. Radiocarbon ages from within the red, siliclastic unit from cores BL96-2 and BL96-3, correlated by depth on the basis of magnetic susceptibility profiles, are labeled as "Anomalous ^{14}C." No polynomial regression is shown for the upper part of BL00-1 because no quadratic equation (as used for the other cores) or other low-order polynomial produced a satisfactory fit to the data.

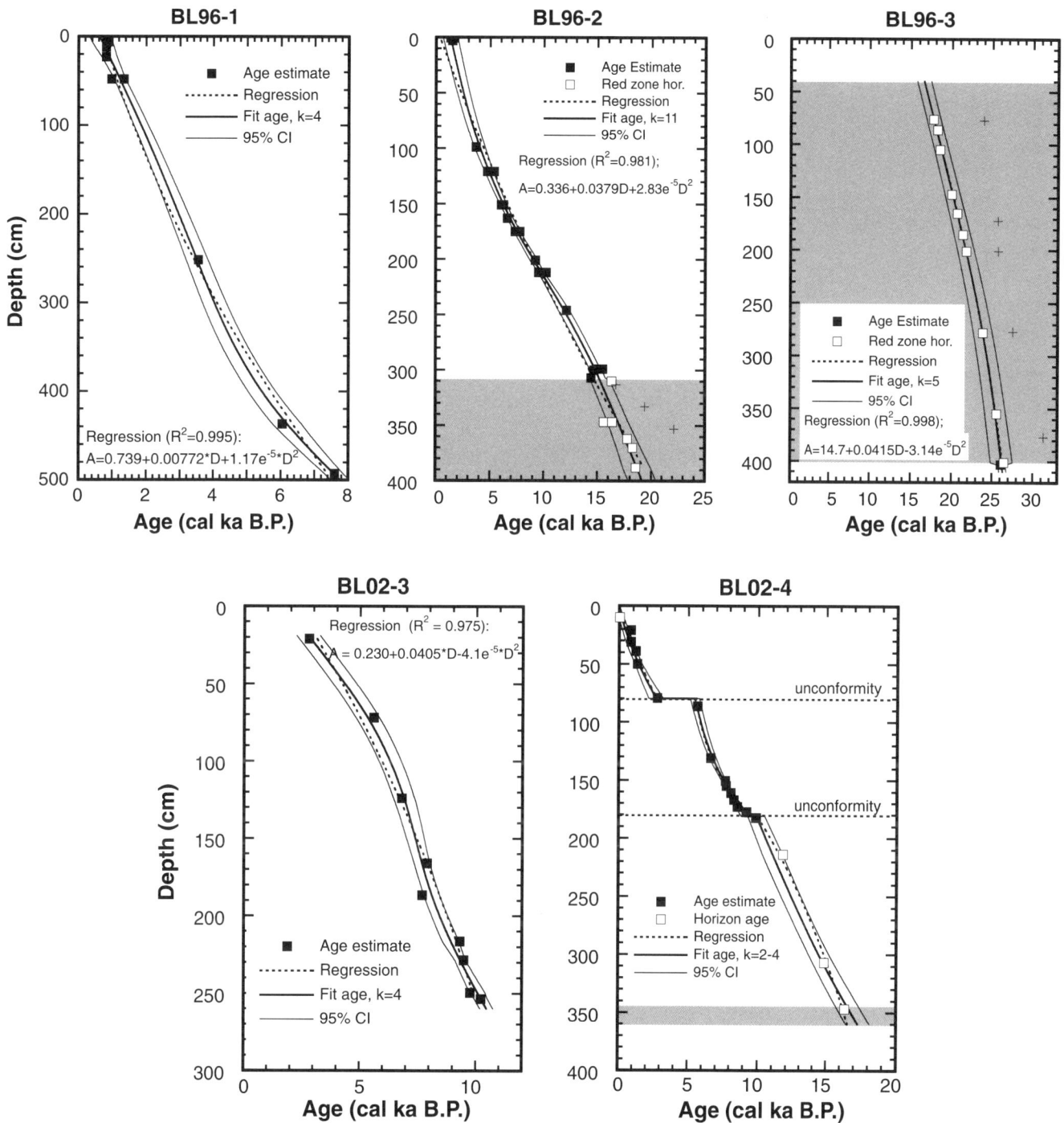

Figure 5. Age models for five Bear Lake cores. Shaded area indicates the red, siliclastic unit of Rosenbaum et al. (this volume). Depth is the independent variable in these plots. Data points outside the red, siliclastic unit are from Figure 2, excluding rejected ages. Data points within the red, siliclastic unit ("Red zone hor.") and in the lowest section of BL02-4 ("Horizon age") are ages of the lithological horizons derived from BL00-1 (Fig. 4), which were transferred to other cores according to the depth correlations in Table 3. Thick dashed line is a quadratic regression through the data; thick solid line is a spline fit (see text). Thin solid lines are the 95% confidence limits for the spline fit. Also shown (pluses) are the anomalously old radiocarbon ages in the red, siliclastic unit in cores BL96-2 and BL96-3 (cf. Fig. 2). The model for BL02-4 was done in three sections for the parts of the core separated by two unconformities at 80 and 181 cm. From the top, these regressions are:

$A = -0.0523 + 0.0263 * D + 1.09e^{-4} * D^2$ ($R^2 = 0.967$)

$A = 7.01 - 0.0415 * D + 2.99e^{-4} * D^2$ ($R^2 = 0.988$)

$A = 8.13 - 0.00561 * D + 8.78e^{-5} * D^2$ ($R^2 = 0.996$).

uniform, mostly between 0.3 and 0.8 mm yr^{-1} in the Holocene, and close to 0.5 mm yr^{-1} for the longer term.

Deriving age models from radiocarbon ages of lake sediments commonly is a difficult process, and Bear Lake proved not to be an exception. Different types of samples each have potential problems, ranging from reservoir effects to terrestrial residence times, all of which can lead to radiocarbon ages that are different than the age of the sediment from which the sample was taken. We experienced most of these problems with samples from Bear Lake, including erroneous ages that were in an attractive stratigraphic progression. However, by analyzing multiple sediment fractions, performing stratigraphic and correlation comparisons, and using multiple consistency checks, we believe that we have generated valid and useful age models for the past 30,000 years of sedimentation in Bear Lake.

ACKNOWLEDGMENTS

We thank the many participants in the Bear Lake project for useful discussions and ideas. Helpful reviews of early versions of this paper were provided by Marith Reheis, Lesleigh Anderson, Eric Grimm, and John Peck.

ARCHIVED DATA

Archived data for this chapter can be obtained from the NOAA World Data Center for Paleoclimatology at http://www.ncdc.noaa.gov/paleo/pubs/gsa2009bearlake/.

REFERENCES CITED

Bard, E., Arnold, M., Hamelin, B., Tisnerat-Laborde, N., and Cabioch, G., 1998, Radiocarbon calibration by means of mass spectrometric ^{230}Th/^{234}U and ^{14}C ages of corals: An updated database including samples from Barbados, Mururoa, and Tahiti: Radiocarbon, v. 40, p. 1085–1092.

Colman, S.M., 2006, Acoustic stratigraphy of Bear Lake, Utah-Idaho–Late Quaternary sedimentation patterns in a simple half-graben: Sedimentary Geology, v. 185, p. 113–125, doi: 10.1016/j.sedgeo.2005.11.022.

Colman, S.M., Jones, G.A., Forester, R.M., and Foster, D.S., 1990, Holocene paleoclimatic evidence and sedimentation rates from a core in southwestern Lake Michigan: Journal of Paleolimnology, v. 4, p. 269–284, doi: 10.1007/BF00239699.

Colman, S.M., Kaufman, D.S., Rosenbaum, J.G., and McGeehin, J.P., 2005, Radiocarbon dating of cores collected in Bear Lake, Utah: U.S. Geological Survey Open-File Report 05-1320, 12 p.

Colman, S.M., Kaufman, D.S., Bright, J., Heil, C., King, J.W., Dean, W.E., Rosenbaum, J.G., Forester, R.M., Bischoff, J.L., and Perkins, M., 2006, Age models for a continuous 250-kyr Quaternary lacustrine record from Bear Lake, Utah-Idaho: Quaternary Science Reviews, v. 25, p. 2271–2282, doi: 10.1016/j.quascirev.2005.10.015.

Dean, W.E., 2009, this volume, Endogenic carbonate sedimentation in Bear Lake, Utah and Idaho, over the last two glacial-interglacial cycles, *in* Rosenbaum, J.G., and Kaufman, D.S., eds., Paleoenvironments of Bear Lake, Utah and Idaho, and its catchment: Geological Society of America Special Paper 450, doi: 10.1130/2009.2450(07).

Dean, W.E., Rosenbaum, J.G., Skipp, G., Colman, S.M., Forester, R.M., Liu, A., Simmons, K.R., and Bischoff, J.L., 2006, Unusual Holocene and late Pleistocene carbonate sedimentation in Bear Lake, Utah-Idaho, USA: Sedimentary Geology, v. 185, p. 93–112, doi: 10.1016/j.sedgeo.2005.11.016.

Dean, W.E., Wurtsbaugh, W., and Lamarra, V., 2009, this volume, Climatic and limnologic setting of Bear Lake, Utah and Idaho, *in* Rosenbaum, J.G., and Kaufman, D.S., eds., Paleoenvironments of Bear Lake, Utah and Idaho, and its catchment: Geological Society of America Special Paper 450, doi: 10.1130/2009.2450(01).

Faegri, K., and Iverson, J., 1975, Textbook of pollen analysis (3rd edition): Copenhagen, Hafner Press.

Heegaard, E., Birks, H.J.B., and Telford, R.J., 2005, Relationships between calibrated ages and depth in stratigraphical sequences: An estimation procedure by mixed-effect regression: The Holocene, v. 15, p. 612–618, doi: 10.1191/0959683605hl836rr.

Jones, G.A., Jull, A.J.T., Linick, T.W., and Donahue, D.J., 1989, Radiocarbon dating of deep-sea sediments—A comparison of accelerator mass spectrometer and beta-decay methods: Radiocarbon, v. 31, p. 104–116.

Kaufman, D.S., 2000, Amino acid racemization in ostracodes, *in* Goodfriend, G., Collins, M., Fogel, M., Macko, S., and Wehmiller, J., eds., Perspectives in amino acid and protein geochemistry: New York, Oxford University Press, p. 145–160.

Kaufman, D.S., Bright, J., Dean, W.E., Moser, K., Rosenbaum, J.G., Anderson, R.S., Colman, S.M., Heil, C.W., Jiménez-Moreno, G., Reheis, M.C., and Simmons, K.R., 2009, this volume, A quarter-million years of paleoenvironmental change at Bear Lake, Utah and Idaho, *in* Rosenbaum, J.G., and Kaufman, D.S., eds., Paleoenvironments of Bear Lake, Utah and Idaho, and its catchment: Geological Society of America Special Paper 450, doi: 10.1130/2009.2450(14).

Kelts, K., Briegel, U., Ghilardi, K., and Hsu, K., 1986, The limnogeology-ETH coring system: Schweizerische Zeitschrift für Hydrologie, v. 48, p. 104–115, doi: 10.1007/BF02544119.

Laabs, B.J.C., Munroe, J.S., Rosenbaum, J.G., Refsnider, K.A., Mickelson, D.M., Singer, B.S., and Chafee, M.W., 2007, Chronology of the last glacial maximum in the upper Bear River Basin, Utah: Arctic, Antarctic, and Alpine Research, v. 39, p. 537–548, doi: 10.1657/1523-0430(06-089)[LAABS]2.0.CO;2.

Licciardi, J.M., Clark, P.U., Brook, E.J., Elmore, D., and Sharma, P., 2004, Variable responses of western U.S. glaciers during the last glaciation: Geology, v. 32, p. 81–84, doi: 10.1130/G19868.1.

Pierce, K.L., 2004, Pleistocene glaciations of the Rocky Mountains, *in* Gillespie, A.R., Porter, S.C., and Atwater, B.F., eds., The Quaternary Period in the United States: Developments in Quaternary Science, v. 1: Amsterdam, Elsevier, p. 63–76.

Rosenbaum, J.G., and Heil, C.W., Jr., 2009, his volume, The glacial/deglacial history of sedimentation in Bear Lake, Utah and Idaho, *in* Rosenbaum, J.G., and Kaufman, D.S., eds., Paleoenvironments of Bear Lake, Utah and Idaho, and its catchment: Geological Society of America Special Paper 450, doi: 10.1130/2009.2450(11).

Rosenbaum, J.G., and Kaufman, D.S., 2009, this volume, Introduction to *Paleoenvironments of Bear Lake, Utah and Idaho, and its catchment*, *in* Rosenbaum, J.G., and Kaufman, D.S., eds., Paleoenvironments of Bear Lake, Utah and Idaho, and its catchment: Geological Society of America Special Paper 450, doi: 10.1130/2009.2450(00).

Rosenbaum, J.G., Dean, W.E., Colman, S.M., and Reynolds, R.L., 2002, Magnetic signature of glacial flour in sediments from Bear Lake, Utah/Idaho: Eos (Transactions, American Geophysical Union), v. 83, no. 47, Abstract GP12B-03.

Rosenbaum, J.G., Dean, W.E., Reynolds, R.L., and Reheis, M.C., 2009, this volume, Allogenic sedimentary components of Bear Lake, Utah and Idaho, *in* Rosenbaum, J.G., and Kaufman, D.S., eds., Paleoenvironments of Bear Lake, Utah and Idaho, and its catchment: Geological Society of America Special Paper 450, doi: 10.1130/2009.2450(06).

Slota, P.J.J., Jull, A.J.T., Linick, T.W., and Toolin, L.J., 1987, Preparation of small samples for ^{14}C accelerator targets by catalytic reduction of CO: Radiocarbon, v. 29, p. 303–306.

Smoak, J.M., and Swarzenski, P.W., 2004, Recent increases in sediment and nutrient accumulation in Bear Lake, Utah/Idaho, USA: Hydrobiologia, v. 525, p. 175, doi: 10.1023/B:HYDR.0000038865.16732.09.

Smoot, J.P., 2009, this volume, Late Quaternary sedimentary features of Bear Lake, Utah and Idaho, *in* Rosenbaum, J.G., and Kaufman, D.S., eds., Paleoenvironments of Bear Lake, Utah and Idaho, and its catchment: Geological Society of America Special Paper 450, doi: 10.1130/2009.2450(03).

Stuiver, M., and Pollach, H.A., 1977, Reporting ^{14}C data: Discussion: Radiocarbon, v. 19, p. 355–363.

Stuiver, M., Reimer, P.J., and Braziunas, T.F., 1998, High-precision radiocarbon age calibration for terrestrial and marine samples: Radiocarbon, v. 40, p. 1127–1151.

Vogel, J.S., Southon, J.R., Nelson, D.E., and Brown, T.A., 1984, Performance of catalytically condensed carbon for use in accelerator mass spectrometry: Nuclear Instruments and Methods in Physics Research B, v. 5, p. 289–293.

MANUSCRIPT ACCEPTED BY THE SOCIETY 15 SEPTEMBER 2008

The Geological Society of America
Special Paper 450
2009

Allogenic sedimentary components of Bear Lake, Utah and Idaho

Joseph G. Rosenbaum
Walter E. Dean
Richard L. Reynolds
Marith C. Reheis
U.S. Geological Survey, Box 25046, Federal Center, Denver, Colorado 80225, USA

ABSTRACT

Bear Lake is a long-lived lake filling a tectonic depression between the Bear River Range to the west and the Bear River Plateau to the east, and straddling the border between Utah and Idaho. Mineralogy, elemental geochemistry, and magnetic properties provide information about variations in provenance of allogenic lithic material in last-glacial-age, quartz-rich sediment in Bear Lake. Grain-size data from the siliciclastic fraction of late-glacial to Holocene carbonate-rich sediments provide information about variations in lake level. For the quartz-rich lower unit, which was deposited while the Bear River flowed into and out of the lake, four source areas are recognized on the basis of modern fluvial samples with contrasting properties that reflect differences in bedrock geology and in magnetite content from dust. One of these areas is underlain by hematite-rich Uinta Mountain Group rocks in the headwaters of the Bear River. Although Uinta Mountain Group rocks make up a small fraction of the catchment, hematite-rich material from this area is an important component of the lower unit. This material is interpreted to be glacial flour. Variations in the input of glacial flour are interpreted as having caused quasi-cyclical variations in mineralogical and elemental concentrations, and in magnetic properties within the lower unit. The carbonate-rich younger unit was deposited under conditions similar to those of the modern lake, with the Bear River largely bypassing the lake. For two cores taken in more than 30 m of water, median grain sizes in this unit range from ~6 μm to more than 30 μm, with the coarsest grain sizes associated with beach or shallow-water deposits. Similar grain-size variations are observed as a function of water depth in the modern lake and provide the basis for interpreting the core grain-size data in terms of lake level.

INTRODUCTION

Bear Lake Valley, on the border between Utah and Idaho, is formed by a tectonically active half-graben (Colman, 2006; Reheis et al., this volume) between the Bear River Range to the west and the Bear Lake Plateau to the east (Fig. 1). The southern part of the valley is occupied by Bear Lake, which is ~32 km long, 6–13 km wide, and 63 m deep. The Bear River flows into the valley north of the lake. During historic times the river did not enter the lake until it was diverted ca. 1912 (Dean et al., this volume) but did flow into the lake at various times during the late Pleistocene (Kaufman et al., this volume; Reheis et al., this volume).

Rosenbaum, J.G., Dean, W.E., Reynolds, R.L., and Reheis, M.C., 2009, Allogenic sedimentary components of Bear Lake, Utah and Idaho, *in* Rosenbaum, J.G., and Kaufman, D.S., eds., Paleoenvironments of Bear Lake, Utah and Idaho, and its catchment: Geological Society of America Special Paper 450, p. 145–168, doi: 10.1130/2009.2450(06). For permission to copy, contact editing@geosociety.org. ©2009 The Geological Society of America. All rights reserved.

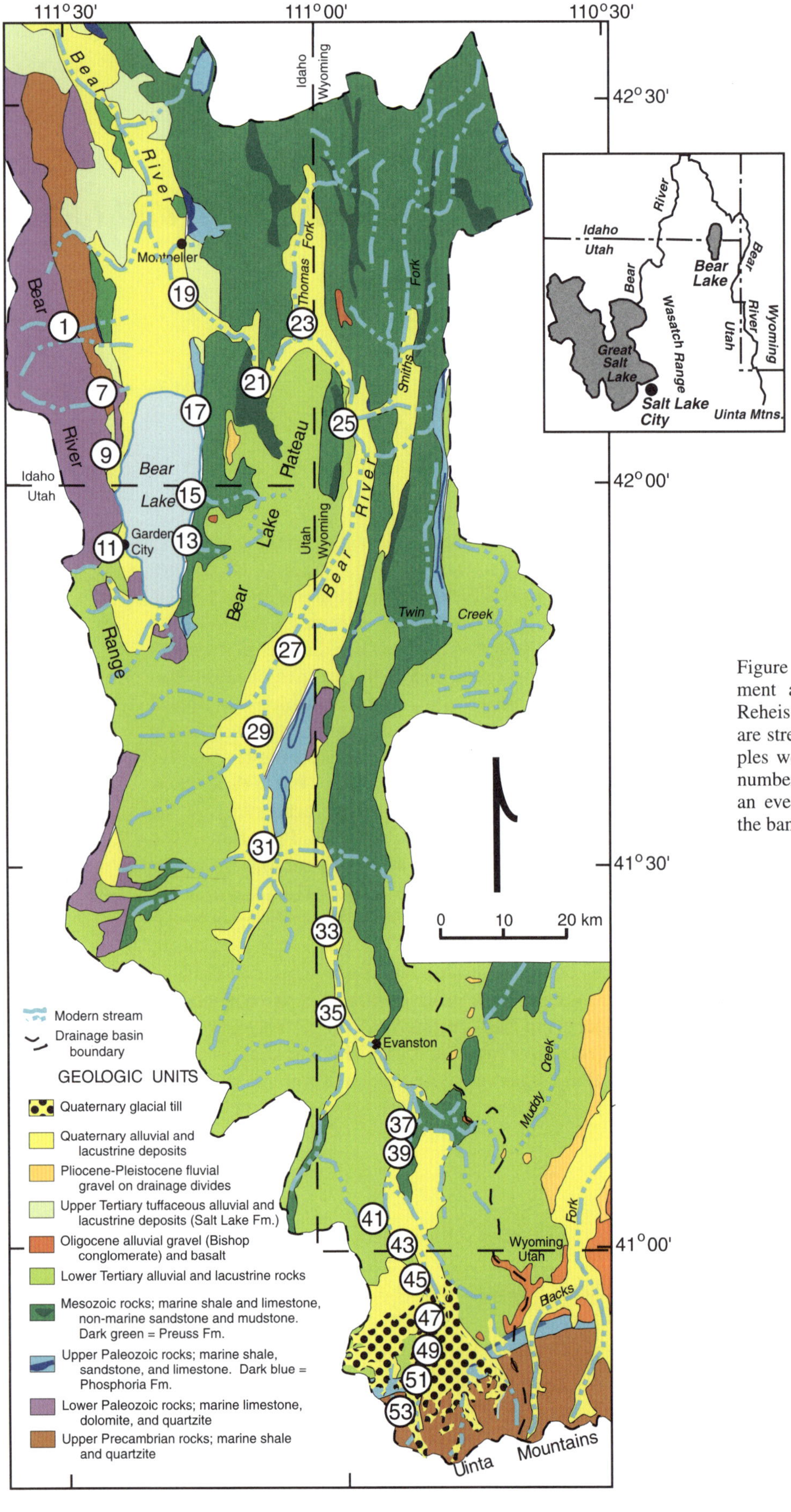

111°30'
111°00'
110°30'
42°30'
42°00'
41°30'
41°00'
Idaho
Wyoming
Utah
Bear River
Montpelier
Thomas Fork
Smiths Fork
Bear Lake
Garden City
Bear Lake Plateau
Bear Range
Twin Creek
Evanston
Muddy Creek
Blacks Fork
Uinta Mountains
Great Salt Lake
Salt Lake City
Wasatch Range
Uinta Mtns.
0 10 20 km
Modern stream
Drainage basin boundary
GEOLOGIC UNITS
Quaternary glacial till
Quaternary alluvial and lacustrine deposits
Pliocene-Pleistocene fluvial gravel on drainage divides
Upper Tertiary tuffaceous alluvial and lacustrine deposits (Salt Lake Fm.)
Oligocene alluvial gravel (Bishop conglomerate) and basalt
Lower Tertiary alluvial and lacustrine rocks
Mesozoic rocks; marine shale and limestone, non-marine sandstone and mudstone. Dark green = Preuss Fm.
Upper Paleozoic rocks; marine shale, sandstone, and limestone. Dark blue = Phosphoria Fm.
Lower Paleozoic rocks; marine limestone, dolomite, and quartzite
Upper Precambrian rocks; marine shale and quartzite

Figure 1. Generalized geologic map of catchment areas for Bear Lake (modified from Reheis et al., this volume). Numbered circles are stream sediment sampling sites. Two samples were collected at each location, an odd-numbered sample from the stream bottom and an even-numbered sample (not shown) from the bank or the overbank deposit.

In the absence of the Bear River, surface flow into Bear Lake is from a number of streams in a small drainage basin. This "local" catchment can be divided into two parts on the basis of bedrock geology (Oriel and Platt, 1980; Dover, 1995). The Bear River Range to the west of the lake is largely underlain by lower Paleozoic formations containing large amounts of dolomite, quartzite, and limestone, with lesser amounts of Precambrian quartzite and shale, as well as clastic Tertiary rocks of the Wasatch Formation. Rocks constituting the Bear Lake Plateau on the east side of the lake consist of abundant Tertiary Wasatch Formation and Mesozoic limestone and clastic sedimentary rocks.

When the Bear River flows into Bear Lake the lake's catchment is much larger. Bedrock along most of the Bear River upstream from Bear Lake is similar to that in the Bear River Plateau, with widespread Tertiary and Mesozoic sedimentary rocks. Bedrock in the river's headwaters, however, comprises Precambrian quartzites and shales of the Uinta Mountain Group (Bryant, 1992).

Bear Lake sediments contain both endogenic and allogenic components. The endogenic component, which consists of minerals formed in the water column and the remains of biota that grew in the lake, directly reflects physical and chemical conditions of the lake. In contrast, allogenic components, which are transported to the lake by water and wind, largely reflect catchment conditions and processes. Some allogenic components may be affected by post-depositional alteration and thereby in part reflect lake conditions rather than catchment conditions.

Here we report data that largely reflect the allogenic component of Bear Lake sediments as well as data from the potential source areas of this component. The data derived by individual techniques, however, do not separate cleanly into groupings that yield information about solely endogenic or solely allogenic materials. For instance, X-ray diffraction data for the lake sediments reported by Dean (this volume), contain information about endogenic carbonate minerals (e.g., calcite and aragonite) as well as information about allogenic minerals (e.g., quartz, calcite, and dolomite). Data from lake sediments presented here (elemental geochemistry on bulk samples, magnetic properties, and grain size of the siliciclastic material) primarily provide information about material derived from the catchment and secondarily about processes within the lake. The bulk-sample elemental data reflect the contents of both carbonate and siliciclastic minerals. Additional elemental data reported by Bischoff et al. (2005) and Dean (this volume), which were acquired from the HCl-soluble component, largely reflect the carbonate minerals of both endogenic and detrital origins. Magnetic properties reflect not only the detrital minerals delivered to the lake but also the effects of post-depositional processes that commonly destroy Fe-oxide minerals or form secondary magnetic phases. Similarly, grain-size distributions of siliciclastic detrital material are affected not only by the sizes of material delivered to the lake but also by sorting during settling and resuspension within the lake.

The large volume of data reported herein contributes to a number of important interpretations presented in other chapters of this volume. For example, (1) variations in mineralogy and magnetic properties are used to correlate stratigraphic horizons among cores (Colman et al., this volume); (2) mineralogical data indicative of provenance help establish the presence or absence of Bear River input to the lake (Kaufman et al., this volume); (3) grain-size data contribute to the interpretation of lake-level history (Smoot and Rosenbaum, this volume); and (4) magnetic property, mineralogical, geochemical, and grain-size data provide the basis for a glacial history based on glacial flour in the lake sediments (Rosenbaum and Heil, this volume).

METHODS

Sampling

Fluvial Sediment

Streams draining the Bear River Range and the Bear River Plateau, as well as the Bear River are potential sources of fluvial input to Bear Lake (Fig. 1). Because environmental change can alter the amounts of material derived from areas of different bedrock lithologies, such change may be reflected in the mineralogical, chemical, and physical properties of lithic materials in the lake sediments. Characterization of materials from different parts of the drainage basins can help constrain interpretations of environmental change based on data obtained from lake-sediment cores.

Samples of fluvial material were collected from (1) streams draining the east and west sides of the local Bear Lake catchment and (2) the Bear River upstream from Bear Lake (Fig. 1). Two samples were taken at each site, an odd-numbered sample from the stream bottom and an even-numbered sample from the stream bank or overbank deposit. The samples, which include all grain sizes up to a few centimeters in diameter, integrate lithologies upstream from the sampling sites. The samples were sieved into four grain-size ranges: pebbles (2 mm to several cm), coarse sand and granules (0.42 mm to ~2 mm), fine and medium sand (0.053 mm to 0.42 mm), and silt and clay (less than 0.053 mm). The pebble fractions were coarsely crushed so that small splits of these fractions represent a mixture of the lithologies present in the samples. Splits were taken from the size fractions for a variety of measurements, which are described below.

Dust

Dust traps were constructed at three sites within 10 m of the shoreline of Bear Lake (Fig. 2) to sample the annual vertical dust deposition to the lake area. Sites were located on the fan of South Eden Creek (BL-1), just south of Garden City (BL-2), and near the Lifton pump station (BL-3). The dust traps were established in 1998 and sampled annually in 1999 and 2000. The dust-trap design is that described by Reheis and Kihl (1995) and yields samples that include both wet and dry dust deposition. Laboratory analyses followed procedure described in Reheis (2003).

Lake Sediment

Lake sediments were cored (Fig. 2) using a variety of devices (Rosenbaum and Kaufman, this volume). Short cores, as much as a

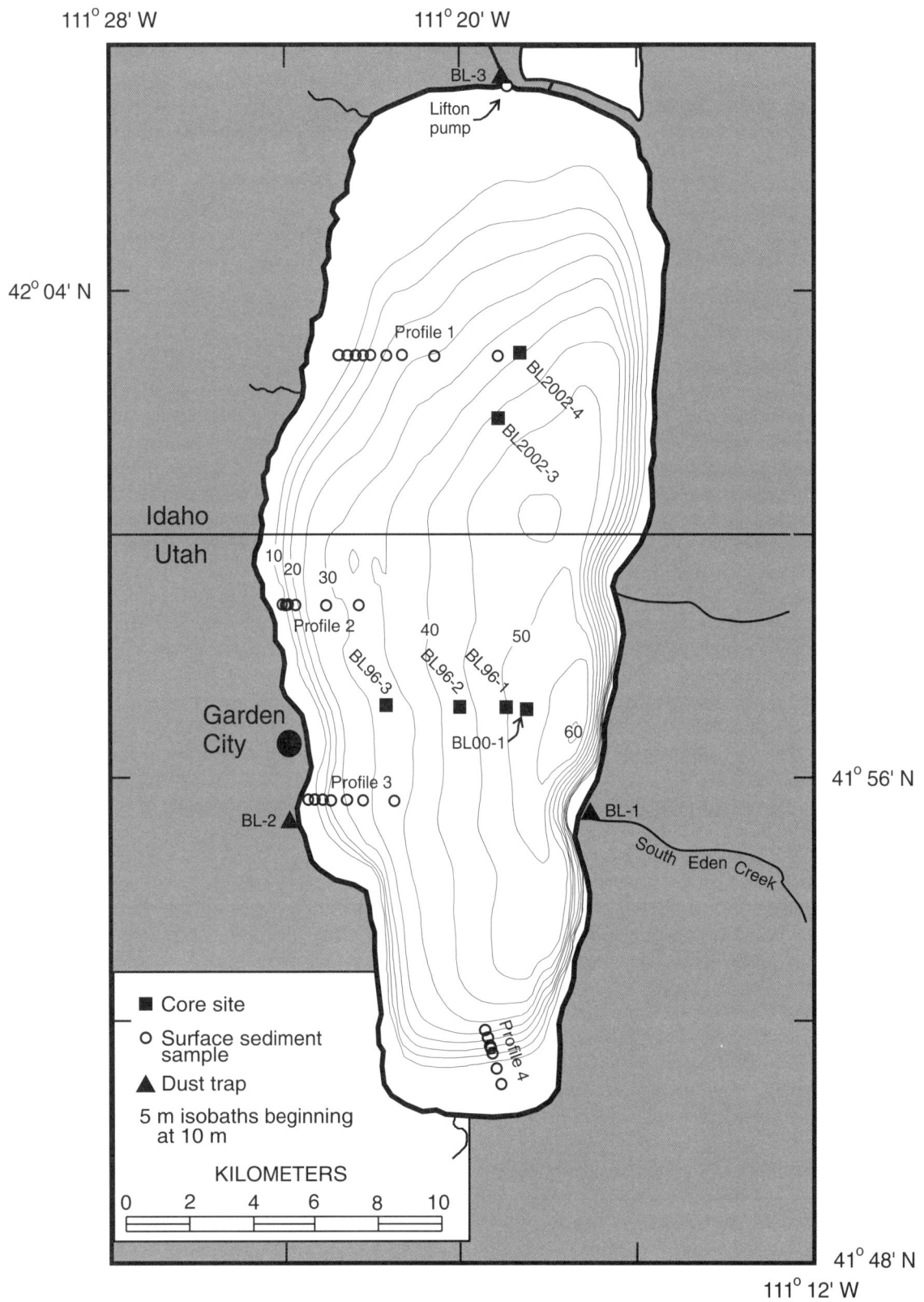

Figure 2. Map of Bear Lake showing bathymetry (Denny and Colman, 2003) and locations of cores, dust traps, and surface sediment samples.

few tens of centimeters in length, preserved the uppermost uncon-solidated sediments. The sampling interval was typically 1 cm, but surface samples, which were taken with the same coring device, are 1.5 cm thick. Longer cores, which penetrated up to 5 m of sediment, were collected in plastic liners. These cores were split lengthwise, with one-half of each core being preserved and logged (Smoot, this volume) and the other half being cut into samples at 1 cm intervals. Samples were then subdivided for a variety of analyses.

Analyses

Major minerals and elemental concentrations were deter-mined for lake-sediment samples and for the silt-and-clay and coarse-sand fractions of fluvial sediment. Standard X-ray diffrac-tion techniques were used for mineral identification (e.g., Moore and Reynolds, 1989). Powdered samples were packed into alu-minum holders and scanned from 15° to 50° using Ni-filtered, Cu-Kα radiation. Semiquantitative mineral contents were deter-mined by normalizing the major peak intensity of a mineral by the sum of major peak intensities for all minerals. This method does not account for differences in the X-ray absorption characteristics of individual minerals; nevertheless, these values provide excel-lent relative values for sample-to-sample comparison as long as the minerals in the samples are similar. The lake-sediment X-ray data are reported by Dean (this volume). Elemental concentra-tions were determined by XRAL Laboratories (Toronto, Canada) using Inductively Coupled Plasma-Atomic Emission Spectrom-etry following digestion of powdered samples in a combination of hydrochloric, nitric, hydrofluoric, and perchloric acids.

For magnetic property analyses, samples of lake sedi-ments and of all four size fractions of fluvial material were packed in nonmagnetic 3.2 cm^3 plastic boxes. The analy-ses include measurements of magnetic susceptibility (MS), anhysteretic remanent magnetization (ARM), and isothermal remanent magnetization (IRM). ARM and IRM were mea-sured with a high-speed spinner magnetometer. ARM was imparted using an alternating field with a peak intensity of 0.1 Tesla (T) and a bias field of 0.1 milliTesla (mT). After mea-surement of ARM, IRM was first imparted in a 1.2 T field (IRM$_{1.2T}$) and then in the opposite direction in a field of 0.3 T (IRM$_{-0.3T}$). Hard isothermal remanent magnetization (HIRM) and the S-parameter (King and Channel, 1991) are given by:

$$HIRM = (IRM_{1.2T} - IRM_{-0.3T})/2, \qquad (1)$$

and

$$S = IRM_{-0.3T}/IRM_{1.2T}. \qquad (2)$$

Samples then were dried and weighed. MS values, which were acquired after drying to eliminate the diamagnetic effects of pore water, were measured in a 600 Hz alternating field with amplitude of ~0.1 mT. The MS readings were corrected for the diamagnetic effect of sample boxes. Dry bulk densities were cal-culated by dividing the dry mass of each sample by the standard volume of the sample boxes (3.2 cm^3).

For selected samples, magnetic minerals were concentrated using a separator described by Reynolds et al. (2001). For cores BL96-1, -2, and -3, samples were chosen to represent magnetic-property variations. In order to obtain enough material for analy-sis, most separates incorporated material from two or more closely spaced samples with similar magnetic properties. The separates were mounted in epoxy, polished, and then observed with reflected-light microscopy (Reynolds and Rosenbaum, 2005). In this man-ner, magnetic minerals were identified from 22 horizons. In addi-tion, five separates were prepared from the silt and clay fraction of fluvial sediments. Most of these separates included material from multiple spatially related sites with similar magnetic properties.

Sample splits for grain-size analyses were treated sequen-tially with HCl, H$_2$O$_2$, and Na$_2$CO$_3$, to remove carbonate, organic matter, and opaline silica, respectively. A laser particle size analyzer, capable of measuring particles ranging from 0.49 to 2000 µm, was used to determine grain-size distributions on the residual siliciclastic material.

RESULTS

Fluvial Sediments

The mineralogical, elemental, and magnetic properties of fluvial sediment vary with grain size and location (Table 1, Figs. 3–5). The samples were divided geographically into four groups. Samples from streams in the local Bear Lake catchment were divided into those from the west side and those from the east side of the lake (Fig. 1). The Bear River samples were assigned to lower (samples 19 through 46) and upper (47 through 54) river groups based on properties described below.

Mineralogy and Elemental Chemistry

In comparison to Bear River sediments, samples from the local catchment are characterized by lower average quartz con-tent and higher contents of carbonate minerals (Table 1, Fig. 3). Calcite is most abundant in samples from the east side of the lake, whereas high concentrations of dolomite are restricted to sam-ples from the west side. These differences, which reflect bedrock geology (Fig. 1), are mirrored in the concentrations of Ca and Mg (Fig. 4). Concentrations of Ca are relatively high within the local catchment, and high values of Mg are restricted to the west-side samples. The between-sample variability of mineral and elemental contents is greater for the local catchment than for the Bear River. The higher variability of the local catchment samples probably reflects contrasting bedrock among the separate areas drained by the small streams. The lower variability of the Bear River samples is attributed to dilution of local bedrock material by a relatively homogeneous mixture of lithologies derived from large areas upstream of each sampling site.

The coarse fraction of Bear River sediments displays little variation in major minerals, although samples 19–29 contain

TABLE 1. AVERAGES OF MINERAL ABUNDANCES, ELEMENTAL ABUNDANCES, MAGNETIC SUSCEPTIBILITY, AND HARD ISOTHERMAL
REMANENT MAGNETIZATION OF SILT-AND-CLAY FRACTION (S&C) AND COARSE-SAND FRACTION (CS) OF FLUVIAL SEDIMENT SAMPLES

	Size fraction	Quartz (%)	Calcite (%)	Dolomite (%)	Feldspar (%)	Al (%)	Ti (%)	Ca (%)	Mg (%)	MS x 10⁻⁷ (m³kg⁻¹)	HIRM x 10⁻⁴ (Am²kg⁻¹)
Bear River	S&C	78	11	4	6	5.2	0.22	4.4	1.14	1.80	4.43
	CS	84	3	1	7	2.3	0.08	2.0	0.36	0.67	3.06
Lower	S&C	77	12	5	5	4.6	0.21	5.4	1.21	1.82	3.38
	CS	88	12	1	7	2.3	0.09	2.5	0.44	0.81	2.64
Upper	S&C	85	0	0	14	7.3	0.25	0.8	0.91	1.71	8.12
	CS	90	1	0	9	2.4	0.04	0.1	0.11	0.21	4.46
Local catchment	S&C	69	10	13	7	3.7	0.21	6.0	1.80	5.63	3.47
	CS	64	22	10	4	2.6	0.12	9.7	1.70	3.03	2.90
West side	S&C	67	4	21	8	3.6	0.21	5.5	2.37	6.09	3.43
	CS	69	13	14	4	2.7	0.13	7.9	2.21	3.23	3.08
East side	S&C	71	20	3	6	3.9	0.21	7.0	0.87	5.03	3.52
	CS	55	40	2	4	2.3	0.11	13.4	0.68	2.76	2.66

more calcite and slightly less quartz than samples from farther upstream (Fig. 3). This downstream increase in calcite, which is clearly reflected in the Ca concentrations, probably reflects greater input from the limestone-bearing Mesozoic bedrock (Fig. 1). The fine fraction of Bear River sediments contains less quartz and more calcite than the coarse fraction, suggesting that the coarse fraction is derived largely from sandstones and quartzites and the fine fraction is derived largely from more friable lithologies. This difference in lithologies is reflected in the contents of Al and Ti.

Moving upstream along the Bear River, both Al and Ti increase in the fine fraction and decrease in the coarse fraction (Fig. 4). In both fractions, Ti:Al decreases upriver, reaching minimum values in the headwaters in the Uinta Mountains.

Very high values of Mg:Ca in the uppermost reaches of the Bear River reflect the absence of carbonate bedrock, whereas elevated values of this ratio from the west side of the local catchment reflect abundant dolomite. The abundance of dolomite is more clearly reflected in the ratio of Mg to the chemically immobile

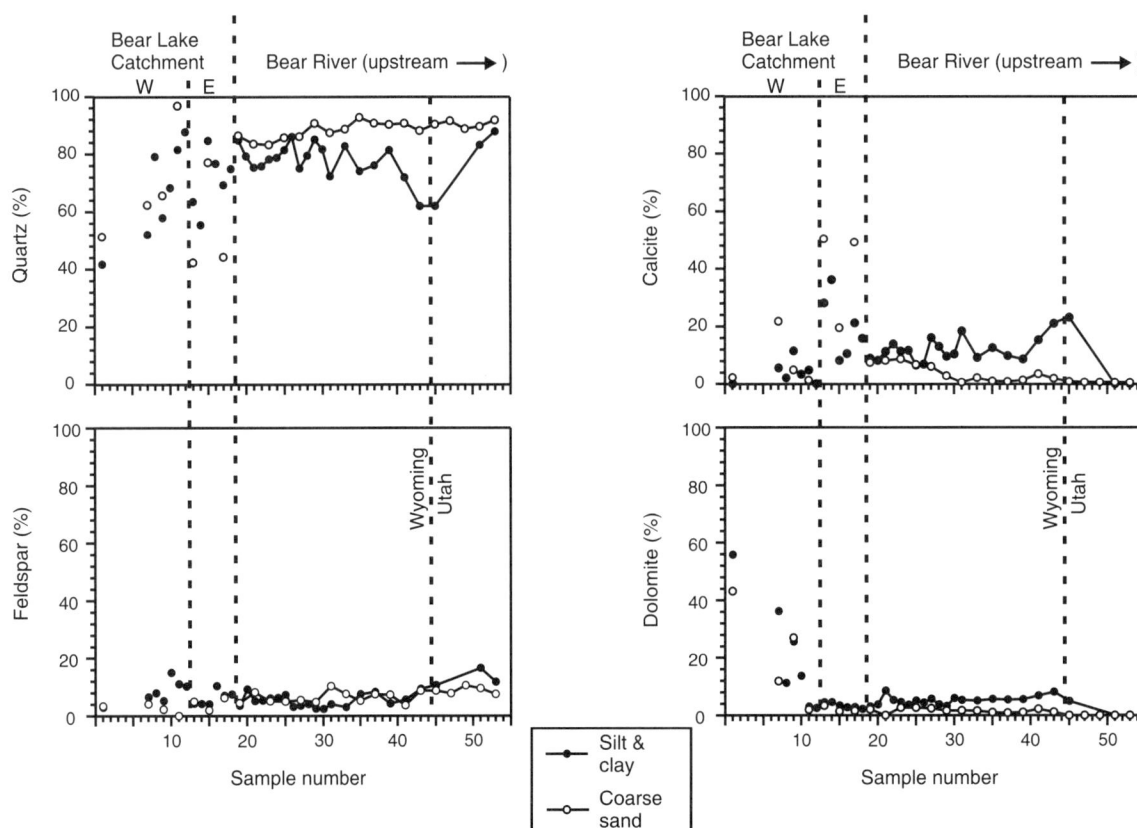

Figure 3. Contents of quartz, calcite, feldspar, and dolomite in two size fractions of fluvial sediment samples as determined by X-ray diffraction. Sample locations shown in Figure 1. Vertical dashed lines separate four catchment areas: west (W) and east (E) sides of the local catchment and the lower and upper Bear River.

element, Al (Fig. 4), which is delivered to the lake in aluminosilicate minerals.

Magnetic Properties

Magnetic properties provide strong contrasts among the four fluvial sediment groups (Table 1, Fig. 5). MS, which provides a measure of the content of ferrimagnetic minerals (e.g., magnetite, titanomagnetite, and maghemite), and HIRM, largely a measure of hematite, do not covary. Values of MS for the silt-and-clay fraction of all sediment groups are higher than corresponding values from the coarser fractions. For the local Bear Lake catchment, MS values are relatively high and HIRM values are relatively low. In comparison to the local catchment, the lower Bear River group has lower MS values and similar HIRM values. Values of MS generally decrease upriver, whereas HIRM values increase upriver into the Uinta Mountains. Values of HIRM for

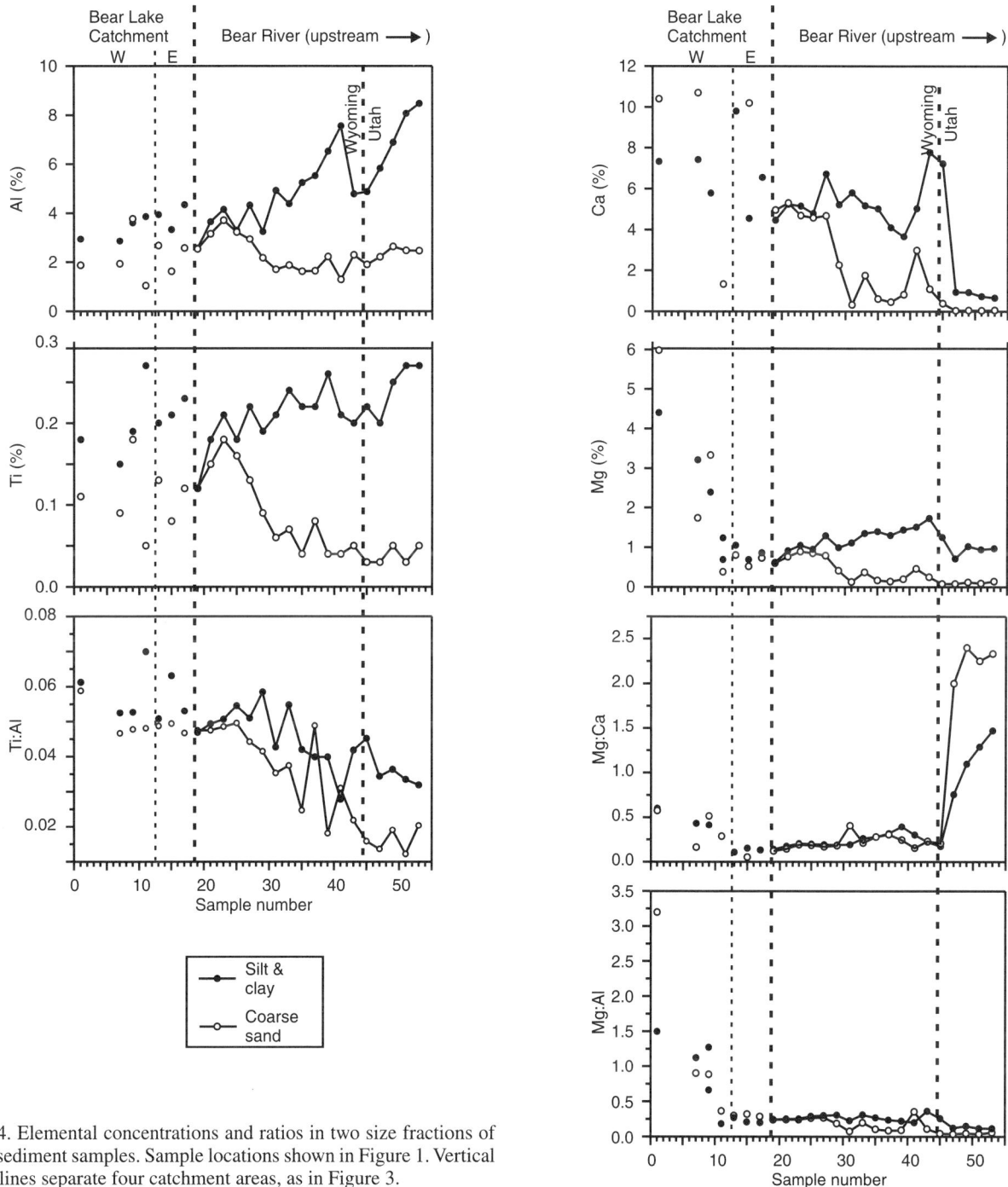

Figure 4. Elemental concentrations and ratios in two size fractions of stream sediment samples. Sample locations shown in Figure 1. Vertical dashed lines separate four catchment areas, as in Figure 3.

the silt-and-clay fraction from the upper Bear River are more than twice the values from the other areas.

Petrographic observations show that magnetite, which occurs commonly as mineral grains and less commonly within rock fragments, is the most abundant mineral in each of the magnetic separates (Reynolds and Rosenbaum, 2005). The magne-tite occurs in a variety of forms, including homogeneous grains and titanomagnetite grains in which the magnetite is subdivided by ilmenite lamellae. Many of these grains formed under high-temperature conditions in igneous or metamorphic rocks. Ferri-magnetic titanohematite and hematite also occur in all samples. Like magnetite, hematite occurs in a variety of forms including

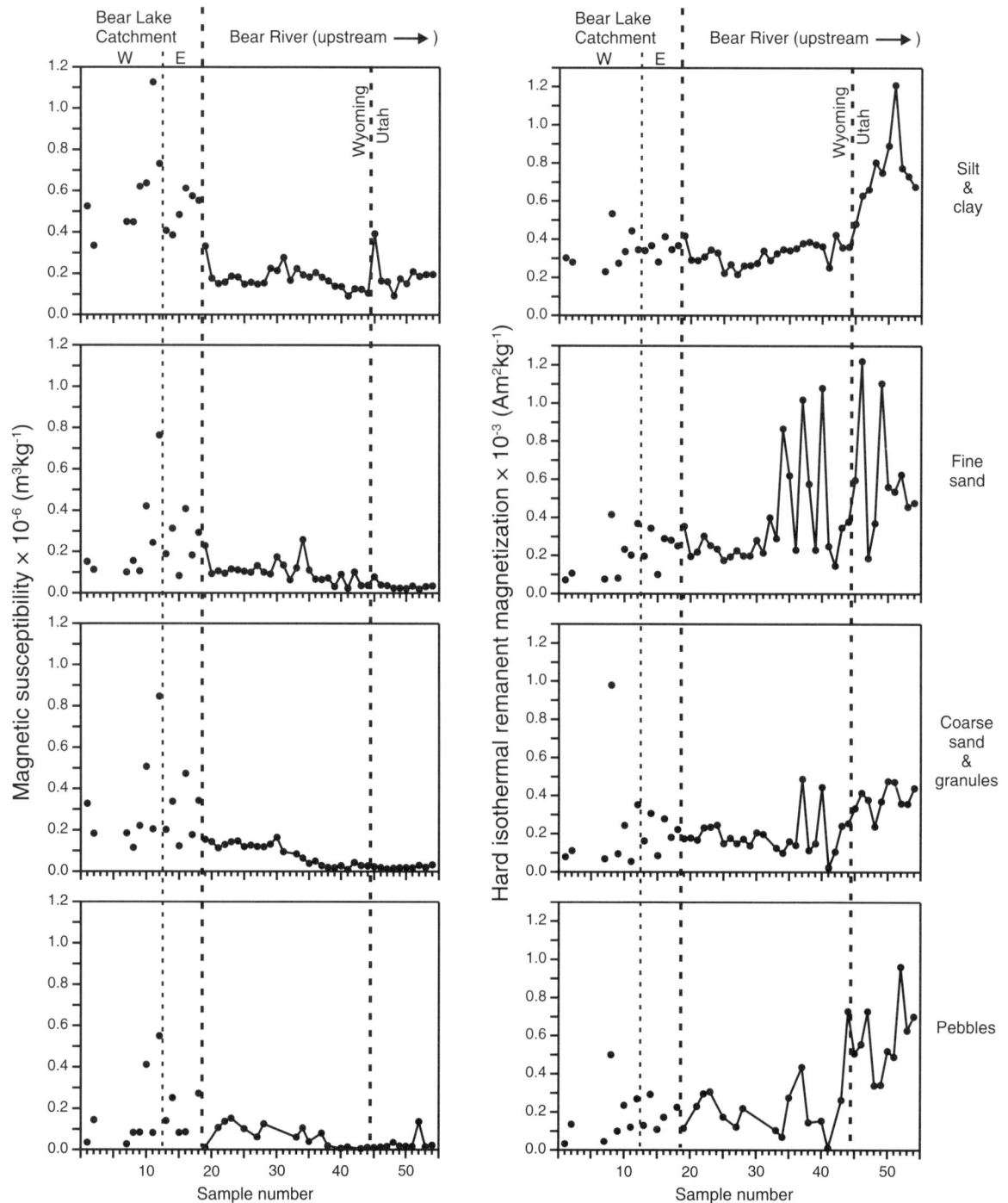

Figure 5. Magnetic susceptibility and hard isothermal remanent magnetization of four size fractions of stream sediment samples. Sample locations shown in Figure 1. Vertical dashed lines separate four catchment areas, as in Figure 3.

specular hematite and fine-grained forms. Some of the specular hematite occurs within titaniferous oxides that formed under high-temperature oxidizing conditions (e.g., pseudobrookite), whereas much of the fine-grained hematite occurs in rock fragments (e.g., red sedimentary rocks) and formed under low-temperature conditions. Red sedimentary rock fragments are particularly abundant in samples from the upper reaches of the Bear River (sites 48 and 51 in Fig. 1). In addition, most of the separates contain small quantities of anthropogenic material including fly ash and steel.

The magnetic grains are characterized by small sizes. Although the silt-and-clay size fraction used for the magnetic separates contains grains as large as 53 μm, few magnetic grains are >20 μm, and most are <10 μm.

Dust

The amounts, composition, and grain size of material collected in dust traps near Bear Lake vary between the two years of collection. The flux of total mineral matter ranges from ~5.5 to >17 g m^{-2} yr^{-1} (Table 2). Much of this variation is produced by variations in sand content. The flux of the <50 μm fraction averages ~5 g m^{-2} yr^{-1} and is less variable. The CaCO$_3$ component of the dust flux is relatively low (<1 g m^{-2} yr^{-1}) and XRD analysis of dust samples indicates that more than 85% of the detritus consists of non-carbonate minerals, mainly quartz (~70%) and plagioclase with minor amounts of clays (G. Skipp, 2004, personal commun.). Concentrations of both salts and carbonate minerals are higher in the 1999–2000 samples than in the 1998–1999 samples (Table 2). These differences are due in large part to the much higher concentrations of sand (which is presumably quartz rich) in the earlier samples. In fact the fluxes of salt are very similar during the two sampling intervals and the fluxes of carbonate minerals are actually slightly lower during the second year.

Lake Sediments

Mineralogy

Sediments deposited over the last 26 k.y. consist of a quartz-rich lower unit and a carbonate-rich upper unit (Dean et al.,

2006; Dean, this volume; Smoot, this volume). Sediments in cores BL96-1 and BL2002-3 are entirely within the upper unit, whereas the transition between the two units occurs at ~3, 0.4, and 3.3 m in cores BL96-2, BL96-3, and BL2002-4, respectively (Figs. 6–10). The carbonate-rich upper unit can be divided into four intervals on the basis of carbonate mineralogy (Dean et al., this volume). In ascending order these are a lower calcite interval, a lower aragonite interval, an upper calcite interval, and an upper aragonite interval.

Quartz content is ~70% in the lower unit and quite uniform. Quartz content in the upper, carbonate-rich unit is generally 20% to 30% with small but well-defined variations. Calcite content in the lower unit is typically between 10% and 20%. Within this unit, the majority of calcite is probably detrital but some may be endogenic. Calcite contents in the aragonite-rich intervals of the upper unit are similar to those in the quartz-rich unit. Within these intervals, much of the calcite may be detrital. The content of dolomite, which is also detrital, is generally between 8% and 9% in the lower unit and between 5% and 8% in the upper unit. Because the sensitivity of the XRD method is limited to a few percent, the dolomite curves are quite noisy; nevertheless, it is apparent that most changes in quartz and dolomite are synchronous. Although the contents of quartz and dolomite generally change at the same time, the dolomite-quartz ratio (Dolo:Qtz) is not constant. Dolo:Qtz increases slightly across the transition from the lower to the upper unit. Within the upper unit, Dolo:Qtz undergoes larger variations that roughly coincide with changes in carbonate mineralogy. Values of Dolo:Qtz are mostly low in the calcite-rich intervals and high in the lower aragonite-rich interval. Values rise in the upper aragonite-rich interval and then fall toward the upper part of the section.

Mass accumulation rates vary widely among the core sites (Table 3). In the quartz-rich lower unit comparison of accumulation rates among sites may not be meaningful because the cores span very different time intervals. For the carbonate-rich upper unit, average mass accumulation rates of both the carbonate and non-carbonate fractions generally increase with water depth. The ratio of carbonate to non-carbonate material also increases with water depth.

TABLE 2. DUST DATA FROM BEAR LAKE, 1998–2000

Trap no.*	OC (%)[†]	CaCO$_3$ (total %)	Salts (total %)	Mineral wt. (g)[§]	Percent of <2 mm fraction				Dust flux (g m^{-2} yr^{-1})	CaCO$_3$ flux (g m^{-2} yr^{-1})	Salt flux (g m^{-2} yr^{-1})	Flux, g m^{-2} yr^{-1} (incl. CaCO$_3$)				
					sand	silt	clay	<20 μm				sand	silt	clay	<20 μm	<50 μm
1998–1999																
BL-1	21.6	4.18	8.15	1.40	15.87	61.70	22.44	48.65	8.77	0.51	1.14	1.21	4.71	1.71	3.71	6.42
BL-2	19.0	4.40	3.64	2.61	72.73	25.13	2.14	9.76	17.60	1.09	0.95	12.11	4.18	0.36	1.62	4.54
BL-3	26.7	4.49	5.10	1.48	31.76	56.10	12.14	33.42	8.01	0.60	0.76	2.30	4.07	0.88	2.42	4.95
1999–2000																
BL-1	11.8	7.17	13.98	0.68	3.31	57.91	38.79	84.27	5.92	0.44	1.04	0.16	2.83	1.89	4.11	4.72
BL-2	18.2	7.25	12.83	0.73	7.45	63.86	28.69	71.82	5.46	0.47	1.02	0.33	2.83	1.27	3.19	4.11
BL-3	21.4	3.33	9.82	1.09	1.65	53.89	44.46	87.50	7.51	0.34	1.17	0.10	3.42	2.82	5.55	6.23

*BL-1, South Eden Creek adjacent to east shore of lake; BL-2, on breakwater at Bear Lake field station on southwest shore of lake; BL-3, on breakwater east of Lifton pump station on north shore of lake.
[†]OC—organic carbon.
[§]Mineral weight excludes organic-matter content.

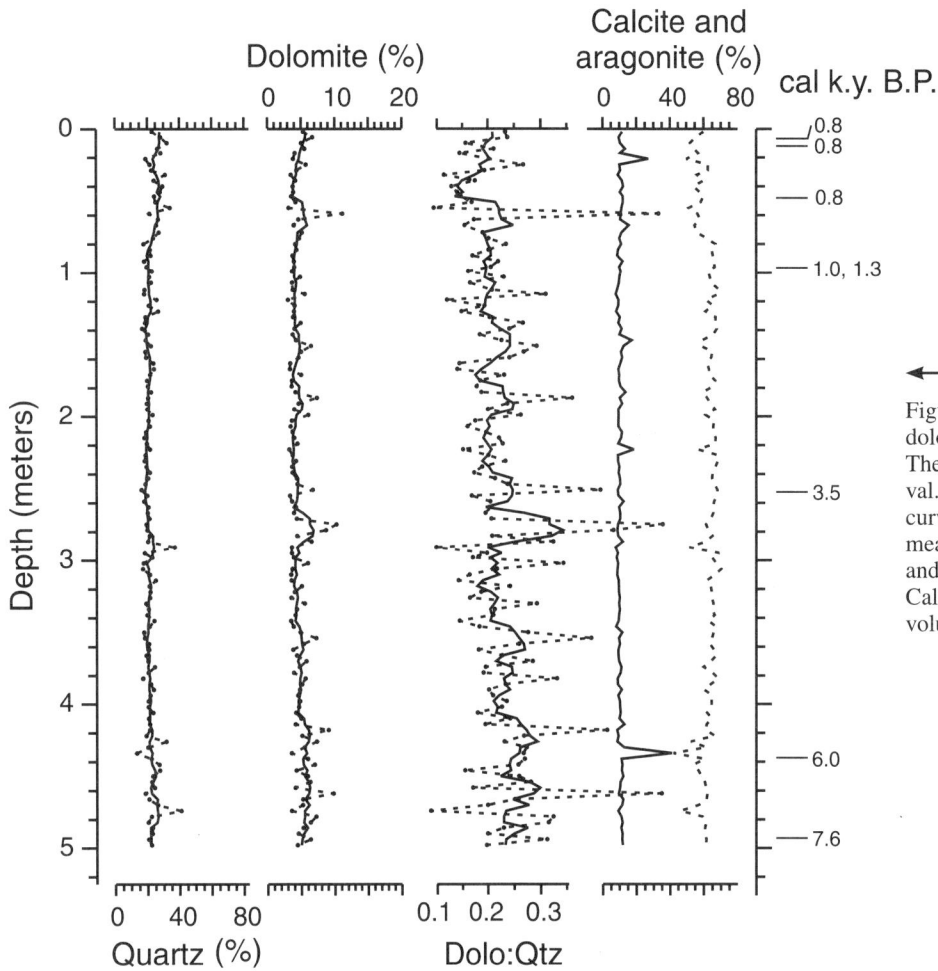

Figure 6. Mineralogic abundances and ratio of dolomite to quartz (Dolo:Qtz) for core BL96-1. The entire core is in the upper aragonite interval. For quartz, dolomite, and Dolo:Qtz, solid curves are drawn through five-point running means. Calcite content is shown by solid line and aragonite content is shown by dashed line. Calibrated ^{14}C ages are from Colman et al. (this volume).

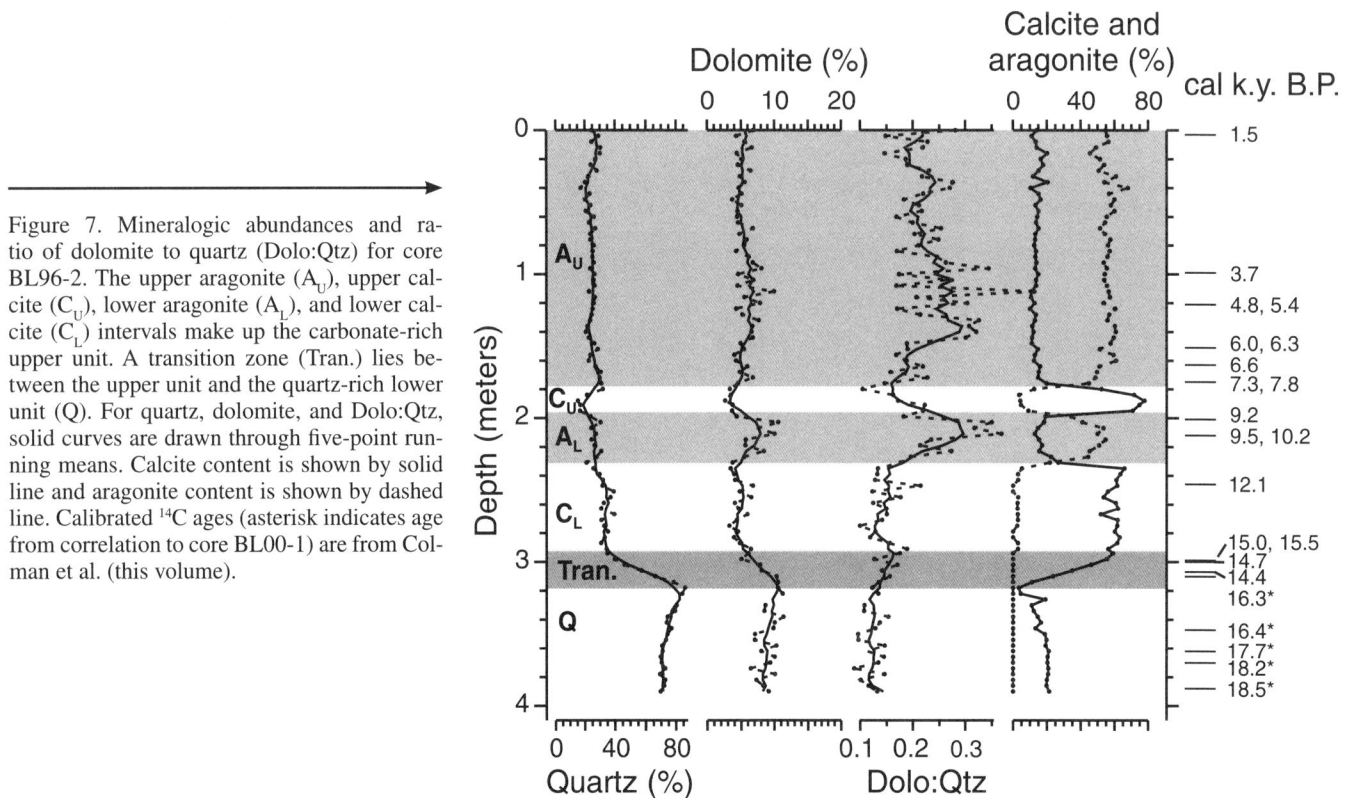

Figure 7. Mineralogic abundances and ratio of dolomite to quartz (Dolo:Qtz) for core BL96-2. The upper aragonite (A_U), upper calcite (C_U), lower aragonite (A_L), and lower calcite (C_L) intervals make up the carbonate-rich upper unit. A transition zone (Tran.) lies between the upper unit and the quartz-rich lower unit (Q). For quartz, dolomite, and Dolo:Qtz, solid curves are drawn through five-point running means. Calcite content is shown by solid line and aragonite content is shown by dashed line. Calibrated ^{14}C ages (asterisk indicates age from correlation to core BL00-1) are from Colman et al. (this volume).

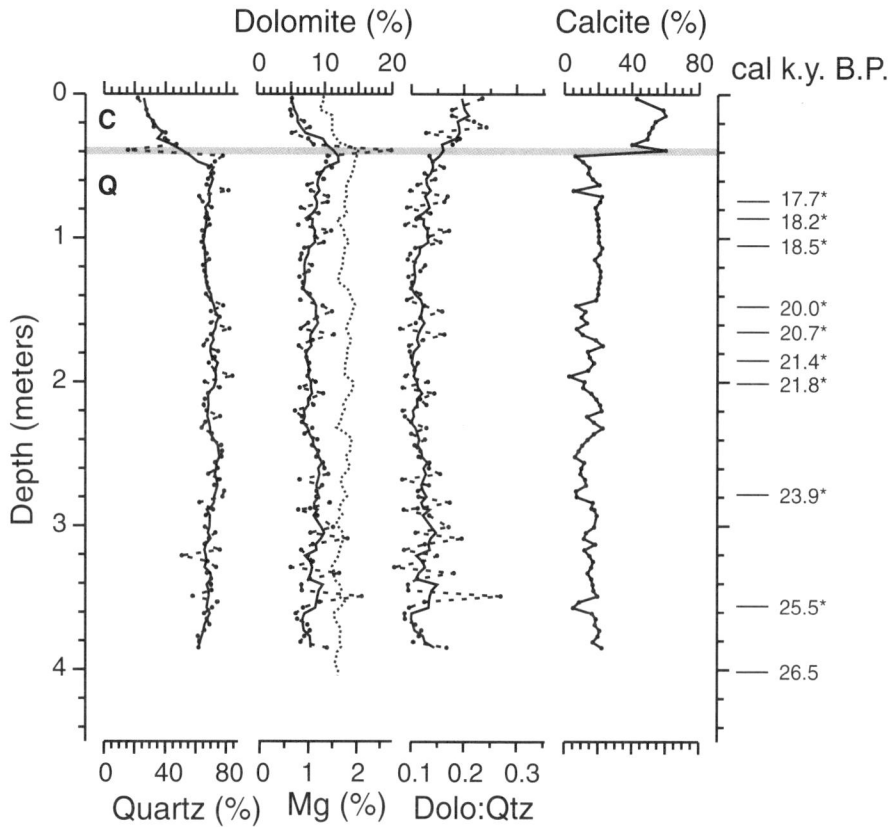

Figure 8. Mineralogic abundances, content of Mg (dotted line), and ratio of dolomite to quartz (Dolo:Qtz) for core BL96-3. The carbonate-rich upper unit (C) and transition zone (horizontal gray band) are thin, relative to other cores. For quartz, dolomite, and Dolo:Qtz, solid curves are drawn through five-point running means. Calibrated ^{14}C ages (asterisk indicates age from correlation to core BL00-1) are from Colman et al. (this volume).

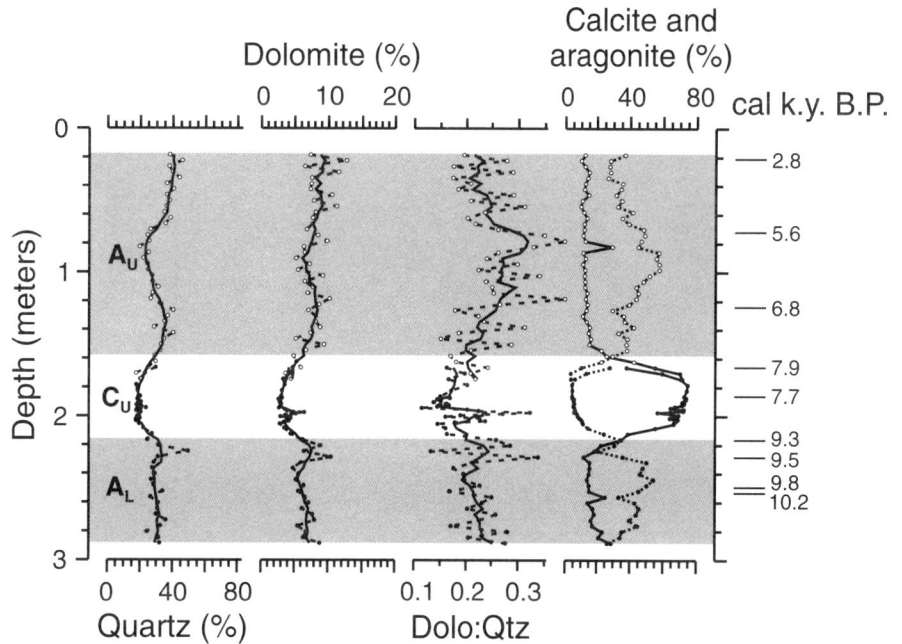

Figure 9. Mineralogic abundances and ratio of dolomite to quartz (Dolo:Qtz) for core BL2002-3. The core penetrated the upper aragonite (A_U), upper calcite (C_U), and lower aragonite (A_L) intervals. Open and closed symbols indicate data from two overlapping core segments. For quartz, dolomite, and Dolo:Qtz, solid curves are drawn through five-point running means. Calcite content is shown by solid line and aragonite content is shown by dashed line. Calibrated ^{14}C ages are from Colman et al. (this volume).

Elemental Chemistry

Variations in contents of Al, Ti, Mg, and Ca within core BL96-3 (Fig. 11) reflect changes in mineralogy like those described above, but because of the high precision of the chemical data some variations are evident in these elements that are not clearly observed in the mineralogy. The chemically immobile elements Al and Ti provide convenient proxies for the noncarbonate detrital material and, like quartz (Fig. 8), they decline upward from the lower unit into the upper unit (Fig. 11). Similarly, across the unit boundary Mg and Ca mimic the upward decrease in dolomite and increase in calcite (Fig. 8), respectively.

In addition to the major changes across the unit boundaries described above, small variations in the contents of Al, Ti, and Mg are evident within the lower unit. These changes are most evident in plots of Mg and of elemental ratios (Fig. 11). Mg from the HCl-soluble carbonate minerals (Bischoff et al., 2005) closely matches the magnitude and variations expected from variations in dolomite. Values of Mg:Ca and Mg:Al have very similar variations throughout the lower unit and diverge abruptly at the unit boundary as Ca content begins to increase due to the precipitation of endogenic calcite. Inspection of Figure 11 reveals that relative maxima and minima in Ti:Al, Mg:Ca, and Mg:Al coincide.

The content of sulfur is uniformly low throughout most of the lower unit, but begins to increase ~30 cm below the transition to the upper unit and peaks at the unit boundary (Fig. 11). Sulfur generally decreases upward through the truncated upper unit sampled at this locality, but remains much higher than throughout most of the lower unit.

Magnetic Properties

Within the quartz-rich lower unit, between the bottom of core BL96-3 and a depth of 1.05 m, magnetic properties vary in a quasi-cyclical manner (Fig. 12). Properties indicative of magnetite content both in an absolute sense (MS, ARM, and IRM) and relative to hematite (S) are strongly correlated. In this interval, there is a negative relation between these properties and HIRM, which indicates hematite content. This part of the lower unit yields nearly identical results in a 120-m-thick cored section (BL00-1, Fig. 2) studied by Heil et al. (this volume). MS features labeled 9–14 and HIRM features labeled 1–10 (Fig. 12) can be unequivocally identified in both records and were used in correlating the two sections and transferring ^{14}C ages to the long core (Colman et al., 2006; Colman et al., this volume).

The relation between the contents of magnetite and hematite changes in the upper part of the lower unit. On the basis of

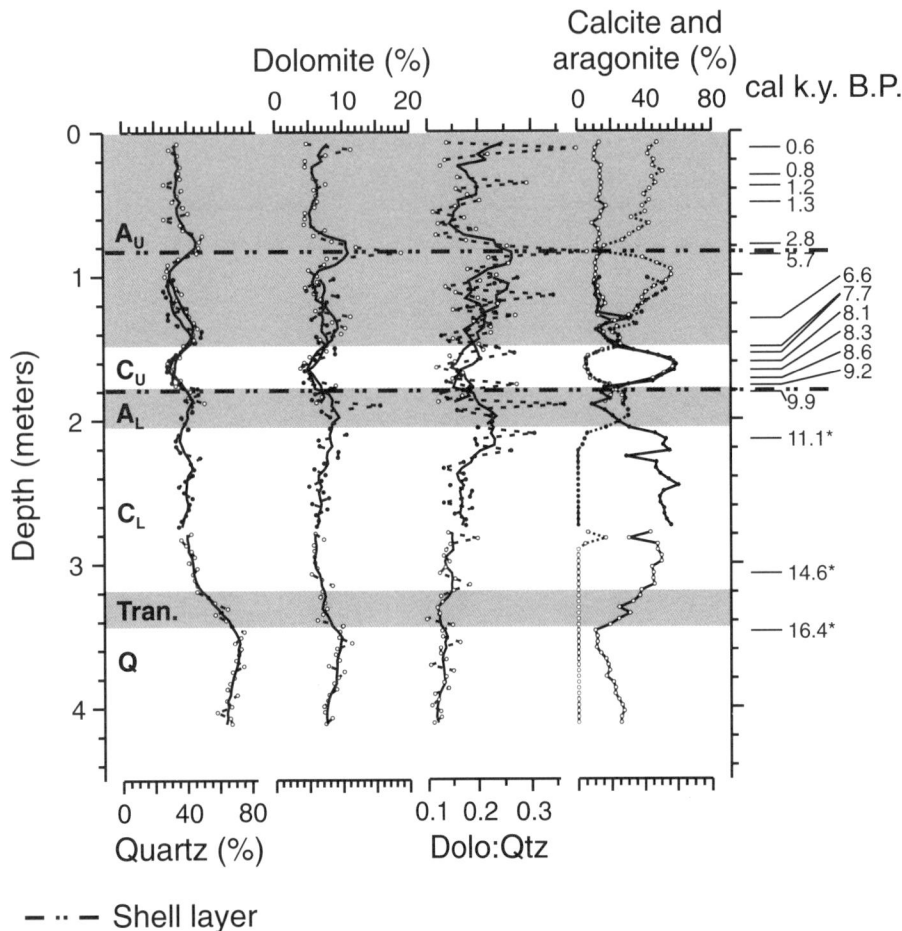

Figure 10. Mineralogic abundances and ratio of dolomite to quartz (Dolo:Qtz) for core BL2002-4. The upper aragonite (A_U), upper calcite (C_U), lower aragonite (A_L), and lower calcite (C_L) intervals make up the carbonate-rich upper unit. A transition zone (Tran.) lies between the upper unit and the quartz-rich lower unit (Q). Open and closed symbols indicate data from three core segments. For quartz, dolomite, and Dolo:Qtz, solid curves are drawn through five-point running means. Calcite content is shown by solid line and aragonite content is shown by dashed line. Calibrated ^{14}C ages (asterisk indicates age from correlation to core BL00-1) are from Colman et al. (this volume).

TABLE 3. MASS ACCUMULATION RATES FOR CARBONATE-RICH UPPER UNIT AND QUARTZ-RICH LOWER UNIT

Core	Water depth (m)	Depth to top of interval (cm)	Depth to bottom of interval (cm)	Age at top of interval (cal yr B.P.)	Age at bottom of interval (cal yr B.P.)	Average density (g cm⁻³)	Average carbonate mineral content (%)	Average quartz content (%)	Average non-carbonate mineral content (%)	Total MAR (g m⁻² yr⁻¹)	Carbonate mineral MAR (g m⁻² yr⁻¹)	Quartz MAR (g m⁻² yr⁻¹)	Non-carbonate mineral MAR (g m⁻² yr⁻¹)
Carbonate-rich upper unit													
BL2002-4	34.9	8	76	110	2500	0.392	60.0	35.7	40.0	107	64	38	43
		80	176	5460	8850	0.759	58.1	36.2	41.9	215	125	78	90
		180	315	9060	15,150	0.746	56.8	38.5	43.2	168	95	64	73
					Weighted average	0.693	57.6	37.5	42.4	169	103	63	72
BL2002-3	42.9	18.5	288.5	2740	11,520	0.794	67.4	27.8	32.6	238	160	66	78
BL96-2	43.0	0	294	1310	14,520	0.724	72.8	26.3	27.2	161	117	42	44
BL96-1	50.0	2	498	650	7710	0.808	78.1	21.9	21.9	567	442	124	124
Quartz-rich lower unit													
BL96-3	33.0	43	385	16,680	25,920	1.090	24.0	69.4	76.0	403	96	280	306
BL2002-4	34.9	345	413	16,570	19,860	1.038	27.4	68.5	72.6	215	59	147	156
BL96-2	43.0	320	396	15,730	18,980	0.971	25.4	74.5	74.0	227	58	169	169

Note: For BL2002-4, mass accumulation rates (MAR) were calculated for intervals between prominent unconformities marked by shell layers and as a weighted average for the three intervals. Mineral content was determined by X-ray diffraction. Carbonate-mineral content is the sum of calcite, aragonite, and dolomite contents. Ages based on age-depth models described in Colman et al. (this volume).

the locations of the transition from the lower to the upper unit, the shape of the HIRM curve, and comparable values of HIRM, the maximum value of HIRM in BL96-3 (9.5×10^{-4} Am² kg⁻¹) at 1.05 m (Fig. 12) correlates with the HIRM peak in BL96-2 (8.6×10^{-4} Am² kg⁻¹) at 3.88 m (Fig. 13), and a secondary peak in BL96-3 (6.4×10^{-4} Am² kg⁻¹) at 0.76 m correlates with a peak in BL96-2 (5.9×10^{-4} Am² kg⁻¹) at 3.62 m. Between these peaks, changes in HIRM are not matched by changes in MS comparable to those that accompany variations in HIRM below 1.05 m in BL96-3. In both BL96-2 and -3, HIRM decreases upward from the secondary peak to the top of the lower unit. However, the magnitude of this decrease differs in the two cores, with HIRM decreasing by factors of 5.1 and 1.7 in BL96-3 and -2, respectively. MS curves in this interval differ markedly. MS values in BL96-3 decrease gradually to the upper boundary of the lower unit, whereas values in BL96-2 increase gradually immediately above the secondary HIRM peak and then more abruptly from 3.32 m to the boundary. This change in rate of increase in MS is also present in the record from BL2002-4 (Fig. 14).

Both MS and HIRM generally decrease upward across the transition between the lower and upper units and continue to decrease in the lower part of the upper unit. Two MS peaks (one labeled 6 in Figs. 13, 14, and 15, and the other labeled 7 in Figs. 13 and 14) interrupt the general decrease in MS. These peaks occur near the lower and upper boundaries of the lower calcite interval. Both MS and HIRM attain very low values in the lower aragonite interval. HIRM then remains low before increasing near the top of the section. The MS records contain several low-amplitude peaks overlying the lower aragonite interval (Figs. 14–16), but only the lowermost of these (labeled 5), which coincides with the upper calcite interval, and an (unnumbered) increase near the top of the section are recognized in more than one core. Individual peaks in MS (1–4), ARM, and IRM in core BL96-1 (Fig. 16) may correlate with zones of slightly elevated values in cores BL2002-3 (Fig. 15) and BL2002-4 (Fig. 14), but similar features do not occur in core BL96-2 (Fig. 13). Magnetic susceptibility peaks 5 and 6 coincide with intervals interpreted as periods of high lake level (Smoot and Rosenbaum, this volume), whereas peak 7 immediately overlies a similar high lake-level interval.

In the lower unit, which is characterized by magnetic properties indicative of relatively abundant magnetite and hematite, petrographic observations of magnetic separates show that detrital magnetic minerals commonly occur as fine-silt-sized (<10 μm), angular particles, or within rock fragments of similar size and shape (Reynolds and Rosenbaum, 2005). Magnetite is the most abundant mineral in eight of 12 magnetic separates (Figs. 12 and 13). Detrital hematite is the second most common mineral in most of the magnetite-rich separates, is subequal to magnetite in one separate, and is the most abundant mineral in three separates. Samples included in these four hematite-rich separates are all from peaks in HIRM. The magnetite occurs in many forms, including titaniferous magnetite, low-titanium magnetite, and as crystals within rock fragments. Hematite occurs mainly in reddened rock fragments and as particles of specular hematite. The reddened rock fragments comprise many different lithic types, but siltstone is the most abundant type in the hematite-rich samples. Iron sulfide minerals are rare. Minor framboidal pyrite is present in two separates and small amounts of griegite are present in four separates at depths greater than 1 m in core BL96-3. The two uppermost separates from the lower unit in BL96-3, which sampled peaks in IRM/MS (Fig. 12), contain more plentiful greigite. There is no petrographic evidence for the dissolution of detrital magnetic grains from the lower unit.

Magnetite is also the most common mineral in the one separate from the transition zone between the lower and upper units in BL96-2 (Fig. 13). The magnetite, which occurs largely as optically homogeneous grains and in rock fragments, shows possible minor dissolution. This is the only magnetic separate studied that contains common pyrite framboids.

Magnetite is the most abundant mineral in five of eight separates from the upper unit (Reynolds and Rosenbaum, 2005). Magnetic properties indicate that this unit has generally low concentrations of both magnetite and hematite (Figs. 13 and 14). The magnetite occurs mostly as small grains and in rock fragments, and, in several separates, shows some evidence of dissolution. Several of the separates are notable. The separate from ~2.80 m in BL96-2, just above the transition zone,

contains relatively abundant material and is the only separate that contains large (~70 μm) grains of magnetite and large (~100 μm) rock fragments with magnetite inclusions. The separate that spans the MS peak at ~2.37 m contains a few magnetite-bearing volcanic glass shards in addition to magnetite as small grains and in rock fragments. The separate at a depth of 4.90 m in BL96-1, within a zone of very low MS (Fig. 16), is unique in that the most abundant mineral is titanohematite (referring to a range of compositions in the hematite-ilmenite solid solution series that are highly magnetic), although titanohematite is present in significant amounts in several other separates. Finally, no individual magnetite grains are present in the uppermost two separates from core BL96-1, which are located in a zone of very low MS. The only Fe-oxides in these separates occur within small rock fragments.

Grain Size

Surface samples from four depth transects (Fig. 2) were subjected to grain-size analyses. Each transect shows a decrease in grain size of the siliciclastic fraction with depth, but the relation between size and depth varies from profile to profile (Fig. 17). The northernmost transect, which has the most constant and lowest bottom slope (Fig. 2), is located close to core sites BL2002-3 and -4 and incorporates the uppermost samples from these cores.

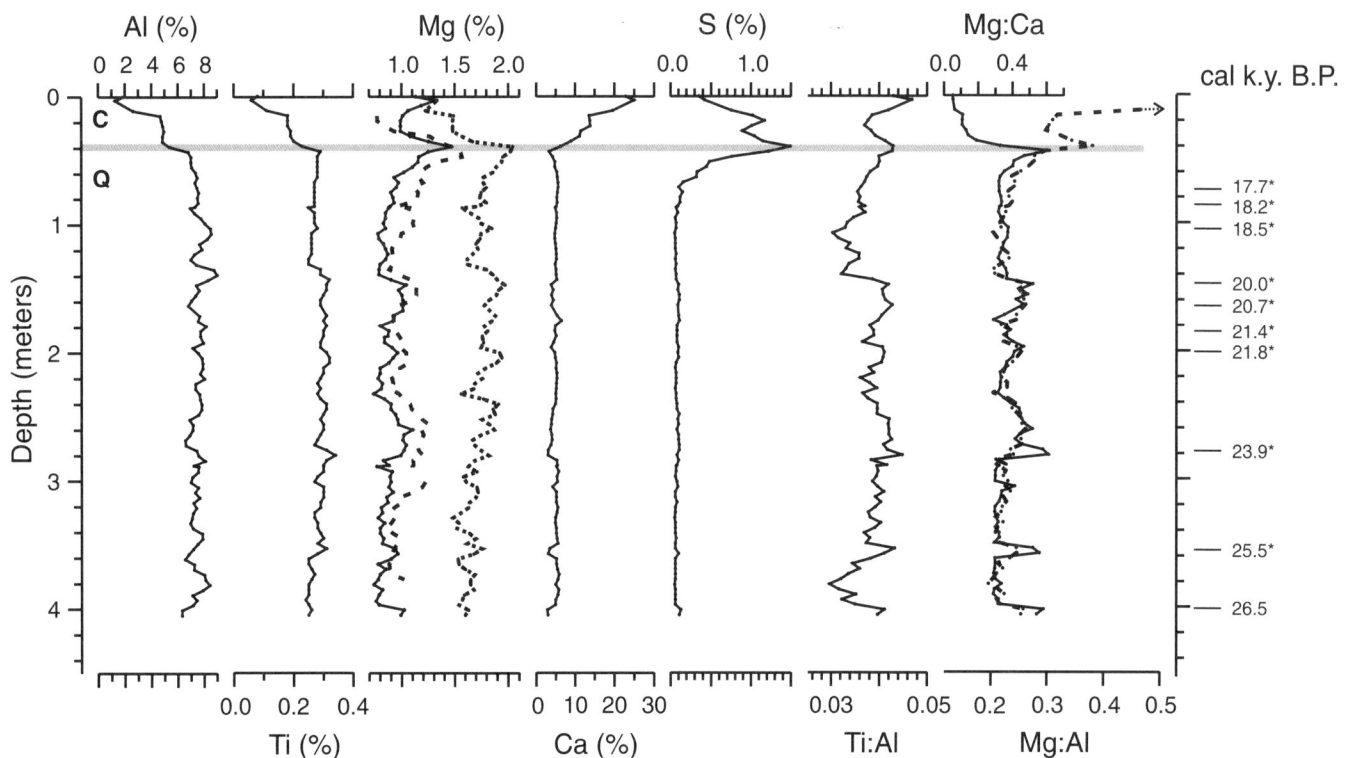

Figure 11. Abundances of selected elements and elemental ratios in core BL96-3. Calibrated ^{14}C ages (asterisk indicates age from correlation to core BL00-1) are from Colman et al. (this volume). Three curves are shown for Mg: (1) Mg from bulk samples (dotted), (2) HCl-extractable Mg (solid; Bischoff et al., 2005), and (3) Mg calculated from five-point running mean values of dolomite content (dashed) under the assumption of stoichiometric dolomite ($CaMg(CO_3)_2$). C and Q denote the carbonate-rich upper unit and quartz-rich lower unit, respectively.

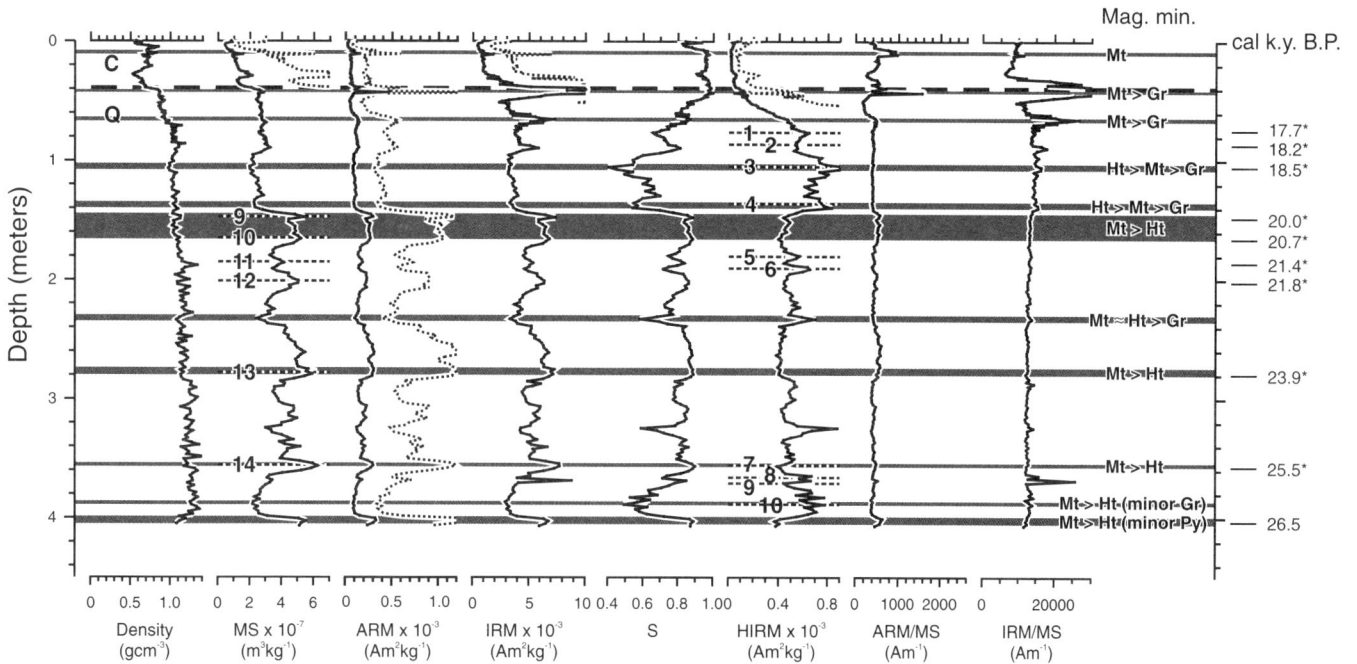

Figure 12. Density and magnetic properties for core BL96-3. In this figure, and in Figures 13–16, magnetic properties are magnetic susceptibility (MS), anhysteretic remanent magnetization (ARM), isothermal remanent magnetization (IRM), the S parameter, hard isothermal remanent magnetization (HIRM), and ratios ARM/MS and IRM/MS. Stratigraphic units (A$_U$—upper aragonite; C$_U$—upper calcite; A$_L$—lower aragonite; C$_L$—lower calcite intervals of the carbonate-rich upper unit, C; Tran.—the transition zone between the upper unit and the quartz-rich lower unit, Q) are delineated by alternating light-gray and white areas or by horizontal dashed lines. For comparison of low-amplitude features among cores, dotted curves are expanded by a factor of 4. Numbered features on the MS and HIRM plots are interpreted to correlate with features of the same numbers on plots of data from other cores. Relative amounts of magnetic minerals (Mag. min.) observed petrographically (Mt—magnetite; Ht—hematite; Ti-ht—titanohematite; Gr—greigite; Py—pyrite) are indicated on horizontal dark-gray bands, which indicate depth intervals sampled for magnetic separates. Calibrated ^{14}C ages (asterisk indicates age from correlation to core BL00-1) are from Colman et al. (this volume).

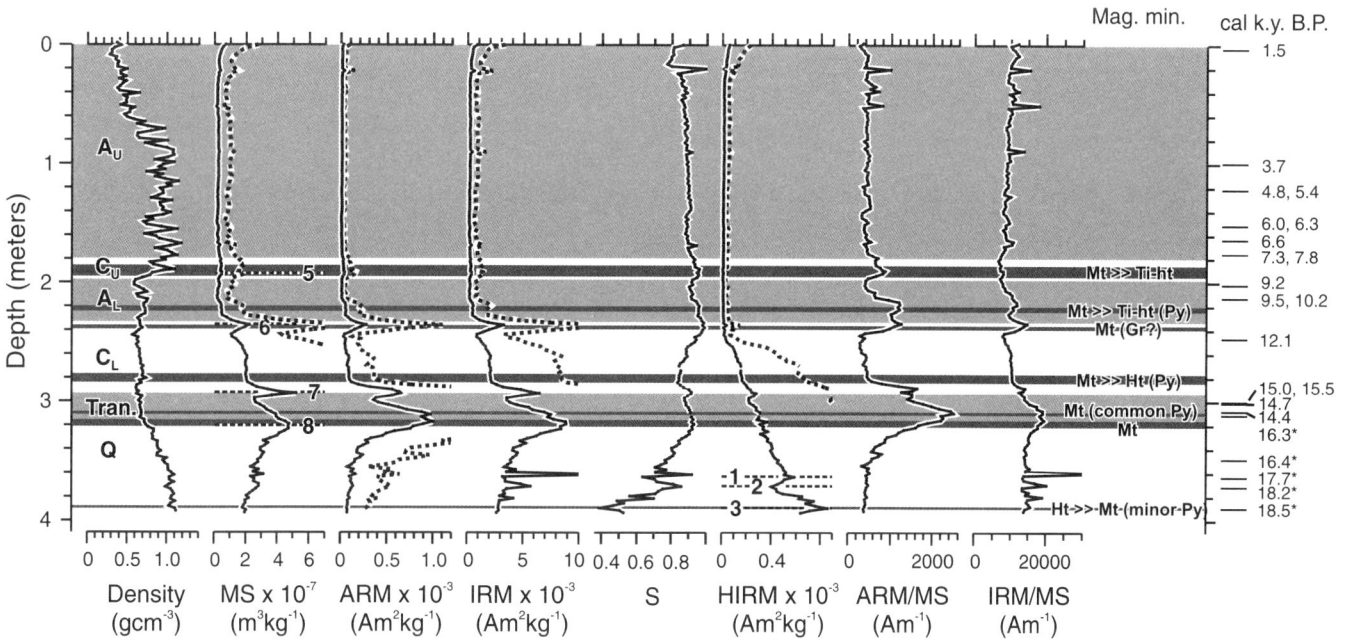

Figure 13. Density and magnetic properties for core BL96-2. See caption to Figure 12 for explanation.

This transect contains median grain sizes ranging from ~6 to 250 μm and yields a simple size-depth relation that is closer to that predicted by a wave model than those for the other transects (Smoot and Rosenbaum, this volume).

In the cored sediment, siliciclastic material from the lower unit is finer grained than that in the upper unit (Fig. 18). Median grain sizes in the lower unit range from <1 μm to 12 μm and aver-age ~3 μm. Sand content averages less than 1%. Below ~0.55 m in core BL96-3, there appear to be well-defined fluctuations in grain size, even though the grain-size range is extremely narrow, with few samples having median grain sizes greater than 4 μm. The coarsest sediment in the lower unit occurs just below the transition to the upper unit. In the upper unit, median grain sizes range from ~3 to 39 μm and average ~15 μm. The average sand

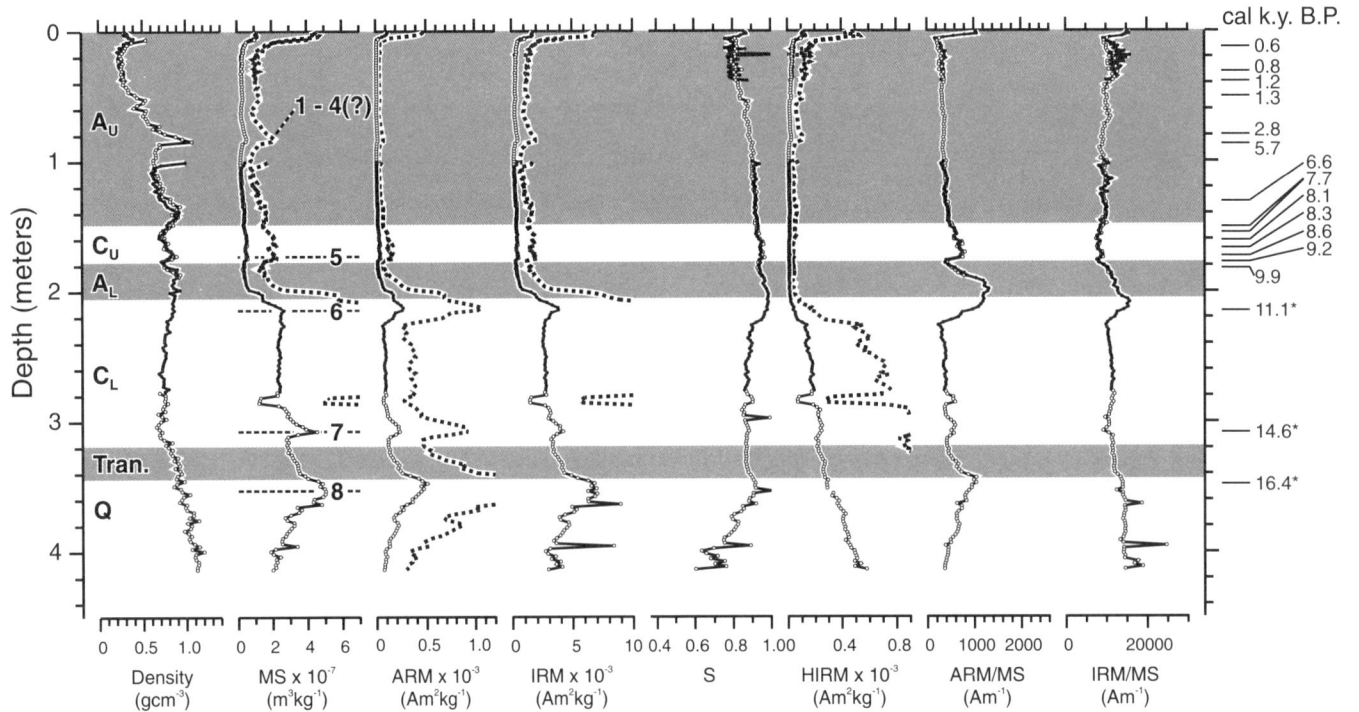

Figure 14. Density and magnetic properties for core BL2002-4. Open and closed symbols indicate data from three core segments. See caption to Figure 12 for explanation.

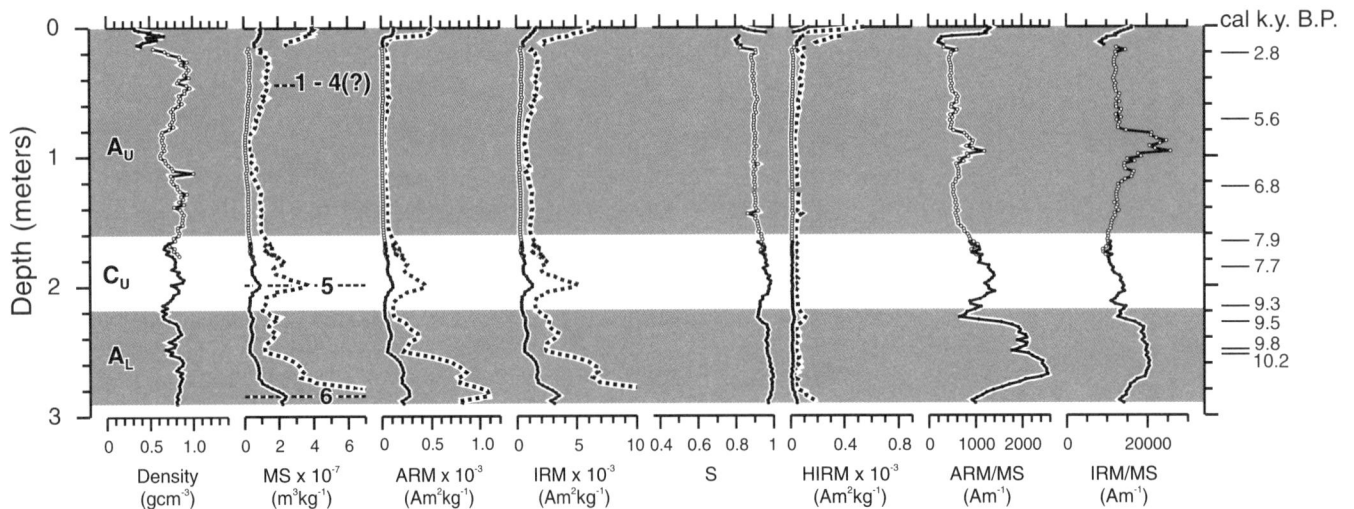

Figure 15. Density and magnetic properties for core BL2002-3. Open and closed symbols indicate data from two core segments. See caption to Figure 12 for explanation.

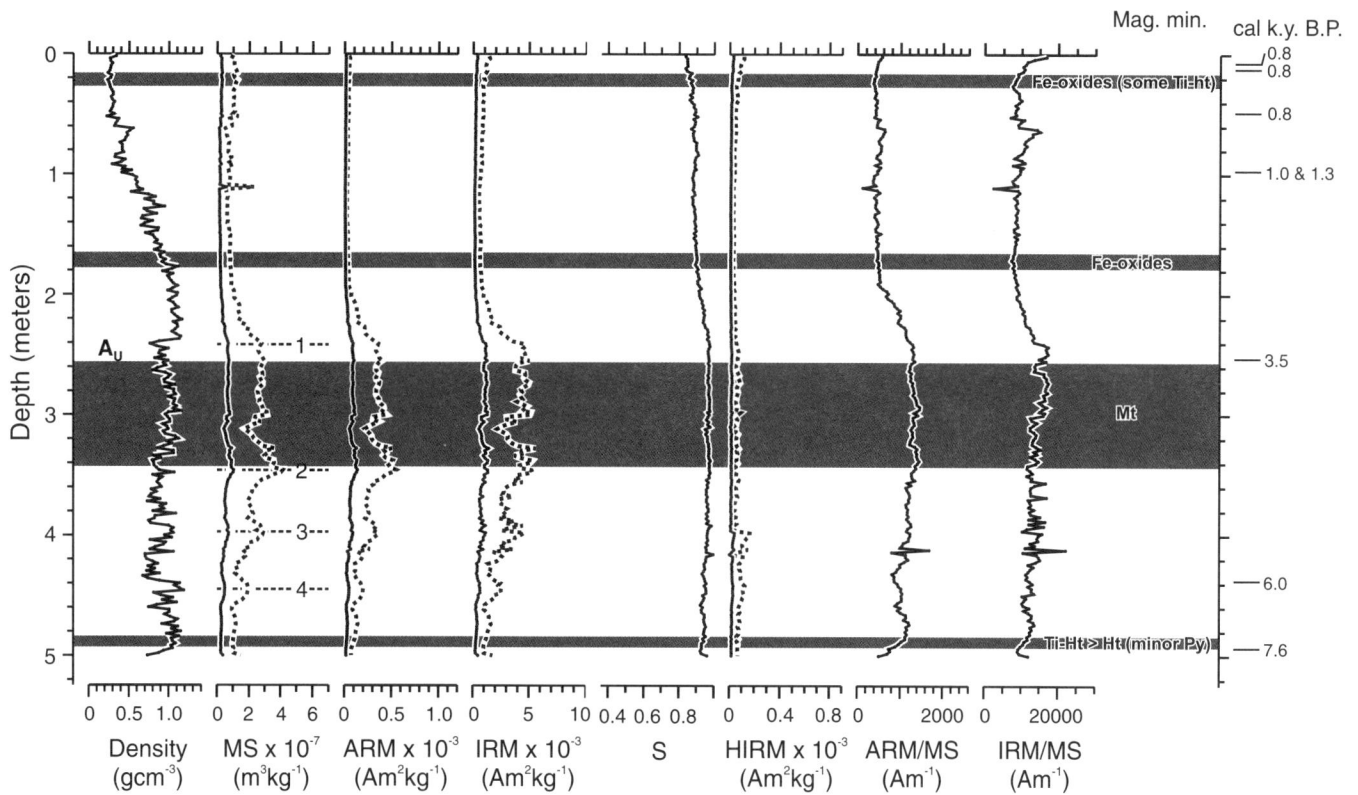

Figure 16. Density and magnetic properties for core BL96-1. See caption to Figure 12 for explanation.

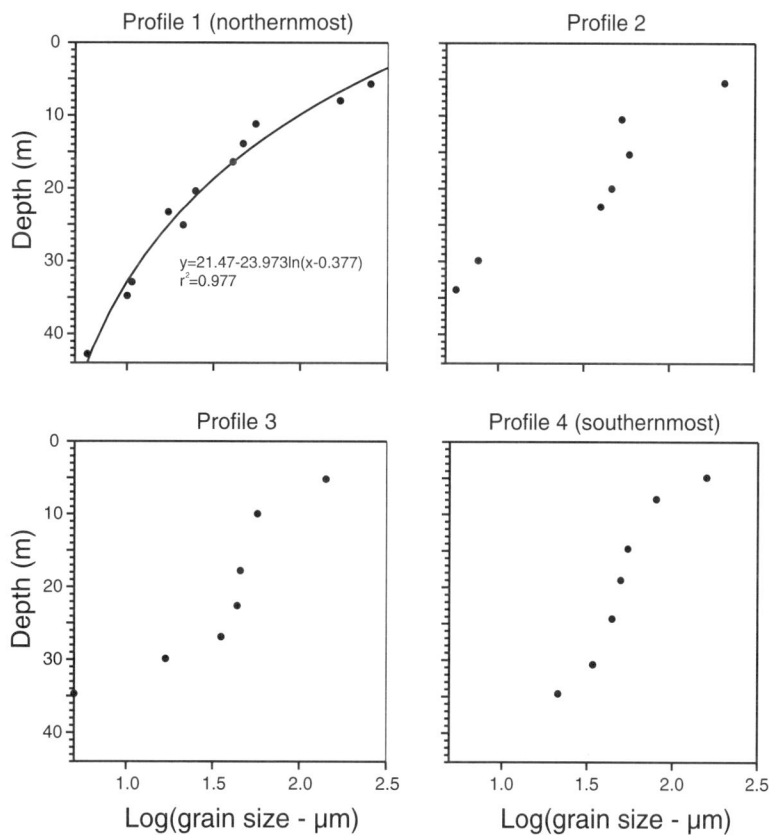

Figure 17. Median-grain-size data for surface samples versus water depth. Profile locations are in Figure 2. Equation and r^2 values are for curve fit to data from profile 1.

content of detrital material in the upper unit is ~9%. Well-defined variations in grain size of siliciclastic material correspond, at least in part, to changes in carbonate mineralogy. The lower calcite interval in BL2002-4 is relatively fine grained, with median grain size averaging 6 μm below 3.30 m and coarsening upward into the lower aragonite interval. Grain size is relatively coarse in the lower aragonite interval, decreases abruptly into the upper calcite interval, and then coarsens again into the upper aragonite interval. Most of the upper aragonite interval is relatively coarse grained, but grain size decreases in the upper 10–15 cm, becoming comparable to that in the lower part of the lower calcite interval. Two zones of coarse sediment in core BL2002-4 enclose shell layers at 0.84 and 1.80 m (Smoot, this volume), which are interpreted as beach or shallow-water deposits (Smoot and Rosenbaum, this volume). Stratigraphic position with respect to mineralogic zones and radiocarbon ages indicate that the lower of these coarse zones in BL2002-4 correlates with a zone from 2.10 to 2.35 m in BL2002-3 that contains two peaks in median grain size.

DISCUSSION

Fluvial Sediment Sources

Data from fluvial sediments indicate significant differences among four potential source areas of detrital sediment in Bear Lake (Table 1): (1) the west side of the local catchment (sites 1–12, Fig. 2), (2) the east side of the local catchment (sites 13–18), (3) the lower Bear River (sites 19–46), and (4) the upper Bear River (sites 47–54). These differences provide the means for interpreting changes in provenance within the cored lake sediment. Some of the differences reflect bedrock geology. Stream sediments from the east and west sides of the local Bear Lake catchment contain less quartz than sediment from the Bear River, reflecting abundant carbonate rocks in the Bear River Range and Bear Lake Plateau. These local catchment areas are differentiated by the abundance of carbonate minerals, with the west side sediments containing more dolomite and having high concentrations of Mg, and the east side sediments containing more calcite. The

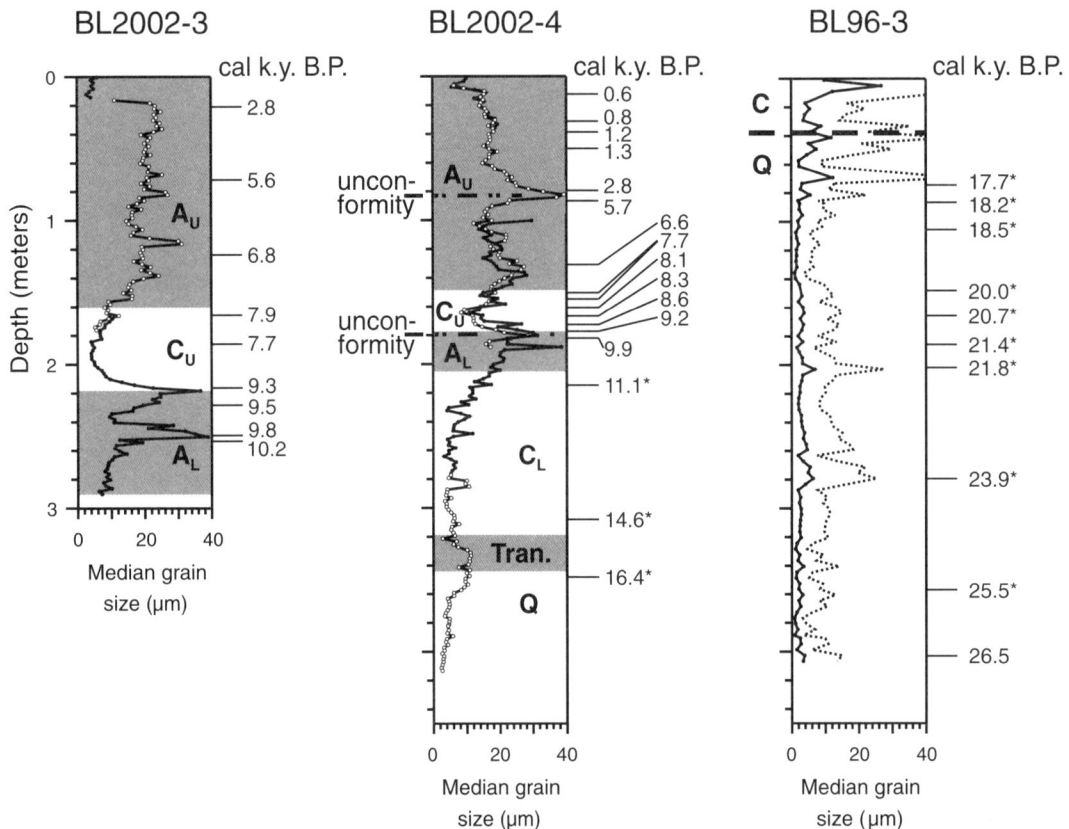

Figure 18. Median grain size versus depth for cores BL2002-3, BL2002-4, and BL96-3. For cores BL2002-3 and -4, open and closed symbols indicate individual core segments. Alternating gray and white zones and horizontal dashed line delineate the upper aragonite (A_U), upper calcite (C_U), lower aragonite (A_L), and lower calcite (C_L) intervals of the carbonate-rich upper unit (C), and the transition zone (Tran.) between the upper unit and the quartz-rich lower unit (Q). Calibrated ^{14}C ages (asterisk indicates age from correlation to core BL00-1) are from Colman et al. (this volume).

Bear River drains areas of widespread clastic sedimentary and metasedimentary rocks. As a result, its sediments contain more quartz and Al, and less Ca than sediments from the local catchment. Hematite content, as indicated by HIRM, is about twice as great in samples from the upper Bear River as in those from either the lower Bear River or the local catchment (Table 1, Fig. 5). The high hematite content in the upper Bear River is probably due to detritus derived from hematite-rich rocks of the Uinta Mountain Group (Ashby et al., 2005). High hematite content in the silt-and-clay fraction corresponds more closely to the extent of till of the last glaciation (Laabs et al., 2007) than to outcrops of the Uinta Mountain Group. Only sites 51–54 are located on Uinta Mountain Group bedrock (Fig. 1), and the maximum extent of last-glacial-aged till, which contains abundant detritus from Uinta Mountain Group rocks, lies between sites 45 and 47. In general, the content of Uinta Mountain Group detritus is diluted within a short distance downstream from the last-glacial limit, although, enhanced hematite content occurs sporadically in coarser fractions somewhat farther downstream, perhaps in outwash from which the fine-grained material has been winnowed.

In contrast to most of the mineralogical and geochemical data, the content of ferrimagnetic minerals, as measured by MS, is not related to bedrock geology. Values of MS for samples from within the local catchment are several times greater than MS values from Bear River sediment, and values from areas of Paleozoic rocks on the west side of the lake are similar to values from areas of Mesozoic rocks on the east side of the lake. Petrographic observations indicate that the stream sediments contain a wide variety of magnetic Fe-Ti-oxide mineral grains, many of which were formed at high temperature in igneous or metamorphic rocks that are lacking in the local Bear Lake drainage basin as well as that of the Bear River. Because the catchment rocks lack sources for many of the Fe-Ti-oxide grains, these grains are interpreted to have been introduced as atmospheric dust (Reynolds and Rosenbaum, 2005). The size of the magnetic grains, mostly <10 µm, is consistent with dust particles that can be transported hundreds of kilometers (Rose et al., 1998; Goudie and Middleton, 2006). These results suggest that the concentration of dust in streams in the local catchment is several times greater than in the sediments of the Bear River. The difference in dust concentration between sediments of the local catchment streams and those of the Bear River could reflect either a difference in the rate of dust deposition or a difference in the amount of dilution of dust by rock material derived from the catchment. Nevertheless, the large contrast in MS provides a potential tool to discriminate between detritus in the lake sediments that was derived from local streams and that derived from the Bear River.

Provenance of Detrital Material

Detrital materials incorporated in the lake sediments include (1) rock material derived through erosion of rocks within the drainage basin and delivered via fluvial transport; (2) dust deposited indirectly via fluvial transport following deposition in the drainage basin; and (3) dust deposited directly in the lake. The limited data (two years) from three dust-trap locations on the north, east, and west sides of Bear Lake indicate relatively low rates of dust deposition (Table 2) compared to dust-fall rates in southwestern deserts (Reheis, 2006). It is not known how well the dust data represent either long-term modern dust deposition or rates of deposition prior to European settlement. Modern dust flux may have been enhanced by anthropogenic sources due to construction and agriculture in the region and even to recreational activities on local beaches. The dust fluxes are also low relative to mass accumulation rates of both carbonate and non-carbonate minerals in the lake sediments (Table 3). In the quartz-rich lower unit, accumulation rates of carbonate and non-carbonate minerals are respectively one and two orders of magnitude higher than in the dust samples. Because of reworking of sediment, especially during periods of low lake level (Smoot and Rosenbaum, this volume), little of the carbonate-rich upper unit exists in water shallower than ~30 m (Colman, 2006; Dean, this volume; Smoot, this volume). Approximately half the area of the full lake is above a depth of 30 m, so sediment focusing may have more or less doubled average mass accumulation rates below this depth. Even assuming a doubling of the observed rate of dust deposition in this unit, accumulation of endogenic carbonate is so rapid that dust contributes little to overall carbonate content of the sediment. Relative to carbonate minerals, fluxes of non-carbonate minerals are lower in the sediment and much higher in dust, so that direct dust deposition in the lake could account for roughly 10%–20% of the non-carbonate minerals in the upper unit.

Variations in mineralogy, elemental concentrations, and magnetic properties with depth in sediment cores may provide information about the source areas of detrital material. In the absence of large amounts of endogenic material, interpretation of mineral and elemental concentrations is straightforward. When detrital material is diluted by abundant endogenic material, such as in the upper unit, interpretations are more complex and more ambiguous. Similarly, magnetic properties provide a powerful tool to discriminate among source areas if post-depositional alteration has not destroyed detrital Fe-oxides or formed secondary magnetic phases.

Similarity between the contents of quartz (70%) and calcite (10%–20%) in the lower unit and the contents of these minerals in the fine-grained fraction of Bear River sediment (Fig. 3) suggests that the lower unit contains little, if any, endogenic calcite. In addition, several observations indicate that magnetic properties in most of the lower unit have been largely unaffected by post-depositional alteration. First, MS values (mostly 0.2–0.5×10^{-6} m^3 kg^{-1}) and HIRM values (mostly 0.3–0.8×10^{-3} Am^2 kg^{-1}) in the lower unit are large in comparison to the upper unit and comparable to values in the fluvial sediments (Fig. 5). Second, post-depositional destruction of Fe-oxides in a reducing environment should affect both magnetite and hematite and, therefore, cannot explain the negative correlation between indicators of magnetite content (i.e., MS, ARM, and IRM) and HIRM. And third, petrographic observations show no evidence of Fe-oxide dissolution and indicate that sulfides are present only in trace amounts except in the uppermost part of the unit in core BL96-3, where magnetic separates contain significant amounts of greigite. Magnetic separates that contain significant amounts of greigite (at 41 and 65 cm in BL96-3, Fig. 12) fall within a zone of elevated sulfur (Fig. 11) and were selected to sample sediment that yielded high values of IRM/MS (Fig. 12), which has been shown to be a good indicator for the presence of this mineral (Reynolds et al., 1994, 1998). Within the sulfur-rich zone in BL96-3, both magnetite and hematite contents decrease upward to the top of the lower unit. In the correlative portion of BL96-2, hematite content decreases, but by a much smaller amount than in BL96-3, and magnetite content increases. Only a few samples near the base of core BL96–2 yield high values of IRM/MS (Fig. 13), suggesting that they contain some greigite. These observations indicate that magnetic minerals in the uppermost 25 cm of the lower unit in BL96-3 have undergone significant post-depositional alteration, and that magnetic minerals in the correlative part of BL96-2 have undergone no alteration or have been altered to a lesser extent.

Interpretation of changes in provenance is relatively straightforward below ~1.4 m in BL96-3 because there is little endogenic material, magnetic properties reflect unaltered detrital minerals, and there are obvious relations among various proxies. The observed variations in HIRM, MS, and dolomite content respectively reflect changes in the proportions of detrital material derived from three sources: Uinta Mountain Group rocks in the headwaters of the Bear River, dust, and bedrock in the Bear River Range. However, determining which source drives the changes is more problematic. One alternative is that variable rates of direct dust input dilute fluvial materials to different extents. Several observations make this alternative unlikely. Given the mass accumulation rates of modern dust and of lower unit sediments (Tables 2 and 3), it seems unlikely that direct dust input to the lake could have been high enough to strongly affect the concentrations of fluvial detritus. Furthermore, dilution of fluvial material by direct deposition of dust cannot explain the positive correlation between MS and dolomite content. Finally, zones with the highest acid-leachable Mg concentrations, 1.45–1.65 m and 2.60–2.80 m (Fig. 19), have concentrations of ~1%. This value is equivalent to ~13% dolomite, indicating that the sediments in these zones comprise largely material from the local catchment (Table 1). MS values in these zones (average ~5 × 10^{-7} m^3 kg^{-1}) are also similar to values of the local catchment samples. It would be highly coincidental for direct dust deposition to the lake and independent fluvial deposition of dolomite to yield values so similar to those observed in modern stream sediments. Therefore, high MS values are interpreted to indicate high concentrations of dust reworked from the land surface. Variations in MS and dolomite are linked because dust and bedrock were delivered to the lake via streams. The negative relations between HIRM and dolomite content (as measured by HCl-extractable Mg) and between HIRM and MS in this part of the section (Fig. 19) are interpreted to arise from millennial-scale variations in the contents of Uinta Mountain Group-rich material delivered by the Bear River and of material from bedrock

and surface of the local catchment, with an increase in one causing dilution of the other. Grain size of siliciclastic material also varies in this part of the section, with finer grain size corresponding to higher content of Uinta Mountain Group material. The variations in content of Uinta Mountain Group material have been interpreted to reflect changes in concentrations of fine-grained glacial flour from the headwaters of the Bear River (Dean et al., 2006; Rosenbaum and Heil, this volume).

Variations in content of glacial flour could have arisen in a variety of ways, including changes in the stream flow within the local catchment, changes in the course of the Bear River that affected the location of its delta or at times allowed the river to bypass the lake, and changes in the flux of glacial flour driven by changes in the extent of glaciation. It is more likely that changes in lake sediment composition were driven by changes in the amount of material delivered by the relatively large sediment-laden Bear River than by changes in the amount of material delivered by the small streams in the local catchment. The nearly identical variations in magnetic properties in cores BL96-3 and BL00-1 (Rosenbaum and Heil, this volume), which are separated by 4.5 km (Fig. 2), suggest that changes in the location of the river and its delta were insignificant, because such changes would be expected to change the distribution of Bear River sediment within the lake. Because Bear Lake appears to produce endogenic carbonate minerals in the absence of the Bear River (Dean et al., 2006; Kaufman et al., this volume), the low, relatively constant content of calcite within the quartz-rich lower unit indicates that Bear River input was probably always present. We therefore favor an interpretation like that of Reynolds et al. (2004) and Rosenbaum and Reynolds (2004) for Upper Klamath Lake (Oregon), in which changes in the content of glacial flour were driven by changes in the extent of glaciers within the catchment.

The relations among proxies for Uinta Mountain Group material and for local catchment material change in the upper part of the lower unit. For descriptive purposes, the upper part of the lower unit has been divided into three zones (Fig. 19). In zone I the correlation between MS and dolomite content (HCl-extractable Mg) in the underlying sediment ceases, with dolomite content becoming high relative to MS. The lack of a close relation between MS and dolomite continues in zone II, and the strong negative relation between MS and HIRM ceases. An abrupt decrease in HIRM at the base of zone II is accompanied by a smaller than "expected" increase in MS, and the HIRM and MS curves diverge across the zone. Within this zone, HIRM generally decreases, whereas dolomite content generally increases. With the exception of one sample, median grain size remains small (mostly ≤5 μm) from the base of the lower unit to the top of zone II. The zone II-III boundary was picked at a point of upward increases in dolomite and grain size in core BL96-3 and at similar increases in MS and dolomite in core BL96-2. (The absence of a similar increase in MS in BL96-3 is attributed to alteration, discussed above.)

In terms of provenance, these observations indicate the following sequence. First (zone I), content of dolomitic bedrock from

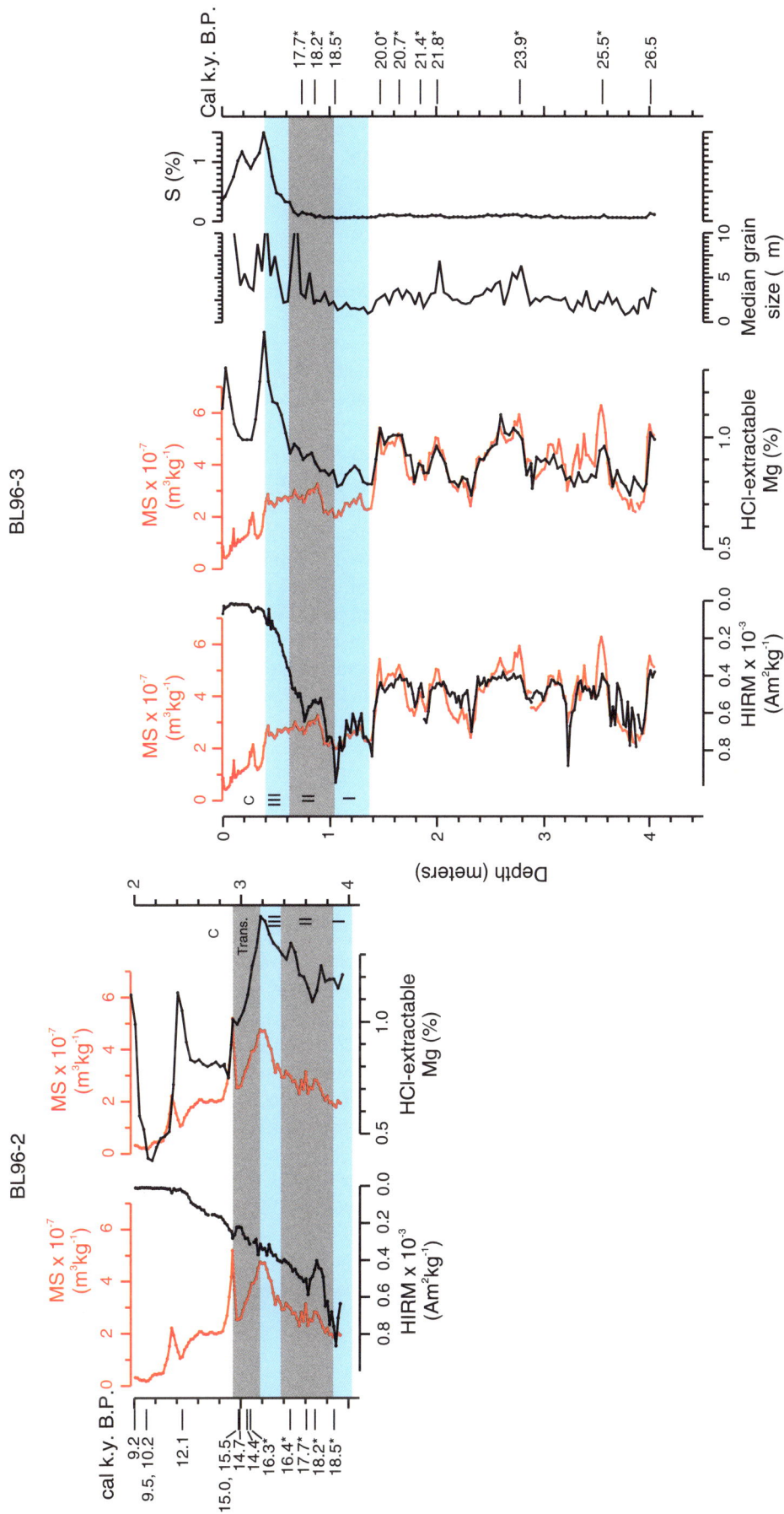

Figure 19. Comparison of magnetic susceptibility (MS), hard isothermal remanent magnetization (HIRM), and content of HCl-extractable Mg (Bischoff et al., 2005) in cores BL96-2 and -3, and median grain size and sulfur content in core BL96-3. Note inverted axis for HIRM. Below 1.2 m, MS is inversely related to HIRM (MS = $7.905 \times 10^{-7} - 7.667 \times 10^{-4} \times$ HIRM, $r^2 = 0.692$). Below 1.4 m, MS is positively correlated with HCl-extractable Mg (MS = $8.602 \times 10^{-7} \times$ Mg-3.529×10^{-7}, $r^2 = 0.681$). These relations were used to scale and position the MS curves with respect to HIRM and Mg. Zones I, II, and III at the top of the quartz-rich lower unit are defined for descriptive purposes (see text). The transition (Trans.) from the lower unit to the carbonate-rich upper unit (C) occurs abruptly at the top of zone III in core BL96-3 and more gradually in core BL96-2. Calibrated ^{14}C ages (asterisk indicates age from correlation to core BL00-1) are from Colman et al. (this volume).

the Bear River Range increased while content of Uinta Mountain Group material remained high. The increase in dolomitic bedrock content was not accompanied by an increase in magnetite-bearing dust from the surface of the local catchment. Second (zone II), decreased content of hematite-rich Uinta Mountain Group material was accompanied by increased content of dolomitic material from the Bear River Range but with little corresponding change in surface material from the local catchment. Third (zone III), both magnetite-rich surface material and dolomitic bedrock from the local catchment increased, whereas the content of hematitic Uinta Mountain Group material declined. Rosenbaum and Heil (this volume) suggest that these variations in the upper portions of the lower unit may record (1) initial input of dolomitic glacial flour produced by growth of glaciers of Pinedale-equivalent age in the Bear River Range (Reheis et al., this volume) without a change in input of magnetite-rich surficial material; (2) growth of glaciers in the headwaters of the Bear River to extend down-valley beyond the extent of hematite-rich Uinta Mountain Group rocks onto less hematite-rich bedrock and concomitant growth of glaciers in the Bear River Range; and (3) decrease in glacial flour content due to glacial retreat or progressive abandonment of Bear Lake by the Bear River, or both.

As the Bear River largely abandoned Bear Lake, endogenic carbonate minerals increased and remained the dominant phases throughout the Holocene. However, the river's influence can be discerned within some intervals of the carbonate-rich upper unit. Dean et al. (2006) interpreted input of Bear River water during deposition of the upper and lower calcite intervals on the basis of geochemistry of the carbonate minerals. In addition, variations in detrital minerals suggest that Bear River sediment entered the lake during deposition of the calcite intervals. Specifically, Dolo:Qtz is generally lower within the calcite intervals than in the aragonite intervals (Figs. 7, 9, and 10) indicating that the local catchment detritus was diluted by quartz-rich, dolomite-poor Bear River sediment.

Above zone III, in the transition zone and the upper unit, magnetic properties as well as mineral and elemental concentrations are not useful indicators of changes in provenance. Low concentrations of Fe-oxide minerals reflect a combination of dilution by endogenic carbonate minerals and post-depositional destruction of detrital Fe-oxide minerals. Magnetic concentration parameters (e.g., MS, ARM, and HIRM) generally decrease upward across the lower calcite interval (Figs. 13 and 14) with little change in detrital material indicated by uniform quartz content (Figs. 7 and 10). This relation indicates a progressive upward increase in alteration of detrital Fe-oxide minerals and suggests an increase in salinity. This interpretation, that greater alteration of Fe-oxide minerals is associated with higher salinity, is consistent with the coincidence or near coincidence of peaks in magnetic concentration parameters (MS features 5–7 in Figs. 13–15) with interpreted high lake levels (Smoot and Rosenbaum, this volume), which presumably would coincide with reduced salinity. Variable Fe-oxide preservation and differences in detrital input probably contribute to low-amplitude variations in magnetic properties (MS features

1–4 in Figs. 14–16). For instance, higher values of MS within the upper aragonite interval in cores BL2002-3 and -4 (Figs. 15 and 14) generally coincide with slightly higher concentrations of detrital quartz (Figs. 9 and 10), whereas magnetic property variations in core BL96-1 (Fig. 16) are not accompanied by similar variations in quartz content (Fig. 6) and are therefore likely caused by differences in preservation of magnetic minerals.

Grain-Size Variations

The observed variation in grain size of modern sediment with water depth has little bearing on interpretation of grain-size variations in the quartz-rich lower unit because conditions during deposition of that unit were very different than at present. During that interval, the Bear River had glaciers in its headwaters and was connected to Bear Lake, the lake probably overflowed continuously, and sedimentation was dominated by fine-grained clastic material. These conditions are similar to those in glacial-aged sediment from Upper Klamath Lake (Oregon), where Reynolds et al. (2004) found variations in bulk sediment grain size, and interpreted finer grain size to reflect higher content of very fine grained glacial flour. Below 1.05 m in core BL96-3, variations in grain size probably reflect variations in glacial-flour content, with finer-grained sediments, which tend to have higher values of HIRM (Fig. 19), containing a higher proportion of glacial flour derived from hematite-rich Uinta Mountain Group rocks. The very fine grain size of less hematite-rich sediment between 1.05 and 0.6 m may also indicate a high content of glacial flour (Rosenbaum and Heil, this volume), but with less Uinta Mountain Group material (i.e., lower HIRM) and more dolomite than in sediments below 1.05 m.

During deposition of the carbonate-rich upper unit, conditions were much closer to those of the modern lake. Grain sizes are generally coarser than in the lower unit, and the coarsest sediment coincides with shell layers, which are interpreted to be beach or shallow-water deposits. The calcite intervals are generally finer grained than the aragonite intervals, suggesting deeper and shallower depositional environments, respectively. For the upper unit, the relation between grain size and water depth for the modern sediments can be used to estimate water depths in the past. Although the depth/grain-size relations differ among the surface profiles (Fig. 17), the location and consistent bottom slope of profile 1 (Fig. 2) make it the obvious choice for modeling paleodepths for cores BL2002-3 and -4. Smoot and Rosenbaum (this volume) combined this type of modeling with detailed sedimentology and paleo-shoreline data to create a lake-level curve.

SUMMARY

Study of modern fluvial sediment and dust provides insights into the origin of variations in the allogenic component of Bear Lake sediment. Two observations that reflect bedrock geology and help determine provenance of the lake sediment are (1) that dolomite content is high in streams draining the Bear River Range on the west side of the lake, and (2) that HIRM values are high in the glaciated headwaters of the Bear River and probably reflect high content of hematite-rich detritus from Uinta Mountain Group rocks. Another observation that contributes to interpretation of provenance is provided by magnetic properties that measure the content of ferrimagnetic minerals (e.g., MS), which largely reflect silt-sized magnetite and titanomagnetite grains that were delivered to the catchment as a component of dust. The concentration of these magnetic grains is about three times higher in streams within the local Bear Lake catchment than in the Bear River.

The above observations are useful in interpreting the origin of mineralogical and magnetic property variations in the siliciclastic lower unit penetrated by cores in Bear Lake because this unit contains little if any endogenic carbonate minerals and its magnetic properties are largely unaffected by post-depositional alteration. Within this unit, quasi-cyclical variations in grain size, HIRM, dolomite content, and MS generally reflect changes in content of very fine grained hematite-rich Uinta Mountain Group detritus delivered to the lake by the Bear River, and of somewhat coarser fluvial material from the local catchment that is rich in dolomite from the bedrock and in magnetite from dust on the land surface. Several factors contribute to an interpretation of the Uinta Mountain Group detritus as glacial flour. First, extensive glaciers were present in the headwaters of the Bear River during the last glacial period. Second, Uinta Mountain Group bedrock is exposed in a small fraction of the catchment. However, Uinta Mountain Group material is abundant in last-glacial-aged till but is largely absent in modern stream sediments downstream from the last glacial limit. Glaciers would have enhanced erosion and provided a source of fine-grained Uinta Mountain Group detritus to the Bear River. And last, below 1.05 m in core BL96-3, the tendency of sediment with a higher content of Uinta Mountain Group material to be finer grained is similar to the relation observed in Upper Klamath Lake (Oregon), where very fine grained, fresh basaltic detritus is interpreted to be glacial flour (Reynolds et al., 2004; Rosenbaum and Reynolds, 2004). In the upper part of the quartz-rich lower unit, the relations among grain size, HIRM, dolomite content, and MS differ from those observed below 1.4 m in core BL96-3. An increase in content of dolomitic bedrock from the local catchment without an increase in dust from the catchment surface or a change in Uinta Mountain Group material is followed by a decrease in Uinta Mountain Group detritus. Although the decrease in Uinta Mountain Group material was previously interpreted to reflect glacial retreat in the Uinta Mountains (Dean et al., 2006), the combined proxies may be more consistent with growth of Uinta glaciers beyond exposures of Uinta Mountain Group rocks and growth of glaciers in the Bear River Range (Rosenbaum and Heil, this volume).

Mineral and elemental concentrations and magnetic properties are less useful for determining provenance of detrital material in the upper unit because of the effects of dilution by large amounts of endogenic carbonate minerals and of post-depositional destruction of detrital Fe-oxide minerals. Nevertheless, relatively low values of Dolo:Qtz in the calcite-rich intervals indicate that

Bear River delivered sediment to the lake during deposition of these intervals. Within this unit, variations in grain size of siliciclastic material largely reflect changes in water depth.

ACKNOWLEDGMENTS

We thank C. Chapman, J. Honke, G. Skipp, and F. Urban for assistance in coring, sampling, and laboratory analyses. L. Brown, H. Goldstein, T. Johnson, M. Rosen, and G. Skipp provided constructive reviews. This work was funded by the Earth Surface Dynamics Program of the U.S. Geological Survey.

ARCHIVED DATA

Archived data for this chapter can be obtained from the NOAA World Data Center for Paleoclimatology at http://www.ncdc.noaa.gov/paleo/pubs/gsa2009bearlake/.

REFERENCES CITED

Ashby, J.M., Geissman, J.W., and Weil, A.B., 2005, Paleomagnetic and fault kinematic assessment of Laramide-age deformation in the eastern Uinta Mountains or, has the eastern end of the Uinta Mountains been bent? *in* Dehler, C.M., Pederson, J.L., Sprinkel, D.A., and Kowallis, B.J., eds., Uinta Mountain geology: Salt Lake City, Utah Geological Association Publication 33, p. 285–320.

Bischoff, J.L., Cummins, K., and Shamp, D.G., 2005, Geochemistry of sediments in cores and sediment traps from Bear Lake, Utah and Idaho: U.S. Geological Survey Open-File Report 2005-1215, http://pubs.usgs.gov/of/2005/1215/ (accessed June 2007).

Bryant, B., 1992, Geologic and structure maps of the Salt Lake City 1° × 2° quadrangle, Utah, and Wyoming: U.S. Geological Survey Miscellaneous Investigations Series Map I-1997, scale 1:250,000.

Colman, S.M., 2006, Acoustic stratigraphy of Bear Lake, Utah-Idaho—Late Quaternary sedimentation patterns in a simple half graben: Sedimentary Geology, v. 185, p. 113–125, doi: 10.1016/j.sedgeo.2005.11.022.

Colman, S.M., Kaufman, D.S., Bright, J., Heil, C., King, J.W., Dean, W.E., Rosenbaum, J.G., Forester, R.M., Bischoff, J.L., Perkins, M., and McGeehin, J.P., 2006, Age models for a continuous 250-kyr Quaternary lacustrine record from Bear Lake, Utah-Idaho: Quaternary Science Reviews, v. 25, p. 2271–2282, doi: 10.1016/j.quascirev.2005.10.015.

Colman, S.M., Rosenbaum, J.G., Kaufman, D., Dean, W.E., and McGeehin, J.P., 2009, this volume, Radiocarbon ages and age models for the last 30,000 years in Bear Lake, Utah-Idaho, *in* Rosenbaum, J.G., and Kaufman, D.S., eds., Paleoenvironments of Bear Lake, Utah and Idaho, and its catchment: Geological Society of America Special Paper 450, doi: 10.1130/2009.2450(05).

Dean, W.E., 2009, this volume, Endogenic carbonate sedimentation in Bear Lake, Utah and Idaho, over the last two glacial-interglacial cycles, *in* Rosenbaum, J.G., and Kaufman, D.S., eds., Paleoenvironments of Bear Lake, Utah and Idaho, and its catchment: Geological Society of America Special Paper 450, doi: 10.1130/2009.2450(07).

Dean, W.E., Rosenbaum, J.G., Skipp, G., Colman, S.M., Forester, R.M., Liu, A., Simmons, K., and Biscoff, J.L., 2006, Unusual Holocene and late Pleistocene carbonate sedimentation in Bear Lake, Utah and Idaho, USA: Sedimentary Geology, v. 185, p. 93–112, doi: 10.1016/j.sedgeo.2005.11.016.

Dean, W.E., Wurtsbaugh, W., and Lamarra, V., 2009, this volume, Climatic and limnologic setting of Bear Lake, Utah and Idaho, *in* Rosenbaum, J.G., and Kaufman, D.S., eds., Paleoenvironments of Bear Lake, Utah and Idaho, and its catchment: Geological Society of America Special Paper 450, doi: 10.1130/2009.2450(01).

Denny, J.F., and Colman, S.M., 2003, Geophysical Surveys of Bear Lake, Utah-Idaho, September 2002: U.S. Geological Survey Open-File Report 03-150.

Dover, J.H., 1995, Geologic map of the Logan 30′ × 60′ quadrangle, Cache and Rich counties, Utah, and Lincoln and Uinta counties, Wyoming: U.S. Geological Survey Miscellaneous Investigations Series Map I-2210 scale, 1:100,000.

Goudie, A.S., and Middleton, N.J., 2006, Desert dust in the global system: New York, Springer, 287 p.

Heil, C.W., Jr., King, J.W., Rosenbaum, J.G., Reynolds, R.L., and Colman, S.M., 2009, this volume, Paleomagnetism and environmental magnetism of GLAD800 sediment cores, Bear Lake, Utah and Idaho, *in* Rosenbaum, J.G., and Kaufman, D.S., eds., Paleoenvironments of Bear Lake, Utah and Idaho, and its catchment: Geological Society of America Special Paper 450, doi: 10.1130/2009.2450(13).

Kaufman, D.S., Bright, J., Dean, W.E., Moser, K., Rosenbaum, J.G., Anderson, R.S., Colman, S.M., Heil, C.W., Jr., Jiménez-Moreno, G., Reheis, M.C., and Simmons, K.R., 2009, this volume, A quarter-million years of paleoenvironmental change at Bear Lake, Utah and Idaho, *in* Rosenbaum, J.G., and Kaufman, D.S., eds., Paleoenvironments of Bear Lake, Utah and Idaho, and its catchment: Geological Society of America Special Paper 450, doi: 10.1130/2009.2450(14).

King, J.W., and Channel, J.E.T., 1991, Sedimentary magnetism, environmental magnetism, and magnetostratigraphy: Reviews of Geophysics, v. 29, p. 358–370.

Laabs, B.J.C., Munroe, J.S., Rosenbaum, J.G., Refsnider, K.A., Mickelson, D.M., Singer, B.S., and Caffee, M.W., 2007, Chronology of the Last Glacial Maximum in the Upper Bear River Basin, Utah: Arctic, Antarctic, and Alpine Research, v. 39, p. 537–548, doi: 10.1657/1523-0430(06-089) [LAABS]2.0.CO;2.

Moore, D.M., and Reynolds, R.C., Jr., 1989, X-ray diffraction and identification and analysis of clay minerals: New York, Oxford University Press, 332 p.

Oriel, S.S., and Platt, L.B., 1980, Geologic map of the Preston 1° × 2° quadrangle, southeastern Idaho and western Wyoming: U.S. Geological Survey, Miscellaneous Investigations Series Map I-1127, scale 1:250,000.

Reheis, M.C., 2003, Dust deposition in Nevada, California, and Utah, 1984–2002: U.S. Geological Survey Open-File Report 03-138, http://pubs.usgs.gov/of/2003/ofr-03-138/ (accessed June 2007).

Reheis, M.C., 2006, 16-year record of dust deposition in southern Nevada and California, U.S.A.: Journal of Arid Environments, v. 67, p. 487–520, doi: 10.1016/j.jaridenv.2006.03.006.

Reheis, M.C., and Kihl, R., 1995, Dust deposition in southern Nevada and California, 1984–1989—Relations to climate, source area, and source lithology: Journal of Geophysical Research (Atmospheres), v. 100, no. D5, p. 8893–8918, doi: 10.1029/94JD03245.

Reheis, M.C., Laabs, J.C., and Kaufman, D.S., 2009, this volume, Geology and geomorphology of Bear Lake Valley and Upper Bear River, Utah and Idaho, *in* Rosenbaum, J.G., and Kaufman, D.S., eds., Paleoenvironments of Bear Lake, Utah and Idaho, and its catchment: Geological Society of America, doi: 10.1130/2009.2450(02).

Reynolds, R.L., and Rosenbaum, J.G., 2005, Magnetic mineralogy of sediments in Bear Lake and its watershed: Support for paleoenvironmental interpretations: U.S. Geological Survey Open-File Report 2005-1406, http://pubs.usgs.gov/of/2005/1406/ (accessed June 2007).

Reynolds, R.L., Tuttle, M.L., Rice, C., Fishman, N.S., Karachewski, J.A., and Sherman, D., 1994, Magnetization and geochemistry of greigite-bearing Cretaceous strata, North Slope basin, Alaska: American Journal of Science, v. 294, p. 485–528.

Reynolds, R.L., Rosenbaum, J.G., Mazza, N., Rivers, W., and Luiszer, F., 1998, Sediment magnetic data (83 to 18 m depth) and XRF geochemical data (83 to 32 m depth) from lacustrine sediment in core OL-92 from Owens Lake, California, *in* Bischoff, J.L., ed., 1998, A high-resolution study of climate proxies in sediments from the last interglaciation at Owens Lake, California: Core OL-92: U.S. Geological Survey Open-File Report 98-132, 20 p.

Reynolds, R.L., Sweetkind, D.S., and Axford, Y., 2001, An inexpensive magnetic mineral separator for fine-grained sediment: U.S. Geological Survey Open-File Report 01-281, 7 p.

Reynolds, R.L., Rosenbaum, J.G., Rapp, J., Kerwin, M.W., Bradbury, J.P., Colman, S.C., and Adam, D., 2004, Record of late Pleistocene glaciation and deglaciation in the southern Cascade Range: I. Petrologic evidence from lacustrine sediment in Upper Klamath Lake, southern Oregon: Journal of Paleolimnology, v. 31, p. 217–233, doi: 10.1023/B:JOPL.0000019230.42575.03.

Rose, N.L., Alliksaar, T., Bowman, J.J., Fott, J., Harlock, S., Punning, J.-M., St. Clair-Gribble, K., Vukic, J., and Watt, J., 1998, The FLAME Project: General discussion and overall conclusions: Water, Air, and Soil Pollution, v. 106, p. 329–351, doi: 10.1023/A:1005093313010.

Rosenbaum, J.G., and Heil, C.W., Jr., 2009, this volume, The glacial/deglacial history of sedimentation in Bear Lake, Utah and Idaho, *in* Rosenbaum, J.G., and Kaufman, D.S., eds., Paleoenvironments of Bear Lake, Utah and Idaho, and its catchment: Geological Society of America Special Paper 450, doi: 10.1130/2009.2450(11).

Rosenbaum, J.G., and Kaufman, D.S., 2009, this volume, Introduction to *Paleoenvironments of Bear Lake, Utah and Idaho, and its catchment, in* Rosenbaum, J.G., and Kaufman, D.S., eds., Paleoenvironments of Bear Lake, Utah and Idaho, and its catchment: Geological Society of America Special Paper 450, doi: 10.1130/2009.2450(00).

Rosenbaum, J.G., and Reynolds, R.L., 2004, Record of late Pleistocene glaciation and deglaciation in the southern Cascade Range: II. Flux of glacial flour in a sediment core from Upper Klamath Lake, Oregon: Journal of Paleolimnology, v. 31, p. 235–252, doi: 10.1023/B:JOPL.0000019229.75336.7a.

Smoot, J.P., 2009, this volume, Late Quaternary sedimentary features of Bear Lake, Utah and Idaho, *in* Rosenbaum, J.G., and Kaufman, D.S., eds., Paleoenvironments of Bear Lake, Utah and Idaho, and its catchment: Geological Society of America Special Paper 450, doi: 10.1130/2009.2450(03).

Smoot, J.P., and Rosenbaum, J.G., 2009, this volume, Sedimentary constraints on late Quaternary lake-level fluctuations at Bear Lake, Utah and Idaho, *in* Rosenbaum, J.G., and Kaufman, D.S., eds., Paleoenvironments of Bear Lake, Utah and Idaho, and its catchment: Geological Society of America Special Paper 450, doi: 10.1130/2009.2450(12).

MANUSCRIPT ACCEPTED BY THE SOCIETY 15 SEPTEMBER 2008

The Geological Society of America
Special Paper 450
2009

Endogenic carbonate sedimentation in Bear Lake, Utah and Idaho, over the last two glacial-interglacial cycles

Walter E. Dean

U.S. Geological Survey, Box 25046, MS 980 Federal Center, Denver, Colorado 80225, USA

ABSTRACT

Sediments deposited over the past 220,000 years in Bear Lake, Utah and Idaho, are predominantly calcareous silty clay, with calcite as the dominant carbonate mineral. The abundance of siliciclastic sediment indicates that the Bear River usually was connected to Bear Lake. However, three marl intervals containing more than 50% $CaCO_3$ were deposited during the Holocene and the last two interglacial intervals, equivalent to marine oxygen isotope stages (MIS) 5 and 7, indicating times when the Bear River was not connected to the lake. Aragonite is the dominant mineral in two of these three high-carbonate intervals. The high-carbonate, aragonitic intervals coincide with warm interglacial continental climates and warm Pacific sea-surface temperatures. Aragonite also is the dominant mineral in a carbonate-cemented microbialite mound that formed in the southwestern part of the lake over the last several thousand years. The history of carbonate sedimentation in Bear Lake is documented through the study of isotopic ratios of oxygen, carbon, and strontium, organic carbon content, $CaCO_3$ content, X-ray diffraction mineralogy, and HCl-leach chemistry on samples from sediment traps, gravity cores, piston cores, drill cores, and microbialites.

Sediment-trap studies show that the carbonate mineral that precipitates in the surface waters of the lake today is high-Mg calcite. The lake began to precipitate high-Mg calcite sometime in the mid–twentieth century after the artificial diversion of Bear River into Bear Lake that began in 1911. This diversion drastically reduced the salinity and $Mg^{2+}:Ca^{2+}$ of the lake water and changed the primary carbonate precipitate from aragonite to high-Mg calcite. However, sediment-trap and core studies show that aragonite is the dominant mineral accumulating on the lake floor today, even though it is not precipitating in surface waters. The isotopic studies show that this aragonite is derived from reworking and redistribution of shallow-water sediment that is at least 50 yr old, and probably older. Apparently, the microbialite mound also stopped forming aragonite cement sometime after Bear River diversion. Because of reworking of old aragonite, the bulk mineralogy of carbonate in bottom sediments has not changed very much since the diversion. However, the diversion is marked by very distinct changes in the chemical and isotopic composition of the bulk carbonate.

After the last glacial interval (LGI), a large amount of endogenic carbonate began to precipitate in Bear Lake when the Pacific moisture that filled the large pluvial lakes of the Great Basin during the LGI diminished, and Bear River apparently abandoned

Dean, W.E., 2009, Endogenic carbonate sedimentation in Bear Lake, Utah and Idaho, over the last two glacial-interglacial cycles, *in* Rosenbaum, J.G., and Kaufman, D.S., eds., Paleoenvironments of Bear Lake, Utah and Idaho, and its catchment: Geological Society of America Special Paper 450, p. 169–196, doi: 10.1130/2009.2450(07). For permission to copy, contact editing@geosociety.org. ©2009 The Geological Society of America. All rights reserved.

Bear Lake. At first, the carbonate that formed was low-Mg calcite, but ~11,000 years ago, salinity and Mg²⁺:Ca²⁺ thresholds must have been crossed because the amount of aragonite gradually increased. Aragonite is the dominant carbonate mineral that has accumulated in the lake for the past 7000 years, with the addition of high-Mg calcite after the diversion of Bear River into the lake at the beginning of the twentieth century.

INTRODUCTION

Bear Lake occupies the southern half of the Bear Lake Valley in northeastern Utah and adjacent Idaho (Fig. 1). The lake is 32 km long and 6–13 km wide with an area of 280 km² at full capacity. Maximum depth is 63 m, and mean depth is 28 m (Birdsey, 1989). The present elevation of the lake when full is 1805 m above sea level, but this level has varied considerably through time (Reheis et al., this volume; Smoot and Rosenbaum, this volume). The natural watershed of the lake is quite small and has a basin-area:lake-area ratio of 4.8 (Wurtsbaugh and Luecke, 1997). Within historical times, the Bear River did not flow into Bear Lake. A series of canals was built beginning in 1911 to bring Bear River water into the lake and a pumping station was completed in 1918 to return Bear Lake water to the river (Birdsey, 1989). Apparently the first Bear River water was diverted into Mud Lake, and presumably overflowed into Bear Lake, in May 1911 (Mitch Poulsen, Bear Lake Regional Commission, 2004, personal commun.). Judging from sparse records, large volumes of Bear River water were not diverted into Bear Lake until at least 1913 (Connely Baldwin, PacifiCorp, 2005, personal commun.). The oldest chemical analysis from Bear Lake is of a water sample collected in 1912 (Kemmerer et al., 1923). I assume that the 1912 analysis is representative of the pre-diversion composition of the lake (Dean et al., 2007). Diversion of Bear River into Bear Lake created a reservoir to supply hydropower and irrigation water. This increased the basin-area:lake-area ratio considerably, to 29.5 (Birdsey, 1989). The mean annual surface hydrologic flux (including precipitation) to the lake is estimated to be 0.48×10^9 m³yr⁻¹ (Lamarra et al., 1986). Outflow is estimated as 0.214×10^9 m³yr⁻¹ which is ~3% of the lake volume (7.86×10^9 m³), giving an average residence time of ~37 yr. The magnitude of groundwater influx may be considerable, but is not known. However, Bright (this volume) concluded that the annual hydrologic balance of the lake is approximately zero. The topographic catchment of Bear Lake is small so that sustaining inflow during persistent dry climate intervals implies that there were (are) large extrabasinal sources of water, necessarily groundwater. The presence of extrabasinal sources of groundwater flow in this setting requires fracture flow along major faults in the highly faulted Bear Lake Valley (Colman et al., 2006; Reheis et al., this volume), and also flow through the highly karstic carbonate rocks of the Bear River Range. The supply of groundwater is primarily from snowmelt (Dean et al., 2007; Bright, this volume).

Acoustic profiling reveals a continuous layered sediment package that is at least 100 m thick and should contain records of at least two or more older glacial cycles (Colman, 2006; Kauf-

man et al., this volume). Such records are out of range of conventional lake coring systems. However, with the development of the Global Lake Drilling to 800 m (GLAD800) system (Dean, et al., 2002), lake cores up to 800 m long are theoretically possible. Preliminary testing of the GLAD800 system was conducted on Bear Lake in September 2000, and two closely spaced cores (holes BL00-1D and BL00-1E) up to 120 m long were collected at site BL00-1 (Fig. 1A; http://www.dosecc.org/html/utah_lakes.html).

The purposes of this paper are to examine the deposition of $CaCO_3$ in Bear Lake on five time scales. First, I examine modern $CaCO_3$ deposition based on evidence from sediment traps. Next, I examine the effect of the diversion of Bear River into Bear Lake on the chemistry, mineralogy, and isotopic composition of $CaCO_3$ deposited in the lake during the twentieth century. The "Rock Pile" (a microbial mound) at the southwest end of the lake documents an unusual development in space and time (past 2000 years) in Bear Lake's history. Then, I examine the initiation of $CaCO_3$ deposition in the lake following the last glacial interval, and changes in the chemistry, mineralogy, and isotopic composition of endogenic carbonate deposited during the latest Pleistocene and the Holocene. Finally, I document the changes in the amount and mineralogy of carbonate deposited in the lake over the last two glacial-interglacial cycles (220,000 cal yr).

Some of this material has been published elsewhere (Dean et al., 2006, 2007) but will be summarized here for a complete overview of all aspects of endogenic carbonate deposition in Bear Lake. However, much of the material is new, and the entire package emphasizes the diversity of the forms and environments of endogenic carbonate deposition that has occurred in Bear Lake over the past 220,000 years.

MATERIALS AND METHODS

Piston cores were collected in 1996 from three localities in Bear Lake (Fig. 1A) using the Kullenberg coring system of the University of Minnesota's Limnological Research Center (UMN-LRC) (Dean et al., 2006). The piston cores recovered a maximum of 5 m of sediment each, but overlapping cores provide a 26,000 yr record. Unknown amounts of sediments were missing from the tops of the piston cores, so surface sediments (up to 50 cm) were collected with a gravity corer in 1998 (Fig. 1A; Dean et al., 2007). Piston cores and gravity cores were collected at five localities at the northern end of the lake in 2002 (Fig. 1A). To assess the seasonality of sedimentation, time-marking sediment traps that dispense Teflon granules in the collection tube every 30 days (Anderson, 1977) were suspended 10 m below the surface of the lake (referred to as surface traps) and 2 m above

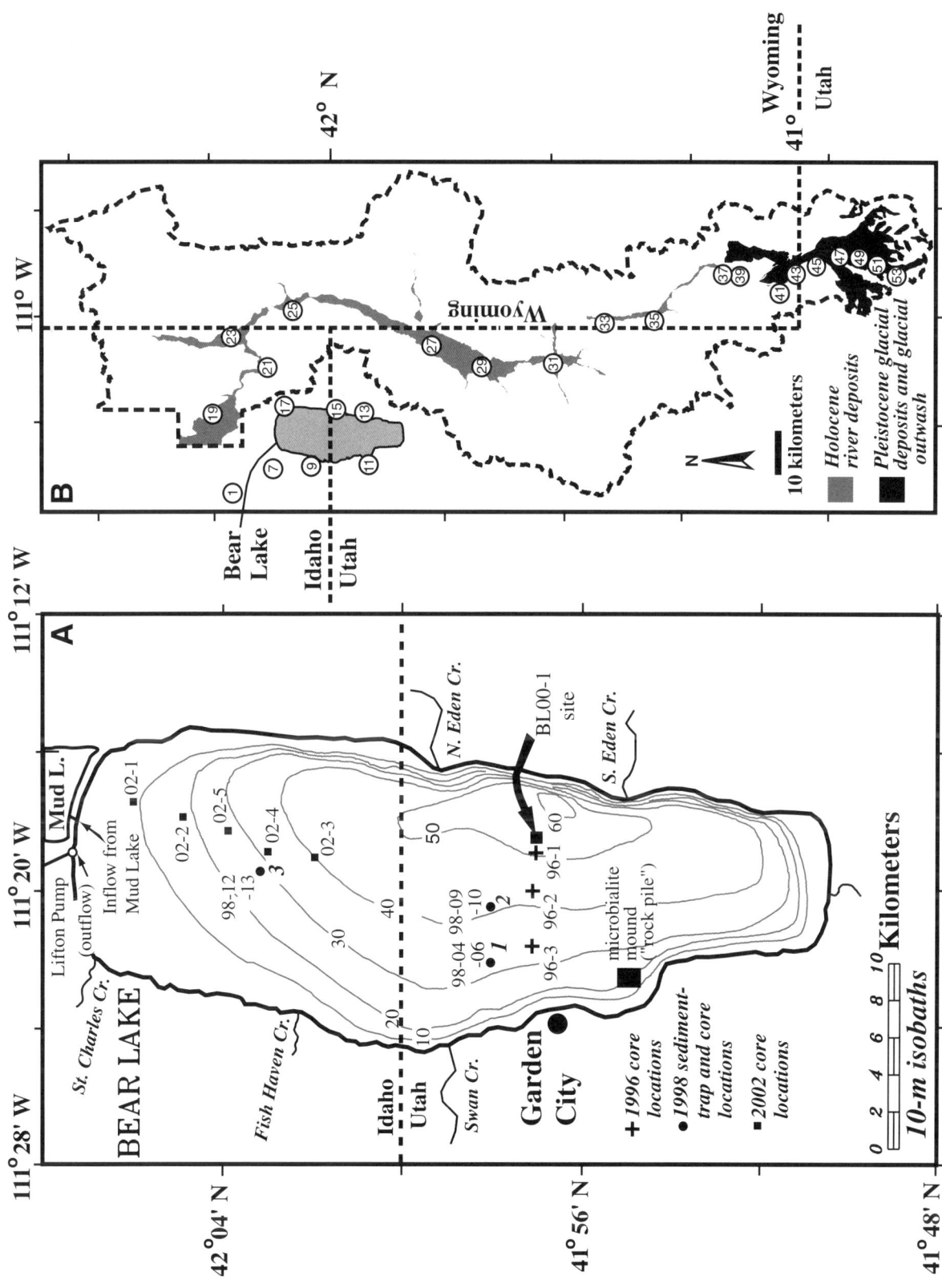

Figure 1. (A) Bathymetric map of Bear Lake, Utah and Idaho, showing the locations of 1996 piston cores, 1998 surface-sediment cores and sediment-trap deployments, 2002 piston and surface-sediment cores, and the microbialite mound ("rock pile"). (B) Locations of stream-sediment cores. Sample numbers in circles correspond to sample numbers in Table 3. Dashed line corresponds to the catchment of the upper reach of the Bear River to its headwaters in the Uinta Mountains. Odd- and even-numbered samples were collected at each site, but only the odd numbers are shown. Odd-numbered samples were taken from the stream bottoms; even-numbered samples were from stream-bank or overbank deposits. See Rosenbaum et al. (this volume) for mineralogy, geochemistry, and magnetic properties of these stream-sediment samples.

the bottom of the lake (bottom traps) at three localities where multiple gravity cores were collected (Fig. 1A) for up to three years (1998–2000).

The GLAD800 cores (from holes BL00-1D and BL00-1E) were shipped under refrigeration to the UMN-LRC. There, the cores were scanned with a Geotek multisensor logger (MSL) that measures porosity, wet bulk density (WBD), magnetic susceptibility (MS), and P-wave velocity. The correlation of sediment recovered in holes BL00-1D and -1E is based on MSL-MS profiles (Colman et al., 2006; Heil et al., this volume). Sampling and analyses followed core curation specifications and protocols that were in place before drilling. The photographs and initial core descriptions (ICD) are available at the National Oceanic and Atmospheric Administration/National Geophysical Data Center, Boulder, Colorado (http://www.ngdc.noaa.gov/mgg/curator/laccore.html). Initial samples for magnetic properties, mineralogy, and geochemistry were taken from the core catcher samples. Additional samples were collected at the same intervals that smear-slide samples were taken at the UMN-LRC in preparing the ICDs.

Concentrations of total carbon (TC) and inorganic carbon (IC) were determined by coulometric titration of CO_2 following extraction from the sediment by combustion at 950 °C and acid volatilization, respectively (Engleman et al., 1985), in U.S. Geological Survey (USGS) laboratories, Denver, Colorado. Weight percent (wt%) IC was converted to wt% $CaCO_3$ by dividing by 0.12, the fraction of carbon in $CaCO_3$. Organic carbon (OC) was determined as the difference between TC and IC. The accuracy and precision for both TC and IC, determined from hundreds of replicate standards usually are better than 0.10 wt%.

Semiquantitative estimates of mineral contents were determined by X-ray diffraction (XRD) techniques (e.g., Moore and Reynolds, 1989) in USGS laboratories, Denver, Colorado. For each sample, raw XRD peak intensities for the main peaks of minerals detected in each sample were converted to semiquantitative percentages by dividing the main peak intensity of a mineral by the sum of the main peak intensities of all minerals. More quantitative estimates of aragonite and calcite, the dominant minerals, were calculated by partitioning the percentage of total $CaCO_3$, determined by coulometry, using the intensity ratios of the main XRD peaks of aragonite to calcite (I-aragonite/I-calcite) and curves of percent aragonite (of total $CaCO_3$) versus I-aragonite/I-calcite determined by Chave (1954) and Lowenstam (1954).

Measurements of ratios of stable isotopes of carbon and oxygen were made on aliquots of the carbon samples (see Dean et al., 2006, 2007 for methods). Isotope measurements on samples from the 1996 cores, 1998 cores, and sediment traps were made in the stable isotope laboratory at the University of Minnesota. Isotope measurements on samples from the 2002 cores were made in the stable isotope laboratory at the University of Arizona. Results of analyses are reported in the usual per mil (‰) δ-notation relative to the Vienna Pee Dee Belemnite (VPDB) marine-carbonate standard for carbon and oxygen:

$$\delta‰ = [(R_{sample}/R_{VPDB})-1] \times 10^3,$$

where R is the ratio (^{13}C:^{12}C) or (^{18}O:^{16}O). Precision is ± 0.1‰ for oxygen and ± 0.06‰ for carbon.

Measurements of dissolved strontium isotope ratios (^{87}Sr/^{86}Sr) were made on filtered, unacidified water samples and on bulk sediment samples leached in 5 mol L^{-1} acetic acid (see Dean et al., 2007, for locations of water samples and details of methods). Precision is usually ±0.00001 for both waters and sediments.

Gas bubbles in Bear Lake were collected with an inverted funnel and water-filled bottle. The chemical composition of the gas was determined by using a Hewlett Packard 6890 series gas chromatograph in the USGS Organic Geochemistry Laboratory, Denver, Colorado. The isotopic composition of gas samples was measured with a Micromass Optima continuous flow Isotope Ratio Mass Spectrometer coupled with an Agilent 6890 Gas Chromatograph and a combustion furnace in the USGS Organic Geochemistry Laboratory, Denver, Colorado. The precision of the method is ±0.2‰.

Chronologies for the 1998 gravity cores are provided by ^{210}Pb dating (Smoak and Swarzenski, 2004). This technique gives approximate ages, at least for the past 100 years. Accelerator mass spectrometer (AMS) ^{14}C ages were obtained from samples of various materials including pollen concentrates, ostracodes, and bulk-sediment OC from piston cores (Colman et al., this volume). Reservoir-corrected and calibrated radiocarbon ages are expressed in kiloannum B.P. (e.g., 26.6 cal ka), and time intervals in thousands of calendar years (cal k.y.). The age model for the 120-m-long sediment sequence in BL00-1 is presented by Colman et al. (2006).

RESULTS AND INTERPRETATION

Sediment Traps

Summer carbonate sediment collected in all surface traps consists of almost pure high-Mg calcite (~10 mol % Mg) (Table 1, Figs. 2 and 3). The bottom trap at the deep site contained aragonite as the dominant mineral, but also had considerable amounts of high-Mg calcite, low-Mg calcite, and quartz, and minor amounts of dolomite (site 2; Table 1, Figs. 1A and 2). In contrast, high-Mg calcite was the dominant mineral in sediments collected in the bottom trap deployed at 27 m below the surface at the north end of the lake, along with considerable amounts of low-Mg calcite, quartz, and aragonite (site 3; Table 1, Figs. 1A and 3). The aragonite occurs as subrounded, needle-shaped crystals ~5 μm long and <1 μm in diameter (Fig. 4A). Most of the calcite occurs as equant, subrounded rhombohedral grains ~4–5 μm in diameter (Fig. 4B).

Three key questions posed by these observations are (1) Why does unstable high-Mg calcite form in the epilimnion? (2) Does high-Mg calcite dissolve and reprecipitate as low-Mg calcite on its trip through the water column? And (3) what is the origin of the aragonite in the hypolimnion?

To my knowledge, the first mention of either high-Mg calcite or aragonite from a freshwater environment is from Lake Balaton, Hungary (Müller, 1970, 1971). In that lake, high-Mg calcite

TABLE 1. BEAR LAKE TRAP SAMPLES FROM LAKE CENTER (SITES 1 AND 2) AND NORTH END (SITE 3).

Site	Surface or Bottom*	Date Deployed	Date Removed	Sample ID	Approximate Date Deposited	XRD Mineralogy#	CaCO$_3$ (%)	OC (%)	^{87}Sr/^{86}Sr	d^{13}C (VPDB) (‰)	d^{18}O (VPDB) (‰)
1	bottom	7/14/1998		0 cm	Summer, 1998		63.2	2.34			
1	bottom	7/14/1998	9/18/1999	50 cm	Summer, 1998	A>HMC>Q>LMC>>D	64.3	2.63			
1	bottom	7/14/1998	9/18/1999	50–100 cm	Summer, 1998	HMC>A>Q>LMC>>D	68.4		0.7093		
2	surface	7/14/1998	9/18/1999	5/7/1999	5/7/1999	HMC>>A-C-Q	80.9	2.34	0.7093	-1.66	-9.24
2	surface	7/14/1998	9/18/1999	7/6/1999	7/6/1999	HMC>>A-C-Q	85.1	2.46			
2	surface	7/14/1998	9/18/1999	8/5/1999	8/5/1999	HMC>>A-C-Q	86.7		0.7092		
2	surface	7/14/1998	9/18/1999	9/4/1999	9/4/1999	HMC>>A-C-Q	87.5			-1.73	-9.72
2	bottom	7/14/1998	9/18/1999	top	Sept., 1999		58.1	5.05			
2	bottom	7/14/1998	9/18/1999	20–30 cm	Summer, 1999	A>HMC>Q>LMC>>D	65.8				
2	bottom	7/14/1998	9/18/1999	30–32 cm	Summer, 1999	A>Q=HMC>LMC>>D	65.3	2.18	0.7094	0.36	-7.94
2	bottom	7/14/1998	9/18/1999	60 cm	Early Summer, 1999	HMC>A>Q>LMC	61.9	4.79			
2	bottom	7/14/1998	9/18/1999	bot. 10 cm	Late Summer, 1998	A>>Q>HMC>LMC>>D	61.8	3.84		-0.19	-7.92
2	bottom	9/22/1999	9/10/2000	1	Sept., 2000		69.7	1.99			
2	bottom	9/22/1999	9/10/2000	2	Summer, 2000		63.3	2.40			
2	bottom	9/22/1999	9/10/2000	3	Summer, 2000	A>HMC>Q>LMC>>D	66.9	1.89	0.7094		
2	bottom	9/22/1999	9/10/2000	4	Spring, 2000		66.9	1.89			
2	bottom	9/22/1999	9/10/2000	5	Summer, 1999		66.8	1.91			
2	bottom	9/22/1999	9/10/2000	6	Summer, 1999	A>HMC>Q>LMC>>D	67.2	1.88			
2	bottom	9/22/1999	9/10/2000	7	Summer, 1999		69.1	1.81			
2	bottom	9/22/1999	9/10/2000	8			68.9	1.85			
2	bottom	9/22/1999	9/10/2000	9		HMC>A>Q>LMC>>D	71.5	1.74			
2	bottom	9/22/1999	9/10/2000	10	Summer, 1999	A=HMC>Q>LMC>>D			0.7094		
2	bottom	9/22/1999	9/10/2000	11	Sept.,1999						
3	surface	7/14/1998	9/8/2000	1	Summer, 2000	all hi-Mg cal			0.7092	-0.50	-8.73
3	surface	7/14/1998	9/8/2000	2	Summer, 2000		74.9	2.53			
3	surface	7/14/1998	9/8/2000	3	Summer, 2000	HMC>>A>Q	69.3	3.12			
3	surface	7/14/1998	9/8/2000	4	Spring, 2000	all hi-Mg cal			0.7093	-1.26	-9.24
3	surface	7/14/1998	9/8/2000	5	Summer, 1999		84.5	1.67			
3	surface	7/14/1998	9/8/2000	6	Summer, 1999	HMC>>A=Q	82.4				
3	surface	7/14/1998	9/8/2000	7	Summer, 1999	all hi-Mg cal			0.7093	-0.83	-8.36
3	surface	7/14/1998	9/8/2000	8	Summer, 1998		83.7	1.72			
3	surface	7/14/1998	9/8/2000	9	Summer, 1998	HMC>>Q>A	78.4	2.20			
3	surface	7/14/1998	9/8/2000	10	July, 1998	all hi-Mg cal			0.7093	-0.38	-7.87
3	bottom	7/14/1998	9/8/2000	1	Late Summer, 2000		64.3	1.79			
3	bottom	7/14/1998	9/8/2000	2	Summer, 2000		63.4	2.03			
3	bottom	7/14/1998	9/8/2000	3	Summer 2000	HMC>A=Q>LMC>>D	64.5				
3	bottom	7/14/1998	9/8/2000	4	Summer, 2000		66.0	1.72			
3	bottom	7/14/1998	9/8/2000	5	Summer, 2000		65.1	1.83			
3	bottom	7/14/1998	9/8/2000	6	Summer, 2000		61.8	2.27	0.7094	-0.27	-8.02
3	bottom	7/14/1998	9/8/2000	7	Spring, 2000		65.1	1.97			
3	bottom	7/14/1998	9/8/2000	8	Dec.,1999	HMC>A>Q>LMC>>D					
3	bottom	7/14/1998	9/8/2000	9	Nov-Dec., 1999		61.3	2.10			
3	bottom	7/14/1998	9/8/2000	10	Nov., 1999	HMC>A>LMC>Q>>D	65.1	2.10			
3	bottom	7/14/1998	9/8/2000	11	Oct.-Nov., 1999		65.8	2.04			
3	bottom	7/14/1998	9/8/2000	12	Oct.? 1999		66.3	1.59	0.7094	-0.50	-8.12
3	bottom	7/14/1998	9/8/2000	13	Sept., 1999		67.1	2.36			
3	bottom	7/14/1998	9/8/2000	14	Aug., 1999		63.0	2.35			
3	bottom	7/14/1998	9/8/2000	15	July, 1999		60.9	1.49			
3	bottom	7/14/1998	9/8/2000	16	July, 1999		71.3	1.08			
3	bottom	7/14/1998	9/8/2000	17	June, 1999		68.3	1.40			
3	bottom	7/14/1998	9/8/2000	18	June, 1999		67.1	2.07			
3	bottom	7/14/1998	9/8/2000	19	Jan-Aprl-99		39.4	5.38			
3	bottom	7/14/1998	9/8/2000	20	Jan.? 1999		70.3	1.45	0.7094	-0.03	-7.80
3	bottom	7/14/1998	9/8/2000	21	Dec., 1998		68.5	1.93			
3	bottom	7/14/1998	9/8/2000	22	Oct.-Nov., 1998		71.1	1.96			
3	bottom	7/14/1998	9/8/2000	23	Sept., 1998		54.3	3.70			
3	bottom	7/14/1998	9/8/2000	24	August, 1998		65.8	2.47			
3	bottom	7/14/1998	9/8/2000	25	August, 1998		71.8	1.88			
3	bottom	7/14/1998	9/8/2000	26	August, 1998	HMC>>Q>A>LMC>>D					
3	bottom	7/14/1998	9/8/2000	27	August, 1998		62.3	3.11			
3	bottom	7/14/1998	9/8/2000	28	August, 1998		56.1	3.33	0.7094	-0.53	-8.18

*Surface—trap 10 m below surface of water. Bottom—trap 2 m above lake bottom.
#XRD mineralogy: HMC—high-Mg calcite; LMC—low-Mg calcite; A—aragonite; D—dolomite; Q—quartz.

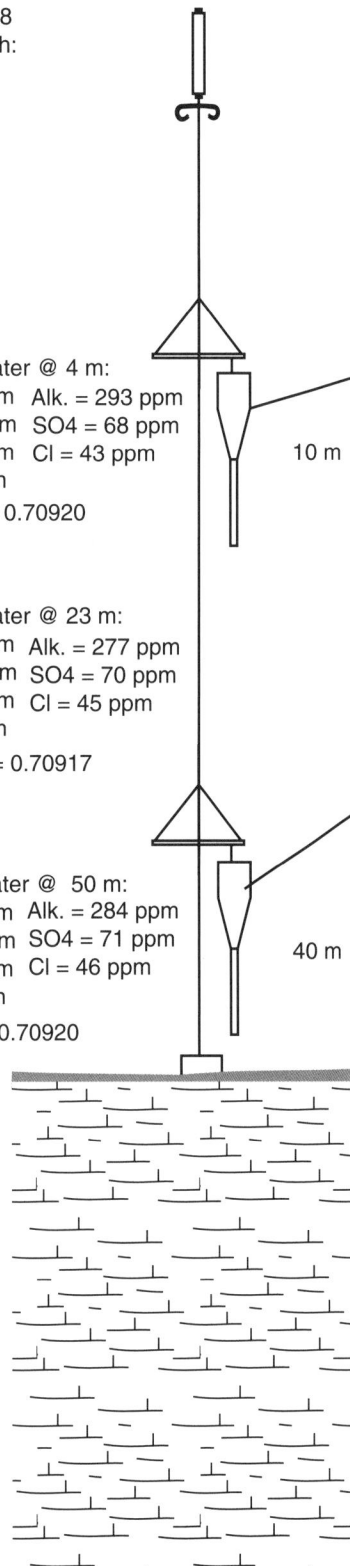

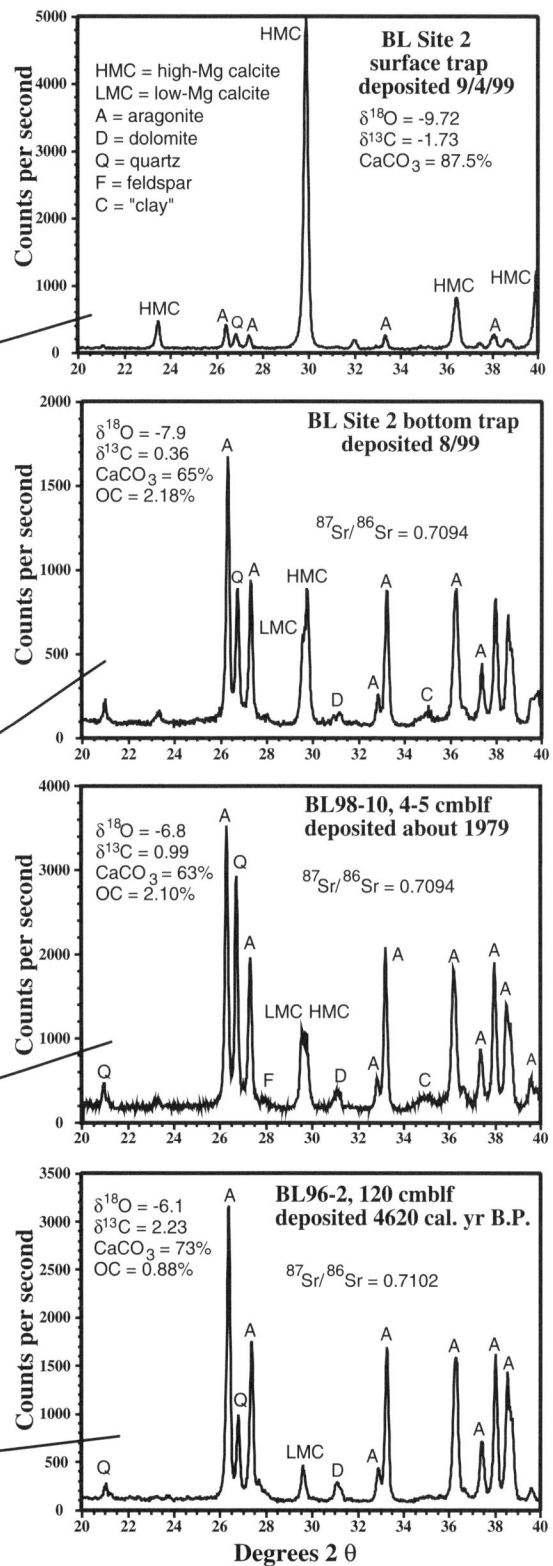

Figure 2. Diagram showing the placement of surface (10 m) and bottom (40 m) sediment traps at site 2 (Fig. 1A) in the center of Bear Lake in a water depth of 43 m. Results of chemical analyses and $^{87}Sr/^{86}Sr$ of water from depths of 4 m, 23 m, and 50 m are shown to the left of the trap diagram (data from Dean et al., 2007). X-ray diffractograms of samples of bulk sediment from the two traps, from 4 to 5 cm in short core BL98-10, and from 150 centimeters below lake floor (cmblf) in piston core BL96-2 are shown to the right of the trap diagram.

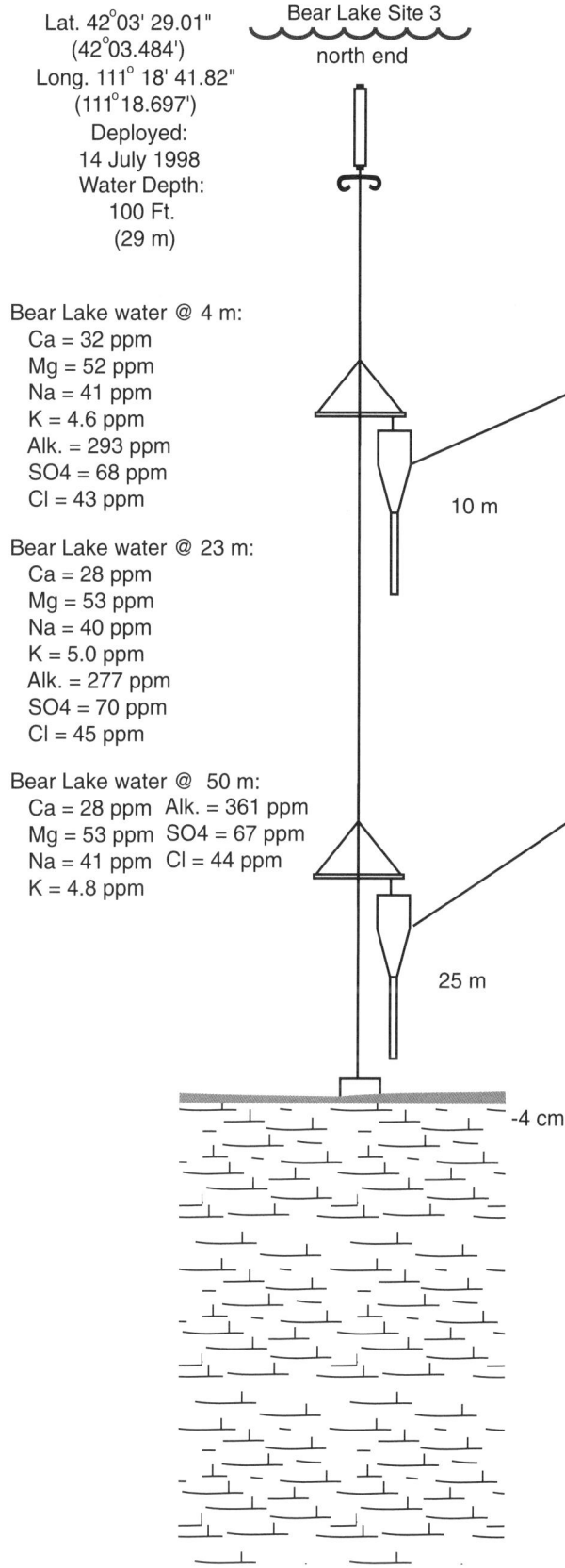

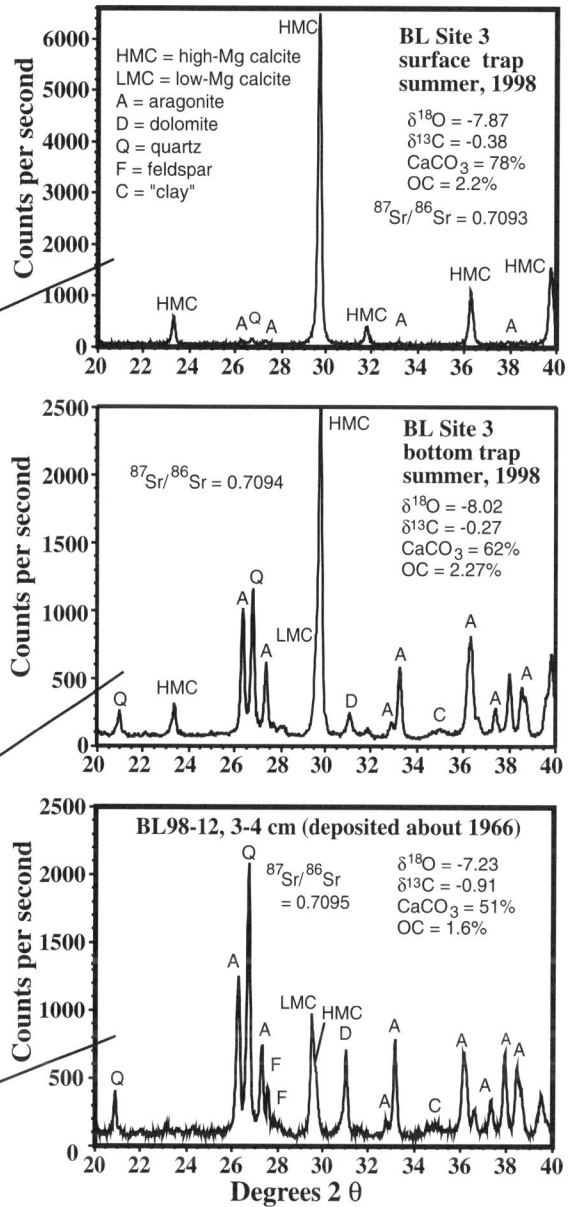

Figure 3. Diagram showing the placement of surface (10 m) and bottom (25 m) sediment traps at site 3 (Fig. 1A) at the north end of Bear Lake in a water depth of 29 m. Results of chemical analyses and $^{87}Sr/^{86}Sr$ of water from depths of 4 m, 23 m, and 50 m are shown to the left of the trap diagram (data from Dean et al., 2007). X-ray diffractograms of samples of bulk sediment from the two traps, and from 3 to 4 cm in short core BL98-12 are shown to the right of the trap diagram.

precipitates in the summer months during strong phytoplankton blooms, and aragonite forms crusts on the leaves of *Potomogeton*, a rooted aquatic plant. The reason for the difference "remains to be solved" (Müller, 1971). On the basis of empirical observations, such as those from Lake Balaton and elsewhere, Müller et al. (1972) suggested that low-Mg calcite forms in lakes having a $Mg^{2+}:Ca^{2+}$ of <2; high-Mg calcite forms in lakes having a $Mg^{2+}:Ca^{2+}$ of 2–12; and aragonite precipitates in lakes hav-

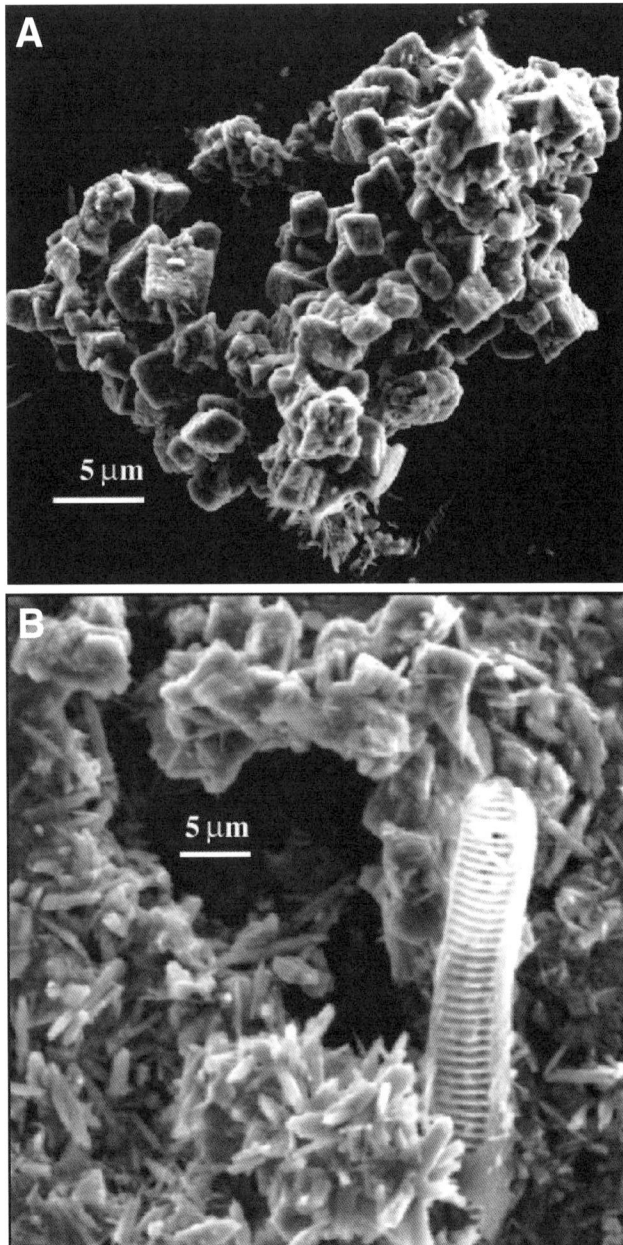

Figure 4. Scanning electron micrographs of bulk sediment showing equant blocks (rounded rhombohedrons) of high-Mg calcite from a surface sediment trap (A), and subrounded needles of aragonite (bottom-left), a pennate diatom (right), and equant blocks of calcite (top-right) from a bottom sediment trap (B).

ing a $Mg^{2+}:Ca^{2+}$ of >12. Last (1982) found that modern bottom sediments in the south basin of Lake Manitoba, Canada, consist mainly of high-Mg calcite with minor amounts of low-Mg calcite, dolomite, and aragonite. He concluded that the low-Mg calcite and dolomite were detrital and that the aragonite came from shell fragments. The high-Mg calcite is derived from precipitation in the water column triggered by photosynthetic removal of CO_2 and concomitant increase of pH. Last (1982) thought that the high $Mg^{2+}:Ca^{2+}$ in the water (1.7) was responsible for generating high-Mg calcite rather than low-Mg calcite.

Positive values of the saturation index (SI; Table 2) show that Bear Lake water is oversaturated with respect to calcite, aragonite, and dolomite at all depths, and $Mg^{2+}:Ca^{2+}$ is ~1.9 (Dean et al., 2007). The 1912 analysis of lake water (Kemmerer et al., 1923; Dean et al., 2007) indicates that the lake was also saturated with respect to all three minerals at that time, but $Mg^{2+}:Ca^{2+}$ was 38, definitely favoring aragonite.

I assume, but have no direct proof, that the precipitation of high-Mg calcite in the epilimnion of Bear Lake is biologically mediated through photosynthetic removal of CO_2 and an increase in pH during the warm summer months (Otsuki and Wetzel, 1974; Kelts and Hsü, 1978; Stabel, 1986; Dean and Megard, 1993; Dean, 1999). The average July temperature of surface water from 1989 to 2004 was 20.4 °C (Dean et al., this volume). The average pH at the surface from 1989 to 2004 was 8.43 ± 0.25 and that at 40 m was 8.36 ± 0.27 (V. Lamarra, 2004, personal commun.). Algal cells can also serve as nuclei for formation of particulate $CaCO_3$ (Stabel, 1986; Wetzel, 2001). Bear Lake is oligotrophic and generally contains low surface concentrations of chlorophyll *a* in summer (Wurtsbaugh and Luecke, 1997; Dean et al., this volume). Diatoms constitute ~80% of the algal population in Bear Lake (Birdsey, 1989), but algal blooms rarely occur (S. Tolentino, C. Luecke, and V. Lamarra, 2000, personal commun.). However, the sediments collected in the surface trap at site 2 in April 1999 included a considerable amount of green algal debris (Fig. 5). Therefore the algal bloom may have mediated the initiation of carbonate precipitation, which continued through the summer.

The dominant carbonate mineral in the sediments deposited in all surface traps during 1998, 1999, and 2000 was high-Mg calcite (Table 1). A layer ~1 cm thick of carbonate sediment accumulated in the trap during late July and early August 1998, but little sediment accumulated in August and September of that year (Fig. 5). In contrast, a much larger amount of carbonate-rich sediment accumulated in July, August, and September 1999, perhaps in response to the April 1999 algal bloom. With a trap amplification ratio of 225 (ratio of the area of the trap opening to the cross sectional area of the collection tube) a 15-cm-thick layer in the collection tube (e.g., 06 July to 05 August 1999, Fig. 5) translates to about a 0.7-mm-thick layer of sediment on the lake floor if all of the carbonate survived as it settled and accumulated on the lake floor.

Calculations using the chemistry of Bear Lake water (Dean et al., 2007) and the stability of Mg-calcites (Bischoff et al., 1987) indicate that at 20 °C the water is supersaturated with respect to aragonite and all Mg-calcites up to 11 mol % Mg, and

TABLE 2. SATURATION INDEX (SI), LOG ION ACTIVITY PRODUCT (IAP), AND LOG EQUILIBRIUM
SOLUBILITY PRODUCT (KT) FOR CALCITE, ARAGONITE, AND DOLOMITE IN WATERS FROM BEAR RIVER AND BEAR LAKE

Location	Calcite			Aragonite			Dolomite		
	SI	log IAP	log KT	SI	log IAP	log KT	SI	log IAP	log KT
Bear River at gauging station, Idaho	0.99	−7.49	−8.48	0.85	−7.49	−8.34	2.03	−15.06	−17.09
Bear Lake, east shore, Utah	1.29	−7.19	−8.48	1.14	−7.19	−8.34	3.21	−13.88	−17.09
Bear Lake at 4 m	0.44	−8.04	−8.48	0.30	−8.04	−8.34	1.45	−15.64	−17.09
Bear Lake at 50 m	0.68	−7.71	−7.87	0.52	−7.71	−8.24	1.64	−14.96	−16.59
Bear Lake, NE of rock pile	0.37	−8.08	−8.45	0.22	−8.08	−8.31	1.27	−15.70	−16.97
Bear Lake, 1979 (Lamarra et al., 1986)	−0.67	−9.12	−8.45	−0.82	−9.12	−8.31	−0.57	−17.55	−16.97
Bear Lake, 1912 (Kemmerer et al., 1923)	−0.21	−8.67	−8.45	−0.36	−8.67	−8.31	1.44	−15.54	−16.97

Note: Data from Dean et al. (2007).

Bear Lake
Utah/Idaho

Shallow Trap
10 m
Below Water Surface

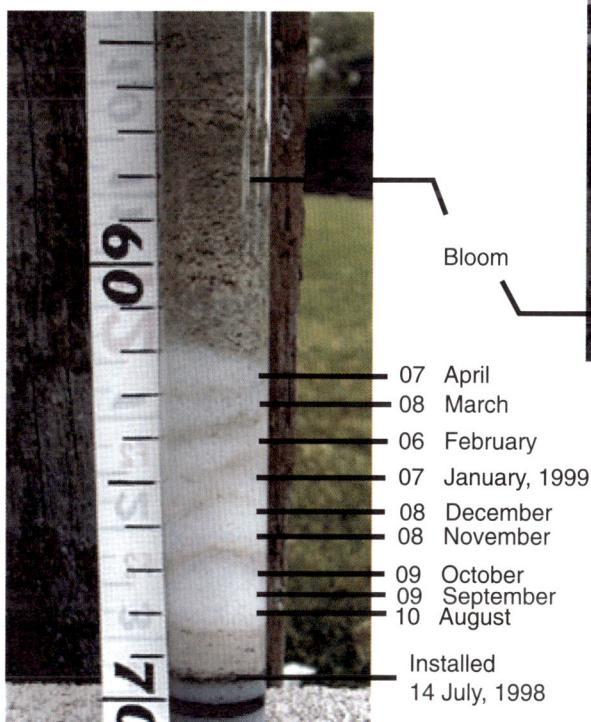

Trap
Debris

Recovery
18 September

04 September

05 August

06 July
06 June
07 May

Bloom

07 April
08 March
06 February
07 January, 1999
08 December
08 November
09 October
09 September
10 August

Installed
14 July, 1998

Trap Amplification
= 225/1

Figure 5. Photograph of the collection tube from a surface sediment trap (10 m) at site 2 (Figs. 1 and 2), deployed on 14 July 1998 and recovered 18 September 1999. White layers are composed of Teflon granules dispensed every 30 days by an automatic timer in the cone of the trap. Note that little sediment accumulated between 10 August 1998 and 7 April 1999. The tan marl deposited between 7 April and 7 May contains abundant phytoplankton debris produced by an algal bloom that was detected in the water column by large increases in chlorophyll *a* (Dean et al., this volume; Fig. 9). The collection tube contained buffered formalin to help preserve any organic matter. All of the CaCO$_3$ deposited in 1999 consists of high-Mg calcite.

that calcite with 4 mol % Mg is the least soluble (i.e., the most stable) calcite composition (Bischoff et al., 2005). At 4 °C, low-Mg calcite is more stable than high-Mg calcite, and aragonite is less stable than either (Bischoff et al., 2005). This, together with the appearance of low-Mg calcite and high-Mg calcite in the bottom traps (Figs. 2 and 3), suggest that metastable high-Mg is being dissolved, and recrystallized as low-Mg calcite, in the cold hypolimnion. Such a stabilization reaction likely occurs as a microsolution-reprecipitation reaction involving calcitization of the original high-Mg calcite (Mackenzie et al., 1983). Thus, high-Mg calcite formed in the epilimnion would dissolve in the water column and reprecipitate as low-Mg calcite. Of course, some of the low-Mg calcite may be detrital.

The above considerations answer the first and second questions posed above, but fail to explain the appearance of aragonite in the hypolimnion (question 3). Calculations by Bischoff et al. (2005) show that in Bear Lake, high-Mg calcite cannot convert to aragonite. In lakes in the United States, aragonite precipitation usually occurs in saline prairie lakes (e.g., Shapley et al., 2005) and does not occur in large, cold, deep, oligotrophic, high-altitude, north temperate lakes fed by snowmelt. Aragonite is a major component in the sediments of Beaver Lake, a marl lake in southeastern Wisconsin, but there the aragonite is derived from mollusk shell fragments (Brown et al., 1992). Dolomite and aragonite are common in the sediments of the saline lakes of the northern Great Plains (North Dakota, Montana, Saskatchewan, and Alberta) where the lake chemistry is dominated by Na^+-Mg^{2+}-SO_4^{2-} waters (Last and Schweyen, 1983). In most lakes where evaporation exceeds precipitation, the volume of groundwater and surface-water inflow is inversely proportional to evaporation. That is, as evaporation increases, the inflow of water decreases, because the lake's hydrologic budget is linked to climate. Therefore, increased evaporation results in less inflow and a saline lake precipitating evaporite minerals or a dry lake. The conditions that form aragonite in saline lakes of the northern Great Plains do not apply to Bear Lake. To help answer the question of where the aragonite in the bottom-trap and post-diversion sediments came from, we need to take a close look at the geochemistry and mineralogy of pre- and post-diversion carbonate, particularly the O, C, and Sr isotopic composition of those sediments.

Surface Sediments

Isotope Geochemistry

Values of $\delta^{18}O$, $\delta^{13}C$, and $^{87}Sr/^{86}Sr$ in bulk carbonate decrease abruptly and markedly at ~12 cm in basin-center core BL98-10 (Fig. 6). I interpret these changes to represent the chemical signature of Bear River diversion early in the twentieth century, reflecting the more ^{18}O-, ^{13}C-, and ^{87}Sr-depleted waters of Bear River (Dean et al., 2007). If this interpretation is correct, the ^{210}Pb ages in years AD in Figure 6 must be too old, particularly the older ones. Smoak and Swarzenski (2004) state that "sediments deposited prior to 100 years ago are assigned artificially 'too old' dates by the CRS [constant rate of supply] model" and that "error

terms are large on the older dates." Sediments deposited prior to diversion have values of $\delta^{18}O$ of about −4‰, whereas the carbonates that accumulated in the less saline waters of the lake after diversion have values of $\delta^{18}O$ of −7‰ to −7.5‰, slightly more enriched in ^{18}O than in sediments from bottom traps and much more enriched than in sediments from surface traps (Table 1, Figs. 2, 3, and 6). Values of $\delta^{13}C$ in pre-diversion sediments are 2.5‰–3‰, whereas those of post-diversion sediments are <1‰ (Figs. 2, 3, and 6), slightly more enriched in ^{13}C than in sediments from the bottom traps and much more enriched than in sediments from surface traps (Table 1, Figs. 2, 3, and 6).

The Sr isotopic ratio of any endogenic carbonate (precipitate or shell) will be the same as in the water from which it formed (e.g., Capo and DePaolo, 1990; Hart et al., 2004). Therefore, the Sr isotopic composition of carbonate that formed in the epilimnion of Bear Lake will be the same as that of the lake water, and we can use $^{87}Sr/^{86}Sr$ to interpret the source of pre-diversion Bear Lake water, and the origin of the mixtures of minerals in bulk carbonate in the trap sediments. Values of $^{87}Sr/^{86}Sr$ in carbonates deposited before diversion are around 0.7102 (Fig. 6), similar to values in stream and spring waters today on the west side of the lake (Fig. 6; Dean et al., 2007). The west-side stream sediments contain a considerable amount of dolomite and calcite (Table 3), presumably derived mainly from Paleozoic carbonate rocks in the Bear River Range to the west of the lake (Oriel and Platt, 1980; Dover, 1985). Values of $^{87}Sr/^{86}Sr$ in carbonates deposited in the lake after diversion are around 0.7094, slightly more enriched in ^{87}Sr than Bear Lake water today ($^{87}Sr/^{86}Sr$ = 0.7092; range 0.70917–0.70922; Dean et al., 2007), and considerably more enriched than Bear River water (~0.7085). Values of $^{87}Sr/^{86}Sr$ in the present lake are intermediate between the pre-diversion lake and Bear River. Following the diversion, values of $^{87}Sr/^{86}Sr$ in Bear Lake (i.e., in post-diversion sediments) started to follow an ^{87}Sr-depletion trend toward values in Bear River (12–8 cm in core BL98-10, Fig. 6), but sediments deposited after AD 1936 (^{210}Pb, Fig. 6) are still enriched in ^{87}Sr ($^{87}Sr/^{86}Sr$ = 0.7094) relative to present lake water ($^{87}Sr/^{86}Sr$ = 0.7092). Therefore, the Sr isotopic composition of the most recent sediments does not reflect the Sr isotopic composition of modern lake water.

Mineralogy

Prior to diversion of Bear River into Bear Lake in the early part of the twentieth century, there is no record of the Bear River entering Bear Lake during historic times. The only information about the water chemistry of Bear Lake prior to diversion is an analysis of water collected in 1912, which showed concentrations of total dissolved solids (TDS), Mg^{2+}, and Ca^{2+}, of 1060, 152, and 4.1 mg L^{-1}, respectively (Kemmerer et al., 1923). Corresponding values in Bear Lake today are 500, 53, and 31 mg L^{-1}, respectively. Such changes should have had a large effect on carbonate minerals precipitating in the lake and on the chemistry of those minerals.

Analyses of cores of surface sediments (up to 30–50 cm long) indicate that prior to diversion the dominant mineral in Bear Lake sediments was aragonite (Dean et al., 2007). After diversion, aragonite continued to be the dominant polymorph of $CaCO_3$ even though $Mg^{2+}:Ca^{2+}$ in the water was much lower. High-Mg calcite began to appear in sediments deposited shortly after Bear River diversion (Fig. 6; Dean et al., 2007).

Source of the Aragonite

The Bear River diversion caused profound changes in the isotopic compositions of C, O, and Sr and the concentrations of

Mg and Sr in carbonate minerals (Fig. 6). The ionic radius of Sr is larger, and that of Mg is smaller, than that of Ca. Therefore, Sr more readily substitutes for Ca in the more open crystal lattice of aragonite, and Mg more readily substitutes for Ca in calcite, so that calcite typically has a lower concentration of Sr and a higher concentration of Mg relative to aragonite. The higher concentration of Sr and lower concentration of Mg in bulk carbonate when aragonite precipitation ceased after the introduction of Bear River water

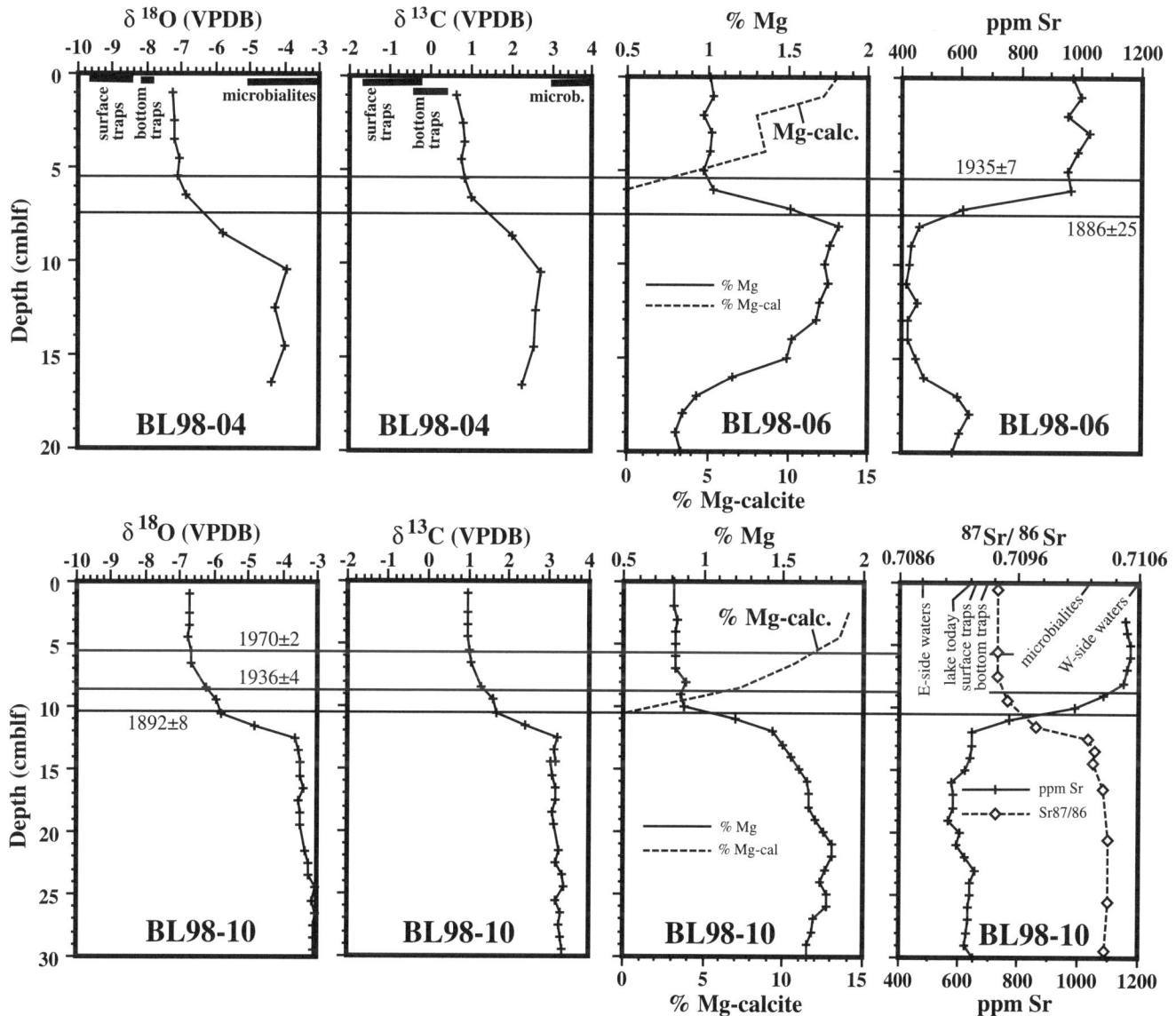

Figure 6. Profiles of values of $\delta^{18}O$ and $\delta^{13}C$ in bulk-sediment samples from cores BL98-04 and BL98-10, concentrations of HCl-soluble Mg and Sr (Bischoff et al., 2005), and abundance of high-Mg calcite calculated from XRD peak intensities (dashed lines) in cores BL98-06 and BL98-10, all versus depth in centimeters below lake floor (cmblf). Values of $^{87}Sr/^{86}Sr$ in carbonate in core BL98-10 also are given (dashed line with diamond symbols). Ranges of values of $\delta^{18}O$ and $\delta^{13}C$ in surface and bottom sediment traps and in microbialites are shown as bars at the tops of the profiles of $\delta^{18}O$ and $\delta^{13}C$ for core BL98-04. Average values of $^{87}Sr/^{86}Sr$ in Bear Lake, east-side springs and creeks, west-side springs and creeks, surface-sediment traps, bottom-sediment traps, and microbialites are shown at the top of the profile of $^{87}Sr/^{86}Sr$ for core BL98-10. Values of $^{87}Sr/^{86}Sr$ for waters are from Dean et al. (2007). Selected ^{210}Pb dates (in years A.D.; Smoak and Swarzenski, 2004) are shown for each core. As discussed in the text, these dates, especially the older ones, are probably too old. XRD—x-ray diffraction; VPDB—Vienna Pee Dee Belemnite.

(Fig. 6) is counterintuitive. This result is due to the much higher concentration of Sr, and much lower concentration of Mg, in Bear River water relative to the pre-diversion lake (Dean et al., 2007).

The sediments were much more enriched in ^{13}C, ^{18}O, and ^{87}Sr prior to diversion. Assuming that $^{87}Sr/^{86}Sr$ in pre-diversion bulk carbonate is representative of the ratio in pre-diversion lake water, then the 1912 Bear Lake had a ratio of ~0.7102 (Fig. 6), which is within the range of west-side waters (0.7101–0.7132), indicating that the west-side waters probably dominated the lake's hydrologic budget. A calculation considering only stream input indicates that ~99% of the 1912 solutes came from west-side waters (Dean et al., 2007).

As pointed out earlier, the Sr isotopic composition of sediments deposited after AD 1936 (by ^{210}Pb) does not reflect the Sr isotopic composition of modern lake water. Any carbonate mineral that precipitates in Bear Lake today should have an $^{87}Sr/^{86}Sr$ of 0.7092, the ratio in lake water today (Fig. 6). The average $^{87}Sr/^{86}Sr$ in bulk carbonate in surface traps is ~0.70925, and that in bottom traps is ~0.70937 (Table 1, Figs. 2, 3, and 6). The bulk carbonate in the trap sediments, especially in the bottom traps, and in the most recently deposited profundal sediments is, therefore, a mixture of carbonate precipitated in the lake today and carbonate more enriched in ^{87}Sr (higher $^{87}Sr/^{86}Sr$), and also more enriched in ^{18}O and ^{13}C (Fig. 6).

X-ray diffraction patterns (Figs. 2 and 3) show that the surface traps contain a small amount of aragonite in addition to the dominant high-Mg calcite. If that enriched aragonite had a $^{87}Sr/^{86}Sr$ of 0.7094 (that of twentieth-century carbonate; Fig. 6), a simple calculation, based on mixing a 0.7094 twentieth-century carbonate

(mostly aragonite) with a 0.7092 endogenic high-Mg calcite, and ignoring differences in Sr contents, shows that the surface traps (with a $^{87}Sr/^{86}Sr$ of 0.70925), on average, contain 75% modern carbonate (high-Mg calcite) and 25% twentieth-century carbonate (mostly aragonite). Because aragonite contains a higher concentration of Sr relative to calcite, this is a liberal estimate for the contribution of high-Mg calcite, which might be less than 75%. The same calculation for the bottom traps shows that, on average, they contain 15% modern carbonate and 85% twentieth-century carbonate. I conclude, therefore, that the aragonite collected in the traps did not precipitate from present-day Bear Lake water, but is reworked aragonite that is at least 50 yr old.

The aragonite in sediments deposited after diversion must have continued to form in the lake for some period of time. If it was reworked pre-diversion aragonite, it would have the isotopic and chemical signatures of pre-diversion aragonite (Fig. 6). Instead, the post-diversion aragonite recorded the changing isotopic and chemical composition of the lake. High-Mg calcite started forming shortly after diversion, and sediments deposited during the latter part of the twentieth century have increasing proportions of high-Mg calcite (Fig. 6). The horizon in the sediments that indicates environmental change due to Bear River diversion is well defined by the C, O, and Sr isotope data (Fig. 6). The differences in isotopic values between the high-Mg calcite in surface traps and the aragonite in pre-diversion sediments represents the actual environmental isotopic change in lake water (Fig. 6). The isotopic values of the twentieth-century surface sediments do not represent the full range of change in the lake water because of mixing of modern endogenic high-Mg calcite and older aragonite that is more enriched in ^{87}Sr (and ^{18}O and ^{13}C).

TABLE 3. X-RAY DIFFRACTION (XRD) MINERALOGY, AND OXYGEN, CARBON, AND STRONTIUM ISOTOPIC COMPOSITION OF THE SILT + CLAY (<63 μm) FRACTION OF STREAM SEDIMENTS IN THE BEAR LAKE CATCHMENT (SEE FIG. 1B FOR LOCATIONS OF SAMPLES)

Sample number	Stream	Quartz (%)*	Dolomite (%)*	Calcite (%)*	Feldspar (%)*	$CaCO_3$ (%)	OC (%)	$\delta^{18}O$-carbonate (per mil, VPDB)	$\delta^{13}C$-carbonate (per mil, VPDB)	$^{87}Sr/^{86}Sr$ (carbonate)	$^{87}Sr/^{86}Sr$ (residue)
West-side creeks											
1	Paris Creek	42	56	0	3	0.0	5.74	−14.3	−6.1	0.71001	0.72098
7	St. Charles Creek	52	36	6	6			−13.0	−5.7	0.71004	0.71663
9	Fish Haven Creek	58	26	11	5					0.71030	0.72283
11	Hodges Canyon	81	3	5	11	0.0	0.81	−14.9	−8.2	0.71025	0.71953
East-side creeks											
13	South Eden Creek	63	4	28	4	25.6	1.85	−13.4	−7.3	0.70892	0.71264
13	South Eden Creek							−11.4	−5.0		
15	North Eden Creek	85	3	8	4	10.8	0.65	−13.5	−4.8	0.70938	0.71489
15	North Eden Creek							−12.4	−5.0		
17	Indian Creek	69	3	21	7			−11.6	−4.5	0.70768	0.71223
Bear River north of Uintas											
19	Bear River	85	3	9	3	9.9	0.49	−12.7	−5.0	0.70907	0.72417
21	Bear River	75	8	11	5					0.70907	0.71737
23	Bear River	78	4	11	6	12.1	1.34	−11.5	−4.0	0.70929	0.71646
25	Bear River	81	5	7	7	11.3	0.41	−11.9	−5.4	0.71006	0.71942
27	Bear River	75	6	16	3					0.71067	0.72131
29	Bear River	85	3	10	2	13.2	0.89	−12.8	−5.7	0.71027	0.72169
31	Bear River	77	6	17	4					0.71078	0.71911
33	Bear River	85	5	9	3	4.5	0.38	−12.2	−5.0	0.71066	0.74241
35	Bear River	80	6	12	8	11.6	2.39	−11.4	−5.9	0.71112	0.72474
37	Bear River	83	6	10	9						
39	Bear River	85	6	8	4	8.0	2.26	−10.0	−5.0	0.71190	0.72881
41	Bear River	77	7	15	6	12.9	1.00	−10.1	−5.3	0.71050	0.73716
43	Bear River	70	9	20	10	22.2	1.64	−9.3	−3.5	0.70990	0.72142
45	Bear River	72	5	22	12						
Bear River in Uintas											
51	Bear River	83	0	0	17						
53	Bear River	88	0	0	12					0.71910	0.74348

Note: OC—organic carbon; VPDB—Vienna Pee Dee Belemnite.
*XRD percentages were calculated from peak intensities.

The Rock Pile

Another part of the endogenic carbonate story in Bear Lake is a mound of carbonate-cemented cobbles and pebbles, known locally as the "rock pile," at the southwestern end of the lake (Fig. 1A), notorious for snagging fishing lines and anchor ropes. This mound rises from a depth of ~18 m below the surface on the north end of the mound to a hummocky surface at a depth of ~12 m. Samples from the rock pile were first provided by Bryce Nielson and Scott Tolentino of the Utah State Division of Wildlife Resources in Garden City, Utah, who brought them up in their gill nets. Each sample had a rounded cobble as a nucleus covered by a thin (<1 cm) layer of carbonate. X-ray diffraction analyses of these samples showed that the carbonate layers were composed of aragonite, and the rounded cobble nuclei were composed of dolomite, suggesting that they came from the Paleozoic carbonate bedrock in the Bear River Range. Limited sampling, underwater video, and SCUBA-based observations indicate that the microbialite mound is made up of many rounded cobbles and pebbles that have been coated and cemented together. Divers described the entire surface of the mound as being coated with a layer of algae. This was subsequently confirmed by underwater video. For this reason I used the term algal mound, or, more correctly, microbialite mound for the rock pile.

Microbialites, or what were formally referred to as "stromatolites," are "organosedimentary deposits that have accreted as a result of a benthic microbial community trapping and binding detrital sediment and/or forming the locus of mineral precipitation" (Burne and Moore, 1987). The role of microbes might simply be through trapping and binding of sediment particles or also through induced or mediated precipitation of $CaCO_3$ by increasing the pH in the micro-environment surrounding the microbes. Once formed, a microbialite may become the locus of passive inorganic cementation. The term microbialite is also used here for individual samples of aragonite-coated cobbles and pebbles.

The largest reported microbialites in lakes are towers up to 40 m high in alkaline (pH>9.7) Lake Van, Turkey (Kempe et al., 1991). The Lake Van microbialites occur where Ca- and CO_2-rich groundwater seeps into the lake, precipitating aragonite. The soft aragonite crusts are then colonized by cyanobacteria, which precipitate more aragonite and stabilize the growing mound.

Although the microbial coating suggests that the microbialite mound in Bear Lake is still forming, radiocarbon ages on four samples of the mound range in age from 3000 to 6500 ^{14}C yr B.P. (Table 4). The difference in age between the bottom of the carbonate layer (just above the dolomite cobble nucleus) and the top of the carbonate layer (outer edge) in the initial three small samples collected in gill nets was ~2000 yr (Table 4, microbialites A, B, and C). A larger sample of the microbialite mound collected by divers in 2002 had a thick rind of aragonite, and a radiocarbon age difference from top to bottom of 3500 yr (Table 4, Fig. 7). The radiocarbon ages suggest that the microbialite mound is not modern but formed during the middle to late Holocene.

Samples of carbonate cement were drilled along transects across the thick carbonate rind of the 2002 microbialite (Fig. 7) for O, C, and Sr isotope analyses. Values of $^{87}Sr/^{86}Sr$ in three samples from the 2002 microbialite (Fig. 8, transect A1–C) and from two other microbialites are remarkably constant, ranging from 0.7101 to 0.7103. These values are similar to those in pre-diversion sediments in cores BL98-10 (Fig. 6) and BL02-4 (Fig. 8). Values of $\delta^{18}O$ and $\delta^{13}C$ range from −3‰ to −5‰, and from 2.5‰ to 4‰, respectively. Analyses of samples drilled from the 2002 microbialite on parallel transects (transects X, R, and G, Fig. 7) produced essentially identical results. Values of $\delta^{18}O$ >−4‰ and values of $\delta^{13}C$ >3‰ (vertical lines on profiles in Fig. 8) occur only in sediments deposited during the past 1500 yr, but before Bear River diversion (Fig. 8). The high values of $\delta^{18}O$ and $\delta^{13}C$ indicate that the microbialite mound, as represented by the 2002 sample (Fig. 7), was not precipitating aragonite during the mid-Holocene because the microbialite samples are much more enriched in ^{18}O, ^{13}C, and ^{87}Sr than mid-Holocene sediments (discussed below). The core and microbialite C and O isotope data presented in Figure 8 indicate that the microbialite mound, at least the top of the mound represented by these few samples, is only several thousand years old, and was not precipitating aragonite cement during the latter part of the twentieth century. Therefore, the radiocarbon ages (Table 4, Fig. 7) must be too old, and this gives some measure of the radiocarbon reservoir effect in Bear Lake.

Values of $\delta^{13}C$ on the outer edge of the 2002 microbialite sample are more than 1‰ enriched in ^{13}C ($\delta^{13}C$ = 4‰–4.5‰, Fig. 8, transects E1–E8 and R1–R6) than Holocene core carbonate ($\delta^{13}C$ = 3‰, Fig. 8, core BL02-4). The higher microbialite values

Sample	Position	[14]C age (yr)	Error (yr)	USGS No.
Microbialite A	Top*	3200	40	3383
Microbialite A	Bottom[†]	5810	40	3384
Microbialite B	Top	5270	40	3385
Microbialite B	Bottom	5400	40	3386
Microbialite C	Top	3920	40	3387
Microbialite C	Bottom	5550	40	3388
Microbialite 2002	Top	3040	40	4145
Microbialite 2002	Middle	5265	40	4143
Microbialite 2002	Bottom	6550	40	4144

TABLE 4. RADIOCARBON AGES OF BULK CARBONATE FROM FOUR MICROBIALITE SAMPLES (A, B, C, AND 2002)

*Top—outer edge of microbialite.
[†]Bottom—just above cobble nucleus.

may represent a slight photosynthetic enrichment by the microbes involved in microbialite formation (Burne and Moore, 1987). All of the carbon isotope results suggest that during the early part of microbialite formation, aragonite cementation was predominantly inorganic, precipitated in isotopic equilibrium with lake water ($\delta^{13}C = 3\text{‰}$), and only in later stages of microbialite formation was there a slight photosynthetic enrichment in ^{13}C in aragonite cement ($\delta^{13}C > 3\text{‰}$ in transects E1–E8 and R1–R6; Fig. 8).

A mound of aragonite-cemented cobbles and pebbles at one particular area of Bear Lake is suspicious. Mounds of inorganically precipitated $CaCO_3$ ("tufa mounds"; e.g., at Mono Lake, California) commonly form where CO_2-charged groundwaters reach the surface as springs, whereupon they degas, increasing the pH of the water, which triggers the precipitation of $CaCO_3$ (e.g., Mono Lake, California; Lake Van, Turkey). This $CaCO_3$ precipitation may or may not be mediated by the photosynthetic activity of cyanobacteria or other microbes. Because we are unable to balance the present lake ionic composition by using that of inflowing surface waters, I suspect that there is a fairly large extrabasinal supply of groundwater, and that groundwater is enriched in ^{18}O, ^{13}C, and ^{87}Sr relative to the lake and surface-water sources (Dean et al., 2007).

Persistent sublacustrine "springs" occur in at least 25 localities in Bear Lake, mostly on the west side and in a north-south line (S. Tolentino and B. Nielson, 2002, personal commun.). They are manifested at the surface of the lake as patches of bubbles. Bubble trains in the water column from some of these springs are strong enough to be seen on sonar images (Bright, this volume). In the spring of 2002, I collected the gas bubbles from three of these springs. Upon analyses, all three proved to be methane with $\delta^{13}C$ values of -68‰ to -70‰ indicating a biogenic origin (Clark and Fritz, 1997). Later that summer, divers observed that there was no water coming from the methane seeps, at least in the summer of 2002.

Methanogenesis is occurring in the sediments of Bear Lake. Coring at site BL00-1 (Fig. 1A) was begun with 3 m drives. However, below ~50 m the sediments were very gassy and expanded in the liner, which necessitated coring with 2 m drives to allow for expansion. Presumably the gas was biogenic methane, although I do not have the appropriate analyses to prove this. However, methane in sediments cannot explain the large volumes of methane originating from vents that are present in the same place year after year and occur in a north-south line parallel to secondary faults that parallel the main bounding fault (East Bear Lake fault; Colman, 2006). Some of these secondary faults cut the youngest sediments and intersect the lake floor. This methane must be coming from a deep source along these secondary faults, and these faults might also be conduits for deep, extrabasinal groundwater.

My current working hypothesis for the origin of the rock pile is that CO_2-charged groundwater entered the lake under the microbialite mound, discharged CO_2, and precipitated $CaCO_3$, perhaps mediated by cyanobacteria and other microbes in much the same way that microbialite mounds are formed in lakes in many parts of the world (Burne and Moore, 1987). Active building of the microbialite mound apparently began about several thousand years ago and apparently ended when Bear River was diverted into Bear Lake. No carbonate is forming on the top of the microbialite mound where the samples discussed here came from, but if my working hypothesis is correct, groundwater may still be entering from below the mound and precipitating carbonate of some unknown mineralogy.

Glacial through Holocene Changes in Carbonate Deposition

The sedimentary sequence in Bear Lake consists of a series of eastward dipping strata, the shallowest of which pinch out to the west (Colman, 2006). Therefore, the three Kullenberg cores that were collected (BL96-1, BL96-2, and BL96-3, Fig. 1A) contain overlapping sequences. This can be seen in the profiles of calcite and aragonite versus depth shown in Figure 9. Core BL96-2 contains all of the lithologic units deposited over the last 19 cal k.y. in Bear Lake condensed into 400 cm of section. Therefore, I will focus most of the discussion on that core, supplemented with some isotope data obtained at higher temporal resolution from core BL02-4 from the north end of the lake (Fig. 1A). According to the age model of Colman et al. (2005; this volume), the section collected in core BL96-2 appears to be nearly continuous and intact. However, pyritized tubes in some horizons in BL96-2 are interpreted as root casts and, if so, would appear to indicate very shallow water or subaerial exposure (Smoot, this volume). Shell layers associated with hiatuses in core BL02-4, collected in shallower water than core BL96-2, suggest subaerial exposure and erosion or nondeposition, and lake levels interpreted from

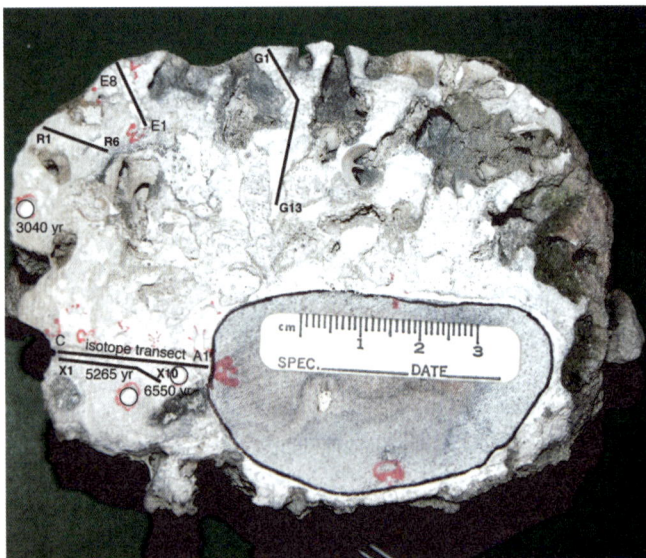

Figure 7. Photograph of a cut half of a 2002 sample of a microbialite containing a rounded dolomite cobble nucleus and a rind of varying thickness of aragonite. Locations of three radiocarbon samples (Table 4) are shown (circles with dates). Letter-number lines are transects of drilled samples for C and O isotope analyses. Results of $\delta^{18}O$ and $\delta^{13}C$ for transects A1–C and E1–E8 are shown in Figure 8.

grain-size data in BL02-4 and -3 indicate times of lower lake level (Smoot and Rosenbaum, this volume; discussed below).

Mineralogy

Mineralogy data for core BL96-2 (Dean et al., 2006) show that the carbonate-poor section deposited prior to 16 cal ka (unit 1) consists of ~75% quartz, 10%–20% $CaCO_3$ as calcite, and minor feldspar and dolomite. Some of the calcite and all of the dolomite deposited during that time probably is detrital. These sediments deposited during the last glacial interval (LGI) consist of red, calcareous, silty clay that are in marked contrast to the light-tan carbonate-rich sediments deposited during the glacial-Holocene transition and the Holocene, regardless of mineralogy.

Several abrupt changes in mineralogy occur within carbonate-rich sediments deposited since 16 cal ka (Fig. 10).

These changes in mineralogy do not always coincide with changes in percentages of $CaCO_3$ (Fig. 10). The percentage of $CaCO_3$, initially as calcite, began to increase ca. 15.5 cal ka (unit 2; Fig. 10), and reached a plateau of ~40% between 15 and 12 cal ka (unit 3). During this interval, minor amounts of aragonite also formed. The percentage of $CaCO_3$ increased to ~70% by 11 cal ka and remained at that level throughout the Holocene and through changes in mineralogy. Between 11 and 9 cal ka, aragonite was the dominant polymorph of $CaCO_3$ formed in the lake (unit 4). A return of calcite precipitation between 9 and 7.5 cal ka (unit 5) indicates a brief freshening event, probably re-entry of Bear River into Bear Lake (see isotope discussion below), which lowered $Mg^{2+}:Ca^{2+}$ out of the aragonite stability field. The sediments deposited over the last 7.5 k.y. (unit 6) consist of ~70% $CaCO_3$, mostly as aragonite with only ~5% calcite (Fig. 10).

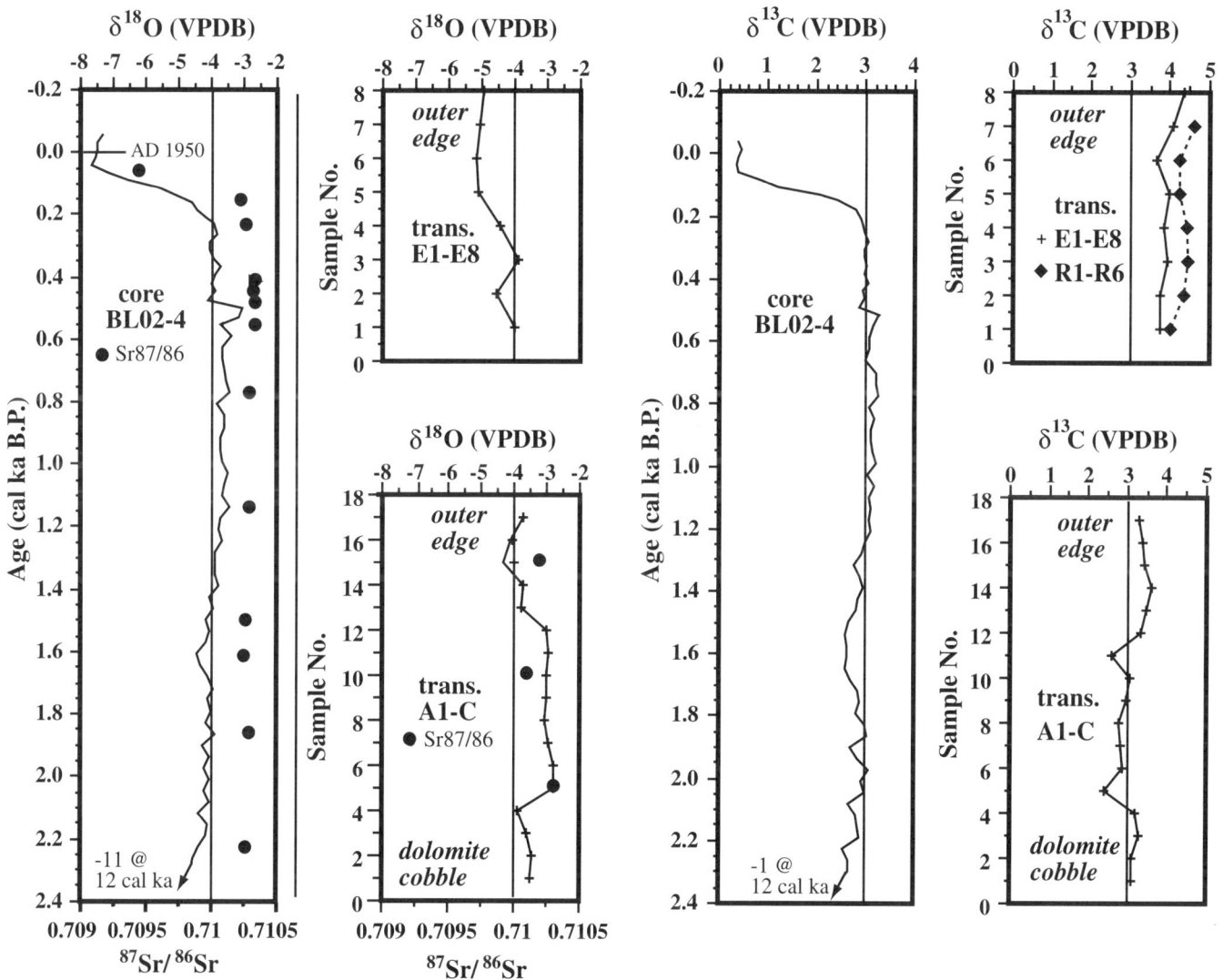

Figure 8. Profiles of $\delta^{18}O$ and $\delta^{13}C$ for core BL02-4 and microbialite transects A1–C and E1–E8 (see Fig. 7 for locations of the transects on the 2002 microbialite sample). Values of $^{87}Sr/^{86}Sr$ for core BL02-4 and microbialite transect A1–C are shown as closed circles on the profiles of $\delta^{18}O$. VPDB—Vienna Pee Dee Belemnite.

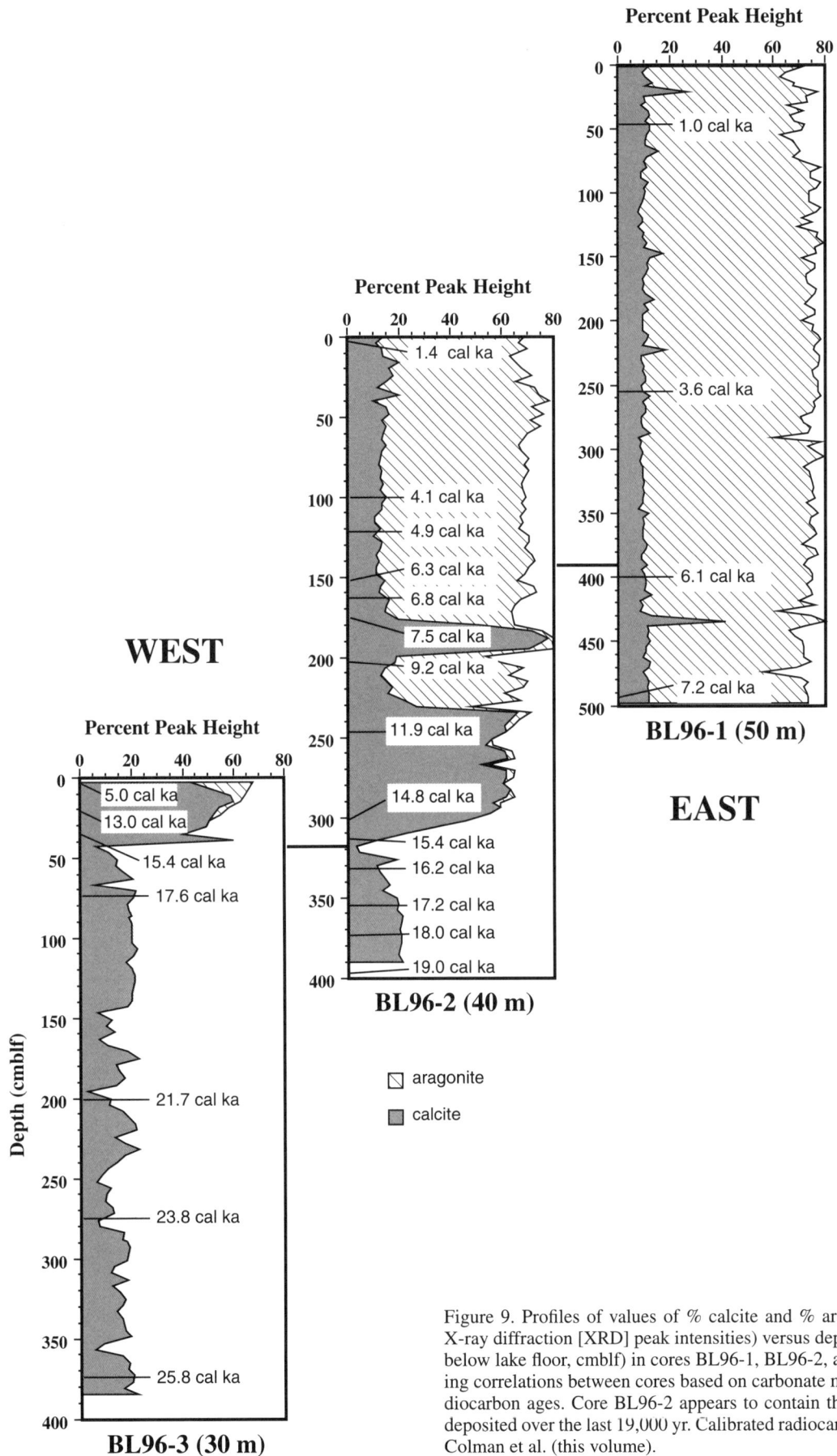

Figure 9. Profiles of values of % calcite and % aragonite (based on X-ray diffraction [XRD] peak intensities) versus depth (in centimeters below lake floor, cmblf) in cores BL96-1, BL96-2, and BL96-3 showing correlations between cores based on carbonate mineralogy and radiocarbon ages. Core BL96-2 appears to contain the entire sequence deposited over the last 19,000 yr. Calibrated radiocarbon ages are from Colman et al. (this volume).

Isotope Geochemistry and Development of Bear Lake over the Last 26 k.y.

According to the strontium isotopic composition of pre- and post-diversion carbonates (Fig. 6), the strontium isotopic composition of twentieth-century carbonates in Bear Lake sediments can be explained by mixing of Bear River water and some pre-diversion lake water having an isotopic composition defined by pre-diversion, late Holocene endogenic carbonate (aragonite; Fig. 6). Therefore, if Bear River entered Bear Lake in the past, the C, O, and Sr isotopic composition of endogenic carbonates should, as a first approximation, provide evidence of such an event. In particular, the very low values of $^{87}Sr/^{86}Sr$ in Bear River water (0.7085) should be an excellent tracer for times when Bear River water entered Bear Lake (Bouchard et al., 1998). When Bear Lake was not connected to Bear River, the hydrology of the lake can be assumed, as a first approximation, to be the main control on isotopic composition. The balance between precipitation and evaporation is perhaps the largest hydrologic factor, but, as discussed earlier, there is an unknown groundwater source that may have varied considerably over time (Dean et al., 2007).

Values of $\delta^{18}O$ and $\delta^{13}C$ in calcite in the calcareous, quartz-rich LGI sediments (unit 1) are lower (Fig. 10) than in Holocene carbonate and in carbonate collected in sediment traps (Fig. 6). The LGI sediments are even more depleted than carbonate rocks (limestones and dolomites) in the Bear River Range to the west of the lake and in the Bear Lake Plateau east of the lake, as judged by an average value of $\delta^{18}O$ of $-8.2 \pm 2.7‰$ and an average value of $\delta^{13}C$ of $0.4 \pm 1.4‰$ in nine samples of those rocks (Bright, this volume; Fig. 11A). However, the LGI sediments are not as depleted as the fine fractions ($<63\mu m$) of stream sediments, particularly those in west-side streams (Table 3, Fig. 11A). Apparently the isotopic composition of bulk carbonate in stream sediments does not reflect the isotopic composition of carbonate bedrock. This is because stream sediments contain grains of endogenic calcite (tufa), which probably formed in the streams. Isotope analyses of several hand-picked grains show that they are depleted in both ^{18}O and ^{13}C (Kaufman et al., this volume; Fig. 11A). It appears that modern stream sediments are mixtures of detritus from carbonate bedrock and endogenic tufa (Fig. 11A). There is no evidence that Bear River and other streams were producing endogenic tufa during the LGI, but the isotopic composition of LGI bulk carbonate (Fig. 10) suggests that some isotopically depleted carbonate was being added to detrital bedrock carbonate, although nine samples are hardly representative of the stratigraphic and geographic distribution of the isotopic composition of carbonate rocks in the Bear River Range and Bear Lake Plateau.

The timing of the isotope excursions might be better defined by the higher resolution isotope data from core BL02-4 from

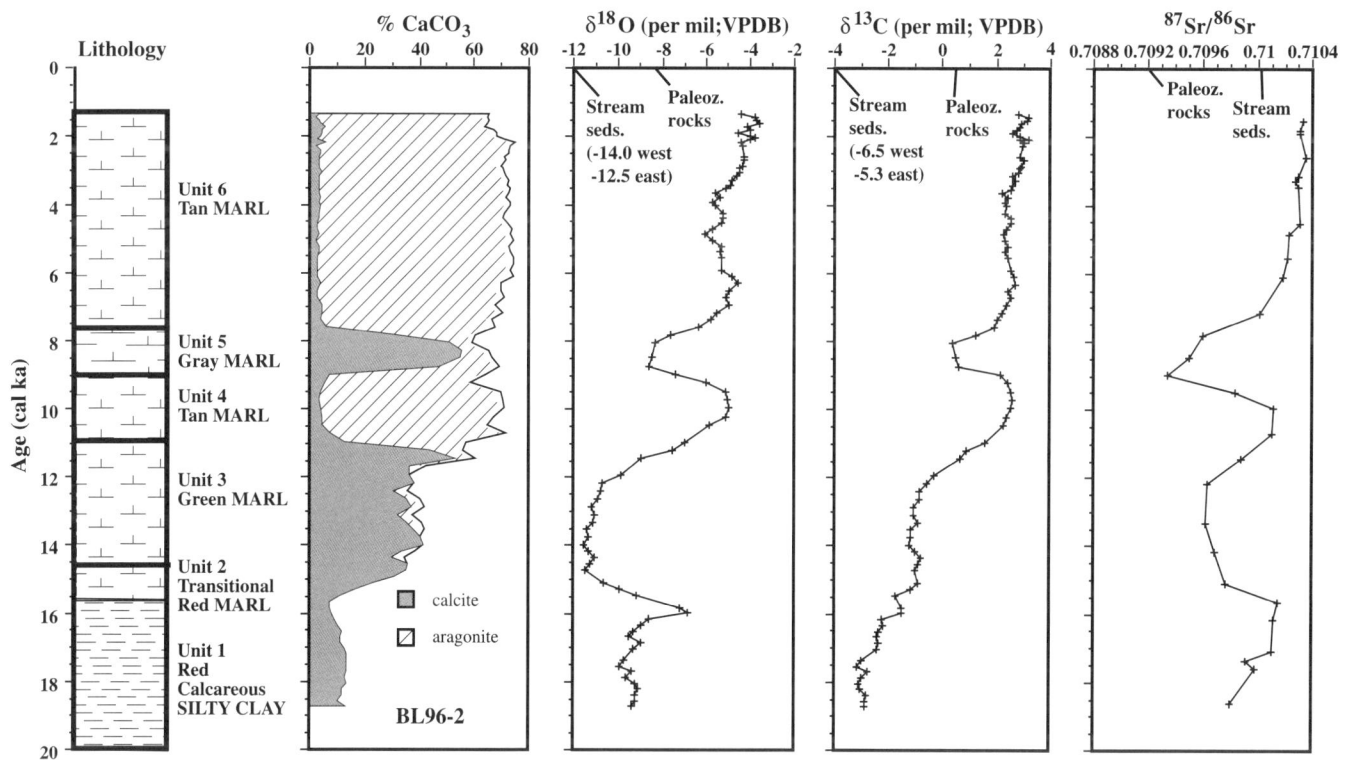

Figure 10. Lithology and profiles of percent $CaCO_3$ as aragonite and calcite (see text for method of partitioning total $CaCO_3$ into aragonite and calcite) and values of $\delta^{18}O$, $\delta^{13}C$, and $^{87}Sr/^{86}Sr$ in bulk carbonate versus age in core BL96-2. Values in stream sediments (Table 3) and Paleozoic carbonate rocks (Bright, this volume) are shown at the top of each profile. The age model was developed by Colman et al. (this volume).

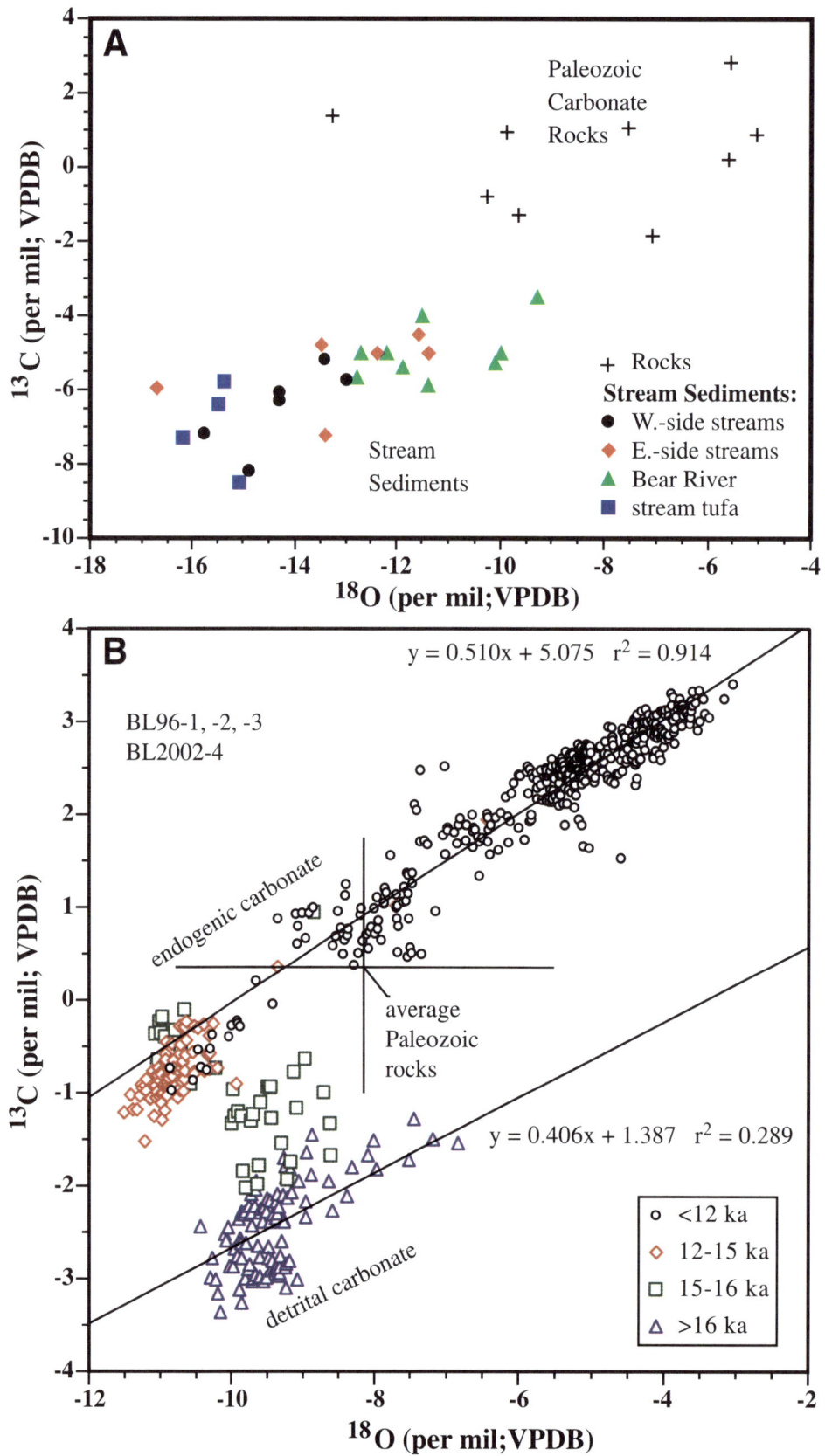

Figure 11. (A) Cross plot of values of δ^{18}O and δ^{13}C in samples from Paleozoic bedrock in the Bear River Range, in the <63 μm fractions of stream sediments, and in hand-picked samples of endogenic stream calcite (tufa). Stream-sediment values are from Table 3. Rock values are from Bright (this volume). (B) Cross plot of values of δ^{18}O and δ^{13}C in samples from cores BL96-1, -2, -3, and BL02-4. Horizontal and vertical bars represent the average and one standard deviation of the carbonate-rock values in A. VPDB—Vienna Pee Dee Belemnite.

shallower water at the north end of the lake (Figs. 1A and 12). Measurements of oxygen and carbon isotopes in this core were made every centimeter, whereas those in core BL96-2 were made every 4 cm (Dean et al., 2006). Unfortunately, the isotope records in core BL02-4 are interrupted by two hiatuses represented by shell layers. The shell layers mark the culmination of lower lake levels of as much as 25 m (Smoot and Rosenbaum, this volume). Nevertheless, the higher-resolution records should help to define the timing of isotope excursions better.

Between 15.5 and 14.5 cal ka (unit 2), values of $\delta^{18}O$ in core BL96-2 declined to -11%, the lowest value in the entire glacial-Holocene sequence, where they remained for ~4 k.y. (Fig. 10). The decline in core BL02-4 was more abrupt, occurring ca. 16 cal ka, and there is no peak in values of $\delta^{18}O$ between 16 and 15 cal ka (Fig. 12) as there is in core BL96-2 (Fig. 10). If the peak in $\delta^{18}O$ in core BL96-2 from the center of the lake is related to the disconnection of Bear River from Bear Lake, and to content of detrital material from Bear Lake catchment streams, this event was not recorded at the north end of the lake. These lower values of $\delta^{18}O$ in sediments deposited between 14.5 and 12 cal ka (unit 3 in Fig. 10;

Fig. 12) might be due to greater influx of Bear River water, but the influx of detrital clastic material was decreasing at that time (Rosenbaum et al., this volume), and there are no corresponding decreases in values of $\delta^{13}C$ (Fig. 10). The low values of $\delta^{18}O$ in calcite deposited between 14.5 and 12 cal ka are therefore most likely due to continued abundance of cold-season precipitation (winter snow).

For most of the sediments in cores BL96-1, -2, -3, and BL02-4, there is a strong covariation between values of $\delta^{18}O$ and $\delta^{13}C$ (Fig. 11B). There is a positive covariation between $\delta^{18}O$ and $\delta^{13}C$ in samples of endogenic carbonate, regardless of mineralogy, younger than ca. 15 cal ka (Fig. 11B), but there are some trends in samples of older sediments that may provide clues for the isotope excursions in the transition between the detrital sediments of the LGI (>16 cal ka) and the endogenic carbonate sediments deposited after 15 cal ka.

A positive covariance between values of $\delta^{18}O$ and $\delta^{13}C$ in lacustrine carbonates often occurs in water bodies with relatively long residence times, and if the correlation is high ($r > 0.7$), this implies that the lake was hydrologically closed (Talbot, 1990, 1994; Talbot and Kelts, 1990; Li and Ku, 1997). Covariant trends may have remarkably long-term persistence through major

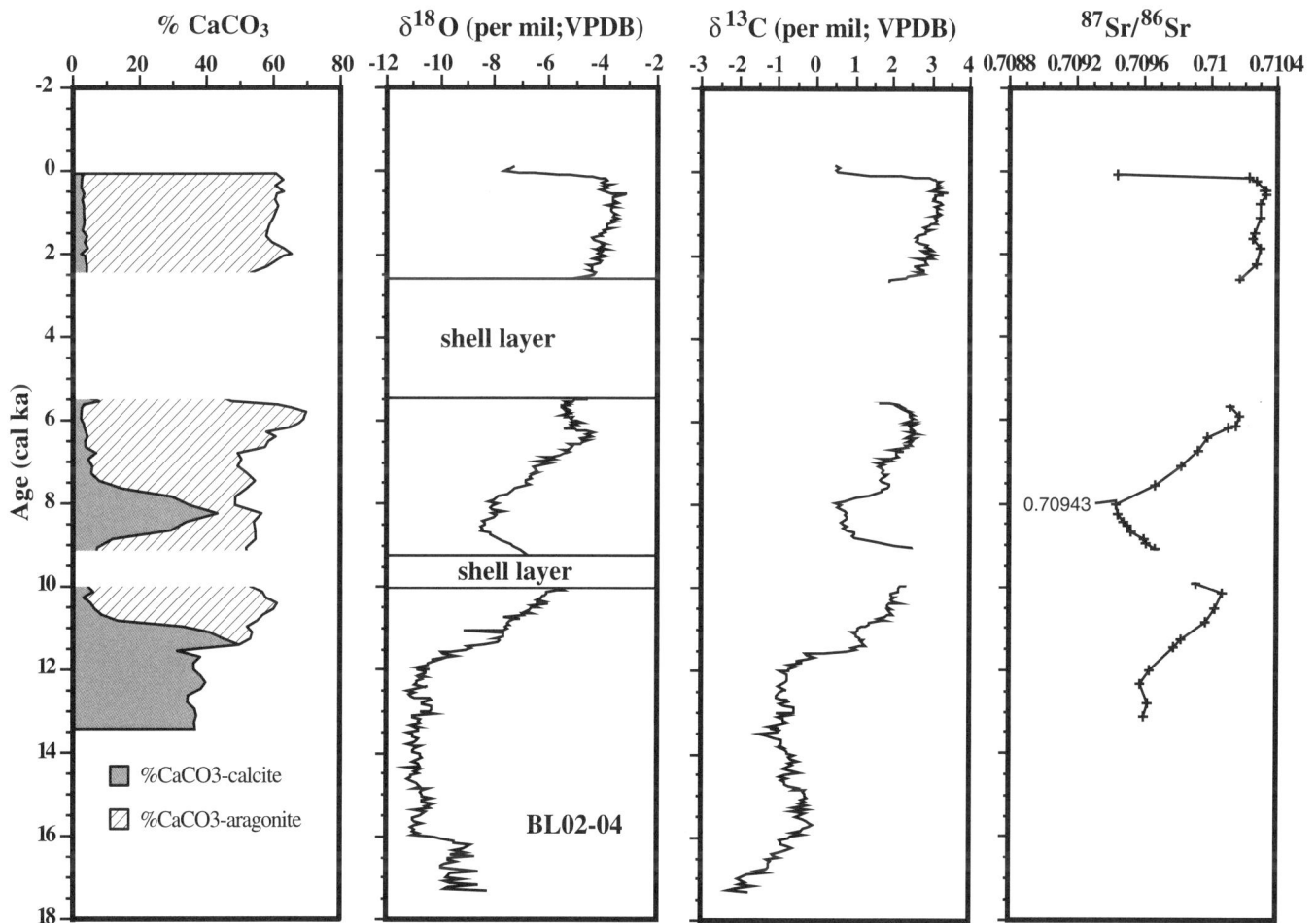

Figure 12. Profiles of percent CaCO$_3$ as aragonite and calcite (see text for method of calculation), and of $\delta^{18}O$, $\delta^{13}C$, and $^{87}Sr/^{86}Sr$ in bulk carbonate versus age (0–20 cal ka) in core BL02-4. The age model was developed by Colman et al. (this volume). VPDB—Vienna Pee Dee Belemnite.

environmental changes (Talbot, 1990), and this certainly applies to the last 15 k.y. in Bear Lake. Sediments deposited before 16 cal ka in BL02-4 and BL96-2, have their own covariant isotope trend, but with much lower values of $\delta^{13}C$ and much weaker isotope covariance than those in the trend for younger sediments (Fig. 11B). These sediments are considered to represent one endmember—sediments that contain glacial flour from the Uinta Mountains (Dean et al., 2006; Rosenbaum and Heil, this volume). However, present-day stream sediments in the Bear River in the Uinta Mountains do not contain carbonate (Table 3), so the carbonate in sediments deposited before 16 cal ka must have been derived from Bear River drainage north of the Uintas in Wyoming and Utah (Fig. 1B) and from the local catchment.

The sediments in BL96-2 and BL2002-4 deposited between 16 and 15 cal ka have a slight negative covariant trend (Fig. 11B). This trend goes from low values of $\delta^{13}C$ (~−2‰) in sediments deposited before 16 cal ka to higher values of $\delta^{13}C$ (~−0.5‰) in sediments deposited between 15.6 and 15.0 cal ka, values that are typical of the endogenic low-Mg calcite deposited between 15 and 12 cal ka (Fig. 11B). Magnetic properties suggest that between 20 and 16 cal ka there was a transition from predominantly glacial flour from the Uinta Mountains in the detrital fraction to predominantly detrital material derived first from Bear River north of the Uintas and then (ca. 16.5 cal ka) from the local Bear Lake catchment (Rosenbaum and Heil, this volume). However, this does not explain the negative isotope covariation between 16 and 15 cal ka, because Paleozoic carbonate rocks in the Bear Lake catchment are more enriched in both ^{18}O and ^{13}C (Fig. 11A), and replacing upper Bear River detrital carbonate (triangles in Fig. 11B) with debris from carbonate rocks would result in a positive covariation. The more likely explanation for the slight negative covariant trend between 16 cal ka and 15 cal ka is a mixing of detrital carbonate in sediments from the upper Bear River drainage north of the Uintas (triangles in Fig. 11B) with endogenic carbonate having higher values of $\delta^{13}C$, (diamonds in Fig. 11B). The endogenic calcite deposited between 15 and 12 cal ka begins to show a positive covariant trend but the variation is small (~1‰ for both $\delta^{18}O$ and $\delta^{13}C$; diamonds in Fig. 11B).

The covariant trend of enrichment in both ^{18}O and ^{13}C in endogenic carbonate younger than 15 cal ka, regardless of mineralogy (low-Mg calcite or aragonite), with a high degree of correlation, is the result of evaporation from a progressively more closed-basin lake with a long residence time as influx of the Bear River greatly diminished. Evaporation and warming of the lake water would lead to loss of CO_2 that is depleted in ^{13}C. Resulting carbon-isotope fractionation between lake water and dissolved inorganic carbon (DIC) and CO_2 would result in ^{13}C-enriched waters and carbonate minerals precipitated from the water (Li and Ku, 1997). In a lake with a long residence time, prolonged burial of ^{13}C-depleted OC could also lead to elevated values of $\delta^{13}C$ in DIC (McKenzie, 1985; Hollander and McKenzie, 1991; Li and Ku, 1997). If this is true, then the slight positive covariant trend in carbonate deposited before 16 cal ka (triangles in Fig. 11B) suggests that the isotopic composition of that carbonate also was influenced by evaporation,

which implies that some of that carbonate is endogenic mixed with carbonate from the upper Bear River drainage.

At about 12 cal ka, values of both $\delta^{18}O$ and $\delta^{13}C$, but especially $\delta^{18}O$, began to increase rapidly (Figs. 10 and 12), most likely due to increased evaporation and/or reduced influx of Bear River water. Between 12 and 10 cal ka, values of $\delta^{18}O$ increased by 6‰ (−11‰ to −5‰). Values of $^{87}Sr/^{86}Sr$ also increased during this interval (Figs. 10 and 12). Evaporation and reduction in Bear River inflow increased the total dissolved solids and $Mg^{2+}:Ca^{2+}$ in the lake water, and by 11 cal ka, a critical value of $Mg^{2+}:Ca^{2+}$ was reached and calcite deposition was rapidly replaced by aragonite deposition (unit 4 in Fig. 10; Fig. 12). As discussed earlier, Müller et al. (1972) suggested that aragonite precipitates only in lakes having an $Mg^{2+}:Ca^{2+}$ of >12. I suggest that by 11 cal ka Bear River was entirely detached from Bear Lake. By 10 cal ka, the mineralogy and isotopic composition of bulk carbonate were approaching values that would be typical of the middle Holocene (6–4 cal ka; Fig. 10). This increase in isotopic values between 12 and 10 cal ka was accompanied by a drop in lake level of more than 25 m below modern lake level, judging from grain-size data and other sedimentological evidence (Smoot and Rosenbaum, this volume). The lower shell layer in core BL02-4 represents the culmination of this lowstand (Fig. 12).

At about 9 cal ka, values of $\delta^{18}O$, $\delta^{13}C$, and $^{87}Sr/^{86}Sr$ decreased considerably (Figs. 10 and 12), suggesting that Bear River water was again entering Bear Lake. Values of $\delta^{18}O$ were at a minimum between 9 and 8 cal ka. Lake level went back up by at least 25 m (Smoot and Rosenbaum, this volume), and the rise was accompanied by a decrease in salinity (decrease in values of $\delta^{18}O$; Figs. 10 and 12) and, presumably, a decrease in $Mg^{2+}:Ca^{2+}$ that resulted in cessation of aragonite precipitation and a return of calcite precipitation (unit 5 in Fig. 10; Fig. 12). The decrease in $^{87}Sr/^{86}Sr$ began a little earlier (9.5 cal ka in both cores) than the decrease in $\delta^{18}O$. The $^{87}Sr/^{86}Sr$ data from core BL02-4 suggest that the Sr isotope decrease may have begun as early as 10 cal ka, below the shell layer in core BL02-04 (Fig. 12), but it is hard to tell because of the shell layer. Values of $^{87}Sr/^{86}Sr$ reached a minimum ca. 9 cal ka in core BL96-2 and ca. 8.2 cal ka in core BL2002-4. The lead of the Sr isotope signal over the O isotope signal probably is because the Sr concentration in the lake today is much lower than that of Bear River, and likely was even lower prior to Bear River diversion. Therefore, changes in Sr isotopic composition of the low-Sr-concentration lake water were not as well buffered as changes in O isotopic composition, which had to affect a much larger oxygen reservoir.

Calcite precipitation lasted ~1000 yr (unit 5 in BL96-2, Fig. 10; upper calcite layer in BL02-4, Fig. 12). Unlike the rapid reversal in the Sr isotope excursion, the well-buffered C and O isotopic composition remained at low values for ~1000 yr before increasing (Figs. 10 and 12). The entire negative Sr and O isotope excursions lasted ~2000 yr. As the river again abandoned the lake, lake level rapidly declined, $Mg^{2+}:Ca^{2+}$ of the lake increased, and by 7.5 cal ka aragonite was again forming (unit 6 in Fig. 10; Fig. 12). The mineralogy of carbonate in Bear Lake sediments has been relatively constant for the last 7 cal k.y. (Fig. 10), but values of $\delta^{18}O$ and $\delta^{13}C$ continued to increase slightly (Figs. 10 and 12), suggesting that the salinity continued to increase after Bear River was again disconnected from Bear Lake.

Carbonate Deposition in Bear Lake during the Last Two Glacial-Interglacial Cycles

The lithology and mineralogy of the sediments collected over the last 26,000 yr, as recovered in piston cores, consists of red, calcareous silty clay deposited between 26 and 16 cal ka overlain by marl deposited during the last 16,000 yr, in which the mineralogy of the carbonate alternates between low-Mg calcite and aragonite with minor amounts of dolomite and quartz. As

discussed earlier, the Holocene environmental history of the Bear Lake is complex (Dean et al., 2006), and the alternation of calcite and aragonite is an important part of that history.

The visual core descriptions of cores recovered from holes BL00-1D and -1E indicate that the top of the recovered section contains the same massive, tan, bioturbated aragonitic marl and underlying red silty clay (Fig. 13, lithologic units 6 and 7; http://www.ngdc.noaa.gov/mgg/curator/laccore.html) that were recovered in piston cores. The overall $CaCO_3$ content of the marl (unit 7) is high, ranging from 60% to 80% with an average of ~70%, mostly in the form of aragonite (Fig. 13). The OC content also is high when compared with the underlying red clay, ranging mostly between 1% and 2%, but higher values commonly occur in older sediments (Fig. 13).

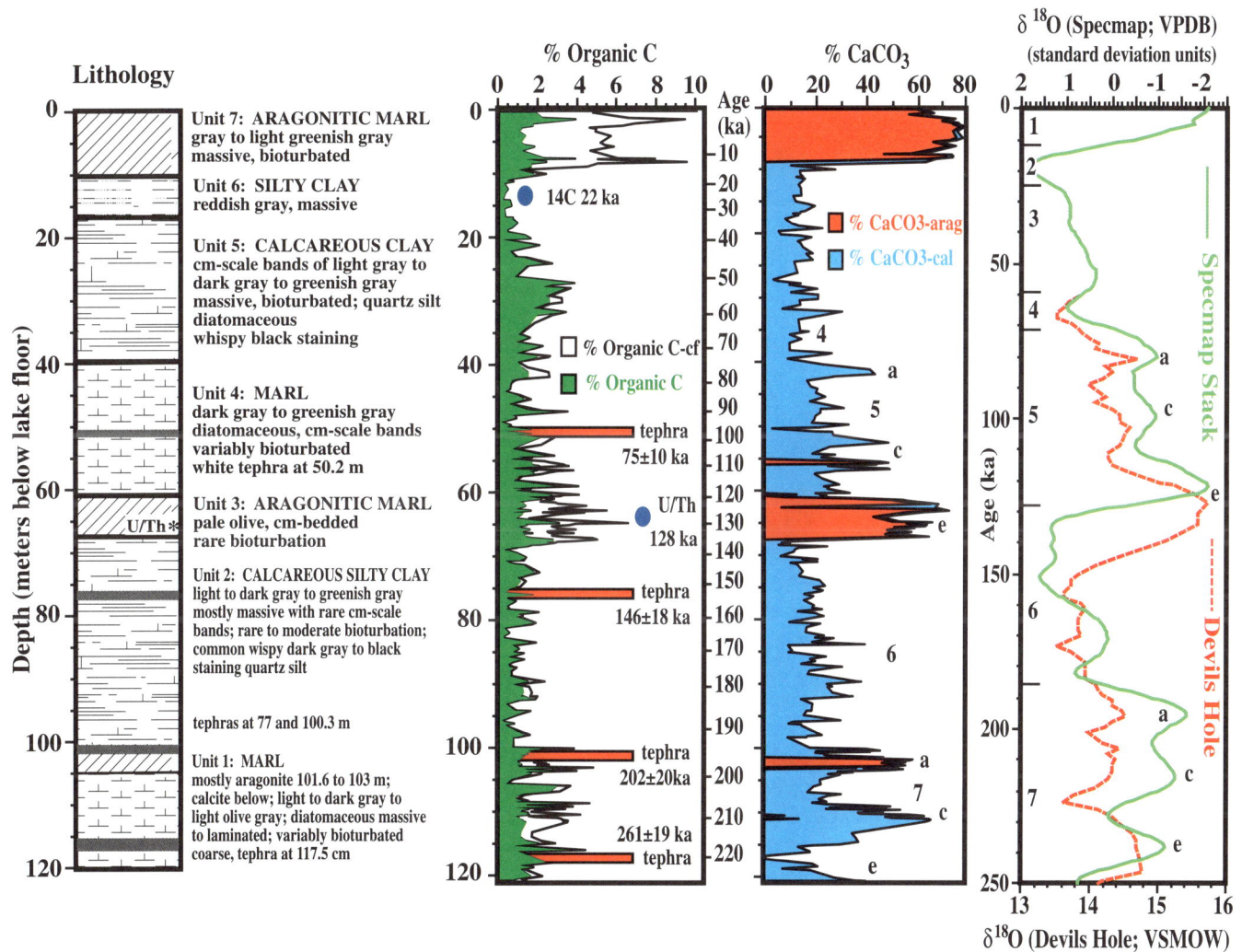

Figure 13. Lithology of sediment; profiles of percentages (calculated from XRD peak intensities) of dolomite and quartz; profiles of percent organic carbon (raw and carbonate-free, cf) and percent $CaCO_3$ as aragonite and calcite (see text for method of calculation) versus depth in samples from cores from holes BL00-1D and -1E. Radiocarbon age, U/Th age, and ages of tephras on the profile for organic carbon are from Colman et al. (2006). Profiles on the right are of values of $\delta^{18}O$ versus age in the SPECMAP stacked marine benthic foraminifera record (Martinson et al., 1987), and in a calcite vein in Devils Hole, Nevada (Winograd et al., 1992). Numbers (1–7) next to the time scale on the $\delta^{18}O$ profile indicate marine oxygen isotope stages 1–7. VPDB—Vienna Pee Dee Belemnite; VSMOW—Vienna Standard Mean Ocean Water. XRD—x-ray diffraction.

On the basis of visual core descriptions and smear-slide analyses, most of the sediments recovered below these top two units were initially described as calcareous clay or silty clay exhibiting some wispy laminations, and some centimeter-scale bedding cycles, distinguished mainly by the degree and style of bioturbation (Fig. 13). Subsequent carbon analyses showed that the clay has highly variable amounts of OC, but generally in the range of 1%–3% (Fig. 13). Variations in amounts of fine-grained sulfides and OC undoubtedly account for most of the variations in color (shades of gray) observed on fresh-cut surfaces. The non-carbonate fraction consists mainly of quartz, with minor amounts of dolomite and feldspar (see Kaufman et al., this volume, for further discussion of the non-carbonate mineralogy). Although several lithologic units contain as much quartz as in the red silty clay deposited during the LGI (~75%), there was no red coloration mentioned in the visual core descriptions, suggesting either that the lake did not receive large quantities of hematite-rich sediment from the Uinta Mountains or that the hematite pigment was destroyed by reducing conditions in the sediments.

The OC concentration in unit 6 is the lowest sustained OC concentration (<0.5%) in the entire record (Fig. 13); few other intervals have OC concentrations this low. In order to determine the effect of $CaCO_3$ dilution on the OC content, OC concentrations were calculated on a carbonate-free (cf) basis (Fig. 13). In the carbonate-rich intervals, the cf concentrations of OC are considerably higher than in bulk sediment. In unit 7, the average bulk-sediment OC concentration is 1.78%, and that on a cf-basis is 5.81%. In units 4 and 3, the average bulk-sediment OC concentration is 1.49%, and that on a cf-basis is 3.18%. In unit 1, the average bulk-sediment OC concentration is 1.36%, and that on a cf-basis is 2.38%. Using OC content as a semiquantitative measure of organic productivity, these results suggest that organic productivity was highest in Bear Lake during the high-carbonate, dry interglacial intervals that have been correlated with marine oxygen isotope stages (MIS) 1, 5, and 7 (Fig. 13; Bright et al., 2006; Colman et al., 2006; Kaufman et al., this volume). The sharpest increase in productivity occurred between units 6 and 7 (Fig. 13).

$CaCO_3$ contents of sediments below lithologic unit 7 are mostly in the range of 20%–40% (Fig. 13), but there are several carbonate-rich (>50% $CaCO_3$) intervals below unit 7. To distinguish high- and low-carbonate intervals, sediments with greater than 20% $CaCO_3$ are here called marls (Fig. 13, lithologic units 1, 3, and 4), and sediments with less than 20% $CaCO_3$ are called calcareous clay or silty clay. The interval from 60 to 67 meters below lake floor (mblf; lithologic unit 3) contains as much as 70% $CaCO_3$, and XRD indicates that most of this is in the form of aragonite (Fig. 13). This interval has been correlated with MIS 5e (Fig. 13; Bright et al., 2006; Colman et al., 2006; Kaufman et al., this volume). Several intervals containing >50% $CaCO_3$ occur between 100 and 115 mblf (Fig. 13, lithologic unit 1) with calcite as the dominant carbonate mineral. This high-carbonate interval has been correlated with MIS 7 (Fig. 13; Bright et al., 2006; Colman et al., 2006; Kaufman et al., this volume). In hole BL00-1E, the high-carbonate peak correlated with MIS 7a contains mostly aragonite (Fig. 13). The peak corre-

lated with MIS 7c contains very minor amounts of aragonite. Below 120 mblf, the $CaCO_3$ content increases rapidly downward (Fig. 13). Perhaps a few more meters of sediment at the base of hole BL00-1E would have obtained a well-defined carbonate peak, possibly aragonite, coincident with MIS 7e.

The striking resemblance of the smoothed $CaCO_3$ carbonate curve to the SPECMAP stacked oxygen isotope curve (Martinson et al., 1987), and to the oxygen isotope curve from a calcite vein in Devils Hole, Nevada (Fig. 13; Winograd et al., 1992) was pointed out by Colman et al. (2006) who used the Devils Hole chronology to "tune" the age model for the BL00-1 cores (see also Kaufman et al., this volume). The correlation of high-carbonate, aragonitic intervals with interglacials has been corroborated by high values of $\delta^{18}O$ and Sr^{87}/Sr^{86} in these intervals (Bright et al., 2006; Kaufman et al., this volume).

Owens Lake, southeastern California, was hydrologically closed, saline, alkaline and highly productive during interglacials when it was disconnected from the Owens River (Bischoff et al., 1997; Menking et al., 1997). The low-carbonate intervals in Bear Lake correlate with global glacial conditions and represent periods when Bear River was directly connected to Bear Lake. Owens Lake was hydrologically open during glacials, when influx from the Owens River was augmented by meltwaters from Sierran glaciers. During the dry interglacials, Bear River was not connected to Bear Lake, but a considerable amount of freshwater influx must have occurred, presumably as groundwater base flow to other streams entering the lake; otherwise, the salinity of the lake would have gone more evaporative, beyond aragonite precipitation, precipitating evaporite minerals (e.g., gypsum and halite; Dean et al., 2007) and producing high values of $\delta^{18}O$. During deposition of aragonitic unit 3 in BL00-1 (Fig. 13), values of $\delta^{18}O$ in bulk carbonate were as high as in aragonite deposited during the late Holocene in BL96-2 and BL02-4 (~–4‰, Figs. 10 and 12; Bright et al., 2006; Kaufman et al., this volume). Values of $\delta^{18}O$ in unit 1 bulk carbonate (MIS-7) only got as high as −8‰ (Bright et al., 2006; Kaufman et al., this volume), suggesting that conditions during deposition of unit 1 were not as evaporative as during deposition of units 3 and 7. Similarly, $^{87}Sr/^{86}Sr$ values in bulk carbonate in unit 3 were as high as in unit 7 (~0.7102), but values in bulk sediment in unit 1 were much lower (~0.7100, Kaufman et al., this volume), indicating more influx of Bear River water with a low $^{87}Sr/^{86}Sr$ (~0.7085). What climatic conditions were ultimately driving wet glacial and dry interglacial conditions in Bear Lake and Owens Lake? The answer has to be in atmospheric circulation over the northeastern Pacific Ocean.

The Bear Lake Record and Glacial/Interglacial Climates of Western United States and NE Pacific

The climate history of the last glacial/interglacial cycle in the western United States and adjacent northeastern Pacific Ocean is fairly well known from both continental and marine records. Most of these records have been obtained from conventional piston coring in lakes and the Pacific margin of the United States, or from outcrops. Continuous marine records of older glacial/interglacial cycles only recently have been revealed by Ocean Drilling Program (ODP) drilling on the Pacific margin (ODP Leg 167; Lyle et al., 2000). The Bear Lake record is the only long, continuous, continental record from the western United States collected from an extant lake.

The presence of red calcareous clay deposited in Bear Lake during the LGI coincides in time with the growth of large lakes in the northern Great Basin, including Bonneville, Lahontan, and Russell (Thompson, 1990; Thompson et al., 1986, 1993; Benson and Thompson, 1987; Currey, 1990). At that time, bristlecone pine occupied the present piñon/juniper life zone implying that the climate was somewhat wetter and summer temperatures were as much as 10 °C below modern. The increased moisture was due to increased winter precipitation produced by enhanced cyclonic flow (i.e., increased domination of the Aleutian lows; Kutzbach et al., 1993). This change in circulation apparently was caused by a splitting of the polar jet stream, and southerly displacement of the southern branch of the jet (Kutzbach, 1987; COHMAP Members, 1988). Precipitation was greatest under the axis of the polar jet, decreased abruptly south of the axis, and decreased less abruptly north of the axis. Consequently, the Great Basin was wetter than present during the LGI, and the Pacific Northwest was drier (Barnosky et al., 1987; Thompson et al., 1993). Sediments deposited during the LGI in Owens Lake, California, at a latitude of 36.5°N contain an ostracode assemblage dominated by *Cytherissa lacustrus*, which indicates very cold, stable limnoclimatic conditions (Bradbury and Forester, 2002). Today *C. lacustrus* lives in cold lakes in boreal forests of Canada and Alaska, and its presence in large numbers implies that polar air masses were present year-round at the latitude of Owens Lake. It is difficult to determine how much wetter the LGI was, but modeling studies by Hostetler et al. (1994) indicate that the presence of the lakes had a considerable influence on the precipitation over these large lake basins. They estimate that at 20 cal ka precipitation in the Bonneville Basin (including the lake) was 3% greater in January and 38% greater in July than it would have been without the lake, due to lake-atmosphere feedbacks.

Although the levels of the large lakes in the northern Great Basin continued to increase during the LGI, they did not reach their maximum levels until after 18 cal ka, followed by precipitous decreases (Benson and Thompson, 1987; Thompson et al., 1986, 1993; Currey, 1990; Oviatt et al., 1992). This suggests that the moisture-delivering storm tracks migrated northward ca. 16 cal ka, when the pluvial lakes of the southern Great Basin began to dry up, and Lakes Bonneville and Lahontan in the northern Great Basin approached maximum levels. By that time (ca. 16 cal ka) the flow in the Bear River had decreased, red clay deposition ceased (top of unit 6), and endogenic carbonate began to increase (increase in $CaCO_3$ in core BL96-2, Fig. 10).

We know from recent changes in levels of lakes that are the remnants of the larger glacial lakes in the Great Basin (e.g., Mono, Pyramid, Walker, and Great Salt Lakes) that lake levels go up when winter precipitation is greater than normal, and go down when winter precipitation is lower than normal (e.g., Benson and Thompson, 1987). The sediments of Pleistocene Lake Manix record major lacustrine phases, interpreted to represent high lake levels, that are coincident with MIS 4 and 6, and lowstands are coincident with MIS 5 and 7 (Jefferson, 2003).

Winter precipitation comes from the Pacific Ocean, and increased winter precipitation occurs by enhanced cyclonic flow around the Aleutian low-pressure system when the polar jet stream moves south of its normal position. Today, this condition exists only during the winter, but during the LGI it may have prevailed throughout the year (e.g., Kutzbach, 1987). When the polar jet stream was in this southerly position, and the Aleutian Low dominated atmospheric circulation over the eastern North Pacific, the southwestern United States was wet (e.g., Van Devender et al., 1987) and the Pacific Northwest was dry (e.g., Barnosky et al., 1987).

A major reconfiguration of Pacific atmospheric circulation and the dominance of cyclonic circulation driven by the Aleutian Low also would have had a major effect on surface oceanic circulation in the North Pacific. Today, surface circulation in the North Pacific is dominated by the North Pacific subtropical gyre. The California Current is the southward flowing, wind-driven, eastern limb of the North Pacific subtropical gyre formed by the divergence of the West Wind Drift on the western margin of North America. This oceanic circulation is controlled by atmospheric-pressure systems of the North Pacific and western North America (the North Pacific High, the Aleutian Low, and the North American Low; Fig. 14). The seasonal strengths and positions of these pressure systems not only generate the weather and climate of the western United States (Huyer and Kosro, 1987; Strub et al., 1987; Thomas and Strub, 1990; Abbott and Barksdale, 1991), but also are part of the atmospheric teleconnections that stretch across the Northern Hemisphere (e.g., Namias et al., 1988). Changes in the position and strength of these atmospheric systems that occurred as a result of the change from global glacial to interglacial conditions had large influences on the climate of the western United States.

Today, the climate of southern Oregon and California is characterized in the spring and summer by strong, persistent, northwesterly winds generated by the juxtaposition of the North Pacific High and North American Low. The average summer position of the subtropical North Pacific High is centered at ~45°N and the average summer position of the strong North American Low, which is maintained by intense heating of the Great Basin, is centered at ~38°N (Fig. 14A). The summer polar jet stream typically is located north of 50°N. Winters are influenced by a weakened North American Low, the migration of the North Pacific High south of 30°N, and the migration of the polar jet stream and associated Aleutian Low to an average position of ~45°N (Fig. 14B). Winters are typically mild, wet, and stormy, with zonal westerly winds.

Temperatures in the continental interior were higher in the Holocene than during the LGI, but what about sea-surface temperatures (SSTs) in the North Pacific? Recent studies suggest that SSTs may have been 3–5 °C cooler than present during the LGI off Oregon and California (e.g., Ortiz et al., 1997; Mix et al.,

1999; Herbert et al., 2001; Pisias et al., 2001; Trend-Staid and Prell, 2002; Barron et al., 2003). Results from analysis of recent ODP cores from the California margin suggest that this glacial-interglacial contrast in SST may have been even greater between MIS 6 and MIS 5 (Lyle et al., 2001).

As discussed earlier, the glacial-to-interglacial (Holocene) transition in Bear Lake in terms of carbonate mineralogy is represented in great detail with excellent radiocarbon dating in cores BL96-2 and BL02-04. The details of this transition are difficult to see in data from site BL00-1 (Fig. 13) because of the coarser sampling interval. I assume that a similar scenario occurred at the transitions to earlier interglacial high-carbonate, aragonitic intervals equivalent to MIS-5 and -7 (Fig. 13), i.e., decreasing precipitation, increasing evaporation:precipitation ratio, increasing salinity leading to carbonate precipitation, first as calcite then, at highest salinity, as aragonite. If this interpretation is correct, it would appear that carbonate precipitation remained more in the calcite-stability environment during the MIS-5 carbonate "event" and even more so during the MIS-7 "event" (Fig. 13).

These high-carbonate, aragonitic interglacial intervals are interpreted to coincide with warm continental climates and warm SSTs recorded in recent ODP cores along the Pacific margin (Lyle et al., 2000; Lyle et al., 2001). This implies that the subtropical North Pacific High and North American Low that today bring warm-dry conditions during the summer (Fig. 14A) were more permanent features of North Pacific circulation during interglacials. During the last glacial interval and other stadials, the Aleutian low-pressure system, presently dominant during the winter (Fig. 14B), was a more permanent feature of North Pacific circulation, producing storms and precipitation that increased the levels of Lakes Bonneville and Lahontan and increased the flow of Bear River. These climatic patterns over the North Pacific are recorded in Bear Lake by river-borne detrital clastic sediments deposited during glacial periods and carbonate sediments, dominated by aragonite, during the brief, more arid interglacial periods.

SUMMARY

1. Time-marking sediment traps placed 10 m below the surface of the lake and 2 m above the bottom of the lake show that high-Mg calcite precipitates from the surface waters of the lake, but aragonite is the dominant mineral that accumulates on the lake floor, along with lesser amounts of high-Mg calcite, low-Mg calcite, dolomite, and quartz.

2. The chemical and isotopic compositions of the high-Mg calcite that forms in the surface waters of the lake today are different from those of the aragonitic sediments deposited in Bear Lake prior to the diversion of Bear River into Bear Lake in the early twentieth century. The aragonite collected in the deep traps did not precipitate from present-day Bear Lake water, but is reworked aragonite from water depths shallower than ~30 m that is at least 50 yr old and possibly older. The diversion of Bear River into Bear Lake created an experiment in which the pre- and post-diversion chemical composition, isotopic composition, and mineralogy of bulk carbonate allows us to distinguish between modern and reworked endogenic carbonate.

3. Radiocarbon ages on samples from the top of an extensive, aragonite-cemented microbialite mound in the southwestern part of the lake suggest that the mound formed during the middle to late Holocene (6500–3000 yr ago). However, O and C isotope values in transects of samples from a thick carbonate rind on one microbialite have values that occur only in sediments

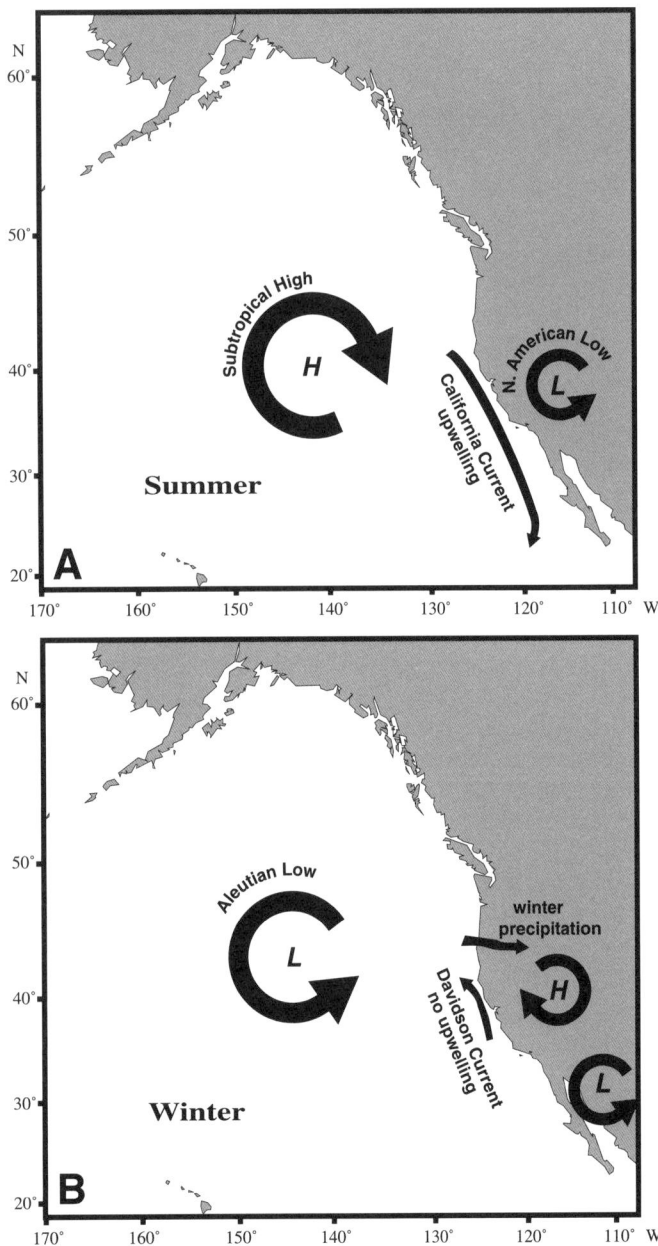

Figure 14. Generalized surface atmospheric pressure cells in the eastern North Pacific and adjacent North America during summer (A) and winter (B).

in cores that were deposited during the last several thousand years. This indicates that the top of the mound began forming several thousand years ago but stopped precipitating aragonite after Bear River diversion. The mound may have originated by CO_2-charged groundwater entering the lake, discharging CO_2, and precipitating $CaCO_3$, mediated by cyanobacteria and other microbes in much the same way that microbialite mounds or tufa mounds form in other lakes today.

4. During the last glacial interval, when increased Pacific moisture filled the large lakes of the Great Basin, the Bear River entered Bear Lake and deposited red, calcareous, silty clay. When Pacific moisture was greatly reduced ~16,000 yr ago, Bear River began to lose its connection to Bear Lake. A combination of warmer water temperatures, increased evaporation, and increased organic productivity triggered the precipitation of $CaCO_3$ as low-Mg calcite. The isotopic evidence shows that the salinity of the lake continued to increase, and by ~11,000 yr ago a threshold was reached when aragonite became the dominate carbonate precipitate. A return to calcite deposition between 8.5 and 7.5 cal ka, accompanied by lower O, C, and Sr isotope ratios, indicates a brief re-entry of Bear River into the lake. The sediments deposited in Bear Lake over the past 7000 years consist of ~75% $CaCO_3$, mostly as aragonite and only ~5% as low-Mg calcite.

5. Most of the time during the past 220,000 years, Bear Lake was connected to Bear River. During these times the Aleutian low-pressure atmospheric circulation system was the dominant feature of the eastern North Pacific, bringing abundant moisture to the western United States and enhancing the flow of Bear River. However, three high-carbonate, aragonitic intervals in sediments deposited over the past 220,000 years indicate times when Bear River was not connected to Bear Lake, and the salinity of the lake increased. During these three intervals the North Pacific subtropical high-pressure system dominated circulation in the eastern North Pacific, bringing dry conditions to the western United States. These three intervals correspond in time to warm, dry interglacials equivalent to marine oxygen isotope stages 1, 5, and 7.

ACKNOWLEDGMENTS

I thank Allen Harrison, Pat Kimball, and Mike Enright for help in collecting the 1998 gravity cores. I thank Roger Anderson for providing the sediment traps and helping to deploy them, and Allen Harrison, Katrina Mosier, and Scott Tolentino for helping to recover the traps. The 1996 Kullenberg piston cores were collected by Dawn Graber, Kerry Kelts, and Bob Thompson. Jeff Honke, Gary Skipp, and Frank Urban were a tremendous help on various coring operations. Gary Skipp, Dave Thornbury, and Kathleen Simmons skillfully performed the X-ray diffraction, carbon, and strontium-isotope analyses. I thank Augusta Warden for providing the gas chromatographic and isotopic analyses of the gas bubbles. Scott Tolentino and Bryce Nielson, Utah Department of Natural Resources, gave freely of their time, boats, and insights on the lake. Eldon Robinson, head ranger at Bear Lake State Park, gave generous use of the park facilities and marina. Allen Riggs (USGS) and Peter Kolesar (Utah State University) contributed their SCUBA diving experience for exploring the methane seeps and the Rock Pile. I thank Mitch Poulsen, Bear Lake Regional Commission, and Connely Baldwin, PacifiCorp, for researching the early records of the Bear River diversion. This and other studies at Bear Lake have benefited enormously from fieldwork by, and discussions with, Darrell Kaufman and Jordon Bright. I acknowledge Dennis Nielson and the DOSECC (Drilling Observation and Sampling of the Earth's Continental Crust) drilling crew for obtaining the Bear Lake GLAD800 cores. I thank Brian Haskell, Doug Schnurrenberger, Blass Valero-Garcés, and others who worked on the drilling platform and performed most of the initial core descriptions. I thank Pete Kolesar, Katrina Moser, Janet Pitman, Joe Rosenbaum, and Mark Shapley for very helpful reviews of earlier drafts of this paper. This research was supported by the USGS Earth Surface Dynamics Program.

ARCHIVED DATA

Archived data for this chapter can be obtained from the NOAA World Data Center for Paleoclimatology at http://www.ncdc.noaa.gov/paleo/pubs/gsa2009bearlake/.

REFERENCES CITED

Abbott, M.R., and Barksdale, B., 1991, Phytoplankton pigment patterns and wind forcing off central California: Journal of Geophysical Research, v. 96(C8), p. 14,649–14,667.

Anderson, R.Y., 1977, Short term sedimentation response in lakes in western United States as measured by automated sampling: Limnology and Oceanography, v. 22, p. 423–433.

Barnosky, C.W., Anderson, P.M., and Bartlein, P.J., 1987, The northwestern U.S. during deglaciation; vegetational history and paleoclimatic implications, *in* Ruddiman, W.F., and Wright, H.E., eds., North America and adjacent oceans during the last deglaciation: Boulder, Colorado, Geological Society of America, Geology of North America, v. K-3, p. 289–321.

Barron, J.A., Heuser, L., Herbert, T., and Lyle, M., 2003, High-resolution climatic evolution of coastal Northern California during the past 16,000 years: Paleoceanography, v. 18, p. 1020, doi: 10.1029/2002PA000768.

Benson, L., and Thompson, R.S., 1987, The physical record of lakes in the Great Basin, in North America and adjacent oceans during the last deglaciation, *in* Ruddiman, W.F., and Wright, H.E., Jr., eds., North America and adjacent oceans during the last deglaciation: Boulder, Colorado, Geological Society of America, Geology of North America, v. K-3, p. 241–260.

Birdsey, P.W., 1989, The limnology of Bear Lake, Utah-Idaho, 1912–1988: A literature review. Utah Department of Natural Resources, Division of Wildlife Resources, Publication no. 89-5, 113 p.

Bischoff, J.L., Fitts, J.P., and Fitzpatrick, J.A., 1997, Responses of sediment geochemistry to climate change in Owens Lake sediment: An 800-ky record of saline/fresh cycles in core OL-92, *in* Smith, G.I., and Bischoff, J.L., eds., An 800,00-year paleoclimatic record from core OL-92, Owens Lake, southeast California: Geological Society of America Special Paper 317, p. 37–48.

Bischoff, J.L., Simmons, K., and Shamp, D., 2005, Geochemistry of sediments in Bear Lake cores and sediment traps: U.S. Geological Survey Open-File Report 2005-1215, http://pubs.usgs.gov/of/2005/1215 (accessed 2005).

Bischoff, W.D., Mackenzie, F.T., and Bishop, F.C., 1987, Stabilities of synthetic magnesian calcites in aqueous solution: Comparison with biogenic materials: Geochimica et Cosmochimica Acta, v. 51, p. 1413–1423, doi: 10.1016/0016-7037(87)90325-5.

Bouchard, D.P., Kaufman, D.S., Hochberg, A., and Quade, J., 1998, Quaternary history of the Thatcher Basin, Idaho, reconstructed from the $^{87}Sr/^{86}Sr$ and amino acid composition of lacustrine fossils—Implications for the diversion of the Bear River into the Bonneville Basin: Palaeogeography, Palaeoclimatology, Palaeoecology, v. 141, p. 95–114, doi: 10.1016/S0031-0182(98)00005-4.

Bradbury, J.P., and Forester, R.M., 2002, Environment and paleolimnology of Owens Lake, California: A record of climate and hydrology for the last 50,000 years, *in* Hershler, R., Madsen, D.B., and Currey, D.R., eds., Great Basin aquatic systems history: Smithsonian Contributions to the Earth Sciences, no. 33, p. 145–173.

Bright, J., 2009, this volume, Isotope and major-ion chemistry of groundwater in Bear Lake Valley, Utah and Idaho, with emphasis on the Bear River Range, *in* Rosenbaum, J.G., and Kaufman, D.S., eds., Paleoenvironments of Bear Lake, Utah and Idaho, and its catchment: Geological Society of America Special Paper 450, doi: 10.1130/2009.2450(04).

Bright, J., Kaufman, D.S., Forester, R.M., and Dean, W.E., 2006, A continuous 250,000 yr record of oxygen and carbon isotopes in ostracode and bulk-sediment carbonate from Bear Lake, Utah-Idaho: Quaternary Science Reviews, v. 25, p. 2258–2270, doi: 10.1016/j.quascirev.2005.12.011.

Brown, B.E., Fassbender, J.L., and Winkler, R., 1992, Carbonate production and sediment transport in a marl lake of southeastern Wisconsin: Limnology and Oceanography, v. 37, p. 184–191.

Burne, R.V., and Moore, L.S., 1987, Microbialites—Organosedimentary deposits of benthic microbial communities: Palaios, v. 2, p. 241–254, doi: 10.2307/3514674.

Capo, R.C., and DePaolo, D.J., 1990, Seawater strontium isotopic variations from 2.5 million years ago to the present: Science, v. 249, p. 51–55, doi: 10.1126/science.249.4964.51.

Chave, K., 1954, Aspects of the biogeochemistry of magnesium: 1. Calcareous marine organisms: Journal of Geology, v. 62, p. 266–283.

Clark, I., and Fritz, P., 1997, Environmental isotopes in hydrogeology: Boca Raton, Florida, Lewis Publishers, 328 p.

COHMAP Members, 1988, Climatic changes in the last 18,000 years: Observations and model simulations: Science, v. 241, p. 1043–1052, doi: 10.1126/science.241.4869.1043.

Colman, S.M., 2006, Acoustic stratigraphy of Bear Lake, Utah-Idaho—Late Quaternary sedimentation in a simple half-graben: Sedimentary Geology, v. 185, p. 1, doi: 10.1016/j.sedgeo.2005.11.022.

Colman, S.M., Kaufman, D., Rosenbaum, J.G., and McGeehin, J.P., 2005, Radiocarbon dating of cores collected in Bear Lake, Utah: U.S. Geological Survey Open-File Report 05-1320, http://pubs.usgs.gov/of/2005/1124 (accessed 2005).

Colman, S.M., Kaufman, D., Heil, C., King, J., Dean, W., Rosenbaum, J.G., Bischoff, J.L., and Perkins, M., 2006, Age models for a continuous 250-ka Quaternary lacustrine record from Bear Lake, Utah-Idaho: Quaternary Science Reviews, v. 25, p. 2271–2282, doi: 10.1016/j.quascirev.2005.10.015.

Colman, S.M., Rosenbaum, J.G., Kaufman, D.S., Dean, W.E., and McGeehin, J.P., 2009, this volume, Radiocarbon ages and age models for the last 30,000 years in Bear Lake, Utah and Idaho, *in* Rosenbaum, J.G., and Kaufman, D.S., eds., Paleoenvironments of Bear Lake, Utah and Idaho, and its catchment: Geological Society of America Special Paper 450, doi: 10.1130/2009.2450(05).

Currey, D.R., 1990, Quaternary paleolakes in the evolution of semidesert basins, with special emphasis on Lake Bonneville and the Great Basin, U.S.A.: Palaeogeography, Palaeoclimatology, Palaeoecology, v. 76, p. 189–214, doi: 10.1016/0031-0182(90)90113-L.

Dean, W.E., 1999, The carbon cycle and biogeochemical dynamics in lake sediments: Journal of Paleolimnology, v. 21, p. 375–393, doi: 10.1023/A:1008066118210.

Dean, W.E., and Megard, R.O., 1993, Environment of deposition of $CaCO_3$ in Elk Lake, *in* Bradbury, J.P., and Dean, W.E., eds., Elk Lake, Minnesota: Evidence for rapid climate change in north-central United States: Geological Society of America Special Paper 276, p. 97–114.

Dean, W., Rosenbaum, J., Haskell, B., Kelts, K., Schnurrenberger, D., Valero-Garcés, B., Cohen, A., Davis, O., Dinter, D., and Nielson, D., 2002, Progress in Global Lake Drilling holds potential for global change research: Eos (Transactions, American Geophysical Union), v. 83, p. 85, 90, 91.

Dean, W.E., Rosenbaum, J.G., Forester, R.M., Colman, S.M., Bischoff, J.L., Liu, A., Skipp, G., and Simmons, K., 2006, Glacial to Holocene evolution

of sedimentation in Bear Lake, Utah-Idaho: Sedimentary Geology, v. 185, p. 93–112, doi: 10.1016/j.sedgeo.2005.11.016.

Dean, W.E., Forester, R.M., Bright, J., and Anderson, R.Y., 2007, Influence of the diversion of Bear River into Bear Lake (Utah and Idaho) on the environment of deposition of carbonate minerals: Evidence from water and sediments: Limnology and Oceanography, v. 53, p. 1094–1111.

Dean, W.E., Wurtsbaugh, W.A., and Lamarra, V.A., 2009, this volume, Climatic and limnologic setting of Bear Lake, Utah and Idaho, *in* Rosenbaum, J.G., and Kaufman, D.S., eds., Paleoenvironments of Bear Lake, Utah and Idaho, and its catchment: Geological Society of America Special Paper 450, doi: 10.1130/2009.2450(01).

Dover, J.H., 1985, Geologic map of the Logan 30′ × 60′ quadrangle, Cache and Rich counties, Utah, and Lincoln and Uinta counties, Wyoming: U.S. Geological Survey Miscellaneous Investigations Series Map I-2210.

Engleman, E.E., Jackson, L.L., Norton, D.R., and Fischer, A.G., 1985, Determination of carbonate carbon in geological materials by coulometric titration: Chemical Geology, v. 53, p. 125–128, doi: 10.1016/0009-2541(85)90025-7.

Hart, W.S., Madsen, D.B., Kaufman, D.S., and Oviatt, C.G., 2004, The $^{87}Sr/^{86}Sr$ ratios of lacustrine carbonates and lake-level history of the Bonneville paleolake system: Geological Society of America Bulletin, v. 116, p. 1107–1119, doi: 10.1130/B25330.1.

Heil, C.W., Jr., King, J.W., Rosenbaum, J.G., Reynolds, R.L., and Colman, S.M., 2009, this volume, Paleomagnetism and environmental magnetism of GLAD800 sediment cores from Bear Lake, Utah and Idaho, *in* Rosenbaum, J.G., and Kaufman, D.S., eds., Paleoenvironments of Bear Lake, Utah and Idaho, and its catchment: Geological Society of America Special Paper 450, doi: 10.1130/2009.2450(13).

Herbert, T.D., Schuffert, J.D., Andreassen, D., Heuser, L., Lyle, M., Mix, A., Ravelo, A.C., Stott, L.D., and Heruera, J.C., 2001, Collapse of the California Current during glacial maximum linked to climate change on land: Science, v. 293, p. 71–76, doi: 10.1126/science.1059209.

Hollander, D.J., and McKenzie, J.A., 1991, CO_2 control on carbon-isotope fractionation during aqueous photosynthesis: A paleo-pCO_2 barometer: Geology, v. 19, p. 929–932, doi: 10.1130/0091-7613(1991)019<0929:CCOCIF>2.3.CO;2.

Hostetler, S.W., Giorgi, F., Bates, G.T., and Bartlein, P.J., 1994, Lake-atmosphere feedbacks associated with paleolakes Bonneville and Lahontan: Science, v. 263, p. 665–668, doi: 10.1126/science.263.5147.665.

Huyer, A., and Kosro, P.M., 1987, Mesoscale surveys over the shelf and slope in the upwelling region near Point Arena, California: Journal of Geophysical Research, v. 92, p. 1655–1681, doi: 10.1029/JC092iC02p01655.

Jefferson, G.T., 2003, Stratigraphy and paleontology of the middle to late Pleistocene Manix Formation, and paleoenvironments of the central Mojave River, southern California, *in* Enzel, Y., Wells, S.G., and Lancaster, N., eds., Paleoenvironments and paleohydrology of the Mojave and southern Great Basin Deserts: Geological Society of America Special Paper 368, p. 43–60.

Kaufman, D.S., Bright, J., Dean, W.E., Rosenbaum, J.G., Moser, K., Anderson, R.S., Colman, S.M., Heil, C.W., Jr., Jiménez-Moreno, G., Reheis, M.C., and Simmons, K.R., 2009, this volume, A quarter-million years of paleoenvironmental change at Bear Lake, Utah and Idaho, *in* Rosenbaum, J.G., and Kaufman, D.S., eds., Paleoenvironments of Bear Lake, Utah and Idaho, and its catchment: Geological Society of America Special Paper 450, doi: 10.1130/2009.2450(14).

Kelts, K., and Hsü, K.J., 1978, Freshwater carbonate sedimentation, *in* Lerman, A., ed., Lakes—Chemistry, geology, physics: Dordrecht, Netherlands, Springer, p. 295–324.

Kemmerer, G., Bovard, J.F., and Boorman, W.R., 1923, Northwestern lakes of the United States; biological and chemical studies with reference to possibilities to production of fish: U.S. Bureau of Fisheries Bulletin, v. 39, p. 51–140.

Kempe, S., Kazmierczak, J., Landmann, G., Konuk, T., Reimer, A., and Lipp, A., 1991, Largest known microbialites discovered in Lake Van, Turkey: Nature, v. 349, p. 605–608, doi: 10.1038/349605a0.

Kutzbach, J.E., 1987, Model simulations of the climatic patterns during the deglaciation of North America, *in* Ruddiman, W.F., and Wright, H.E., eds., North America and adjacent oceans during the last deglaciation: Boulder, Colorado, Geological Society of America, Geology of North America, v. K-3, p. 425–446.

Kutzbach, J.E., Guetter, P.J., Behling, P.J., and Selin, R., 1993, Simulated climate changes: Results of the COHMAP climate-model experiments, *in* Wright, H.E., Jr., Kutzbach, J.E., Webb, T., III, Ruddiman, W.F., Street-Perrott,

F.A., and Bartlein, P.J., eds., Global climates since the Last Glacial Maximum: Minneapolis, University of Minnesota Press, p. 24–93.

Lamarra, V., Liff, C., and Carter, J., 1986, Hydrology of Bear Lake Basin and its impact on the trophic state of Bear Lake, Utah-Idaho: The Great Basin Naturalist, v. 46, p. 690–705.

Last, W.M., 1982, Holocene carbonate sedimentation in Lake Manitoba, Canada: Sedimentology, v. 29, p. 691–704, doi: 10.1111/j.1365-3091.1982.tb00074.x.

Last, W.M., and Schweyen, T.H., 1983, Sedimentology and geochemistry of saline lakes of the Great Plains: Hydrobiologia, v. 105, p. 245–263, doi: 10.1007/BF00025192.

Lowenstam, H.A., 1954, Factors affecting the aragonite:calcite ratios in carbonate-secreting marine organisms: Journal of Geology, v. 62, p. 284–322.

Li, H.-C., and Ku, T.-L., 1997, $\delta^{13}C$-$\delta^{18}O$ covariance as a paleohydrological indicator for closed-basin lakes: Palaeogeography, Palaeoclimatology, Palaeoecology, v. 133, p. 69–80, doi: 10.1016/S0031-0182(96)00153-8.

Lyle, M., Koizumi, Richter C., and Moore, T.C., Jr., eds., 2000, Proceeding of the Ocean Drilling Program, Scientific Results, Volume 167: College Station, Texas, Ocean Drilling Program.

Lyle, M., Heuser, L., Herbert, T., Mix, A., and Barron, J., 2001, Interglacial theme and variations: 500 k.y. of orbital forcing and associated responses from the terrestrial and marine biosphere, U.S.: Pacific Northwest: Geology, v. 29, p. 1115–1118, doi: 1001130/0091-7613(2001)029<1115:itavky>2.0.CO;2.

Mackenzie, F.T., Bischoff, W.B., Bishop, F.C., Loijens, M., Schoonmaker, J., and Wollast, R., 1983, Magnesian calcites: Low-temperature occurrence, solubility and solid-solution behavior, *in* Reeder, R.J., ed., Carbonates: Mineralogy and chemistry: Chelsea, Michigan, Mineralogical Society of America, p. 97–144.

Martinson, D.G., Pisias, N.G., Hays, J.D., Imbrie, J., and Moore, T.C., Jr., 1987, Age dating and the orbital theory of the ice ages: Development of a high-resolution 0 to 3,000,000-year chronostratigraphy: Quaternary Research, v. 27, p. 1–29, doi: 10.1016/0033-5894(87)90046-9.

McKenzie, J.A., 1985, Carbon isotopes and productivity in the lacustrine and marine environment, *in* Stumm, W., ed., Chemical processes in lakes: New York, John Wiley and Sons, p. 99–118.

Menking, K.M., Bischoff, J.L., Fitzpatrick, J.A., Burdette, J.W., and Rye, R.O., 1997, Climatic/hydrologic oscillations since 155,000 yr B.P. at Owens Lake, California, reflected in abundance and stable isotope composition of sediment carbonate: Quaternary Research, v. 48, p. 58–68, doi: 10.1006/qres.1997.1898.

Mix, A.C., Lund, D.C., Pisias, N.G., Boden, P., Bornmalm, L., Lyle, M., and Pike, J., 1999, Rapid climate oscillations in the Northeast Pacific during the last deglaciation reflect northern and southern hemisphere sources, *in* Clark, P.U., Webb, R., and Keigwin, L.D., eds., Mechanisms of global climate change at millennial time scales: American Geophysical Union, Geophysical Monograph 112, p. 127–148.

Moore, D.M., and Reynolds, R.C., Jr., 1997, X-ray diffraction and identification and analysis of clay minerals (second edition): New York, Oxford University Press, 378 p.

Moser, K.A., and Kimball, J.P., 2009, this volume, A 19,000-year record of hydrologic and climatic change inferred from diatoms from Bear Lake, Utah and Idaho, *in* Rosenbaum, J.G., and Kaufman, D.S., eds., Paleoenvironments of Bear Lake, Utah and Idaho, and its catchment: Geological Society of America Special Paper 450, doi: 10.1130/2009.2450(10).

Müller, G., 1970, High-magnesian calcite and protodolomite in Lake Balaton (Hungary) sediments: Nature, v. 226, p. 749–750, doi: 10.1038/226749a0.

Müller, G., 1971, Aragonite inorganic precipitation in a freshwater lake: Nature, v. 226, p. 18.

Müller, G., Irion, G., and Forstner, U., 1972, Formation and diagenesis of inorganic Ca-Mg carbonates in the lacustrine environment: Naturwissenschaften, v. 59, p. 158–164, doi: 10.1007/BF00637354.

Namias, J., Yuan, X., and Cayan, D.R., 1988, Persistence of North Pacific sea surface temperature and atmospheric flow patterns: Journal of Climate, v. 1, p. 682–703.

Oriel, S.S., and Platt, L.B., 1980, Geologic map of the Preston 1° × 2° quadrangle, southeastern Idaho and western Wyoming: U.S. Geological Survey Miscellaneous Investigations Series Map I-1127.

Ortiz, J., Mix, A., Hostetler, S., and Kashgarian, M., 1997, The northern California Current at 42°N of the last glacial maximum: Reconstruction based on planktonic foraminifera: Paleoceanography, v. 12, p. 191–206, doi: 10.1029/96PA03165.

Otsuki, A., and Wetzel, R.G., 1974, Calcium and total alkalinity budgets and calcium carbonate precipitation of a small hard-water lake: Archiv für Hydrobiologie, v. 73, p. 14–30.

Oviatt, C.G., Curry, D.R., and Sack, D., 1992, Radiocarbon chronology of Lake Bonneville, eastern Great Basin, USA: Palaeogeography, Palaeoclimatology, Palaeoecology, v. 99, p. 225–241, doi: 10.1016/0031-0182(92)90017-Y.

Pisias, N.G., Mix, A.C., and Hueser, L., 2001, Millennial scale climate variability of the northeast Pacific Ocean and northwest North America based on radiolaria and pollen: Quaternary Science Reviews, v. 20, p. 1561–1576, doi: 10.1016/S0277-3791(01)00018-X.

Reheis, M.C., Laabs, B.J.C., and Kaufman, D.S., 2009, this volume, Geology and geomorphology of Bear Lake Valley and upper Bear River, Utah and Idaho, *in* Rosenbaum, J.G., and Kaufman, D.S., eds., Paleoenvironments of Bear Lake, Utah and Idaho, and its catchment: Geological Society of America Special Paper 450, doi: 10.1130/2009.2450(02).

Rosenbaum, J.G., and Heil, C.W., Jr., 2009, this volume, The glacial/deglacial history of sedimentation in Bear Lake, Utah and Idaho, *in* Rosenbaum, J.G., and Kaufman, D.S., eds., Paleoenvironments of Bear Lake, Utah and Idaho, and its catchment: Geological Society of America Special Paper 450, doi: 10.1130/2009.2450(11).

Rosenbaum, J.G., Dean, W.E., Reynolds, R.L., and Reheis, M.C., 2009, this volume, Allogenic sedimentary components of Bear Lake, Utah and Idaho, *in* Rosenbaum, J.G., and Kaufman, D.S., eds., Paleoenvironments of Bear Lake, Utah and Idaho, and its catchment: Geological Society of America Special Paper 450, doi: 10.1130/2009.2450(06).

Shapley, M.D., Johnson, W.C., Engstrom, D.R., and Ostercamp, W.R., 2005, Late-Holocene flooding and drought in the Northern Great Plains, USA, reconstructed from tree rings, lake sediments, and ancient shorelines: The Holocene, v. 15, p. 29–41, doi: 10.1191/0959683605hl781rp.

Smoak, J.M., and Swarzenski, P.W., 2004, Recent increases in sediment and nutrient accumulation in Bear Lake, Utah/Idaho, USA: Hydrobiologia, v. 525, p. 175–184, doi: 10.1023/B:HYDR.0000038865.16732.09.

Smoot, J.P., 2009, this volume, Late Quaternary sedimentary features of Bear Lake, Utah and Idaho, *in* Rosenbaum, J.G., and Kaufman, D.S., eds., Paleoenvironments of Bear Lake, Utah and Idaho: Geological Society of America Special Paper 450, doi: 10.1130/2009.2450(03).

Smoot, J.P., and Rosenbaum, J.G., 2009, this volume, Sedimentary constraints on late Quaternary lake-level fluctuations at Bear Lake, Utah and Idaho, *in* Rosenbaum, J.G., and Kaufman, D.S., eds., Paleoenvironments of Bear Lake, Utah and Idaho: Geological Society of America Special Paper 450, doi: 10.1130/2009.2450(12).

Stabel, H.-H., 1986, Calcite precipitation in Lake Constance—Chemical equilibrium, sedimentation, and nucleation by algae: Limnology and Oceanography, v. 31, p. 1081–1093.

Strub, P.T., Allen, J.S., Huyer, A., and Smith, R.L., 1987, Seasonal cycles of currents, temperatures, winds, and sea level over the northeast Pacific continental shelf: 35°N to 48°N: Journal of Geophysical Research, v. 92, p. 1507–1526, doi: 10.1029/JC092iC02p01507.

Talbot, M.R., 1990, A review of the palaeohydrological interpretations of carbon and oxygen isotope ratios in primary lacustrine carbonates: Chemical Geology, v. 80, p. 261–279.

Talbot, M.R., 1994, Paleohydrology of the late Miocene Ridge Basin lake, California: Geological Society of America Bulletin, v. 106, p. 1121–1129, doi: 10.1130/0016-7606(1994)106<1121:POTLMR>2.3.CO;2.

Talbot, M.R., and Kelts, K., 1990, Paleolimnological signatures from carbon and oxygen isotopic ratios in carbonates from organic-rich lacustrine sediments, *in* Katz, B., ed., Lacustrine basin exploration—Case studies and modern analogs: American Association of Petroleum Geologists Memoir 50, p. 99–111.

Thomas, A.C., and Strub, P.T., 1990, Seasonal and interannual variability of pigment concentrations across a California Current frontal zone: Journal of Geophysical Research, v. 95, p. 13,023–13,042, doi: 10.1029/JC095iC08p13023.

Thompson, R.S., 1990, Late Quaternary vegetation and climate in the Great Basin, *in* Betancourt, J.L., Van Devender, T.R., and Martin, P.S., eds., Packrat middens—The last 40,000 years of biotic change: Tucson, University of Arizona Press, p. 200–239.

Thompson, R.S., Benson, L., and Hattori, E.M., 1986, A revised chronology for the last Pleistocene lake cycle in the Lahontan Basin: Quaternary Research, v. 25, p. 1–9, doi: 10.1016/0033-5894(86)90039-6.

Thompson, R.S., Whitlock, C., Bartlein, P.J., Harrison, S.P., and Spaulding, W.G., 1993, Climatic changes in the western United States since 18,000 yr B.P., *in* Wright, H.E., Jr., Kutzbach, J.E., Webb, T., Ruddiman, W.F., Street-Perrott, F.A., and Bartlein, P.J., eds., Global climates since the last glacial maximum: Minneapolis, University of Minnesota Press, p. 468–513.

Trend-Staid, M., and Prell, W.L., 2002, Sea surface temperature at the Last Glacial Maximum: A reconstruction using the modern analog technique: Paleoceanography, v. 17, p. 1065, doi: 10.1029/2000PA000506.

Van Devender, T.R., Thompson, R.S., and Betancourt, J.L., 1987, Vegetation history of the deserts of southwestern North America; The nature and timing of Late Wisconsin-Holocene transition, *in* Ruddiman, W.F., and Wright, H.E., eds., North America and adjacent oceans during the last deglaciation: Boulder, Colorado, Geological Society of America, Geology of North America, v. K-3, p. 323–352.

Wetzel, R.G., 2001, Limnology—Lake and river ecosystems (3rd edition): Academic Press, 1006 p.

Winograd, I.J., Coplen, T.B., Landwehr, J.M., Riggs, A.C., Ludwig, K.R., Szabo, B.J., Kolesar, P.T., and Revesz, K.M., 1992, Continuous 500,000-year climate record from vein calcite in Devils Hole, Nevada: Science, v. 258, p. 255–260, doi: 10.1126/science.258.5080.255.

Wurtsbaugh, W., and Luecke, C., 1997, Examination of the abundance and spatial distribution of forage fish in Bear Lake (Utah/Idaho): Salt Lake City, Final Report of Project F-47-R, Study 5, to the Utah Division of Wildlife Resources, 217 p.

MANUSCRIPT ACCEPTED BY THE SOCIETY 15 SEPTEMBER 2008

The Geological Society of America
Special Paper 450
2009

Ostracode endemism in Bear Lake, Utah and Idaho

Jordon Bright

Department of Geology, Box 4099, Northern Arizona University, Flagstaff, Arizona 86011, USA

ABSTRACT

Bear Lake, Utah and Idaho, is one of only a few lakes worldwide with endemic ostracode species. In most lakes, ostracode species distributions vary systematically with depth, but in Bear Lake, there is a distinct boundary in the abundances of cosmopolitan and endemic valves in surface sediments at ~7 m water depth. This boundary seems to coincide with the depth distribution of endemic fish, indicating a biological rather than environmental control on ostracode species distributions. The cosmopolitan versus endemic ostracode species distribution persisted through time in Bear Lake and in a neighboring wetland.

The endemic ostracode fauna in Bear Lake implies a complex ecosystem that evolved in a hydrologically stable, but not invariant, environmental setting that was long lived. Long-lived (geologic time scale) hydrologic stability implies the lake persisted for hundreds of thousands of years despite climate variability that likely involved times when effective moisture and lake levels were lower than today. The hydrologic budget of the lake is dominated by snowpack meltwater, as it likely was during past climates. The fractured and karstic bedrock in the Bear Lake catchment sustains local stream flow through the dry summer and sustains stream and groundwater flow to the lake during dry years, buffering the lake hydrology from climate variability and providing a stable environment for the evolution of endemic species.

INTRODUCTION

Bear Lake, Utah and Idaho, is an interesting lake for several reasons. It is one of the most long-lived extant lakes in North America, if not *the* most long-lived extant lake on the continent (Bright et al., 2006; Colman, 2006). The limestone- and dolomite-rich watershed surrounding the lake generates an unusual water chemistry within the lake (Dean et al., 2007, this volume). Throughout much of the Holocene (and other relatively arid, Holocene-like climates) Bear Lake has precipitated aragonite as the dominant carbonate mineral (Bright et al., 2006; Dean et al., 2006, Dean, this volume), which is also unusual for a high-altitude, northern temperate lake (Dean et al., 2007). Bear Lake

also contains four endemic fish species (Sigler and Sigler, 1996), which, excluding the Great Lakes basin (Smith, 1981; Smith and Todd, 1984; Reed et al., 1998), is the largest number of endemic fishes in any extant North American lake. Recent studies have suggested that the speciation of the Bear Lake whitefish is a recent event, and may still be occurring today (Vuorinen et al., 1998; Toline et al., 1999; Miller, 2006).

Bear Lake also contains an endemic deep-water ostracode fauna. Cosmopolitan ostracodes are uncommon in the lake and are primarily restricted to the shallowest littoral zone or lake margins, which again is unusual. Lakes normally exhibit an ostracode distribution associated with environmental change along a depth gradient from the lake margin through the littoral zone and into the hypolimnion (e.g., Kitchell and Clark, 1979; Mourguiart and Montenegro, 2002). These biofacies owe their existence to changes in species productivity along the depth gradient. Segments of the gradient may include wetlands, springs, and the upper, middle, sublittoral, and profundal zones. Rather than having a continuum of ostracode species assemblages from the lake margins into the profundal zone, modern Bear Lake exhibits an apparent and relatively abrupt change in the ostracode species assemblage at a water depth of ~7 m. This paper investigates the ostracode distributions, both modern and ancient, within Bear Lake and the surrounding area, and provides two hypotheses to explain those distributions.

BACKGROUND

Bear Lake

Bear Lake is an oligotrophic, alkaline (pH ~8.4; Dean et al., 2007, this volume) lake straddling the Utah-Idaho border (Fig. 1). The lake resides in a half-graben situated between the Bear River Range and the Bear Lake Plateau. The lake's elevation is ~1805 m above sea level and it has a surface area of ~287 km^2, a volume of ~8.1 × 10^9 m^3, and a maximum depth of ~65 m. Geophysical studies show a thick sedimentary sequence below the lake (Denny and Colman, 2003), with sediments possibly as old as 6 Ma (Colman, 2006).

Bear Lake's outline is smooth, without any prominent coves or bays (Smart, 1958). The littoral zone is rocky and exposed to persistent wind and wave action. Rooted plants are able to grow only in the most protected shore areas (Smart, 1958; Lamarra et al., 1986). The deepest area of the lake (~65 m) is on the eastern side between North and South Eden Canyons (Fig. 1). The lake bottom slopes 0.3° to 0.5° toward this deepest point. The bottom of Bear Lake is very flat and regular with distinctive sediment characteristics. Sandy lake-bottom sediments extend out to a water depth of ~12 m. A sand-silt bottom is prevalent between water depths of ~12 and ~30 m, and a silt-marl bottom is prevalent at water depths >~30 m (Smart, 1958).

Local climate is characterized by a mean annual temperature (MAT) near the lake of ~6 °C and a mean annual precipitation (MAP) of ~316 mm (MAT and MAP are long-term averages (16–95 yr) from Lifton, Idaho, Bear Lake State Park, and Laketown, Utah; http://wrcc.sage.dri.edu/summary/climsmid and /climsmut.html). Annual precipitation throughout the region is dominated by snow. Summers are warm and dry. Evaporation rates at Bear Lake are poorly known. Type A pan measurements at Lifton, Idaho, place the evaporation rate near 1000 mm yr^{-1} (www.wrcc.dri.edu/CLIMATEDATA.html), but a more intensive evaporation study at the lake suggests that the rate may be closer to ~600 mm yr^{-1} (Amayreh, 1995).

The catchment area of Bear Lake contains numerous flowing springs, wetlands, small to moderate-sized streams, and the Bear River. The streams and springs on the west side of Bear Lake are cold and dilute (total dissolved solids [TDS] ~250–350 mg

L^{-1}), whereas those on the east side are primarily cold and more saline (TDS ~400–2000 mg L^{-1}) (Bright, this volume). A small hot spring complex (47 °C) discharges at the northeast corner of the lake (Fig. 1). The TDS of Bear River base flow was 550 mg L^{-1} during the summer of 2000.

Since ca. A.D. 1912, flow from the Bear River has been seasonally diverted into Mud Lake and from there into Bear Lake for storage and then withdrawn for irrigation and power generation. The diversion of Bear River water into Bear Lake both reduced the lake's TDS and changed its solute composition by dramatically increasing the alk:Ca ratio (Dean et al., 2007, this volume). Bear Lake's TDS was ~1100 mg L^{-1} ca. 1912 (Birdsey, 1989) and is ~550 mg L^{-1} today. Prior to Bear River diversion, local inflow, including direct precipitation, was sufficient to maintain limited outflow (McConnell et al., 1957). A hydrologic balance model developed by Lamarra et al. (1986) suggests that Bear Lake would have overflowed throughout the 1970s and into the early 1980s if the lake had been left in its natural state.

The Mud Lake wetlands were sustained by local stream flow as well as by spring and diffuse groundwater discharge from the surrounding highlands prior to Bear River diversion. In the past, Mud Lake water may have entered Bear Lake through the sandbar that divides the two water bodies (McConnell et al., 1957). The sandbar was reinforced during the 1912 Bear River diversion and Mud Lake water now enters Bear Lake though an inlet gate near Lifton. Groundwater exchange through the sandbar between Mud Lake and Bear Lake probably still occurs, however.

Ostracode Ecology

Ostracodes are microscopic, aquatic, bivalved crustaceans. Their life cycles depend, in part, on environmental parameters that link species occurrences to climate and hydrology. They are sensitive to three primary environmental factors: (1) chemical hydrology (hydrochemistry), including both the general solute composition, but especially the total carbonate alkalinity (alk) to calcium ratio (alk/Ca), and, secondarily, the total dissolved solids (TDS) (Forester, 1983; Smith, 1993; Curry, 1999); (2) physical hydrology, including the general setting, e.g., lakes, springs, and streams (Curry, 1999), as well as the nature and variability of the setting's water properties (e.g., permanence), and of its ground-, surface-, and atmospheric-water interchange; and (3) water temperature as determined by the annual water-temperature profile derived from the resident air masses and/or by the groundwater flow path(s) (Forester, 1985, 1991a; Roca and Wansard, 1997). A study by Curry (1999) did not show any significant correlation between water temperature and the distributions of several common North American ostracode species, however.

Ostracode occurrences in lakes are primarily linked to climate through their coupled water-air-temperature dependent biogeographic distributions and their effective-moisture driven hydrochemical requirements (Delorme, 1969; DeDeckker, 1981; Forester, 1987) and are secondarily linked to physical hydrology. A lake's physical hydrology can significantly modify the climate

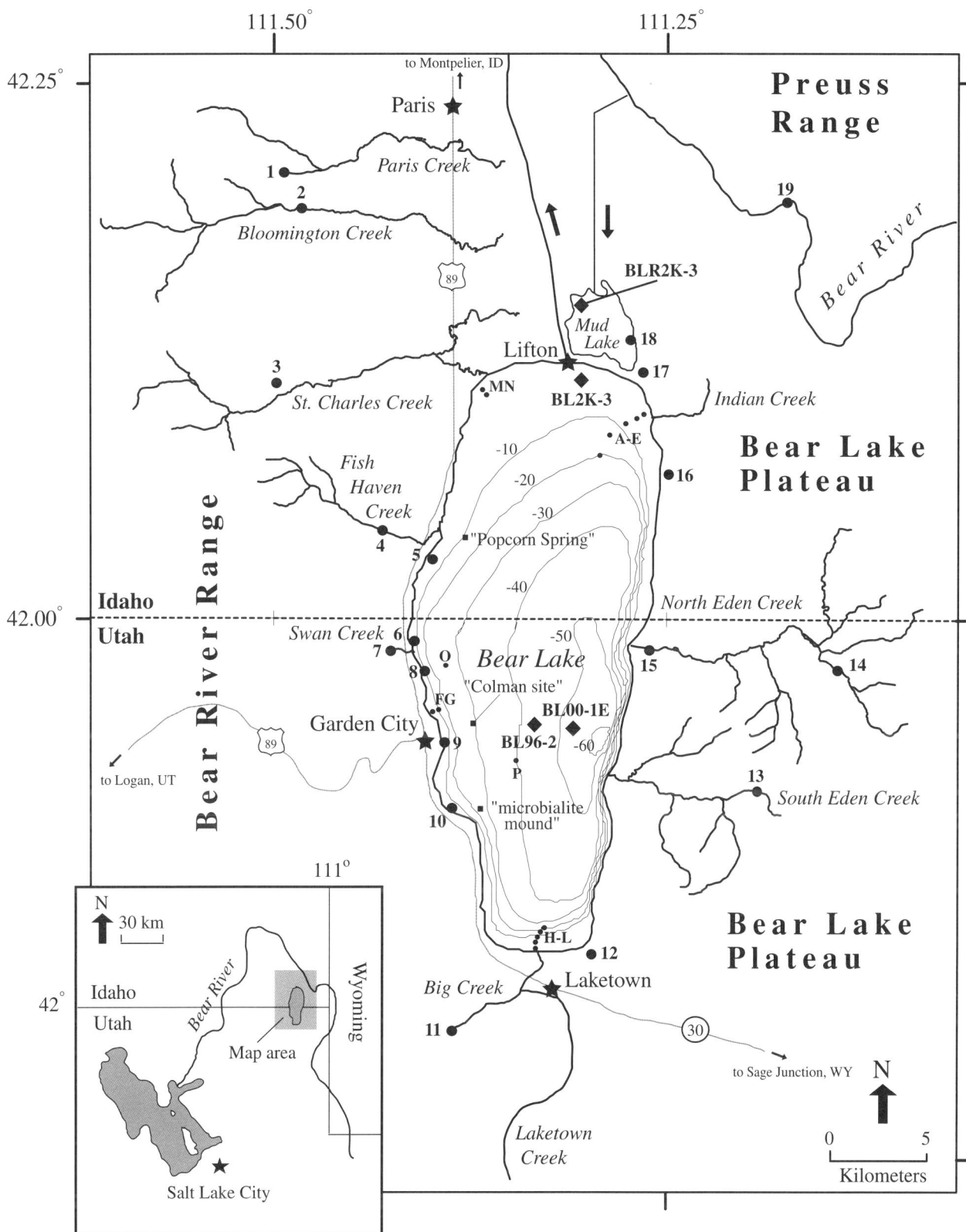

Figure 1. Bear Lake locality map. Bathymetry lines are in 10 m increments. Numbered sites and site P were sampled for ostracodes from 1999 to 2004. Lake sites A–O were sampled for ostracodes in 1999. Core localities labeled BLR2K-3, BL2K-3, BL96-2, and BL00-1E. Sites P, "Colman site", and "Popcorn Spring" are methane seeps. "Microbialite mound" is a probable sublacustrine spring.

signature by, for example, groundwater through-flow or changes in the relative contributions of multiple groundwater sources (Smith et al., 2002a).

MATERIALS AND METHODS

The ostracode data come from numerous grab samples from springs and wetlands around the lake, ~30 Eckman surface grab samples from the lake floor, and four lake sediment cores, BLR2K-3, BL2K-3, BL96-2, and BL00-1 (Fig. 1). The lake-bottom sediment recovered in the Eckman sampler was typically stiff gray mud overlain by a thin (up to a few cm thick) layer of light-tan oozy mud. The tan mud was collected into Ziplock® baggies for further processing. Core BL96-2 was taken with a Kullenberg corer (Kullenberg, 1947), whereas BLR2K-3 and BL2002-4 were taken with a UWITEC piston corer (Mondsee, Austria), with a core diameter of 5 cm. Core BL00-1 was taken with the GLAD800 (Global Lake Drilling to 800 m) drill rig (http//www.dosecc.org/html/glad800.html). Radiocarbon ages for cores BLR2K-3, BL2K-3, and BL96-2 are presented in Colman et al. (this volume). The chronology for core BL00-1 is presented in Colman et al. (2006, 2007) and Kaufman et al. (this volume). Cores BLR2K-3, BL2K-3, and BL96-2 were sampled at ~3–4 cm intervals and each sample was ~1 cm thick. Core BL00-1 was sampled at ~1 m intervals and each sample was ~3 cm thick. All sediment samples were processed to concentrate ostracodes using a standard protocol developed for the calcareous microfossil laboratory at the U.S. Geological Survey in Denver, Colorado (Forester, 1988). The process involved disaggregating the sample by soaking for ~1 week in a weak detergent (Calgon®) solution. Solutions containing stubborn sediments were frozen and thawed repeatedly. Once processed, the sand-sized residue (>150 µm) was size sorted and adult ostracode valves were identified to species when possible and counted. Surface samples were processed following the same protocol and species presence was noted.

RESULTS

Cosmopolitan ostracode species are common and diverse in springs and wetlands throughout North America and are generally well known with an established taxonomy (e.g., Furtos, 1933; Delorme, 1971, 1970a, 1970b, 1970c, 1970d; Forester et al., 2006). Cosmopolitan ostracode species inhabit the springs and wetlands in the Bear Lake catchment (Table 1). These habitats contain *Cypridopsis vidua* and *Limnocythere itasca* in the wetlands, *Candona acuminata* around the orifices of springs, and *Cavernocypris wardi* in cold springs and groundwater settings. As expected, sites with composite hydrologic sources, for example springs discharging into wetlands, have diverse assemblages reflecting the hydrologic heterogeneity.

Cosmopolitan ostracode taxa are also common in the upper littoral zone of Bear Lake. For example, at the Bear Lake Training Center site (site 10; Table 1, Fig. 1), the ostracode species

assemblage is diverse, with 11 species living in a small spring that emerged ~3 m lakeward from the shoreline during recent low lake levels. In 2003, the lake regressed farther and the spring discharge expanded lakeward across the former littoral zone. *Herpetocypris brevicaudata* appeared in the 2003 collections, even though it was absent in prior year collections, and it is now common in several exposed shoreline and former littoral zone springs (Table 1). Small numbers of empty cosmopolitan ostracode valves and empty endemic species valves occur together in lake sediment samples out to ~7 m water depth (Table 2). Valves of cosmopolitan species are also rarely found at deeper depths in the lake, but they all appear to be transported shells (poor preservation, single, often juvenile valves).

Empty valves of the endemic species (Fig. 2) are extremely abundant in the surface-sediment samples at water depths >7 m (Table 2), often constituting most of the sand-sized fraction. The ostracode fauna in the most recent lake-bottom sediments is composed of eight candonid species, two limnocytherid species, and one unidentified taxon (Fig. 2). The unknown taxon is a cyprid, but its relationship to other cyprids is not understood. The endemic ostracode species diversity is highest in deeper water (Table 2). It is unclear if the increased diversity is an accurate representation of live distributions, or if empty ostracode valves have simply been reworked to deeper parts of the lake, however. The lack of modern sediment in water depths <30 m (Colman, 2006; Dean et al., 2006) indicates that sediment reworking is an important process in this lake and likely alters the postmortem distribution of ostracode valves. The rarity or absence of ostracode soft body parts in the modern lake sediment variously implies limited present-day productivity (perhaps due to extinction associated with the Bear River diversion), rapid decomposition of soft parts, and/or endemic ostracode populations that normally have a patchy distribution. The possibility of a patchy distribution is supported by abundant soft-part-bearing ostracode carapaces in a sediment sample taken from a large methane seep at 40 m water depth (site P; Fig. 1). Most of the bottom samples reported here were collected in <20 m water depth in an attempt to find live cosmopolitan ostracodes in the littoral zone. Sampling was limited (*n* = 4) in the 24–38 m depth zone where Smart (1958) reported a peak in ostracode concentrations. With the one exception, no live ostracodes were recovered at the few deep-water sample sites. Live ostracodes were recovered at the 40-m-deep methane seep on the first attempt, however. Other possible methane seep locations within the lake are known and further sampling is necessary to determine if the endemic ostracode distributions are linked to these seeps.

The present-day spatial relation between the cosmopolitan and endemic ostracodes is also evident in the fossil record. The stratigraphic distribution of the most common cosmopolitan ostracodes, those with an abundance of 1% or more, and all occurrences of the endemic ostracodes are shown in Figures 3, 4, 5, and 6 for cores BLR2K-3, BL2K-3, BL96-2, and BL00-1, respectively.

Core BLR2K-3 from Mud Lake (Fig. 1) is dominated by cosmopolitan ostracode taxa (Fig. 3). Approximately 75% of the ostracodes counted in the samples from BLR2K-3 are *Limnocythere itasca* and *Physocypria globula*, and ~15% are *Candona caudata* and *Potamocypris* sp. Core BL2K-3 was taken from the north end of Bear Lake in ~4 m of water (Fig. 1). Ostracode composition fluctuates between being cosmopolitan-dominated and endemic-dominated, with the exception of *Physocypria globula*, which occurs throughout the samples (Fig. 4). Core BL96-2 from the central part of the lake (~40 m water depth; Fig. 1) contains endemic ostracode species only (Fig. 5). Seven of the endemic candonids, the two species of limnocytherids, and the unknown taxon from the modern lake-bottom samples are present (Fig. 5). Seventy-eight percent of the valves in BL96-2 are from just two candonid species (*Candona* sp. 1 and 2; Fig. 2), and 96% of the valves are from four candonids and one limnocytherid species. Core BL00-1, taken in ~50 m of water near the deepest part of

the lake (Fig. 1), contains primarily endemic species. Six of the endemic candonids, one of the limnocytherids, and the unknown taxon are present. The cosmopolitan species *Cytherissa lacustris* is also intermittently present (Fig. 6).

DISCUSSION

Continental surface water (e.g., lakes, wetlands) is ephemeral over geologic time scales. Pluvial lakes of late Pleistocene age in the western and southwestern United States persisted for few tens of thousands of years (e.g., Benson et al., 1990; Lowenstein et al., 1999; Cohen et al., 2000; Garcia and Stokes, 2006). The Great Lakes of the Midwest have been in their current configuration for less than ~18,000 yr, since the retreat of the last

TABLE 1. OSTRACODE SPECIES SURROUNDING BEAR LAKE

Site (Fig. 1)	Lat. (°N)	Long. (°W)	Ostracodes Genus-species
Paris Spring			
1	42.206	111.498	*Cavernocypris wardi, Prionocypris canadensis*
Jarvis Spring			
2	42.191	111.483	*Cavernocypris wardi, Cypria ophthalmica, Candona sigmoides, Strandesia* sp.
Blue Pond Spring			
3	42.105	111.495	*Candona acuminata, Candona sigmoides, Cavernocypris wardi, Cypria ophthalmica, Strandesia deltoidea*
Sadducee Spring			
4	42.051	111.460	*Cavernocypris wardi, Strandesia* sp.
"South Fish Haven" littoral zone spring			
5	42.026	111.402	*Herpetocypris brevicaudata, Ilyocypris bradyi, Cypridopsis vidua, Heterocypris incongruens, Strandesia meadensis, Cavernocypris wardi, Cypridopsis okeechobei*
"Swan Creek north" littoral zone spring			
6	41.985	111.406	*Herpetocypris brevicaudata, Cypridopsis vidua, Ilyocypris bradyi, Limnocythere itasca, Strandesia meadensis, Candona stagnalis, Physocypria* sp.
Swan Creek Spring			
7	42.985	111.427	*Candocyprinotus ovatus*
Littoral zone beach seeps near Swan Creek			
8	41.978	111.402	*Herpetocypris brevicaudata, Ilyocypris bradyi, Candona* sp., *Heterocypris incongruens, Limnocythere itasca, Physocypria* sp., *Cypridopsis vidua*
Spring discharge onto beach at Garden City			
9	41.944	111.390	*Herpetocypris brevicaudata, Ilyocypris bradyi, Candona* sp., *Heterocypris incongruens, Limnocythere itasca, Physocypria* sp., *Cypridopsis vidua*
"Bear Lake Training Center" littoral zone spring			
10	41.915	111.389	*Candona acuminata, Candona stagnalis, Candona caudata, Candona candida, Cavernocypris wardi, Cypridopsis vidua, Physocypria globula, Strandesia meadensis, Potamocypris unicaudata, Limnocythere itasca, Herpetocypris brevicaudata*
Big Spring			
11	41.809	111.389	*Ilyocypris bradyi, Cypria ophthalmica, Cypridopsis okeechobei, Candona acuminata, Candona sigmoides, Candona* sp., *Strandesia* sp.
Falula Spring			
12	41.842	111.302	*Cavernocypris wardi, Candona acuminata, Cypria ophthalmica, Ilyocypris bradyi*
Seep, South Eden Canyon			
13	41.921	111.192	*Cavernocypris wardi, Ilyocypris bradyi, Heterocypris fretensis, Heterocypris incongruens, Candona sigmoides, Candona stagnalis, Candona acuminata, Cypridopsis vidua, Potamocypris* sp., *Strandesia* sp., *Cyclocypris* spp.
Spring, North Eden Canyon			
14	41.997	111.140	*Cavernocypris wardi, Strandesia* sp., *Candona sigmoides*
North Eden Creek			
15	41.986	111.255	*Cavernocypris wardi, Ilyocypris bradyi, Cyclocypris ampla, Candona stagnalis, Candona acuminata, Cypria ophthalmica*
"Cedars and Shade" spring			
16	42.075	111.250	*Cavernocypris wardi, Ilyocypris bradyi, Candona sigmoides*
Mud Lake Hot Spring			
17	42.115	111.264	*Candona compressa, Ilyocypris bradyi, Heterocypris fretensis, Heterocypris incongruens, Darwinula stevensoni, Darwinula* sp.
Mud Lake			
18	42.125	111.264	*Candona acuminata, Candona stagnalis, Candona caudata, Cypridopsis vidua, Physocypria globula, Cyclocypris serena, Heterocypris fretensis, Potamocypris unicaudata, Potamocypris* sp., *Limnocythere itasca*
Bear River at Harer gauge, ID			
19	42.195	111.166	*Candona acuminata, Candona distincta, Candona stagnalis, Candona caudata, Cavernocypris wardi, Cypria ophthalmica, Ilyocypris bradyi, Cypridopsis vidua, Limnocythere inopinata, Limnocythere paraornata, Pelocypris albomaculata, Physocypria globula, Cyclocypris serena, Cyclocypris laevis, Strandesia* sp.

Note: Lat.—latitude; Long.—longitude. Locations in quotations are common names.

TABLE 2. OSTRACODES FROM BEAR LAKE SURFACE SEDIMENT

Site (Fig. 1)	Lat. (°N)	Long. (°W)	Depth (m)	Endemic ostracodes (see Fig. 2) Genus-species	Cosmopolitan ostracodes Genus-species
Bear Lake off Indian Creek					
A	42.094	111.263	1.0	*Candona* sp. 2	*Ilyocypris bradyi*
B	42.094	111.264	4.0	*Candona* sp. 2	*Physocypria* sp.
C	42.090	111.275	7.0	*Candona* spp. 1–7, *Limnocythere* spp. 1 and 2, unidentified genus	*Ilyocypris bradyi*
D	42.087	111.280	14.5	*Candona* spp. 1, 2, and 4, *Limnocythere* spp. 1 and 2	none
E	42.083	111.284	20.0	*Candona* spp. 1–8, *Limnocythere* spp. 1 and 2, unidentified genus	rare *Ilyocypris bradyi*, *Physocypria* sp.
Bear Lake, bay south of marina					
F	N.D.	N.D.	2.0	*Candona* sp. 2, *Limnocythere* sp. 2	*Ilyocypris bradyi*
G	N.D.	N.D.	5.0	*Candona* spp. 1 and 2, *Limnocythere* sp. 2	*Ilyocypris bradyi*, *Cypridopsis vidua*
Bear Lake off Big Creek					
H	41.846	111.337	1.5	none	fragments of ostracodes
I	41.848	111.337	3.5	none	fragments of ostracodes
J	41.852	111.334	6.5	*Candona* spp. 1–4, *Candona* spp. 6–8, *Limnocythere* spp. 1 and 2	*Physocypria* sp., *Cypridopsis vidua*, *Cyclocypris* sp.
K	41.857	111.329	15.0	*Candona* spp. 1–4, *Candona* sp. 8, *Limnocythere* spp. 1 and 2, unidentified genus	none
L	41.860	111.326	25.0	*Candona* spp. 1–5, *Candona* sp. 7, *Limnocythere* spp. 1 and 2, unidentified genus	none
Bear Lake northwest corner					
M	42.105	111.369	2.0	rare *Candona* sp. 2	*Physocypria* sp., *Ilyocypris bradyi*, *Cypridopsis vidua*, *Cyclocypris* sp.
N	42.098	111.362	4.5	*Candona* spp. 1, 2, and 5, *Limnocythere* sp. 2	
Bear Lake east of Swan Creek					
O	N.D.	N.D.	24.0	*Candona* spp. 1–7, *Limnocythere* spp. 1 and 2	none
Sublacustrine methane seep					
P	41.937	111.342	40.0	*Candona* spp. 1–7, *Limnocythere* sp. 1, unidentified genus	none

Note: Lat.—latitude; Long.—longitude; N.D.—no data; all ostracode valves at sites A–O were empty, with preservation ranging from pristine to encrusted. Encrusted valves were more common in shallow water. Live ostracodes were recovered only at site P.

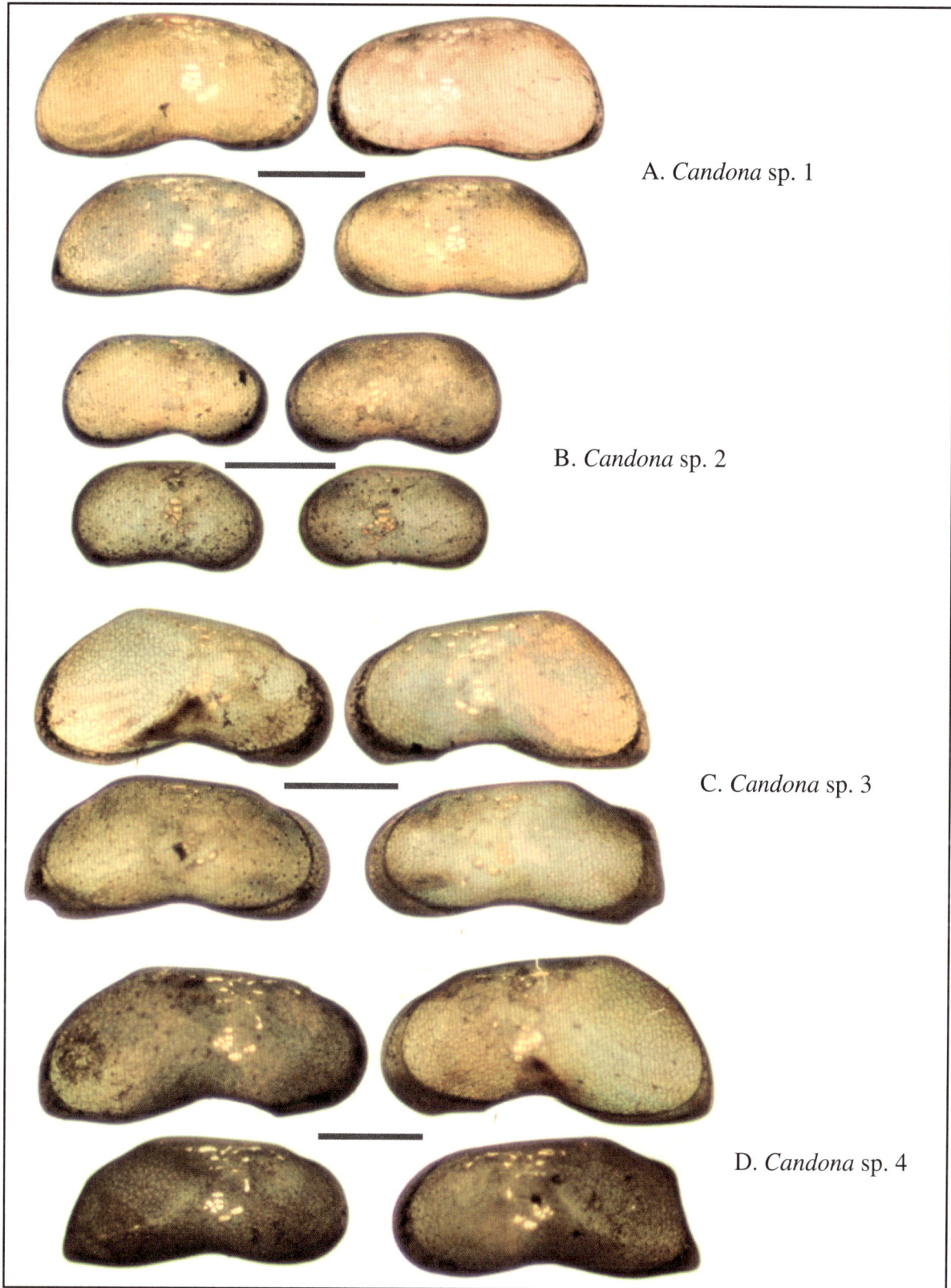

Figure 2 (*continued on following three pages*). Endemic Bear Lake ostracode fauna and two cosmopolitan ostracodes photographed under transmitted light. Groupings A–K are from an Eckman lake-bottom sample taken in ~30 m of water, and are endemic to the lake. Grouping L is from a grab sample taken at Jarvis Spring (Site 2; Fig. 1). Grouping M is from a late Pleistocene aged deposit from Térapa, Sonora, Mexico. Valve placement is identical for each image grouping. Upper left—male right valve; upper right—male left valve; lower left—female right valve; lower right—female left valve. All images are external lateral views taken on a Leica DMLP microscope using a Cannon EOS Rebel XT digital camera. Scale bar in each grouping—0.5 mm.

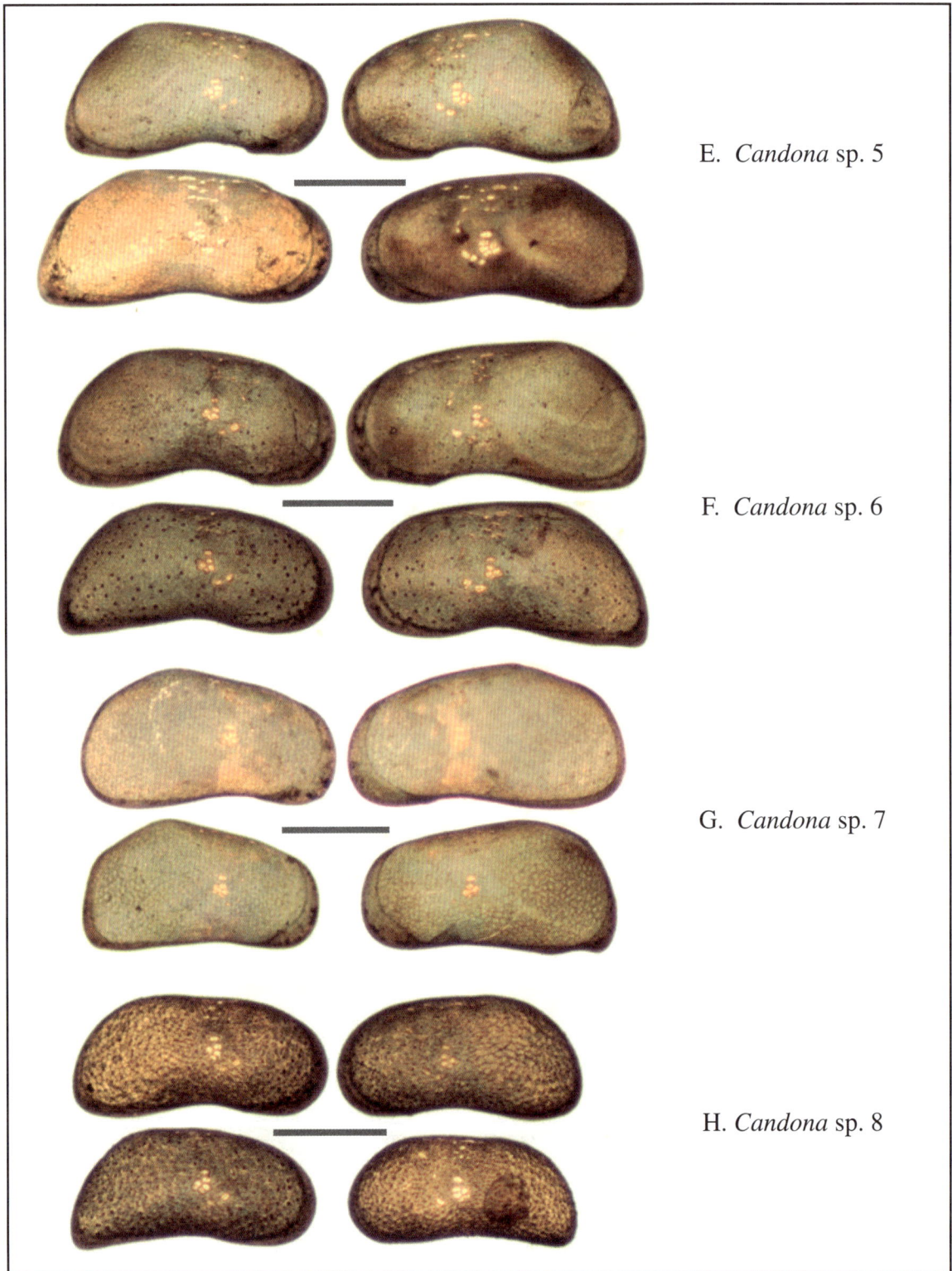

E. *Candona* sp. 5

F. *Candona* sp. 6

G. *Candona* sp. 7

H. *Candona* sp. 8

Figure 2 (*continued*).

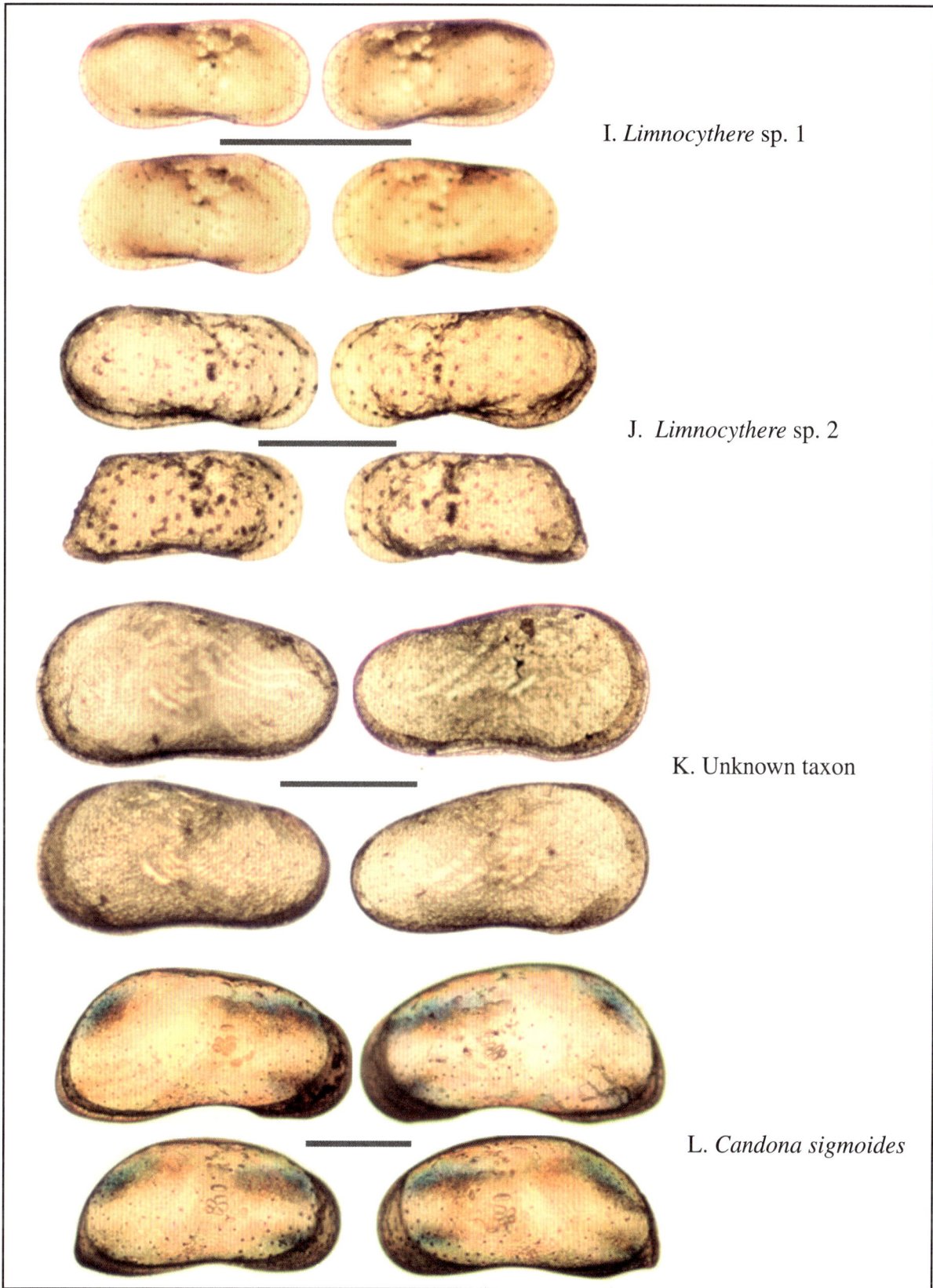

I. *Limnocythere* sp. 1

J. *Limnocythere* sp. 2

K. Unknown taxon

L. *Candona sigmoides*

Figure 2 (*continued*).

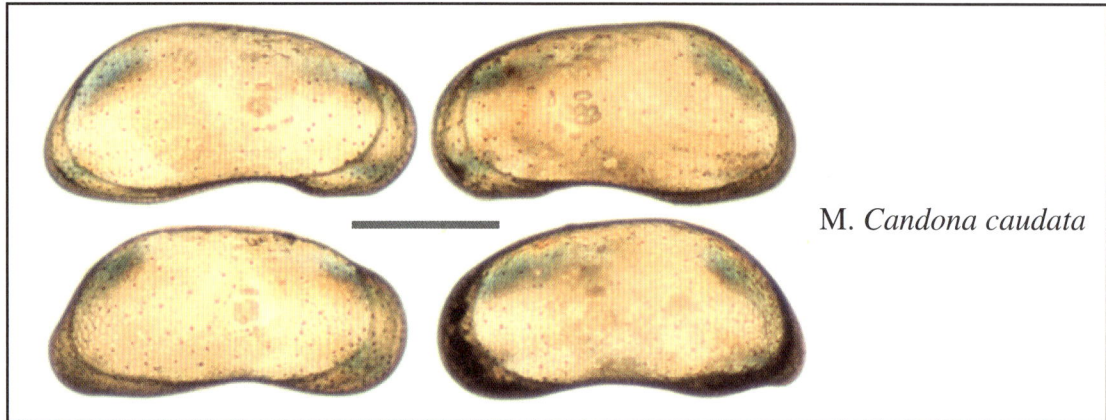

M. *Candona caudata*

Figure 2 (*continued*).

continental glaciers (e.g., Calkin and Feenstra, 1985; Colman et al., 1994; Lewis et al., 1994; Breckenridge, 2007). Groundwater discharge habitats (e.g., springs, wetlands) are susceptible to changes in the elevation of the local water table, which in turn is controlled by wet and dry cycles that operate on annual, decadal, century, and millennial scales (e.g., Quade et al., 1995, 1998; Fritz et al., 2000; St. George and Nielson, 2002). Cosmopolitan ostracodes (and other) species are those that are widespread and have adapted to ephemeral or variable habitats, giving them an advantage over species that have not. Important adaptive traits might include the ability of an individual to withstand desiccation, or the ability to generate large numbers of eggs or offspring that are easily transported from one location to another or that are capable of surviving adverse conditions (e.g., McLay, 1978a, 1978b). Ostracode (and other) species with poor dispersal mechanisms, low reproductive rates, or a limited home range would be more susceptible to extinction during adverse conditions (Cohen and Johnson, 1987). It seems plausible that once evolved, cosmopolitan species may *require* habitat variability in order to trigger key life phases (e.g., mating or egg hatching). For example, the spring thaw may trigger the hatching of *Candona candida* eggs, and seasonal fluctuations in salinity appear to be important in the life cycle of *Candona rawsoni*, because it is not found in lakes or ponds that lack that particular characteristic (Forester, 1987).

At the other extreme of continental aquatic habitats are geologically long-lived lakes. These lakes are unique in that they persist for hundreds of thousands to millions of years and experience large-scale changes in climate, lake chemistries, and lake levels (Frogley et al., 2002). Extant lakes that fall into this category are Lake Baikal, Lake Tanganyika, Lake Malawi, and several others (Frogley et al., 2002). These ancient lakes harbor incredible species diversity and are host to many endemic species (Martens and Schön, 1999).

Leading theories pertaining to the development of endemic species in these lakes include long-term stability (e.g., Cohen and Johnson, 1987), repeated speciation and extinction within the lake system (McCune, 1987), and repeated immigration events (Martens and Schön, 1999). The development of complex ecosystems and associated endemic species may be a time-dependent phenomenon, with a few thousand years of stability likely insufficient to produce endemism (e.g., Forester, 1991b; Wells et al., 1999; Smith et al., 2002b). Stability for many tens to hundreds of thousands of years and gene pool isolation are thought to be necessary to evolve endemic populations (e.g., Martens, 1997). Large pluvial lakes in western North America (Forester, 1987) and the Great Lakes in the Midwest (Forester et al., 1994) were (are) presumably stable environments for a few thousand years, yet, with the exception of Lake Bonneville, did (do) not contain endemic ostracode species. Lake Bonneville is interesting in that several of the fossil ostracode taxa found in its sediments are very similar to those from the profundal zone of Bear Lake (Spencer et al., 1984). Similarly, several of the fossil fish species found in Lake Bonneville sediments are presently living in, and are endemic to, Bear Lake (Sigler and Sigler, 1987, 1996; Broughton, 2000). Bear Lake, with its continuous history (cf. Balch et al., 2005, and Bright et al., 2006) is likely the source of the endemic species (e.g., Miller, 2006).

The necessity of long-term stability in the development of endemism has been countered by debate on the origin of endemic cichlid fish species in Lakes Victoria and Malawi of East Africa, the cisco species flock of the Great Lakes, and the whitefish complex of Bear Lake, Utah and Idaho. Lake Victoria contains over 500 endemic cichlid fish species (Verheyen et al., 2003). Complete desiccation of the lake ~15,000 yr ago implies that the lake's endemic diversity has evolved since that time (Johnson et al., 1996; Stager et al., 2004), although competing theories do exist (Verheyen et al., 2003; Rutaisire et al., 2004). In Lake Malawi, the >200 endemic cichlid fish species that inhabit the rocky shore zone and islands of the lake may have developed very recently, as little as 200–300 yr ago (Owen et al., 1990). In recent history the Great Lakes contained as many as eight endemic cisco (*Coregonus*) species (Smith, 1981; Smith and Todd, 1984; Todd and Smith, 1992). The evolution of the Great Lakes ciscoes is thought to have occurred recently, since the last retreat of continental glaciers from the Great Lakes basins ~15,000 yr ago (Smith, 1981; Bailey and Smith, 1981; Reed et al., 1998). And at Bear Lake, genetic studies on its endemic whitefish suggest a recent divergence between the three whitefish species (Vuorinen et al., 1998; Miller, 2006). Bones of two of the Bear Lake endemic whitefish (*Prosopium gemmifer* [Bonneville cisco] and *P. spilonotus* [Bonneville whitefish]) have been found in the earliest deposits of Lake Bonneville (Stansbury phase; Smith et al., 1968) and are likely ~20,000 yr old (Broughton, 2000), indicating speciation by at least that time. Bones from all three endemic Bear Lake whitefish have been identified from Lake Bonneville sediments that were deposited 10,160–11,270 ^{14}C yr

B.P. (~11,700–13,500 cal yr B.P.; Broughton, 2000), indicating that all three species were established by at least that time. The apparent lack of *Prosopium abyssicola* (Bear Lake whitefish) in the older Lake Bonneville sediments may be due to several factors, the most simplistic being that specimens of *P. abyssicola* have yet to be found (or correctly identified) in the older sediments, or more controversially, that *P. abyssicola* had not yet speciated and arrived later in the Lake Bonneville sequence.

Forester (1991b) suggested that endemic lacustrine ostracodes evolve because biologic selection pressures exceed physical selection pressures in lakes where stable environmental conditions persist over long time intervals. Biologic selection pressures involve

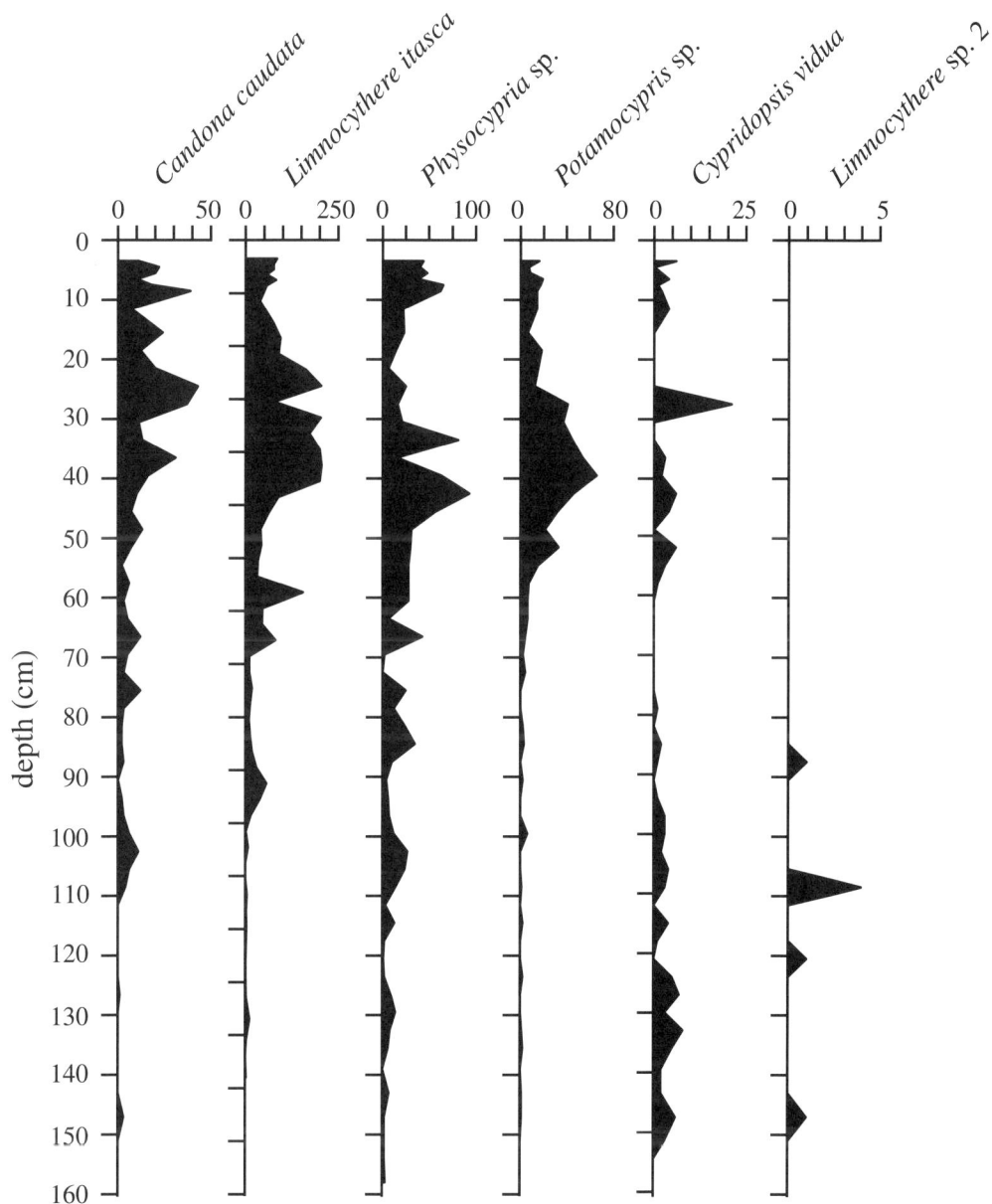

Figure 3. Stratigraphic distribution of common cosmopolitan and all endemic ostracodes in core BLR2K-3. Abundance is reported in valves per gram of sediment.

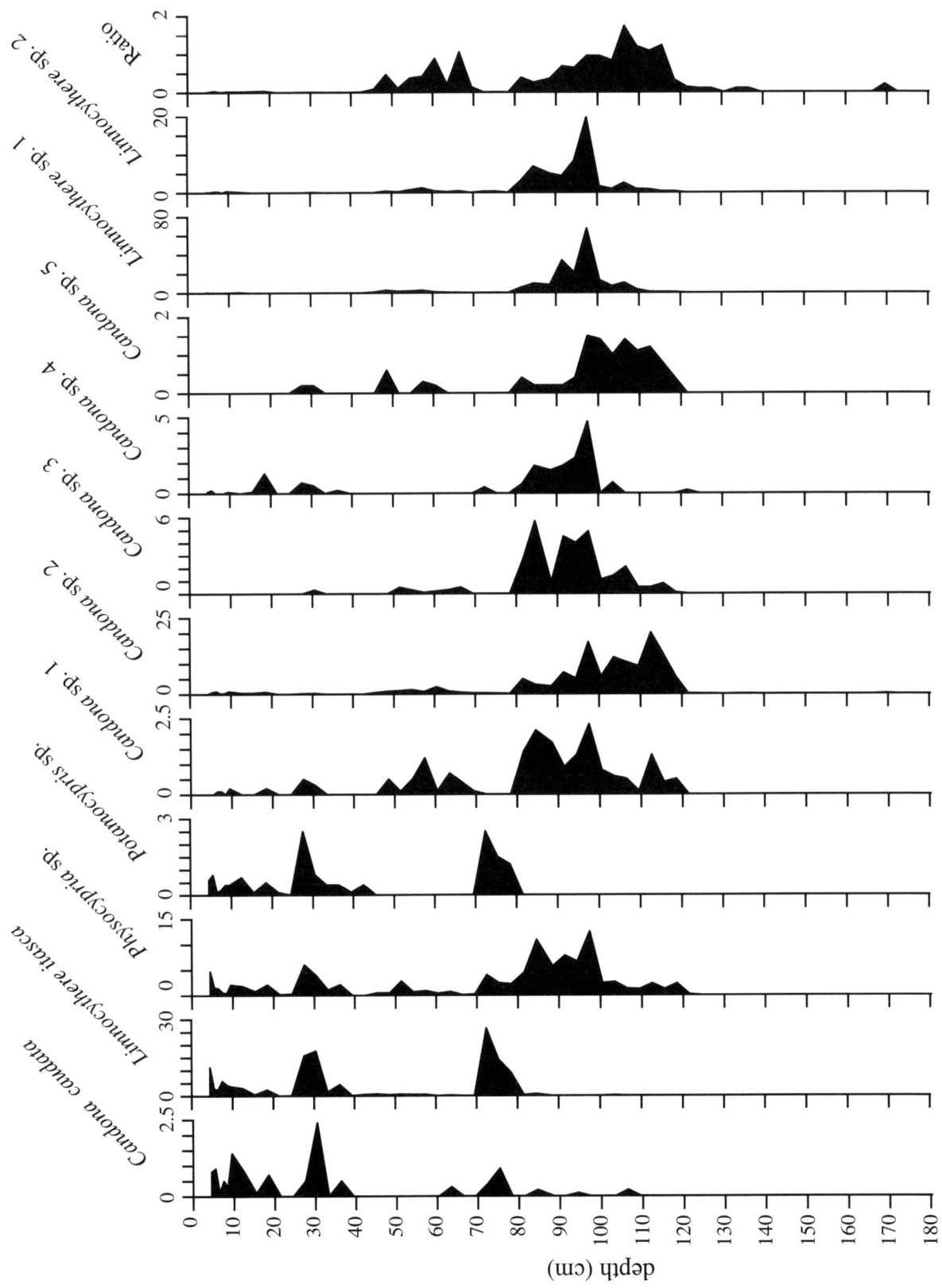

Figure 4. Stratigraphic distribution of common cosmopolitan and all endemic ostracodes in core BL2K-3. Abundance is reported as one-tenth of the actual valves per gram of sediment.

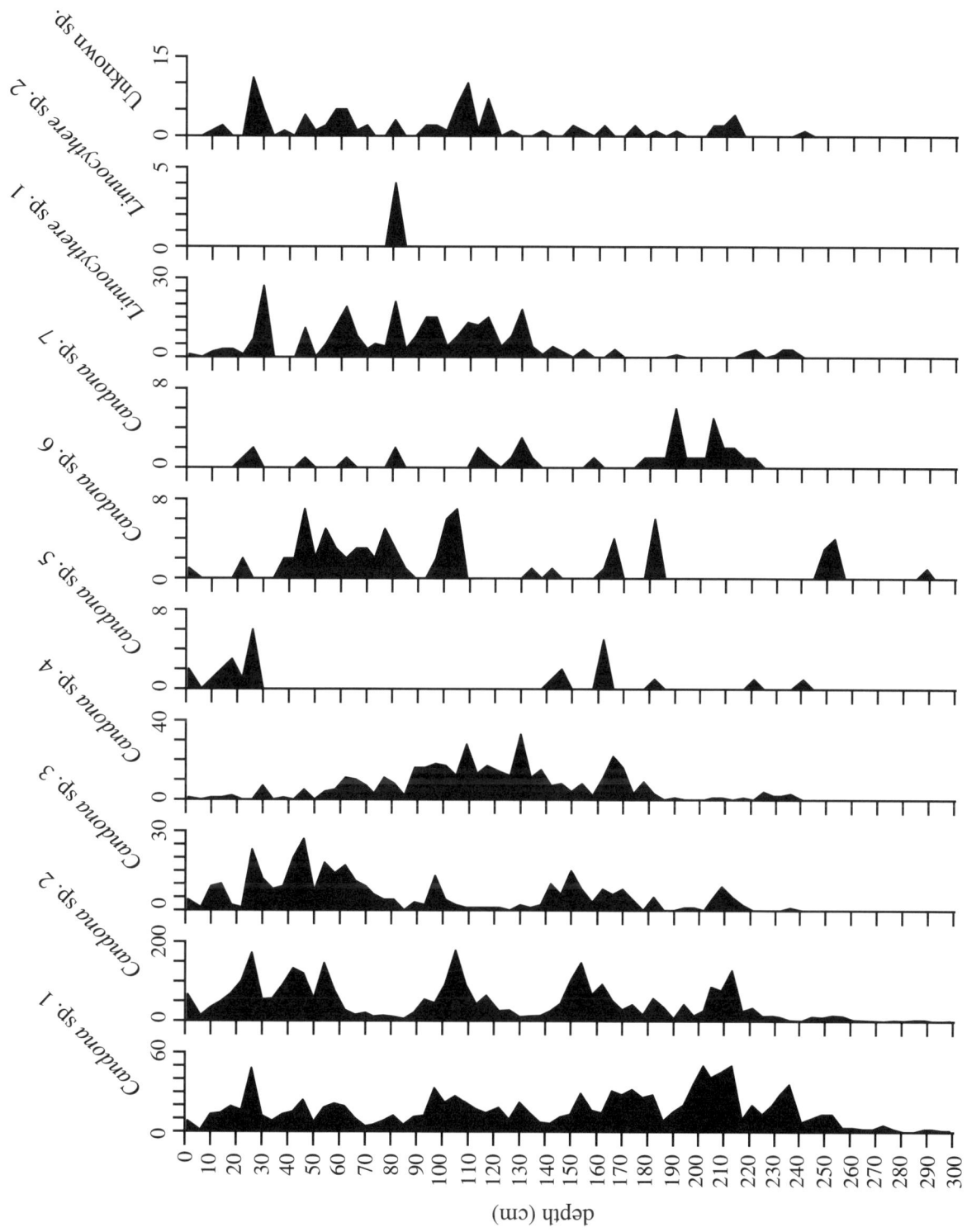

Figure 5. Stratigraphic distribution of endemic ostracodes in core BL96-2; no cosmopolitan ostracodes were found in this core. Abundance is reported in valves per gram of sediment.

adaptive strategies related to surviving inter- and intra-species competition (e.g., reproduction, predation). The development of thick shells or spines and other ornamentation, brooding behavior, and extended parental care are examples of adaptations in response to biologic pressures such as predator-prey relationships and mate recognition (e.g., Martens and Schön, 1999). Physical selection pressures involve adaptive strategies and physiologies suited to changing physical and chemical environments. The development of desiccation-resistant eggs and eggs that hatch at random intervals after production (Angell and Hancock, 1989), short life spans that guarantee several populations and egg clutches per year (e.g., Delorme, 1978, 1982), and the ability for some ostracodes to enter into a state of torpidity when environmental conditions are less than favorable (Delorme and Donald, 1969) are all adaptations to physical selection pressures.

Endemism is presumably favored when biological pressures are at a maximum and when physical selection pressures, such as environmental changes, are relatively invariant and play a secondary role in the survival of the species. Biological and physical selection pressures are interdependent, however. For example, the environmental variability that requires special physical adaptations for success (physical selection) also places limitations on the survival potential for the predators (biological selection) of a given species.

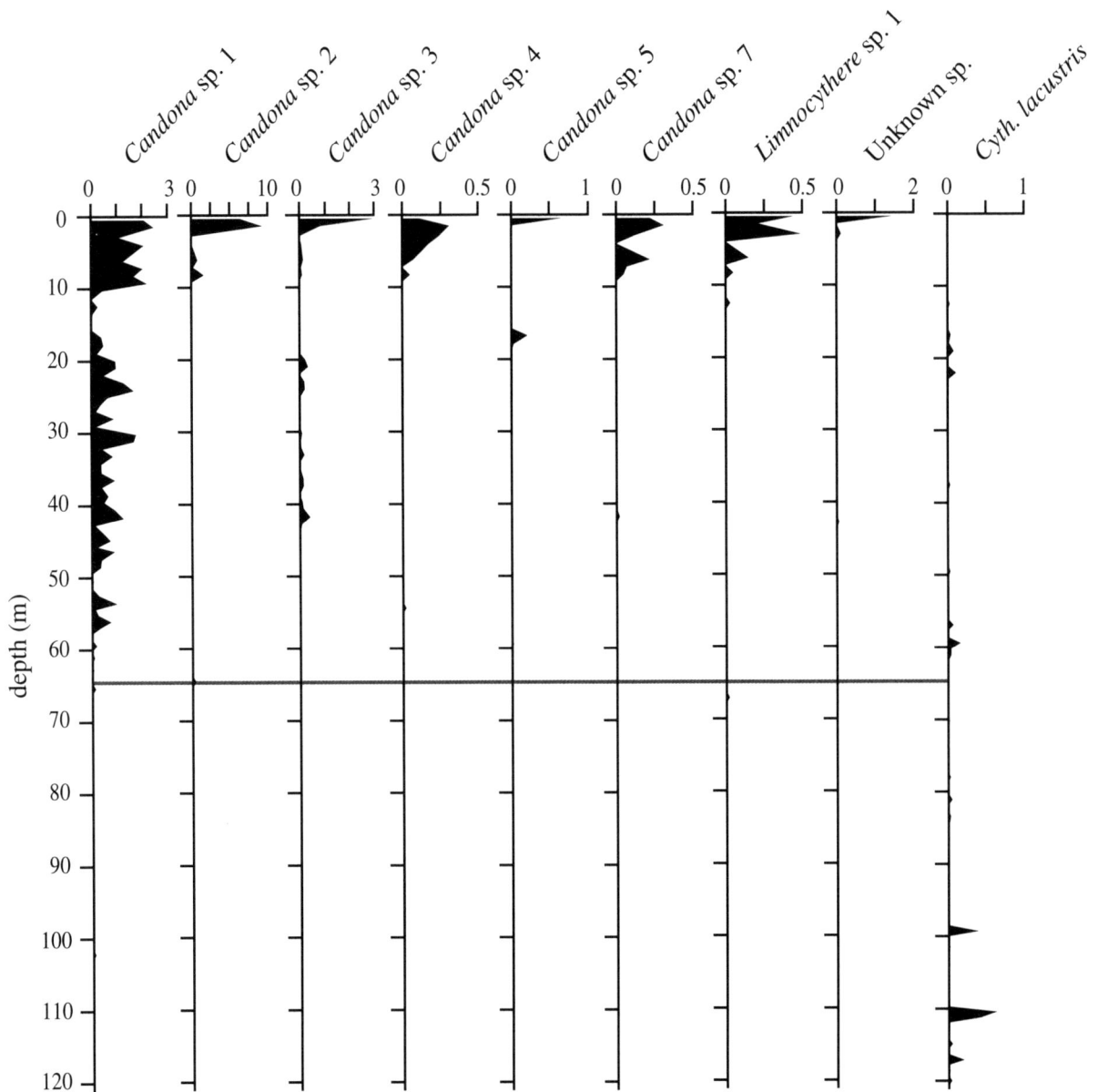

Figure 6. Stratigraphic distribution of endemic ostracodes and *Cytherissa lacustris* in core BL00 1E. Abundance is reported in valves per gram of sediment. Aragonite from ~65 m depth (gray line) generated a U-Th age of ca. 128 ka (Colman et al., 2006, 2007).

An example of an ancient, geologically long lived, lacustrine environment is the Pliocene-aged Glenns Ferry Formation of the western United States. The deep-lake phase of the Glenns Ferry Formation lasted roughly 1 million years (ca. 3.5 to ca. 2.5 Ma; Thompson, 1996; and references therein) and was populated by both endemic and cosmopolitan ostracodes. The endemic ostracodes of the Glenns Ferry Formation resemble contemporaneous cosmopolitan species except for unusual ornamentation such as a large mid-valve spine. Others were highly ornamented relative to typical cosmopolitan continental ostracodes (Forester, 1991b). The ornamented Glenns Ferry Formation endemic taxa populated only the deep lake sediment, whereas cosmopolitan taxa were common in all marginal hydrologic settings. A similar cosmopolitan-endemic ostracode distribution occurs at Bear Lake.

Ostracode Distributions in Bear Lake

Bear Lake is an oligotrophic lake (Wurtsbaugh and Hawkins, 1990). Mean annual whole-lake benthic invertebrate biomass estimates are extremely low, only 0.34 g dry weight per square meter (g dry wt m^{-2}), which is the lowest value recorded for a temperate-zone lake (Wurtsbaugh and Hawkins, 1990). Invertebrate biomass estimates for the more productive littoral zone are only slightly higher, but are always less than 1 g dry wt m^{-2} (Wurtsbaugh and Hawkins, 1990). Ostracodes, although rare, do constitute a large part of the littoral zone biomass, however (Wurtsbaugh and Hawkins, 1990). The ostracodes reported by Wurtsbaugh and Hawkins (1990) were not identified to species, so it is unknown whether the ostracodes were cosmopolitan or endemic species. Smart (1958) conducted an extensive biologic survey of the bottom sediments of Bear Lake and reported that ostracodes (simply reported as "*Candona* species") were most abundant in water depths of 24–38 m. The decrease in ostracode abundance at depths greater than ~38 m may be an apparent decrease resulting from sediment resuspension and focusing in the deeper portions of the lake, however (Colman, 2006; Dean et al., 2006). The "*Candona* species" reported by Smart (1958) were not identified to species level and here are assumed to represent the endemic, candonid-rich, profundal assemblage.

It is unclear why cosmopolitan ostracode valves are not found in water depths greater than ~7 m. Live cosmopolitan ostracodes have been clearly identified only from lake marginal springs and the marsh complexes that have developed on the exposed lake floor during the recent low-water years (Table 2). Live endemic ostracodes, and identified as such, have been recovered only from one deep-water methane seep, although they are apparently abundant at water depths of 24–38 m (Smart, 1958). One hypothesis is that cosmopolitan ostracodes do not actively live in the lake at any depth and the lake is inhabited entirely by the endemic fauna. In this scenario the cosmopolitan ostracode valves found in Bear Lake's littoral zone are reworked from the springs along the lake margins and from marsh sediments that were deposited on the exposed lake floor during previous low lake levels. An alternative hypothesis is that cosmopolitan ostracodes do inhabit the littoral zone of the lake, but only to a water depth less than 7 m (with empty valves likely being reworked out to ~7 m depth). In either scenario cosmopolitan ostracodes are unable to inhabit the deeper parts of the lake.

The fish distributions within Bear Lake may partially explain the cosmopolitan-endemic ostracode distributions in the lake. Fish densities in Bear Lake are low in water <5 m deep, with native Utah chub and Utah suckers being most numerous throughout the year and endemic juvenile Bear Lake sculpin (*Cottus extensus*) being numerous only during the summer (Wurtsbaugh and Hawkins, 1990). Bear Lake sculpin and Bear Lake whitefish (*Prosopium abyssicola*) are most abundant in water depths >30 m, with adult Bear Lake sculpin abundances increasing in shallower water (~15 m) during the summer. Bear Lake whitefish (*P. spilonotus*) abundances are highest in water depths of ~15–30 m. (Wurtsbaugh and Hawkins, 1990; Kennedy et al., 2006).

Analyses of the gut contents of Bear Lake fish indicate that ostracodes (unidentified taxa) are a large part of the diet of several fish species. Ostracodes constitute 59%–99% of the diet of Bear Lake whitefish and Bear Lake sculpin, both of which are endemic to the lake. The next two most active ostracode predators are juvenile Bonneville whitefish and the Utah sucker (Table 3). The Bonneville whitefish is also endemic to the lake, but the Utah sucker is not (Wurtsbaugh and Hawkins, 1990; Thompson, 2003; Tolentino and Thompson, 2004; Kennedy, 2005; Kennedy et al., 2006). Because the ostracodes from the fish gut analyses were not identified to species, however, the relative abundance of endemic or cosmopolitan ostracode species consumed by the various fish species is unknown. Bear Lake whitefish are substantially more numerous in water depths greater than ~40 m (Kennedy et al., 2006), which implies that the ostracodes they consume are exclusively endemics.

The ecology of the endemic ostracode fauna must involve adaptive strategies that allow them to survive in spite of the oligotrophic nature of the lake and the heavy predation pressure of the endemic fish. One theoretical possibility is that the endemic ostracodes exploit the fish predation and, for example, use the endemic fish as a dispersal mechanism (e.g., Kornicker and Sohn, 1971; Mellors, 1975; Vinyard, 1979; Smith, 1985; Bartholmé et al., 2005). Finding new food resources in a large oligotrophic lake, especially if that food resource has a patchy distribution, would be key to survival. Fish are highly mobile and migrate over large areas of the lake. Exploiting that mobility (actively or passively) would allow the endemic ostracodes (or their eggs) to be dispersed throughout the lake, and from food patch to food patch. Maximum endemic ostracode and fish densities overlap in water depths of ~20–38 m. If fish are a dispersal mechanism for the endemic ostracodes, then it is reasonable to expect that endemic ostracode densities would coincide with high fish densities. Cosmopolitan ostracode species attempting to invade this endemic ecosystem may be unable to cope with the intense predation pressure, or might not have a reproductive strategy compatible with the endemic fish behavior, and would be restricted to lake marginal springs or the shallowest portions of the littoral zone (<5 m deep) where fish densities are lowest.

The morphology of Bear Lake is likely another factor that contributes to the confinement of cosmopolitan ostracodes to the shallow littoral zone. The large expanses of gently sloping lake bottom means that any fluctuation in lake levels would expose and flood large expanses of Bear Lake's littoral zone. For example, a 4 m decrease in lake level (from full lake) would expose ~30 km² of lake bottom. Cosmopolitan ostracodes adapted to variable habitats (strong colonizers) would likely be restricted to this zone (in the emergent marshes). The endemic deep-water ostracode fauna are apparently unable to exploit the shallow littoral zone and are restricted to the deeper parts of the lake where the impacts of lake-level fluctuations are minimal to nonexistent. In addition to the fish predation pressure, cosmopolitan ostracode species may not be able to invade the deeper parts of the lake due to its comparative stability. As a result of these variables, and probably others, the cosmopolitan-endemic ostracode distribution in Bear Lake mirrors that of the Glenns Ferry Formation.

Ostracode Distribution in Bear Lake Sediment Cores

The same species assemblage distinctions noted in the modern lake sediment also existed in the past. Core BLR2K-3 from Mud Lake contains sediments deposited over the past ~8000 yr, but primarily over the past ~2000 yr (Colman et al., this volume), that are dominated by cosmopolitan ostracodes (Fig. 3). These taxa suggest that Mud Lake was principally a shallow lake and wetland, similar to today. The chemical composition of Mud Lake prior to the 1912 Bear River diversion is unknown, but the near absence of endemic ostracodes in the Mud Lake core indi-

cates that it has been incompatible with their requirements for the last several thousand years.

Core BL2K-3 contains sediments deposited at the north end of Bear Lake over the past 3500 yr (Colman et al., this volume). Radiocarbon ages from this core are not in stratigraphic order, suggesting sediment reworking or complications with some of the radiocarbon samples. The problematic radiocarbon data limit the usefulness of this core, but several trends in the ostracode data are apparent (Fig. 4). First, the antithetic behavior of the endemic and cosmopolitan ostracode faunas at this site suggests a species distribution pattern that was similar to the present. Second, any given sediment sample from this core was dominated by either cosmopolitan or endemic ostracodes, but never both, with the exception of *Physocypria* sp. And third, there does not appear to be a faunal gradation between the cosmopolitan and endemic dominated samples. The difference between the two sample types is distinct. The cosmopolitan-rich zones may represent times when the core site was situated in less than ~7 m of water and endemic fish predation was minimal or nonexistent.

Core BL96-2 contains entirely endemic-dominated ostracode assemblages (Fig. 5) from sediments that were deposited over the last ~26,000 yr (Colman et al., this volume). The absence of cosmopolitan taxa in the deep-water setting is atypical of lakes throughout North America with oxygenated hypolimnia. The lack of cosmopolitan species indicates that the lake depth at the core site has never been shallower than ~5 m. This suggests that Bear Lake has never dried out, or never exceeded the environmental tolerances of the endemic ostracode species from the late Wisconsin to the present day.

Core BL00-1 contains sediments that were deposited near the deepest part of the lake over the last 220,000 yr (Kaufman et al., this volume). With one exception, core BL00-1 contains

TABLE 3. OSTRACODE CONSUMPTION BY BEAR LAKE FISH SPECIES

Fish species (size, season)	% Diet	Reference
Bear Lake whitefish (100–150 mm, spring)	99	Thompson, 2003
Bear Lake whitefish (150–200 mm, spring)	90	Thompson, 2003
Bear Lake whitefish (200–250 mm, spring)	73	Thompson, 2003
Bear Lake whitefish (>250 mm, spring)	75	Thompson, 2003
Bear Lake whitefish (100–150 mm, summer)	59	Thompson, 2003
Bear Lake whitefish (150–200 mm, summer)	75	Thompson, 2003
Bear Lake whitefish (200–250 mm, summer)	83	Thompson, 2003
Bear Lake whitefish (>250 mm, summer)	99	Thompson, 2003
Bear Lake sculpin (small)	60	Wurtsbaugh and Hawkins, 1990
Bear Lake sculpin (medium)	68	Wurtsbaugh and Hawkins, 1990
Bear Lake sculpin (large)	73	Wurtsbaugh and Hawkins, 1990
Bonneville whitefish (100–150 mm, spring)	33	Thompson, 2003
Bonneville whitefish (150–200 mm, spring)	14	Thompson, 2003
Bonneville whitefish (200–250 mm, spring)	3	Thompson, 2003
Bonneville whitefish (250–300 mm, spring)	1	Thompson, 2003
Bonneville whitefish (300–350 mm, spring)	3	Thompson, 2003
Bonneville whitefish (>350 mm, spring)	0	Thompson, 2003
Bonneville whitefish (all sizes, summer)	0	Thompson, 2003
Utah sucker (small)	19	Wurtsbaugh and Hawkins, 1990
Utah sucker (large)	38	Wurtsbaugh and Hawkins, 1990
Utah chub	16	Wurtsbaugh and Hawkins, 1990
Carp	6	Wurtsbaugh and Hawkins, 1990
Dace	0	Wurtsbaugh and Hawkins, 1990
Redsides	0	Wurtsbaugh and Hawkins, 1990

Note: Bear Lake whitefish, Bonneville whitefish, and Bear Lake sculpin are endemic to Bear Lake.

entirely endemic species (Fig. 6). Some of the endemic species, but not all (primarily *Candona* sp. 1; Fig. 2A), are found in sediments as old as ca. 130 ka (MIS 5). The core extends through a second glacial-interglacial cycle (to MIS 7), but few whole ostracodes are preserved. Identifiable fragments of *Candona* sp. 1 (Fig. 2A) exist to the bottom of the core, however.

Endemic ostracode species diversity is greatest in the Holocene aragonitic sediments in the upper 10 m of core BL00-1E and in the upper 3 m of core BL96-2 (Figs. 5 and 6). It is unknown whether the higher endemic ostracode diversity in the Holocene sediment is an example of relatively recent ostracode speciation (younger than 15 ka) or whether the lower species diversity in pre-Holocene-aged sediment is due to a preservational bias or some other mechanism.

Cause of Endemism in Bear Lake

The Bear Lake ostracode assemblage contains 11 endemic species, although some of the candonids are morphologically similar to local cosmopolitan taxa and could be ecophenotypes or subspecies of local or other cosmopolitan species (compare Fig. 2A and Figs. 2L and 2M, and compare Fig. 2C with *Candona mendotaensis* from Kitchell and Clark, 1979). The profundal zone of Bear Lake supports more ostracode species than is observed in other large North American lakes. For example, Lake Michigan contains only two ostracode species in the deepest parts of the lake (91–244 m; Buckley, 1975), and rarely more throughout the Holocene (Forester et al., 1994). The endemic Bear Lake ostracodes also exhibit some unusual morphologic features, such as hooked caudal processes, sharp valve margins, and unusual ornamentation (Figs. 2A, 2C, and 2G). However, the morphology and species diversity is less than in the Glenns Ferry Formation sediments (Forester, 1991b) and in other ancient extant lakes such as Lake Baikal or Lake Tanganyika (Mazepova, 1994; Park and Downing, 2000). Bear Lake also supports four species of endemic fish (Sigler and Sigler, 1987, 1996) and perhaps other taxa, but overall the lake does not have a large number of species as is observed in other lakes with endemic faunas. The reduced morphological expression and lower endemic diversity at Bear Lake may result from its oligotrophic state and possibly less complex and variable ecosystems than in other larger endemic-containing lakes (e.g., Lake Baikal or Lake Tanganyika). The reduced endemic diversity at Bear Lake may also be due to its young age and small size relative to lakes like Baikal or Tanganyika.

Bear Lake has been able to produce endemic (ostracode) faunas because of its unique hydrology and chemistry. A possible paleoshoreline is presently submerged under ~22 m of water (Colman, 2006), suggesting that Bear Lake periodically experiences substantial decreases in lake level, decreases that may or may not be climate related (Smoot and Rosenbaum, this volume). Intuitively, Bear Lake should become saline or possibly even go dry during periods of low effective moisture (e.g., the Great Salt Lake; Oviatt et al., 1999; Balch et al., 2005), resulting in the extinction of the endemic fauna. There are no evaporite deposits or any known high-salinity-tolerant ostracode species in core BL2K-3, BL96-2, or the 120-m-long core to suggest elevated lake TDS during times of maximum aridity (e.g., the middle Holocene, OIS 5e; Dean et al., 2006). Bear Lake may never develop a high TDS water mass during arid climates because the majority of local solute inflow is calcium, magnesium, and bicarbonate (Dean et al., 2007; Bright, this volume). Carbonate precipitation (as aragonite or calcite) removes a large portion of the calcium and bicarbonate from the water column. The degree and rate of solute enrichment within the lake are then limited because the majority of the solute load is lost to carbonate precipitation. Groundwater leakage from the lake removes solutes as well. Additionally, the oxygen isotope ($\delta^{18}O$) values on bulk sediment from core BL96-2 are lower during the arid middle to late Holocene (ca. 3–5 ka) than during the latest Holocene (younger than 3 ka; Dean et al., 2006). Isotopic depletion during arid climates has been noted in other settings where groundwater is a large component of a lake's hydrologic budget (e.g., Smith et al., 1997; 2002a). A large groundwater influx derived predominantly from the Bear River Range (Bright, this volume) sustains the lake through arid climates and provides a stable and persistent habitat for the biota inhabiting the lake.

Further Research

The ostracode faunas and distribution patterns in Bear Lake provide several intriguing avenues for additional research. Further research should focus on the following: (1) Determining if, or to what extent, the methane seeps within the lake contribute to the health of the endemic ostracode fauna. Ostracodes are a key component in the diets of three of Bear Lake's endemic fish and are an integral part of the lake's food chain. The methane seeps may provide a food source or a refuge for the endemic ostracodes during times of unfavorable chemical and physical conditions. (2) Conducting a thorough sampling of the modern lake sediments, with the ostracodes identified to species level, noting which species are currently living in the lake and their distributions. (3) Identifying which species of ostracodes are being consumed by the various fish species in Bear Lake and determining if there are seasonal or spatial variations in those predation patterns. (4) Documenting soft-part anatomy and studying the genetic material of the endemic ostracodes to determine their evolutionary relationships. Do they represent a recent radiation, or are they more ancient lineages? Some of the species may be ecophenotypes or subspecies of extant local cosmopolitan forms (Fig. 2).

CONCLUSIONS

Bear Lake is contained in a tectonically active basin with a sedimentary record that may extend back nearly 6 m.y. As many as 11 species of endemic ostracodes inhabit Bear Lake, yet cosmopolitan ostracodes have been unable to successfully colonize the lake. The evolutionary reason for the genesis of the endemic ostracodes in Bear Lake is unknown, but may be related to a combination of environmental stability, fish predation pressure, and the numerous isolated methane seeps within the lake. Extensive fish predation may also play a key role in the endemic-cosmopolitan ostracode distribution in the lake. Bear Lake contains one of the most diverse endemic ecosystems in North America, in spite of its relatively small size and oligotrophic nature. The endemic ostracodes suggest that, in addition to being a long-lived lake, Bear Lake has an environment that has remained relatively stable in spite of large fluctuations in climate, lake chemistry, and lake level. Groundwater that discharges in Bear River Range streams plays a key role in the modern hydrologic balance of Bear Lake and likely has done so through most of its existence. This persistent groundwater variable buffers the lake, to some extent, from the effects of climate change and generates a permanent lacustrine habitat suitable for the generation of endemic species.

ACKNOWLEDGMENTS

Funding for this study was provided by the U.S. Geological Survey's Earth Surface Dynamics Program. Countless conversations with Rick Forester (retired) and Scott Tolentino (Utah Division of Wildlife Resources, Bear Lake Station) have been greatly appreciated. A conversation with Dennis Shiozawa (Brigham Young University) on the development of the endemic Bear Lake whitefish complex was appreciated. Scott Tolentino was instrumental in locating and sampling the sublacustrine methane seep and lake-marginal springs. Reviews by Alison Smith and Brandon Curry contributed greatly to the improvement of this paper.

REFERENCES CITED

Amayreh, J., 1995, Lake evaporation: A model study [Ph.D. thesis]: Logan, Utah State University, 178 p.

Angell, R.W., and Hancock, J.W., 1989, Response of eggs of *Heterocypris incongruens* (Ostracoda) to experimental stress: Journal of Crustacean Biology, v. 9, p. 381–386, doi: 10.2307/1548561.

Bailey, R.M., and Smith, G.R., 1981, Origin and geography of the fish fauna of the Laurentian Great Lakes basin: Canadian Journal of Fisheries and Aquatic Sciences, v. 38, p. 1539–1561, doi: 10.1139/f81-206.

Balch, D.P., Cohen, A.S., Schnurrenberger, D.W., Haskell, B.J., Valero Garces, B.L., Beck, J.W., Cheng, H., and Edwards, R.L., 2005, Ecosystem and paleohydrological response to Quaternary climate change in the Bonneville Basin, Utah: Palaeogeography, Palaeoclimatology, Palaeoecology, v. 221, p. 99–122, doi: 10.1016/j.palaeo.2005.01.013.

Bartholmé, S., Samchyshyna, L., Santer, B., and Lampert, W., 2005, Subitaneous eggs of freshwater copepods pass through fish guts: Survivability, hatchability, and potential ecological implications: Limnology and Oceanography, v. 50, p. 923–929.

Benson, L.V., Currey, D.R., Dorn, R.I., Lajoie, K.R., Oviatt, C.G., Robinson, S.W., Smith, G.I., and Stine, S., 1990, Chronology of expansion and contraction of four Great Basin lake systems during the past 35,000 years: Palaeogeography, Palaeoclimatology, Palaeoecology, v. 78, p. 241–286, doi: 10.1016/0031-0182(90)90217-U.

Birdsey, P.W., Jr., 1989, The limnology of Bear Lake: A literature review: Salt Lake City, Utah Department of Natural Resources, Publication no. 89-5, 113 p.

Breckenridge, A., 2007, Lake Superior varve stratigraphy and implications for eastern Lake Agassiz outflow from 10,700 to 8900 cal ybp (9.5–8.0 ¹⁴C ka): Palaeogeography, Palaeoclimatology, Palaeoecology, v. 246, p. 45–61, doi: 10.1016/j.palaeo.2006.10.026.

Bright, J., 2009, this volume, Isotope and major-ion chemistry of groundwater in Bear Lake Valley, Utah and Idaho, with emphasis on the Bear River Range, *in* Rosenbaum, J.G., and Kaufman, D.S., eds., Paleoenvironments of Bear Lake, Utah and Idaho, and its catchment: Geological Society of America Special Paper 450, doi: 10.1130/2009.2450(04).

Bright, J., Kaufman, D., Forester, R., and Dean, W., 2006, A continuous 250,000 yr record of oxygen and carbon isotopes in ostracode and bulk-sediment carbonate from Bear Lake, Utah-Idaho: Quaternary Science Reviews, v. 25, p. 2258–2270, doi: 10.1016/j.quascirev.2005.12.011.

Broughton, J.M., 2000, Terminal Pleistocene fish remains from Homestead Cave, Utah, and implications for fish biogeography in the Bonneville Basin: Copeia, v. 2000, p. 645–656, doi: 10.1643/0045-8511(2000)000[0645:TPFRFH]2.0.CO;2.

Buckley, S.B., 1975, Study of post-Pleistocene ostracod distribution in the soft sediments of southern Lake Michigan [Ph.D. thesis]: Urbana-Champaign, University of Illinois, 293 p.

Calkin, P.E., and Feenstra, B.H., 1985, Evolution of the Erie-Basin Great Lakes, *in* Karrwo, P.F., and Calkin, P.E., eds., Quaternary evolution of the Great Lakes: St. John's, Newfoundland, Geological Association of Canada Special Publication 30, p. 149–170.

Cohen, A.S., and Johnson, M.R., 1987, Speciation in brooding and poorly dispersing lacustrine organisms: Palaios, v. 2, p. 426–435, doi: 10.2307/3514614.

Cohen, A.S., Palacios-Fest, M.R., Negrini, R.M., Wigand, P.E., and Erbes, D.B., 2000, A paleoclimate record for the past 250,000 years from Summer Lake, Oregon, USA: II. Sedimentology, paleontology and geochemistry: Journal of Paleolimnology, v. 24, p. 151–182, doi: 10.1023/A:1008165326401.

Colman, S.M., 2006, Acoustic stratigraphy of Bear Lake, Utah-Idaho—Late Quaternary sedimentation patterns in a simple half-graben: Sedimentary Geology, v. 185, p. 113–125, doi: 10.1016/j.sedgeo.2005.11.022.

Colman, S.M., Clark, J.A., Clayton, L., Hansel, A.K., and Larsen, C.E., 1994, Deglaciation, lake levels, and meltwater discharge in the Lake Michigan basin: Quaternary Science Reviews, v. 13, p. 879–890, doi: 10.1016/0277-3791(94)90007-8.

Colman, S.M., Kaufman, D.S., Bright, J., Heil, C., King, J.W., Dean, W.E., Rosenbaum, J.G., Forester, R.M., Bischoff, J.L., Perkins, M., and McGeehin, J.P., 2006, Age model for a continuous, ca. 250-ky Quaternary lacustrine record from Bear Lake, Utah-Idaho: Quaternary Science Reviews, v. 25, p. 2271–2282, doi: 10.1016/j.quascirev.2005.10.015.

Colman, S.M., Kaufman, D.S., Bright, J., Heil, C., King, J.W., Dean, W.E., Rosenbaum, J.G., Forester, R.M., Bischoff, J.L., Perkins, M., and McGeehin, J.P., 2007, Corrigendum to "Age model for a continuous, ca. 250-ky Quaternary lacustrine record from Bear Lake, Utah-Idaho": Quaternary Science Reviews, v. 26, p. 1192, doi: 10.1016/j.quascirev.2007.02.006.

Curry, B.B., 1999, An environmental tolerance index for ostracodes as indicators of physical and chemical factors in aquatic habitats: Palaeogeography, Palaeoclimatology, Palaeoecology, v. 148, p. 51–63, doi: 10.1016/S0031-0182(98)00175-8.

Dean, W.E., 2009, this volume, Endogenic carbonate sedimentation in Bear Lake, Utah and Idaho, over the last two glacial-interglacial cycles, *in* Rosenbaum, J.G., and Kaufman, D.S., eds., Paleoenvironments of Bear Lake, Utah and Idaho, and its catchment: Geological Society of America Special Paper 450, doi: 10.1130/2009.2450(07).

Dean, W., Rosenbaum, J., Skipp, G., Colman, S., Forester, R., Liu, A., Simmons, K., and Bischoff, J., 2006, Unusual Holocene and late Pleistocene carbonate sedimentation in Bear Lake, Utah and Idaho, USA: Sedimentary Geology, v. 185, p. 93–112, doi: 10.1016/j.sedgeo.2005.11.016.

Dean, W., Forester, R., Anderson, R., Bright, J., and Simmons, K., 2007, Influence of the diversion of Bear River into Bear Lake (Utah and Idaho) on the environment of deposition of carbonate minerals: Evidence from water and sediments: Limnology and Oceanography, v. 52, p. 1094–1111.

Dean, W., Wurtsbaugh, W., and Lamarra, V., 2009, this volume, Climatic and limnologic setting of Bear Lake, Utah and Idaho, *in* Rosenbaum, J.G., and Kaufman, D.S., eds., Paleoenvironments of Bear Lake, Utah and Idaho, and its catchment: Geological Society of America Special Paper 450, doi: 10.1130/2009.2450(01).

DeDeckker, P., 1981, Ostracodes of athalassic saline lakes: Hydrobiologia, v. 81, p. 131–144, doi: 10.1007/BF00048710.

Delorme, L.D., 1969, Ostracodes as Quaternary paleoecological indicators: Canadian Journal of Earth Sciences, v. 6, p. 1471–1476.

Delorme, L.D., 1970a, Freshwater ostracodes of Canada: Part I: Subfamily Cypridinae: Canadian Journal of Zoology, v. 48, p. 153–168.

Delorme, L.D., 1970b, Freshwater ostracodes of Canada: Part II: Subfamily Cypridopsinae and Herpetocypridinae, and family Cyclocyprididae: Canadian Journal of Zoology, v. 48, p. 253–266, doi: 10.1139/z70-042.

Delorme, L.D., 1970c, Freshwater ostracodes of Canada: Part III: Family Candonidae: Canadian Journal of Zoology, v. 48, p. 1099–1127, doi: 10.1139/z70-194.

Delorme, L.D., 1970d, Freshwater ostracodes of Canada: Part IV: Families Ilyocyprididae, Notodromadidae, Darwinulidae, Cytherideidae, and Entocytheridae: Canadian Journal of Zoology, v. 48, p. 1251–1259, doi: 10.1139/z70-214.

Delorme, L.D., 1971, Freshwater ostracodes of Canada: Part V: Families Limnocytheridae, Loxoconchidae: Canadian Journal of Zoology, v. 49, p. 43–64, doi: 10.1139/z71-009.

Delorme, L.D., 1978, Distribution of freshwater ostracodes in Lake Erie: Journal of Great Lakes Research, v. 4, p. 216–220.

Delorme, L.D., 1982, Lake Erie oxygen: The prehistoric record: Canadian Journal of Fisheries and Aquatic Sciences, v. 39, p. 1021–1029, doi: 10.1139/f82-137.

Delorme, L.D., and Donald, D., 1969, Torpidity of freshwater ostracodes: Canadian Journal of Zoology, v. 47, p. 997–999, doi: 10.1139/z69-160.

Denny, J.F., and Colman, S.M., 2003, Geophysical survey of Bear Lake, Utah-Idaho, September 2002: U.S. Geological Survey Open-File Report 03-150, CD-ROM.

Forester, R.M., 1983, Relationship of two lacustrine ostracode species to solute composition and salinity: Implications for paleohydrochemistry: Geology, v. 11, p. 435–439, doi: 10.1130/0091-7613(1983)11<435:ROTLOS>2.0.CO;2.

Forester, R.M., 1985, *Limnocythere bradburyi* n. sp.: A modern ostracode from central Mexico and a possible Quaternary paleoclimatic indicator: Journal of Paleontology, v. 59, p. 8–20.

Forester, R.M., 1987, Late Quaternary paleoclimate records from lacustrine ostracodes, *in* Ruddiman, W.F., and Wright, H.E., Jr., eds., North America and adjacent oceans during the last deglaciation: Boulder, Colorado, Geological Society of America, Geology of North America, v. K-3, p. 261–276.

Forester, R.M., 1988, Nonmarine calcareous microfossil sample preparation and data acquisition procedures: U.S. Geological Survey Technical Procedure HP-78 R1, p. 1–9.

Forester, R.M., 1991a, Ostracode assemblages from springs in the western United States: Implications for paleohydrology: Memoirs of the Entomological Society of Canada, v. 155, p. 181–201.

Forester, R.M., 1991b, Pliocene-climate history of the western United States derived from lacustrine ostracodes: Quaternary Science Reviews, v. 10, p. 133–146, doi: 10.1016/0277-3791(91)90014-L.

Forester, R.M., Colman, S.M., Reynolds, R.L., and Keigwin, L.D., 1994, Lake Michigan's Late Quaternary limnological and climate history from ostracode, oxygen isotope, and magnetic susceptibility records: Journal of Great Lakes Research, v. 20, p. 93–107.

Forester, R.M., Smith, A.J., Palmer, D.F., and Curry, B.B., 2006, North American non-marine ostracode database NANODe, Version 1: Kent, Ohio, Kent State University, http://www.kent.edu/nanode (accessed January 2008).

Fritz, S.C., Ito, E., Yu, Z., Laird, K.R., and Engstrom, D.R., 2000, Hydrologic variation in the northern Great Plains during the last two millennia: Quaternary Research, v. 53, p. 175–184, doi: 10.1006/qres.1999.2115.

Frogley, M.R., Griffiths, H.I., and Martens, K., 2002, Modern and fossil ostracods from ancient lakes, *in* Holmes, J.A., and Chivas, A.R., eds., The Ostracoda: Applications in Quaternary research: Washington, D.C., American Geophysical Union, Geophysical Monograph 131, p. 167–184.

Furtos, N.C., 1933, The Ostracoda of Ohio: Columbus, Ohio State University, Ohio Biological Survey 5, Bulletin 29, p. 411–524.

Garcia, A.F., and Stokes, M., 2006, Late Pleistocene highstand and regression of a small, high altitude pluvial lake, Jakes Valley, central Great Basin, USA: Quaternary Research, v. 65, p. 179–186, doi: 10.1016/j.yqres.2005.08.025.

Johnson, T.C., Scholz, C.A., Talbot, M.R., Kelts, K., Ricketts, R.D., Ngobi, G., Beuning, K., Ssemmanda, I., and McGill, J.W., 1996, Late Pleistocene desiccation of Lake Victoria and rapid evolution of cichlid fishes: Science, v. 273, p. 1091–1093, doi: 10.1126/science.273.5278.1091.

Kaufman, D.S., Bright, J., Dean, W.E., Moser, K., Rosenbaum, J.G., Anderson, R.S., Colman, S.M., Heil, C.W., Jr., Jiménez-Moreno, G., Reheis, M.C., and Simmons, K.R., 2009, this volume, A quarter-million years of paleoenvironmental change at Bear Lake, Utah and Idaho, *in* Rosenbaum, J.G., and Kaufman, D.S., eds., Paleoenvironments of Bear Lake, Utah and

Idaho, and its catchment: Geological Society of America Special Paper 450, doi: 10.1130/2009.2450(14).

Kennedy, B.M., 2005, Trade-offs in environmental growth conditions and predation risk relate to observed ecological separation between two closely related endemic whitefishes in Bear Lake, Utah/Idaho [M.S. thesis]: Logan, Utah State University, 84 p.

Kennedy, B.M., Thompson, B.W., and Luecke, C., 2006, Ecological differences between two closely related morphologically similar benthic whitefish (*Prosopium spilonotus* and *Prosopium abyssicola*) in an endemic whitefish complex: Canadian Journal of Fisheries and Aquatic Sciences, v. 63, p. 1700–1709, doi: 10.1139/F06-065.

Kitchell, J.A., and Clark, D.L., 1979, Distribution, ecology, and taxonomy of recent freshwater Ostracoda of Lake Mendota, Wisconsin: Madison, University of Wisconsin–Madison Natural History Series, no. 1, 24 p.

Kornicker, L.S., and Sohn, I.G., 1971, Viability of ostracode eggs egested by fish and effect of digestive fluids on ostracode shells—Ecologic and paleoecologic implications, *in* Oertli, H.J., ed., Colloque sur la paléoécologie des ostracodes, Pau, 20 July 1970: Centre Recherches Pau-SNPA Bulletin, v. 5 (Suppl.), p. 125–135.

Kullenberg, B., 1947, The piston core sampler: Svenska Hydrografisk-Biologiska Kommissionens Skrifter, v. 3, p. 1–40.

Lamarra, V., Liff, C., and Carter, J., 1986, Hydrology of Bear Lake basin and its impact on the trophic state of Bear Lake, Utah-Idaho: The Great Basin Naturalist, v. 46, p. 690–705.

Lewis, C.F.M., Moore, T.C., Jr., Rea, D.K., Dettman, D.L., Smith, A.M., and Mayer, L.A., 1994, Lakes of the Huron Basin: Their record of runoff from the Laurentide ice sheet: Quaternary Science Reviews, v. 13, p. 891–922, doi: 10.1016/0277-3791(94)90008-6.

Lowenstein, T.K., Li, J., Brown, C., Roberts, S.M., Ku, T.-L., Luo, S., and Yang, W., 1999, 200 k.y. paleoclimate record from Death Valley salt core: Geology, v. 27, p. 3–6, doi: 10.1130/0091-7613(1999)027<0003:KYPRFD>2.3.CO;2.

Martens, K., 1997, Speciation in ancient lakes: Trends in Ecology and Evolution, v. 12, p. 177–182, doi: 10.1016/S0169-5347(97)01039-2.

Martens, K., and Schön, I., 1999, Crustacean biodiversity in ancient lakes: A review: Crustaceana, v. 72, p. 899–910, doi: 10.1163/156854099503807.

Mazepova, G., 1994, On comparative aspects of ostracod diversity in the Baikalian fauna, *in* Martens, K., Goddeeris, B., and Coulter, G., eds., Speciation in ancient lakes: Archiv für Hydrobiologie Ergebnisse der Limnologie, v. 44, p. 197–202.

McConnell, W.J., Clark, W.J., and Sigler, W.F., 1957, Bear Lake: Its fish and fishing: Utah State Department of Fish and Game, Idaho Department of Fish and Game, Wildlife Management Department of Utah State Agricultural College, 76 p.

McCune, A.R., 1987, Lakes as laboratories of evolution: Endemic fishes and environmental cyclicity: Palaios, v. 2, p. 446–454, doi: 10.2307/3514616.

McLay, C.L., 1978a, Comparative observations on the ecology of four species of ostracodes living in a temporary freshwater puddle: Canadian Journal of Zoology, v. 56, p. 663–675, doi: 10.1139/z78-094.

McLay, C.L., 1978b, The population biology of *Cyprinotus carolinensis* and *Herpetocypris reptans* (Crustacea, Ostracoda): Canadian Journal of Zoology, v. 56, p. 1170–1179, doi: 10.1139/z78-161.

Mellors, W.K., 1975, Selective predation of ephippal *Daphnia* and the resistance of ephippal eggs to digestion: Ecology, v. 56, p. 974–980, doi: 10.2307/1936308.

Miller, B.A., 2006, The phylogeography of *Prosopium* in western North America [M.S. thesis]: Provo, Brigham Young University, 164 p.

Mourguiart, P., and Montenegro, M.E., 2002, Climate changes in the Lake Titicaca area: Evidence from ostracode ecology, *in* Holmes, J.A., and Chivas, A.R., eds., The Ostracoda: Applications in Quaternary Research: Washington, D.C., American Geophysical Union, Geophysical Monograph 131, p. 151–165.

Oviatt, J.G., Thompson, R.S., Kaufman, D.S., Bright, J., and Forester, R.M., 1999, Reinterpretation of the Burmester core, Bonneville Basin, Utah: Quaternary Research, v. 52, p. 180–184, doi: 10.1006/qres.1999.2058.

Owen, R.B., Crossley, R., Johnson, T.C., Tweddle, D., Kornfield, I., Davison, S., Eccles, D.H., and Engstrom, D.E., 1990, Major low levels of Lake Malawi and their implications for speciation rates in cichlid fishes: Proceedings of the Royal Society of London, ser. B, Biological Sciences, v. 240, p. 519–553.

Park, L.E., and Downing, K.F., 2000, Implications of phylogeny reconstruction for ostracod speciation modes in Lake Tanganyika: Advances in Ecological Research, v. 31, p. 303–330, doi: 10.1016/S0065-2504(00)31017-0.

Quade, J., Mifflin, M.D., Pratt, W.L., McCoy, W., and Burckle, L., 1995, Fossil spring deposits in the southern Great Basin and their implications for changes in water-table levels near Yucca Mountain, Nevada, during Quaternary time: Geological Survey of America Bulletin, v. 107, p. 213–230, doi: 10.1130/0016-7606(1995)107<0213:FSDITS>2.3.CO;2.

Quade, J., Forester, R.M., Pratt, W.L., and Carter, C., 1998, Black mats, spring-fed streams, and late-Glacial-age recharge in the southern Great Basin: Quaternary Research, v. 49, p. 129–148, doi: 10.1006/qres.1997.1959.

Reed, K.M., Dorschner, M.O., Todd, T.N., and Phillips, R.B., 1998, Sequence analysis of the mitochondrial DNA control region of ciscoes (genus *Coregonus*): Taxonomic implications for the Great Lakes species flock: Molecular Biology, v. 7, p. 1091–1096.

Roca, J., and Wansard, G., 1997, Temperature influence on development and calcification of *Herpetocypris brevicaudata* Kaufmann, 1900 (Crustacea: Ostracoda) under experimental conditions: Hydrobiologia, v. 347, p. 91–95, doi: 10.1023/A:1003067218024.

Rutaisire, J., Booth, A.J., Masembe, C., Nyakaana, S., and Muwanika, V.B., 2004, Evolution of *Labeo victorianus* predates the Pleistocene desiccation of Lake Victoria: Evidence from mitochondrial DNA sequence variation: South African Journal of Science, v. 100, p. 607–608.

Sigler, W.F., and Sigler, J.W., 1987, Fishes of the Great Basin: A natural history: Reno, University of Nevada Press, 425 p.

Sigler, W.F., and Sigler, J.W., 1996, Fishes of Utah: A natural history: Salt Lake City, University of Utah Press, 375 p.

Smart, E.W., 1958, An ecological study of the bottom fauna of Bear Lake, Idaho and Utah [Ph.D. thesis]: Logan, Utah State University, 88 p.

Smith, A.J., 1993, Lacustrine ostracodes as hydrochemical indicators in lakes of the north-central United States: Journal of Paleolimnology, v. 8, p. 121–134, doi: 10.1007/BF00119785.

Smith, A.J., Donovan, J.J., Ito, E., and Engstrom, D.R., 1997, Ground-water processes controlling a prairie lake's response to mid-Holocene drought: Geology, v. 25, p. 391–394, doi: 10.1130/0091-7613(1997)025<0391:GWPCAP>2.3.CO;2.

Smith, A.J., Donovan, J.J., Ito, E., Engstrom, D.R., and Panek, V.A., 2002a, Climate-driven hydrologic transients in lake sediment records: Multiproxy record of mid-Holocene drought: Quaternary Science Reviews, v. 21, p. 625–646, doi: 10.1016/S0277-3791(01)00041-5.

Smith, D.G., 1985, Recent range expansion of the freshwater mussel *Anodonta implicata* and its relationship to clupeid fish restoration in the Connecticut River system: Freshwater Invertebrate Biology, v. 4, p. 105–108, doi: 10.2307/1467182.

Smith, G.R., 1981, Late Cenozoic freshwater fishes of North America: Annual Review of Ecology and Systematics, v. 12, p. 163–193, doi: 10.1146/annurev.es.12.110181.001115.

Smith, G.R., and Todd, T.N., 1984, Evolution of species flocks of fishes in north temperate lakes, *in* Echelle, A., and Kornfield, I., eds., Evolution of fish species flocks: Orono, University of Maine, Orono Press, p. 45–68.

Smith, G.R., Stokes, W.L., and Horn, K.F., 1968, Some Late Pleistocene fishes of Lake Bonneville: Copeia, v. 1968, p. 807–816, doi: 10.2307/1441848.

Smith, G.R., Dowling, T.E., Gobalet, K.W., Lugaski, T., Shiozawa, D.K., and Evans, R.P., 2002b, Biogeography and timing of evolutionary events among Great Basin fishes, *in* Hershler, R., Madsen, D.B., and Currey, D.R., eds., Great Basin aquatic systems history: Smithsonian Contributions to the Earth Sciences, no. 33, p. 175–234.

Smoot, J.P., and Rosebaum, J.G., 2009, this volume, Sedimentary constraints on late Quaternary lake-level fluctuations at Bear Lake, Utah and Idaho, *in* Rosenbaum, J.G., and Kaufman, D.S., eds., Paleoenvironments of Bear Lake, Utah and Idaho, and its catchment: Geological Society of America Special Paper 450, doi: 10.1130/2009.2450(12).

Spencer, R.J., Baedecker, M.J., Eugster, H.P., Forester, R.M., Goldhaber, M.B., Jones, B.F., Kelts, K.F., McKenzie, J., Madsen, D.B., Rettig, S.L., and Bowser, C.J., 1984, Great Salt Lake, and precursors, Utah: The last 30,000 years: Contributions to Mineralogy and Petrology, v. 86, p. 321–334, doi: 10.1007/BF01187137.

St. George, S., and Nielson, E., 2002, Hydroclimatic change in southern Manitoba since A.D. 1409 inferred from tree rings: Quaternary Research, v. 58, p. 103–111, doi: 10.1006/qres.2002.2343.

Stager, J.C., Day, J.J., and Santini, S., 2004, Origin of the superflock of cichlid fishes from Lake Victoria, East Africa: Comment: Science, v. 304, p. 963, doi: 10.1126/science.1091978.

Thompson, B.W., 2003, An ecological comparison of two endemic species of whitefish in Bear Lake, Utah/Idaho [M.S. thesis]: Logan, Utah State University, 131 p.

Thompson, R.S., 1996, Pliocene and early Pleistocene environments and climates of the western Snake River Plain, Idaho: Marine Micropaleontology, v. 27, p. 141–156, doi: 10.1016/0377-8398(95)00056-9.

Thompson, R.S., Whitlock, C., Bartlein, P.J., Harrison, S.P., and Spaulding, W.G., 1993, Climatic changes in the western United States since 18,000 yr B.P., *in* Wright, H.E., Jr., Kutzbach, J.E., Webb, T., III, Ruddiman, W.F., Street-Perrott, F.A., and Bartlein, P.J., eds., Global climates since the Last Glacial Maximum: Minneapolis, University of Minnesota Press, p. 468–513.

Todd, T., and Smith, G.R., 1992, A review of differentiation in Great Lakes ciscoes: Polskie Archiwum Hydrobiologii, v. 39, p. 261–267.

Tolentino, S.A., and Thompson, B.W., 2004, Meristic differences, habitat selectivity and diet separation of *Prosopium spilonotus* and *P. abyssicola*: Annales Zoologici Fennici, v. 41, p. 309–317.

Toline, C.A., Seamons, T.R., and Davis, C., 1999, Quantification of molecular and morphological differentiation of whitefish taxa in Bear Lake: Salt Lake City, Final Report to the Utah Division of Natural Resources, Project F-47-R, Study 5, 53 p.

Verheyen, E., Salzburger, W., Snoeks, J., and Meyer, A., 2003, Origin of the superflock of cichlid fishes from Lake Victoria, East Africa: Science, v. 300, p. 325–329, doi: 10.1126/science.1080699.

Vinyard, G., 1979, An ostracode (*Cypridopsis vidua*) can reduce predation from fish by resisting digestion: American Midland Naturalist, v. 102, p. 188–190, doi: 10.2307/2425084.

Vuorinen, J.A., Bodaly, R.A., Reist, J.D., and Luczynski, M., 1998, Phylogeny of five *Prosopium* species with comparisons with other Coregonine fishes based on isozyme electrophoresis: Journal of Fish Biology, v. 53, p. 917–927.

Wells, T.M., Cohen, A.S., Park, L.E., Dettman, D.L., and McKee, B.A., 1999, Ostracode stratigraphy and paleoecology from surficial sediments of Lake Tanganyika, Africa: Journal of Paleolimnology, v. 22, p. 259–276, doi: 10.1023/A:1008046417660.

Wurtsbaugh, W.A., and Hawkins, C., 1990, Trophic interactions of fish and invertebrates in Bear Lake, Utah/Idaho: Logan, Ecology Center Special Publication, 167 p.

MANUSCRIPT ACCEPTED BY THE SOCIETY 15 SEPTEMBER 2008

The Geological Society of America
Special Paper 450
2009

A 19,000-year vegetation and climate record for Bear Lake, Utah and Idaho

Lisa A. Doner*

Center for the Environment, Plymouth State University, Plymouth, New Hampshire 03264, USA

ABSTRACT

Pollen analysis of sediments from core BL96-2 at Bear Lake (42°N, 111°20′W), located on the Utah-Idaho border in America's western cordillera, provides a record of regional vegetation changes from full glacial to the late Holocene. The reconstructed vegetation records are mostly independent of Bear Lake's hydrologic state and are therefore useful for identifying times when climate forcing contributed to lake changes. The Bear Lake pollen results indicate that significant changes in the Bear Lake vegetation occurred during the intervals 15,300–13,900, 12,000–10,000, 7500–6700, 6700–5300, 3800–3600, and 2200–1300 cal yr B.P. These intervals coincide with regional shifts in vegetation and climate, documented in pollen, isotope and biogeographic records in the Basin and Range region, suggesting that large-scale climate was the primary forcing factor for these intervals of change. Maximum aridity and warmth is indicated from 12,000 to 7500 cal yr B.P., followed by intervals of generally more mesic and cool conditions, especially after 7500 cal yr B.P.

INTRODUCTION

Paleoenvironmental histories are often derived from multiproxy studies of lake and wetland sediments. The aquatic environment tends to preserve characteristics in the sediments that are otherwise lost through oxidation or erosion. Deep lakes, in particular, can provide high-resolution records because of reduced wind-induced sediment mixing and profundal anoxia. Environmental proxies include diatoms, ostracodes, mollusks, oxygen and carbon isotopes, carbon content, particle-size, geochemistry, sedimentary magnetism, and pollen. In highly variable lake environments, the climatic signals of these proxies can be masked or confounded by internal forcing factors or strongly mitigated by within-lake environmental conditions. For example, in the past 19,000 years, Bear Lake's chemistry has varied widely as a result of changes in Bear River influx, resulting in relatively

large magnitude changes in oxygen isotope ratios (Dean, this volume). The influence of temperature on these isotopic ratios is mixed with, and hidden by, these hydrologic events. Although pollen grains are subject to current-driven sorting and preservation conditions in the lake, most of the source area is terrestrial and dominantly extra-local, especially for large lakes (Tauber, 1977; Bradshaw, 1994; Jackson, 1994). Long-distance transport of pollen in river systems, or redeposition of pollen that has been exposed to an oxygenated environment, is easily distinguished from airborne, rapidly deposited pollen by its preservation state (Cushing, 1967). Pollen data can thus serve as a bellwether for distinguishing local-lake hydrologic changes from regional climate changes.

One of a series of papers in this volume examining Quaternary sediment records from Bear Lake, Utah and Idaho, this paper focuses on post-glacial vegetation reconstructions from

*E-mail: donerl@mac.com

Doner, L.A., 2009, A 19,000-year vegetation and climate record for Bear Lake, Utah and Idaho, *in* Rosenbaum, J.G., and Kaufman, D.S., eds., Paleoenvironments of Bear Lake, Utah and Idaho, and its catchment: Geological Society of America Special Paper 450, p. 217–227, doi: 10.1130/2009.2450(09). For permission to copy, contact editing@geosociety.org. ©2009 The Geological Society of America. All rights reserved.

pollen analyses and identification of climate forcing of sedimentary changes. The overall aim of the Bear Lake project is to reconstruct local and regional changes associated with tectonic basin subsidence, Bear River migration, and climate change. Few continuous vegetation records from the glacial maximum to the late Holocene in the Basin and Range region of the western United States exist because of erosion of the long-term sediment record at low elevations from lake desiccation and at high elevations from mountain glaciations. Proximal to Bear Lake, pre-Holocene paleovegetation records have been developed from the Great Salt Lake/Bonneville Basin, covering the last 15 m.y. (Davis and Moutoux, 1998), from a 70,000 yr vegetation reconstruction in a modern marsh at Grays Lake, Idaho, ~80 km northeast of Bear Lake (Beiswenger, 1991), and from a deglacial sequence at Rapid Lake, Wyoming (Fall et al., 1995). Holocene records in the region also come from these records plus peat deposits in the Uinta Mountains, southeast of Bear Lake (Munroe, 2003). A new, multi-glacial pollen record from the GLAD800 drilling project at Bear Lake reveals regional vegetation sensitivity to summer insolation, global ice volume, and some Heinrich events (Jiménez-Moreno et al., 2007).

Although Bear Lake sediments show clear responses to the dominant climate changes of glacial-interglacial cycles and transition intervals (Jiménez-Moreno et al., 2007; Kaufman et al., this volume), interpretation of Bear Lake's response to the milder climate changes of the late Quaternary is complicated. Tectonic activity along the graben, diversion of the Bear River out of the lake watershed, and varying inputs from multiple groundwater sources caused strong variations in lake level and water source during the last 25 k.y. (Reheis et al., this volume). These hydrologic changes in the late Pleistocene confuse climate interpretations from diatoms, ostracodes, isotopes, and carbonates (Moser and Kimball, this volume; Bright, this volume, Chapter 8; Dean, this volume). Sedimentary magnetism also shows a strong glacial-interglacial component from transport of sediments to Bear Lake by the Bear River but after diversion of the river, this climate signal is lost by post-depositional destruction of Fe-oxide minerals (Rosenbaum and Heil, this volume). In this situation, where changes in water and sediment sources and chemistries create uncertainties in identification of forcing factors by alternative proxies, the onus to demonstrate climate forcing of late Quaternary changes in the sediments is on the pollen record.

SITE DESCRIPTION

Bear Lake (42°N, 111°20′W) occupies the southern reaches of Bear Lake Valley, a north-south oriented half-graben located between the Bear River Range to the west and the Bear Lake Plateau to the east (Fig. 1). The modern lake has a maximum elevation of 1805 m above sea level, constrained by the Lifton Bar, a natural beach barrier to the north, with a maximum lake extent of 32 km by 12 km, 282 km² surface area, and 63 m maximum depth (Birdsey, 1989). Dingle Swamp dominates the northern half of Bear Lake Valley and extends from Mud Lake, adjoining

Lifton Bar, to the Bear River confluence with the Rainbow Canal. The lake today is fed primarily from spring-fed streams to the west, south, and east (Bright, this volume, Chapter 4). The Bear River has predominantly fed into the lake over the past 250 k.y., although this connection is periodically lost during warmer interglacial intervals (Kaufman et al., this volume). During the late Pleistocene, the lake was a maximum of 8 m above the modern level and the lake extent included the connection to the Bear River (Reheis et al., this volume).

Bear Lake's present climate is generally continental, with hot, dry summers, and winters cold enough to freeze the lake most years. Wind-blown ice accumulates in push ridges on the shores (Wurtsbaugh and Luecke, 1998) and probably contributes to the

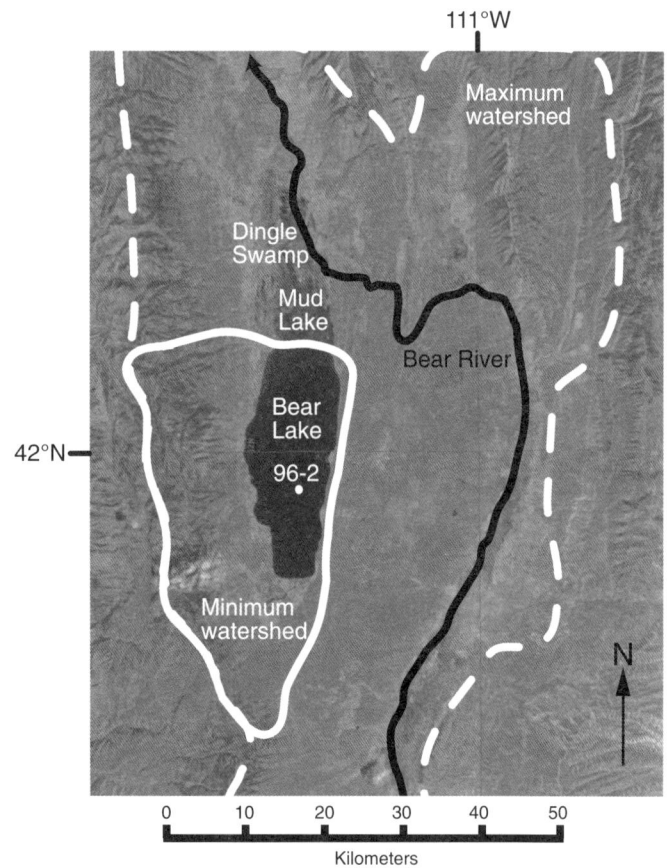

Figure 1. Satellite image of the Bear Lake region showing approximate boundaries of the Bear Lake watershed with Bear River influx (maximum) and without (minimum). The Bear River travels parallel to the lake on the far side of the eastern highlands, then turns sharply west and enters Dingle Swamp in the northern Bear Lake Valley. The swamp currently drains to the north, away from the lake. Prior to 1912, Mud Lake was isolated from Bear Lake by a barrier beach. At low lake levels, the lake is physically isolated from the swamp and the Bear River is not part of the lake's hydrology. At high lake levels, lake and swamp waters commingle and the Bear River's influence on the lake's hydrology is significant.

weakly vegetated lake margin. Bear Lake's climate differs from that of Great Salt Lake, ~120 km to the southwest, by Bear Lake's higher elevation and bounding ranges that create much larger orographic effects and potential for significant direct inputs from snowmelt. Although not well documented, evaporation potential at Bear Lake is expected to be much less than at Great Salt Lake because of cooler temperatures, topographic interference with wind patterns, and reduced hours of direct daylight (Kaliser, 1972; Amayreh, 1995). Because of these differences, Great Salt Lake and Bear Lake vegetation and paleoclimate records cannot be directly compared, although both should experience similar regional-scale trends in precipitation and drought that influence lake level and regional vegetation.

Modern vegetation around Bear Lake is mixed woodland, farmland, and rangeland. The valley is not heavily developed but is extensively farmed, beginning with settlement in the early 1800s. The native vegetation is dominated by shrubs, mostly sagebrush (*Artemisia* spp.) and juniper (*Juniperus* spp.) in the lower elevations of the watershed, with cottonwood (*Populus angustifolia* and *P. trichocarpa*) and willow (*Salix* spp.) near drainages. A mixed conifer forest dominates the upper elevations of the watershed, with lodgepole (*Pinus contorta*) and limber (*Pinus flexilis*) pine, Gambel oak (*Quercus gambelii*), maple (*Acer* spp.), Engelmann (*Picea engelmannii*) and blue (*Picea pungens*) spruce, and Douglas (*Psuedotsuga menziesii*), white (*Abies concolor*), and subalpine (*Abies lasiocarpa*) fir in the highlands. Government-sponsored Web sites provide GIS-based maps of modern vegetation distributions in the Bear Lake region (Albee et al., 1988; Thompson et al., 2000).

METHODS

One of three long sediment cores collected with a Kullenberg corer at Bear Lake in 1996, the 4-m-long BL96-2 core was determined by ^{14}C AMS (accelerator mass spectrometry) dating to have nearly complete representation of the Holocene (Colman et al., this volume). This core was subsampled for pollen analysis at irregular intervals, with closer intervals near lithologic changes. Pollen preparation followed standard techniques (Faegri and Iversen, 1989). Two tablets containing spores of *Lycopodium clavatum* from Lund University, batch 307862 with an average concentration of 13,500 ± 308 spores, were added to each sample to provide a control basis for concentration estimates. Prepared pollen samples are stored with glycerin in capped glass vials. Pollen keys and reference slides in collections kept by the U.S. Geological Survey (USGS) Earth Surface Processes Team in Denver, Colorado, aided in identification of unfamiliar grains (McAndrews et al., 1973; Heusser and Peteet, 1988; Faegri and Iversen, 1989; Moore et al., 1991).

Pollen identifications and counts were completed using an Olympus BH2 microscope with a combined ocular and dry-lens magnification factor of 500. Pollen reference collections at the USGS Earth Surface Processes Team (Denver) and the Institute of Arctic and Alpine Research, University of Colorado, aided

identification of pollen types. USGS photographs and reference slides of pine pollen types were especially useful in recognizing species-specific variations in Quaternary and pre-Quaternary pollen morphology. Pine species were split out conservatively, with uncertain grains being grouped as diploxylon or haploxylon types, or as *Pinus*-type. To avoid problems of differential sorting, pollen grains were identified along a minimum of five evenly spaced transects, covering either half the slide from midline to outer edge, or the full slide, until a minimum of 300 grains were identified, excluding spores and aquatic types. The pollen types were grouped as trees, shrubs, herbs, aquatic, spores, and indeterminate. Indeterminate types were categorized as broken, corroded, crumpled, degraded, hidden, thinned and crumpled, or unknown, as suggested by Cushing (1967). Pollen percentages for each group were calculated relative to a total sum of trees, shrubs, herbs, and indeterminate types. Total pollen concentrations (in grains per gram of dry sediment) include all vascular pollen and spore types. Cyperaceae (sedges) in the western United States occur commonly as aquatic species and so were grouped as aquatic pollen. Pollen data calculations were completed in the program TILIA, version 2.0.b.4 (Grimm, 1990), with pollen diagrams from TILIAGRAPH, version 2.0.5.b (Grimm, 1998). Types with less than 1% maximum occurrence (rare types) are presented as numbers of grains rather than percentages.

The multivariate analysis program CONISS (Grimm, 1987) generated Euclidean-distance dissimilarity values for terrestrial pollen and spores with more than 1% maximum occurrence (Overpeck et al., 1985). Hierarchical, polythetic agglomerative dendrograms with minimum-variance clustering are used to indicate which samples are most alike (least dissimilar), with these appearing as tight clusters close to the axis. The distance off the axis (dissimilarity distance) provided an estimate of total sample variability and was useful in determining the relative differences between clusters (Gaugh, 1982). Because the presence and absence of sporadically occurring types is statistically weaker than for more abundant types, these rare types were excluded from the dissimilarity calculations. Horizontal "zone" lines indicate high dissimilarity between adjacent samples.

The age model used here is the polynomial (y = 0.336 + 0.0379x + 2.83e^{-5}x^2, R^2 = 0.981) that Colman et al. (this volume) fit to calibrated ^{14}C ages using the midpoints of 1σ intervals. Although the sampling resolution is low, the age basis for each of the samples is well constrained by this age model. The timing of intrasample changes is most accurately defined by the age of bounding samples. Thus, intervals of change in this record are designated by the two nearest neighbors.

RESULTS

Pollen counts were completed on 16 samples with identification of 110 pollen types, including 21 tree, 11 shrub, 58 herb, 9 spore, and 11 aquatic types. The pollen grains were generally well preserved and identifiable. Indeterminate pollen reached a maximum of 40% of the total pollen sum in the oldest sample, but after 15,500 cal yr B.P. was never more than 6.5%. Degraded and broken grains dominate the indeterminate fraction with hidden grains never more than 35% of this sub-sum.

All total pollen sums exceeded 400 grains, except in the two glacial samples that had pollen sums over 300 grains. Concentration values ranged from 5900 grains•g^{-1}, ca. 18,500 cal yr B.P., to 154,000 grains•g^{-1}, around 7000 cal yr B.P. A relatively large increase in concentration before 14,000 cal yr B.P. followed distinct decreases in indeterminate types. These results suggest that surface-transported pollen came from less distant sources, with less dilution of the pollen by sediment, by 15,500 cal yr B.P. Both of these changes are indicative of waning inputs of the Bear River's sediment load. Although herbaceous types dominate throughout the record, with herb:tree ratios of 1.1–2.6, periods of high total pollen concentration are generally associated with higher tree percentages. This is not true for the late Holocene, when the percentage of tree pollen was at a maximum but concentration was low. This concentration decrease in the uppermost sediments might result from increased dilution due to sedimentation rate increases. Alternatively, long-distance transport of tree pollen could enhance the representation of trees when local pollen inputs are low, such as occurs in alpine and arctic settings (Bourgeois, 1990; Fall, 1992; Markgraf, 1980).

In the dissimilarity analyses, 46 terrestrial pollen types with less than 1% maximum representation (rare types) were excluded, whereas ten rare types, originally identified to genus, were recategorized to genus-type or family to raise those representations above 1%. Dissimilarity analyses were run on 31 types, with nine identified to family, two to order, 15 to genera, and five to species (all *Pinus* spp.). Of these 31 types, 12 are tree types from six different genera. Four pollen zones, based on maximum dissimilarity values, are identified by hierarchical clustering with periods of greatest change in the pollen composition during the following intervals: 15,300–13,900, 12,000–10,000, 6700–5300, and 2200–1300 cal yr B.P. Maximum dissimilarity occurred during the deglacial interval (ca. 15,300–13,900 cal yr B.P.); the remaining clusters have progressively smaller difference values toward the present. Within-cluster dissimilarities are 50%–100% less than with their neighboring zones, with highest variability in within-cluster distances occurring prior to 10,000 cal yr B.P.

Relative percentages of the major pollen types, and raw counts of the rare types, are plotted in groups according their zone of maximum representation and shown with dissimilarity clustering results. These reveal a progression of dominant pollen types through time (Figs. 2 and 3). Most of the vegetation types identified, even those from the earliest parts of the record, occur in the Bear River watershed today (Albee et al., 1988; Burns and Honkala, 1990; Thompson et al., 2000; Welsh et al., 2003). The remaining non-local types occur within the neighboring regions of northern Arizona, southern Utah, and Wyoming. Their presence in the sediments of Bear Lake could be the result of long-distance pollen transport or biome shifts. The assemblages and proportional representation of these pollen types differ greatly over the last 19,000 cal yr, however, and these changes form the

basis of the zone interpretations. All ages are rounded to the nearest 100 yr. Uncertainty in the timing of vegetation changes is high because of the low sampling resolution.

Zone 1 (19,000–15,300 cal yr B.P., full glacial) includes the full glacial and initial deglacial period. The most abundant pollen in this zone is sage (*Artemisia*). Pollen types reaching dominance in this interval are birch (*Betula*) and willow (*Salix*) and a variety of alpine meadow types common today in the Basin and Range. Families with steppe affiliations, such as Chenopodiaceae and Caryophyllaceae, are also well represented although they reach maxima later. Spruce (*Picea*), pine (*Pinus*), and fir (*Abies*) pollen occur in low abundances and may have grown vegetatively most of the time, producing pollen infrequently because of unfavorable climate conditions for reproduction. Spore types occurring in highest abundance here include ferns common to woodlands and rocky soils. These results are all consistent with the cold but generally ice-free environment associated with last glacial interval in the lower elevations of the Basin and Range region (Anderson et al., 1999; Barnosky et al., 1987; Benson et al., 1990; Madsen and Currey, 1979; Thompson et al., 1993).

Zone 2 (13,900–12,000 cal yr B.P., deglacial) includes the last stages of deglaciation and the Younger Dryas stage. This zone is defined by just two samples, so environmental responses are only coarsely resolved. Three species of pine, piñon (*P. edulis*), limber (*P. flexilis*), and ponderosa (*P. ponderosa*), reach their maxima in the interval, as do the rarer types poplar (*Populus*) and chestnut/chinquapin (*Castanea*). Many herbaceous types typical of woodlands and open forests occur in this zone, some reaching their maximum here. Percentages of sage and ragweed (*Ambrosia*), and other shrubs and herbs associated with arid steppe, are lower for both samples than in Zone 1. The pollen composition in this zone suggests an expansion of woodlands and mixed forests within the Bear Lake watershed.

Zone 3 (10,000–6700 cal yr B.P., early to middle Holocene) contains a rich tree assemblage with maximum percentages of juniper (Cupressaceae), Douglas fir (*Pseudotsuga menziesii*), lodgepole pine (*Pinus contorta*), singleleaf piñon pine (*Pinus monophylla*), oak (*Quercus*), and maximum occurrences of the rare tree types of maple (*Acer*), ash (*Fraxinus*), and undifferentiated haploxylon pines. Alpine, woodland and wet meadow shrubs and herb types are also well represented in this zone, including alder (*Alnus*), Moschatel (*Adoxa*), *Dryas*, *Filipendula*, buckbean (*Menyanthes*), Saxifragaceae, and *Valeriana*. Steppe types increase as well, with maximum levels of sage, ragweed, shadbush (*Amelanchier*), goosefoot (Chenopodiaceae), pinks (Caryophyllaceae), and greasewood (*Sarcobatus*). There is a time-transgressive change in the pollen assemblage in this zone, with the dominant steppe plants, sage and ragweed, reaching their maxima before 8000 cal yr B.P. This is followed at 7500–7000 cal yr B.P., with maxima in tree and alpine types. This, in turn, is followed by a distinct assemblage comprising mostly Ponderosa and lodgepole pines, shadbush, sage, pinks, and greasewood ca. 6700 cal yr B.P.

Zone 4 (5300–2200 cal yr B.P., middle to late Holocene) has the highest percentages of the rose family (Rosaceae), followed

Bear Lake Pollen Major Taxa

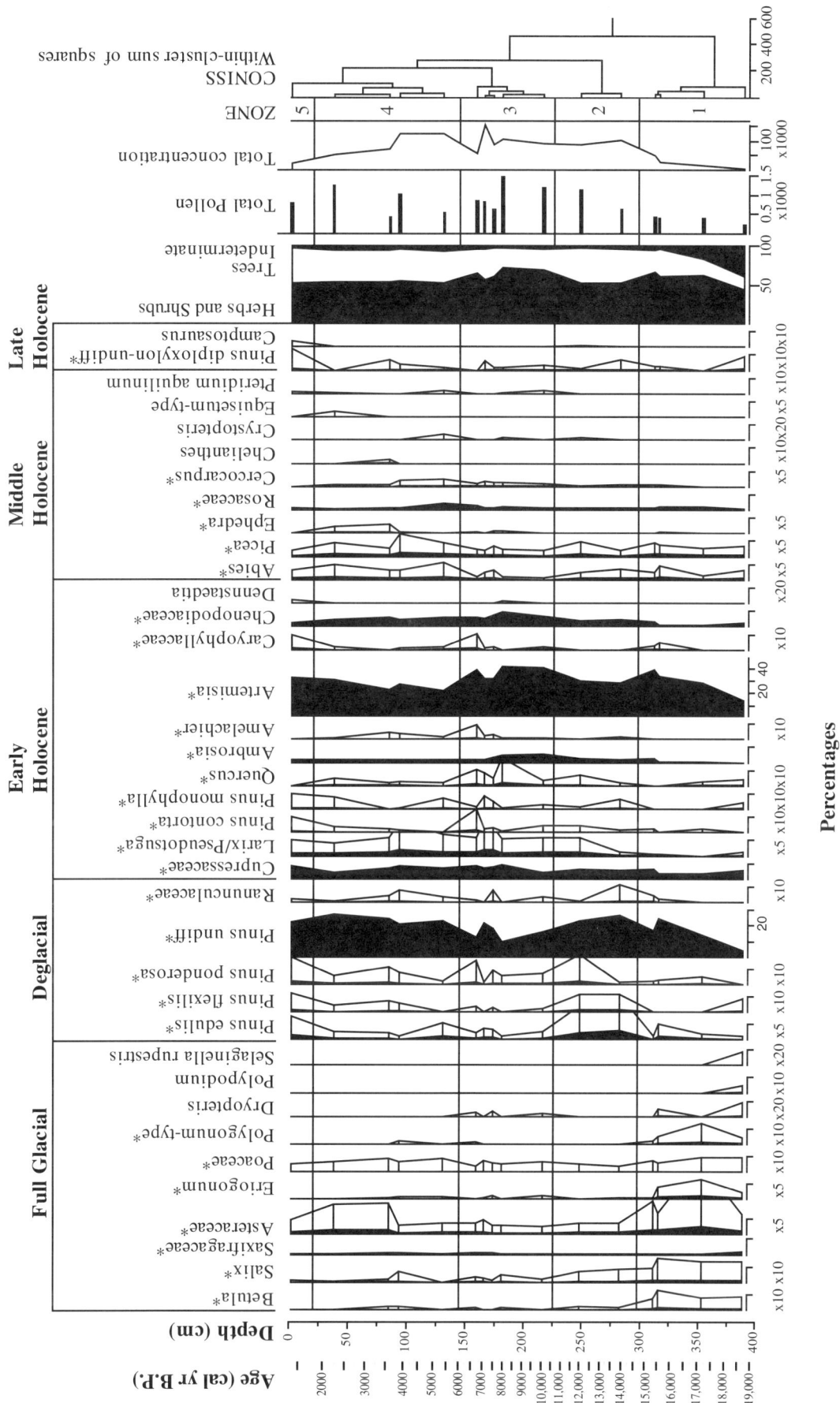

Figure 2. Major pollen taxa from the Bear Lake core 96-2, shown as percentages. Note that most of the x-axis curves are exaggerated by a factor of 5–20 (shown as hollow curves with histograms at sample points) and that the x-axis scaling (width) varies. Unexaggerated types are plotted as solid black curves. X-axis tick marks are every 10% with labels every 20%. Asterisks indicate pollen types used to create the CONISS dendrogram that, in turn, was used to define the zone boundaries (horizontal black lines). Pollen types are grouped according to the timing of their maximum zone values as dominant in the full glacial (before 15,300 cal yr B.P.), deglacial (ca. 13,900–12,000 cal yr B.P.), early Holocene (ca. 10,000–6700 cal yr B.P.), middle Holocene (ca. 5300–2200 cal yr B.P.), or late Holocene (1300 cal yr B.P.). Ages assigned to the pollen zone boundaries are approximations, calculated as the midpoints between the pollen samples in the neighboring zones.

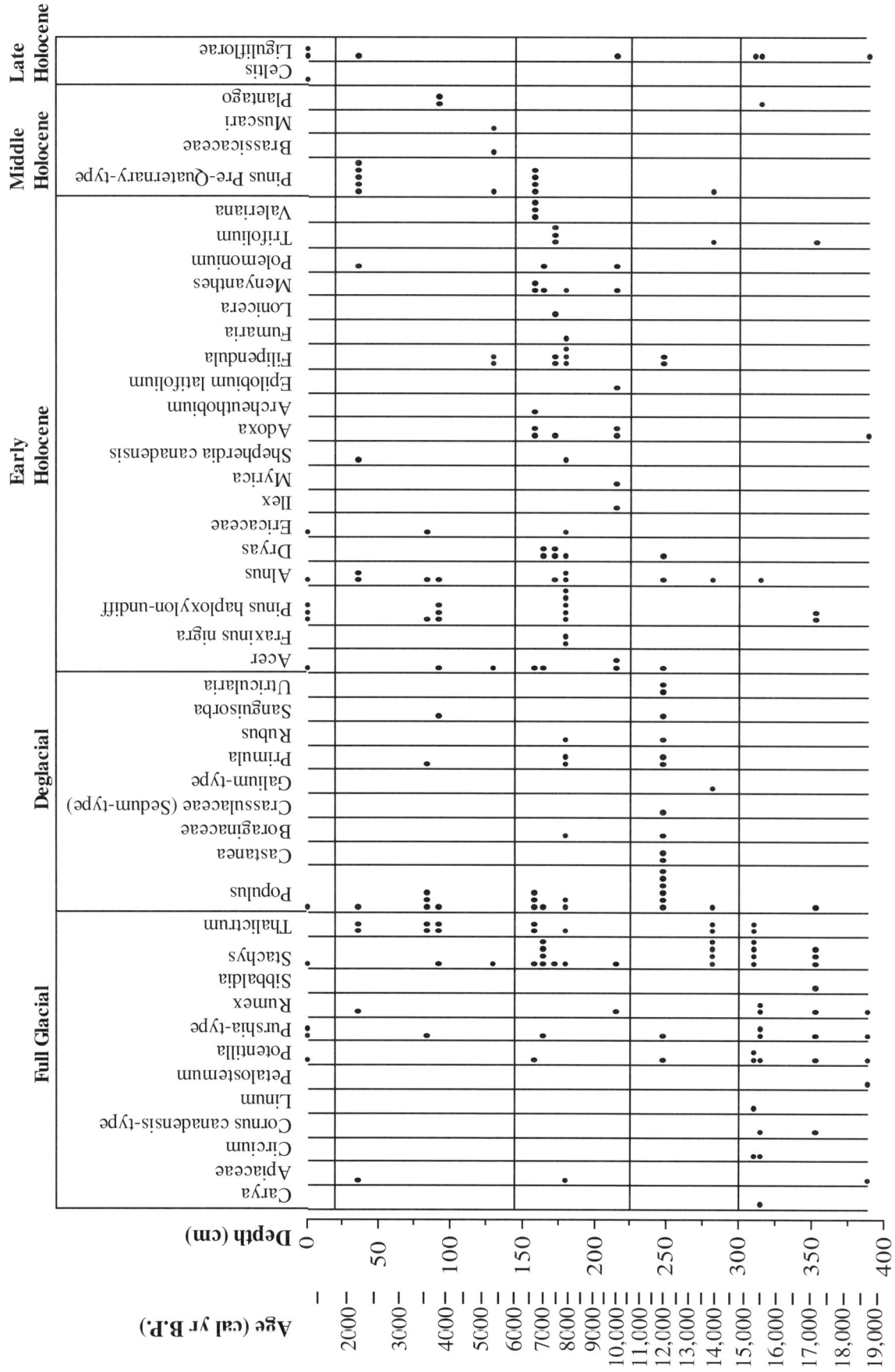

Figure 3. Minor (rare) pollen taxa from Bear Lake core BL96-2, shown as individual grain counts with one black dot per grain. Pollen types are grouped as in Figure 2. The horizontal zone lines are shown for reference although none of these rare types were included in the CONISS result.

by maximum values of the forest alpine types fir (*Abies*) and spruce (*Picea*), as well as mountain mahogany (*Cercocarpus*). Douglas fir and poplar continue to be relatively well represented in this zone. These moist forest types all reach maxima before 3800 cal yr B.P. After that, steppe and dry woodland types including pine, composites (Asteraceae), and Mormon tea (*Ephedra*) increase abruptly, while willow, buttercup (Ranunculaceae), and shadbush decrease (by 3600 cal yr B.P.). Total pollen concentration decreases by ~50% over this transition, but the pollen results offer no apparent explanation for this decrease. Moisture-loving horsetail (*Equisetum*) and pre-Quaternary pine types increase to their maxima ca. 2200 cal yr B.P., but decrease again after. These pre-Quaternary types, identified from pollen reference collections of Tertiary and Cretaceous pollen, indicate erosion and transport of pollen from sediments of these ages. The peak in pre-Quaternary types at the same time as peak aquatic types suggests an increase in precipitation leading to erosional down-cutting by streams tributary to the lake. Higher sediment inputs would cause pollen concentration decreases through dilution.

Zone 5 (ca. 1300 cal yr B.P., late Holocene) is defined by only one sample but it is distinct from the previous four samples of Zone 4. Upland types with maximum occurrence in this zone are the dandelion tribe (Liguliflorae) and three-needle pines (*Pinus* diploxylon). Pines of all identified types, except pre-Quaternary, and juniper increase slightly, while all the other tree types either decrease or remain at Zone 4 levels. Pinks and rose family types also increase. Many types present elsewhere in the core are not represented in this sample and only the oldest full-glacial sample has fewer types included in the CONISS statistics. Pollen concentration is lower than at any time since 16,350 cal yr B.P.

Amounts of aquatic pollen in sediments often correspond to changes in lake dimensions, including length of shoreline, and efficiency of pollen transport from shallow regions to the core site (Digerfeldt, 2003). The basin morphology of Bear Lake is such that increases in the size of Bear Lake beyond the Lifton Bar would greatly increase the habitat for rooting aquatic plants, providing an additional source of aquatic pollen to the core site during lake highstands. In the aquatic pollen types (Fig. 4), several distinct changes can be recognized. Because of their strong linkages with lake level, these aquatic pollen changes are detailed:

1. Between 19,000 and 15,500 cal yr B.P. the sedge family (Cyperaceae) reaches its greatest percentages. This sedge may be more tundra and cold meadow species than fully aquatic ones. Only two definitively aquatic types, pondweed (*Potamogeton natans*) and the green algae *Pediastrum*, occur during this interval.

2. Between ca. 15,500 and 15,400 cal yr B.P., sedges decrease by 50%, green algae increase and two new aquatic types, quillwort (*Isoetes*) and cattail (*Typha latifolia)* make their first occurrence in the record.

3. Circa 13,700 cal yr B.P., sedge is high again, but below pre-15,500 cal yr B.P. levels, and no other aquatic types occur besides green algae.

4. After 12,000 cal yr B.P., aquatic types become more diverse with simultaneous occurrence of sedges, cattail, pondweed and green algae. An interval of even higher aquatic diversification follows, with a maximum around 7800 cal yr B.P., with continuing representation by cattail, pondweed, and green algae and introduction of bur-reed (Sparganiaceae), duckweed (Lemnaceae), watermilfoil (*Myriophyllum alterniflorum*), water lily (*Nymphaea*), and arrowhead (*Sagittaria*).

5. Circa 7500 cal yr B.P., all but three aquatic types disappear. Sedge occurs at reduced levels, but one cattail species and green algae are distinctly higher in this sample. By 7000 cal yr B.P., this pattern is changed, with reduced levels of cattail and green algae but reappearance of pondweed. By 6700 cal yr B.P., cattail, sedge, and green algae are at local minima, and pondweed is the only other type represented.

6. At 5300 cal yr B.P., sedge and pondweed are relatively high and cattail and bur-reed reach maximum representation. Green algae are at their minimum at this time. Between 5300 and 3800 cal yr B.P., aquatic types are represented by green algae, sedge, cattail, and pondweed, but by 3500 cal yr B.P., only green algae and sedge occur.

7. Aquatic diversity remains low after 3500 cal yr B.P. with only cattail, sedge, and pondweed present at 2200 cal yr B.P. The youngest sample at 1300 cal yr B.P., is similarly depauperate, but contains the first occurrence of another cattail species (*T. angustifolium*), relatively strong representation by bur-reed, and low levels of sedge.

DISCUSSION

The Bear Lake paleovegetation record is largely in agreement with long-term regional paleoclimate reconstructions, many of which are based on pollen data. Records extending back several million years from sediments beneath modern Great Salt Lake (Davis and Moutoux, 1998), and 225 k.y. at Bear Lake (Jiménez-Moreno et al., 2007), indicate that sage-dominated steppe vegetation is regionally persistent over multiple glacial cycles, with increases in juniper, ragweed, and shadbush during interglacials, and in sage and conifers during glacials. Vegetation assemblages in the region are generally continuous through these glacial cycles, without the local extinctions that mark climate transitions along the tundra-forest boundary, for instance. Despite this, distinct climate-driven changes are identifiable at many sites by changes in relative abundances of steppe versus woodland/forest pollen (e.g., Thompson, 1990; Beiswenger, 1991; Fall et al., 1995; Davis and Moutoux, 1998; Quade et al., 1998; Munroe, 2003).

The most significant change in the Bear Lake pollen record, based on the Euclidean distances in the cluster analysis, is during deglaciation, the onset of which has been dated in various sites around the Great Basin between 17.5 and 16.0 ka (Licciardi et al., 2001. This deglacial interval has been tied to Heinrich Event 1 (Clark and Bartlein, 1995; Phillips et al., 1996). Although Heinrich events are thought to be cold periods with high moisture balance (Denton et al., 1999), Great Basin regional vegetation records indicate a trend toward warmer and moister conditions between 17,000 and 14,000 cal yr B.P. (Cole and Arundel, 2005;

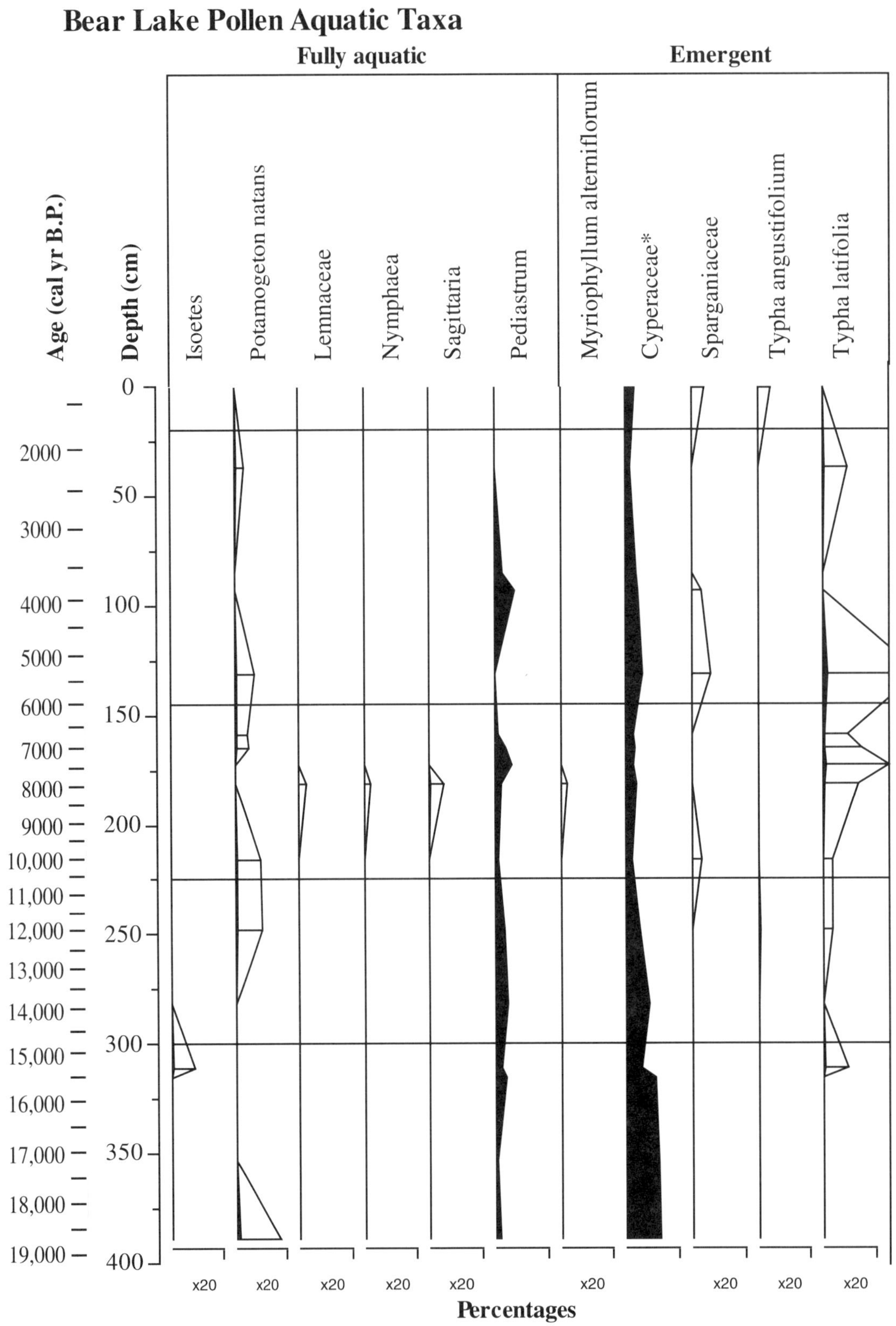

Bear Lake Pollen Aquatic Taxa

Figure 4. Aquatic pollen taxa from Bear Lake core BL96-2, shown as percentages. Tick marks and exaggeration are as in Figure 2. Horizontal zone lines are shown for reference although only one of these types (Cyperaceae) was included in the CONISS result. Cyperaceae (sedge) can occur as an upland plant in tundra biomes, or as an aquatic plant. Although its period of maximum representation is during the full glacial interval, it is shown here to highlight periods when the aquatic type might be present. The aquatic pollen are grouped into fully aquatic (submerged) and emergent types. Fully aquatic types can live in deep-water conditions (>20 m) if the water clarity is high, whereas emergent types live in shallow water (<3 m), along shorelines and in marshes.

Betancourt, 1990). Isotopes from northern Arizona packrat middens, and the biogeographic distribution of Utah agave found in the middens, indicate that minimum temperatures during this time were 1.5–3.5 °C warmer than before 17,000 yr B.P. (Cole and Arundel, 2005). Eolian records in the Canyonlands area suggest this warm and moist trend did not extend to the southeastern Colorado Plateau region (Reheis et al., 2005). At some time between 15,300 and 13,900 cal yr B.P., Bear Lake's vegetation switched from the cold-tolerant steppe/tundra plants that occurred in the full glacial interval to pine-woodland vegetation, agreeing in timing and level of response with the regional records but suggesting that the climate transition began after 15,300 cal yr B.P.

The deglacial vegetation at Bear Lake persisted from 13,900 to 12,000 cal yr B.P. The pine woodland vegetation of this period is consistent with regional records (Fall et al., 1995; Madsen et al., 2001). Although the Younger Dryas (ca. 12,900–11,600 cal yr B.P.) is documented in various records in western North America (Doerner and Carrera, 2001; Reasoner and Jodry, 2000; Polyak et al., 2004; Oviatt et al., 2005), it cannot be distinguished in the Bear Lake pollen data from the single sample located within the interval. Despite this, the second largest transition in the Bear Lake vegetation record overlaps part of the Younger Dryas, with the end of the deglacial and start of the early Holocene pollen zones (12,000 and 10,000 cal yr B.P.). Over this transition, conifers and cold-tolerant trees and herbs around Bear Lake gave way to sage-steppe plants.

Faunal data suggest that the early Holocene in the Great Basin region was moist and cool (Grayson, 2000; Madsen et al., 2001). Regional vegetation reconstructions are poorly dated for this interval, but mesic conditions are also suggested for Grays Lake, Idaho, by approx. 7900 cal yr B.P. (7100 ^{14}C yr B.P.) (Beiswenger, 1991) and for marsh deposits created by increased levels of Great Salt Lake, dated at 7650 ^{14}C yr B.P. (~8400 cal yr B.P.) (Murchison and Mulvey, 2000). Other records for the Great Basin also indicate greater effective moisture ca. 6400–6000 ^{14}C yr B.P. (7400–6800 cal yr B.P.) (summarized by Madsen et al., 2001). This interpretation of a cool, wet early Holocene is complicated by high-elevation vegetation records from the Uinta Mountains of northeastern Utah that indicate higher-than-modern temperatures by 9400 cal yr B.P. (Munroe, 2003), and by Grand Canyon midden records of Utah agave that suggest temperatures were above modern by 8500 cal yr B.P. (Cole and Arundel, 2005). In the Bear Lake record, the early Holocene interval is represented by just two samples, at 10,000 and 8000 cal yr B.P., in which sage reaches a maximum and tree pollen a minimum. Aquatic vegetation reaches highest diversity in the sample at 8000 cal yr B.P., suggesting that Bear Lake had a large extent of shallow water, consistent with flooding of the Bear Lake Valley. In the early Holocene pollen zone, diversity of plant types increases between 7500 and 6700 cal yr B.P. This high-diversity period includes an oscillation in vegetation (7500–7000 cal yr B.P.), with higher levels of cold-tolerant trees, shrubs, and herbs and green algae, while arid steppe plants, drought-tolerant trees, and emergent aquatic plants decreased. By the end of this oscillation, many types of pollen had returned to their earlier levels.

The Bear Lake vegetation changed to a conifer-dominated forest by 5300 cal yr B.P., with distinctly less sage and a wide variety of tree types. The vegetation then closely resembled that of the 7500–7000 cal yr B.P. oscillation, except that sage is lower in the post-5300 cal yr B.P. interval. Circa 3600 cal yr B.P. a smaller magnitude change occurred in the vegetation, with increased ponderosa pine, asters, poplar, and *Ephedra*, and decreased spruce, fir, willow and sage pollen. This is also seen in the 2200 cal yr B.P. sample. A return to cooler temperatures after 5000 cal yr B.P. is documented in other records with higher conifer representation at lower elevations, fauna consistent with cooler intervals, and increased levels of Great Salt Lake and Ruby Marsh (Madsen et al., 2001). The final transition in the Bear Lake vegetation record occurs between 2200 and 1300 cal yr B.P. Unfortunately the core-top is absent and the past 2000 yr is represented by only one sample. The vegetation represented in this sample differs from the middle Holocene ones in having a very low diversity and a higher proportion of pine. Pollen concentration in this sample is lower than in any other post-glacial sample.

Comparison of the Bear Lake vegetation record with other records from Bear Lake shows periods of concordance and discordance. The vegetation records support isotopic and magnetic indicators of Bear River influx and deglaciation after 17,000 cal yr B.P. (Rosenbaum and Heil, this volume). In addition, a sharp increase in $\delta^{18}O$ ca. 12,000 cal yr B.P., followed by an increase in Mg in endogenic calcite from 11,500 to 11,000 cal yr B.P. (Dean, this volume), coincides with the transition from deglacial to early Holocene vegetation. High lake levels postulated from isotope, diatom, and sedimentary records at 9000–7500 cal yr B.P. (Laabs and Kaufman, 2003; Dean, this volume; Moser and Kimball, this volume; Reheis et al., this volume; Smoot and Rosenbaum, this volume) are poorly resolved by the pollen record with samples only at 10,000 and 8000 cal yr B.P. The upland record in these two samples indicates arid conditions, with minima in tree types and maxima in steppe plants until 7500 cal yr B.P., yet the aquatic vegetation record concurs with a lake highstand interpretation at 8000 cal yr B.P. Following this evidence of a highstand, the maximum in tree vegetation between 7500 and 7000 cal yr B.P. suggests cooler, moister climates in the Bear Lake watershed. At the same time, a minor increase in quartz and magnetic susceptibility values in the Bear Lake sediments points to Bear River influences in the lake (Rosenbaum and Heil, this volume).

The offset in timing between the interpreted lake highstand in the lake geochemistry records (9000–7500 cal yr B.P.) and corroborating evidence in the upland pollen record (7500–7000 cal yr B.P.) is ~1000 yr. Although there is not sufficient data to explain the cause of this offset, it could be due to lags in overland transport of pollen from higher-elevation regions in the watershed to the lake. During periods of higher effective moisture, upland erosion may be minimized by vegetation cover. Vegetation die-offs in subsequent arid intervals would allow erosional downcutting that would transport both pollen and minerogenic matter to the lake. This hypothesis would also explain the increased influx of quartz at 7500–7000 cal yr B.P. It does not explain, however, why indeterminate and pre-Quaternary types, both indicators of erosional down-cutting, did not increase until the end of the tree-rich interval, at 6700 cal yr B.P. Alternatively, increased winter precipitation could create a highstand from 9000 to 7500 cal yr B.P. without significantly affecting upland vegetation if climate conditions during the growing season remained relatively dry. In the later Holocene, peaks in calcite mass accumulation rates and minor fluctuations in $\delta^{18}O$, ca. 4500–2800 cal yr B.P. (Dean et al., 2006), seem to be in phase with the 3800–3600 cal yr B.P. transition in the Bear Lake vegetation, and in regional vegetation records, toward cooler, more mesic conditions.

CONCLUSIONS

In summary, the Bear Lake vegetation record corresponds well, although with low temporal resolution, to regional records of change in vegetation and climate, documented in pollen, isotope, and biogeographic records in the Basin and Range region. Significant changes in the Bear Lake vegetation occurred during the intervals 15,300–13,900, 12,000–10,000, 7500–6700, 6700–5300, 3800–3600, and 2200–1300 cal yr B.P. All of these intervals, except for the last, fit within periods of change detected in other Bear Lake paleoenvironmental proxies, suggesting that those changes at Bear Lake were due to climatic forcing. In the pollen record, maximum aridity and warmth are indicated from 12,000 to 7500 cal yr B.P., but this is based on three samples separated by millennia. Data are insufficient to show whether the apparent aridity in the early Holocene was continuous or interrupted by a mesic interval, 9000–7500 cal yr B.P., as suggested by other Bear Lake environmental proxies, although aquatic pollen do support evidence of a highstand by 8000 cal yr B.P. The Bear Lake vegetation record indicates that several intervals change toward a more mesic and cool environment around the lake after 7500 cal yr B.P., compared to early Holocene conditions. The past 2000 yr are represented by only one sample, ca. 1300 cal. yr B.P., that is distinctly different from any other in the preceding 5000 years.

ACKNOWLEDGMENTS

The U.S. Geological Survey, Earth Surface Processes Team in Denver, Colorado, provided the samples, equipment, reference materials, and support services for this work, including helpful advice on the editing and scientific content.

ARCHIVED DATA

Archived data for this chapter can be obtained from the NOAA World Data Center for Paleoclimatology at http://www.ncdc.noaa.gov/paleo/pubs/gsa2009bearlake/.

REFERENCES CITED

Albee, B.J., Shultz, L.M., and Goodrich, S., 1988, Atlas of the vascular plants: Digital version: Utah Museum of Natural History, http://www.gis.usu.edu/Geography-Department/utgeog/utvatlas/ (accessed December 2005).

Amayreh, J., 1995, Lake evaporation: A model study [Ph.D. thesis]: Logan, Utah State University, 178 p.

Anderson, R.S., Hasbargen, J., Koehler, P.A., and Feiler, E.J., 1999, Late Wisconsin and Holocene subalpine forests of the Markagunt Plateau of Utah, southwestern Colorado Plateau, U.S.A.: Arctic, Antarctic, and Alpine Research, v. 31, p. 366–378, doi: 10.2307/1552585.

Barnosky, C.W., Anderson, P.M., and Bartlein, P.J., 1987, The northwestern U.S. during deglaciation; Vegetational history and paleoclimatic implications, *in* Ruddiman, W.F., and Wright, H.E., eds., North America and adjacent oceans during the last deglaciation: Boulder, Colorado, Geological Society of America, Geology of North America, v. K-3, p. 289–321.

Beiswenger, J.M., 1991, Late Quaternary vegetational history of Grays Lake, Idaho: Ecological Monographs, v. 61, p. 165–182, doi: 10.2307/1943006.

Benson, L.V., Currey, D.R., Dorn, R.I., Lajoie, K.R., Oviatt, C.G., Robinson, S.W., Smith, G.I., and Stine, S., 1990, Chronology of expansion and contraction of four Great Basin lake systems during the past 35,000 years: Palaeogeography, Palaeoclimatology, Palaeoecology, v. 78, p. 241–286, doi: 10.1016/0031-0182(90)90217-U.

Betancourt, J.L., 1990, Late Quaternary biogeography of the Colorado Plateau, *in* Betancourt, J.L., Devender, T.R.V., and Martin, P.S., eds., Packrat middens: The last 40,000 years of biotic change: Tuscon, Arizona, University of Arizona Press, p. 259–293.

Birdsey, P.W., 1989, The limnology of Bear Lake, Utah-Idaho, 1912–1988: A literature review: Utah Department of Natural Resources, Division of Wildlife Resources, Publication no. 89–5, 113 p.

Bourgeois, J.C., 1990, Seasonal and annual variation of pollen content in the snow of a Canadian high Arctic ice cap: Boreas, v. 19, p. 313–322.

Bradshaw, R.H.W., 1994, Quaternary terrestrial sediments and spatial scale: The limits to interpretation, *in* Traverse, A., ed., Sedimentation of organic particles: Cambridge, UK, Cambridge University Press, p. 239–252.

Bright, J., 2009, this volume, Chapter 4, Isotope and major-ion chemistry of groundwater in Bear Lake Valley, Utah and Idaho, with emphasis on the Bear River Range, *in* Rosenbaum, J.G., and Kaufman, D.S., eds., Paleoenvironments of Bear Lake, Utah and Idaho, and its catchment: Geological Society of America Special Paper 450, doi: 10.1130/2009.2450(04).

Bright, J., 2009, this volume, Chapter 8, Ostracode endemism in Bear Lake, Utah and Idaho, *in* Rosenbaum, J.G., and Kaufman, D.S., eds., Paleoenvironments of Bear Lake, Utah and Idaho, and its catchment: Geological Society of America Special Paper 450, doi: 10.1130/2009.2450(08).

Burns, R.M., and Honkala, B.H., technical coordinators, 1990, Silvics of North America: 1. Conifers; 2. Hardwoods: Washington, D.C., U.S. Department of Agriculture, Forest Service, Agriculture Handbook 654, v. 2, 877 p.

Clark, P.U., and Bartlein, P.J., 1995, Correlation of late Pleistocene glaciation in the western United States with North American Heinrich Events: Geology, v. 23, p. 483–486, doi: 10.1130/0091-7613(1995)023<0483:COLPGI>2.3.CO;2.

Cole, K.L., and Arundel, S.T., 2005, Carbon isotopes from fossil packrat pellets and elevational movements of Utah agave plants reveal the Younger Dryas cold period in Grand Canyon, Arizona: Geology, v. 33, p. 713–716, doi: 10.1130/G21769.1.

Colman, S.M., Rosenbaum, J.G., Kaufman, D.S., Dean, W.E., and McGeehin, J.P., 2009, this volume, Radiocarbon ages and age models for the last 30,000 years in Bear Lake, Utah and Idaho, *in* Rosenbaum, J.G., and Kaufman, D.S., eds., Paleoenvironments of Bear Lake, Utah and Idaho, and its catchment: Geological Society of America Special Paper 450, doi: 10.1130/2009.2450(05).

Cushing, E.J., 1967, Evidence for differential pollen preservation in late Quaternary sediments in Minnesota: Review of Palaeobotany and Palynology, v. 4, p. 87–101, doi: 10.1016/0034-6667(67)90175-3.

Davis, O.K., and Moutoux, T.E., 1998, Tertiary and Quaternary vegetation history of the Great Salt Lake, Utah, USA: Journal of Paleolimnology, v. 19, p. 417–427, doi: 10.1023/A:1007959203433.

Dean, W.E., 2009, this volume, Endogenic carbonate sedimentation in Bear Lake, Utah and Idaho, over the last two glacial-interglacial cycles, *in* Rosenbaum, J.G., and Kaufman, D.S., eds., Paleoenvironments of Bear Lake, Utah and Idaho, and its catchment: Geological Society of America Special Paper 450, doi: 10.1130/2009.2450(07).

Dean, W.E., Rosenbaum, J., Skipp, G., Colman, S., Forester, R., Liu, A., Simmons, K., and Bischoff, J., 2006, Unusual Holocene and late Pleistocene carbonate sedimentation in Bear Lake, Utah and Idaho, U.S.A.: Sedimentary Geology, v. 185, p. 93–112, doi: 10.1016/j.sedgeo.2005.11.016.

Denton, G.H., Heusser, C.J., Lowell, T.V., Moreno, P.I., Anderson, B.G., Heusser, L.E., Schlüchter, C., and Marchant, D.R., 1999, Interhemispheric linkage of paleoclimate during the last glaciation: Geografiska Annaler, v. 18A, p. 107–153, doi: 10.1111/j.0435-3676.1999.00055.x.

Digerfeldt, G., 2003, Studies on past lake-level fluctuations, *in* Berglund, B.E., ed., Handbook of Holocene palaeoecology and palaeohydrology: Caldwell, New Jersey, Blackburn Press, p. 127–143.

Doerner, J.P., and Carrera, P.E., 2001, Late Quaternary vegetation and climatic history of the Long Valley area, west-central Idaho, U.S.A.: Quaternary Research, v. 56, p. 103–111, doi: 10.1006/qres.2001.2247.

Faegri, K., and Iversen, J., 1989, Textbook of pollen analysis (4ᵗʰ edition, Faegri, K., Kaland, P.E., and Krzywinski, K., eds.): New York, John Wiley and Sons, 328 p.

Fall, P.L., 1992, Spatial patterns of atmospheric pollen dispersal in the Colorado Rocky Mountains, USA: Review of Palaeobotany and Palynology, v. 74, p. 293–313, doi: 10.1016/0034-6667(92)90013-7.

Fall, P.L., Davis, P.T., and Zielinski, G.A., 1995, Late Quaternary vegetation and climate of the Wind River Range, Wyoming: Quaternary Research, v. 43, p. 393–404, doi: 10.1006/qres.1995.1045.

Gaugh, H.G., Jr., 1982, Multivariate analysis in community ecology: New York, Cambridge University Press, 298 p.

Grayson, D.K., 2000, Mammalian responses to middle Holocene climatic change in the Great Basin of the western United States: Journal of Biogeography, v. 27, p. 181–192, doi: 10.1046/j.1365-2699.2000.00383.x.

Grimm, E., 1987, CONISS: A Fortran 77 program for stratigraphically constrained cluster analysis by the method of incremental sum of squares: Computers and Geosciences, v. 13, p. 13–35, doi: 10.1016/0098-3004(87)90022-7.

Grimm, E., 1990, TILIAGRAPH v.1.2 pollen graphics program: Springfield, Illinois State Museum.

Grimm, E., 1998, TILIAGRAPH v. 2.0.5.b pollen graphics program: Springfield, Illinois State Museum.

Heusser, C.J., and Peteet, D.M., 1988, Spores of *Lycopodium* and *Selaginella* of North Pacific America: Canadian Journal of Botany, v. 66, p. 508–525.

Jackson, S.T., 1994, Pollen and spores in Quaternary lake sediments as sensors of vegetation composition: Theoretical models and empirical evidence, *in* Traverse, A., ed., Sedimentation of organic particles: Cambridge, UK, Cambridge University Press, p. 253–286.

Jiménez-Moreno, G., Anderson, R.S., and Fawcett, P.J., 2007, Orbital- and millennial-scale vegetation and climate changes of the past 225 ka from Bear Lake, Utah-Idaho (USA): Quaternary Science Reviews, v. 26, p. 1713–1724, doi: 10.1016/j.quascirev.2007.05.001.

Kaliser, B.N., 1972, Environmental geology of Bear Lake area, Rich County, Utah: Utah Geological and Mineralogical Survey Bulletin, v. 96, 32 p.

Kaufman, D.S., Bright, J., Dean, W.E., Rosenbaum, J.G., Moser, K., Anderson, R.S., Colman, S.M., Heil, C.W., Jr., Jiménez-Moreno, G., Reheis, M.C., and Simmons, K.R., 2009, this volume, A quarter-million years of paleoenvironmental change at Bear Lake, Utah and Idaho, *in* Rosenbaum, J.G., and Kaufman, D.S., eds., Paleoenvironments of Bear Lake, Utah and Idaho, and its catchment: Geological Society of America Special Paper 450, doi: 10.1130/2009.2450(14).

Laabs, B.J.C., and Kaufman, D.S., 2003, Quaternary highstands in Bear Lake Valley, Utah and Idaho: Geological Society of America Bulletin, v. 115, p. 463–478, doi: 10.1130/0016-7606(2003)115<0463:QHIBLV>2.0.CO;2.

Licciardi, J.M., Clark, P.U., Brook, E.J., Pierce, K.L., Kurz, M.D., Elmore, D., and Sharma, P., 2001, Cosmogenic ³He and ¹⁰Be chronologies of the late Pinedale northern Yellowstone ice cap, Montana, USA: Geology, v. 29, p. 1095–1098, doi: 10.1130/0091-7613(2001)029<1095:CHABCO>2.0.CO;2.

Madsen, D.B., and Currey, D.R., 1979, Late Quaternary glacial and vegetation changes, Little Cottonwood Canyon area, Wasatch Mountains, Utah: Quaternary Research, v. 12, p. 254–270, doi: 10.1016/0033-5894(79)90061-9.

Madsen, D.B., Rhode, D., Grayson, D.K., Broughton, J.M., Livingston, S.D., Hunt, J., Quade, J., Schmitt, D.N., and Shaver, M.W., 2001, Late Quaternary environmental change in the Bonneville basin, western USA: Palaeogeography, Palaeoclimatology, Palaeoecology, v. 167, p. 243–271, doi: 10.1016/S0031-0182(00)00240-6.

Markgraf, V., 1980, Pollen dispersal in a mountain area: Grana, v. 19, p. 127–146.

McAndrews, J.H., Berti, A.A., and Norris, G., 1973, Key to the Quaternary pollen and spores of the Great Lakes region: Royal Ontario Museum, Life Sciences Miscellaneous Publication, v. 64.

Moore, P.D., Webb, J.A., and Collinson, M.E., 1991, Pollen analysis: London, Blackwell Scientific Publications, 216 p.

Moser, K.A., and Kimball, J.P., 2009, this volume, A 19,000-year record of hydrologic and climatic change inferred from diatoms from Bear Lake, Utah and Idaho, *in* Rosenbaum, J.G., and Kaufman, D.S., eds., Paleoen-

vironments of Bear Lake, Utah and Idaho, and its catchment: Geological Society of America Special Paper 450, doi: 10.1130/2009.2450(10).

Munroe, J.S., 2003, Holocene timberline and palaeoclimate of the northern Uinta Mountains, northeastern Utah, USA: The Holocene, v. 13, p. 175–185, doi: 10.1191/0959683603hl600rp.

Murchison, S.B., and Mulvey, W.E., 2000, Late Pleistocene and Holocene shoreline stratigraphy on Antelope Island, *in* King, J., ed., Geology of Antelope Island: Utah Geological Survey Miscellaneous Publication 00-1, p. 77–83.

Overpeck, J.T., Webb, T., III, and Prentice, I.C., 1985, Quantitative interpretation of fossil pollen spectra: Dissimilarity coefficients and the method of modern analogs: Quaternary Research, v. 23, p. 87–108, doi: 10.1016/0033-5894(85)90074-2.

Oviatt, C.G., Miller, D., McGeehin, J., Zachary, C., and Mahan, S., 2005, The Younger Dryas phase of Great Salt Lake, Utah, USA: Palaeogeography, Palaeoclimatology, Palaeoecology, v. 219, p. 263–284, doi: 10.1016/j.palaeo.2004.12.029.

Phillips, F.M., Zreda, M.G., Benson, L.V., Plummer, M.A., Elmore, D., and Sharma, P., 1996, Chronology for fluctuations in late Pleistocene Sierra Nevada glaciers and lakes: Science, v. 274, p. 749–751, doi: 10.1126/science.274.5288.749.

Polyak, V.J., Rasmussen, J.B.T., and Asmerom, Y., 2004, Prolonged wet period in the southwestern United States through the Younger Dryas: Geology, v. 32, p. 5–8, doi: 10.1130/G19957.1.

Quade, J., Forester, R.M., Pratt, W.L., and Carter, C., 1998, Black mats, spring-fed streams, and late-glacial-age recharge in the Southern Great Basin: Quaternary Research, v. 49, p. 129–148, doi: 10.1006/qres.1997.1959.

Reasoner, M.A., and Jodry, M.A., 2000, Rapid response of alpine timberline vegetation to the Younger Dryas climate oscillation in the Colorado Rocky Mountains, USA: Geology, v. 28, p. 51–54, doi: 10.1130/0091-7613(2000)28<51:RROATV>2.0.CO;2.

Reheis, M.C., Reynolds, R.L., Goldstein, H., Roberts, H.M., Yount, J.C., Axford, Y., Cummings, L.S., and Shearin, N., 2005, Late Quaternary eolian and alluvial response to paleoclimate, Canyonlands, southeastern Utah: Geological Society of America Bulletin, v. 117, p. 1051–1069, doi: 10.1130/B25631.1.

Reheis, M.C., Laabs, B.J.C., and Kaufman, D.S., 2009, this volume, Geology and geomorphology of Bear Lake Valley and upper Bear River, Utah and Idaho, *in* Rosenbaum, J.G., and Kaufman, D.S., eds., Paleoenvironments of Bear Lake, Utah and Idaho, and its catchment: Geological Society of America Special Paper 450, doi: 10.1130/2009.2450(02).

Rosenbaum, J.G., and Heil, C.W., Jr., 2009, this volume, The glacial/deglacial history of sedimentation in Bear Lake, Utah and Idaho, *in* Rosenbaum, J.G., and Kaufman, D.S., eds., Paleoenvironments of Bear Lake, Utah and Idaho, and its catchment: Geological Society of America Special Paper 450, doi: 10.1130/2009.2450(11).

Smoot, J.P., and Rosenbaum, J.G., 2009, this volume, Sedimentary constraints on late Quaternary lake-level fluctuations at Bear Lake, Utah and Idaho, *in* Rosenbaum, J.G., and Kaufman, D.S., eds., Paleoenvironments of Bear Lake, Utah and Idaho, and its catchment: Geological Society of America Special Paper 450, doi: 10.1130/2009.2450(12).

Tauber, H., 1977, Investigations of aerial pollen transport in a forested area: Dansk Botanisk Arkiv, v. 32, p. 1–121.

Thompson, R.S., 1990, Late Quaternary vegetation and climate in the Great Basin, *in* Betancourt, J.L., et al., eds., Packrat middens—The last 40,000 years of biotic change: Tucson, University of Arizona Press, p. 200–239.

Thompson, R.S., Whitlock, C., Bartlein, P.J., Harrison, S.P., and Spaulding, W.G., 1993, Climatic changes in the western United States since 18,000 yr B.P., *in* Wright, H.E., Jr., Kutzbach, J.E., Webb, T., III, Ruddiman, W.F., Street-Perrott, F.A., and Bartlein, P.J., eds., Global climates since the last glacial maximum: Minneapolis, University of Minnesota Press, p. 468–513.

Thompson, R.S., Anderson, K.H., and Bartlein, P.J., 2000, Atlas of relations between climatic parameters and distributions of important trees and shrubs in North America—Introduction and conifers: U.S. Geological Survey Professional Paper 1650-A, p. 1–269.

Welsh, S.L., Atwood, N.D., Goodrich, S., and Higgins, L.C., 2003, A Utah flora, 3ʳᵈ edition: Provo, Brigham Young University Press, 912 p.

Wurtsbaugh, W., and Luecke, C., 1998, Limnological relationships and population dynamics of fishes in Bear Lake (Utah/Idaho): Salt Lake City, Utah Division of Wildlife Resources, Report of Project F-47-R, Study 5, 73 p.

MANUSCRIPT ACCEPTED BY THE SOCIETY 15 SEPTEMBER 2008

The Geological Society of America
Special Paper 450
2009

A 19,000-year record of hydrologic and climatic change inferred from diatoms from Bear Lake, Utah and Idaho

Katrina A. Moser
Department of Geography, University of Western Ontario, 1151 Richmond Street North, London, Ontario N6A 5C2, Canada

James P. Kimball
Department of Geography, University of Utah, Salt Lake City, Utah 84112-9155, USA

ABSTRACT

Changes in diatom fossil assemblages from lake sediment cores indicate variations in hydrologic and climatic conditions at Bear Lake (Utah-Idaho) during the late glacial and Holocene. From 19.1 to 13.8 cal ka there is an absence of well-preserved diatoms because prolonged ice cover and increased turbidity from glacier-fed Bear River reduced light and limited diatom growth. The first well-preserved diatoms appear at 13.8 cal ka. Results of principal components analysis (PCA) of the fossil diatom assemblages from 13.8 cal ka to the present track changes related to fluctuations of river inputs and variations of lake levels. Diatom abundance data indicate that the hydrologic balance between 13.8 and 7.6 cal ka is strongly tied to river inputs, whereas after 7.6 cal ka the hydrologic balance is more influenced by changes in lake evaporation. Wet conditions maintained high river inputs from 13.8 to 10.8 cal ka and from 9.2 to 7.6 cal ka, with a dry interval between 10.8 and 9.2 cal ka. After 9.2 cal ka until 2.9 cal ka lake levels were high except for two periods, one between 7.6 and 5.8 cal ka and one between 4.3 and 3.8 cal ka, as a result of decreased effective moisture. After 2.9 cal ka, fossil diatom assemblages suggest drier conditions until 1.6 cal ka to the present, when fragments of large, pennate diatoms appear, possibly the result of a rapid lake transgression. Although similarities exist between the Bear Lake records and other western hydrologic and climatic records, the covariations are not strong. Our data suggest that climatic regimes at Bear Lake have changed frequently over time, perhaps as a consequence of the position of several important climatic boundaries near Bear Lake.

Moser, K.A., and Kimball, J.P., 2009, A 19,000-year record of hydrologic and climatic change inferred from diatoms from Bear Lake, Utah and Idaho, *in* Rosenbaum, J.G., and Kaufman, D.S., eds., Paleoenvironments of Bear Lake, Utah and Idaho, and its catchment: Geological Society of America Special Paper 450, p. 229–246, doi: 10.1130/2009.2450(10). For permission to copy, contact editing@geosociety.org. ©2009 The Geological Society of America. All rights reserved.

INTRODUCTION

Bear Lake is located on the border between Utah and Idaho, close to the eastern edge of Utah (Fig. 1). The lake is situated beneath the present-day mean winter position of the polar front and is just northwest of the region most strongly influenced by monsoonal precipitation (Mock, 1996; Adams and Comrie, 1997). As a result of its location, Bear Lake receives maximum precipitation during January and May (WRCC, 2007). The January peak results from Pacific airstreams flowing through the low-elevation gap of the Snake River Plain (Bryson and Hare, 1974), whereas the peak in May is due to meridional troughs and cutoff lows, which can draw moisture from the Gulf of Mexico and produce upslope precipitation (Hirschboeck, 1991). Owing to its position, Bear Lake is sensitive to climatic change, and small shifts in the atmospheric boundaries may result in significant changes to the hydrologic balance of Bear Lake. Tectonics, stream diversions, and changes in groundwater input could also affect the Bear Lake hydrologic balance.

Bear Lake, which is 32 km long and 6–13 km wide with an area of 280 km², contains sediments extending back hundreds of thousands and perhaps millions of years (Dean et al., 2006). The lake lies in a tectonically active half-graben underlain by Paleozoic and Mesozoic sedimentary rocks that include limestones, dolostones, shales, sandstones, and quartzites (Davidson, 1969). The maximum depth is 63 m, and the mean depth is 28 m (Birdsey, 1989). When full, the surface elevation is 1805 m above sea level (asl), but lake levels may have been as much as 11 m higher at times during the late Pleistocene (Laabs and Kaufman, 2003). Historically, Bear River bypassed the lake until it was diverted into the lake between 1911 and 1918, thereby transforming Bear Lake into a reservoir (Birdsey, 1989). The diversion increased the ratio of watershed area to lake area from 4.8 to 29.5 (Wurtsbaugh and Luecke, 1997). Prior to diversion, the lake was often topographically closed, overflowing intermittently. Since diversion, lake level has fluctuated, with minima in 1935 and 2004 ~5.5 m below full.

Bear Lake is an alkaline (pH = 8.4–8.6), oligotrophic system (Table 1; Birdsey, 1989; Wurtsbaugh and Hawkins, 1990). Although

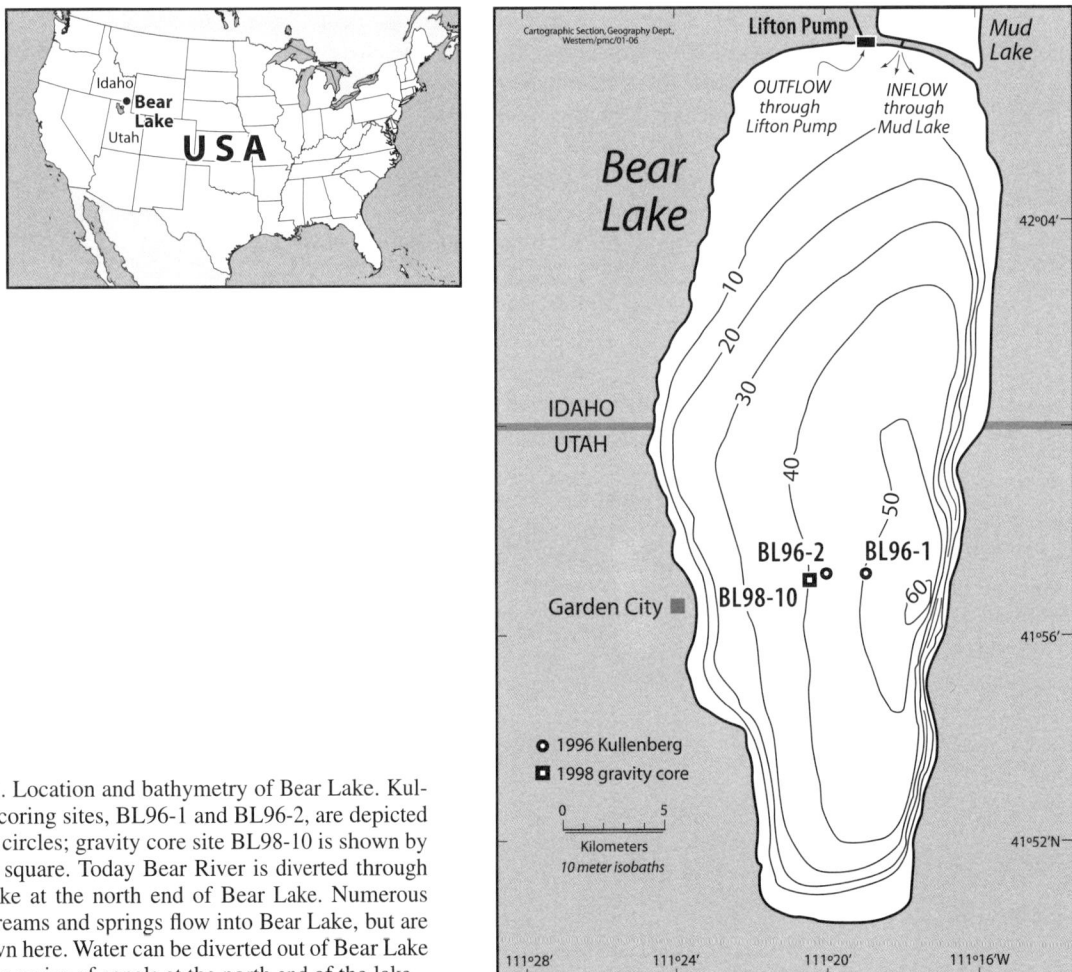

Figure 1. Location and bathymetry of Bear Lake. Kullenberg coring sites, BL96-1 and BL96-2, are depicted by open circles; gravity core site BL98-10 is shown by an open square. Today Bear River is diverted through Mud Lake at the north end of Bear Lake. Numerous small streams and springs flow into Bear Lake, but are not shown here. Water can be diverted out of Bear Lake through a series of canals at the north end of the lake.

TABLE 1. LIMNOLOGICAL VARIABLES OF BEAR LAKE PRE- AND POST-DIVERSION

Limnological variable	1912[a]	1979[b]	1980s	1997–1998[g]
pH	N/A	8.3–9.0[c]	7.4–9.1[d]	7.9–8.5
Total alkalinity (mg L^{-1})	586	265	194–267[d]	N/A
Ca (mg L^{-1})	4.1	69	25–76[e]	N/A
Mg (mg L^{-1})	152	41	28–52[e]	N/A
Na (mg L^{-1})	66.3	39	N/A	N/A
K (mg L^{-1})	10.5	3	N/A	N/A
Cl (mg L^{-1})	78.5	46	N/A	N/A
SO$_4$ (mg L^{-1})	96.8	16	N/A	N/A
Secchi depth (m)	N/A	N/A	~5[d]	4.1–11.2
Chlorophyll a (µg L^{-1})	N/A	0.25	0.25[d]	0.60
Total nitrogen (mg L^{-1})	N/A	N/A	0.25–1.1[f]	0.25
Total phosphorus (µg/L^{-1})	N/A	N/A	<20, except in spring[d]	Max. = 14–25
Total organic carbon	N/A	N/A	4[f]	N/A

Note: Data from Birdsey, 1989.
[a]Kemmerer et al.,1923;
[b]Werner, 1982;
[c]Lamarra et al., 1979;
[d]BLRC, 1986;
[e]Birdsey, 1982;
[f]Birdsey, 1985;
[g]Ecosystems Research Institute, 1998.

the lake does not freeze every year, it is dimictic (Wurtsbaugh and Hawkins, 1990). A deep chlorophyll maximum (20–30 m) can occur during the summer months (Birdsey, 1989). The diversion of Bear River into Bear Lake ca. 1912 markedly changed the chemistry of Bear Lake, increasing nutrient loading and decreasing alkalinity (Table 1; Birdsey, 1989; Dean et al., 2007).

Diatoms, single-celled algae that are characterized by cell walls composed of opaline silica, commonly record climatic (e.g., Smol and Cumming, 2000) and hydrologic (e.g., Wolin and Duthie, 1999) change. Diatoms have frequently been used to determine changes in lake level (Wolin and Duthie, 1999). For example, variations in lake depth result in changes in the relative areas of pelagic and littoral zones, and thus the ratio of planktic to benthic diatoms. It is, therefore, possible to infer past lake levels from the relative proportion of planktic to benthic diatoms (e.g., Bradbury, 1997). Lake-water salinity in closed basin lakes is also strongly linked to effective moisture (Fritz et al., 1991, 1999). In closed-basin lakes, decreases in effective moisture lead to ionic concentration (salinity), which results in changes in diatom community composition (Fritz et al., 1993, 1999). Diatoms have also been used to track changes in fluvial inputs (Ludlam et al., 1996).

In this research we use diatom fossil assemblages to interpret a 19,000-year record of environmental change. This record is compared with other paleolimnological evidence from Bear Lake in order to determine hydrologic and climatic change.

METHODS

Diatom analyses were performed on three cores, BL96-1, BL96-2, and BL98-10, which are respectively 5.00, 3.92, and 0.36 m long (Rosenbaum and Kaufman, this volume). Radiocarbon dating indicates that BL96-1 and BL96-2 span the past 7.6

and 19.1 cal ka, respectively (Colman et al., this volume), and ^{210}Pb dating indicates that BL98-10 spans the past few hundred years (Smoak and Swarzenski, 2004). The same cores that were used in Dean et al. (2006) are used in this manuscript; however, an improved chronological model is applied (Colman et al., this volume). In order to compare the data presented in Dean et al. (2006) with the data presented here, some of the data from the Dean et al. (2006) paper were re-plotted here using the new age model.

Preparation and analysis of 50 samples from BL96-1, 39 samples from BL96-2, and 11 samples from BL98-10 were made using standard procedures (Battarbee et al., 2001). First, a 10% HCl solution was used to remove carbonate minerals. After two or three washes of the slurry with deionized water, a mixture of concentrated nitric and sulfuric acid was applied to remove organic material. Residual acids were removed with a series of deionized water washes. Coverslips were prepared by evaporating a small aliquot of the resulting slurries, and were attached to glass slides using Naphrax® (refractive index > 1.74). Approximately 600 diatoms were identified and enumerated on each slide. Identification was done using oil immersion at 1000× on a Nikon Eclipse E600 microscope equipped with differential interface contrast optics. Diatom taxonomy was based mainly on Krammer and Lange-Bertalot (1986–1991) and Cummings et al. (1995).

A computer program for canonical community ordination (CANOCO; ter Braak and Šmilauer, 1999) was used to perform principal components analysis (PCA) to explore relationships among diatom taxa, and to compare diatom community composition changes with other paleolimnological proxies. For our analyses of the diatom data, we included only diatom taxa that occurred in at least two samples as well as in one sample in abundances greater than 1% (Table 2 lists these taxa and provides authority names).

TABLE 2. SPECIES NAMES AND CODES

Species Number	Species	Authority
1	*Pseudostaurosira brevistriata* (total)	(Grunow in Van Heurck) D.M. Williams and Round
2	*P. brevistriata*	(Grunow in Van Heurck) D.M. Williams and Round
3	*P. brevistriata*/rhomboid	
4	*P. brevistriata*/oval	
5	*Staurosira elliptica*	(Schumann) D.M. Williams and Round
6	*Staurosirella pinnata*	(Ehrenberg) D.M. Williams and Round
7	*S. pinnata* var. *accumunata*	(A. Mayer) Regenbogen D.M. Williams and Round
8	*Staurosira construens*	(Ehrenberg) Grunow D.M. Williams and Round
9	*Staurosirella leptostauron*	(Ehrenberg) Hustedt D.M. Williams and Round
10	*Nitzschia fonticola*	Grunow
11	*N. gracilis*	Hantzsch
12	*N.* sp.	
13	*Achnanthes curtissima*	Carter
14	*A. lanceolata*	(Brébisson) Grunow
15	*A. saccula*	Carter
16	*Navicula rhyncocephala*	Kützing
17	*N. pupula*	Kützing
18	*N. capitata* var. *lueneburgensis*	(Grunow) Patrick
19	*N. oblonga*	Kützing
20	*N. tuscula*	(Ehrenberg) Grunow
21	*N. cryptotenella*	Lange-Bertalot
22	*N.* sp.	
23	*Stephanodiscus minutulus*	(Kützing) Cleve and Moller
24	*S. niagarae*	Ehrenberg
25	*S. medius*	Håkansson
26	*Cyclotella michiganiana*	Skvortzow
27	*C. rossii*	Håkansson
28	*C. meneghiniana*	Kützing
29	*C.* sp.	
30	*C. bodanica* var. *affinis*	(Grunow) Cleve-Euler
31	*C. ocellata*	Pantocsek
32	*Diploneis elliptica*	(Kützing) Cleve
33	*Cocconeis placentula* var. *lineata*	(Ehrenberg) Van Heurck
34	*Diatoma tenue* var. *elongatum*	Lyngbye
35	*Caloneis* sp. Bear Lake	
36	*C. schumanniana*	(Grunow) Cleve
37	*Amphora pediculus*	(Kützing) Grunow
38	*A. inariensis*	Krammer
39	*A. libyca*	Ehrenberg
40	*A. ovalis*	(Kützing) Kützing
41	*Cymbella mesiana*	Cholnoky
42	*C. aspera*	(Ehrenberg) Cleve
43	*Surirella minuta*	Brébisson in Kützing
44	*Surirella ovalis*	Brébisson
45	*Pinnularia viridis*	(Nitzsch) Ehrenberg

RESULTS

Diatom Stratigraphy

The cored section was divided into five zones by visual inspection of the diatom stratigraphies (Figs. 2–4). Core BL96-2 includes zones 1, 2, 3, and 4, which were further divided into subzones 1a, 1b, 2a, 2b, 2c, 3a, 3b, and 3c. BL96-1 comprises zones 3 and 4, and BL98-10 contains zones 4 and 5. Diatom zone and subzone boundaries are sometimes coincident with the lithologic unit boundaries described by Dean et al. (2006) and Dean (this volume) (Figs. 2 and 3). The boundary between diatom zones 1a and 1b is roughly coincident with the boundary between lithologic unit 1, a red calcareous silty clay, and unit 2, a transitional red marl; the boundary between diatom zones 2a and 2b is approximately correlative with the boundary between unit 3, a green marl, and unit 4, a tan marl; the boundary between

diatom zones 2c and 3a is coincident with the boundary between unit 5, a gray marl, and unit 6, a tan marl. Diatom zones 3 and 4, and therefore all of core BL96-1, are composed only of lithologic unit 6.

Comparison of results from cores BL96-1 and BL96-2 indicates good agreement (i.e., within 100 yr) in the ages of the onset of subzone 3b (Table 3). For more recent zone boundaries, however, there are greater differences in the ages between the two cores. The discrepancies are ~300 yr and 500 yr for the onset of zones 3c and 4, respectively. Also, distinct spikes in *Navicula oblonga*, which probably record the same event, occur at 1.9 cal ka and 1.6 cal ka in core BL96-2 and BL96-1, respectively. These discrepancies may be due to errors in the age models caused by small, unrecognized unconformities in the cored sections (Smoot and Rosenbaum, this volume). For the discussion below, ages of correlative boundaries in BL96-1 and BL96-2 have been averaged. The age of the boundary between zones 4 and 5, recorded

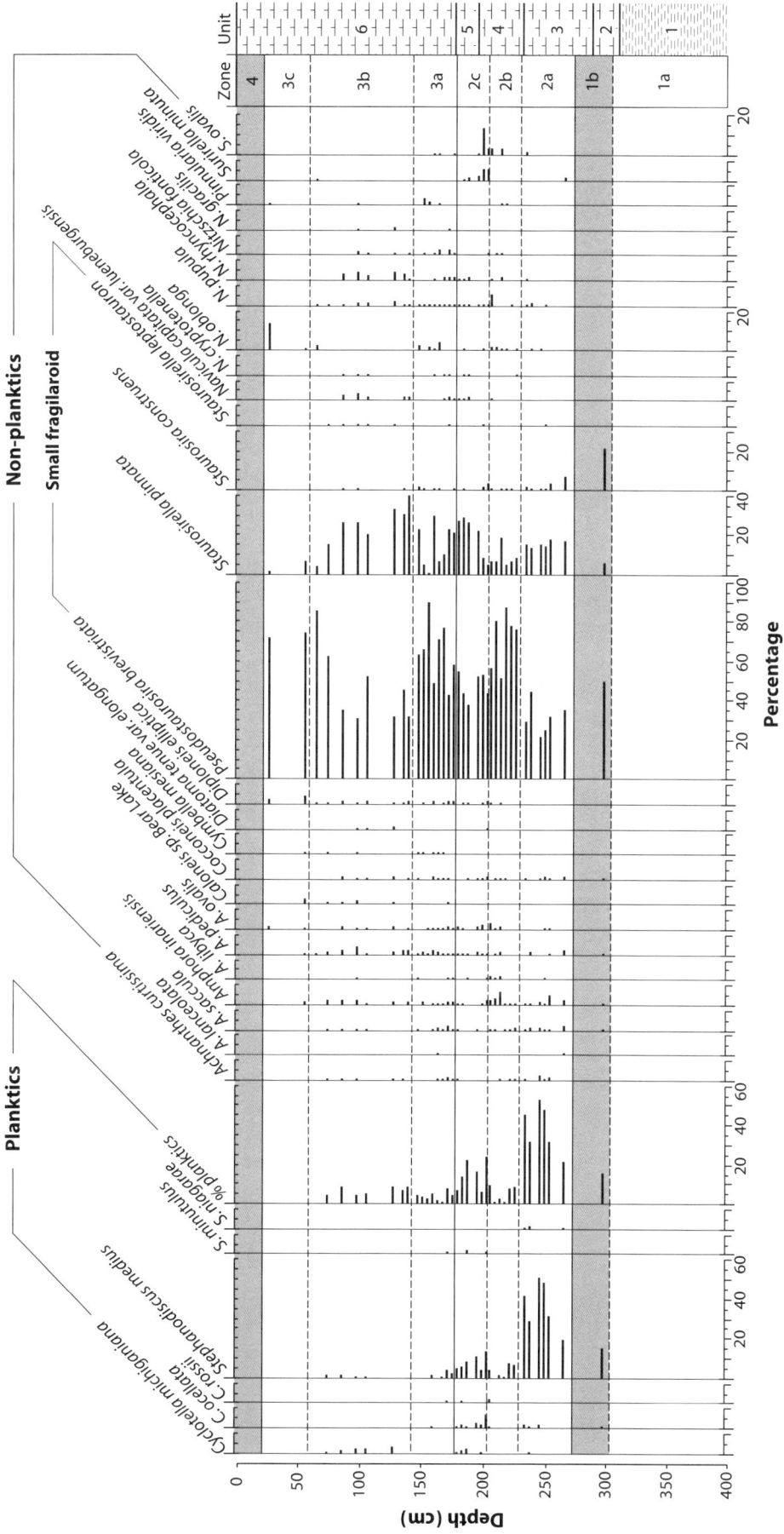

Figure 2. Diatom stratigraphy of BL96-2. Core BL96-2 extends to 19.1 cal ka and includes only the main diatom taxa. Zones depicted were determined by visual inspection and corroborated by principal components analysis (PCA). Zones 1 and 4 are distinguished from the other zones by the absence of well-preserved diatoms. Zone 1a contains no evidence of diatoms, whereas zones 1b and 4 contain dissolved and broken diatoms (shown by shading). PCA (Figs. 5–7) shows that zone 2 is distinct from zone 3 on the basis of diatom community composition, and this difference is attributed to changes in the hydrology of Bear Lake. During the deposition of zone 2, changes in river inputs affected the hydrology and chemistry of Bear Lake, whereas during zone 3, evaporation was more important in determining the hydrology and chemistry. Lithologic units are based on Dean et al. (2006) and are described in the text.

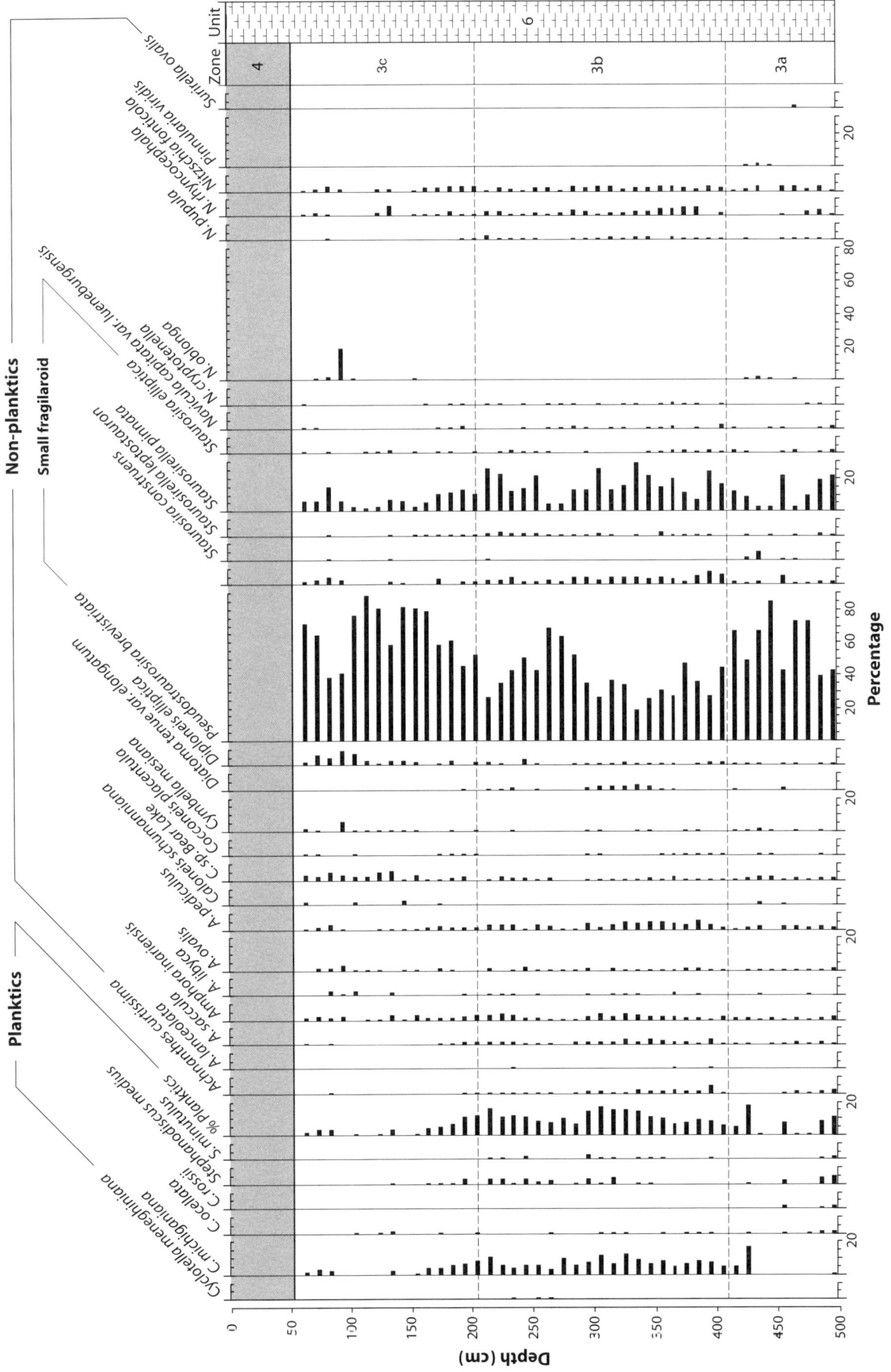

Figure 3. Diatom stratigraphy of BL96-1. Core BL96-1 spans the last 7.6 cal k.y. and includes only the main diatom taxa. Zones were determined by visual inspection and were corroborated by principal components analysis (PCA). The diatom community composition of these zones is virtually the same as in BL96-2.

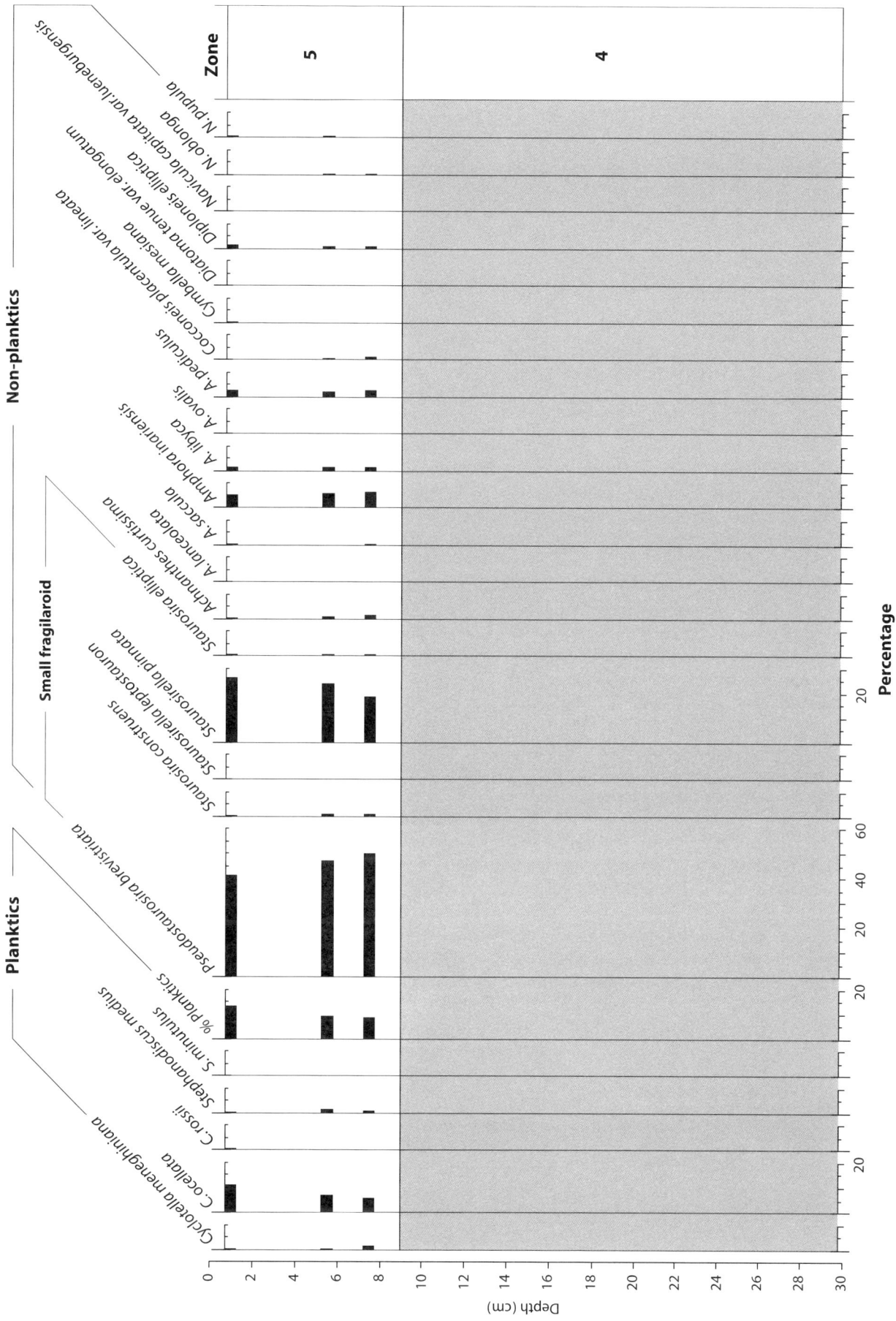

Figure 4. Diatom stratigraphy of BL98-10. The base of core BL98-10 is not dated, but the chronology for the upper part of the core is provided by ^{210}Pb (Smoak and Swarzenski, 2004). Only the main diatom taxa were included. BL98-10 contains zones 4 and 5, and ^{210}Pb indicates that the top of zone 4 is ca. A.D. 1936, although Dean et al. (2007) suggest that this age is too old. Zone 4 contains mainly diatom fragments, and zone 5 represents modern diatom assemblages of Bear Lake.

in core BL96-1 and BL98-10, is uncertain. Zone 4 comprises the top 50 cm of BL96-1, and the uppermost sediments have an age of 0.7 cal ka; however, the most recent sediments are missing (Colman et al., this volume). In core BL98-10 the top of zone 4 occurs between 8 and 9 cm, which is equivalent to ca. A.D. 1936 according to ^{210}Pb ages (Smoak and Swarzenski, 2004); however, Dean (this volume) suggested that the ^{210}Pb ages are too old. For this paper, we use the ca. A.D. 1936 date to refer to the top of zone 4, recognizing that this age is likely too old. All ages are in calendar years before 1950.

Zone 1 (19.1–13.8 cal ka)

Zone 1a (19.1–14.6 cal ka), which contains no evidence of diatoms, is distinguished from zone 1b (14.6–13.8 cal ka) by the appearance of broken and dissolved diatoms.

Zone 2 (13.8–7.6 cal ka)

Zone 2 marks the appearance of well-preserved diatoms. Relative to the overlying sediments, zone 2 contains abundant planktic diatoms, indicating large proportions of pelagic area relative to littoral area resulting from high lake levels (Wolin and Duthie, 1999). Zone 2a (13.8–10.8 cal ka) contains the greatest percentage of planktic species in the stratigraphy (an average of 38.4%). Of the planktic diatoms, *Stephanodiscus medius* is most common, and *Cyclotella ocellata* is common in lower percentages. *Stephanodiscus medius* is often associated with eutrophic conditions and is common during the transition from spring circulation to summer stratification (Kienel et al., 2005). The trophic status of *Cyclotella ocellata* is more ambiguous (Baier et al., 2004), but it has been found in the deep chlorophyll maximum (Stoermer et al., 1996) suggesting its presence may be indicative of deeper light penetration and/or strong stratification during late spring and early summer (Stoermer, 1993; Stoermer et al., 1996). The presence of these two diatoms in zone 2a suggests ameliorating conditions, with warmer temperatures and less turbidity than in zone 1, which would lead to thermal stratification.

The benthic diatom *Pseudostaurosira brevistriata*, which is the most common diatom throughout the core, is least abundant in zone 2a, whereas *Staurosirella pinnata* and *Staurosira*

construens are relatively abundant. All three of these diatoms are epipelic and often found in association in alkaline waters. *Staurosirella pinnata* and *Staurosira construens* are indicative of mesotrophic waters, whereas *Pseudostaurosira brevistriata* are indicative of more oligotrophic waters (Cummings et al., 1995; Moser et al., 2004).

Zone 2b (10.8–9.2 cal ka) is distinguished from zone 2a by an abrupt decrease in the abundance of planktic species (average = 4.16%) and an increase in the abundance of the benthic species *Pseudostaurosira brevistriata*. The abundance of another benthic species, *Staurosirella pinnata*, declines.

Proportions of planktic taxa (average = 13% in BL96-2)—primarily *Stephanodiscus medius*, but also *Cyclotella ocellata* and *C. michiganiana*—in zone 2c (9.2–7.6 cal ka) are greater than in zone 2b, but less than in zone 2a. The abundance of *Pseudostaurosira brevistriata* declines relative to subzone 2b, whereas the abundance of *Staurosirella pinnata* increases. The base of subzone 2c is marked by a peak in *Surirella ovalis* and *Surirella minuta*. In samples from Bear Lake and inflowing streams, *Surirella ovalis* was only found in appreciable amounts (>1%) in Bear River sediments (Kimball, 2001). *Surirella minuta* is associated with fresh waters (Risberg et al., 1999) and has been found in eutrophic waters (Krammer and Lange-Bertalot, 1988).

Zone 3 (7.6–1.8 cal ka)

The boundary between zones 2 and 3 is defined by an abrupt decrease in the abundances of planktic diatoms (average = 3.1% in BL96-2 and 7.3% in BL96-1) and a substantial increase in the abundance of small, benthic/tychoplanktic fragilarioid taxa (average = 76.5% in BL96-2 and 69.2% in BL96-1). These shifts indicate a marked change in the hydrology of Bear Lake.

Zone 3a (7.6–5.9 cal ka) is delimited by the near disappearance of planktic species, a decrease of the abundance of the benthic diatom *Staurosirella pinnata*, and an increase of the abundance of *Pseudostaurosira brevistriata*.

Zone 3b (5.9–2.9 cal ka) is distinguished from zone 3a by an increase in the abundance of planktic species, mainly *Cyclotella michiganiana* and *Stephanodiscus medius*. The epiphytic species *Diatoma tenue* var. *elongatum*, which is a eutrophic diatom, reached its greatest abundance (1.6%) at Bear Lake in the middle of this zone. The benthic diatom *Pseudostaurosira brevistriata* shows an overall decline, whereas *Staurosirella pinnata* increases.

Zone 3c (2.9 to1.6 cal ka) is marked by an abrupt decrease in the abundance of planktic species and the benthic diatom *Staurosirella pinnata*. *Pseudostaurosira brevistriata* increases and *Caloneis* sp. Bear Lake reaches its greatest abundance in this zone. We were not able to identify *Caloneis* sp. Bear Lake as a previously described diatom, and recording the taxonomy is beyond the scope of this paper. This diatom has not been not observed in Bear Lake today (Kimball, 2001), so we have no ecological information for this taxon, although it is likely a epipelic diatom. A large spike in the abundance of *Navicula oblonga*, a epipelic, halophilous species, occurs in this zone.

TABLE 3. COMPARISON OF THE TIMING OF
THE ONSET OF ZONES BETWEEN CORES

Zone	BL96-2	BL96-1	BL98-10
1a	19.1	N/A	N/A
1b	14.6	N/A	N/A
2a	13.8	N/A	N/A
2b	10.8	N/A	N/A
2c	9.2	N/A	N/A
3a	7.6	N/A	N/A
3b	5.8	5.9	N/A
3c	3.0	2.7	N/A
4	1.8	1.3	N/A
5	N/A	N/A	1998–present

Note: all ages are in cal ka years

Zone 4 (1.6 cal ka to ca. A.D. 1936)

Zone 4 is characterized by broken and uncountable diatoms. Most of the diatom pieces observed were from large, pennate diatoms, including *Navicula oblonga*, *Diploneis elliptica*, *Pinnularia viridis*, and *Cymbella mesiana*, as well as the small, benthic/tychoplanktic fragilarioid taxa. *Pinnularia viridis* is well adapted to drying conditions; when conditions become dry this diatom moves deeper into the sediment (Evans, 1958, 1959). Diatom breakage and the presence of mainly littoral diatoms are suggestive of a shallow, high-energy environment (Flower, 1993), which would be consistent with reworking of littoral sediments during a lake transgression. Grazing can also result in broken diatoms, and it is possible that this zone represents a change in grazing intensity in the littoral zone at Bear Lake. However, the conditions that would result in a dramatic increase in grazing at Bear Lake during this time are unknown.

Zone 5 (ca. A.D. 1936–present)

Zone 5 represents present-day conditions, with Bear River water flowing into Bear Lake. Although the ^{210}Pb age, which indicates that the 1912 diversion is represented by sediments deposited at ~10.5 cm (Smoak and Swarzenski, 2004), may be correct, Dean (this volume) suggested that striking changes in the isotopic composition of bulk carbonate at 12.5 cm in BL98-10 mark the diversion. It is thus likely that there is a lag between the diversion and the change from diatom zone 4 to zone 5. Benthic diatoms, including *Pseudostaurosira brevistriata* and *Staurosirella pinnata* dominate this zone. A large pennate diatom, *Diploneis elliptica*, is

also present. This shows that, despite the large size of Bear Lake, the littoral diatom community is well represented in the deep sediments. Planktic diatoms are relatively common and are dominated by *Cyclotella ocellata* and *Stephanodiscus medius*.

Principal Components Analysis (PCA)

Principal components analysis (PCA) was used to summarize diatom data in BL96-2 and BL96-1 (Figs. 5–7). Prior to analyses, the diatom percentage data were square-root transformed to increase the importance of less abundant diatom taxa. In Figure 5, results from a standardized PCA (i.e., species scores were standardized to have a mean of 0 and a variance of 1) were scaled so that distances between samples approximated Euclidean distances. Therefore, species scores in Figure 5 are regression coefficients of the standardized species data onto the sample scores. The results presented in Table 4 were determined using a different scaling, so that the species scores are the correlation of the species to the ordination axis defined by the sample scores. Only taxa with a correlation of greater than the absolute value of 0.6 are included in the table.

Owing to the absence of fossil diatoms between 19.1 and 14.6 cal ka, samples from this interval were not included in the PCA. The top three samples of BL98-10 were the only samples that contained diatoms representing modern-day conditions. These samples were plotted passively (i.e., the samples did not influence the position of the axes) on the PCA biplots for comparison with BL96-2 and BL96-1 samples.

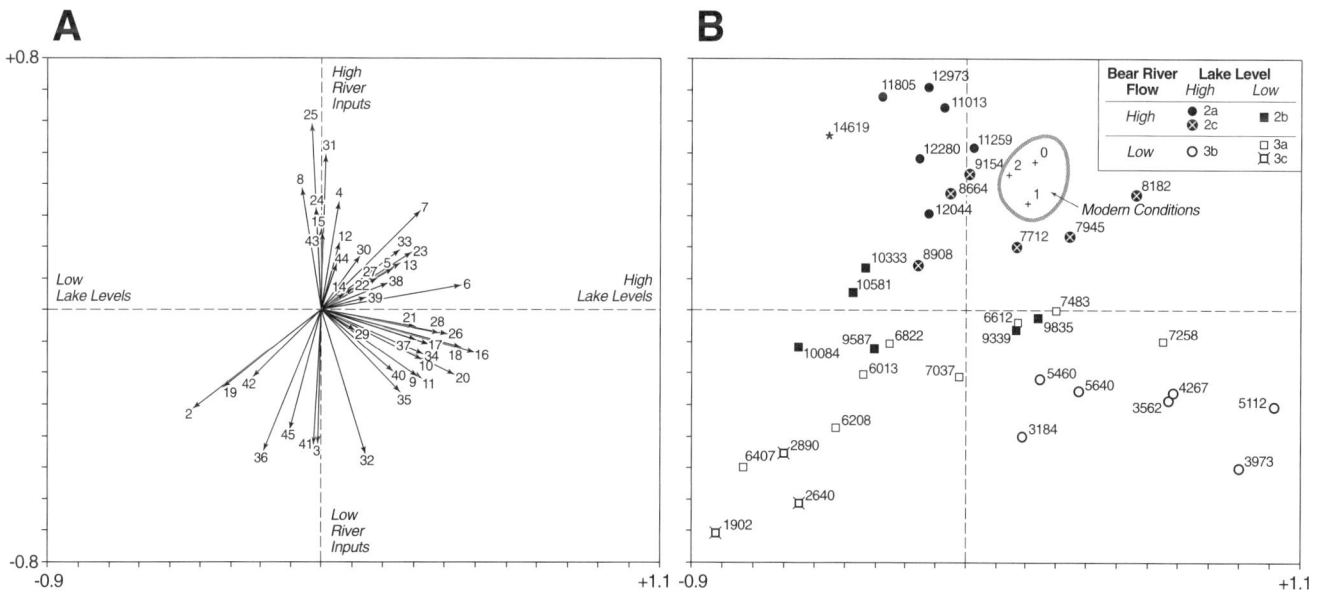

Figure 5. Principal components analysis biplot of diatom data from BL96-2. Eigenvalues are $\lambda_1 = 0.20$; $\lambda_2 = 0.14$. (A) Species are denoted by arrows, and the numbers refer to species names listed in Table 2. The position of the arrow is determined by the regression coefficients of the standardized species data onto the sample score. The interpretation of the axes is based on the ecology of diatom species that are most strongly correlated with the axes, and is explained in detail in the text. (B) Points represent different diatom samples, and the number assigned to a point represents the age of that sample. The closer two points are the more similar those samples are in terms of diatom community composition. Zone 1b is represented by a single sample, denoted by *. Zone 3 is depicted by open symbols and zone 2 is shown by black symbols.

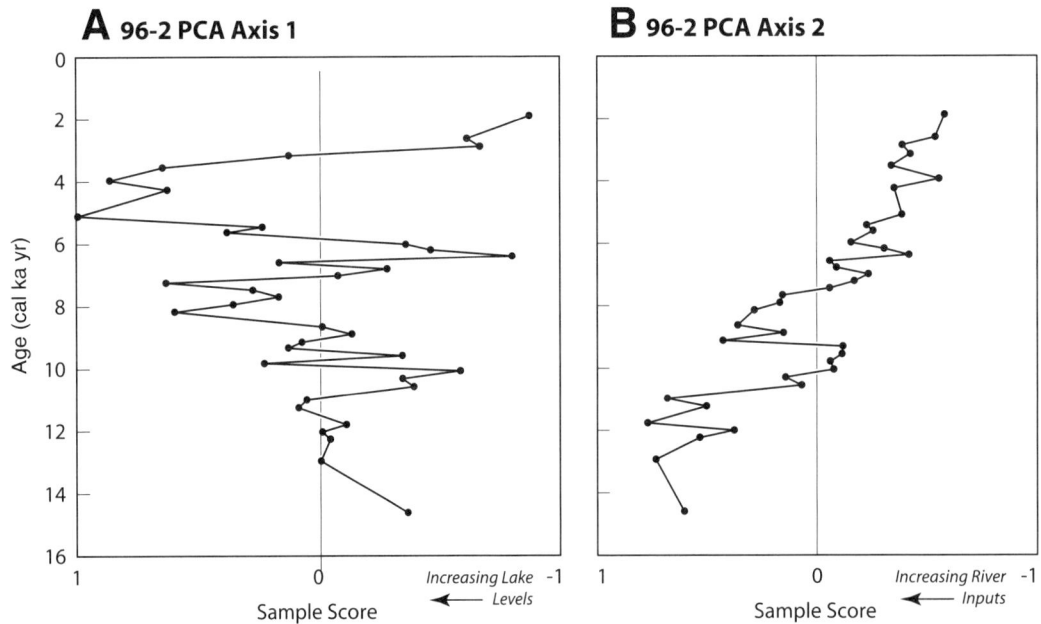

Figure 6. (A) Principal components analysis (PCA) axis 1 sample scores plotted against time for core BL96-2. (B) PCA axis 2 sample scores plotted against time for core BL96-2. The x-axis for both plots is reversed, so that wet conditions are to the left to be consistent with later plots. Note that prior to 7.6 cal ka river inputs and lake level change in the same direction, reflecting the importance of river and stream inflows in determining Bear River lake levels prior to this time. After 7.6 cal ka, lake levels change independently of river inputs, indicating the greater importance of other variables, such as evaporation, in determining lake levels.

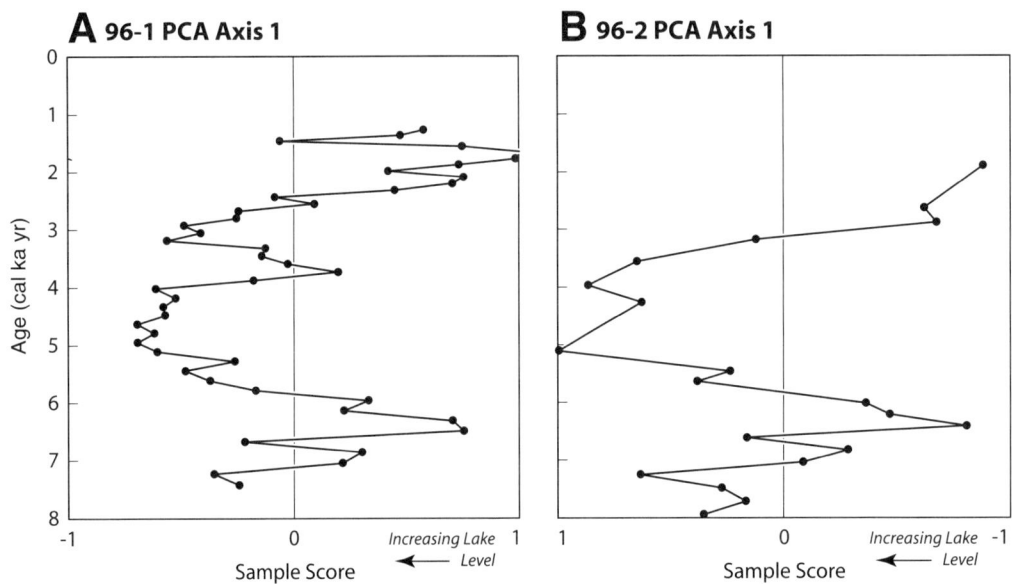

Figure 7. Principal components analysis (PCA) axis 1 sample scores plotted against age for core BL96-1 (A) and for core BL96-2 (B). Note that the BL96-1 axis 1 is reversed compared to the BL96-2 axis 1. When comparing two different PCA biplots, axes can be reversed, so that positive values in one biplot mean the same thing as negative values in the other plot. In this case increasing lake levels are shown to the left of both plots. Note the strong similarities between these two plots. The eigenvalue for BL96-1 for axis 1 is 0.26.

The eigenvalues (i.e., the dispersion of the species scores on the ordination axis) for the first two axes of the PCA of BL96-2 were $\lambda_1 = 0.20$ and $\lambda_2 = 0.14$, and of BL96-1 were $\lambda_1 = 0.26$ and $\lambda_2 = 0.08$. The total variance explained by the first four axes in the PCA of BL96-2 was 50%, and for BL96-1 was 46%. Because the eigenvalue for the second PCA axis of the BL96-1 analysis was low, it will not be considered further in this paper. Also, because the trends observed in the PCA biplot for BL96-1 and BL96-2 were similar, only the biplot for BL96-2 is depicted (Fig. 5).

PCA BL96-2 Biplots

The ecology of the diatom taxa that determine the interpretation of the first axis of the PCA biplot for BL96-2 indicate that this axis represents lake-level changes at Bear Lake, with greater values indicating rising lake levels and lower values falling lake levels. Planktic taxa, including *Cyclotella michiganiana* (28) and *C. meneghiana* (21), and periphytic diatoms, including *Staurosirella pinnata* (6), *Navicula tuscula* (20), *N. rhyncocephala* (16), and *N. capitata* var. *lueneburgensis* (18), have high scores (>0.6) (Table 4) and are, therefore, positively correlated with axis 1 and important in interpreting axis 1. The high positive planktic diatom scores suggest that positive sample scores are indicative of times of increased lake level (Wolin and Duthie, 1999). This is supported by the ecology of the other diatom taxa that are strongly correlated with axis 1. According to Rühland et al. (2003) most of these taxa have large optimal values for dissolved organic carbon (DOC) and total nitrogen (TN) (Table 4). Allochthonous carbon is composed mainly of DOC, whereas autochthonous carbon is mainly particulate organic carbon (POC) (Wetzel, 2001). In Bear Lake today the ratio of allochthonous to autochthonous carbon is ~50:1, likely as a result of the large organic carbon loading from Bear River to Bear Lake (Birdsey, 1985, 1989). Although no measurements are available for Bear Lake, other research suggests that DOC inputs to Bear Lake would increase during wetter years (Schindler et al., 1997). Today nitrogen loading to Bear

Lake is predominately from Bear River, and levels of nitrogen increase during wetter years (Birdsey, 1989). In addition, rising lake levels during wet years would flood shores of the lake and release DOC and nitrogen from these areas to the lake.

Pseudostaurosira brevistriata (2) has a large absolute score (>0.60) and is negatively correlated with PCA axis 1 (Table 4, Fig. 5). *P. brevistriata* has low TN and DOC optima (Table 4; Rühland et al., 2003). *Navicula oblonga* (19), which is also negatively correlated with axis 1 (species score = −0.36), is a halophilous species (Risberg et al., 1999) and *Cymbella aspera* (42) (axis 1 species score = −0.31) has been described as aerophilous (Zong, 1998). The ecology of these diatoms is consistent with drier conditions causing reduced TN and DOC, and with falling lake levels, which would expose shorelines and increase concentrations of ions. As well, no planktic diatoms are negatively correlated with axis 1 and have large absolute scores, suggesting that negative values indicate shallower waters.

Three lines of evidence indicate that the second PCA axis of the BL96-2 analysis represents river inputs, with positive values indicating increased fluvial inputs and negative values decreased inputs. First, taxa strongly correlated with axis 1, *Stephanodiscus medius* (25) and *Cyclotella ocellata* (31), are planktic and eutrophic, whereas all taxa strongly negatively correlated with axis 2 are periphytic (Table 4).

Second, post-diversion BL98-10 samples, which represent a time when Bear River water flowed into Bear Lake, have relatively high and positive scores on the second axis. The present nutrient budget to Bear Lake is largely composed of inputs from Bear River. In fact, Birdsey (1989) estimated that 60%–80% of post-diversion phosphorus delivery to Bear Lake is from Bear River. Moreover, nutrients increased significantly to Bear Lake when Bear River was connected to Bear Lake (Table 1; Birdsey, 1989). Although nutrient loading to Bear River is probably greater in historical times than it would have been prior to agriculture and other human activities, greater catchment size would

TABLE 4. DIATOM ECOLOGICAL INFORMATION

Species Number	Species	Species Score on Axis 1 of 96-2	Species Scores on Axis 2 of 96-2	DOC Optimum ug/L	TN Optimum mg/L
Positively Correlated to Axis 1					
16	*Navicula rhyncocephala*	0.87		N/A	N/A
6	*Staurosirella pinnata*	0.79		13.30	473.20
18	*N. capitata* var. *lueneburgensis*	0.79		N/A	N/A
20	*N. tuscula*	0.75		N/A	N/A
26	*Cyclotella michiganiana*	0.7		23.50	882.10
28	*C. meneghiniana*	0.66		N/A	N/A
Mean Value of Optima				18.40	677.65
Negatively Correalted to Axis 1					
2	*Pseudostaurosira brevistriata*	−0.74		8.50	344.10
Mean Value of Optima				8.50	344.10
Positively Correlated to Axis 2					
25	*Stephanodiscus medius*		0.88	N/A	N/A
31	*Cyclotella ocellata*		0.73	10.50	268.80
Mean Value of Optima				10.50	268.80
Negatively Correlated to Axis 2					
32	*Diploneis elliptica*		−0.68	N/A	N/A
36	*Caloneis schumannii*		−0.66	N/A	N/A
3	*Pseudostaurosira brevistriata* form rhomboid		−0.63	N/A	N/A
41	*Cymbella mesiana*		−0.62	N/A	N/A

Note: DOC—dissolved organic carbon; TN—total nitrogen.

have increased terrestrial inputs and hence nutrient delivery (Schindler, 1971; Prairie and Kalff, 1986).

Third, several species found only in modern samples collected from rivers and streams flowing into Bear Lake today, including *Cocconeis placentula* (33), *Amphora inariensis* (38), *Surirella ovalis* (44), *Achnanthes curtissima* (13), and *Achnanthes lanceolata*, have positive scores on the second PCA axis (Kimball, 2001).

PCA Sample Scores

Prior to 7.6 cal ka the direction of change shown in the plot of PCA axis 1 sample scores is similar to the direction of change shown in the plot of PCA axis 2, although the values of the scores and magnitude of change between scores are different (Fig. 6). Assuming that axis 1 represents lake levels and axis 2 river inputs, prior to 7.6 cal ka lake level was strongly linked to Bear River inflows, but after this time lake level was more strongly controlled by other factors, probably evaporation. Increases in saline and aerophilic diatoms after 7.6 cal ka suggest that effective moisture is an important driver of lake-level fluctuations.

From 13.8 to 10.8 cal ka inferred lake levels were moderate and relatively stable, whereas inferred river inputs were high. At 10.8 cal ka, lake levels fell rapidly in response to decreased river and stream inputs. River inputs remained low until 9.2 cal ka, and although lake levels fluctuated between 10.8 and 9.2 cal ka, they also remained generally low. Beginning ca. 9.2 cal ka, river inputs increased rapidly and remained high until 7.6 cal ka.

Beginning ca. 9.2 cal ka, river inputs increased rapidly. Lake levels began to rise ca. 8.5 cal ka and remained high until 7 cal ka. As river inputs slowly declined following 7.6 ka, however, lake levels fluctuated rapidly, likely in response to changes in evaporation.

Sample scores for BL96-2 on axis 1 show similar trends for BL96-1 on axis 1 (Fig. 7), and the diatom species that most influence the position of axis 1 are similar in analyses of the two cores. Because BL96-1 has better temporal resolution, we will focus on it to examine lake-level fluctuations. Lake levels were low from 7.6 to 5.9 cal ka, but increased rapidly at 5.9 cal ka and remained high until 2.9 cal ka. This generally wet period was interrupted by a dry period between 4.3 and 3.8 cal ka. Lake levels dropped rapidly at 2.9 cal ka and remained low until 1.6 cal ka.

Discussion

Late Pleistocene (19.1–13.8 cal ka): Poor Diatom Preservation

The absence of diatoms from 19.1 to 14.6 cal ka in zone 1a is probably related to the harsh conditions at Bear Lake during the last glacial interval (24–15 cal ka) (Dean et al., 2006). Cold temperatures during this time would have increased ice cover, which would have decreased transmission of light. Turbidity would have been high as a result of increased sediment load in Bear River originating from the Uinta Mountains (Dean et al., 2006; Rosenbaum and Kaufman, this volume), which would

have further reduced light and limited diatom growth (Bradshaw et al., 2000). Detrital quartz dominated sediment deposition until ca. 18 cal ka, when it decreased rapidly and was replaced with endogenic calcite (Dean, this volume). By ca. 15 cal ka, approximately the same time that diatoms first appear, low-Mg calcite dominated sediments deposited at Bear Lake. Diatoms deposited between 14.6–13.8 cal ka are poorly preserved as a result of dissolution. Dissolution of diatoms is affected by pH (Lewin, 1961; Iler, 1979), major-ion composition and concentration (Flower 1993; Barker et al., 1994), temperature (Marshall, 1980), and silica saturation levels (Flower, 1993; van Cappellen and Qiu, 1997). It is unlikely that poor diatom preservation between 14.6 and 13.8 cal ka occurred as a result of changes in pH or major-ion composition or concentration because (1) the diatom community composition suggests that Bear Lake has always been alkaline, (2) endogenic calcite dominated sediment deposition, and (3) Mg:Ca ratios remained relatively unchanged between 16 and 11 cal ka (Dean et al., 2006). More likely, diatom dissolution during this period was related to continued cold temperatures and limited silica. Increasing temperatures increase silica dissolution (Marshall, 1980), and as a result increase silica concentrations in streams. Temperatures in the Great Basin during the last glacial maximum are estimated to have been 10 °C lower than present (Thompson, 1988). Such low temperatures would decrease weathering rates, and thereby reduce silica concentrations in inflowing streams (Vesley et al., 2005). Reduced temperatures would also reduce thermal stratification and productivity (Sorvari et al., 2002), and might decrease internal lake cycling of silica, although silica cycling in lakes is complicated and affected by many processes (Wetzel, 2001).

Latest Pleistocene to Middle Holocene (13.8–7.6 cal ka): Bear River Influence on Bear Lake

The appearance of well-preserved diatoms and the presence of *Stephanodiscus medius* and *Cyclotella ocellata* in zone 2 are indicative of warmer temperatures and reduced turbidity relative to zone 1. These changes would have led to increased thermal stratification and increased nutrient cycling (Interlandi et al., 1999). The pollen record, which records a switch from conifers and cold-tolerant trees and herbs to sage-steppe plants, also indicates warmer temperatures (Doner, this volume). Warmer temperatures in the Great Basin at this time are attributed to greater-than-present summer insolation as a result of orbital variations (Thompson et al., 1993).

The presence of mesotrophic and eutrophic diatoms indicates high nutrient availability, which would result from increased runoff. PCA indicates that river inputs were high during deposition of zone 2a (13.8–10.8 cal ka), but that lake levels were moderate (Fig. 6). Increased values of isotope ratios for Sr, C, and O after ca. 12 ka suggest that Bear River was not connected to Bear Lake after this time (Fig. 8; Dean et al., 2006). The difference in the timing of the Bear River disconnection suggested by the diatom evidence compared to the isotope and carbonate chemistry may be related to the greater sensitivity of the latter variables to Bear

River inflows. The geochemical signature of Bear River is distinct from that of Bear Lake and other streams flowing into Bear Lake (Dean et al., 2007; Dean, this volume); however, although the Bear River diatom assemblage is distinct from Bear Lake diatoms, it is not distinct from that of other streams entering the lake (Kimball, 2001). Taken together, the evidence suggests that Bear River disconnected by 12 cal ka, but that other fluvial inputs continued to be relatively high until 10.8 cal ka. Between 13.8 and 10.8 cal ka, PCA of diatom data indicates that, despite continued high river inputs, Bear Lake levels were moderate, perhaps reflecting increasing evaporation during the summer (Fig. 6). High inflows from rivers accompanied by relatively high evaporation suggest that winters were wet and summers were warm and dry.

Increasingly arid conditions are reported for the Bonneville Basin beginning ca. 13.9–13.3 cal ka (Godsey et al., 2005; Madsen et al., 2001), which is earlier than what is inferred for Bear Lake. The delay in the onset of arid conditions at Bear Lake could be because Bear Lake's hydrologic balance was more influenced by glaciers, which may have persisted in the Uinta Mountains until 10 cal ka (Munroe, 2003), or could indicate a different climatic control for Bear Lake and the Bonneville Basin at this time. More humid conditions existed between 13.6 and 10.7 cal ka on the Great Plains than during the middle Holocene (Valero-Garcés et al., 1997), suggesting a common climatic control between this region and Bear Lake that may be related to the position of the jet stream and to storms tracking eastward from the Pacific (Thompson et al., 1993).

Planktic diatoms decrease rapidly at 10.8 cal ka and remain low until 9.2 cal ka, indicating either a decrease in nutrients due to reduced river inputs or a decrease in lake levels (Figs. 2 and 6). The period between 10.8 and 9.2 cal ka is also marked by a switch from calcite-rich to aragonite-rich sediments (Fig. 8; Dean et al., 2006). Although aragonite deposition in a cold, oligotrophic lake is unusual, it likely was in response to increasing temperatures and salinity (Dean et al., 2006; Dean, this volume). Records from the northern Great Plains similarly indicate slowly and then rapidly increasing aridity beginning ca. 10.5 cal ka and 7.9 cal ka, respectively (e.g., Laird et al., 1996; Valero-Garcés et al., 1997).

In contrast to the northern Great Plains, evidence suggests that conditions at Bear Lake became wetter between 9.2 and 7.6 cal ka. Planktic diatoms increase in tandem with a decrease of oxygen isotope values and a switch from aragonite to calcite deposition (Fig. 8). Stable isotope evidence, including O, C, and Sr, indicates that this period was a time of increased lake levels as a result of reconnection of Bear River to Bear Lake (Dean et al., 2006). Peaks in *Surirella ovalis* and *S. minuta*, found by Kimball (2001) to occur mainly in Bear River today, suggest greater river input (Fig. 2). PCA of the diatom data similarly point to greater river inputs as well as higher lake levels (Fig. 6), thereby corroborating Dean et al.'s (2006) interpretation. According to Smoot and Rosenbaum (this volume), lake levels increased ~10 m above modern limit, and the timing of this shift is roughly coincident with the Willis Ranch shoreline, which is at ~1814 m asl (8 m above modern lake levels) (Laabs and Kaufman, 2003).

The lack of correlation between the Bear Lake record and records from the northern Great Plains and much of the western United States (Fritz et al., 2001) between 9 and 7.5 cal ka

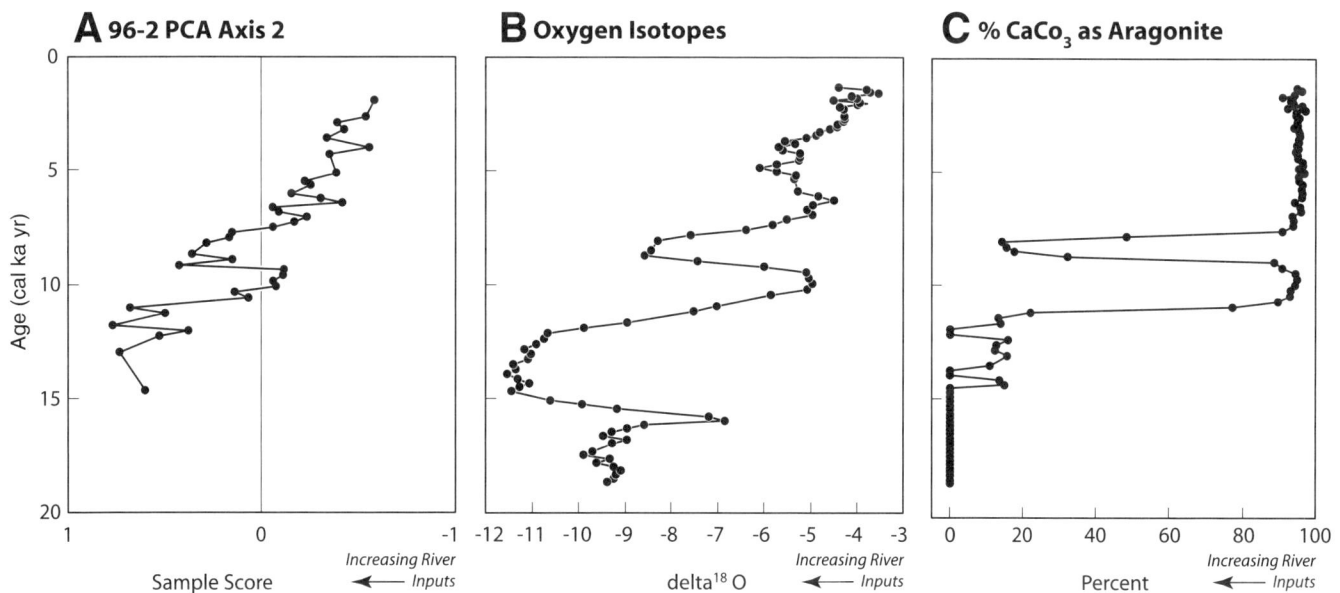

Figure 8. (A) Principal components analysis (PCA) axis 2 scores plotted against age for core BL96-2. (B) δ¹⁸O in bulk carbonate versus age in BL96-2 (after Dean et al., 2006). (C) Percent CaCO₃ as aragonite in BL96-2 (after Dean et al., 2006). The x-axis for the diatom PCA score plot has been reversed to match the direction of inferred river inputs in the plots of oxygen isotopes and % CaCO₃ as aragonite. For all three plots, scores to the right of the figure represent lower river inputs.

implies that either these two regions were affected by two different climatic regimes or that non-climatic factors were controlling Bear Lake's hydrologic balance. It is possible that the reconnection between Bear River and Bear Lake was the result of tectonic or geomorphic processes, although no evidence has been found to support this. However, it is also difficult to explain this event climatically.

Wetter conditions at Bear Lake are consistent with pollen evidence from adjacent regions, which indicates greater summer precipitation during the early Holocene than today (Whitlock and Bartlein, 1993). Wetter conditions could be linked to intensified monsoonal circulation during the middle Holocene. Bear Lake climate during this time is similar to those described for regions in the southwest, which were wetter and warmer between 9 and 4 cal ka (Betancourt et al., 1993; Poore et al., 2005). Several researchers have hypothesized that wetter conditions in the southwest beginning 9 cal ka were the result of intensified monsoonal circulation as a result of greater seasonal differences in insolation. However, as will be discussed below, the timing of increased monsoonal circulation is contentious (Barron et al., 2005).

Middle to Late Holocene (7.6 cal ka to ca. A.D. 1918): Recording Lake-Level Variations

Changes in diatom community composition, which are reflected in the PCA, indicate reduced river inflow to Bear Lake following 7.6 cal ka. This is supported by a switch from calcite to aragonite deposition and increasing Sr isotope values, which point to a final disconnection between Bear River and Bear Lake at this time (Fig. 8) (Dean et al., 2006). PCA also shows that there is reduced correlation between river inputs and lake-level variation following 7.6 cal ka, indicating that after this time Bear Lake levels were more influenced by changes in factors other than stream flow, most likely summer precipitation and evaporation.

PCA suggests lower lake levels, and increased isotope values indicate greater evaporation, between 7.6 and 5.9 cal ka, most likely due to increased aridity at this time. Beginning ca. 5.9 cal ka until 2.9 cal ka, PCA suggests that lake levels increased rapidly and isotope values suggest decreased evaporation, which together indicate a wetter climate. Following 2.9 cal ka, diatom data show falling lake levels and increasing isotope values record increasing evaporation, indicating that conditions became drier again (Fig. 9). A brief dry period occurred between 4.3 and 3.8 cal ka. A lake-level model applied to other Bear Lake cores indicates lake-level changes similar to those inferred from diatoms, although the timing of correlative lake-level highs and lows between cores is not exact (Fig. 8; Smoot and Rosenbaum, this volume). For example, Smoot and Rosenbaum (this volume) found that lake level remained low until 3.5 cal ka.

From the inferred lake-level changes, the Bear Lake record could be interpreted as a generally wet period extending from 9.2 to 2.9 cal ka that is interrupted by two dry intervals, 7.6–5.9 cal ka and 4.3–3.8 cal ka. Many records from the southwest United States indicate that conditions were wetter during the middle Holocene (post 9–7 cal ka) as a result of intensified monsoonal flow (e.g., Spaulding, 1991; Ely et al., 1993; Waters and Haynes, 2001). Today most precipitation at Bear Lake occurs during January; however, years with increased monsoonal circulation result in greater August precipitation in the area of Bear

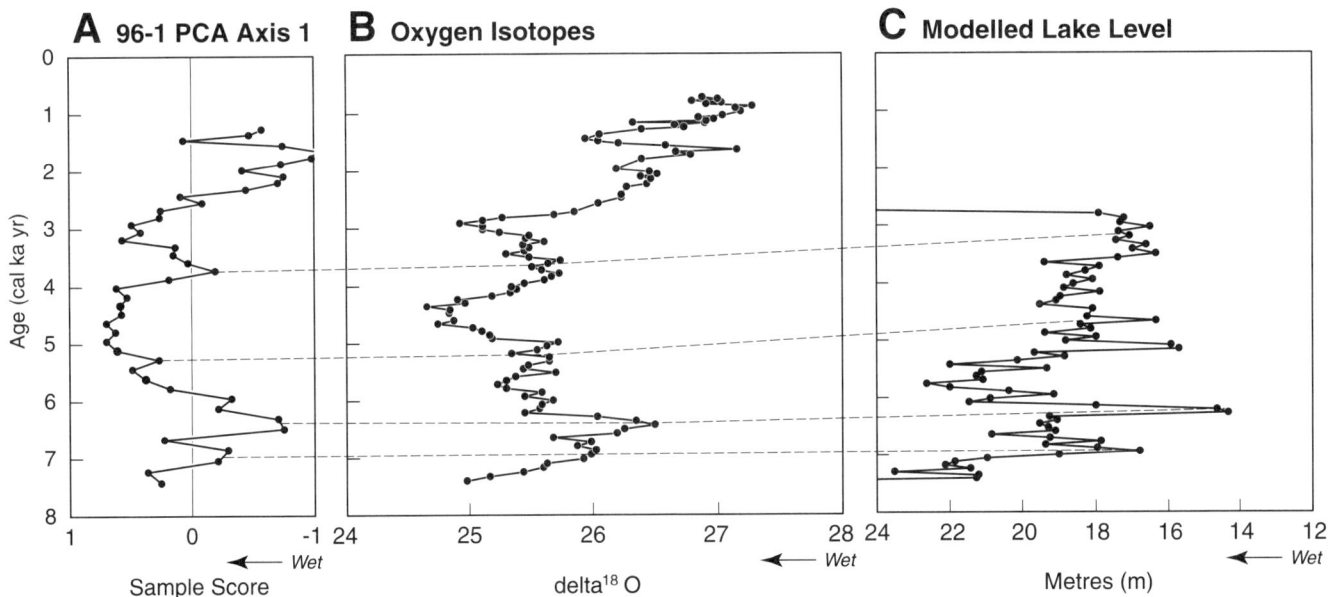

Figure 9. (A) Principal components analysis (PCA) axis 1 sample scores plotted against age for core BL96-1. (B) $\delta^{18}O$ in bulk carbonate versus age in BL96-1 (Dean, this volume). (C) Modeled lake level based on grain-size analysis (Smoot and Rosenbaum, this volume). Lake levels are plotted in reverse (i.e., from highest to lowest) on the x-axis so that in all plots, points on the right-hand side of the figure represent drier conditions. Dotted lines connect what are believed to be times of similar high or low lake levels.

Lake (Mock and Brunelle-Daines, 1999). Increased monsoonal flow during the middle Holocene has been attributed to altered latitudinal and seasonal distribution of insolation caused by variations in Earth's orbital parameters at that time (Thompson et al., 1993; Mock and Brunelle-Daines, 1999). An increase in monsoonal moisture in the southwest United States should occur in tandem with higher sea-surface temperature (SST) records from the Gulf of California and Gulf of Mexico, but SST in the Gulf of California began to increase at 6 cal ka (Barron et al., 2005), and SSTs in the Gulf of Mexico are highest between 7.0 and 4.7 cal ka (Poore et al., 2003, 2005). Moreover, modeling has suggested that orbital variations alone cannot explain regional-scale moisture patterns (Diffenbaugh and Sloan, 2004), and that ocean-atmosphere feedbacks are important forcing mechanisms for determining mid-Holocene moisture balance in the western United States (Vettoretti et al., 1998; Harrison et al., 2003).

On the Great Plains conditions were dry from 10 to 4.5 cal ka, with maximum aridity between 7.5 and 5.5 cal ka (Fritz et al., 2001). This pattern of a wet Southwest and dry Midcontinent is similar to present-day patterns of moisture/aridity resulting from the formation of a crescent-shaped region of enhanced subsidence bordering the area of the enhanced monsoon (Higgins et al. 1997, 1998; Higgins and Shi, 2000). Modeling has also demonstrated that mid-Holocene aridity in North America is dynamically linked to an orbitally induced enhancement of the summer monsoon in the American Southwest (Harrison et al., 2003). These experiments indicate that Bear Lake is on the border between increased moisture due to enhanced monsoonal flow versus increased aridity due to associated subsidence (Harrison et al., 2003). We hypothesize that subtle changes in monsoonal strength and related subsidence result in shifts in moisture at Bear Lake during the Holocene. Therefore, the dry periods at Bear Lake ca. 7.6–5.9 cal ka and 4.3–3.8 cal ka may indicate monsoon core shifts when Bear Lake fell within the area of subsidence (aridity). These periods of increased aridity at Bear Lake also coincide with dry conditions in other parts of the Great Basin including (1) the driest conditions of the Holocene at Pyramid Lake between 7.6 and 6.3 cal ka (Mensing et al., 2004), (2) dry conditions at Mono Lake between 7.4 and 4.5 cal ka (Davis, 1999), (3) desiccation of Walker Lake at 5 cal ka (Benson et al., 1991), and (4) persistent drought at Owens Lake from 6.5 to 3.8 cal ka (Benson et al., 2002).

The timing of the second dry event could also be related to a widespread drought that occurred between 4.3 and 4.1 cal ka in mid-continental North America (Booth et al., 2005). This event may have been global; however, the forcing mechanism remains uncertain (Booth et al., 2005).

The period from 2.9 to 1.6 cal ka is characterized by decreasing lake levels and increased evaporation. This zone is capped by an interval of broken pennate diatoms, which may be indicative of reworking of littoral sediments during the past 700 years as lake levels increased. The decreasing lake levels beginning at 2.9 cal ka may be related to reduced monsoonal circulation. Marine records from the Gulf of Mexico showing a decrease in the planktic foraminifera, *Globigerinoides sacculifera* following 3 cal ka (Poore et al., 2005) indicate decreased monsoonal circulation, and Barron et al. (2004, 2005) report a decrease in upwelling in the Gulf of California, which they attribute to reduced monsoonal circulation beginning at 2.8 cal ka.

Many paleoenvironmental records from the United States indicate that at 2.8 cal ka modern conditions were established. For example, lake sediment records from the continental United States indicate more humid and less variable conditions beginning ca. 2.8 cal ka (Clark et al., 2002), and in the Sierra Nevada, wetter but more variable conditions were recorded (Brunelle and Anderson, 2003; Bloom, 2006). Records from the Great Basin are in general agreement with these records suggesting that the period following 3 cal ka was wetter than the middle Holocene (Benson et al., 2002).

CONCLUSIONS

Diatoms at Bear Lake record changes in hydrologic and climatic condition for the last 19.1 cal ka.

- From 19.1 to 14.6 cal ka the complete absence of diatoms indicates that Bear Lake was characterized by cold, turbid, and silica-poor conditions. The presence of diatoms between 14.6 and 13.8 cal ka is linked to increased light due to decreased turbidity as the glaciers disappeared from the Uinta Mountains and the sediment load of inflowing streams was reduced. The poor preservation of the frustules during this time, however, suggests that conditions continued to be relatively cold.

- Between 13.8 and 7.6 cal ka, the PCA of the diatom data generally indicates greater river inputs than following 7.6 cal ka. The period between 10.8 and 9.2 cal ka, however, was characterized by diatoms indicative of low river inputs and low lake levels. Other geochemical evidence suggests that Bear River was not flowing into Bear Lake at this time. After 7.6 cal ka diatoms indicate low river inputs until ca. A.D. 1912, when the Bear River was diverted by humans into Bear Lake.

- In general, between 9.2 and 2.9 cal ka, diatom community composition and stable isotope ratios suggest that lake levels were generally high and evaporation low, but that at least two prolonged periods of lower lake levels and greater evaporation occurred: 7.6–5.9 cal ka and 4.3–3.8 cal ka. Increased monsoonal circulation could have increased summer precipitation and reduced evaporation at Bear Lake following 9.2 cal ka. The two prolonged dry periods recorded at Bear Lake during this time may indicate subtle changes in monsoonal strength although other climatic mechanisms remain to be considered.

- Following 2.9 cal ka, Bear Lake shows marked changes that are characterized by falling lake levels, indicative of drier conditions linked to reduced monsoonal circulation as recorded in marine sediment from the Gulf of Mexico and the Gulf of California.

- Because the location of Bear Lake is at the junction of several major climatic boundaries in the western United States (Mock, 1996), it is expected that climatic, and therefore hydrologic, conditions would change frequently as the boundaries shift in response to large-scale climatic forcings. We suggest, therefore, that the shifts from periods when effective moisture at Bear Lake was similar to the Great Plains and Great Basin to periods when effective moisture at Bear Lake was similar to the monsoon-dominated Southwest track changes in the relative mean positions of important air-mass boundaries.

ACKNOWLEDGMENTS

This research was funded by the U.S. Geological Survey. Special thanks are given to the team of scientists who worked on this project and provided many interesting discussions at workshops and meetings. This manuscript greatly benefited from comments from Joe Rosenbaum, Walt Dean, Pat Bartlein, and Jeffery Stone, who read earlier versions of this paper.

ARCHIVED DATA

Archived data for this chapter can be obtained from the NOAA World Data Center for Paleoclimatology at http://www.ncdc.noaa.gov/paleo/pubs/gsa2009bearlake/.

REFERENCES CITED

Adams, D.K., and Comrie, A.C., 1997, The North American monsoon: Bulletin of the American Meteorological Society, v. 78, p. 2197–2213, doi: 10.1175/1520-0477(1997)078<2197:TNAM>2.0.CO;2.

Baier, J., Lucke, A., Negendank, J.F.W., Schleser, G.W., and Zolitschka, B., 2004, Diatom and geochemical evidence of mid- to late-Holocene climatic changes at Lake Holzmaar, West-Eifel (Germany): Quaternary International, v. 113, p. 81–96, doi: 10.1016/S1040-6182(03)00081-8.

Barker, P., Fontes, J.C., Gasse, F., and Druart, J.C., 1994, Experimental dissolution of diatom silica in concentrated salt solutions and implications for paleoenvironmental reconstruction: Limnology and Oceanography, v. 39, p. 99–110.

Barron, J.A., Bukry, D., and Bischoff, J.L., 2004, High resolution paleoceanography of the Guaymas Basin, Gulf of California, during the past 15,000 years: Micropaleontology, v. 50, p. 185–207, doi: 10.1016/S0377-8398(03)00071-9.

Barron, J.A., Bukry, D., and Dean, W.E., 2005, Paleoceanographic history of the Guaymas Basin, Gulf of California, during the past 15,000 years based on diatoms, silicoflagellates, and biogenic sediments: Marine Micropaleontology, v. 56, p. 81–102, doi: 10.1016/j.marmicro.2005.04.001.

Battarbee, R.W., Jones, V.J., Flower, R.J., Cameron, N.G., Bennion, H., Carvalho, L., and Juggins, S., 2001, Diatoms, in Smol, J.P., Birks, H.J., and Last, W.M., eds., Terrestrial, algal and siliceous indicators, v. 2 of Tracking environmental change using lake sediments: Dordrecht, Kluwer Academic Publishers, p. 155–202.

Benson, L.V., Meyers, P.A., and Spencer, R.J., 1991, Change in the size of Walker Lake during the past 5000 years: Palaeogeography, Palaeoclimatology, Palaeoecology, v. 81, p. 189–214, doi: 10.1016/0031-0182(91)90147-J.

Benson, L., Kashgarian, M., Rye, R., Lund, S., Paillet, F., Smoot, J., Kester, C., Mensing, S., Meko, D., and Linstrom, S., 2002, Holocene multidecadal and multicentennial droughts affecting northern California and Nevada: Quaternary Science Reviews, v. 21, p. 659–682, doi: 10.1016/S0277-3791(01)00048-8.

Betancourt, J.L., Pierson, E.A., Rylander, K.A., Fairchild-Parks, J.A., and Dean, J.S., 1993, Influence of history and climate on New Mexico piñon-juniper woodlands: USDA Forest Service General Technical Report RM-236, p. 42–62.

Birdsey, P.W., 1985, Coprecipitation of phosphorus with calcium carbonate in Bear Lake, Utah [M.S. thesis]: Logan, Utah State University, 122 p.

Birdsey, P.W., 1989, The limnology of Bear Lake, Utah-Idaho, 1912–1988: A literature review: Utah Department of Natural Resources, Division of Wildlife Resources, Publication no. 89-5, 113 p.

Bloom, A.M., 2006, A Paleolimnological investigation of climatic and hydrological conditions during the Late Pleistocene and Holocene in the Sierra Nevada, California, USA [Ph.D. thesis]: Salt Lake City, University of Utah, 175 p.

Booth, R.K., Jackson, S.T., Forman, S.L., Kutzbach, J.E., Bettis, E.A., Kreig, J., and Wright, D.K., 2005, A severe centennial-scale drought in mid-continental North America 4200 years ago and apparent global linkages: The Holocene, v. 15, p. 321–328, doi: 10.1191/0959683605hl825ft.

Bradbury, J.P., 1997, A diatom record of climate and hydrology for the past 200 ka from Owens Lake, California with comparison to other great basin records: Quaternary Science Reviews, v. 16, p. 203–219, doi: 10.1016/S0277-3791(96)00054-6.

Bradshaw, E.G., Jones, V.J., Birks, H.J.B., and Birks, H.H., 2000, Diatom responses to late-glacial and early-Holocene environmental changes at Krakenes, western Norway: Journal of Paleolimnology, v. 23, p. 21–34, doi: 10.1023/A:1008021016027.

Brunelle, A., and Anderson, R.S., 2003, Sedimentary charcoal as an indicator of late-Holocene drought in the Sierra Nevada, California, and its relevance to the future: The Holocene, v. 13, p. 21–28, doi: 10.1191/0959683603hl591rp.

Bryson, R.A., and Hare, F.K., 1974, The climates of North America, in Bryson, R.A., and Hare, F.K., eds., Climates of North America: Amsterdam, Elsevier, World survey of climatology, v. 11, p. 1–47.

Clark, J.S., Grimm, E.C., Donovan, D.D., Fritz, S.C., Engstrom, D.R., and Almendinger, J.E., 2002, Drought cycles and landscape response to past aridity on prairies of the northern Great Plains, USA: Ecology, v. 83, p. 595–601.

Colman, S.M., Rosenbaum, J.G., Kaufman, D.S., Dean, W.E., and McGeehin, J.P., 2009, this volume, Radiocarbon ages and age models for the last 30,000 years in Bear Lake, Utah and Idaho, in Rosenbaum, J.G., and Kaufman, D.S., eds., Paleoenvironments of Bear Lake, Utah and Idaho, and its catchment: Geological Society of America Special Paper 450, doi: 10.1130/2009.2450(05).

Cummings, B.F., Wilson, S.E., Hall, R.I., and Smol, J.P., 1995, Diatoms from British Columbia (Canada) lakes and their relationship to salinity, nutrients and other limnological variables, in Lange-Bertalot, L.H., ed., Bibliotheca Diatomologica, Band 31: Berlin/Stuttgart, J. Cramer, p. 207.

Davidson, D.F., 1969, Some aspects of geochemistry and mineralogy of Bear Lake sediments, Utah-Idaho [M.S. thesis]: Logan, Utah State University, 67 p.

Davis, O.K., 1999, Pollen analysis of a late-glacial and Holocene sediment core from Mono Lake, Mono County, California: Quaternary Research, v. 52, p. 243–249, doi: 10.1006/qres.1999.2063.

Dean, W.E., 2009, this volume, Endogenic carbonate sedimentation in Bear Lake, Utah and Idaho, over the last two glacial-interglacial cycles, in Rosenbaum, J.G., and Kaufman, D.S., eds., Paleoenvironments of Bear Lake, Utah and Idaho, and its catchment: Geological Society of America Special Paper 450, doi: 10.1130/2009.2450(07).

Dean, W.E., Rosenbaum, J., Skipp, G., Colman, S., Forester, R., Liu, A., Simmons, K., and Bischoff, J., 2006, Unusual Holocene and late Pleistocene carbonate sedimentation in Bear Lake, Utah and Idaho, USA: Sedimentary Geology, v. 185, p. 93–112, doi: 10.1016/j.sedgeo.2005.11.016.

Dean, W.E., Forester, R.M., Bright, J., and Anderson, R.Y., 2007, Influence of the diversion of Bear River into Bear Lake (Utah and Idaho) on the environment of deposition of carbonate minerals: Evidence from water and sediments: Limnology and Oceanography, v. 52, p. 1094–1111.

Diffenbaugh, N.S., and Sloan, L.C., 2004, Mid-Holocene orbital forcing of regional-scale climate: A case study of western North America using a high-resolution RCM: American Meteorological Society, v. 17, p. 2927–2937.

Doner, L.A., 2009, this volume, A 19,000-year vegetation and climate record for Bear Lake, Utah and Idaho, in Rosenbaum, J.G., and Kaufman, D.S., eds., Paleoenvironments of Bear Lake, Utah and Idaho, and its catchment: Geological Society of America Special Paper 450, doi: 10.1130/2009.2450(09).

Ely, L.L., Enzel, Y., Baker, V.R., and Cayan, D.R., 1993, A 5000-year record of extreme floods and climate change in the Southwestern United States: Science, v. 262, p. 410–412, doi: 10.1126/science.262.5132.410.

Evans, J.H., 1958, The survival of fresh water algae during dry periods. Part I, An investigation of five small ponds: Journal of Ecology, v. 46, p. 149–168, doi: 10.2307/2256910.

Evans, J.H., 1959, The survival of fresh water algae during dry periods. Part II, Drying experiments. Part III, Stratification of algae in pond margin litter and mud: Journal of Ecology, v. 47, p. 55–81, doi: 10.2307/2257248.

Flower, R.J., 1993, Diatom preservation: Experiments and observations on dissolution and breakage in modern and fossil material: Hydrobiologia, v. 269/270, p. 473–484, doi: 10.1007/BF00028045.

Fritz, S.C., Juggins, S., Battarbee, R.W., and Engstrom, D.R., 1991, Reconstruction of past changes in salinity and climate using a diatom-based transfer function: Nature, v. 352, p. 706–708, doi: 10.1038/352706a0.

Fritz, S.C., Juggins, S., and Battarbee, R.W., 1993, Diatom assemblages and ionic characterization of lakes in the northern Great Plains, North America: A tool for reconstructing past salinity and climate fluctuations: Canadian Journal of Fisheries and Aquatic Sciences, v. 50, p. 1844–1856.

Fritz, S.C., Cumming, B.F., Gasse, F., and Laird, K.R., 1999, Diatoms as indicators of hydrologic and climatic change in saline lakes, *in* Stoermer, E.F., and Smol, J.P., eds., The diatoms: Applications for the environmental and earth sciences: Cambridge, UK, Cambridge University Press, p. 41–72.

Fritz, S.C., Metcalfe, S.E., and Dean, W.E., 2001, Holocene climate patterns in the Americas inferred from paleolimnological records, *in* Markgraf, V., ed., Interhemispheric climate linkages: London, Academic Press, p. 241–263.

Godsey, H.S., Currey, D.R., and Chan, M.A., 2005, New evidence for an extended occupation of the Provo shoreline and implications for regional climate change, Pleistocene Lake Bonneville, Utah, USA: Quaternary Research, v. 63, p. 212–223, doi: 10.1016/j.yqres.2005.01.002.

Harrison, S.P., Kutzbach, J.E., Liu, Z., Bartlein, P.J., Otto-Bliesner, B., Muhs, D., Prentice, I.C., and Thompson, R.S., 2003, Mid-Holocene climates of the Americas: A dynamical response to changed seasonality: Climate Dynamics, v. 20, p. 663–688.

Higgins, R.W., and Shi, W., 2000, Dominant factors responsible for interannual variability of the summer monsoon in the southwestern United States: Journal of Climate, v. 13, p. 759–776, doi: 10.1175/1520-0442(2000)013 <0759:DFRFIV>2.0.CO;2.

Higgins, R.W., Yao, Y., and Wang, X.L., 1997, Influence of the North American monsoon system on the US summer precipitation regime: Journal of Climate, v. 10, p. 2600–2622, doi: 10.1175/1520-0442(1997)010<2600: IOTNAM>2.0.CO;2.

Higgins, R.W., Mo, K.C., and Yao, Y., 1998, Interannual variability of the US summer precipitation regime with emphasis on the south-western monsoon: Journal of Climate, v. 11, p. 2582–2606, doi: 10.1175/1520-0442(1998)011 <2582:IVOTUS>2.0.CO;2.

Hirschboeck, K.K., 1991, Climate and floods: National water summary 1988–89—Floods and droughts: Hydrologic Perspectives on Water Issues: U.S. Geological Survey Water-Supply Paper, v. 2375, p. 67–88.

Iler, R., 1979. The chemistry of silica: Solubility, polymerization, colloid and surface properties, and biochemistry: New York, John Wiley and Sons, 866 p.

Interlandi, S.J., Kilham, S.S., and Theriot, E.C., 1999, Responses of phytoplankton to varied resource availability in large lakes of the greater Yellowstone Ecosystem: Limnology and Oceanography, v. 44, p. 668–682.

Kienel, U., Schwab, M.J., and Schettler, G., 2005, Distinguishing climatic from direct anthropogenic influences during the past 400 years in varved sediments from Lake Holzmaar (Eifel, Germany): Journal of Paleolimnology, v. 33, p. 327–347, doi: 10.1007/s10933-004-6311-z.

Kimball, J.P., 2001, Late Quaternary environmental change as inferred from diatoms of the sediments of Bear Lake, Utah/Idaho [M.S. thesis]: Salt Lake City, University of Utah Department of Geography, 109 p.

Krammer, K., and Lange-Bertalot, H., 1986, Bacillariophyceae, Teil 1: Naviculaceae: New York, Gustav Fischer, 876 p.

Krammer, K., and Lange-Bertalot, H., 1988, Bacillariophyceae, Teil 2: Bacillariaceae, Epithemiaceae, Surirellaceae: New York, Gustav Fischer, 610 p.

Krammer, K., and Lange-Bertalot, H., 1991a, Bacillariophyceae, Teil 1: Achnanthaceae, Kritische Erganzungen zu Navicula (Lineoltae) und Gomphonema Gestamtliteraturverzeichnis Teil 1–4: New York, Gustav Fischer, 437 p.

Krammer, K., and Lange-Bertalot, H., 1991b, Bacillariophyceae, Teil 3: Centrales, Fragilariaceae, Eunotiaceae: New York, Gustav Fischer, 576 p.

Laabs, B.J.C., and Kaufman, D.S., 2003, Quaternary highstands in Bear Lake Valley, Utah and Idaho: Geological Society of America Bulletin, v. 115, p. 463–478, doi: 10.1130/0016-7606(2003)115<0463:QHIBLV>2.0.CO;2.

Laird, K.R., Fritz, S.C., Grimm, E.C., and Mueller, P.C., 1996, Century-scale paleoclimatic reconstruction from Moon Lake, a closed basin lake in the Northern Great Plains: Limnology and Oceanography, v. 41, p. 890–902.

Lewin, J., 1961, The dissolution of silica from diatom walls: Geochimica et Cosmochimica, v. 21, p. 182–198, doi: 10.1016/S0016-7037(61)80054-9.

Ludlam, S.D., Feeney, S., and Douglas, M.S.V., 1996, Changes in the importance of lotic and littoral diatoms in a high arctic lake over the last 191 yrs: Journal of Paleolimnology, v. 16, p. 184–204.

Madsen, D.B., Rhode, D., Grayson, D.K., Broughton, J.M., Livingston, S.D., Hunt, J., Quade, J., Schmitt, D.N., and Shaver, M.W., III, 2001, Late Quaternary environmental change in the Bonneville basin, western USA: Palaeogeography, Palaeogeography, Palaeoclimatology, v. 167, p. 243–271, doi: 10.1016/S0031-0182(00)00240-6.

Marshall, W.L., 1980, Amorphous silica solubilities. 1. Behaviour in aqueous sodium nitrate solution; 25–300 °C, 0–6 molar. 3. Activity coefficient relations and predictions of solubility behaviour in salt solutions 0–350 °C: Geochimica et Cosmochimica, v. 44, p. 907–913 and 925–931.

Mensing, S.A., Benson, L.V., Kashgarian, M., and Lund, S., 2004, A Holocene pollen record of persistent droughts from Pyramid Lake, Nevada, USA: Quaternary Research, v. 62, p. 29–38, doi: 10.1016/j.yqres.2004.04.002.

Mock, C.J., 1996, Climatic controls and spatial variations of precipitation in the western United States: Journal of Climate, v. 9, p. 1111–1125, doi: 10.1175/ 1520-0442(1996)009<1111:CCASVO>2.0.CO;2.

Mock, C.J., and Brunelle-Daines, A.R., 1999, A modern analogue of western United States summer palaeoclimate at 6000 years before present: The Holocene, v. 9, p. 541–545, doi: 10.1191/095968399668724603.

Moser, K.A., Smol, J.P., and MacDonald, G.M., 2004, Ecology and distribution of diatoms from boreal forest lakes in Wood Buffalo National Park, northern Alberta and the Northwest Territories, Canada: Academy of Natural Sciences of Philadelphia Special Publication 22, 59 p.

Munroe, J.S., 2003, Holocene timberline and paleoclimate of the northern Uinta Mountain, northeastern Utah, USA: The Holocene, v. 13, p. 175–185, doi: 10.1191/0959683603hl600rp.

Poore, R.Z., Dowsett, R.J., Verardo, S., and Quinn, T.M., 2003, Millennial- to century-scale variability in Gulf of Mexico Holocene climate records: Paleoceanography, v. 18, p. 1048, doi: 10.1029/2002PA000868.

Poore, R.Z., Pavich, M.J., and Grissino-Mayer, H.D., 2005, Record of the North American southwest monsoon from Gulf of Mexico sediment cores: Geology, v. 33, p. 209–212, doi: 10.1130/G21040.1.

Prairie, Y.T., and Kalff, J., 1986, Effect of catchment size on phosphorus export: Water Resources Bulletin, v. 22, p. 465–470.

Risberg, J., Sandgren, P., Teller, J.T., and Last, W.M., 1999, Siliceous microfossils and mineral magnetic characteristics in a sediment core from Lake Manitoba, Canada: A remnant of glacial Lake Agassiz: Canadian Journal of Earth Sciences, v. 36, p. 1299–1314, doi: 10.1139/cjes-36-8-1299.

Rosenbaum, J.G., and Kaufman, D.S., 2009, this volume, Introduction to *Paleoenvironments of Bear Lake, Utah and Idaho, and its catchment*, *in* Rosenbaum, J.G., and Kaufman D.S., eds., Paleoenvironments of Bear Lake, Utah and Idaho, and its catchment, Geological Society of America Special Paper 450, doi: 10.1130/2009.2450(00).

Rühland, K.M., Smol, J.P., and Pienitz, R., 2003, Ecology and spatial distributions of surface-sediment diatoms from 77 lakes in the subarctic Canadian treeline region: Canadian Journal of Botany, v. 81, p. 57–73, doi: 10.1139/b03-005.

Schindler, D.E., 1971, A hypothesis to explain differences and similarities among lakes in the experimental lakes area, northwestern Ontario: Journal of the Fisheries Research Board of Canada, v. 28, p. 295–301.

Schindler, D.W., Jefferson, C.P., Bayley, S.E., Parker, B.R., Beaty, K.G., and Stainton, M.P., 1997, Climate-induced changes in the dissolved organic carbon budgets of boreal lakes: Biogeochemistry, v. 36, p. 9–28, doi: 10.1023/A:1005792014547.

Smoak, J.M., and Swarzenski, P.W., 2004, Recent increases in sediment and nutrient accumulation in Bear Lake, Utah/Idaho, USA: Hydrobiologia, v. 525, p. 175–184, doi: 10.1023/B:HYDR.0000038865.16732.09.

Smol, J.P., and Cumming, B.F., 2000, Tracking long-term changes in climate using algal indicators in lake sediments: Journal of Phycology, v. 36, p. 986–1011, doi: 10.1046/j.1529-8817.2000.00049.x.

Smoot, J.P., and Rosenbaum, J.G., 2009, this volume, Sedimentary constraints on late Quaternary lake-level fluctuations at Bear Lake, Utah and Idaho, *in* Rosenbaum, J.G., and Kaufman, D.S., eds., Paleoenvironments of Bear Lake, Utah and Idaho, and its catchment: Geological Society of America Special Paper 450, doi: 10.1130/2009.2450(12).

Sorvari, S., Korhola, A., and Thompson, R., 2002, Lake diatom response to recent Arctic warming in Finnish Lapland: Global Change Biology, v. 8, p. 171–181, doi: 10.1046/j.1365-2486.2002.00463.x.

Spaulding, W.G., 1991, A middle Holocene vegetation record from the Mojave Desert of North America and its paleoclimatic significance: Quaternary Research, v. 35, p. 427–437, doi: 10.1016/0033-5894(91)90055-A.

Stoermer, E.F., 1993, Evaluating diatom succession: Some peculiarities of the Great Lakes case: Journal of Paleolimnology, v. 8, p. 71–83, doi: 10.1007/BF00210058.

Stoermer, E.F., Emmert, G., Julius, M.L., and Schelske, C.L., 1996, Paleolimnologic evidence of rapid recent change in Lake Erie's trophic status: Canadian Journal of Fisheries and Aquatic Sciences, v. 53, p. 1451–1458, doi: 10.1139/cjfas-53-6-1451.

ter Braak, C.J.F., and Šmilauer, P., 1999, CANOCO for Windows, version 4.02: Wageningen, Netherlands, Center for Biometry Wageningen (CPRO-DLO).

Thompson, R.S., 1988, Western North America vegetation dynamics in the western United States: Modes of response to climatic fluctuations, *in* Huntley, B., and Webb, T., III, eds., Vegetation history: Dordrecht, Kluwer Academic Publishers, p. 415–457.

Thompson, R.S., Whitlock, C., Bartlein, P.J., Harrison, S.P., and Spaulding, W.G., 1993, Climatic changes in the western United States since 18,000 years B.P., *in* Wright, H.E., Kutzbach, J.E., Webb, III, T., Street-Perrot, F.A., and Bartlein, P.J., eds., Global climates since the Last Glacial Maximum: Minneapolis, University of Minnesota Press, p. 468–513.

Valero-Garcés, B.L., Laird, K.R., Fritz, S.C., Kelts, K., Ito, E., and Grimm, E.C., 1997, Holocene climate in the northern Great Plains inferred from sediment stratigraphy, stable isotopes, carbonate geochemistry, diatoms, and pollen at Moon Lake, North Dakota: Quaternary Research, v. 48, p. 359–369, doi: 10.1006/qres.1997.1930.

van Cappellen, P., and Qiu, L., 1997, Biogenic silica dissolution in sediments of the southern ocean. II. Kinetics: Deep-Sea Research, v. 44, p. 1129–1149, doi: 10.1016/S0967-0645(96)00112-9.

Vesley, J., Maier, V., Kopacek, J., Safanda, J., and Norton, S.A., 2005, Increasing silicon concentrations in Bohemian forest lakes: Hydrology and Earth System Sciences, v. 9, p. 699–706.

Vettoretti, G., Peltier, W.R., and McFarlane, N.A., 1998, Simulations of mid-Holocene climate using an atmospheric general circulation model: Journal of Climate, v. 11, p. 2607–2627, doi: 10.1175/1520-0442(1998)011<2607:SOMHCU>2.0.CO;2.

Waters, M.R., and Haynes, C.V., 2001, Late Quaternary arroyo formation and climate change in the American Southwest: Geology, v. 29, p. 399–402, doi: 10.1130/0091-7613(2001)029<0399:LQAFAC>2.0.CO;2.

Wetzel, R.G., 2001, Limnology: Lakes and river ecosystems (3rd edition): New York, Academic Press, 1006 p.

Whitlock, C., and Bartlein, P.J., 1993, Spatial variations of Holocene climate change in the Yellowstone region: Quaternary Research, v. 39, p. 231–238, doi: 10.1006/qres.1993.1026.

Wolin, J.A., and Duthie, H.C., 1999, Diatoms as indicators of water level change in freshwater lakes, *in* Stoermer, E.F., and Smol, J.P., eds., The diatoms: Applications for the environmental and earth sciences: Cambridge, UK, Cambridge University Press, p. 183–202.

WRCC (Western Regional Climate Center), 2007, http://www.wrcc.dri.edu/ (accessed 10 April 2007).

Wurtsbaugh, W., and Hawkins, C., 1990, Trophic interactions between fish and invertebrates in Bear Lake, Utah-Idaho: Salt Lake City, Report of Project F-26-R, to the Utah Division of Wildlife Resources, 167 p.

Wurtsbaugh, W., and Luecke, C., 1997, Examination of the abundance and spatial distribution of forage fish in Bear Lake (Utah/Idaho): Salt Lake City, Report of Project F-47-R, Study 5, to the Utah Division of Wildlife Resources, 217 p.

Zong, Y., 1998, Diatom records and sedimentary responses to sea-level change during the last 8000 years in Roudsea Wood, northwest England: The Holocene, v. 8, p. 219–228, doi: 10.1191/095968398671096338.

Manuscript Accepted by the Society 15 September 2008

The Geological Society of America
Special Paper 450
2009

The glacial/deglacial history of sedimentation in Bear Lake, Utah and Idaho

Joseph G. Rosenbaum

U.S. Geological Survey, Box 25046, Federal Center, Denver, Colorado 80225, USA

Clifford W. Heil Jr.

Graduate School of Oceanography, University of Rhode Island, Narragansett, Rhode Island 02882, USA

ABSTRACT

Bear Lake, in northeastern Utah and southern Idaho, lies in a large valley formed by an active half-graben. Bear River, the largest river in the Great Basin, enters Bear Lake Valley ~15 km north of the lake. Two 4-m-long cores provide a lake sediment record extending back ~26 cal k.y. The penetrated section can be divided into a lower unit composed of quartz-rich clastic sediments and an upper unit composed largely of endogenic carbonate. Data from modern fluvial sediments provide the basis for interpreting changes in provenance of detrital material in the lake cores. Sediments from small streams draining elevated topography on the east and west sides of the lake are characterized by abundant dolomite, high magnetic susceptibility (MS) related to eolian magnetite, and low values of hard isothermal remanent magnetization (HIRM, indicative of hematite content). In contrast, sediments from the headwaters of the Bear River in the Uinta Mountains lack carbonate and have high HIRM and low MS. Sediments from lower reaches of the Bear River contain calcite but little dolomite and have low values of MS and HIRM. These contrasts in catchment properties allow interpretation of the following sequence from variations in properties of the lake sediment: (1) ca. 26 cal ka—onset of glaciation; (2) ca. 26–20 cal ka—quasi-cyclical, millennial-scale variations in the concentrations of hematite-rich glacial flour derived from the Uinta Mountains, and dolomite- and magnetite-rich material derived from the local Bear Lake catchment (reflecting variations in glacial extent); (3) ca. 20–19 cal ka—maximum content of glacial flour; (4) ca. 19–17 cal ka—constant content of Bear River sediment but declining content of glacial flour from the Uinta Mountains; (5) ca. 17–15.5 cal ka—decline in Bear River sediment and increase in content of sediment from the local catchment; and (6) ca. 15.5–14.5 cal ka—increase in content of endogenic calcite at the expense of detrital material. The onset of glaciation indicated in the Bear Lake record postdates the initial rise of Lake Bonneville and roughly corresponds to the Stansbury shoreline. The lake record indicates that maximum glaciation occurred as Lake Bonneville reached its maximum extent ca. 20 cal ka and that deglaciation was under way while Lake Bonneville remained at its

Rosenbaum, J.G., and Heil, C.W., Jr., 2009, The glacial/deglacial history of sedimentation in Bear Lake, Utah and Idaho, *in* Rosenbaum, J.G., and Kaufman, D.S., eds., Paleoenvironments of Bear Lake, Utah and Idaho, and its catchment: Geological Society of America Special Paper 450, p. 247–261, doi: 10.1130/2009.2450(11). For permission to copy, contact editing@geosociety.org. ©2009 The Geological Society of America. All rights reserved.

peak. The transition from siliciclastic to carbonate sedimentation probably indicates increasingly evaporative conditions and may coincide with the climatically driven fall of Lake Bonneville from the Provo shoreline. Although lake levels fluctuated during the Younger Dryas, the Bear Lake record for this period is more consistent with drier conditions, rather than cooler, moister conditions interpreted from many studies from western North America.

INTRODUCTION

Bear Lake is a large, long-lived lake that lies in the southern end of a fault-bounded valley astride the Utah-Idaho border (Fig. 1), and contains a thick sequence of Quaternary sediments (Colman, 2006). The upper 120 m of these sediments, which were penetrated near the depocenter by core BL00-1 during testing of the GLAD800 coring platform (Dean et al., 2002), provide a record of at least the last two glacial–interglacial cycles (Kaufman et al., this volume).

The Bear River arises in glaciated terrain in the Uinta Mountains, flows generally northward, and then crosses an outwash fan north of the lake as it enters the Bear Lake Valley (Reheis et al., this volume). During historic times and throughout much of the Holocene the river continued to the north (Fig. 1) without entering the lake, exiting the valley before bending to the south on its way to Great Salt Lake. At times in the past, however, the river flowed into the lake (Kaufman et al., this volume). Many factors may influence the relation between the river and the lake, including surficial processes such as migration of the river on the outwash fan, active tectonism that may have changed the river's course, and changes in climatic conditions that caused the lake to rise or fall, thereby capturing or abandoning the river (Reheis et al., this volume).

The endogenic and allogenic materials in Bear Lake provide a complex record of past environmental conditions. Endogenic minerals (Dean et al., 2006; Dean, this volume) and biologic materials yield a record of changing hydrologic conditions, whereas allogenic materials provide information about erosional processes and fluvial transport within the catchment areas. The presence or absence of the Bear River has a large influence on both the endogenic and allogenic components. When the Bear River did not enter Bear Lake, the lake was fed by short local streams and sublacustrine springs and was probably topographically closed or intermittently overflowing. During such times, the lake precipitated abundant carbonate minerals; lake level (Smoot and Rosenbaum, this

Figure 1. Shaded relief digital elevation model of Bear Lake area (view to the northeast) and location of Bear Lake and the Bear River. The dashed arrow indicates the approximate path of the river when it entered the lake.

volume), mineralogy, and stable isotopes (Dean, this volume) were sensitive to evaporative conditions; and all fluvial sediment was derived from the local catchment. When the Bear River flowed into Bear Lake, the lake was more likely to be fresh and overflowing. At such times the lake precipitated little if any carbonate minerals and fluvial sediments were derived from the Bear River above Bear Lake Valley as well as from the local catchment.

In this paper we interpret the paleoenvironmental record from the onset of the last period of extensive local glaciation to the beginning of the Holocene. The analysis is based on multi-proxy data sets from two 4-m-long cores, BL96-2 and BL96-3, located in ~40 m and 30 m of water respectively (Fig. 2). Core BL96-2 contains ~3 m of carbonate-rich sediment overlying ~1 m of siliciclastic material (Dean et al., 2006). Core BL96-3

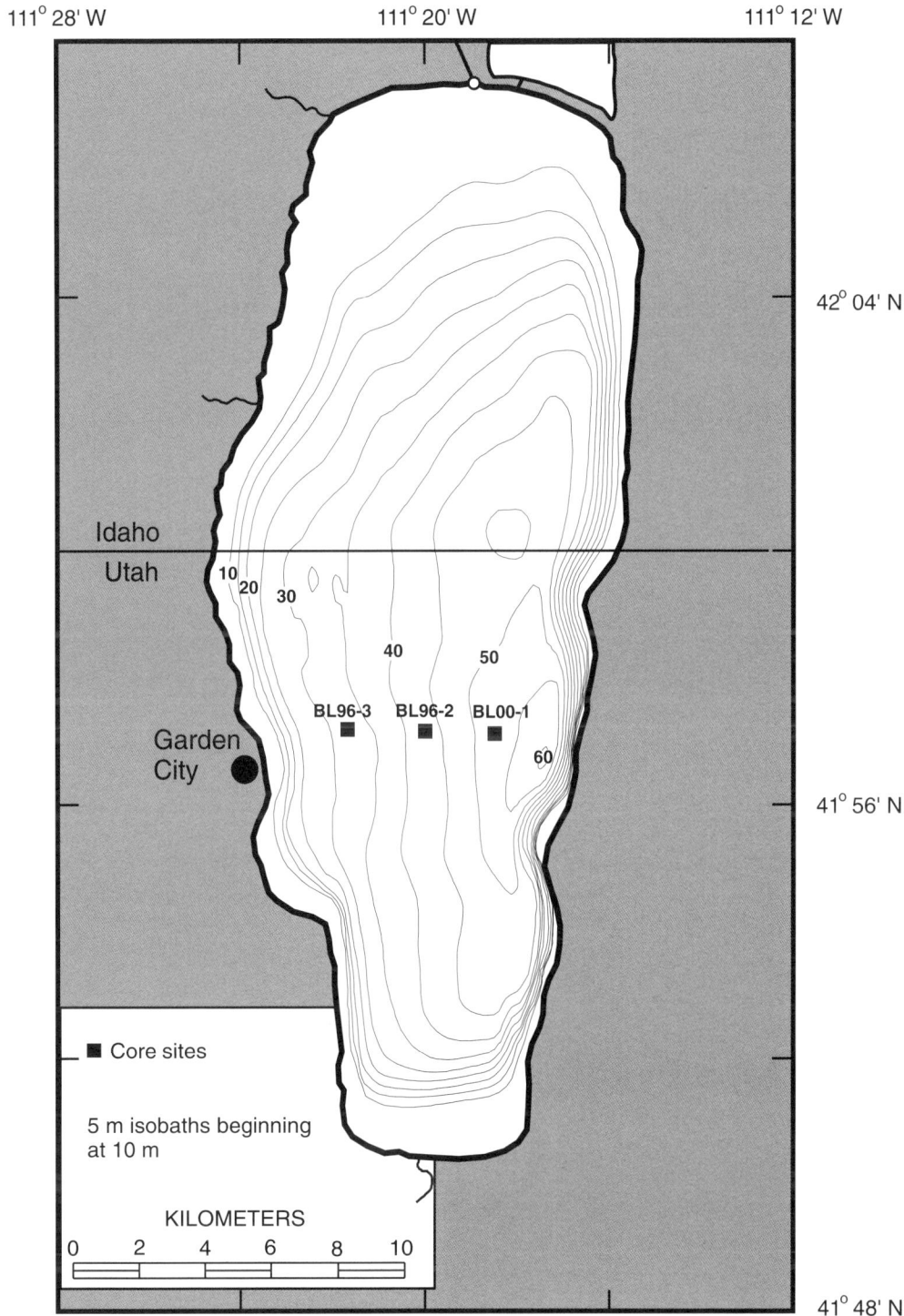

Figure 2. Coring locations BL96-2, BL96-3, and BL00-1. Bathymetry was mapped in 2002 by Denny and Colman (2003).

contains a highly attenuated, incomplete section of the carbonate-rich sediment overlying more than 3.5 m of the older siliciclastic unit. Several distinct horizons (e.g., changes in mineralogy and magnetic properties) allow the sections penetrated by these and other cores to be precisely correlated (Colman et al., this volume; Dean, this volume; Rosenbaum et al., this volume). Together the two cores provide a nearly complete composite record extending from ca. 26 cal ka to the late Holocene. Chronology is provided by age models based on radiocarbon ages obtained mostly from pollen concentrates (Colman et al., this volume). These concentrates contain pollen and other refractory organic material (e.g., charcoal) that survived dissolution of carbonate and silicate minerals during standard pollen preparation procedures. For the carbonate-rich sediment in core BL96-2, the age model is based on 17 radiocarbon ages. Because of the discontinuous nature of the record, no age model was created for the attenuated carbonate-rich section in core BL96-3. Earlier studies (Colman et al., 2005; Dean et al., 2006; Laabs et al., 2007) accepted ages from sediments interpreted herein to contain glacial flour, but Colman et al. (this volume) determined that all such ages are thousands of years too old. Therefore, for glacial flour-rich sediments, an age model using ages from above and below the glacial flour-bearing sediments was constructed for core BL00-1 (Fig. 2). Interpolated ages for distinct horizons were then transferred to cores BL96-2 and BL96-3. Six such horizons were used to provide age control for core BL96-2 and ten were used for BL96-3 (Colman et al., this volume). The radiocarbon analyses provide an apparently reliable chronology, although potential systematic errors of a few hundred years from possible reworking of sediment (Smoot and Rosenbaum, this volume) and storage in the catchment cannot be fully evaluated. Although actual dating uncertainty is probably several hundred years, we give calibrated ages to the nearest 100 years provided by the cubic spline models of Colman et al. (this volume) for the purpose of describing variations in proxy data and environmental change with age.

The record of environmental change derived from the siliciclastic sediments provides an interpreted glacial history. This record is largely based on changes in provenance of detrital material in the Bear Lake sediments. Therefore, we first describe some properties of fluvial sediments from the Bear River and from streams entering the lake that allow us to infer such changes.

PROPERTIES OF FLUVIAL SEDIMENTS

As shown by Rosenbaum et al. (this volume), there are significant differences in the mineralogy, elemental composition, and magnetic properties of fluvial sediments among the various areas that have supplied detrital material to Bear Lake. In part, these differences reflect differences in bedrock geology (Fig. 3), but, in the case of some magnetic properties, differences reflect variable concentrations of dust in the fluvial material (Reynolds and Rosenbaum, 2005). Although the ferrimagnetic Fe-oxide grains in dust are mostly not derived from the local bedrock, Rosenbaum et al. (this volume) demonstrate that such grains are delivered to the lake largely through fluvial input and therefore provide a powerful tool to assess changes in provenance for the Bear Lake sediments. For the purpose of understanding the provenance of detrital material, the catchment was divided into three areas: (1) the local Bear Lake catchment (i.e., stream sampling sites 1 and 7–17 on Fig. 3), (2) the lower Bear River (sites 19–45), and (3) the headwaters of the Bear River in the Uinta Mountains (sites 47–53).

Although there are significant differences among the properties of the different size fractions of the alluvial sediment, only data from the finest-sized fraction (<63 µm) from the local catchment and the lower Bear River are considered here because Bear Lake sediments contain little detrital material coarser than silt, except in very shallow water (Rosenbaum et al., this volume). For the headwaters of the Bear River, data for both the <63 µm fraction and the coarse-sand-sized fraction are considered, because late Pleistocene glaciers in that area would have generated fine-grained material from lithologies now largely represented in the coarser fractions of the river sediments. There is little difference among the magnetic properties of the fine-sand, coarse-sand, and pebble fractions, and mineralogical and chemical data were acquired from only the <63 µm and coarse-sand-sized fractions (Rosenbaum et al., this volume).

Local Bear Lake Catchment

On average, samples of <63 µm fluvial material from the local catchment contain more dolomite and more ferrimagnetic minerals than other potential sediment sources (Fig. 4). Outcrops on the west side of Bear Lake are largely lower Paleozoic formations containing large amounts of dolomite, quartzite, and limestone. The west-side exposures contain lesser amounts of Precambrian quartzite and shale, and fluvial clastic rocks of the Tertiary Wasatch Formation (Fig. 3). In contrast, rocks on the east side of the lake consist of Mesozoic limestone and clastic sedimentary rocks (with little dolomite) as well as widespread Wasatch Formation (Oriel and Platt, 1980; Dover, 1995). The widespread dolomitic rocks on the west side of the lake are reflected in the mineralogy and elemental chemistry of fluvial sediments in the area (Table 1). The average dolomite and Mg contents of the local catchment samples are much higher than averages from other potential sediment sources (Fig. 4). Differences in bedrock between east-side and west-side sources contribute to the high dispersion of dolomite, calcite, and Mg contents.

The average magnetic susceptibility of samples from the local catchment is more than 2.5 times greater than that from either of the other areas (Table 1, Figs. 4 and 5). Samples from both the east and west sides of the lake have similar magnetic properties despite differences in bedrock. The stream sediments, both from the local catchment and from along the Bear River, contain a wide variety of silt-sized Fe-oxide mineral grains (including magnetite, titanomagnetite, and ilmenohematite), most of which must have been derived from lithologies not present in the watersheds and are therefore interpreted to be a component of dust (Reynolds and Rosenbaum, 2005). The high susceptibilities of stream

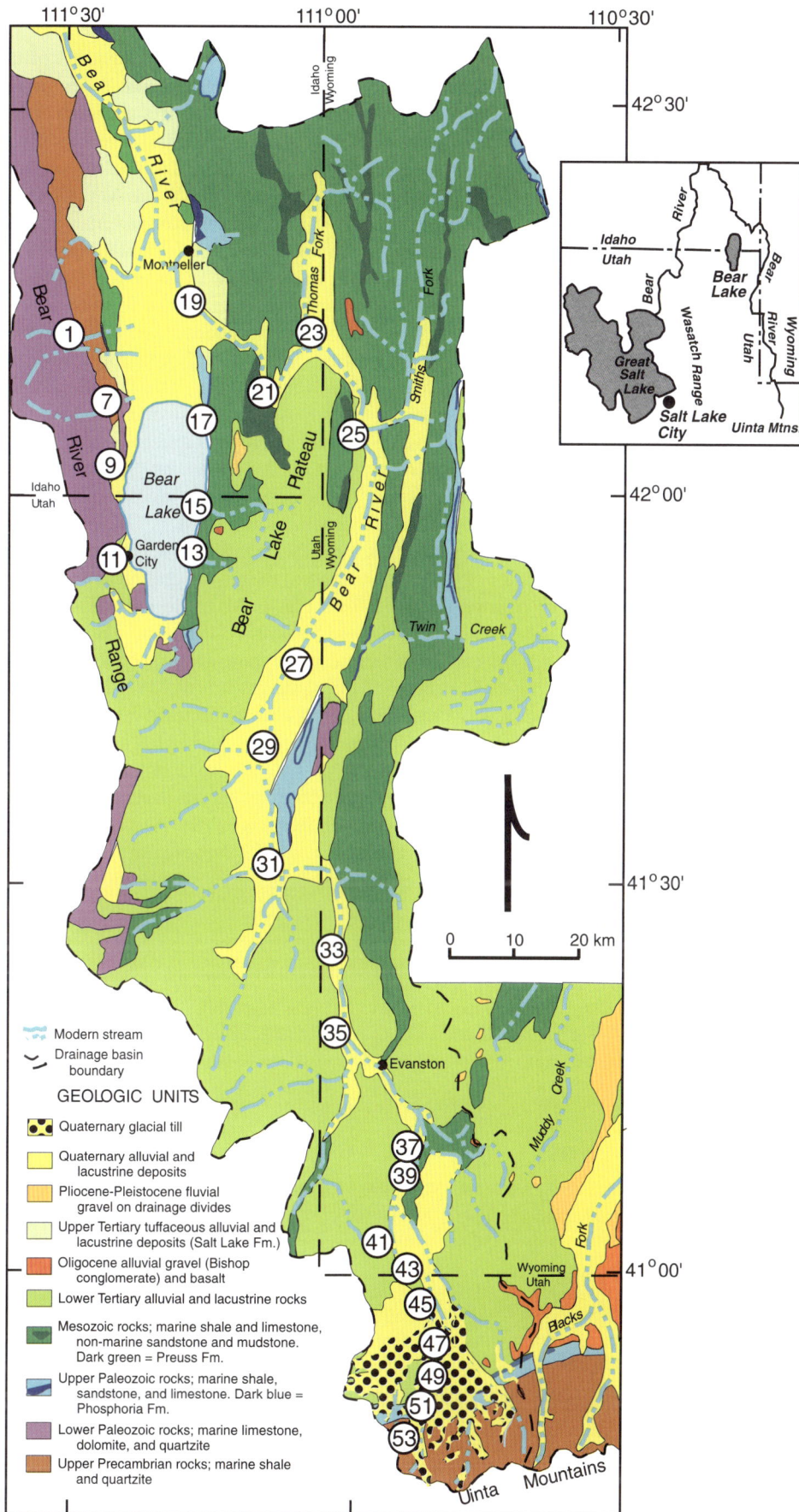

Figure 3. Generalized geologic map of the Bear Lake area and the Bear River drainage above Bear Lake Valley (modified from Reheis et al., this volume). Numbered circles are stream sample locations. Odd-numbered samples were collected from stream bottoms. Even-numbered samples (not shown) were collected from bank or over-bank deposits near locations of the odd-numbered samples.

GEOLOGIC UNITS

Modern stream

Drainage basin boundary

Quaternary glacial till

Quaternary alluvial and lacustrine deposits

Pliocene-Pleistocene fluvial gravel on drainage divides

Upper Tertiary tuffaceous alluvial and lacustrine deposits (Salt Lake Fm.)

Oligocene alluvial gravel (Bishop conglomerate) and basalt

Lower Tertiary alluvial and lacustrine rocks

Mesozoic rocks; marine shale and limestone, non-marine sandstone and mudstone. Dark green = Preuss Fm.

Upper Paleozoic rocks; marine shale, sandstone, and limestone. Dark blue = Phosphoria Fm.

Lower Paleozoic rocks; marine limestone, dolomite, and quartzite

Upper Precambrian rocks; marine shale and quartzite

samples from the local catchment as compared to susceptibilities from along the Bear River reflect either greater amounts of dust deposition or less dilution by locally derived rock material.

Headwaters of the Bear River

The mineralogy and elemental chemistry of samples from the headwaters of the Bear River differ markedly from those of

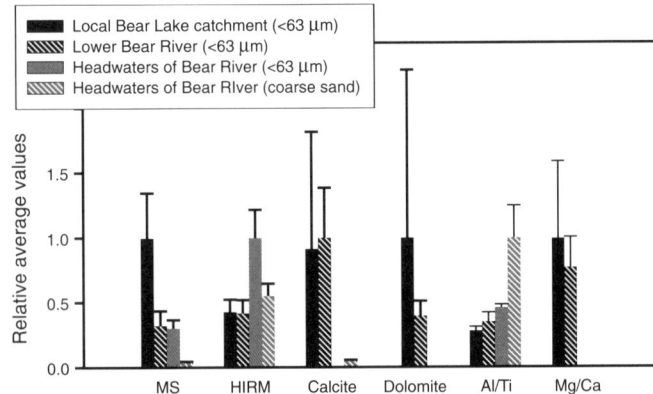

Figure 4. Relative average values and standard deviations of magnetic susceptibility (MS), hard isothermal remanent magnetization (HIRM), minerals, and elemental ratios for fluvial sediments. Mg/Ca values are not shown for samples from the headwaters because these samples contain very small quantities of carbonate minerals, and for these samples Mg and Ca reflect silicate minerals. In the lower reaches of the Bear River and in the Bear Lake catchment Mg and Ca probably reflect dolomite and calcite.

the local Bear Lake catchment (Fig. 4). The headwaters of the Bear River contain Precambrian quartzites and shales of the Uinta Mountain Group (Bryant, 1992), which commonly contain abundant hematite cement (Ashby et al., 2005). The Bear River sediments in this area contain little calcite and no dolomite and, as a consequence, have low contents of Ca and Mg (Table 1). Al and Ti contents of the <63 μm and coarse-sand fractions differ greatly. Values for the finer fraction are slightly higher than those from the local Bear Lake catchment and lower reaches of the Bear River, whereas values for the coarser fraction are significantly lower than from either of these areas. In comparison to these other source areas, values of Al/Ti from the headwaters of the Bear River are significantly higher, with the fine-fraction yielding values ~1.5 times those from these other areas and the coarse fraction values ~3 times as high (Table 1).

Magnetic properties of the headwater samples are also distinct from other areas. The hematite content (as measured by hard isothermal remanent magnetization [HIRM]) of the coarse-sand fraction is greater than the contents from the other source areas by a factor of ~1.3 and the content of the <63 μm fraction is greater by a factor of ~2.4 (Table 1, Fig. 4). Magnetic susceptibility (MS) values for the two size fractions also differ greatly. Both indicate that the content of ferrimagnetic minerals is much lower than in samples from the local Bear Lake catchment. MS of the finer fraction is about the same as for samples from the lower reaches of the Bear River, whereas MS for the coarser fraction is an order of magnitude lower. For the two size fractions of the headwater samples, differences in hematite content and in elemental concentrations probably reflect differences in the lithologies contained in the fractions. Differences in MS, however, are probably due

TABLE 1. MEAN VALUES AND STANDARD DEVIATIONS FOR MAGNETIC, MINERALOGICAL, AND ELEMENTAL DATA FROM STREAM SAMPLES IN DIFFERENT CATCHMENT AREAS

	Local Bear Lake Catchment <63 μm	Lower Bear River <63 μm	Upper Bear River <63 μm	Upper Bear River coarse sand
MS x 10^{-7} (m^3kg^{-1})	5.63 ± 1.96 (14)	1.82 ± 0.64 (28)	1.71 ± 0.37 (8)	0.21 ± 0.07 (8)
HIRM x 10^{-4} (Am^2kg^{-1})	3.47 ± 0.81 (13)	3.38 ± 0.84 (26)	8.12 ± 1.76 (8)	4.46 ± 0.78 (8)
Quartz (%)	69 ± 14 (13) 71 ± 10 east 67 ± 17 west	77 ± 6 (20)	82 (2)	90 ± 1 (4)
Calcite (%)	11 ± 11 (13) 20 ± 11 east 4 ± 4 west	12 ± 5 (20)	0 (2)	<1 (4)
Dolomite (%)	12 ± 17 (13) 3 ± 1 east 21 ± 19 west	5 ± 1 (20)	0 (2)	0 (4)
Ca (%)	6.03 ± 2.28 (8)	5.38 ± 1.15 (14)	0.81 ± 0.14 (4)	0.05 ± 0.01 (4)
Mg (%)	1.80 ± 1.39 (8)	1.21 ± 0.29 (14)	0.91 ± 0.14 (4)	0.11 ± 0.03 (4)
Mg/Ca	0.30 ± 0.17 (8)	0.23 ± 0.07 (14)	1.15 ± 0.031 (4)	2.25 ± 0.18 (4)
Al (%)	3.72 ± 0.68 (8)	4.64 ± 1.32 (14)	7.31 ± 1.20 (4)	2.44 ± 0.17 (4)
Ti (%)	0.21 ± 0.04 (8)	0.21 ± 0.03 (14)	0.25 ± 0.03 (4)	0.04 ± 0.01 (4)
Al/Ti	18 ± 2 (8)	22 ± 5 (14)	30 ± 2 (4)	65 ± 16 (4)

Note: Number of measurements given in parentheses. Values for quartz, calcite, and dolomite in the local Bear Lake catchment are given for all 13 samples from around the lake and for samples subdivided into those from the east side (6 samples) and those from the west side (7 samples). MS—magnetic susceptibility; HIRM—hard isothermal remanent magnetization.

in large part to origin of most of the ferrimagnetic Fe-oxides as a component of dust. The silt-sized dust particles are highly concentrated in the <63 μm fraction of the alluvial sediment.

Lower Bear River

Properties of alluvial sediments from the lower reaches of the Bear River are similar to those of sediments from the local Bear Lake catchment, but differ in several important ways. Sediments from the two areas have similar contents of calcite, Al, Ti, and Ca, and nearly identical values of HIRM (Table 1, Fig. 4). The content of dolomite in the lower Bear River sediment is about the same as in the streams on the east side of Bear Lake and much lower than in streams on the west side (Table 1), so that overall the Bear River sediments have less dolomite (and therefore less

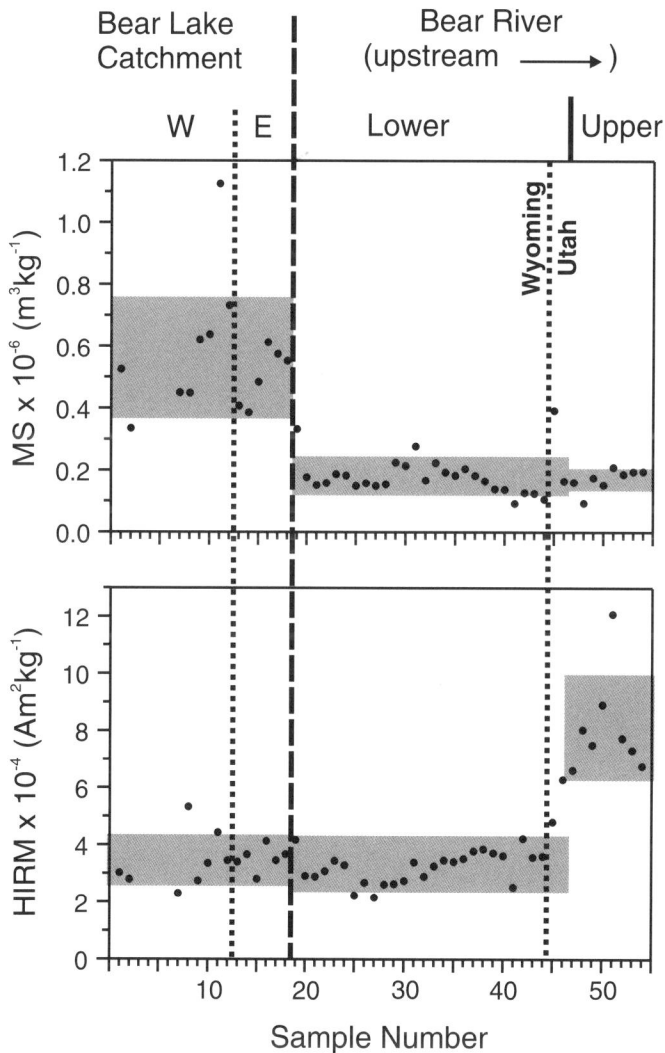

Figure 5. Magnetic susceptibility (MS) and hard isothermal remanent magnetization (HIRM) of catchment samples. Sample locations are shown on Figure 3. Shaded boxes indicate one standard deviation about the mean.

Mg) than average sediment from streams in the local catchment. In addition, MS values are much lower for the Bear River sediments than for those from the local catchment, indicating that the Bear River sediment contains lower concentrations of dust.

In contrast to samples from the headwaters, the lower Bear River sediments have lower values of Al/Ti and HIRM as well as higher concentrations of carbonate minerals (Table 1). Importantly, sediments derived from the small catchment area in the Uinta Mountains are quickly diluted by material from other parts of the catchment so that under present (interglacial) conditions Precambrian Uinta Mountain Group material is a minor component of Bear River sediment (see HIRM in Fig. 5).

PROPERTIES OF LAKE SEDIMENTS

Stratigraphic Zones

The composite section sampled by cores BL96-2 and BL96-3 was divided into seven zones based on changes in carbonate and quartz contents, magnetic properties (MS and HIRM), and elemental ratios (Fig. 6). Zones 1a–1d are characterized by high quartz content, low carbonate content, and intermediate to high values of MS and HIRM relative to the full range of values observed in the section. Zone 1a extends from the base of core BL96-3 to the top of a sharp decrease in MS and coincident increase in HIRM. This zone is distinguished by well-defined variations in MS, HIRM, Mg/Ca, and Al/Ti. Within zone 1a, MS is negatively correlated with HIRM and positively correlated with Mg/Ca (Fig. 6 and 7). In addition, Al/Ti varies with HIRM. As noted by Rosenbaum et al. (this volume), these relations begin to change in zone 1b (equivalent to their zone I). The top of zone 1b occurs at maxima in HIRM and Al/Ti (Fig. 6). Within this zone, variations in MS are unmatched by variations in Mg/Ca, which remains essentially constant (Fig. 7). Values of Mg/Ca are consistently higher than those predicted by the relation between Mg/Ca and MS in zone 1a. Zone 1c extends from the top of zone 1b to an upward increase in Mg/Ca. Values of Mg/Ca in zone 1c are similar to those in zone 1b and do not vary with MS. Relative to zones 1a and 1b, MS in zone 1c varies less with HIRM. The upper boundary of zone 1d is defined by the beginning of a decrease in quartz content and concomitant increase in carbonate content. Within this zone, Mg/Ca varies with MS and HIRM varies inversely with MS, but with trends that are offset from those in zone 1a (Fig. 7). Zone 2a marks a major transition in sedimentation. Within this zone, the sediment becomes progressively enriched in endogenic carbonate at the expense of detrital material (e.g., quartz). MS, HIRM, and Mg/Ca decrease upward across zone 2a. The lower boundary of zone 2b is the beginning of an interval of nearly constant calcite content (~40%). The calcite content increases abruptly in the upper part of zone 2b. MS and HIRM continue to decrease across this zone (although there are two spikes in MS of unknown origin) and Mg/Ca is uniformly low. The boundary between zones 2b and 2c is defined by a sharp increase in aragonite content. MS, HIRM, and Mg/Ca are low throughout zone 2c.

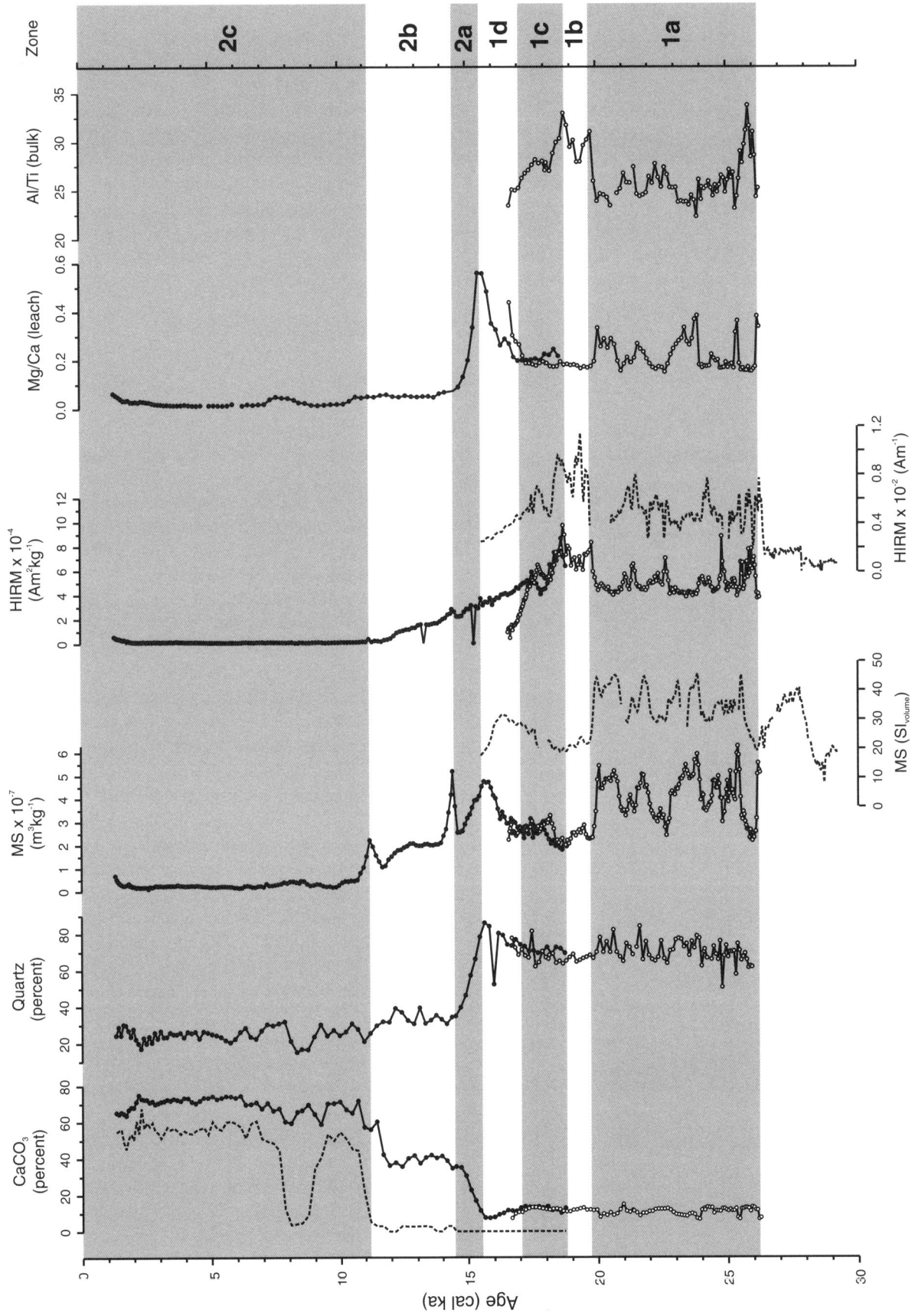

Figure 6. Contents of CaCO₃ and quartz (Dean, this volume), magnetic susceptibility (MS), hard isothermal remanent magnetization (HIRM; Rosenbaum et al., this volume), and elemental ratios versus age for cores BL96-2 (solid circles) and BL96-3 (open circles). Ages are from cubic spline models of Colman et al. (this volume). Data from the uppermost 40 cm of core BL96-3 are not plotted because the age control for that part of the core is limited. The dashed line on the carbonate plot shows the content of aragonite. Dashed curves for MS and HIRM are from core BL00-1 (Heil et al., this volume). Mg/Ca values were determined from acid-leached samples (Bischoff et al., 2005); Al/Ti values were determined for bulk samples (Rosenbaum et al., this volume). Stratigraphic zones are described in the text.

Changes in Provenance

Between 26 and 15.5 cal ka, zones 1a–1d, the sediments contain little, if any, endogenic carbonate. These sediments consist largely of quartz with lesser amounts of calcite, dolomite, and small quantities of other minerals (Dean et al., 2006). The quartz and carbonate-mineral contents remain relatively constant (Fig. 6), and detrital Fe-oxide minerals appear largely unaffected by post-depositional alteration (Reynolds and Rosenbaum, 2005; Rosenbaum et al., this volume). Within this interval, the properties of stream sediments, described above, provide a qualitative basis for interpreting changes in sources of the lithogenic material (Table 2).

The quasi-cyclical variations in magnetic properties and elemental ratios in zone 1a (Fig. 6) are interpreted to reflect variations in the concentrations of detrital material derived from Uinta Mountain Group rocks and of that derived from local Bear Lake catchment. HIRM and Al/Ti are highly correlated (Fig. 7) and high values of these parameters indicate high concentrations of sediment derived from headwaters of the Bear River. Similarly, MS and Mg/Ca are highly correlated and high values of these parameters indicate elevated concentration of sediment from the local Bear Lake catchment, with high values of MS indicating high input of dust from surficial material, and high values of Mg/Ca indicating high input from dolomitic bedrock. The strong

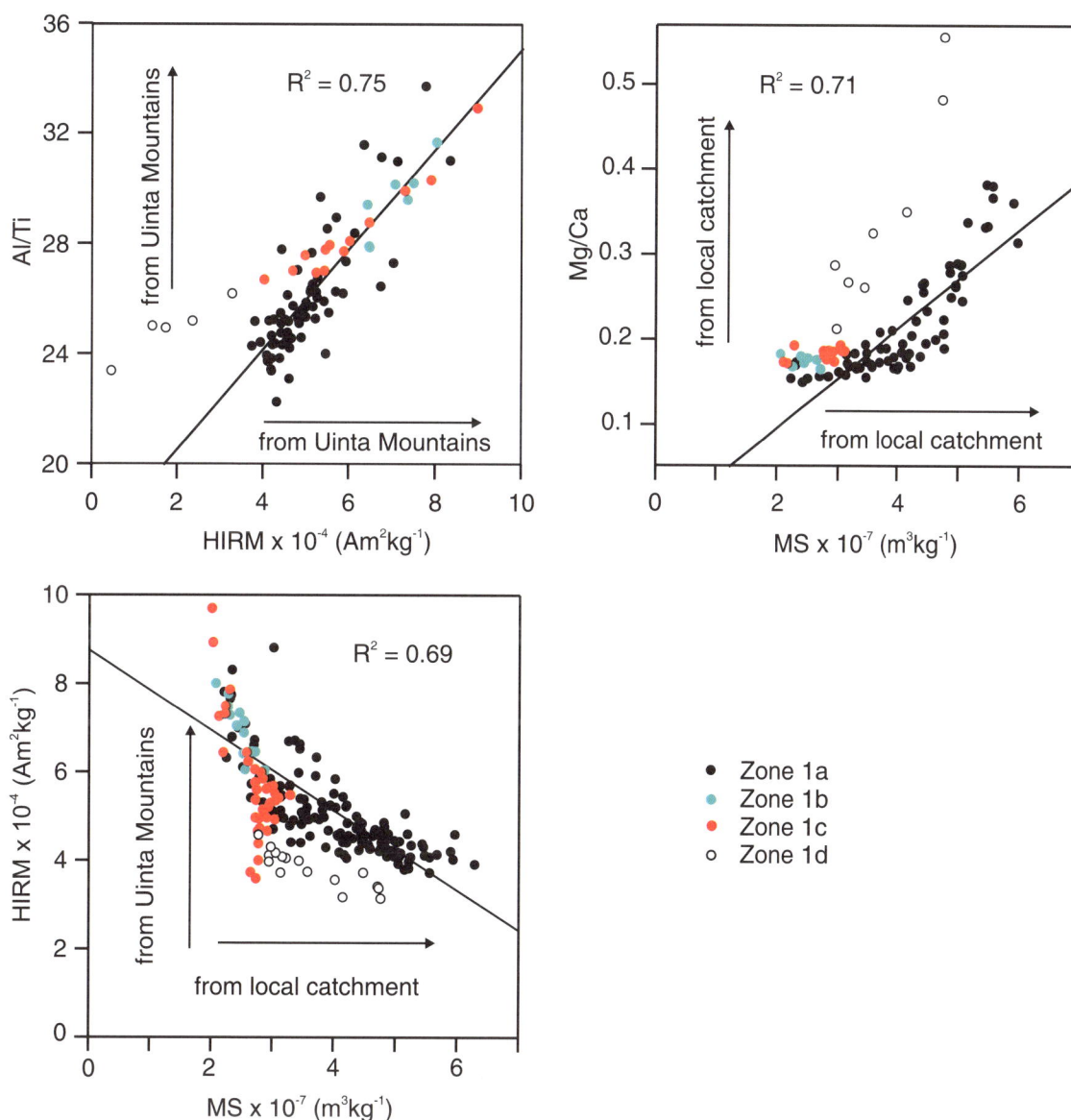

Figure 7. Plots of magnetic properties and elemental ratios for samples from zones 1a–1d, showing: (1) the positive relations between HIRM and Al/Ti (indicative of material from the Uinta Mountains), and MS and Mg/Ca (indicative of the local Bear Lake catchment), and (2) the negative relation between MS and HIRM. For Al/Ti versus HIRM, linear fit is for zones 1a, 1b, and 1c. For HIRM versus MS, linear fit is for zones 1a and 1b. For Mg/Ca versus MS, linear fit is for zone 1a.

TABLE 2. QUALITATIVE CHARACTERISTICS OF SEDIMENT SOURCE AREAS

	Local Bear Lake catchment <63 μm	Lower Bear River <63 μm	Upper Bear River <63 μm	Upper Bear River coarse-sand
Dolomite/Calcite (leachable Mg/Ca)	High	Low	None*	None*
Al/Ti	Low	Low	Intermediate	High
MS	High	Intermediate	Intermediate	Low
HIRM	Low	Low	High	Intermediate

Note: MS—magnetic susceptibility; HIRM—hard isothermal remanent magnetization.
*Because Upper Bear River rocks contain very little carbonate and no dolomite, Mg in these sediments is not acid leachable.

negative relation between concentrations of material from the Bear Lake catchment and from the Uinta Mountains (Fig. 7) suggests that the variations are driven by the same mechanism. As discussed below, the Uinta Mountain detritus is thought to be largely glacial flour. As the supply of glacial flour changed in response to waxing and waning of glaciers, the flux of Uinta Mountain-derived material into Bear Lake varied and variably diluted the locally derived sediment.

In zone 1b, high values of HIRM and Al/Ti and low values of MS and Mg/Ca indicate high concentrations of material from the headwaters of the Bear River and low concentrations of material from the local catchment. However, the relation between dust-derived material and bedrock from the local catchment differs from that in zone 1a. In zone 1b, Mg/Ca values do not vary with MS and, as noted previously, are high with respect to MS (Fig. 7). This relation suggests that, in comparison to zone 1a, material from the local catchment is enriched in detritus from the local bedrock relative to that from surficial material.

Decreasing values of HIRM and Al/Ti across zone 1c indicate decreasing concentration of sediment from the Uinta Mountain Group rocks (Fig. 6). This decrease is not offset by an increase in MS comparable to the relation in the underlying zones (Fig. 7). The concentration of dolomitic detritus is similar to that in zone 1b. These relations indicate that the decrease in content of Uinta Mountain Group material is not fully offset by an increase in content of material from the local catchment and, therefore, must be increasingly replaced by material derived from lower reaches of the Bear River.

The content of detritus from the Uinta Mountain Group (i.e., HIRM and Al/Ti) continues to decrease upward across zone 1d. In this zone, the content of sediment derived from the local Bear Lake catchment increases. Both MS and Mg/Ca increase, although the relation of these components to each other and of MS to HIRM differ from that in zone 1a (Fig. 7). These changes could be produced by changes in input from the local catchment or by changes in the input from the Bear River, or both.

Zone 2a represents the transition from sediment dominated by detrital material to sediment dominated by endogenic carbonate; within this zone and zones 2b and 2c, the magnetic properties and elemental data provide little information about the sources of detrital material. Decreases in HIRM and MS across zone 2a are consistent with dilution of detrital material by increasing amounts of endogenic carbonate, whereas decreasing values of Mg/Ca

reflect large increases in the content of Ca due to calcite precipitation rather than changes in the proportions of detrital dolomite and calcite. HIRM and MS continue to decline through zone 2b and remain at very low values through zone 2c. Carbonate content (which is nearly constant across zone 2b and increases from ~40% to 70% from the top of zone 2b to the base of zone 2c) is not high enough to fully account for the low values of HIRM and MS by dilution. Therefore, increasing post-depositional destruction of detrital Fe-oxide minerals is thought to produce the low values of HIRM and MS within these zones.

INTERPRETATIONS OF ENVIRONMENTAL CHANGE

Glacial History

Glacially derived material in lake sediments has been used to infer variations in glacial extent (Leonard, 1985; Karlén and Matthews, 1992; Bischoff and Cummins, 2001; Rosenbaum and Reynolds, 2004). Although there is considerable uncertainty about the timing of sediment discharge and its relation to subglacial storage and subsequent flushing (Hicks et al., 1990; Harbor and Warburton, 1993; Hallet et al., 1996), Leonard (1997) argues that glacial flour output largely reflects glacial extent when averaged over periods of several tens to hundreds of years or longer. This position is supported by a study of glacial flour in sediments of Upper Klamath Lake, Oregon (Rosenbaum and Reynolds, 2004).

The interpretations of glacial history from Bear Lake sediments presented below is much more detailed than those presented by Dean et al. (2006) and Laabs et al. (2007). In addition, because of revisions to the Bear Lake chronology (Colman et al., this volume), the timing of the onset of glaciation, maximum glaciation, and subsequent glacial retreat differs significantly from that presented in the earlier studies.

ΔHIRM as Rock Flour Proxy

Magnetic properties and geochemical data indicate that sediments in zones 1a–1c contain large quantities of material derived from Uinta Mountain Group rocks. This material is interpreted to be largely glacial rock flour for several reasons. (1) The age of the sediments coincides with a period in which glaciers were widespread in the Rocky Mountains including the Uinta Mountains (Pierce, 2004; Laabs et al., 2007). (2) The glaciated area in the headwaters of the Bear River makes up a small fraction of the drainage basin above Bear Lake (Fig. 3) so that in the absence of enhanced erosion, such as that provided by glaciers, detritus from the Uinta Mountains would be expected to be a minor component of Bear Lake sediment. This expectation is corroborated by the observation that HIRM values for modern stream sediments are elevated within the glacial limits in the Bear River Valley (Fig. 3) but abruptly decrease downstream due to dilution (Fig. 5). (3) In general, the proxy indicators of Uinta Mountain Group-derived material (i.e., HIRM and Al/Ti) in zones 1a–1c covary with content of fine-silt- and clay-sized siliciclastic material (Fig. 8). This relation is similar to that observed for Upper Klamath Lake where glacially derived rock flour in the lake sediment is finer than other detrital rock material (Reynolds et al., 2004; Rosenbaum and Reynolds, 2004).

Variations in glacial-flour content could reflect (1) changes in the location of the Bear River and its delta, (2) changes in the course of the Bear River that caused the river to alternately enter and bypass the lake, or (3) changes in the flux of glacial flour that reflect changes in the extent of glaciation. As discussed by Rosenbaum et al. (this volume), changes in the location of the river and its delta are probably insignificant because such changes would alter the distribution of sediment within the lake and, therefore, cannot account for the nearly identical variations

observed in cores BL96-3 and BL00-1 (Fig. 6), which are separated by 4.5 km (Fig. 2). In the absence of the Bear River, Bear Lake produces large amounts of endogenic carbonate minerals (Dean et al., 2006). The low uniform content of calcite within zones 1a–1d indicates that the Bear River was connected to the lake throughout this interval. We therefore interpret variations in glacial flour to reflect changes in the extent of glaciation in the headwaters of the Bear River.

High concentrations of hematite (i.e., high values of HIRM) indicate high concentrations of glacial flour derived from the Uinta Mountain Group, but not all hematite is derived from these rocks. The initial precipitation of endogenic carbonate at the base of zone 2a, following the increasing dominance of sediment derived from the local catchment, suggests that at this point little if any sediment was being delivered by the Bear River. The value of HIRM at this point ($\sim$3.3 $\times$ 10^{-4} Am^2 kg^{-1}) is close to the average value of HIRM for stream samples from the local catchment (Table 1). Fortuitously, the average HIRM value for the lower Bear River samples is close to that of the local catchment. Therefore, 3.3 $\times$ 10^{-4} Am^2 kg^{-1} is taken as an estimate of HIRM for siliciclastic sediment lacking hematite-rich detritus from the Uinta Mountains.

In core BL96-3, magnetic properties were determined for $\sim$190 samples spanning the $\sim$10,500 yr interval of zones 1a–1d, an average of $\sim$55 yr per sample. Within these zones, we interpret variations of HIRM that are defined by three or more values (Fig. 6) to reflect changes in glacial extent. ΔHIRM (i.e., HIRM - 3.3 $\times$ 10^{-4} Am^2 kg^{-1}) is therefore a measure of the content of glacial flour derived from the Uinta Mountain Group. Values of ΔHIRM are >0 through all but the uppermost part of zones 1a–1d, indicating significant glaciation during most of the period. This proxy record of late Wisconsin glaciation of the Uinta Mountains (derived from the Bear Lake sediments) and a similar record for the southern Cascade Range (derived from Upper Klamath Lake) both identify a number of millennial-scale variations in glacial extent prior to the maximum advance of ice (Fig. 9).

Glacial flour derived from rocks other than those of the Uinta Mountain Group will not contribute to ΔHIRM. Such glacial flour would have been produced by portions of glaciers extending down the Bear River Valley beyond the extent of Precambrian rocks of the Uinta Mountain Group (Fig. 3). In addition, small glaciers along the crest of the Bear River Range west of Bear Lake (Reheis et al., this volume) would have contributed hematite-poor dolomitic glacial flour to the lake.

Onset of Glaciation

The lowest content of glacial flour in zone 1a occurs at the base of core BL96-3. Magnetic properties from a longer record in core BL00-1 (Heil et al., this volume) are precisely correlated to the BL96-3 data and demonstrate that core BL96-3 captures the entire interval of elevated hematite content (Fig. 6). Therefore, the onset of extensive glaciation in the northwestern Uinta Mountains (e.g., the headwaters of the Bear River) apparently occurred ca. 26 cal ka, similar to the timing of the 1965 radiocarbon-based

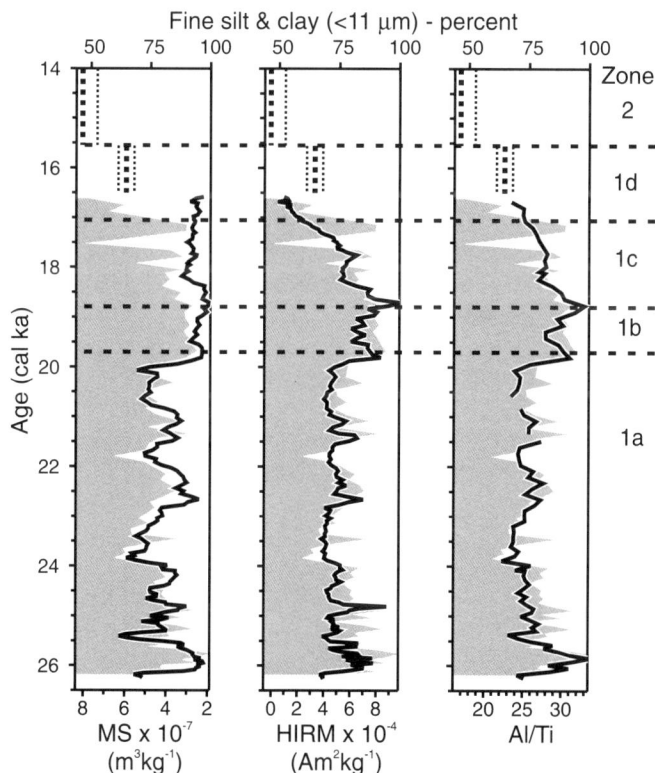

Figure 8. Content of <11 μm material (shaded curve) in core BL96-3 plotted with MS, HIRM, and Al/Ti. Stratigraphic zones are described in the text. Vertical dashed lines show means and standard deviations of <11 μm material for groups of samples outside the range of the age model for the core.

model of glacial extent for the Rocky Mountains presented by Porter et al. (1983), which is shown in Figure 9. There is no evidence in the Bear Lake record for extensive glaciation prior to ca. 26 cal ka, like that shown in the 1983 model. Changes in vegetation within the Bonneville Basin ca. 32 cal ka indicate that a cold dry climatic regime that had existed for ~10,000 years began to give way to moister but still cool conditions (Madsen et al., 2001). The onset of glaciation indicated in the Bear Lake record postdates this vegetation change and the initial growth of Lake Bonneville by ~6000 years (Fig. 10), and apparently coincides with formation of the Stansbury shoreline.

Maximum Glaciation

The Bear Lake record indicates that maximum content of glacial flour derived from the Uinta Mountain Group, interpreted from ΔHIRM (Fig. 9), was attained ca. 19.7 cal ka and persisted for ~800 years (i.e., zone 1b). During the last glacial interval, the maximum ice extent in the Bear River Valley extended well beyond outcrops of Precambrian Uinta Mountain Group rocks (Fig. 3). The distal portions of glaciers would have produced glacial flour largely from Tertiary sedimentary rocks and would not have contributed to high values of ΔHIRM. Therefore, ΔHIRM probably underestimates the maximum content of glacial flour from the Uinta Mountains. Although, it is possible that hematite-

poor glacial flour diluted hematite-rich glacial flour to such an extent that peak glacial-flour content occurred after the maximum in ΔHIRM, the peak could not have been more than a few hundred years later given the cosmogenic ages, 18.1–18.7 cal ka, from terminal moraines in the Bear River Valley (Fig. 9; Laabs et al., 2007). Lacking a definitive proxy for glacial flour derived from Tertiary sedimentary rocks and recognizing the above uncertainty, we nevertheless interpret ΔHIRM to represent the extent of glaciation in the Bear River drainage.

During the period of maximum glaciation, zone 1b, the content of material from dolomitic bedrock (i.e., Mg/Ca) no longer varied with material derived from surficial deposits (i.e., MS) in the same manner as in zone 1a (Fig. 7). In comparison to zone 1a, the content of dolomitic material in zone 1b is consistently high with respect to surficial material. We suggest that this relative enrichment of dolomitic material reflects glacial flour produced by small glaciers on the crest of the Bear River Range (Reheis et al., this volume).

The timing of maximum glacial extent derived from the Bear Lake record coincides with the maximum glacial extent in the southern Cascade Range derived from Upper Klamath Lake, Oregon (Rosenbaum and Reynolds, 2004). The timing also coincides with the late stages of widespread glaciation according to radiocarbon-based age models for the Rocky Mountains (Porter

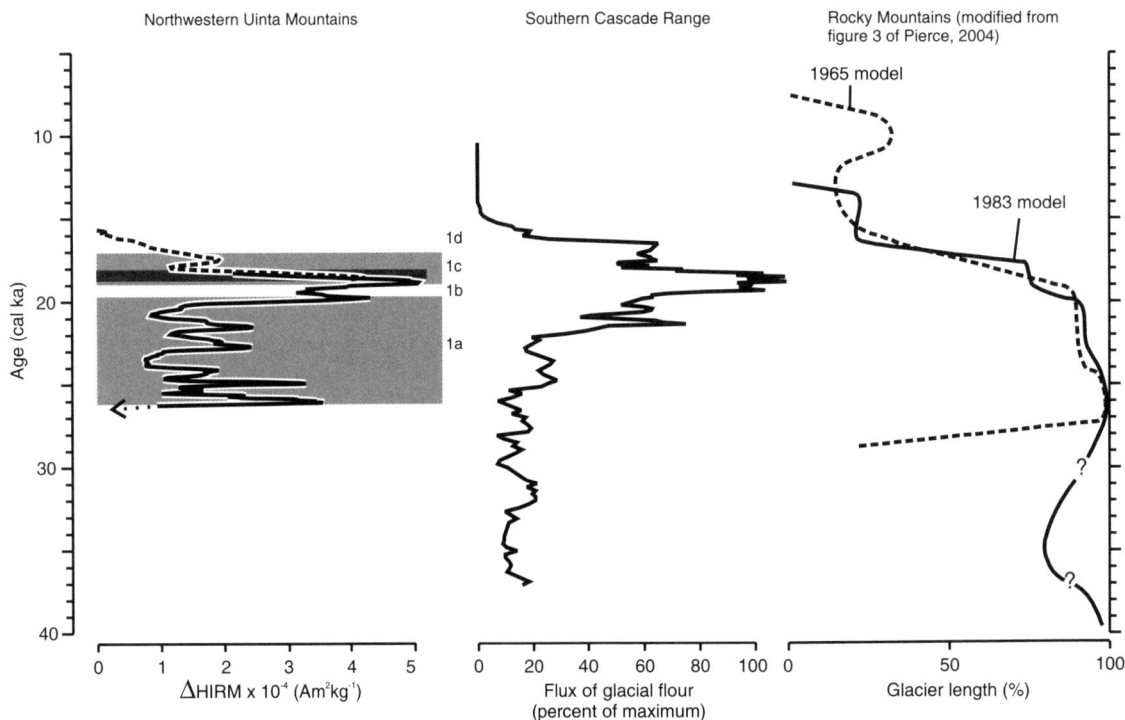

Figure 9. Comparison of the Bear Lake glacial record for the northwestern Uinta Mountains with a record from the southern Cascades derived from glacial flour in Upper Klamath Lake (Rosenbaum and Reynolds, 2004), and with models of glacial length for the Rocky Mountains (Porter et al., 1983). Zones 1a–1d are as shown in Figure 6. For the Bear Lake record, the solid (dashed) line is the portion of the record from core BL96-3 (BL96-2). The dark horizontal band in the Bear Lake record indicates the range of cosmogenic ages (18.1–18.7 ka) from terminal moraines in the Bear River Valley (Laabs et al., 2007).

et al., 1983; Pierce, 2004). An age for the maximum extent of glaciation of 19.7–18.9 cal ka, as indicated by the Bear Lake record, implies that glaciers attained their greatest size as Lake Bonneville grew to its maximum depth (Fig. 10; Oviatt et al., 1992; Oviatt, 1997).

Glacial Recession

In zone 1c, glacial extent in the headwaters of the Bear River decreased abruptly at 18.9 cal ka (Fig. 9). The initial recession was followed by a short re-advance that culminated shortly before 17.5 cal ka. Within this interval, the content of fine silt and clay remained high (Fig. 8), and the relation between the content of dolomitic material and surficial material was like that in zone 1b, indicating probable continued input of glacially derived material from the Bear River Range. Glacial extent then decreased continuously in the top of zone 1c and through zone 1d. No glacial input from the Bear River Range is apparent in zone 1d and glacial flour from the Uinta Mountains became undetectable at 15.7 cal ka.

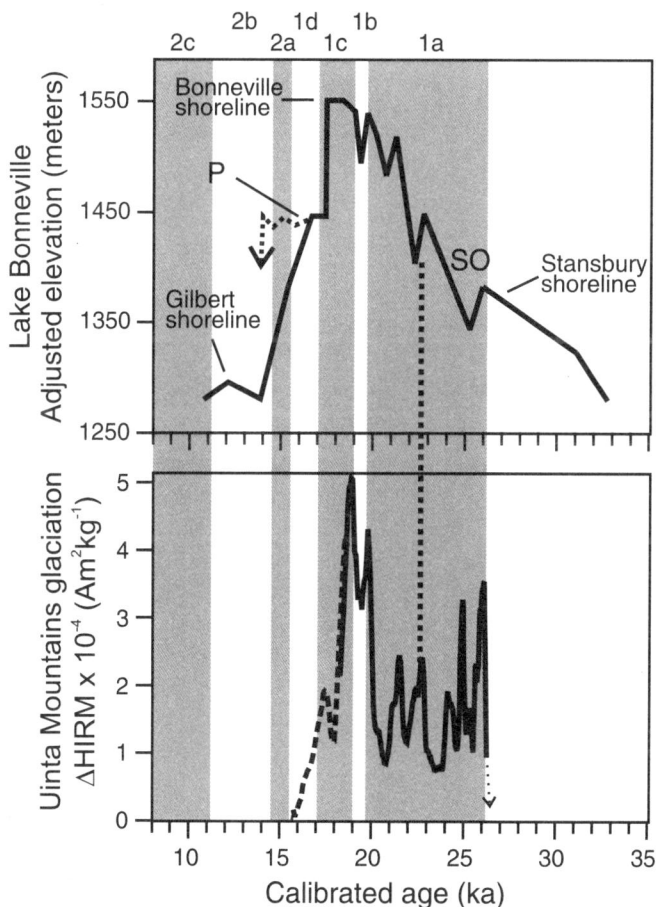

Figure 10. Comparison of the Bear Lake glacial flour record with the Lake Bonneville hydrograph (modified from Oviatt, 1997). The dashed arrow indicates an alternative interpretation of the timing of regression following formation of the Provo shoreline (Godsey et al., 2005). SO indicates the Stansbury oscillation. Zones 1a–2c are those shown in Figure 6.

These features closely match variations in the record from Upper Klamath Lake (Fig. 9). The initial rapid recession following the glacial maximum occurred during the highstand of Lake Bonneville, and the beginning of the final retreat apparently occurred during or after formation of the Provo shoreline (Fig. 10).

Post-Glacial Record

Although glaciers persisted at high elevations in the Uinta Mountains until ca. 10 cal ka (Munroe, 2003), Bear Lake received little glacial flour after ca. 15.7 cal ka. Increasing content of endogenic calcite beginning ca. 15.5 cal ka (zone 2a) indicates increased salinity of the lake and probable abandonment of the lake by the Bear River (Dean et al., 2006). The period of increasing carbonate content occurs within the period in which Oviatt (1997) interpreted climatic conditions to produce the major regression of Lake Bonneville from the Provo shoreline (Fig. 10). Godsey et al. (2005), however, suggest that Lake Bonneville remained at or near the Provo shoreline until 13.0 cal ka and then regressed rapidly.

Lake-level reconstructions based on sedimentological evidence indicate an abrupt ~20 m drop in the level of Bear Lake ca. 12.8 cal ka (Smoot and Rosenbaum, this volume). Over the next 1000 years, a period roughly coinciding with the Younger Dryas, the lake declined to more than 30 m below modern level, rose to near the modern level, fell again by more than 30 m, and rose once again. Other evidence suggests increasingly evaporative conditions during this period and the subsequent several hundred years. Such evidence includes a sharp increase in $\delta^{18}O$ ca. 12.0 cal ka (Dean et al., 2006) and an increase in the content of Mg in endogenic calcite from ~4 mol percent to ~8 mol percent beginning ca. 11.5 cal ka (based on X-ray diffraction data; G. Skipp, 2005, personal commun.). At 11.0 cal ka, salinity reached a threshold and aragonite began to precipitate (Fig. 6). Many studies indicate that climatic conditions in western North America were cool and moist during the Younger Dryas (Thompson et al., 1993; Phillips et al., 1994; Rhode and Madsen, 1995; Quade et al., 1998; Allen and Anderson, 2000; Reasoner and Jodry, 2000; Currey et al., 2001; Madsen et al., 2001; Polyak et al., 2004), and recent work at Great Salt Lake indicates that lake level rose to the Gilbert shoreline during this period (Oviatt et al., 2005). However, the Bear Lake results are more consistent with studies that indicate low lake levels in the Great Basin during this period (Benson et al., 1997; Licciardi, 2001).

SUMMARY

Sediments deposited in Bear Lake after ca. 26 cal ka consist of a lower siliciclastic unit, which was deposited until 15.5 cal ka, and overlying carbonate-rich sediments. For the siliciclastic material, differences in magnetic and elemental properties of fluvial sediments document changes in detrital input from three sediment source areas: the headwaters of the Bear River in the Uinta Mountains, the Bear River below the Uinta Mountains and above Bear Lake, and the local Bear Lake watershed. Under present conditions, hematite-rich material from Uinta Mountain Group rocks is a minor constituent of Bear River sediment because it is rapidly diluted below the headwaters. It is a major component, however, in the siliciclastic Bear Lake sediments, indicating not only that the Bear River was connected directly to Bear Lake but also that the river was transporting much more detritus derived from Uinta Mountain Group rocks than at present. We interpret that glaciation was the cause of enhanced transport of this material, that the Uinta Mountain Group material is largely glacial flour, and that the quantity of glacial flour is related to the extent of glaciation. The onset of extensive glaciation at 26 cal ka postdates moistening of conditions in the region indicated by the initial growth of Lake Bonneville and by changes in vegetation. The Bear Lake record indicates that maximum glaciation occurred between ca. 19 and 20 cal ka as Lake Bonneville attained its maximum depth and that rapid deglaciation began ca. 19 cal ka while Lake Bonneville remained high. The onset of deglaciation indicated in the Bear Lake record is in good agreement with cosmogenic ages from terminal moraines in the Bear River Valley (Laabs et al., 2007). The content of glacial flour in the Bear Lake sediments continued to decrease until ca. 15.5 cal ka. Shortly thereafter, the Bear River abandoned Bear Lake and endogenic carbonate minerals replaced clastic sediments. The properties used to interpret changes in provenance for sediments older than 15.5 cal ka cannot be used in the younger sediments because of the effects of the endogenic minerals and the post-depositional destruction of detrital Fe-oxide minerals.

From 15.5 to 10 cal ka, evidence of environmental change is largely based on changes in the types and compositions of endogenic carbonate minerals (Dean, this volume), and sedimentary indicators of lake-level history (Smoot and Rosenbaum, this volume). This evidence includes increasing concentration of calcite from 15.5 to 14.5 cal ka, indicating withdrawal of the Bear River from Bear Lake and increasing evaporative conditions. This interval in the Bear Lake record coincides with a period when Lake Bonneville may have been declining due to climatic conditions. Although lake levels in the Bonneville Basin apparently rose during the Younger Dryas, evidence from Bear Lake suggests generally low but fluctuating lake levels and increasingly evaporative conditions during this period.

ACKNOWLEDGMENTS

We thank M. Hudson, H. Goldstein, S. Zimmerman, R. Negrini, and D. Clark for constructive reviews. Funding was provided by the Earth Surface Dynamics Program of the U.S. Geological Survey.

ARCHIVED DATA

Archived data for this chapter can be obtained from the NOAA World Data Center for Paleoclimatology at http://www.ncdc.noaa.gov/paleo/pubs/gsa2009bearlake/.

REFERENCES CITED

Allen, B.D., and Anderson, R.Y., 2000, A continuous, high-resolution record of late Pleistocene climate variability from the Estancia basin, New Mexico: Geological Society of America Bulletin, v. 112, p. 1444–1458, doi: 10.1130/0016-7606(2000)112<1444:ACHRRO>2.0.CO;2.

Ashby, J.M., Geissman, J.W., and Weil, A.B., 2005, Paleomagnetic and fault kinematic assessment of Laramide-age deformation in the eastern Uinta Mountains or, has the eastern end of the Uinta Mountains been bent? *in* Dehler, C.M., Pederson, J.L., Sprinkel, D.A., and Kowallis, B.J., eds., Uinta Mountain geology: Utah Geological Association Publication 33, p. 285–320.

Benson, L., Burdett, J., Lund, S., Kashgarian, M., and Mensing, S., 1997, Nearly synchronous climate change in the Northern Hemisphere during the last glacial termination: Nature, v. 388, p. 263–265, doi: 10.1038/40838.

Bischoff, J.L., and Cummins, K., 2001, Wisconsin glaciation of the Sierra Nevada (79,000–15,000 yr B.P.), as recorded by rock flour in sediments of Owens Lake, California: Quaternary Research, v. 55, p. 14–24, doi: 10.1006/qres.2000.2183.

Bischoff, J.L., Cummins, K., and Shamp, D.G., 2005, Geochemistry of sediments in cores and sediment traps from Bear Lake, Utah and Idaho: U.S. Geological Survey Open-File Report 2005-1215, http://pubs.usgs.gov/of/2005/1215/ (accessed June 2007).

Bryant, B., 1992, Geologic and structure maps of the Salt Lake City 1° × 2° quadrangle, Utah and Wyoming: U.S. Geological Survey Miscellaneous Investigations Series Map I-1997, 1:250,000.

Colman, S.M., 2006, Acoustic stratigraphy of Bear Lake, Utah–Idaho—Late Quaternary sedimentation patterns in a simple half-graben: Sedimentary Geology, v. 185, p. 113–125, doi: 10.1016/j.sedgeo.2005.11.022.

Colman, S.M., Kaufman, D.S., Rosenbaum, J.G., and McGeehin, J.P., 2005, Radiocarbon dating of cores collected from Bear Lake, Utah and Idaho: U.S. Geological Survey Open-File Report 2005-1320, http://pubs.usgs.gov/of/2005/1320/ (accessed June 2007).

Colman, S.M., Rosenbaum, J.G., Kaufman, D.S., Dean, W.E., and McGeehin, J.P., 2009, this volume, Radiocarbon ages and age models for the last 30,000 years in Bear Lake, Utah and Idaho, *in* Rosenbaum, J.G., and Kaufman, D.S., eds., Paleoenvironments of Bear Lake, Utah and Idaho, and its catchment: Geological Society of America Special Paper 450, doi: 10.1130/2009.2450(05).

Currey, D.R., Lips, E., Thein, B., Wambeam, T., and Nishazawa, S., 2001, Elevated Younger Dryas lake levels in the Great Basin, western U.S.A.: Geological Society of America Abstracts with Programs, v. 33, p. A-217.

Dean, W.E., 2009, this volume, Endogenic carbonate sedimentation in Bear Lake, Utah and Idaho, over the last two glacial-interglacial cycles, *in* Rosenbaum, J.G., and Kaufman, D.S., eds., Paleoenvironments of Bear Lake, Utah and Idaho, and its catchment: Geological Society of America Special Paper 450, doi: 10.1130/2009.2450(07).

Dean, W., Rosenbaum, J., Haskell, B., Kelts, K., Schnurrenberger, D., Valero-Garces, B., Cohen, A., Davis, O., Dinter, D., and Nielson, D., 2002, Progress in global lake drilling holds potential for global change research: Eos (Transactions, American Geophysical Union), v. 83, p. 85, 90–91.

Dean, W.E., Rosenbaum, J.G., Skipp, G., Colman, S.M., Forester, R.M., Liu, A., Simmons, K., and Biscoff, J.L., 2006, Unusual Holocene and late Pleistocene carbonate sedimentation in Bear Lake, Utah and Idaho, USA: Sedimentary Geology, v. 185, p. 93–112, doi: 10.1016/j.sedgeo.2005.11.016.

Denny, J.F., and Colman, S.M., 2003, Geophysical surveys of Bear Lake, Utah-Idaho, September 2002: U.S. Geological Survey Open-File Report 03-150.

Dover, J.H., 1995, Geologic map of the Logan 30′ × 60′ quadrangle, Cache and Rich counties, Utah, and Lincoln and Uinta counties, Wyoming: U.S. Geological Survey Miscellaneous Investigations Series Map I-2210, scale 1:100,000.

Godsey, H.S., Currey, D.R., and Chan, M.A., 2005, New evidence for an extended occupation of the Provo shoreline and implications for regional climate change, Pleistocene Lake Bonneville, Utah, USA: Quaternary Research, v. 63, p. 212–223, doi: 10.1016/j.yqres.2005.01.002.

Hallet, B., Hunter, L., and Bogen, J., 1996, Rates of erosion and sediment evacuation by glaciers: A review of field data and their implications: Global and Planetary Change, v. 12, p. 213–235, doi: 10.1016/0921-8181(95)00021-6.

Harbor, J., and Warburton, J., 1993, Relative rates of glacial and nonglacial erosion in alpine environments: Arctic and Alpine Research, v. 25, p. 1–7, doi: 10.2307/1551473.

Heil, C.W., Jr., King, J.W., Rosenbaum, J.G., Reynolds, R.L., and Colman, S.M., 2009, this volume, Paleomagnetism and environmental magnetism of GLAD800 sediment cores from Bear Lake, Utah and Idaho, *in* Rosenbaum, J.G., and Kaufman, D.S., eds., Paleoenvironments of Bear Lake, Utah and Idaho, and its catchment: Geological Society of America Special Paper 450, doi: 10.1130/2009.2450(13).

Hicks, D.M., McSaveney, M.J., and Chinn, T.J.H., 1990, Sedimentation in proglacial Ivory Lake, Southern Alps, New Zealand: Arctic and Alpine Research, v. 22, p. 26–42, doi: 10.2307/1551718.

Karlén, W., and Matthews, J.A., 1992, Reconstructing Holocene glacier variations from glacial lake sediments: Studies from Nordvestlandet and Jostedalsbreen-jotunheimen, southern Norway: Geografiska Annaler, v. 74A, p. 327–348, doi: 10.2307/521430.

Kaufman, D.S., Bright, J., Dean, W.E., Moser, K., Rosenbaum, J.G., Anderson, R.S., Colman, S.M., Heil, C.W., Jr., Jiménez-Moreno, G., Reheis, M.C., and Simmons, K.R., 2009, this volume, A quarter-million years of paleoenvironmental change at Bear Lake, Utah and Idaho, *in* Rosenbaum, J.G., and Kaufman, D.S., eds., Paleoenvironments of Bear Lake, Utah and Idaho, and its catchment: Geological Society of America Special Paper 450, doi: 10.1130/2009.2450(14).

Laabs, B.J.C., Munroe, J.S., Rosenbaum, J.G., Refsnider, K.A., Mickelson, D.M., and Cafee, M.W., 2007, Chronology of the last glacial maximum in the upper Bear River basin, Utah: Arctic, Antarctic, and Alpine Research, v. 39, p. 537–548, doi: 10.1657/1523-0430(06-089)[LAABS]2.0.CO;2.

Leonard, E.M., 1985, Glaciological and climatic controls on lake sedimentation, Canadian Rocky Mountains: Zeitschrift für Gletscherkunde und Glazialgeologie, v. 21, p. 35–42.

Leonard, E.M., 1997, The relationship between glacial activity and sediment production: Evidence from a 4450-year varve record of neoglacial sedimentation in Hector Lake, Alberta, Canada: Journal of Paleolimnology, v. 17, p. 319–330, doi: 10.1023/A:1007948327654.

Licciardi, J.M., 2001, Chronology of latest Pleistocene lake-level fluctuations in the pluvial Lake Chewaucan basin, Oregon, USA: Journal of Quaternary Science, v. 16, p. 545–553, doi: 10.1002/jqs.619.

Madsen, D.B., Rhode, D., Grayson, D.K., Broughton, J.M., Livingston, S.D., Hunt, J., Quade, J., Schmitt, D.N., and Shaver, M.W., 2001, Late Quaternary environmental change in the Bonneville basin, western USA: Palaeogeography, Palaeoclimatology, Palaeoecology, v. 167, p. 243–271, doi: 10.1016/S0031-0182(00)00240-6.

Munroe, J.S., 2003, Holocene timberline and palaeoclimate of the northern Uinta Mountains, northeastern Utah, USA: The Holocene, v. 13, p. 175–185, doi: 10.1191/0959683603hl600rp.

Oriel, S.S., and Platt, L.B., 1980, Geologic map of the Preston 1° × 2° quadrangle, southeastern Idaho and western Wyoming: U.S. Geological Survey Miscellaneous Investigations Series Map I-1127, scale 1:250,000.

Oviatt, C.G., 1997, Lake Bonneville fluctuations and global climate change: Geology, v. 25, p. 155–158, doi: 10.1130/0091-7613(1997)025<0155:LBFAGC>2.3.CO;2.

Oviatt, C.G., Currey, D.R., and Sack, D., 1992, Radiocarbon chronology of Lake Bonneville, eastern Great basin, USA: Palaeogeography, Palaeoclimatology, Palaeoecology, v. 99, p. 225–241, doi: 10.1016/0031-0182(92)90017-Y.

Oviatt, C.G., Miller, D.M., McGeehin, J.P., Zachary, C., and Mahan, S., 2005, The Younger Dryas phase of Great Salt Lake, Utah, USA: Palaeogeography, Palaeoclimatology, Palaeoecology, v. 219, p. 263–284, doi: 10.1016/j.palaeo.2004.12.029.

Pierce, K.L., 2004, Pleistocene glaciations of the Rocky Mountains, *in* Gillespie, A.R., Porter, S.C., and Atwater, B.F., eds., The Quaternary Period in the United States: Amsterdam, Elsevier, p. 63–76.

Phillips, F.M., Campbell, A.R., Smith, G.I., and Bischoff, J.L., 1994, Interstadial climate cycles: A link between western North America and Greenland: Geology, v. 22, p. 1115–1118, doi: 10.1130/0091-7613(1994)022<1115:ICCALB>2.3.CO;2.

Polyak, V.J., Rasmussen, J.B.T., and Asmerom, Y., 2004, Prolonged wet period in the southwestern United States through the Younger Dryas: Geology, v. 32, p. 5–8, doi: 10.1130/G19957.1.

Porter, S.C., Pierce, K.L., and Hamilton, T.D., 1983, Late Pleistocene glaciation in the Western United States, *in* Porter, S.C., ed., The Late Pleistocene, v. 1 of Wright, H.E., Jr., ed., Late Quaternary environments of the United States: Minneapolis, Minnesota University of Minnesota Press, p. 71–111.

Quade, J., Forester, R.M., Pratt, W.L., and Carter, C., 1998, Black mats, spring-fed streams, and late-glacial-age recharge in the southern Great Basin: Quaternary Research, v. 49, p. 129–148, doi: 10.1006/qres.1997.1959.

Reasoner, M.A., and Jodry, M.A., 2000, Rapid response of alpine timberline vegetation to the Younger Dryas climate oscillation in the Colorado Rocky Mountains, USA: Geology, v. 28, p. 51–54, doi: 10.1130/0091-7613(2000)28<51:RROATV>2.0.CO;2.

Reheis, M.C., Laabs, B.J.C., and Kaufman, D.S., 2009, this volume, Geology and geomorphology of the Bear Lake Valley and upper Bear River, Utah and Idaho, *in* Rosenbaum, J.G., and Kaufman, D.S., eds., Paleoenvironments of Bear Lake, Utah and Idaho, and its catchment: Geological Society of America Special Paper 450, doi: 10.1130/2009.2450(02).

Reynolds, R.L., and Rosenbaum, J.G., 2005, Magnetic mineralogy of sediments in Bear Lake and its watershed: Support for paleoenvironmental interpretations: U.S. Geological Survey Open-File Report 2005-1406, http://pubs.usgs.gov/of/2005/1406/ (accessed June 2007).

Reynolds, R.L., Rosenbaum, J.G., Rapp, J., Kerwin, M.W., Bradbury, J.P., Colman, S., and Adam, D., 2004, Record of late Pleistocene glaciation and deglaciation in the southern Cascade Range: I. Petrologic evidence from lacustrine sediment in Upper Klamath Lake, southern Oregon: Journal of Paleolimnology, v. 31, p. 217–233, doi: 10.1023/B:JOPL.0000019230.42575.03.

Rhode, D., and Madsen, D.B., 1995, Late Wisconsin vegetation in the northern Bonneville basin: Quaternary Research, v. 44, p. 246–256, doi: 10.1006/qres.1995.1069.

Rosenbaum, J.G., and Reynolds, R.L., 2004, Record of late Pleistocene glaciation and deglaciation in the southern Cascade Range: II. Flux of glacial flour in a sediment core from Upper Klamath Lake, Oregon: Journal of Paleolimnology, v. 31, p. 235–252, doi: 10.1023/B:JOPL.0000019229.75336.7a.

Rosenbaum, J.G., Dean, W.E., Reynolds, R.L., and Reheis, M.C., 2009, this volume, Allogenic sedimentary components of Bear Lake, Utah and Idaho, *in* Rosenbaum, J.G., and Kaufman, D.S., eds., Paleoenvironments of Bear Lake, Utah and Idaho, and its catchment: Geological Society of America Special Paper 450, doi: 10.1130/2009.2450(06).

Smoot, J.P., and Rosenbaum, J.G., 2009, this volume, Sedimentary constraints on Late Quaternary lake-level fluctuations at Bear Lake, Utah and Idaho, *in* Rosenbaum, J.G., and Kaufman, D.S., eds., Paleoenvironments of Bear Lake, Utah and Idaho, and its catchment: Geological Society of America Special Paper 450, doi: 10.1130/2009.2450(12).

Thompson, R.S., Whitlock, C., Bartlein, P.J., Harrison, S.P., and Spaulding, W.G., 1993, Climatic changes in the western United States, since 18,000 yr B.P., *in* Wright, H.E., Jr., Kutzbach, J.E., Webb, T., III, Ruddiman, W.F., Street-Perott, F.A., and Bartlein, P.J., eds., Global climates since the Last Glacial Maximum: Minneapolis, University of Minnesota Press, p. 468–513.

MANUSCRIPT ACCEPTED BY THE SOCIETY 15 SEPTEMBER 2008

The Geological Society of America
Special Paper 450
2009

Sedimentary constraints on late Quaternary lake-level fluctuations at Bear Lake, Utah and Idaho

Joseph P. Smoot
U.S. Geological Survey, MS 926A, National Center, Reston, Virginia 20192, USA

Joseph G. Rosenbaum
U.S. Geological Survey, Box 25046, MS 980, Denver Federal Center, Denver, Colorado 80225, USA

ABSTRACT

A variety of sedimentological evidence was used to construct the lake-level history for Bear Lake, Utah and Idaho, for the past ~25,000 years. Shorelines provide evidence of precise lake levels, but they are infrequently preserved and are poorly dated. For cored sediment similar to that in the modern lake, grain-size distributions provide estimates of past lake depths. Sedimentary textures provide a highly sensitive, continuous record of lake-level changes, but the modern distribution of fabrics is poorly constrained, and many ancient features have no modern analog. Combining the three types of data yields a more robust lake-level history than can be obtained from any one type alone. When smooth age-depth models are used, lake-level curves from multiple cores contain inconsistent intervals (i.e., one record indicates a rising lake level while another record indicates a falling lake level). These discrepancies were removed and the multiple records were combined into a single lake-level curve by developing age-depth relations that contain changes in deposition rate (i.e., gaps) where indicated by sedimentological evidence. The resultant curve shows that, prior to 18 ka, lake level was stable near the modern level, probably because the lake was overflowing. Between ca. 17.5 and 15.5 ka, lake level was ~40 m below the modern level, then fluctuated rapidly throughout the post-glacial interval. Following a brief rise centered ca. 15 ka (= Raspberry Square phase), lake level lowered again to 15–20 m below modern from ca. 14.8–11.8 ka. This regression culminated in a lowstand to 40 m below modern ca. 12.5 ka, before a rapid rise to levels above modern ca. 11.5 ka. Lake level was typically lower than present throughout the Holocene, with pronounced lowstands 15–20 m below the modern level ca. 10–9, 7.0, 6.5–4.5, 3.5, 3.0–2.5, 2.0, and 1.5 ka. High lake levels near or above the modern lake occurred ca. 8.5–8.0, 7.0–6.5, 4.5–3.5, 2.5, and 0.7 ka. This lake-level history is more similar to records from Pyramid Lake, Nevada, and Owens Lake, California, than to those from Lake Bonneville, Utah.

Smoot, J.P., and Rosenbaum, J.G, 2009, Sedimentary constraints on late Quaternary lake-level fluctuations at Bear Lake, Utah and Idaho, *in* Rosenbaum, J.G., and Kaufman, D.S., eds., Paleoenvironments of Bear Lake, Utah and Idaho, and its catchment: Geological Society of America Special Paper 450, p. 263–290, doi: 10.1130/2009.2450(12). For permission to copy, contact editing@geosociety.org. ©2009 The Geological Society of America. All rights reserved.

INTRODUCTION

Bear Lake in Utah and Idaho (Fig. 1) has fluctuated in size frequently in response to changing conditions of inflow and evaporation. Historically, the lake has fluctuated from a highstand at 1805.5 m above sea level (asl) numerous times since 1920 to a low level of 1799 m asl in 1936 (Dean et al., this volume). These fluctuations are a response to regional climate change (Dean et al., this volume), although some pre-historical fluctuations may also reflect physiographic changes due to stream capture or tectonism (Reheis et al., this volume). During much of the late Pleistocene, Bear Lake was fed directly by the Bear River (Dean et al., 2006; Kaufman et al., this volume), which entered the basin from the east at a location north of the present lake (Reheis et al., this volume). During that time, the lake drained back into the Bear River via channels just east of the Bear River Range (Reheis et al., this volume).

This paper uses sedimentological data to derive a radiocarbon-dated record of lake-level fluctuations for the past ~30,000 years. The data include shoreline elevations, grain-size changes, and textural comparisons. Each data type has it strengths and weaknesses. The largest obstacle to evaluating lake-level changes is establishing the ages of different types of deposits. The density and reliability of radiocarbon ages are highly variable in the various cores and outcrops. For cored sediments, ages used in this paper are radiocarbon ages that have been converted to calendar years before 1950 (cal yr B.P.) using the terrestrial calibration set of Stuiver et al. (1998). The age estimates are median probabilities (1σ errors) reported by Colman et al. (this volume).

SHORELINE DEPOSITS

Shoreline deposits provide the most direct evidence of past lake levels. These include a characteristic suite of structures or grain-size distributions that can be inferred to have been deposited within a meter or two of the water surface at the lake edge. Shoreline deposits include wave-formed beach and bar deposits and delta topsets. Such deposits range in thickness from several meters to a thin veneer overlying a wave-cut terrace.

Modern wave-formed shoreline deposits at Bear Lake range from boulder-cobble deposits to sand (Fig. 2). Unlike the shoreline deposits of oceans or very large lakes, many of the Bear Lake shoreline deposits are not very well sorted. The relatively small size of the lake results in highly variable wave energy with the coarsest grain sizes moved during storms. Former boulder shorelines are recognized by open-framework packing with layers or patches of grains sorted by size and shape (e.g., Smoot and Lowenstein, 1991) (Fig. 3A). Sandy beaches range in character from well-sorted shoreface deposits (Figs. 3B and 3C) to poorly sorted shell gravel sheets (Fig. 3D; Smoot, this volume). Gravel sheets form where the lake-floor slope is very low. Storm waves break hundreds of meters offshore in a meter or two of water, allowing only low-energy waves to reach the shoreline (Smoot, this volume). Wave-formed bars are ridges that may form many tens of meters from shore (Smoot and Lowenstein, 1991). By comparison

to wave-formed bars in marine settings, the bars are formed at depths where the crests are less than a meter or two below the surface (e.g., Komar, 1998, p. 292–302). Wave-formed bars occur where the lake-floor slope is steep enough to allow larger waves to move closer to shore. The bars are pushed shoreward by storms until they become attached to the shoreline. The lakeward sides of the bars are subsequently eroded and redistributed, or they act as a new shoreface for beach deposits. The relatively small surface area of Bear Lake produces only small-scale offshore bars (tens of centimeters thick). Wave-cut terraces are erosional surfaces. Wave erosion is an initial response to rising water levels but long-term erosion is a function of sediment supply (see Komar, 1998, p. 121–129). Large terraces are commonly attributed to lake still stands (e.g., Benson, 1994; Adams and Wesnousky, 1998), but stepped terraces would be expected in a rising lake. Sediments deposited over a terrace range from a single layer of imbricated pebbles to a sediment platform several meters thick that is built into the lake. A sediment platform is deposited as a series of lakeward-dipping foresets composed of wave-sorted sediment (see Smoot and Lowenstein, 1991). Such deposits are thin within the Holocene section at Bear Lake, but some of the meters-thick Pleistocene "fan delta" deposits noted in Laabs and Kaufman (2003) may have originated in this manner.

Deltas are lakeward-thickening wedges of sediment formed at the intersection of streams or rivers and standing bodies of water (e.g., Coleman, 1981). Delta deposits in Bear Lake have two styles. Birdfoot distributaries with mouth bar deposits occur where the Bear River previously intersected the lake. Gilbert-type delta deposits (Gilbert, 1885) with well-defined delta foresets and topsets occur where sediment-laden streams debouched from canyons directly into the lake. Robertson (1978) examined the distribution of fluvial and deltaic sediments in the north end of Bear Lake Valley, and some data on delta distributions were provided by Laabs and Kaufman (2003). Colman (2006) presented seismic evidence of Pleistocene deltaic deposits in the eastern side and north end of the lake. The sedimentary characteristics of delta deposits around Bear Lake have not been studied in any detail.

Laabs and Kaufman (2003) and Reheis et al. (2005, this volume) documented evidence of shorelines in outcrops around Bear Lake. The study of Laabs and Kaufman (2003) included data from Williams et al. (1962), Robertson (1978), and McCalpin (1993). Most of the shoreline features noted in these studies are beach gravel deposits. They mention delta and fan-delta deposits, but do not indicate recognition of the topset-foreset transition, which is necessary to accurately ascertain water depth. Additional shoreline data include sandy beach deposits and shell gravels in sediment cores within the lake (Smoot, this volume). Colman (2006) recognized an erosional terrace in seismic records of the lake floor that was interpreted as a shoreline feature. The top of the terrace is at 1798 m and the base is at 1784 m elevation. Dean (this volume) described an aragonite-cemented, boulder-cobble deposit called the "rock pile" whose base is at 1786 m and top is at 1792 m. Dean (this volume) interpreted this deposit as a microbialite mound probably related to spring activity. The

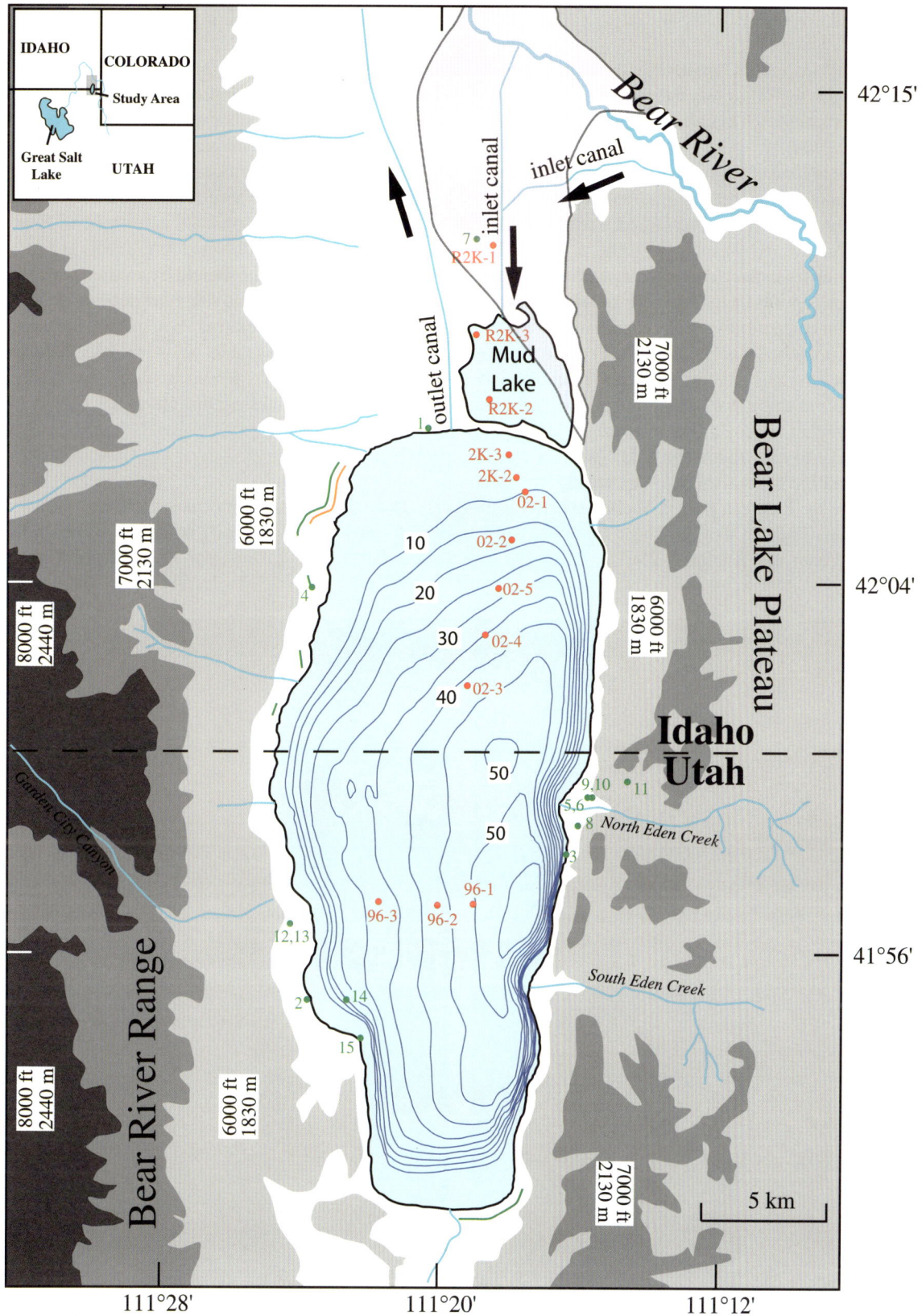

Figure 1. Map of Bear Lake at the modern highstand level (1805.5 m). Isobaths are in meters (note the absence of a 5 m contour). Land elevation contours are rounded off to the nearest 10 m. Red dots are core localities. Green dots are locations of dated shorelines, and numbers refer to samples in Table 1. Green lines show the Willis Ranch shoreline (1814 m) and the orange line shows the Garden City shoreline (1811 m). Pink area to north shows extent of Pleistocene deltaic sands from Robertson (1978). Map is modified from Laabs and Kaufman (2003). Bathymetry is from Denny and Colman (2003).

boulder descriptions are more consistent with a wave-formed shoreline or bar deposit with algal tufa buildups, similar to shoreline deposits in other lakes, such as Lake Lahontan in Nevada and California (Benson, 1994).

Wave-formed shoreline deposits are characterized by a mixture of erosion and aggregation features reflecting sediment availability, accommodation space, duration of a lake level, and changing wave energy (Adams and Wesnousky, 1998, p. 1327). The sediment that accumulates in these deposits may include material eroded from previous deposits including shells and organic material. Furthermore, the contact between the shoreline deposit and underlying deposits may be an erosional surface, and a shift of lake level may cause a former shoreline feature to be completely eroded. The net effect of these characteristics is that the age of a shoreline deposit is often uncertain. For instance, a shell gravel in core BL02-5 (Fig. 1) has white snail shells with an age of 2860 cal yr B.P. and black snail shells of apparently the same species with an age of 3520 cal yr B.P. Shells collected from the Lifton shoreline (Fig. 1) show a range of ages (8700–6400 cal yr B.P.) (Laabs and Kaufman, 2003) that suggest re-

working and possible reoccupation of the same shoreline. Most shoreline ages referred to in this paper were derived from shelly material within the deposit or within adjacent finer-grained material. Similar results from multiple shells provide more confidence in the shoreline ages. Additional problems with dating shorelines are discussed later in the section on chronology.

Abundant evidence of fault movement in Bear Lake Valley (McCalpin, 1993; Colman, 2006; Reheis et al., this volume) adds to the uncertainty of shoreline elevations. McCalpin (1993) argued that 12,000-year-old shoreline deposits on the east side of the lake were probably uplifted ~8 m by fault movement. Colman (2006) noted numerous faults below the lake floor, some cutting the youngest sediment. Movement along these faults may have changed lake-floor and shoreline elevations.

Shoreline Results

The subaerial shoreline record for the past 35,000 years is relatively sparse. Deposits at 1801–1806 m asl reflect historical fluctuations of the lake. Older deposits occur between

Figure 2. Modern shoreline features. (A) Boulder-cobble strandlines on west side of lake. Background shows small sand bars exposed by a recent drop in lake level. (B) Sandy shoreline on east side of lake. Note small, wave-cut terrace in background. (C) Shell-rich sandy shoreline on west side of lake. The steep shoreface is producing lakeward-dipping foresets. Note ripples underwater in front of shore. (D) Shell gravel on northeastern shore of lake. A sheet of shell gravel extends a couple of kilometers inland on the flat shoreline.

1808 and 1814 m asl and yield ages ranging from ca. 6390 to 16,300 cal yr B.P. (sources of observations and ages are in Table 1). A well-defined shoreline on the northwest side of the lake (1811 m asl), called the Garden City shoreline (Williams et al., 1962), is undated. On the west side of the lake, the deposits of the Raspberry Square highstand and the Willis Ranch shoreline occur at 1814 m asl, indicating that the lake rose to this elevation ca. 16,000 cal yr B.P. and again ca. 9200 cal yr B.P. (Table 1). The ages for these and other shorelines may be too old because of reservoir effects or reworking of shells. A number of other shoreline features on the east side of the lake occur at this elevation and higher, but correlating the features is confounded by differential uplift along faults bounding the east side of the valley (McCalpin,

1993; Laabs and Kaufman, 2003). No shoreline features above the modern lake level have been observed for the period from ca. 16,000 to 35,000 cal yr B.P. Fluvial and marsh deposits dated at 13,000–9000 cal yr B.P. in the area north of the lake suggest that the lake did not exceed the 1808 m asl shoreline and probably did not exceed 1806 m asl during that period (Laabs, 2001). Emerged beach gravel exposed on the east side of Bear Lake at Cisco Beach is dated ca. 12,000 cal yr B.P. (Laabs and Kaufman, 2003), although the extent to which the gravel has been uplifted is not known for certain.

Sublacustrine geomorphic expressions of shorelines include (1) an undated erosional bench at 1784–1798 m asl; (2) the upper surface of a delta deposit located at the northern end of the lake

Figure 3. Cross sections of shoreline sediment. (A) Exposure of boulder-cobble strandline deposit at North Eden Canyon. Note characteristic open-framework packing of small boulders with similar sizes and shapes and abrupt transition to well-sorted gravel (matrix is infiltrated around clasts). Scale bar is 20 cm. (B) Wave-rippled sand with thin muddy partings (dark) in core BL02-5. This represents deposition in less than 5 m of water. Scale is in centimeters. (C) Beach sand in trench on west side of lake near Garden City. Shoreface sand forms lakeward-dipping planar sand beds. These overlie steep, shell-rich, lakeward-dipping foresets. Base of trench is muddy sand with wave ripples. Knife is 25 cm long. (D) Shell gravel beach deposit in core BL2K-3. The gravel is overlain and underlain by muddy sand. Scale is in centimeters.

observed on seismic profiles at 1798 m asl and having an estimated age for the overlying reflector of 35,000 yr (Colman, 2006); and (3) the boulder deposit at 1786–1792 m asl (about the same elevation as the erosional bench) known as the "rock pile" (Fig. 1). Aragonite coatings on a cobble from the boulder deposit yield ages ranging from 7470 to 3300 cal yr B.P., but these ages may be too old (Dean, this volume). Also, the boulder deposit could be much older than the aragonite coating. During a period of low lake level, bones of a mastodon were recovered from the area of a sublacustrine spring at an elevation of 1804 m asl. The bones, which yielded an age of 22,000 cal yr B.P., indicate that the spring was subaerial at that time (Laabs and Kaufman, 2003).

Cores taken in ~35 m of water or less (i.e., at or above 1770.5 m asl) contain shoreline deposits or deposits that formed at lake depths less than 5 m (Table 1). The shoreline deposits are typically shell gravels that formed as gravel sheet deposits. In the deeper cores (BL02-5, BL96-3, and BL02-4), graded sand and shell gravel layers may not be true shorelines, but probably formed in less than 5 m of water, shallow enough for waves to move the grain sizes in the layers (Smoot, this volume). Ages of these deposits are better constrained by radiocarbon dating of pollen concentrates from adjacent sediments rather than by ages from shells within the deposits (Colman et al., this volume). Core shoreline data indicate that the lake dropped as low as 1776 m asl sometime between 9800 and 9200 cal yr B.P. and ca. 2800 cal yr B.P. Lake level may also have been as low as 1779 m asl sometime between 5000 and 7000 cal yr B.P., according to shell gravel age constraints in BL96-3, BL02-4, and BL02-5 (Table 1).

TABLE 1. SHORELINE AGES AND ELEVATIONS

Location	Elevation (m asl)	Age (cal yr B.P.)	Dated material	Comments
1	1808	8090 ± 70 <6390 ± 60	mollusk shell charcoal	(1) Lifton shoreline
2	1808	8540 ± 100	mollusk shell	(2) Lifton shoreline
3	1808	8780 ± 270	mollusk shells	(2) Lifton shoreline
4	1814	9220 ± 360	mollusk shells	(2) Willis Ranch shoreline
5	1814	9840 ± 130	charcoal in overlying marsh deposits	(1) North Eden Canyon fault uplift
6	1814	10,400 ± 130	charcoal in overlying marsh deposits	(3) North Eden Canyon fault uplift
7	1810	11,260 ± 50	mollusk shells	(1) Fault uplift?
8	1814	12,510 ±170	mollusk shell	(4) Cisco Beach fault uplift
9	1814	12,540 ±170	*Discus* shell	(1) North Eden Canyon fault uplift
10	1814	12,830 ± 90	*Discus* shell	(1) North Eden Canyon fault uplift
11	1830	15,150 ± 760	mollusk shell	(3) North Eden Canyon fault uplift
12	1814	16,310 ± 240	*Limnaea* shell	(1) Raspberry Square phase deposit
13	1814	16,010 ± 270	*Limnaea* and *Discus* shells	(1) Raspberry Square phase deposit
14	1786–1792	>7475 ± 40 >2000	aragonite crust comparison to δ^{18}O of lake sediment	(5) Rock pile
15	1804	22,000 ± 375	mastodon bones in spring	(4)
BL2K-3	1800	3540 ± 80	shells	(6)
BL02-1	1796	>1910 ± 60 <6180 ± 50 7810 ± 80	pollen pollen gastropod	(6)
BL02-1	1796	>9250 ± 120 <11,540 ± 140	pollen pollen	(6)
BL02-1	1796	<13,240 ± 50 12,080 ± 240	pollen gastropod	(6)
BL02-2	1788	>1880 ± 50 <25,390 ± 190 8360 ± 30	pollen pollen gastropod	(6)
BL02-5	1779	<23,170 ± 310 7000 ± 100	pollen gastropod	(6)
BL96-3	1773	>5050 ± 80	pollen	(6)
BL02-4	1771	>2780 + 50 <5660 ± 50 2860 ± 70 3520 ± 40	pollen pollen gastropod gastropod	(6)
BL02-4	1771	>9200 + 70 <9870 ± 40 10,000 ± 100	pollen pollen gastropod	(6)

Note: Locations are keyed to Figure 1. ">" and "<" indicate ages in overlying and underlying sediment, respectively. Numbers in parentheses indicate source of data: 1—Reheis et al. (2005); 2—Williams et al. (1962); 3—McCalpin (1993); 4—Laabs and Kaufman (2003); 5—Dean (this volume); 6—Colman et al. (this volume). asl—above sea level.

GRAIN SIZE

The classic description of sediment distribution indicates a decrease of grain size from the margins to the center of a lake (see Hakanson and Jansson, 1983). Both stream inflow and wave-induced processes operate to segregate grain sizes in this manner. In a delta setting, the stream jet loses momentum upon intersecting the lake, which causes saltating grains to cease moving and sediment suspended by turbulence to settle (see Smith and Ashley, 1985; Nemec, 1995). The depth of wave influence on sediment size and the level of energy available are dependent upon the height and period of the waves, which are in turn dependent upon the wind stress and fetch (Johnson, 1980, Hakanson and Jansson, 1983; Rowan et al., 1992). Greater water depth requires unusually large waves to move bottom sediment, whereas in shallow water, even small, gentle waves can suspend and transport silt and clay. At a given site, changes in grain size may reflect changes in water depth. Other factors, such as changes in grain shape, variations in the lake's thermal structure, or differences in the initial concentration of suspended sediment may affect grain settling (Sturm, 1979). The correlation of decreasing grain size to depth applies only to mechanically transported sediments. Sizes of grains formed biologically (e.g., shells), or chemically (i.e., endogenic and authigenic minerals), can be independent of lake depth. For this study, grain sizes of siliciclastic material in surface samples were examined as a function of water depth and then compared with similar data from core samples.

Methods

Samples of surface sediment (upper 1.5 cm) were collected along four depth transects (Fig. 4A) and 1-cm-thick core samples were collected at 2 cm intervals from cores from sites BL02-3 and BL02-4 (Fig. 1), and at 4 cm intervals from core BL96-3 (Rosenbaum et al., this volume). The surface samples were collected with a piston corer and extruded on site to prevent mixing during transport. Each sample was treated to remove carbonate, organic material, and biogenic silica (Rosenbaum et al., this volume). The residue, consisting of siliciclastic sediment, was analyzed using a laser particle-size analyzer.

Approach and Application of Grain-Size Comparisons

Each transect of surface samples shows a progressive fining away from the shoreline (Figs. 4 and 5); however, the variation of grain size with depth is different for each transect. This implies that there is no simple relationship between water depth and grain size. Because delta input is virtually absent in the modern lake, these grain-size data were compared with mathematical models based on wave influence.

Johnson (1980) and Rowan et al. (1992) attempted to quantify the relation between lake depth and grain size using Airy wave models that relate wave height to wind speed and fetch. The wave height, wavelength, and wave celerity of an Airy wave

are mutually constrained; by determining one parameter the others can be derived. Rowan et al. (1992) provided equations where the maximum possible wave height of a deep-water Airy wave (wavelength is less than four times water depth) is constrained by the maximum effective fetch (Hakanson and Jansson, 1983, p. 188–191) at any given point. For a given water depth, the bottom shear stress of that wave can be derived. Komar and Miller (1973, 1975) established formulas relating the minimum wave bottom-shear stress to movement of different grain sizes. The transport formulas assume the grains approximate quartz spheres, so they are not applicable to grains less than silt size. Combining these relationships, one can constrain the maximum depths for waves to move different grain sizes for a lake. This technique was applied to Bear Lake using a 1 km grid over the modern highstand surface area. The results were extended to shallow water using equations provided by Johnson (1980) under the assumption that the adjacent deep-water wave periods remain constant.

The results of the analysis—the maximum grain size that can be moved by the largest possible wave at each grid point—were contoured (Fig. 4A) and compared with grain sizes from each of the surface transects. For all transects, the observed median grain size in the area of sand-sized grain transport is well below the maximum, suggesting that maximum-size waves have not occurred recently. The finer grain sizes, however, are consistently coarser than the grain size predicted by the wave model. This difference is least pronounced in the northernmost transect, which is the area with lowest bottom slope. In comparison to the northernmost transect, transects over steeper bottoms show different grain-size relations to depth, with a tendency for coarser grain sizes to occur farther offshore. This observation suggests that a mechanism other than wave transport influences grain size. Gravity flows are the most likely candidate for bottom transport, explaining the association with steeper slopes. Gravity flows can be initiated by a variety of mechanisms including floods, earthquakes, and high wave activity. The evidence for this mechanism in Bear Lake is discussed below.

Sediment traps deployed in the lake during this study (Dean, this volume) collected high-Mg calcite near the surface, but mostly aragonite near the lake floor. This is best explained by the presence of subsurface gravity transport of sediment containing aragonite eroded by waves from nearshore deposits. The transport of shallow-water sediment into deeper areas of a lake is called sediment focusing (Lehman, 1975; Davis and Ford, 1982; Hilton, 1985; Blais and Kalff, 1995). The bottom slopes over many areas of the lake are well within values conducive to gravity flows (Rowan et al., 1992; Hakanson, 1995). The erosion and transport of sediments from the lake edge to the lake center is also indicated because there is little Holocene sediment in water shallower than 30 m compared with much thicker deep-water Holocene deposits (Colman, 2006; Dean et al., 2006; Smoot, this volume). Smear slides (mm-scale sample) from some sediment intervals are dominated by a single carbonate mineral phase with a characteristic crystal habit (such as 3–5 µm aragonite needles versus 7–10 µm aragonite needles) and by diatoms of one or two genera (Smoot, this volume). In contrast, smear slides of other sediment intervals contain mixtures of different carbonate mineral phases and crystal sizes as well as highly variable mixtures of benthic and pelagic diatoms. Smoot (this volume) interpreted these latter smear slides as evidence of sediment mixing due to erosion and transport. In some closed-basin lakes, the effect of sediment focusing is exag-gerated by falling lake levels (Smoot, 2003; Smoot and Benson, 2004). During the period that sediment traps were deployed and surface sampling occurred, the water level in Bear Lake was fall-ing (Dean, this volume; Rosenbaum et al., this volume).

The depth-grain size relation of gravity-flow deposits is not easily defined and is site dependent. Therefore, those transects

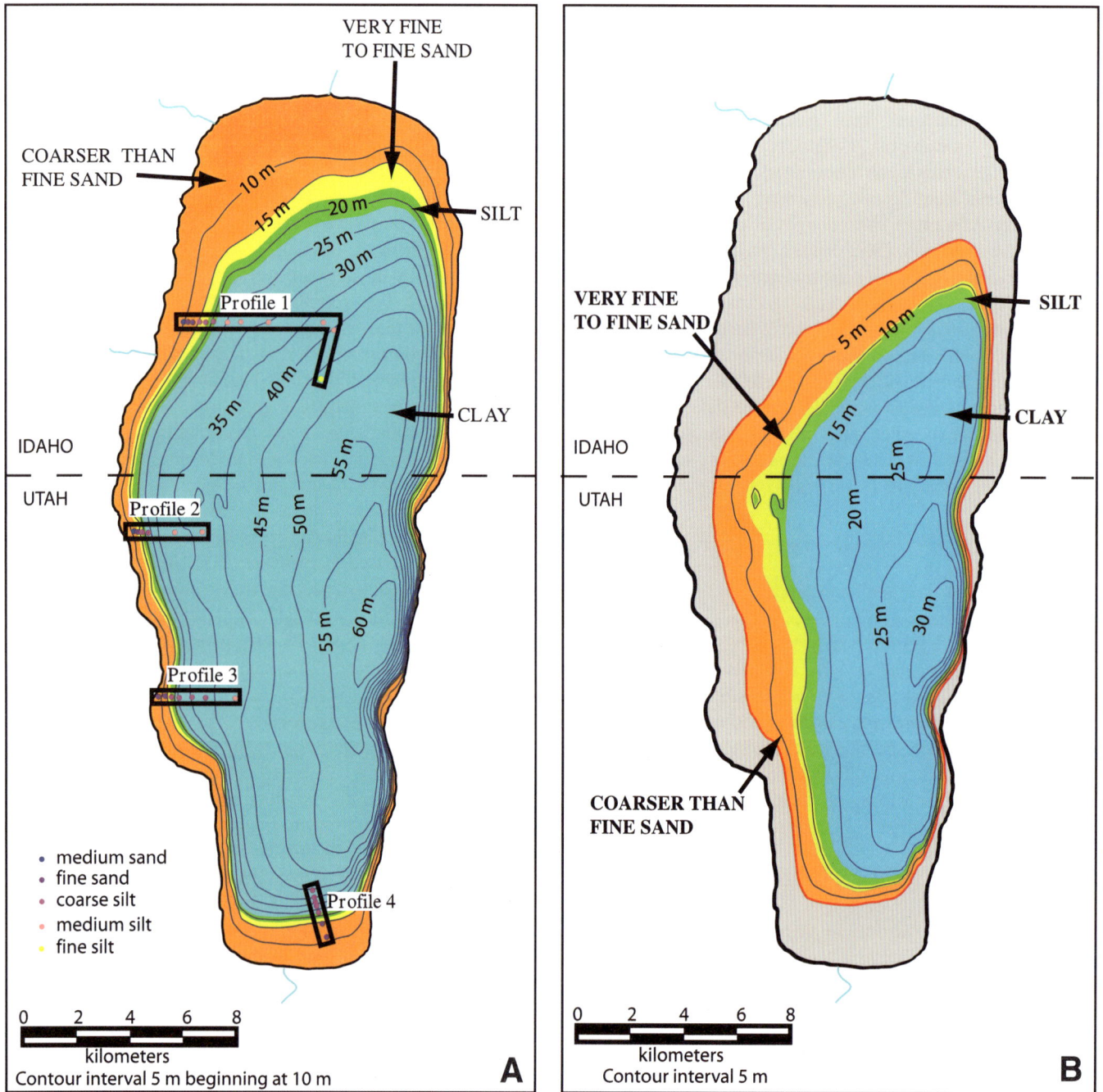

Figure 4. Maps showing maximum depth of wave transport for different grain sizes at Bear Lake. Bathymetry is from Denny and Colman (2003). (A) Model for Bear Lake at the maximum historical elevation. The collection sites of four transects of surface samples are shown with the median grain size at each locality. (B) Model for Bear Lake at a level 25 m below the historical maximum. Gray shows area of modern highstand above this lake model.

with strong indications of sediment focusing (profiles 2, 3, and 4) were considered poor candidates for modeling lake-depth changes. Median grain sizes for the northernmost transect (profile 1) are the closest to the wave model (Fig. 4A), and this transect has a similar slope to that of the core transect from the northern edge of the lake to the center. Therefore, a curve was fit to the median grain size versus depth data for profile 1 (Fig. 5). This curve was used to calculate a modeled depth for each grain-size sample from core sites BL02-3, BL02-4, and BL96-3.

The strength of this approach is that it relates past grain-size distributions to those from modern lake-bottom sediment, which can be directly related to a physical model. As long as the boundary conditions for sediment transport are similar, the technique should provide reasonable approximations of the lake depth through time. There will always be an uncertainty with respect to the influence of gravity flows in disrupting the pattern, but the slope of surface profile 1 is similar to the slopes from which the cores were taken (Fig. 4A), and the surface samples incorporate modern sediment focusing. A more subtle problem is related to changes in the wave model in response to changes in the lake size. When water level was lower than the modern level, Bear Lake was smaller and the maximum waves were smaller. Under such a scenario, the minimum depth for accumulation of sediment finer than silt is much less than for the modern Bear Lake (Fig. 4B). Therefore, modeled water depths based on modern surface samples overestimate the depth of water in a smaller lake. For a larger lake, the opposite is true, but shoreline constraints suggest that this is not a significant problem at Bear Lake because it did not have a substantially larger fetch at higher surface elevations.

The grain-size-distribution model is not applicable to times when the Bear River was transporting sediment into Bear Lake. The influx of sediment from the delta setting changes the surface-sediment distributions, particularly those of the finer sediment fractions. It is also likely that the grain-size-distribution approach will provide spurious results during periods when siliciclastic sediment not available today was reworked and mixed with the car-bonate sediment. The siliciclastic residue from Holocene aragonite ranges from 20% to 25% (Dean et al., 2006; Dean, this volume); siliciclastic fractions significantly greater than this range will probably distort the modeled depths. A very small Bear Lake will be more removed from the immediate mountain drainages than a larger lake. It is possible that the relative siliciclastic input from the mountains will always be finer grained in the smaller lake than in the larger lake, skewing model depths toward deeper values.

Another potential source of error in the grain-size method results from differences between surface samples and core samples. The time interval represented by a water-rich surface sample is shorter than that for a more compacted core sample of similar thickness. This disparity could create a significant problem in sediment intervals where depth changed rapidly (Fig. 6). Burrowing may have caused greater mixing in core samples, because more time was available for mixing, and mixing could include both older and younger deposits. The 1-cm-thick core samples may also include material from two very different layers (e.g., when a sample spans an erosional contact), thereby averaging contrasting grain sizes.

Grain-Size Results

The results from site BL02-3 (modern depth 43 m) indicate that lake levels were below modern lake (bml) for all but the interval from ~175 to 200 cm and another short interval at ~290 cm (Fig. 7). The model indicates that lake levels were on average 20 m bml and reached 25 m bml twice (~220 and ~250 cm). At site BL02-4 (modern depth 35 m), the model indicates that lake levels were below the modern level for all of the upper 225 cm except for intervals at ~5–10 cm and ~160–165 cm (Fig. 7). The model indicates that lake level averaged ~10 m bml and reached ~17 m bml twice (~85 and ~190 cm). The model-derived depths for the interval from ~225 cm to the base of BL02-4 are mostly at or above the modern lake level, but these results are questionable because the siliciclastic content in this interval is significantly

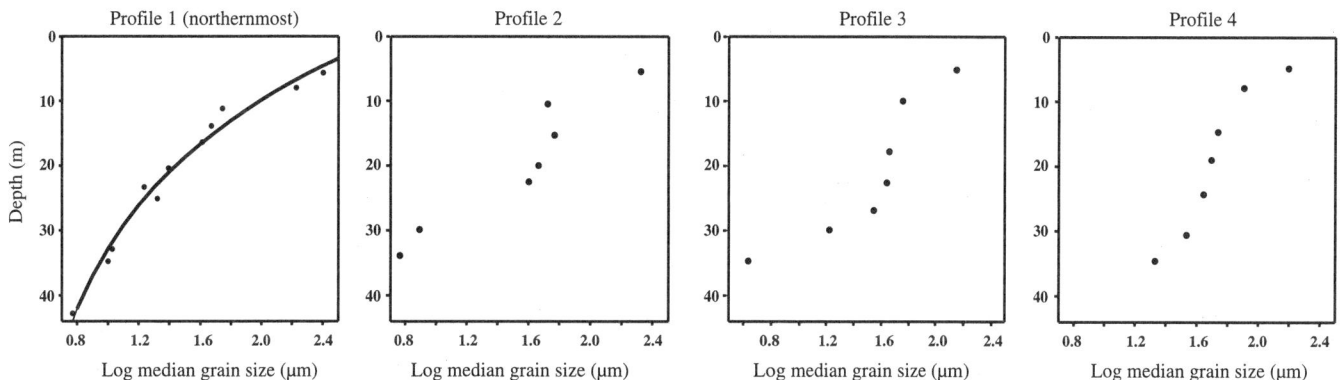

Figure 5. Log of median grain size for surface samples along the four sampling transects shown in Figure 4A. Graphs show median grain size for each sample. The curve in profile 1 was used to compute modeled lake depths shown in Figure 7. The equation for the curve is $Y = 21.4664 - 23.9725 * \ln(X - 0.3772)$ and $R^2 = 0.9800$.

Smoot and Rosenbaum

higher than that of the modern lake sediments (Dean, this volume). The time significance of these intervals will be discussed in more detail following a discussion of chronology.

The siliciclastic-rich portions of BL02-4 (Fig. 7) and BL96-3 are not suitable for the lake-depth reconstruction model based on profile 1, but they do show important variability in grain size. Smoot (this volume) noted that most of the clay-sized siliciclastic sediment is rock flour rather than clay minerals throughout the siliciclastic-rich intervals. Rosenbaum et al. (this volume) and Rosenbaum and Heil (this volume) showed a correspondence between finer grain size and the content of rock material derived from the Uinta Mountains, which they interpret to be glacial flour. They interpret variations in grain size, as well as concomitant variations in magnetic properties and geochemistry, to reflect varying proportions of fine-grained glacial flour from the Uinta Mountains and coarser-grained material from the

local catchment. Such variations could have arisen either from changes in input of glacial flour (due to changes in glacial activity or changes in the Bear River), or to changes in the input of local material (due to processes in and around the lake). Rosenbaum et al. (this volume), and Rosenbaum and Heil (this volume) favor an interpretation that glacial activity in the Uinta Mountains is largely responsible for the observed variations.

SEDIMENTARY TEXTURES

Sedimentary textures are manifestations of bulk-sediment grain size (as opposed to the grain size of the siliciclastic fraction described in the previous section), sorting, bedding features, other sedimentary structures, and biological or chemical components that collectively define the sediment. The sedimentary features used in this study were defined by hand-lens-scale, continuous analysis of

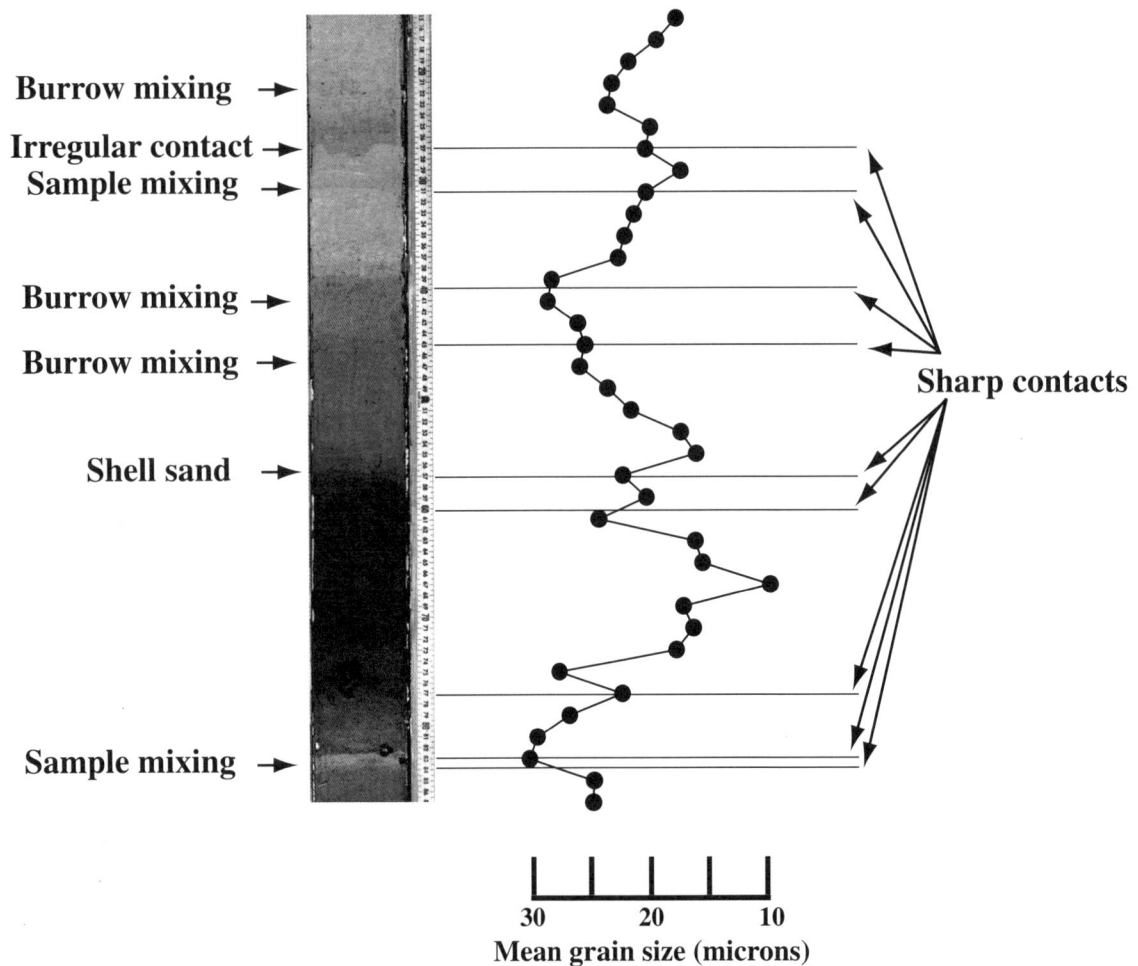

Figure 6. Median-grain-size data of siliciclastic residue plotted against sedimentary features in a segment of core BL02-4. Finer material is overlain by coarser material at numerous sharp contacts. Samples spanning these contacts average contrasting grain sizes, thereby reducing variability. Such averaging is most significant at irregular contacts and across very thin beds. Burrow mixing across sedimentary contacts also changes grain sizes from depositional values. Note that a layer composed largely of ostracode shells is medium-sand-sized, but its siliciclastic residue is medium-silt-sized. Scale is in centimeters.

all cores, X-radiography of selected cores, smear-slide petrography at irregular intervals from most cores, and SEM image analysis of selected intervals (Smoot, this volume). These analyses provide a more continuous depiction of variability than most other techniques that rely on discrete sampling, which averages properties over centimeter-thick samples. Although few sedimentary textures are absolute depth indicators, the composite distribution of textures allows one to interpret relative depth changes. Key features for constraining lake levels in Bear Lake include wave-formed structures, erosional features, grain types, sorting, root structures, and mineralogy. Additional lake-level information can be inferred from bioturbation styles and bulk grain-size distributions.

Like grain size, the formation of different types of wave-formed ripples (Figs. 8A and 8B) is linked to quantifiable wave

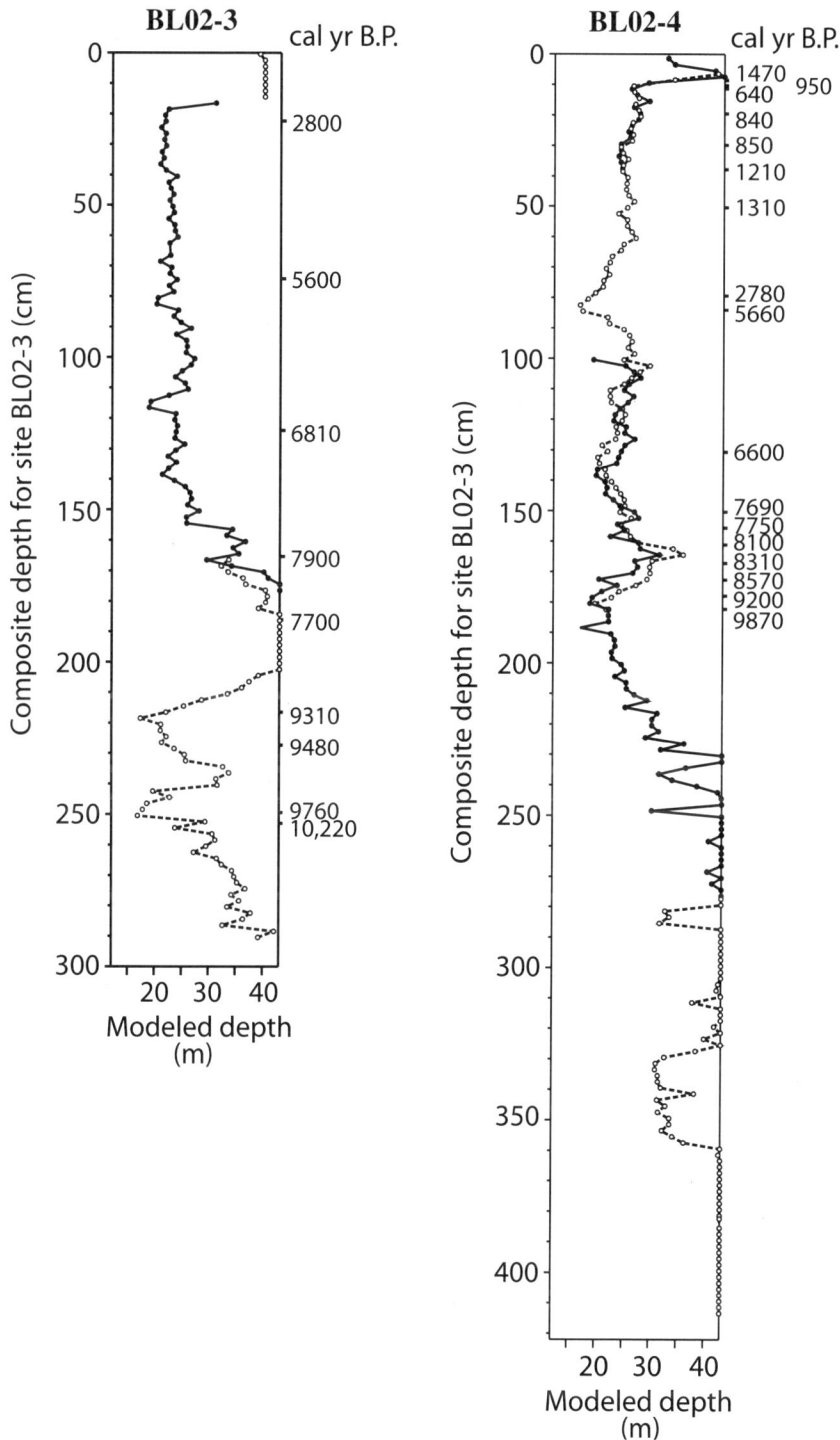

Figure 7. Modeled water depths for core sites BL02-3 and BL02-4 (Fig. 1). Depths were computed using median grain sizes of 1-cm-thick core samples and the polynomial based on the depth distribution of grain sizes of surface samples on profile 1 (Fig. 5). Samples with grain sizes that indicated depths greater than 43 m (the maximum depth for profile 1) were assigned model depths of 43 m. The sampled sections were acquired in overlapping core segments that are indicated by alternating open and closed symbols. Numbers to the right of each graph are calibrated radiocarbon ages (Colman et al., this volume).

RIPPLES

A

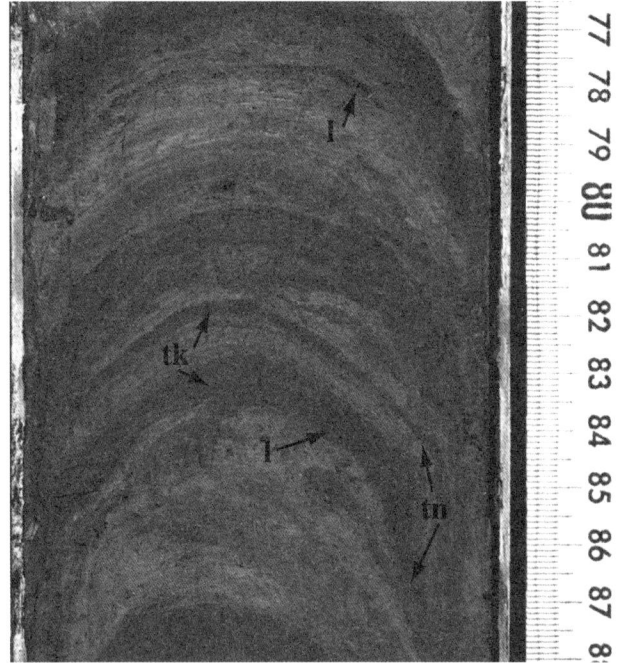

B

BURROWS

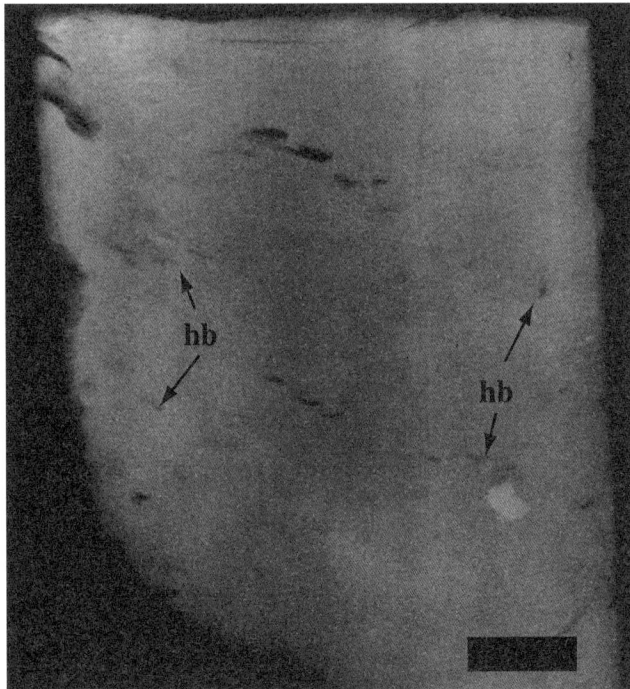

C

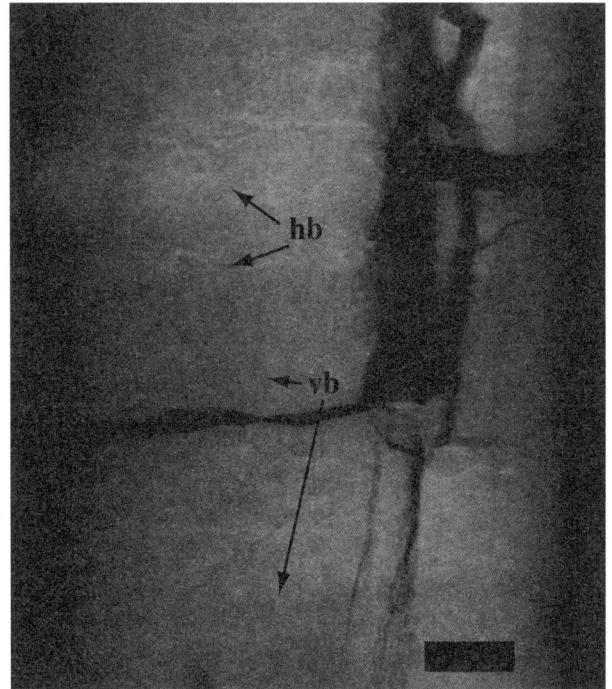

D

Figure 8. Depth-sensitive sedimentary structures. (A) Medium to fine sand with wave-ripple structures at site BL02-1. Ripples are recognized as lenses (l) and layers that thicken (tk) and thin (tn) laterally. Some ripples show internal lamination (il) that indicates vortex ripple forms. Aragonite mud interbeds are lighter colored. Scale is in centimeters. (B) Wave-ripple structures in siliciclastic sediment of core BL2K-2-1. Fine to medium sand forming lenses (l) and laminae that thicken (tk) and thin (tn) laterally represent rolling-grain ripples. Scale is in centimeters. (C) Aragonite mud with small horizontal burrows (hb) viewed in an X-ray image from core BL96-1. The limited size and horizontal orientation suggest low oxygen availability with limited infauna. Scale bar is 1 cm. (D) Siliciclastic mud with abundant randomly oriented burrows including vertical (vb) and horizontal (hb) orientations viewed in an X-ray image from core BL96-3. The variable sizes and orientations suggest a well-oxygenated environment with varied infauna. Scale bar is 1 cm.

properties (Clifton, 1976; Harms et al., 1982). For instance, rolling-grain ripples form at the initiation of sediment movement under wave bottom-shear stress, whereas vortex ripples require higher shear stress under the same depth and grain-size conditions. Rolling-grain ripples have broad crests and large wavelength-to-height ratios in contrast to vortex ripples, which have sharper crests and smaller wavelength-to-height ratios. In cross section, rolling-grain ripples appear as thin, elongate lenses with no internal stratification, whereas vortex ripples comprise thicker, shorter lenses with internal cross-lamination (i.e., Harms et al., 1982). Smoot and Benson (1998) extended this concept to include wave-formed lag deposits, and noted that the relative abundance of locally derived shell material incorporated in wave-formed deposits increases with distance from external source areas such as mountain fronts or river mouths. The Bear Lake sediment cores provide only a few opportunities to use wave-formed ripple types as depth constraints. Wave-formed deposits in cores taken many kilometers from the modern shoreline clearly show the impact of distance from external source areas by the predominance of shell material over clastic grains. These deposits appear to be less apparent in grain-size analyses of siliciclastic residue.

As discussed in the section on grain size, sediment focusing occurs frequently regardless of lake size. Basin geometry controls the amount of sediment transported lakeward with steeper basin floors promoting more transport (Hakanson, 1977, 1982; Blais and Kalff, 1995). Falling lake levels change the dynamics of this relationship by exposing more fine-grained sediment initially deposited in deep water to wave conditions, shifting the location of proximal wave-formed turbidites basinward, and decreasing the surface area for distribution of wave-reworked sediment (Smoot and Benson, 2004). In contrast, rising lake levels engender the reverse of these conditions. The net effect is that the sedimentary record of sediment focusing is more pronounced in falling lakes than in rising lakes. In Bear Lake sediment cores, zones with pronounced evidence of sediment focusing are interpreted as indicative of falling lake levels. The intervals of increased sediment focusing are recognized by smear-slide analysis, and by distinct intervals with more scattered coarse grains or graded intervals with sharp, erosional bases (Smoot, this volume). The latter are interpreted as wave-formed turbidite deposits that increase in size and frequency with falling lake levels.

A variety of burrowing styles occur in Bear Lake sediment (Smoot, this volume). The nature of burrows depends upon the type of organism, sediment type, the rate of sediment accumulation, water chemistry, and availability of food or oxygen (e.g., Hasiotis, 2002). Shifts in burrow style within a vertical sediment record indicate changes in one or more of these parameters. In Bear Lake, shifts from small, bedding-plane-parallel burrows to larger, more randomly oriented burrows (Figs. 8C and 8D) may indicate changes in water depth. Smoot and Benson (1998) argued that the degree of lake stratification controls available oxygen on the lake floor, which changes the assemblage and strategies of burrowing organisms. When the water column is well stratified, low-oxygen bottom conditions limit burrowing to a small fauna

that mine only the upper sediment surface. In contrast, when the water column is well mixed, well-oxygenated bottom conditions allow a richer fauna with more vertical mixing of sediment. Chemical stratification is more likely to occur during rising lake levels when freshwater inflow overlies more concentrated water from the lowstand. At Bear Lake, small bedding-plane-parallel burrows are associated with core intervals with little evidence of sediment focusing, whereas large random burrows are associated with intervals that are dominated by features indicating sediment focusing. This strengthens the interpretation that the amount and type of burrowing reflect changes in lake level.

The presence of vertical root structures requires either subaerial exposure or very shallow water. Subaqueous plants produce tiny "holdfast" roots, which are restricted to the upper 1–2 cm of sediment (Hutchinson, 1975). Vertical roots that are 1 mm in diameter or larger probably represent subaerial vegetation. Root casts in the Bear Lake cores are distinguished from burrows by tapering diameters, bifurcation into smaller diameters, and carbonaceous residues (Fig. 9). The distinction between root structures and burrows is not always straightforward, particularly with large sediment-filled root casts. In the Bear Lake cores, root structures are commonly associated with other soil-like features such as cutans, carbonate concretions, and evaporites (Smoot, this volume). Root structures penetrate the sediment, so they may occur within subaqueous deposits, reflecting later subaerial exposure.

The use of sedimentary textures as lake-level indicators has both strengths and weaknesses. Sedimentary textures are measured continuously, are independent of vertical scale, and use multiple criteria to identify even small changes in depositional environment. The variability of sediment characteristics, however, also makes the categorization of sediment packets more subjective than the straightforward physical measurements of grain size or color. Thus, a subtle indicator of a lake-level change may be lost in the sediment categorization. In addition, the distributions of modern sedimentary textures are poorly constrained because the uppermost water-rich deposits were not preserved as undisturbed core. Furthermore, some of the sedimentary textures do not have modern analogs. Therefore, sedimentary textures provide a high probability for recognizing relative lake-level changes, but low precision for quantifying absolute lake-level changes.

Sedimentary Texture Results

Bear Lake sediments can be broadly divided into several sediment categories: aragonite mud, calcite mud, shell-rich sand and gravel, mixed calcitic-siliciclastic mud, and siliciclastic mud (Fig. 10; Smoot, this volume). These sediments define a basic stratigraphy. Upper and lower aragonite zones are separated by a calcite zone that is easily correlated among the cores. A second calcite zone underlies the lower aragonite and overlies a mixed calcitic-siliciclastic mud interval. A third calcite zone underlies this mixed interval and overlies a zone of siliciclastic sediment that extends to the base of the studied section.

The upper and lower aragonite zones are subdivided into three types, which are distinguished by sedimentary textures, predominately grain-size distribution and diatom characteristics (Smoot, this volume). Aragonite I, which has an intermediate grain size, is the dominant type found in the upper part of deep-water cores (i.e., BL 96-1, BL96-2, BL02-3, and BL02-4), whereas the coarser-grained Aragonite III is more common in the upper parts of shallow-water cores (BL02-1, BL02-2, BL2K-2, BL02-5, and BL96-3). Aragonite II is finer grained than the other two types, contains small pelagic diatoms (Moser and Kimball, this volume), and is restricted to the two deepest-water cores (BL96-1 and BL96-2). Aragonite II was not identified in BL02-3, which was taken at the same depth as BL96-2 (43 m), suggesting either that they were taken at the shallowest depth boundary for Aragonite II or that some factor other than depth controls the distribution of the fabric. The distribution of the different types of aragonite mud and their relative grain sizes are interpreted as indicating relative lake depths. Aragonite III is the shallowest-water type and Aragonite II as the deepest-water type. In con-

trast to the quantitative grain-size data discussed previously, the grain sizes discussed here include shelly material, crystals and grains of carbonate, as well as siliciclastic material. Changes in grain size of siliciclastic material roughly coincide with boundaries between Aragonite I and Aragonite III, but do not distinguish between Aragonite I and Aragonite II.

Calcite mud, which has virtually no aragonite or clastic sediment in smear slides, is interpreted as occurring in a rising lake or deep lake. The calcite intervals, which, in part, also include mixed carbonate and siliciclastic components, are believed to represent the highest Holocene lake levels because they are continuous across a range of water depths with little change in character, they are very fine grained, and they have a predominance of small pelagic diatoms (Moser and Kimball, this volume).

Sediment composed of shell-rich sand and gravel indicates shallow water or shoreline deposits. These deposits commonly overlie erosional surfaces.

Sediment composed of mixed calcite and siliciclastic material occurs in two settings. In the Holocene record, such deposits

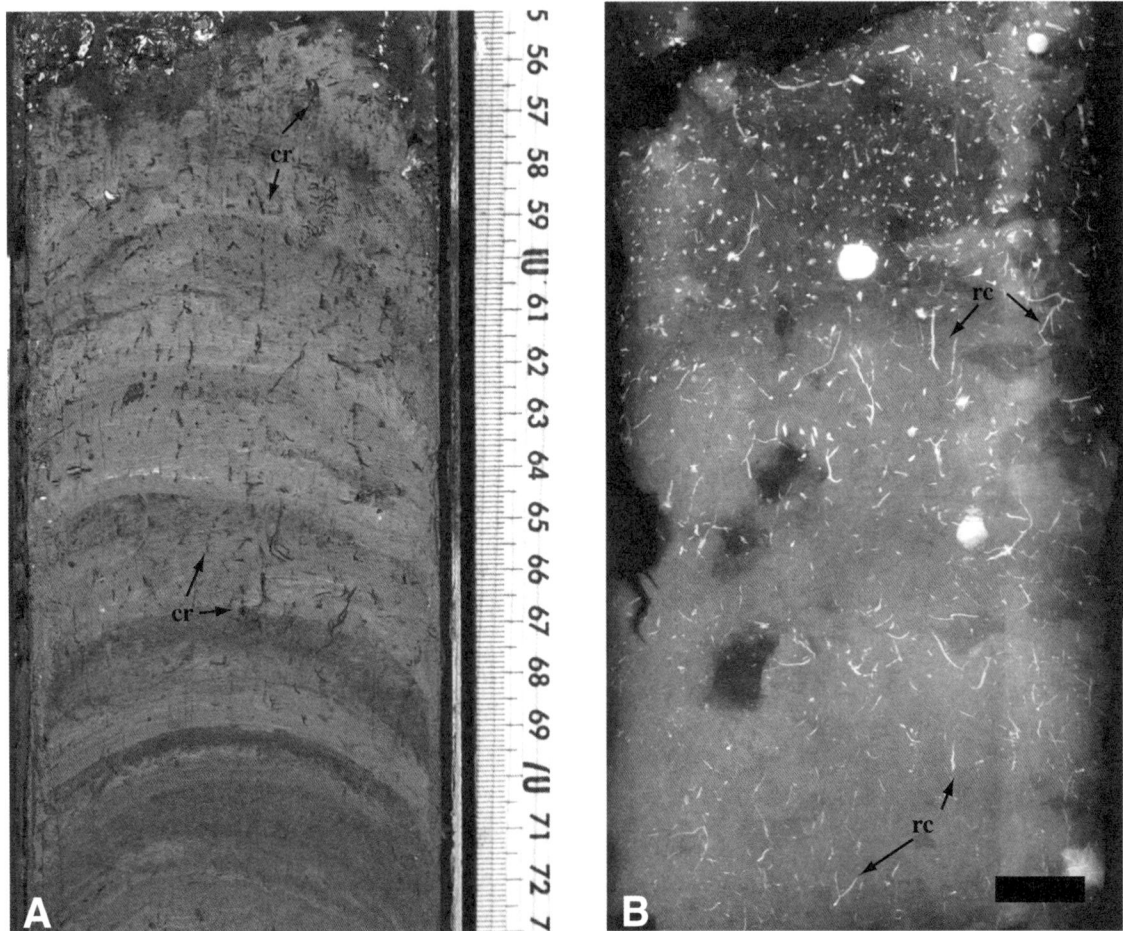

Figure 9. Root-like structures in cores. (A) Carbonized root-like structures (cr) with pyrite coatings in core BL2K-1. Note how the features decrease in size and abundance downward. Scale is in centimeters. (B) Root-like casts lined with framboidal pyrite (rc) viewed in an X-ray image of a section of BL96-3. Note how features decrease in size and abundance downward. Large white circular features are carbonate concretions interpreted as soil features. Scale bar is 1 cm.

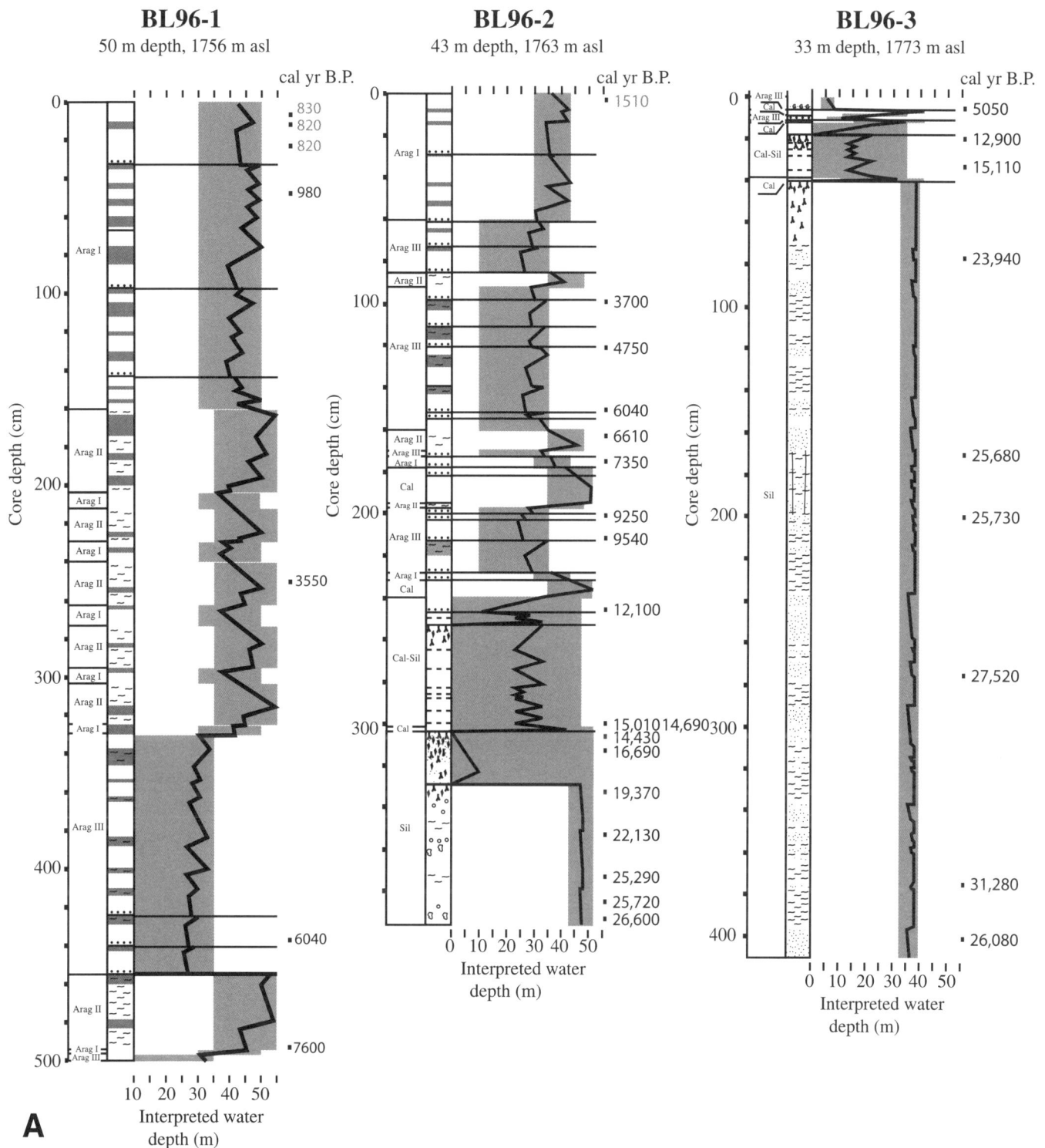

Figure 10 (*continued on next page*). Depth models based on sedimentary textures in cores BL96-1, BL96-2, and BL96-3 (A) and at core sites BL02-3 and BL02-4 (B). Each core shows the sediment type as defined by Smoot (this volume) and the significant sedimentary structures. Interpreted depths are shown by the black lines, with gray areas signifying the range of possible depths based on this technique. Black horizontal lines indicate sharp bedding surfaces (with coarser-grained sediment overlying finer-grained material) that may indicate erosion. Numbers to right of column are calibrated radiocarbon dates (Colman et al., this volume). Scale to left of each core is in centimeters. Legend appears on following page. In Figure 10B, photographs of core segments are left of the schematic drawings, and curves composed of dashed lines represent modeled depths from grain-size data (Fig. 7). asl—above sea level.

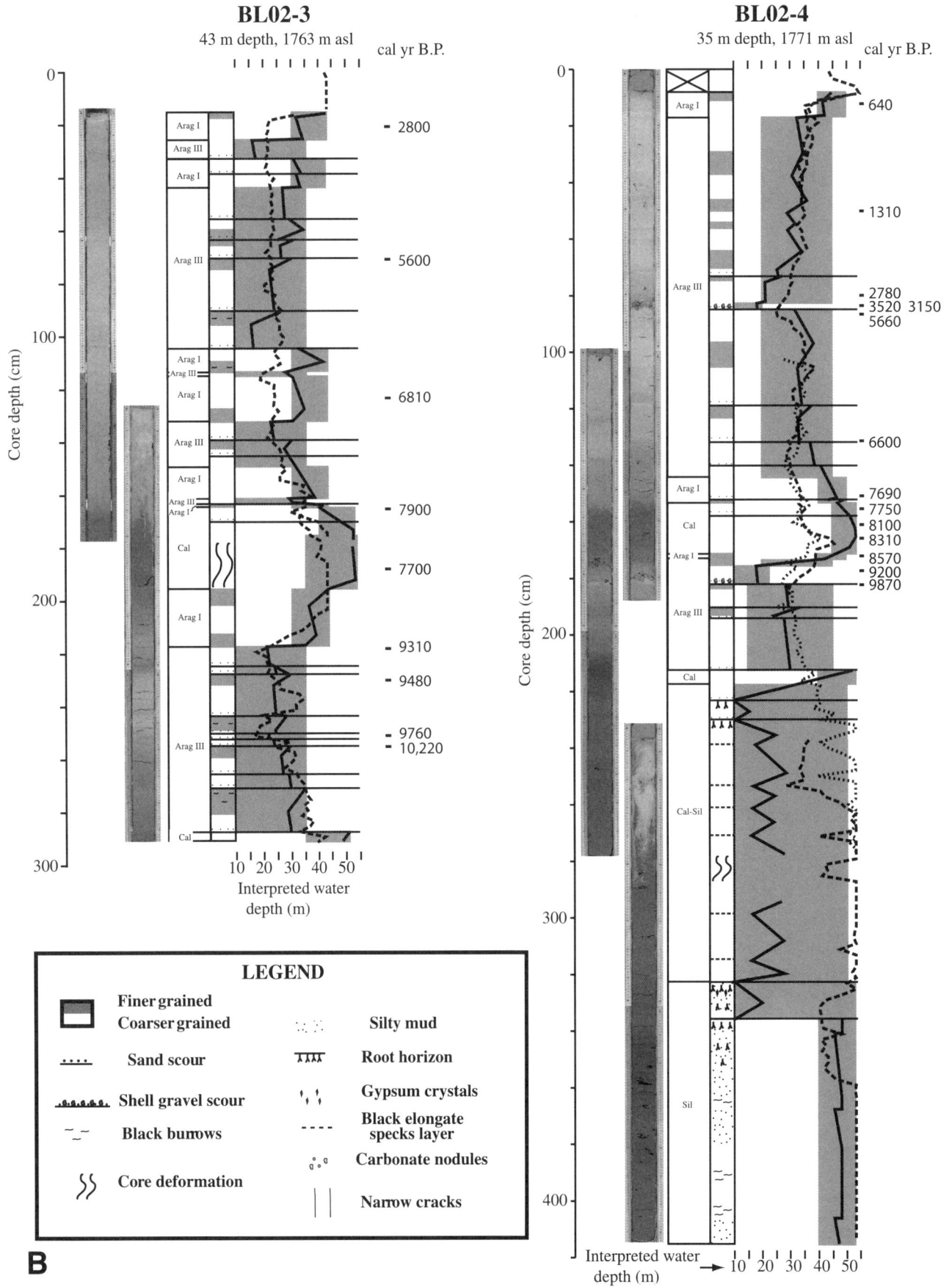

Figure 10 (*continued*).

reflect marsh sedimentation north of the lake (Smoot, this volume). Deposits with similar root features, organic-rich bands, and tiny gastropods occur in core BL02-1 (100–170 cm depth). In the Pleistocene section, mixed calcite and siliciclastic mud deposits comprise an interval between the aragonite deposits and the underlying siliciclastic mud. The water-depth interpretations for these deposits are uncertain, because they have no modern analog. Their association with horizons of root casts suggests they were formed in shallow water, but their fine grain size is more consistent with deposition in deep water.

Siliciclastic mud intervals were deposited when the Bear River was a major source of sediment and Bear Lake was spilling (Dean et al., 2006; Kaufman et al., this volume). Siliciclastic mud in the shallow part of the lake (cores BL2K-2, BL02-1, and BL02-2) is sandier than correlative sediment from the deeper parts of the lake, suggesting a depth not much greater than the modern lake. Wave-formed ripples in BL2K-2 constrain the depth at least intermittently to less than 5 m above the modern lake. This interpretation is consistent with the dearth of shoreline features in the same age range (discussed earlier) as the siliciclastic mud intervals in the cores.

In addition to this broad division of sediment types, there are other indications of lake-level fluctuation. These indications largely reflect differences in sediment focusing and occur in each type of aragonite mud. The variability is most pronounced in Aragonite I, whereas in Aragonite III, greater sediment focusing is characterized by more abundant rock fragments and broken diatoms. In Aragonite II intervals, greater sediment focusing is indicated by small-scale transitions into Aragonite I. Sediment focusing is also indicated in the upper portions of calcite mud intervals. Intervals that display more sediment focusing are interpreted as deposits from shallower water than intervals of the same mud type that lack evidence of sediment focusing. This interpretation is supported by shifts in bioturbation patterns. Small bedding-plane-parallel burrows are typical of the finer-grained intervals interpreted as rising lake levels, and random burrows are associated with increased sediment focusing. The magnitude of lake-level change represented by these variations is not well established, but, by comparison to modern lake fluctuations (Dean et al., this volume), is thought to be on the scale of ~5 m in most cases.

Horizons from which root-cast sequences extend downward are found in BL96-2, BL96-3, BL02-3, BL02-4, BL02-5, BL02-2, BL02-1, and BL2K-2 (Smoot, this volume). These horizons are interpreted to have formed when lake level regressed below the core site. There are at least three such horizons in core BL96-2, each indicating that lake level was below 1763 m asl. Rooted horizons occur in similar stratigraphic positions in the other cores, and a fourth rooted horizon in BL02-4 occurs in a position equivalent to a sandy layer in BL96-2 (Smoot, this volume). Two of the root-cast sequences are within the siliciclastic mud interval, including its upper boundary, and two of them are within the mixed calcitic-siliciclastic mud interval. As mentioned earlier, the bulk of the siliciclastic mud interval was probably deposited at depths at or slightly greater than that of the modern lake. The two

rooted horizons are interpreted as indicating a rapid drop in lake level, with a slight transgression within the lowstand.

Water-depth indicators associated with the mixed calcitic-siliciclastic mud are equivocal. Horizons that contain small horizontally elongated sulfide coatings that appear to be tubes may represent development of subaqueous vegetation, but the features could also be burrows or some other tube-shaped feature. Subaqueous vegetation would be consistent with shallow water (around 10–20 m maximum depth). The mixed calcitic-siliciclastic mud interval thins from 60 cm in BL96-2 to ~25 cm in BL96-3 (1773 m asl), and is missing in BL02-5 (1783 m asl) and shallower cores. BL02-1 (1796 m asl) contains a marsh deposit (100–170 cm depth) that is at least partly correlative to the mixed calcitic-siliciclastic mud interval, suggesting two possible depth scenarios for the lake (Fig. 11). Nearly pure calcite mud intervals occur at the base and top of the mixed calcitic-siliciclastic mud interval. These suggest deep-water conditions, so the associated root-cast horizons may have formed during lowstands and may have no bearing on depth at the time of deposition.

LAKE-DEPTH MODELS

Sedimentary textures and shorelines were used to construct models of lake depth changes (Fig. 10). For these models, sediment types constrain the probable range of depths. The maximum depth of Aragonite I is 50 m, the depth of core BL96-1, and the minimum depth is 35 m, the depth of BL02-4. Aragonite II represents lake depth greater than 50 m. The maximum depth of Aragonite III is taken as 33 m, the depth of BL96-3, and the minimum depth is 18 m, the depth of BL02-2. The assemblage of sedimentary features in calcitic mud and their continuity even in cores from modern shallow-water environments suggest they represent the greatest depths. Sediments of the Willis Ranch and Raspberry Square highstands are the highest features (1814 m asl) within the age range of these deposits. Lake elevation during deposition of much of the siliciclastic mud is assumed to be near the elevation of the Lifton shoreline (1808 m asl) given the constraints for forming wave vortex ripples in fine sand in cores BL2K-2 (1797 m asl), BL02-1 (1796 m asl), and BL02-2 (1787 m asl), and the absence of shorelines of similar age above or below the Lifton shoreline. Two additional assumptions provide further constraints on the model. First, within a given sediment type, coarser grains and erosional surfaces indicate shallower water. For instance, an increase in evidence of sediment focusing is modeled by a 5 m decrease in water depth. Second, within the calcitic-siliciclastic mud, small horizontal tubes are taken to indicate shallower water than sediment lacking such tubes. These assumptions are reasonable, but the upper-limit assumptions for all but the calcite mud are largely unsubstantiated.

Comparison of results for the upper aragonite mud interval among the different cores shows some variability in the interpreted lake-depth histories (Fig. 10). Modeled depths for core BL96-1 are consistently greater in the upper 330 cm and lowest 45 cm than in the intervening 115 cm, where modeled depths fluctuate

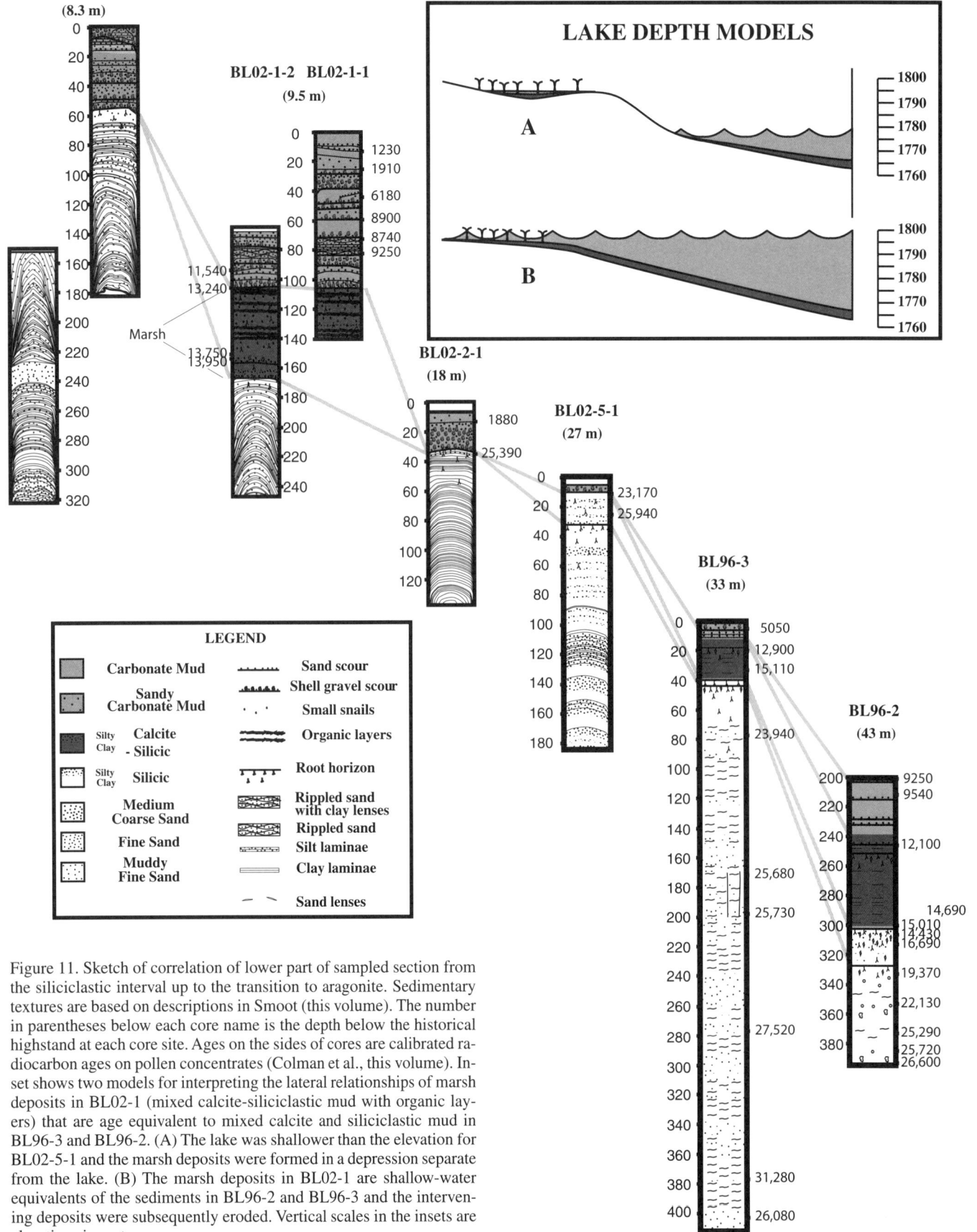

Figure 11. Sketch of correlation of lower part of sampled section from the siliciclastic interval up to the transition to aragonite. Sedimentary textures are based on descriptions in Smoot (this volume). The number in parentheses below each core name is the depth below the historical highstand at each core site. Ages on the sides of cores are calibrated radiocarbon ages on pollen concentrates (Colman et al., this volume). Inset shows two models for interpreting the lateral relationships of marsh deposits in BL02-1 (mixed calcite-siliciclastic mud with organic layers) that are age equivalent to mixed calcite and siliciclastic mud in BL96-3 and BL96-2. (A) The lake was shallower than the elevation for BL02-5-1 and the marsh deposits were formed in a depression separate from the lake. (B) The marsh deposits in BL02-1 are shallow-water equivalents of the sediments in BL96-2 and BL96-3 and the intervening deposits were subsequently eroded. Vertical scales in the insets are elevations in meters.

around ~25 m bml. The uppermost 160 cm varies between the modern level and ~10 m bml, whereas the interval from 160 to 330 cm shows greater fluctuations, varying from ~15 m bml to ~3 m above modern level. Modeled depths for the upper 60 cm in BL96-2 fluctuate within 10 m of the modern level. Depths from the underlying 100 cm are mostly 10–20 m bml but rise to near modern lake levels in a 10 cm interval at a core depth of 90 cm. Underlying this lowstand sequence is a 15 cm highstand sequence that contains a sharp peak that rises above the modern level. In the upper 40 cm of BL02-3, modeled depths indicate two periods when lake levels were ~10 m bml, separated by a period when the lake was 15 m lower. Modeled depths from the underlying 65 cm are mostly ~10–15 m bml beginning following a lowstand ~25 m bml at a core depth of 95–105 cm. Underlying this lowstand sequence is a highly fluctuating sequence from 105 to 160 cm that includes two peaks to near modern lake level and a lowstand interval at ~20 m bml. Near the base of the upper arago-nite zone, modeled lake-level falls from above the modern level in the calcite zone to ~15 m bml. In the upper 15 cm of BL02-4, mod-eled depths are within 5 m of the modern lake. Modeled depths in the underlying 120-cm-thick interval are mostly between 10 and 15 m bml, with an interval ~25 m bml at ~85 cm. Within the interval 135–170 cm (overlying the calcite mud), modeled depths range from greater than modern depths to 10 m bml. The abbrevi-ated section in BL96-3 indicates depths ~25 m bml.

Interpreted histories for the lower aragonite zone also show core-to-core variations. For BL96-2, modeled depths are mostly 15 m bml and indicate one peak within 5 m of the modern level. For BL02-3, modeled depths show a peak within 5 m of modern before falling ~15 m then rising gradually to the contact with the overlying calcite zone. In BL02-4, the lower aragonite interval is only 35 cm thick. The modeled depth history indicates mostly lake levels ~15 m bml with one minimum 10 m lower followed by an abrupt rise to the overlying calcite zone. In BL96-3, the 2-cm-thick aragonite mud interval indicates lake levels ~20 m bml.

The mixed calcitic-siliciclastic mud interval varies in thick-ness among the three cores (Fig. 10). Seven horizons with small horizontal tubes occur below the root-cast horizon in BL96-2. Six such horizons occur below the lower root-cast horizon in BL02-4, and four occur below the root-cast horizon in BL96-3. It is difficult to compare this interval in cores BL96-2 and BL02-4 because sedi-mentary structures in this part of BL02-4 are not as well defined as in the other cores. X-radiography was not used to study BL02-4 (Smoot, this volume) and the top of the deepest core segment from this site was highly disturbed (Fig. 10B). The three closely spaced horizontal tube horizons in BL96-3 and BL96-2 may be correla-tive. If so, a large part of the BL96-3 mixed calcitic-siliciclastic mud interval below the root-cast horizon was eroded.

The siliciclastic mud interval is characterized by a pattern of upward-coarsening sequences at each site where it was sampled (BL96-2, BL96-3, and BL02-4). Two complete sequences were penetrated in both BL96-2 and BL02-4. In BL96-3, the sequences are much more layered and coarser grained. The upward-coarsening successions vary from ~40-cm to ~80-cm thick, and

there is a suggestion of a larger-scale upward coarsening sequence ~130-cm thick. The sequences are shown as shallowing succes-sions, but they may have more to do with shifts in abundance of source material (Rosenbaum and Heil, this volume) or changes in the inflow (Smoot, this volume). The siliciclastic mud inter-val has two root-cast sequences in the uppermost few decimeters including the contact with the mixed calcitic-siliciclastic mud. These indicate that lake levels fell at least 43 m below the mod-ern lake level. In BL96-2 and BL96-3, the immediately overly-ing material is a calcite mud that is interpreted as a transgressive deposit. The absence of this calcite mud in BL02-4 suggests that it may not indicate conditions as deep as the younger calcite lay-ers, although it is possible that the layer was missed in BL02-4.

Comparison of Depth Models from Grain-Size Data and from Textural Features

Modeled lake levels determined from grain-size data and independently from sedimentary textures can be directly compared for cores BL02-3 and BL02-4 (Fig. 10B). The grain-size model for the aragonite and calcite mud intervals generally indicates shallower-lake conditions than the sedimentary-texture model, but well within the ranges of possible depths for each sedimentary texture. The two methods generally produce the same sense of change, but the magnitude of change is frequently different. The onset of changes is also slightly different in many places.

Several factors contribute to the differences produced by the two techniques. The textural features include chemical and biogenic sediments that are excluded from the grain-size data. Aragonite III is distinguished from Aragonite I by the abundance of broken and abraded diatoms and ostracodes in the former. The small portions of siliciclastic material in the aragonites may not have enough variability to distinguish between these fabrics. Sandy beds overlying sharp and sometimes irregular contacts are not always recognized in grain-size data because samples either missed these beds or averaged coarse-grained material with adjacent fine-grained sediment. Burrow mixing of sediment between layers may also contribute to smoothing of the grain-size data. For instance, thin, very-fine-grained carbonate layers are not apparent in the grain-size data because of burrow mixing across boundaries. The assessment of near modern lake condi-tions for Aragonite I, however, is not verifiable due to a lack of surface sediment for comparison, and the grain-size data indicate that this may not be a good assumption. If lower lake levels were assumed for this fabric, the texture-based model would indicate levels more akin to the grain-size model.

The grain-size model for calcitic-siliciclastic mud in BL02-4 suggests depths at the model maximum for most of the thickness and minor intervals with drops to modern depths or 5 m below modern depths. The textural evidence for shallow lake conditions in the mixed calcitic-siliciclastic mud intervals is equivocal. If the lake was shallow, the absence of wave features is difficult to explain. If the tiny horizontal tube features are correctly inter-preted as root hairs, perhaps they formed under tens of meters of clear water. Marsh deposits in BL02-1 (Fig. 11) provide the strongest constraint on lake depth during deposition of the mixed calcitic-siliciclastic mud interval, thus establishing that as the maximum lake elevation (1796 m asl) for that interval. The conflicting values with the grain-size model are attributed to the effect of siliciclastic sediment input, which is not present in the modern depth model. The minor coarser-grained intervals within the calcitic-siliciclastic mud appear to be coincident with the presence of sulfide-coated egg casings and carapaces (Smoot, this volume) that may have skewed the results.

As noted earlier, the grain-size model is not applicable to the siliciclastic mud interval. The best lake-level constraints for the siliciclastic mud interval are the absence of beach deposits of similar age and the presence of marsh and fluvial deposits of similar age in the area north of the lake at elevations lower than the Lifton shoreline deposits. Increases of the silt content of BL96-2 and BL96-3 are interpreted as indicators of shallower lake depths, which are correlative to shifts from predominately silt and clay to sand-rich intervals in cores from sites BL2K-2, BL02-1, and BL02-2.

CHRONOLOGY

In order to combine the records from the various cores into a single lake-depth history, chronologies for the different data sets must be consistent. Most of the radiocarbon ages for the cores are from pollen concentrates (Colman et al., this volume). A few ages were obtained from ostracode shells, total organic carbon, and mollusk shells. The pollen concentrates contain small amounts of other carbon-bearing material that may be detrital in character. Colman et al. (this volume) interpret a reservoir effect (Broecker and Walton, 1959; Olsson, 1986; Bjorck and Wohlfarth, 2001) of 370 years for the ostracode ages by comparing ages of ostracodes and pollen from the same horizons. A similar comparison between pollen and total organic carbon ages suggests that ages from total organic carbon are too old by ~480 years. There has been no attempt to establish a reservoir effect on the shells of snails or clams, although ages of shell layers and pollen-derived ages from adjacent layers suggest that a reservoir effect of ~400 years is reasonable. There are no data available to determine if the reservoir effect varied with lake depth or species.

Smoothed Age Models

Colman et al. (this volume) developed age models for cores BL96-1, BL96-2, BL96-3, BL02-3, and BL02-4 using both polynomial and cubic spline techniques. With the exception of core BL02-4, each core was assumed to be continuous with no significant breaks or abrupt variations in sedimentation rate. The age model for core BL02-4 consisted of three curves separated by age breaks at shell gravel layers thought to indicate unconformities. Within the siliciclastic mud interval that Rosenbaum and Heil (this volume) interpreted as dominated by glacial flour, Colman et al. (this volume) rejected five radiocarbon ages (from

pollen concentrates) from each of cores BL96-2 and BL96-3. They noted that pollen within this interval is sparse and poorly preserved in contrast to the underlying and overlying strata. Their age models for BL96-2, BL96-3, and BL02-4 through the siliciclastic interval are based on ages above the interval dominated by glacial flour, a calibrated age of 26,080 yr B.P. from the bottom of BL96-3 (just below the interval dominated by glacial four), two radiocarbon ages from BL00-1 (below the interval dominated by glacial flour) that are consistent with the age from the base of BL96-3, and correlation of horizons among the cores. The two curve-fitting techniques discussed by Colman et al. (this volume) yield nearly identical results within the limits of error, and correlations among cores based on geochemistry and magnetic properties are consistent with these age models.

The initial step in construction of a combined lake-level curve is to convert lake depths to surface elevations. This was accomplished by assuming that the elevation of each core site was constant through the time of sediment accumulation. There was no attempt to correct for variations in accumulation rate versus subsidence. Comparison of interpreted lake-elevation curves for individual cores (BL96-1, BL96-2, BL96-3, BL02-3, and BL02-4) shows some scatter at any time. General trends of rising and falling lake level are more or less synchronous. However, taking the age models at face value, there are significant intervals with conflicting senses of lake-level change (examples shown in Figs. 12A and 13A). Such conflicts occur ca. 1000, 2500, 3400, 3800, 4300, 5200, 5700, 6900, 9000, 9500, 10,000, 10,500, 10,900, 11,300, and 11,700 cal yr B.P. The models also indicate that a root-forming event at 1763 m asl coincides with a lake at 1806 m asl (ca. 18,300 cal yr B.P.), whereas a root-forming event at 1773 m asl coincided with lake level at 1810 m asl (ca. 23,700 cal yr B.P.). Lake-level variations must be synchronous at each location. Therefore, if the sedimentary evidence is interpreted correctly, the smoothed age models have incorrectly presented the rates of deposition in at least some of the cores.

Alternative Chronology

Sedimentological evidence of erosional surfaces overlain by coarse sediment indicative of breaks in sedimentation (like those in the age model for core BL02-4) occurs in all of the cores, but commonly with more subtle textural expressions. The variety of textural styles recognized in the cores also suggests that sedimentation rates may be highly variable. Therefore, an alternative approach to chronology is required. In this approach, the emphasis is on observable differences in the sediment and the correlation of depositional conditions rather than on mathematical simplicity.

The alternative approach employs an iterative procedure (Fig. 14). The core from the greatest depth (BL96-1) was plotted first. Ages were linearly interpolated between dated horizons and extrapolated to the core ends (Fig. 14A). The second deepest-water core (BL96-2) was then placed on the same plot with the radiocarbon dates constraining the ages (Fig. 14B). Using the same depth proxies, the intervals between constraining radiocarbon ages on both cores were adjusted so that the sense of lake-level change is synchronous (Fig. 14C). If possible, the relative magnitudes of lake-level change were also aligned. Erosional surfaces were then aligned, so that each erosional surface in the deeper-water core is equivalent to a similar surface in the shallower-water core. If insertion of an erosional gap allowed a better alignment of the records, the gap was constrained by the following requirements: that erosion was greater in the shallower core, that the erosion began earlier in the shallower core, and that sedimentation resumed earlier in the deeper-water core. This procedure was repeated for all well-dated cores with overlapping records (Fig. 14D), and for dated portions of cores lacking extensive age control. Where present, grain-size-based depth curves were adjusted in the same manner as curves based on sedimentary features from the same core. Shoreline data were placed on the same plot. The subaerial shorelines appear to be completely out of phase with any of the depth indicators in the core. A reasonable argument—assuming the shoreline data are the most reliable—would be to discard all core data. As previously discussed, however, subaerial shoreline data have poor age control due to the reliance on ages

from shells, but are the best indicators of true depth. Therefore, shoreline deposits were used to constrain lake depth for relatively broad time intervals. For instance, the Willis Ranch and Raspberry Square highstand deposits are the highest features with age ranges that overlap the cored intervals. The calcite layers in the cores have the grain-size range, textural characteristics, and distribution that indicate the deepest-water conditions. If the two most continuous calcite layers are correlated with the two highest shorelines, the other shoreline depth constraints fall roughly in phase with the core data. Although this approach requires an assumption that the shoreline ages are too old, the difference is consistent with observed problems in the cores and reasonable processes in the shoreline environment.

Most of the variability in the composite lake-level history (Figs. 12B and 13B) results from differences between depth estimates based on grain-size data and those based on sedimentary features. In some cases, relative lake-level changes could not be reconciled with the radiocarbon ages. In these situations, one of the ages must be assumed to be incorrect, if the assumptions about synchronous lake-level rise and fall are valid.

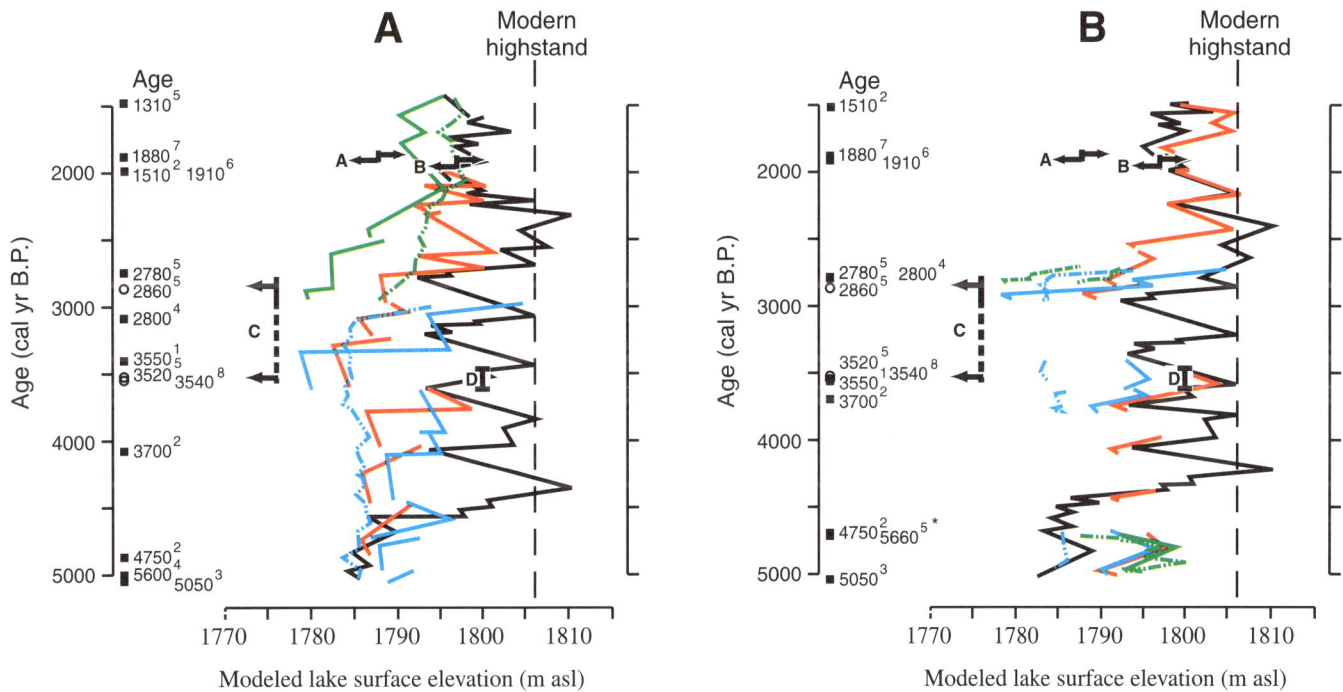

Figure 12. For the period 5000–1500 cal yr B.P., interpretations of lake-surface elevation history for individual cores and different types of data using (A) the smoothed age model of Colman et al. (this volume) and (B) the alternative chronology described in the text. Lake-surface elevation curves utilize sedimentary texture models (solid lines) for cores BL96-1 (black), BL96-2 (red), BL02-3 (blue), and BL02-4 (green), and also grain-size models (dashed lines) for sites BL02-3 and BL02-4. The surface elevations at each core site were determined by adding modeled water depths to the elevation of the sediment surface at that site. Vertical bars indicate shoreline features listed in Table 1, and bars with arrows are depth-significant features from other cores. Arrows indicate the relative depth with respect to the core depth (shallower left and deeper right), and arrows in both directions indicate the relative transition in depth. The vertical lengths of bars indicate the analytical confidence of ages. A—Transition from gravel sheet to Aragonite III in BL02-2; B—Transition from gravel sheet to Aragonite III in BL02-1; C—Shell ages from gravel sheet in BL02-4; D—Shell age from gravel sheet in BL2K-3. Calibrated radiocarbon ages on pollen concentrates (filled boxes) and on shells (open circles) are from Colman et al. (this volume). Shell ages are reduced by 400 years. Superscripts for ages refer to cores: 1—BL96-1; 2—BL96-2; 3—BL96-3; 4—BL02-3; 5—BL02-4; 6—BL02-1; 7—BL02-2; 8—BL2K-3. Asterisk indicates age rejected in creating the alternative chronology. asl—above sea level.

Where possible, such conflicts were resolved by accepting the calibrated age that is most consistent with the greatest number of cores. When necessary, ages from sandy intervals were rejected because such intervals are likely to be contaminated by older material. We devised two age models using this technique, one that assumed the ages within the siliciclastic interval with poor pollen preservation are too old, and one using the same criteria for acceptance of ages as the rest of the core. The lake-level history for the deposits overlying the siliciclastic interval is unaffected by the two models. Only a few ages do not fall within their confidence intervals (Colman et al., this volume) on the composite age scale. Six ages were discarded for the following reasons: (1) An age of 640 cal yr B.P. in core BL02-4 is one of several out-of-order ages that are believed to be a result of sediment reworking following diversion of the Bear River into the lake ca.

1912. (2) Sediment characteristics of a sample yielding an age of 5660 cal yr B.P. in core BL02-4 are more consistent with deposits younger than 5600 cal yr B.P. in core BL02-3, and the sample may also contain reworked materials from the overlying sand. (3) The youngest age in BL96-1 (7600 cal yr B.P.) is in an interval that probably contains reworked material and the character of the deposits is inconsistent with material with similar ages in BL02-3 and BL02-4. (4) In BL02-3, an age of 7900 cal yr B.P. is plotted ~100 years younger than its confidence limits in order to maintain consistency with similar materials in BL02-4. The alternative is to assume that two ages within BL02-4 are incorrect. (5) An age of 7700 cal yr B.P. in BL02-3 is out of sequence and within fluidized sediment. (6) An age of 14,690 cal yr B.P. in BL96-2 is inconsistent with a second age of 15,010 cal yr B.P. in the same core and an identical age in BL96-3.

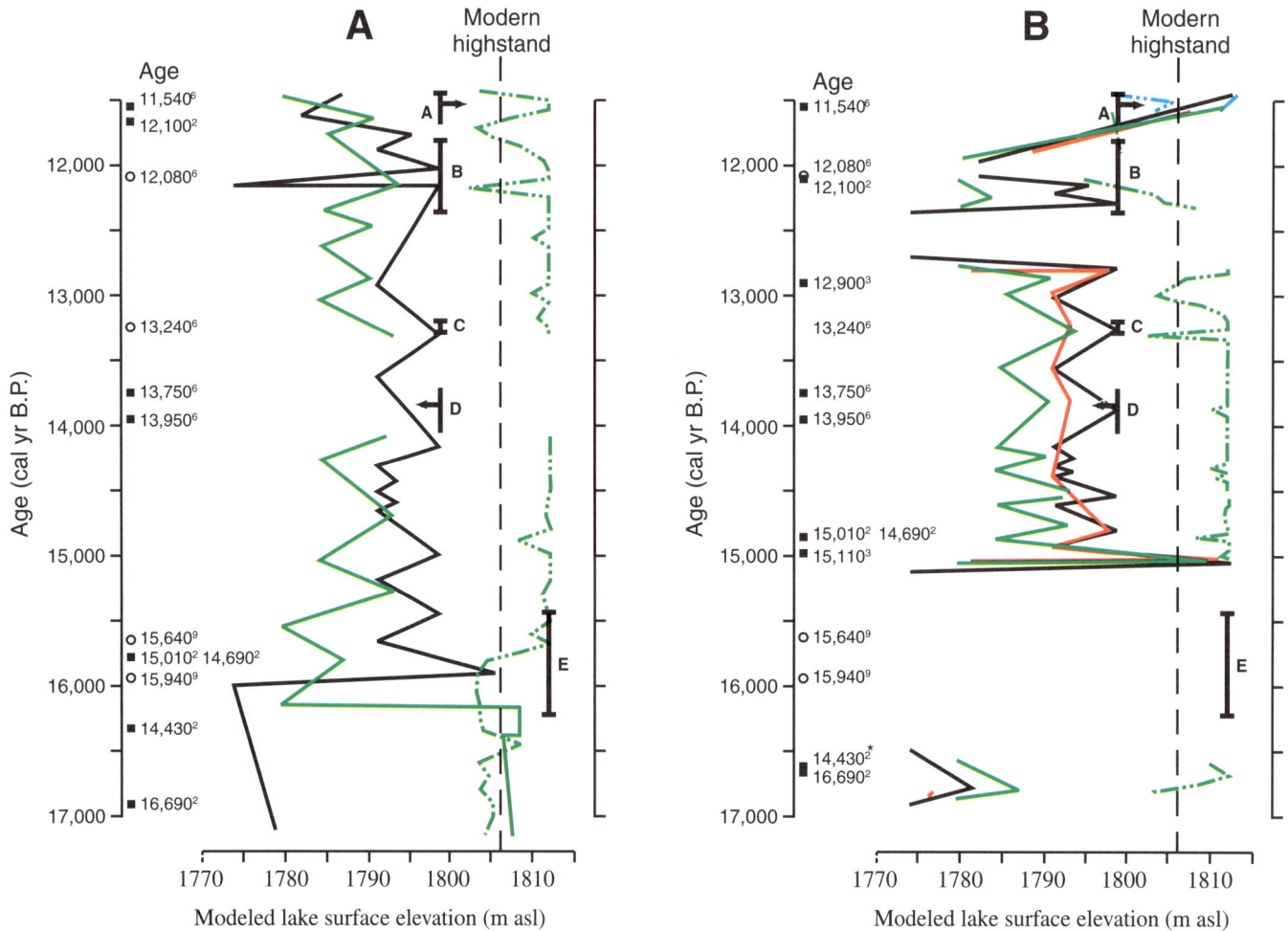

Figure 13. For the period 17,000–11,500 cal yr B.P., interpretations of lake-surface elevation history for individual cores and different types of data using (A) the smoothed age model of Colman et al. (this volume) and (B) the alternative chronology described in the text. Lake-surface elevation curves utilize sedimentary texture models (solid lines) for cores BL96-2 (black), BL96-3 (red), BL02-3 (blue), and BL02-4 (green), and also grain-size models (dashed lines) for sites BL02-3 and BL02-4. See caption for Figure 12 for explanation of numbers and symbols used in this figure. A—Aragonite overlying gravel sheet in BL02-1; B—Shell age from gravel sheet in BL02-1; C—Marsh deposit directly below gravel sheet in BL02-1; D—Marsh deposit in BL02-1; E—Shell ages from the deposits of the Raspberry Square phase. Ages with the superscript 9 from Laabs and Kaufman (2003). Shell ages are reduced by 400 years. asl—above sea level.

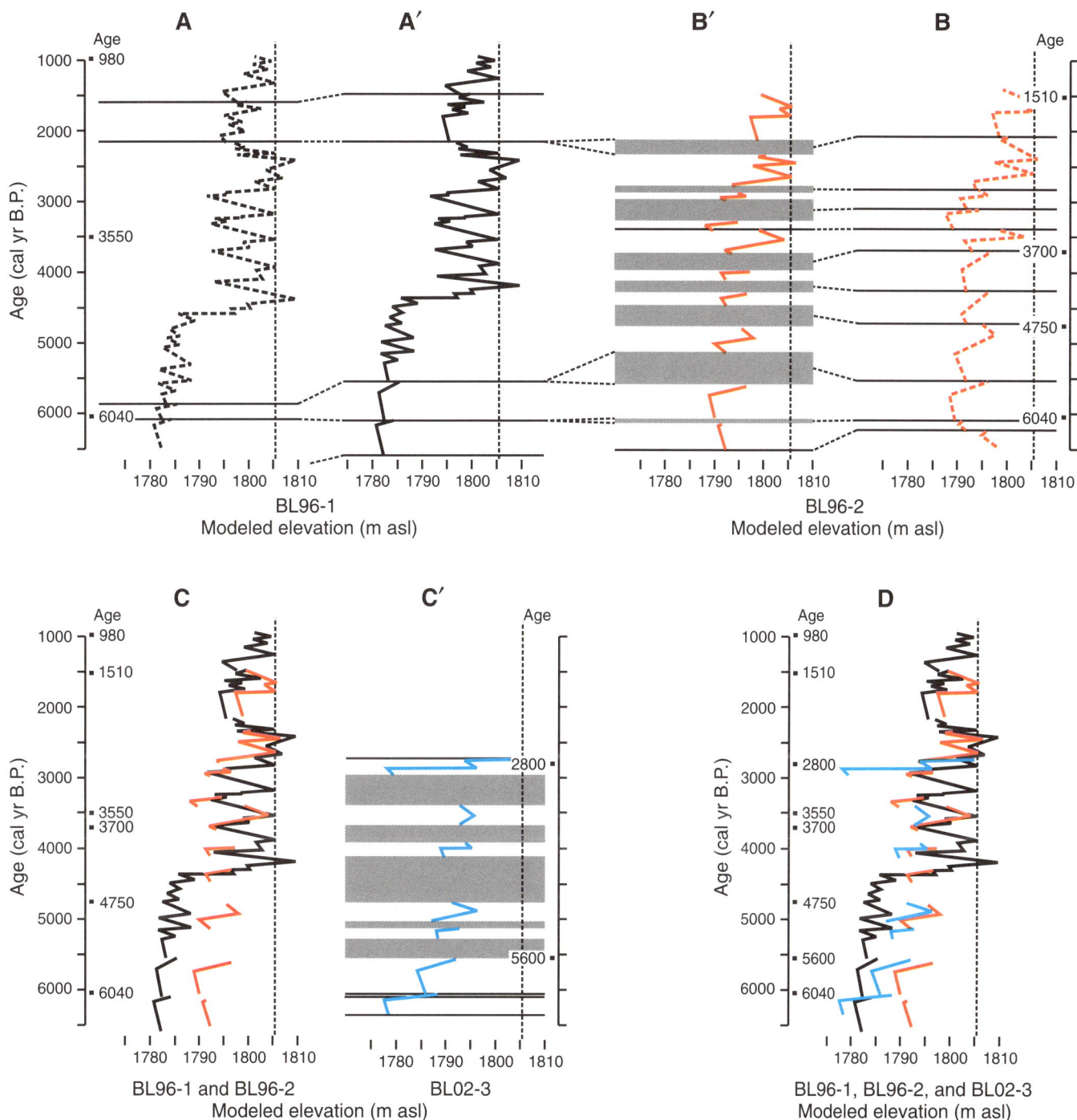

Figure 14. Illustration of method used to produce the alternative chronology for construction of the composite lake-level history (Fig. 16A). Modeled water depths (Fig. 10) have been converted to lake elevation. Portions of the Holocene records (based on sedimentary features) are shown for three cores. Calibrated radiocarbon ages (cal yr B.P.) from Colman et al. (this volume) and their locations are shown to the left or right of age axes. Vertical dashed lines indicate the modern highstand. Horizontal lines signify sharp bedding surfaces (with coarser-grained sediment overlying finer-grained material) that may indicate erosion. (A, B). Unadjusted results for the deepest water site (BL96-1) and the second deepest water site (BL96-2). (A′, B′) Adjusted results for BL96-1 and BL96-2. The curves were adjusted so that lake-level changes are in phase without shifting dated horizons. Erosional gaps (gray bars) were introduced in the shallower record to bring the records into phase and to account for missing fluctuations. Note that erosional surfaces in the deeper record have correlative surfaces in the shallower record. (C, C′). The combined records for BL96-1 and BL96-2 and the adjusted record for next shallower core (BL02-3). The record for BL02-3 was adjusted in the same manner described above. (D) The combined records for BL96-1, BL96-2, and BL02-3. Similar adjustments were made for site BL02-4. asl—above sea level.

The two age models for the siliciclastic interval differ considerably. If the radiocarbon ages are treated in a manner similar to the rest of the core, only two ages fall out of the confidence limits of Colman et al. (this volume): (1) An age of 14,430 cal yr B.P. in BL96-2 is out of order with the other ages. (2) An age of 26,080 cal yr B.P. in BL96-3 is out of order with other dates in that core. Using this correlation, magnetic feature 10 of Colman et al. (this volume) is not the same age in BL96-2 and BL96-3. Because Colman et al. (this volume) discard most of the ages from the siliciclastic interval, it is constrained only by the one age at the base of BL96-3, and by ages overlying rooted horizons that are interpreted as depositional gaps. For the second model, we assumed that the magnetic correlation between cores BL96-3 and BL00-1 by Rosenbaum and Heil (this volume) is correct. We assumed that the sedimentation rates in BL96-3 and BL00-1 were proportional, so that a linear sedimentation trend could be projected through the correlative magnetic features and the two radiocarbon ages in BL00-1 and the one radiocarbon age in BL96-3. Using this approach, the age of the oldest rooted horizon in BL96-3 projects to ca. 20,000 cal yr B.P. We then aligned magnetic feature 10 in BL96-2 to BL96-3, and then matched the silty intervals that are interpreted as indicators of lake-level fluctuations. Assuming a proportional accumulation rate in BL96-2 to the projection of the ages in BL00-1 and BL96-3 produces an age of ca. 19,100 cal yr B.P. for the lowest rooted horizon in BL96-2. The oldest rooted horizon in BL02-4 was assumed to be between the ages at BL96-2 and BL96-3 and the silty intervals were then aligned. The radiocarbon age of 19,370 cal yr B.P. in BL96-2 falls within the confidence interval using this approach. Like Colman et al. (this volume), we rejected the age of 14,430 cal yr B.P. in BL96-2 because it is younger than accepted ages from overlying sediment. If we reject the age of 16,690 cal yr B.P. in BL96-2, the sedimentation interval between the two root horizons is completely unconstrained. Therefore, we accepted this age for lack of any other information.

A comparison of the smoothed age curves of Colman et al. (this volume) to the inferred ages based on the alternative chronology (Fig. 15) shows minor differences. The largest differences are due to the incorporation of erosional breaks and to associated changes in sedimentation rate. In general, the coarser-grained intervals that are attributed to lake-level drops have higher sedimentation rates than the fine-grained intervals attributed to lake-level rises. The largest deviations from the models of Colman et al. (this volume) are in the siliciclastic intervals, even when using the same magnetic correlation and rejected radiocarbon ages. The difference is due to the projection of accumulation rates across rooted horizons by Colman et al. (this volume). One magnetic feature in BL02-4 was not correlated to its presumed match in BL96-2 because the rooted horizons would occur out of sequence.

Composite Lake Level

Based on the alternative chronology, we used the individual core water-depth curves (Fig. 10) to create a composite lake-level curve (Fig. 16A). For the carbonate-rich sediments, grain-size data were used to establish water depth and sedimentary features to indicate changes in lake level. Interpretations based on sedimentary features were favored where grain-size data miss obvious textural changes. In otherwise unconstrained intervals, nonlinear sedimentation rates were used so that similar sediment types had similar relative accumulation rates. For example, transgressive aragonite layers dominated by precipitated crystals had lower accumulation rates than regressive layers dominated by sediment focusing. At Bear Lake, subaerial paleoshorelines are excellent indicators of past lake levels, but the reported ages are inconsistent with ages and interpreted depths from the cores. We suggest that the shoreline ages are too old due to reworking of older shell material and unknown reservoir effects. We interpret the Willis Ranch shoreline (9220 ± 360 cal yr B.P., Table 1) to be equivalent to the upper calcite layer, which is dated several hundred to a thousand years younger. If the published age of the Raspberry Square phase (Table 1) is too old by a similar amount, then it is equivalent to the lowest calcite layer. Similar adjustments in shell ages from Lifton shoreline deposits (Table 1) make the elevation of this shoreline correspond to that of the lake-level rise at 6800 cal yr B.P. The siliciclastic mud intervals are mostly constrained by a spilling lake with little change of depth and by maximum depths suggested by wave-rippled sand in BL2K-2. It was assumed that coarser-grained intervals accumulated more quickly than finer-grained intervals, as suggested by the intervals constrained by radiocarbon ages. The mixed calcitic-siliciclastic mud interval is constrained by the depth of the marsh deposits in BL02-1 and the rooted horizons.

DISCUSSION

The lake-level curve (Fig. 16A) indicates that, prior to ca. 18,000 cal yr B.P., the lake level was stable and subsequently varied radically. This difference is attributed to a change from an older spilling lake to one that was primarily closed. The timing of the first major drop in Bear Lake appears to coincide with the time Lake Bonneville was rising to the Bonneville shoreline (Fig. 16B). If the differences are not due to dating problems in one or both of the records, the diversion of Bear River from Bear Lake may have allowed more water to reach Lake Bonneville without the leakage and evaporation inherent with a period of storage in a lake basin. Lake Bonneville was spilling over its threshold when Bear Lake first became topographically closed. The post-Provo drop in the level of Lake Bonneville roughly corresponds to the 15,000–12,000 cal yr B.P. low lake stand in Bear Lake, and the rise of Lake Bonneville to the Gilbert shoreline at 11,500 cal yr B.P. matches a calcite layer overlying a rooted horizon in Bear Lake. These events occurred after the Bonneville flood, when the lake rapidly dropped from the Bonneville shoreline to the Provo shoreline. The lowstand between ca. 14.8 and 11.8 ka was punctuated by a drop to ~40 m below modern level from ca. 12.8 to 11.8 ka.

For the most part, lake levels at Bear Lake during the Holocene were lower than the historical range of lake levels. Peaks near or above the modern lake occurred at 8500–8000,

7000–6500, 4500–3500, 2500, and 700 cal yr B.P. Rises in the surface of Owens Lake (California) ca. 9000–7500, 4250–3000, and 500 cal yr B.P. (Bacon et al., 2006) roughly coincide with rises of Bear Lake. In addition, Bear Lake high-water periods, except the earliest, roughly correspond to "wetter" intervals at Pyramid Lake (Fig. 16C), although their relative magnitudes differ. In the Owens Lake record (Fig. 16F), there are intervals of increased wetness at 8500, 7700–6500, 2500, and 1000 cal yr B.P. At Bear Lake, levels 15–20 m below the modern highstand occurred ca. 10,000–9000, 7000, 6500–4500, 3500, 3000–2500,

2000, and 1500 cal yr B.P. Each of these dry periods has roughly synchronous "dryer" intervals in the Pyramid Lake and Owens Lake records. The interval of 6500–4500 cal yr B.P. corresponds to a major erosional break with soils at Owens Lake, tree stumps of that age occur below lake level at Lake Tahoe, and during that time the Truckee River stopped flowing to Pyramid Lake (Benson, 2004). The 3000 and 7000 cal yr B.P. lowstands at Bear Lake roughly coincide with dry intervals at Pyramid Lake indicated by a sandy erosional surface in the deep part of the lake (Benson, 2004) and by an isotopically heavy interval, respectively.

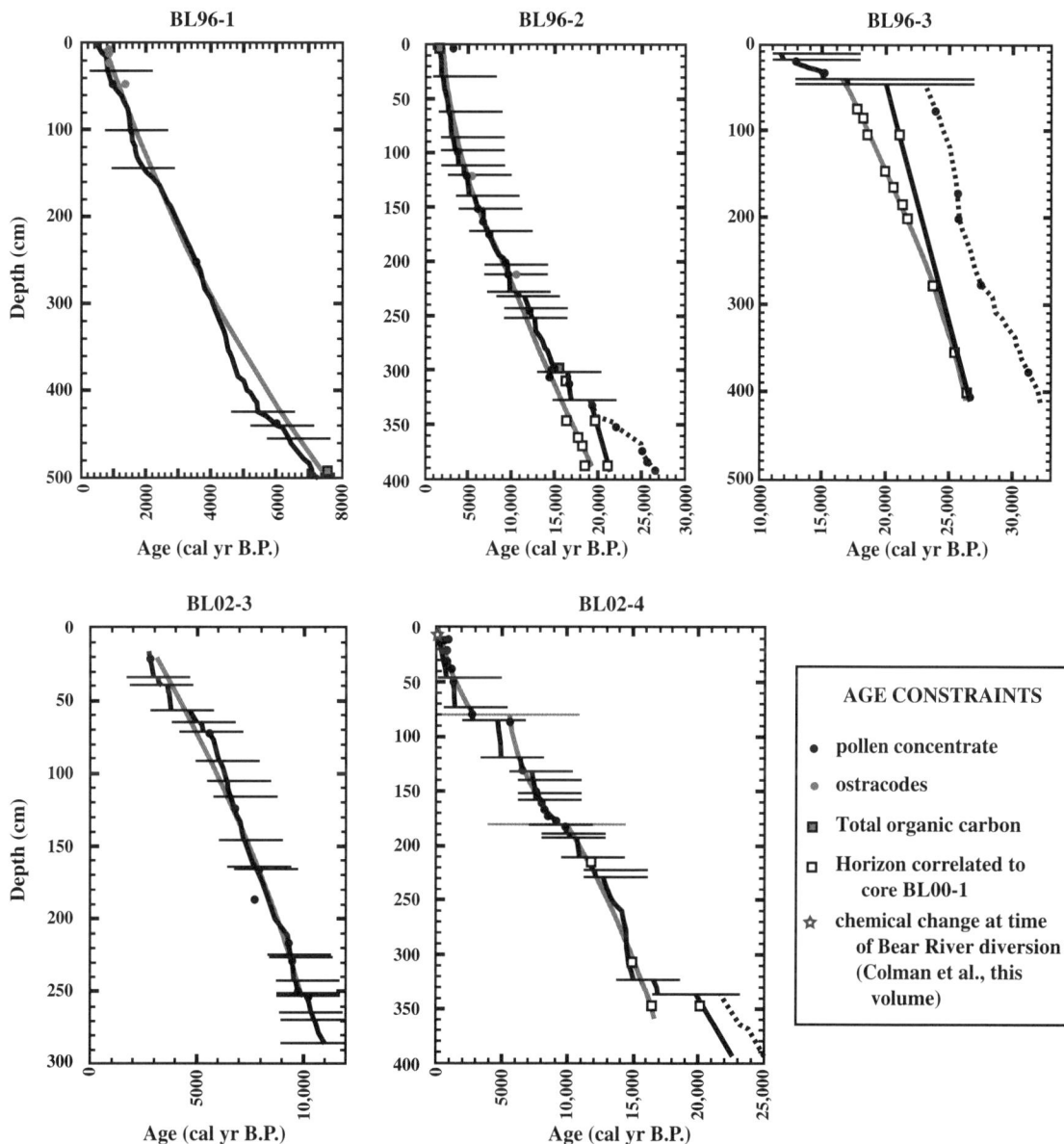

Figure 15. Comparison of age-versus-depth plots using the smoothed age models of Colman et al. (this volume) (heavy gray lines) and the alternative chronology described in the text (heavy black lines). Thin horizontal lines indicate erosional surfaces used in the smoothed age models (gray) and the alternative chronology (black). Dashed black lines indicate the chronology for the siliciclastic interval using the radiocarbon ages versus the solid black lines that assume the radiocarbon ages are too old as in Colman et al. (this volume). Symbols indicate age constraints.

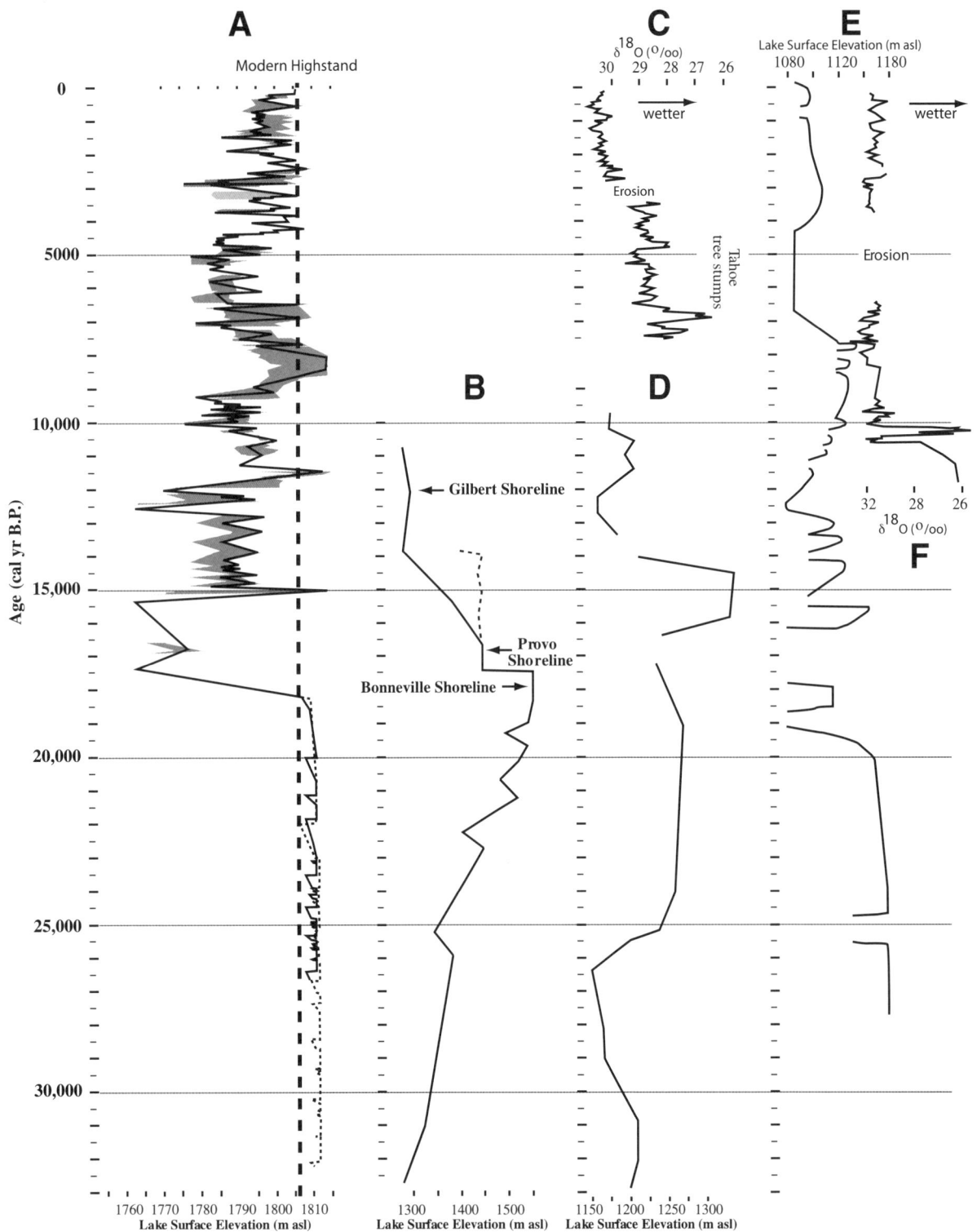

Figure 16. Lake-surface elevation history of Bear Lake compared with surface elevation changes and climate proxies from other lakes. (A) Lake-elevation curve using the alternative chronology. Lake elevations were chosen from the combined data sets using criteria given in text. Gray area shows range of interpretations in the combined data sets. The grain-size data for siliciclastic-rich intervals are not included in that range. Dashed line shows model if radiocarbon ages in the siliciclastic interval that were rejected by Colman et al. (this volume) are used. (B) Lake surface elevations for Lake Bonneville, Utah, based primarily on shoreline dates (modified from Oviatt [1997]). The dashed line indicates an alternative interpretation by Godsey et al. (2005). (C) Oxygen isotope measurements from sediments of Pyramid Lake, Nevada (modified from Benson [2004]). Lighter values are interpreted as higher inflow (wetter). (D) Lake surface elevations for Pyramid Lake, Nevada, based primarily on tufa elevations and ages (modified from Benson [2004]). (E) Lake-surface elevations for Owens Lake. Most ages are from tufa and shells with no reservoir-effect correction (modified from Bacon et al. [2006]). (F) Oxygen isotope measurements from sediments of Owens Lake, California (modified from Benson [2004]). Lower values are interpreted as higher inflow (wetter). asl—above sea level.

The Pleistocene lake-level record at Bear Lake after 18,000 cal yr B.P. is similar to the Pyramid Lake (Fig. 16D) and Owens Lake (Fig. 16E) records. The lowstand at 17,500–15,500 cal yr B.P. at Bear Lake coincides with a lowstand at Pyramid Lake and to a period of erosion and soil development at Owens Lake (Benson, 2004; Bacon et al., 2006). The Owens Lake record indicates that the drop in lake level began earlier than indicated in the Bear Lake record, and that a brief rise in lake level was also earlier. Some of this offset could be due to the uncorrected reservoir effect in dating tufa and shells at Owens Lake. The highstand at 15,000 cal yr B.P. at Bear Lake, correlated with the Raspberry Square phase, roughly equates to the Lahontan highstand at Pyramid Lake; spilling-lake conditions at Owens Lake appear earlier according to the data of Bacon et al. (2006). The lowstand at Bear Lake culminating ca. 12,500 cal yr B.P. corresponds to the drop of Lake Lahontan at Pyramid Lake to the level of the modern closed lake and to a lowstand at Owens Lake. The Bear Lake highstand at 11,500 cal yr B.P. is roughly synchronous with a lake-level rise at Pyramid Lake, although a similar rise is missing in the Owens Lake elevation curve of Bacon et al. (2006). The isotopic data of Benson (2004), however, suggest a period of spilling conditions at that time. The lake-level record at Bear Lake before 18,000 cal yr B.P. does not appear to show any of the variability described from Pyramid Lake or Owens Lake.

SUMMARY

The Bear Lake sedimentary record provides a detailed history of lake-level fluctuations, but some of the age constraints of these changes are less clear. The history of fluctuations was reconstructed by combining three techniques for estimating past water depth. The most clear-cut indicators of lake elevation are shoreline deposits, but they have relatively low preservation potential and are difficult to date. A quantitative lake-level proxy, based on grain-size data from cored sediment, was developed by comparison to modern grain-size distributions. This proxy is best applied to the aragonite sequences most similar to the modern, pre-diversion sediments, and is inappropriate for siliciclastic-rich Pleistocene deposits. Sedimentary textures are sensitive indicators of lake-level fluctuations, but are poorly constrained with respect to actual lake levels. Interpretations of lake-level from sedimentary textures are weakest for deposits with no modern analogs, whose depths of deposition are a matter of speculation. Derivation of a single lake-level history using these three types of data from multiple cores and subaerial exposures required development of an alternative chronology. Smooth age-depth models for Bear Lake cores (Colman et al., this volume) are inadequate because they assume no abrupt changes in sedimentation rate and no gaps in the record. The alternative chronology assumes that the pollen-based radiocarbon ages are mostly correct, that erosional gaps exist in the record, and that gaps are larger in shallower-water deposits than in deeper-water deposits. The lake-level curve produced using the alternative chronology shows some coincidence of lake-level drops and rises with wet and dry intervals at Pyramid Lake and Owens Lake, particularly during the past 8000 years. The Lake Bonneville record is less coincident with that of Bear Lake. The Bonneville highstand occurred during a Bear Lake lowstand that followed diversion of the Bear River away from the basin. The Bear Lake lowstand may have occurred during a major drought that was not recorded in the Bonneville Basin because it was a spilling lake.

The approach of using combined sedimentological constraints on lake levels should be applicable to most lakes. Shoreline data will be most useful in settings with high clastic sediment input, varied sources of material suitable for dating, and a range of slopes within the basin. Grain-size data will be most useful if the basin floor is very flat. Best results will come from lakes with large areas well removed from river mouths and a continuous record with no changes in provenance. The range of sedimentary textures will vary with each lake basin and will have to be evaluated for that basin. Closed-basin drainages are much more variable in depth and area, providing more contrast in sedimentary features than is found in spilling lakes.

ARCHIVED DATA

Archived data for this chapter can be obtained from the NOAA World Data Center for Paleoclimatology at http://www.ncdc.noaa.gov/paleo/pubs/gsa2009bearlake/.

REFERENCES CITED

Adams, K.D., and Wesnousky, S.G., 1998, Shoreline processes and the age of the Lake Lahontan highstand in the Jessup Embayment, Nevada: Geological Society of America Bulletin, v. 110, p. 1318–1332, doi: 10.1130/0016-7606(1998)110<1318:SPATAO>2.3.CO;2.

Bacon, S.N., Burke, R.M., Pezzopane, S.K., and Jayko, A.S., 2006, Last glacial maximum and Holocene lake levels of Owens Lake, eastern California, USA: Quaternary Science Reviews, v. 25, p. 1264–1282, doi: 10.1016/j.quascirev.2005.10.014.

Benson, L.V., 1994, Carbonate deposition, Pyramid Lake Subbasin, Nevada: 1. Sequence of formation and elevational distribution of carbonate deposits (tufas): Palaeogeography, Palaeoclimatology, Palaeoecology, v. 109, p. 55–87, doi: 10.1016/0031-0182(94)90118-X.

Benson, L.V., 2004, Western Lakes, in Gillespie, A.R., Porter, S.C., and Atwater, B.F., eds., The Quaternary Period in the United States: Amsterdam, Elsevier, Developments in Quaternary Science, v. 1, p. 185–204.

Bjorck, S., and Wohlfarth, B., 2001, 14C chronostratigraphic techniques in paleolimnology, in Last, W.M., and Smol, J.P., eds., Tracking environmental change using lake sediments, Volume 1: Basin analysis, coring, and chronological techniques: Dordrecht, Kluwer Academic Publishers, p. 205–245.

Blais, J.M., and Kalff, J., 1995, The influence of lake morphometry on sediment focusing: Limnology and Oceanography, v. 40, p. 582–588.

Broecker, W.S., and Walton, A.F., 1959, The geochemistry of 14C in freshwater systems: Geochimica et Cosmochimica Acta, v. 16, p. 15–38, doi: 10.1016/0016-7037(59)90044-4.

Clifton, H.E., 1976, Wave-formed sedimentary structures: A conceptual model, in Davis, R.A., Jr., and Ethington, R.L., eds., Beach and nearshore sedimentation: Tulsa, Oklahoma, SEPM Special Publication 24, p. 126–148.

Coleman, J.M., 1981, Deltas: Processes and Models for Exploration: Minneapolis, Burgess Publication Co., 124 p.

Colman, S.M., 2006, Acoustic stratigraphy of Bear Lake, Utah-Idaho—Late Quaternary sedimentation patterns in a simple half-graben: Sedimentary Geology, v. 185, p. 113–125, doi: 10.1016/j.sedgeo.2005.11.022.

Colman, S.M., Rosenbaum, J.G., Kaufman, D.S., Dean, W.E., and McGeehin, J.P., 2009, this volume, Radiocarbon ages and age models for the past 30,000 years in Bear Lake, Utah and Idaho, in Rosenbaum, J.G., and

Kaufman, D.S., eds., Paleoenvironments of Bear Lake, Utah and Idaho, and its catchment: Geological Society of America Special Paper 450, doi: 10.1130/2009.2450(05).

Davis, M.B., and Ford, M.S., 1982, Sediment focusing in Mirror Lake, New Hampshire: Limnology and Oceanography, v. 27, p. 137–150.

Dean, W.E., 2009, this volume, Endogenic carbonate sedimentation in Bear Lake, Utah and Idaho, over the last two glacial-interglacial cycles, *in* Rosenbaum, J.G., and Kaufman, D.S., eds., Paleoenvironments of Bear Lake, Utah and Idaho, and its catchment: Geological Society of America Special Paper 450, doi: 10.1130/2009.2450(07).

Dean, W., Rosenbaum, J., Skipp, G., Colman, S., Forester, R., Liu, A., Simmons, K., and Bischoff, J., 2006, Unusual Holocene and late Pleistocene carbonate sedimentation in Bear Lake, Utah and Idaho, USA: Sedimentary Geology, v. 185, p. 93–112, doi: 10.1016/j.sedgeo.2005.11.016.

Dean, W.E., Wurtsbaugh, W.A., and Lamarra, V.A., 2009, this volume, Climatic and limnologic setting of Bear Lake, Utah and Idaho, *in* Rosenbaum, J.G., and Kaufman, D.S., eds., Paleoenvironments of Bear Lake, Utah and Idaho, and its catchment: Geological Society of America Special Paper 450, doi: 10.1130/2009.2450(01).

Denny, J.F., and Colman, S.M., 2003, Geophysical surveys of Bear Lake, Utah-Idaho, September 2002: U.S. Geological Survey Open-File Report 03-150.

Gilbert, G.K., 1885, The topographic features of lake shores: U.S. Geological Survey, Firth Annual Report, p. 69–123.

Godsey, H.S., Currey, D.R., and Chan, M.A., 2005, New evidence for an extended occupation of the Provo Shoreline and implications for regional climate change, Pleistocene Lake Bonneville, Utah, USA: Quaternary Research, v. 63, p. 212–223, doi: 10.1016/j.yqres.2005.01.002.

Hakanson, L., 1977, The influence of wind, fetch, and water depth on distribution of sediments in Lake Vanern, Sweden: Canadian Journal of Earth Sciences, v. 14, p. 397–412.

Hakanson, L., 1982, Lake bottom dynamics and morphometry: The dynamic ratio: Water Resources Research, v. 18, p. 1444–1450, doi: 10.1029/WR018i005p01444.

Hakanson, L., 1995, Models to predict net and gross sedimentation in lakes: Marine and Freshwater Research, v. 46, p. 305–319.

Hakanson, L., and Jansson, M., 1983, Principles of lake sedimentology: Berlin, Springer-Verlag, 316 p.

Harms, J.C., Southard, J.B., and Walker, R.G., 1982, Structures and sequences in clastic rocks: Tulsa, Oklahoma, SEPM Short Course no. 9, 249 p.

Hasiotis, S.T., 2002, Continental trace fossils: Tulsa, Oklahoma, SEPM Short Course Notes no. 51, 132 p.

Hilton, J., 1985, A conceptual framework for predicting the occurrence of sediment focusing and sediment redistribution in lakes: Limnology and Oceanography, v. 30, p. 1131–1143.

Hutchinson, G.E., 1975, Limnological botany, v. 3 of A treatise on limnology: New York, John Wiley and Sons, 660 p.

Johnson, T.C., 1980, Sediment redistribution by waves in lakes, reservoirs, and embayments, *in* Proceedings, Symposium on Surface Water Impoundments, Minneapolis, June 1980: Minneapolis, American Society of Civil Engineers, p. 1307–1317.

Kaufman, D.S., Bright, J., Dean, W.E., Rosenbaum, J.G., Moser, K., Anderson, R.S., Colman, S.M., Heil, C.W., Jr., Jiménez-Moreno, G., Reheis, M.C., and Simmons, K.R., 2009, this volume, A quarter-million years of paleoenvironmental change at Bear Lake, Utah and Idaho, *in* Rosenbaum, J.G., and Kaufman, D.S., eds., Paleoenvironments of Bear Lake, Utah and Idaho, and its catchment: Geological Society of America Special Paper 450, doi: 10.1130/2009.2450(14).

Komar, P.D., 1998, Beach processes and sedimentation (2nd edition): Upper Saddle River, New Jersey, Prentice Hall, 544 p.

Komar, P.D., and Miller, M.C., 1973, The threshold of sediment movement under oscillatory water waves: Journal of Sedimentary Petrology, v. 43, p. 1101–1110.

Komar, P.D., and Miller, M.C., 1975, On the comparison between the threshold of sediment motion under waves and unidirectional currents with a discussion on the practical evaluation of the threshold: Journal of Sedimentary Petrology, v. 45, p. 362–367.

Laabs, B.J.C., 2001, Quaternary lake-level history and tectonic geomorphology of Bear Lake Valley, Utah and Idaho [M.S. thesis]: Flagstaff, Northern Arizona University, 132 p.

Laabs, B.J.C., and Kaufman, D.S., 2003, Quaternary highstands in Bear Lake Valley, Utah and Idaho: Geological Society of America Bulletin, v. 115, p. 463–478, doi: 10.1130/0016-7606(2003)115<0463:QHIBLV>2.0.CO;2.

Lehman, J.T., 1975, Reconstructing the rate of accumulation of lake sediment: The effect of sediment focussing: Quaternary Research, v. 5, p. 541–550, doi: 10.1016/0033-5894(75)90015-0.

McCalpin, J.P., 1993, Neotectonics of the northeastern Basin and Range margin, western U.S.A.: Zeitschrift für Geomorphologie, N.F., Suppl. Bd., v. 94, p. 137–157.

Moser, K.A., and Kimball, J.P., 2009, this volume, A 19,000-year record of hydrologic and climatic change inferred from diatoms from Bear Lake, Utah and Idaho, *in* Rosenbaum, J.G., and Kaufman, D.S., eds., Paleoenvironments of Bear Lake, Utah and Idaho, and its catchment: Geological Society of America Special Paper 450, doi: 10.1130/2009.2450(10).

Nemec, W., 1995, The dynamics of deltaic suspension plumes, *in* Ori, M.N., and Postma, G., eds., Geology of deltas: Rotterdam, A.A. Balkema, p. 31–93.

Olsson, I., 1986, Radiometric dating, *in* Berglund, B.E., ed., Handbook of Holocene palaeoecology and palaeohydrology: New York, John Wiley and Sons, p. 273–312.

Oviatt, C.G., 1997, Lake Bonneville fluctuations and global climate change: Geology, v. 25, p. 155–158, doi: 10.1130/0091-7613(1997)025<0155: LBFAGC>2.3.CO;2.

Reheis, M.C., Laabs, B.J.C., Forester, R.M., McGeehin, J.P., Kaufman, D.S., and Bright, J., 2005, Surficial deposits in the Bear Lake Basin: U.S. Geological Survey Open-File Report 2005-1088, 30 p.

Reheis, M.C., Laabs, B.J.C., and Kaufman, D.S., 2009, this volume, Geology and geomorphology of Bear Lake Valley and upper Bear River, Utah and Idaho, *in* Rosenbaum, J.G., and Kaufman, D.S., eds., Paleoenvironments of Bear Lake, Utah and Idaho, and its catchment: Geological Society of America Special Paper 450, doi: 10.1130/2009.2450(02).

Robertson, G.C., 1978, Surficial deposits and geologic history, northern Bear Lake, Idaho [M.S. thesis]: Logan, Utah State University, 162 p.

Rosenbaum, J.G., and Heil, C.W., Jr., 2009, this volume, The glacial/deglacial history of sedimentation in Bear Lake, Utah and Idaho, *in* Rosenbaum, J.G., and Kaufman, D.S., eds., Paleoenvironments of Bear Lake, Utah and Idaho, and its catchment: Geological Society of America Special Paper 450, doi: 10.1130/2009.2450(11).

Rosenbaum, J.G., Dean, W.E., Reynolds, R.L., and Reheis, M.C., 2009, this volume, Allogenic sedimentary components of Bear Lake, Utah and Idaho, *in* Rosenbaum, J.G., and Kaufman, D.S., eds., Paleoenvironments of Bear Lake, Utah and Idaho, and its catchment: Geological Society of America Special Paper 450, doi: 10.1130/2009.2450(06).

Rowan, D.J., Kalff, J., and Rasmussen, J.B., 1992, Estimating the mud deposition boundary depth in lakes from wave theory: Canadian Journal of Fisheries and Aquatic Sciences, v. 49, p. 2490–2497, doi: 10.1139/f92-275.

Smith, N.D., and Ashley, G., 1985, Proglacial lacustrine environment, *in* Ashley, G., Shaw, J., and Smith, N.D., eds., Glacial sedimentary environments: Tulsa, Oklahoma, SEPM Special Publication 16, p. 135–216.

Smoot, J.P., 2003, Impact of sedimentation styles on paleoclimate proxies in late Pleistocene through Holocene lakes in the western U.S.: International Limnogeology Congress, 3rd, Tucson, Abstracts, p. 273.

Smoot, J.P., 2009, this volume, Late Quaternary sedimentary features of Bear Lake, Utah and Idaho, *in* Rosenbaum, J.G., and Kaufman, D.S., eds., Paleoenvironments of Bear Lake, Utah and Idaho, and its catchment: Geological Society of America Special Paper 450, doi: 10.1130/2009.2450(03).

Smoot, J.P., and Lowenstein, T.K., 1991, Depositional environments of non-marine evaporites, *in* Melvin, J.L., ed., Evaporites, petroleum and mineral resources: Elsevier, Amsterdam, Developments in Sedimentology, v. 50, p. 189–347.

Smoot, J.P., and Benson, L.V., 1998, Sedimentary structures as indicators of paleoclimatic fluctuations, Pyramid Lake, Nevada, *in* Pitman, J.K., and Carroll, A.R., eds., Modern and ancient lake systems: Salt Lake City, Utah Geological Association, Guidebook 26, p. 131–161.

Smoot, J.P., and Benson, L.V., 2004, Mechanical mixing of climate proxies by sediment focusing in Pyramid Lake, Nevada: A cautionary tale: Geological Society of America Abstracts with Programs, v. 36, no. 5, p. 473.

Stuiver, M., Reimer, P.J., and Braziunas, T.F., 1998, High-precision radiocarbon age calibration for terrestrial and marine samples: Radiocarbon, v. 40, p. 1127–1151.

Sturm, M., 1979, Origin and composition of clastic varves, *in* Schluchter, C., ed., Moraines and varves: Rotterdam, A.A. Balkema, p. 281–285.

Williams, J.S., Willard, A.D., and Parker, V., 1962, Recent history of Bear Lake Valley, Utah-Idaho: American Journal of Science, v. 260, p. 24–36.

MANUSCRIPT ACCEPTED BY THE SOCIETY 15 SEPTEMBER 2008

The Geological Society of America
Special Paper 450
2009

Paleomagnetism and environmental magnetism of GLAD800 sediment cores from Bear Lake, Utah and Idaho

Clifford W. Heil Jr.
John W. King
Graduate School of Oceanography, University of Rhode Island, Narragansett, Rhode Island 02882, USA

Joseph G. Rosenbaum
Richard L. Reynolds
U.S. Geological Survey, Box 25046, Federal Center, Denver, Colorado 80225, USA

Steven M. Colman
Large Lakes Observatory and Department of Geological Sciences, University of Minnesota Duluth, Duluth, Minnesota 55812, USA

ABSTRACT

A ~220,000-year record recovered in a 120-m-long sediment core from Bear Lake, Utah and Idaho, provides an opportunity to reconstruct climate change in the Great Basin and compare it with global climate records. Paleomagnetic data exhibit a geomagnetic feature that possibly occurred during the Laschamp excursion (ca. 40 ka). Although the feature does not exhibit excursional behavior (≥40° departure from the expected value), it might provide an additional age constraint for the sequence. Temporal changes in salinity, which are likely related to changes in freshwater input (mainly through the Bear River) or evaporation, are indicated by variations in mineral magnetic properties. These changes are represented by intervals with preserved detrital Fe-oxide minerals and with varying degrees of diagenetic alteration, including sulfidization. On the basis of these changes, the Bear Lake sequence is divided into seven mineral magnetic zones. The differing magnetic mineralogies among these zones reflect changes in deposition, preservation, and formation of magnetic phases related to factors such as lake level, river input, and water chemistry. The occurrence of greigite and pyrite in the lake sediments corresponds to periods of higher salinity. Pyrite is most abundant in intervals of highest salinity, suggesting that the extent of sulfidization is limited by the availability of SO_4^{2-}. During MIS 2 (zone II), Bear Lake transgressed to capture the Bear River, resulting in deposition of glacially derived hematite-rich detritus from the Uinta Mountains. Millennial-scale variations in the hematite content of Bear Lake sediments during the last glacial maximum (zone II) resemble Dansgaard-Oeschger (D-O) oscillations and Heinrich events (within dating uncertainties), suggesting that the influence of millennial-scale climate oscillations

Heil, C.W., Jr., King, J.W., Rosenbaum, J.G., Reynolds, R.L., and Colman, S.M., 2009, Paleomagnetism and environmental magnetism of GLAD800 sediment cores from Bear Lake, Utah and Idaho, *in* Rosenbaum, J.G., and Kaufman D.S., eds., Paleoenvironments of Bear Lake, Utah and Idaho, and its catchment: Geological Society of America Special Paper 450, p. 291–310, doi: 10.1130/2009.2450(13).

can extend beyond the North Atlantic and influence climate of the Great Basin. The magnetic mineralogy of zones IV–VII (MIS 5, 6, and 7) indicates varying degrees of post-depositional alteration between cold and warm substages, with greigite forming in fresher conditions and pyrite in the more saline conditions.

INTRODUCTION

Lakes occupying tectonic depressions in the Great Basin have accumulated nearly continuous Quaternary sediment records following their late Tertiary subsidence. These lakes, and the sedimentary processes within them, are sensitive to changes in temperature and precipitation (Benson et al., 1990; Morrison, 1991; Oviatt et al., 1992; Grayson, 1993). As a result, our understanding of Quaternary climate change and its influence on the western United States can be improved by studying the sediments within them. A key requirement for interpreting and correlating any sedimentary record in the context of climate change is a robust age model. Accurate dating is challenging because many of the dating techniques have varying reliability (Colman et al., 2006) and the magnitude of ^{14}C reservoir effects for these lakes is not well known (Benson, 1999). Studies of the past direction and strength of Earth's magnetic field have the potential to provide another method to address some of the questions about timing of climatic events in the Great Basin, with respect to records elsewhere, including the North Atlantic. For sedimentary records shorter than 0.78 m.y. (like our Bear Lake record), there are three types of geomagnetic behavior that can be used to help constrain the chronology. The first type, excursions, are brief (i.e., 10^3 years) but significant departures from the geocentric axial dipole, which describes the behavior of Earth's magnetic field as a "bar magnet" centered on the spin axis of Earth. During these departures, the magnetic field approaches (and commonly attains) a polarity reversal and then returns to its previous state (King and Peck, 2001). The second type of geomagnetic behavior is secular variation. Paleomagnetic secular variation (PSV) is described as short-term ($10^2–10^4$ years) changes in the non-dipole component of Earth's magnetic field. This secular variability can result in as much as a 30–40° drift of the geomagnetic pole away from the geographic pole and have a regional influence of 3000–5000 km at Earth's surface (King, 1983; Lund, 1996; King and Peck, 2001). The third form of geomagnetic behavior commonly used in constructing an age model is based on changes in the strength (intensity) of Earth's magnetic field over time. The intensity of Earth's magnetic field has both a non-dipole and a dipole component and has changed through time, making it useful for regional correlations on secular variation (SV) time scales and global correlations on longer time scales (10^3-10^6 years) (King, 1983; King and Peck, 2001). Sedimentary records characterizing changes in all three of these geomagnetic behaviors have been used successfully for correlating and dating lacustrine sequences (Liddicoat and Coe, 1979; King, 1983; King et al., 1983; Negrini et al., 1984; Lund et al., 1988; Liddicoat, 1992; Lund, 1996; Brachfeld and Banerjee, 2000; Lewis et al., 2007).

The reliability of paleomagnetic results may be compromised by physical deformation of sediments during or after coring. Tests for sediment deformation can be made by multiple sampling of the sections (Thompson, 1984) and by measurement of magnetic fabric (Rosenbaum et al., 2000). Paleomagnetic reliability can also be compromised by post-depositional chemical alteration of original detrital magnetic minerals (Snowball and Torii, 1999; Roberts et al., 2005; Sagnotti et al., 2005). The degree and effects of such alteration can usually be recognized by a combination of mineral magnetic, petrographic, and geochemical analyses (Reynolds et al., 1994). Although some post-depositional authigenic minerals (e.g., greigite) can record the directional characteristics of Earth's magnetic field at the time of mineral formation, it is often impossible to determine when such minerals formed relative to deposition of the surrounding sediments (Roberts et al., 2005; Sagnotti et al., 2005).

Many of the approaches used to evaluate paleomagnetic reliability also form the backbone of environmental magnetism applied to interpreting the depositional and post-depositional processes (physical, chemical, and biological) responsible for magnetic signals (Thompson and Oldfield, 1986; Reynolds and King, 1995; Verosub and Roberts, 1995). Such magnetic records may provide valuable information about a range of paleoenvironmental factors, such as (1) conditions of weathering, erosion, and sediment transport in the catchment; (2) conditions of sediment transport and deposition in the lake; and (3) chemical conditions in lake water and sediments. The goal of this study is to examine these factors to assess the paleomagnetic directional signal recorded in the sediments of Bear Lake and to interpret changes in the depositional and post-depositional conditions as they relate to glacial and interglacial conditions within the Bear Lake catchment during the past ~220,000 years.

GEOGRAPHIC AND HYDROLOGIC SETTING

Bear Lake is located in northeastern Utah and southeastern Idaho (Fig. 1). The lake is contained within the half-graben Bear Lake Valley (McCalpin, 1993) and has a maximum depth of 63 m, with a mean depth of 28 m (Birdsey, 1989). The present lake surface is 1805 m above sea level. The Bear River Range (Fig. 1) captures most of the moisture from the prevailing westerlies, yielding average annual precipitation of 125 cm yr^{-1} in the western portion of the catchment (http://wcc.nrcs.usda.gov/snow/), mostly occurring in the winter. In contrast, the eastern portion of the catchment receives an average of 30.5 cm yr^{-1}, slightly more during the winter (http://wrcc.sage.dri.edu/summary/climsmid.html). Prior to ca. 1918, when a series of canals was constructed to divert the Bear River into the lake, Bear Lake was topographically closed and evaporation dominated the hydrologic system throughout the Holocene (Dean, this volume). However, geomorphic and stratigraphic evidence indicates that the level of Bear Lake exceeded its present level several times during the Quaternary (Laabs and Kaufman, 2003). When the lake expanded, it may have captured the natural channel of the Bear River whose headwaters are in the Uinta Mountains located ~150 km south of Bear Lake.

METHODS

Two holes, separated by a few tens of meters, were continuously cored to depths of 100 m (BL00-1D) and 120 m (BL00-1E). The holes were located at 41°57'06"N, 111°18'30"W, at a water

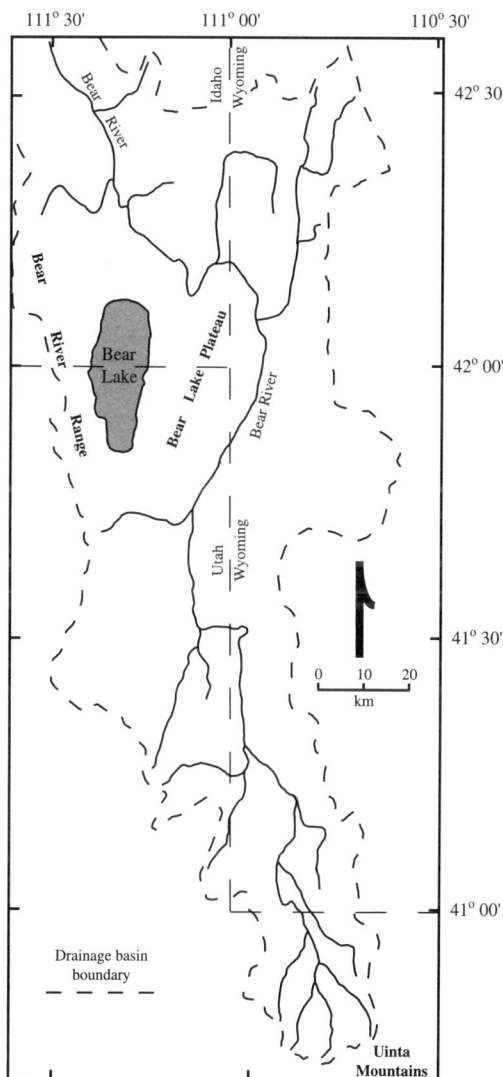

Figure 1. Location map of Bear Lake on the Utah and Idaho border. The curved dashed line delineates the drainage basin of the Bear River (solid black line). The Uinta Mountains are located in the southeastern corner of the map.

depth of 54.8 m. The core sampling tools used were developed by DOSECC (Drilling, Observation and Sampling of the Earth's Continental Crust), based on ODP and commercial designs, to recover-high quality core samples. They include a hydraulic piston corer (HPC), a non-rotating, extended shoe corer, and a rotating, extended core bit corer (the Alien). Both the HPC and extended shoe corer recover 6.6 cm diameter cores, whereas the Alien corer recovers 6.1 cm diameter cores. The HPC was used to recover the uppermost 45 m of sediment in both holes. In hole BL00-1D, the Alien corer was used from 45 to 69 m depth. The extended shoe corer was used for the remaining 31 m of hole BL00-1D (69–100 m) and for the lower 75 m of hole BL00-1E (45–120 m). During the initial core description process, the magnetic susceptibility (K) was measured at 2 cm intervals with a Bartington Instruments loop sensor (80-mm diameter) attached to a GEOTEK® multi-sensor core logger, and ~250 unoriented samples were collected in 3.2 cm³ non-magnetic plastic boxes. After the initial core description phase, u-channel samples were taken from the split core surface.

Natural remanent magnetization (NRM) was measured at 2 cm intervals along the u-channel samples using a 2-G® Enterprises small-access cryogenic magnetometer (model 755–1.65 UC) at the paleomagnetic laboratory of the Graduate School of Oceanography at the University of Rhode Island. The samples were subjected to stepwise alternating field demagnetization at seven fields (0, 5, 10, 15, 20, 30, and 40 mT). After the 40 mT step, the magnetization was typically less than 40% of its initial value and no further demagnetization was performed. Following demagnetization, the characteristic remanent magnetization (ChRM) and maximum angular deviation (MAD) values were calculated using the software created and described by Mazaud (2005).

Nineteen extra samples were taken directly from the u-channels where a distinct low in inclination was identified. These samples were placed in plastic boxes and used for measuring anisotropy of magnetic susceptibility (AMS) to determine whether the directional data were the result of core deformation or whether these recorded a geomagnetic feature. The AMS measurements were made with a KLY-2 Kappa bridge.

Following NRM measurements, anhysteretic remanent magnetization (ARM) was imparted in a 100 mT alternating field with a bias field of 100 μT (79.6 Am⁻¹). The u-channels were then demagnetized at the same levels as the NRM in order to generate NRM/ARM ratios for paleointensity reconstructions (Meynadier et al., 1992; Yamazaki and Ioka, 1994; Lehman et al., 1996). Also, the susceptibility of anhysteretic remanent magnetization (K_{ARM}) was determined by dividing the initial ARM intensity by the bias field (79.6 Am⁻¹).

Isothermal remanent magnetizations (IRM) were imparted to the u-channel samples, following measurement of ARM. Samples were first subjected to a steady field of 1.2 T using a direct-field CENCO electromagnet, from which the resulting magnetization is referred to as the saturation IRM (SIRM). Samples were then given a backfield IRM (BIRM) in an oppositely directed steady field of 0.3 T. The S parameter and "hard" IRM

(HIRM) were calculated using the SIRM and BIRM as described in King and Channell (1991). Measurement of SIRM was problematic for u-channel samples from depths greater than 30 m, because the imparted magnetization approached the upper limits of the system's sensitivity. For this reason, we adopted the method described by Roberts (2006), who suggested imparting an IRM at 1.2 T and then demagnetizing the resulting IRM at 100 mT before measurement. The measured value is here called IRM_{100mT}. In addition to the u-channel measurements, the mineral magnetic properties of the ~250 discrete magnetic samples were characterized. For the discrete samples, IRM was measured with a AGICO JR-5 high-speed spinner magnetometer.

The magnetic properties reflect types, amounts, and magnetic grain sizes of magnetic minerals in the sediments. For sediments lacking in ferrimagnetic minerals (e.g., titanomagnetite and greigite) K reflects the content of paramagnetic minerals and of diamagnetic materials (e.g., calcite, quartz, and water). Magnetic Susceptibility (K), ARM, and IRM commonly reflect the content of ferrimagnetic minerals (for example, magnetite, titanomagnetite, titanohematite, and greigite) but differ in their sensitivity to variations in magnetic grain size (Dunlop and Özdemir, 1997). The ferrimagnetic minerals are characteristically strongly magnetic and have relatively low coercivities. The HIRM parameter is a measure of the concentration of high-coercivity minerals (e.g., hematite and goethite). For ferrimagnetic grains large enough to carry remanent magnetization (≥0.03 μm), K has a weak dependence on magnetic grain size, increasing somewhat as grains increase from single domain to multi-domain. For such remanence-carrying grains, both IRM and ARM decrease with increasing grain size so that large multi-domain grains contribute relatively little in comparison to smaller grains. In comparison to IRM, ARM varies more strongly for very small grains (<1 μm), with ARM being particularly strong for magnetite grains with diameters on the order of 0.1 μm or smaller. Because of the differing dependence of these magnetic properties on the size of ferrimagnetic grains, the ratios ARM/K, IRM/K, and ARM/IRM provide convenient qualitative measures of "magnetic grain size" (with higher values indicating finer sizes). Grain-size interpretations based on these ratios assume that K, ARM, and IRM reflect the same ferrimagnetic minerals. This assumption is commonly violated when K is significantly affected by paramagnetic and or diamagnetic material, when IRM contains a large hematite component, and when there are major changes in ferrimagnetic minerals.

Both low- and high-coercivity minerals (e.g., magnetite and hematite, respectively) acquire significant magnetization at 1.2 T. Demagnetization at 100 mT usually removes most of the magnetization from low-coercivity, ferrimagnetic phases, but has little effect on high-coercivity phases. When a significant amount of single-domain ferrimagnetic material (e.g., greigite) is present, however, the ferrimagnetic component dominates IRM_{100mT} because the 100 mT field is not large enough to demagnetize these extremely stable, highly magnetic grains. Greigite, an authigenic ferrimagnetic Fe-sulfide mineral, typically occurs as dominantly single-domain size grains (Roberts, 1995; Snowball, 1997). Therefore, IRM_{100mT} either provides a measure of the content of high-coercivity minerals such as hematite or is an indicator for the presence of greigite. The ambiguity in the interpretation of IRM_{100mT} can be largely avoided by using an additional screen for the presence of greigite described by Reynolds et al. (1994, 1998). Single-domain greigite acquires a large IRM in a 1.2 T field but is largely unaffected by the 100 mT alternating field used to impart ARM. Therefore, samples characterized by high values of SIRM/K and relatively low values of K_{ARM}/K are likely to contain significant amounts of greigite.

First-order reversal curves (FORCs) were generated to further characterize some of the magnetic grain size and mineralogical changes. Following the methods of Roberts et al. (2000) and Pike et al. (2001), a set of 200 partial hysteresis curves was generated from 39 samples from the Bear Lake section using a Princeton Measurements Corporation, Alternating Gradient Magnetometer. Samples were chosen on the basis of magnetic property variations and petrographic observations (Reynolds and Rosenbaum, 2005). FORC diagrams were generated using the FORCOBELLO program of Winklhofer and Zimanyi (2005) with a smoothing factor of five. In sediments from Bear Lake, with a complex mixture of magnetic minerals, the usual hysteresis parameters can be ambiguous because of the complexity of grain interactions. FORC diagrams can help identify the presence or absence of magnetostatic interactions and characterize the contributions of different magnetic grain sizes (Roberts et al., 2000; Pike et al., 2001). Magnetostatic grain interactions cause spreading of contours parallel to the H_b axis. The shape of the contours is indicative of magnetic grain size, with single-domain grains producing closed contours and multi-domain grains producing divergent contours (Roberts et al., 2000). The median switching field, a measure of the coercivities of the different magnetic grain sizes and mineralogies in the sample, is plotted on the H_c axis (Roberts et al., 2000; Pike et al., 2001). The FORC diagrams and hysteresis data were used in conjunction with the other magnetic properties and petrographic observations to characterize variations in magnetic mineralogy.

Identification of magnetic minerals can be unambiguously made through direct reflected-light petrographic observations when the mineral grains exceed ~5 μm (Petersen et al., 1986). Less precise identification (e.g., distinguishing Fe-oxides from Fe-sulfides) can be made for grain sizes down to ~1 μm. Reflected-light petrographic analyses were made on magnetic minerals that had been concentrated from bulk sediment with a pumped-slurry separator (Reynolds et al., 2001) and then mounted in epoxy and polished.

RESULTS

Magnetic Properties

Mineral magnetic parameters for the Bear Lake sequence show intervals of distinctive variations related to the input, formation, and preservation of magnetic minerals. Values of K range from near zero in the upper few meters to greater than 60×10^{-5} (SI) in several intervals down-core (Fig. 2). Values of K_{ARM}/K, SIRM/K, and ARM/SIRM display large-magnitude variations with some intervals having consistently low values of one or more ratios, and other zones having higher, more variable values. Similarly, IRM_{100mT} displays large-scale variations with intervals of low values separated by intervals of higher, more variable values. Based on the variations in these magnetic properties, the Bear Lake record was divided into seven zones (Table 1, Fig. 2). Dean (this volume) and Kaufman et al. (this volume) describe the sedimentology of the GLAD800 cores from Bear Lake and identify seven zones that roughly coincide with our mineral magnetic zones (Fig. 2), illustrating the need to consider the bulk sediment variations when interpreting the mineral magnetic signal. Although each peak and trough of the magnetic properties do not correspond exactly to sedimentologic variations, the relationship between carbonate content and the amount of magnetic material is obvious. Intervals containing aragonitic marl have the lowest K values due to dilution of detrital material, dissolution of magnetic Fe-oxides, or reduced input of terrestrially derived silt. Conversely, the intervals with higher silt content (e.g., Zone II) and lower carbonate have the highest K values. The mineral magnetic zones and a brief description of the bulk sedimentology are given below.

Zone I

Zone I spans the uppermost 9 m. This interval has low values of K, IRM_{100mT}, and SIRM/K, and high values of K_{ARM}/K and ARM/SIRM. The low values of K and IRM_{100mT} are due in part to dilution of detrital material by endogenic carbonate and in part to post-depositional destruction of magnetite and hematite (Rosenbaum and Heil, this volume). In addition to dilution and post-depositional alteration, the low silt content of this interval suggests a diminished input of detrital, Fe-oxide-bearing material. The diamagnetic properties of the abundant carbonate contribute to the extremely low values of K. High values of ARM/SIRM and FORC data (Fig. 3A), however, indicate that the small quantity of ferrimagnetic material in this interval has a fine magnetic grain size, single domain (SD) or pseudo–single domain (PSD). We speculate that this fraction is protected from alteration within larger detrital silicate grains and rock fragments, and that the quantity of these protected grains is so small that they affect the grain-size indicators only in the absence of other ferrimagnetic minerals. The upward decrease in ARM/SIRM associated with extremely low values of K may indicate that even these protected grains have been destroyed in the uppermost sediments. Petrographic observations of magnetic separates from Bear Lake cores BL96-1, -2, and -3, located a short distance from the BL00 drill site, show that sediments correlative with zone I contain small quantities of magnetite and hematite, and that pyrite and greigite are either absent or present in minor quantities (Reynolds and Rosenbaum, 2005). Many of the Fe-oxides observed in these samples, including both magnetite and hematite, occur in rock fragments, lending support to the speculation that such inclusions are protected from alteration. It

should be noted, however, that grains large enough to be optically identified are too large to explain the fine magnetic grain size discussed above.

Zone II

This zone, which extends from 9 to 18 m (Fig. 2), differs markedly from overlying and underlying sediments. This zone consists of massive, reddish-gray silty clay and contains less than 20% $CaCO_3$ (Dean, this volume; Kaufman et al., this volume). High values of both K and IRM_{100mT} indicate an abundance of ferrimagnetic material. The grain-size-sensitive ratios and FORC data (Fig. 3B) indicate the presence of coarse (multi-domain, MD) magnetic grains, and the FORC data also indicate the subordinate presence of SD, high-coercivity minerals as well. The contribution of the high-coercivity minerals is overwhelmed by the more magnetic and coarser low-coercivity minerals in the grain-size-sensitive ratios. Petrographic observations in correlative sediments from other cores (Reynolds and Rosenbaum, 2005) indicate that magnetite, titanomagnetite, and hematite control magnetic properties. The magnetic minerals represent a wide variety of compositions, textures, and sizes including homogeneous magnetite grains tens of micrometers in diameter. Little dilution of detrital material by endogenic carbonate (Fig. 2) and good preservation of magnetite account for high values of K. These factors also contribute to high hematite content (indicated by the high IRM_{100mT} values) having multiple origins, including input of hematite-rich rock fragments from the Uinta Mountains (Reynolds and Rosenbaum, 2005; Rosenbaum and Heil, this volume).

Zone III

This zone, which extends from 18 to 41 m, is characterized by K and ARM/SIRM values intermediate to those of zones I and II (Fig. 2), and low values of IRM_{100mT}. The sediment is calcareous clay with centimeter-scale gray to greenish-gray bands that are massive and show signs of bioturbation. In addition, there is silt-sized quartz, diatoms, and >3% organic carbon (Dean, this volume; Kaufman et al., this volume). The most abundant mineral in a magnetic separate from a depth of 35.07 m is titanohematite (Reynolds and Rosenbaum, 2005). The sample contains a small amount of magnetite, but most magnetite has been dissolved or replaced by pyrite. It is difficult to compare the degree of alteration between zones I and III using the petrographic data, but the difference in carbonate content between the two zones largely accounts for the difference in K, suggesting that the degree of alteration is similar. FORC data (Fig. 3C) indicate the presence of coarser (MD), low-coercivity minerals with a contribution of finer (SD), slightly higher coercivity minerals (indicated by the less divergent contours extending to higher fields). These "higher coercivity," SD minerals should not be confused with the high-coercivity, coarser minerals identified in zone II (e.g., hematite). They are low-coercivity minerals but have a slightly higher coercivity than the other low-coercivity minerals present (note: for the remainder of this paper, the phrase "higher coercivity" is used in reference to these minerals). It is likely that this mixture of SD,

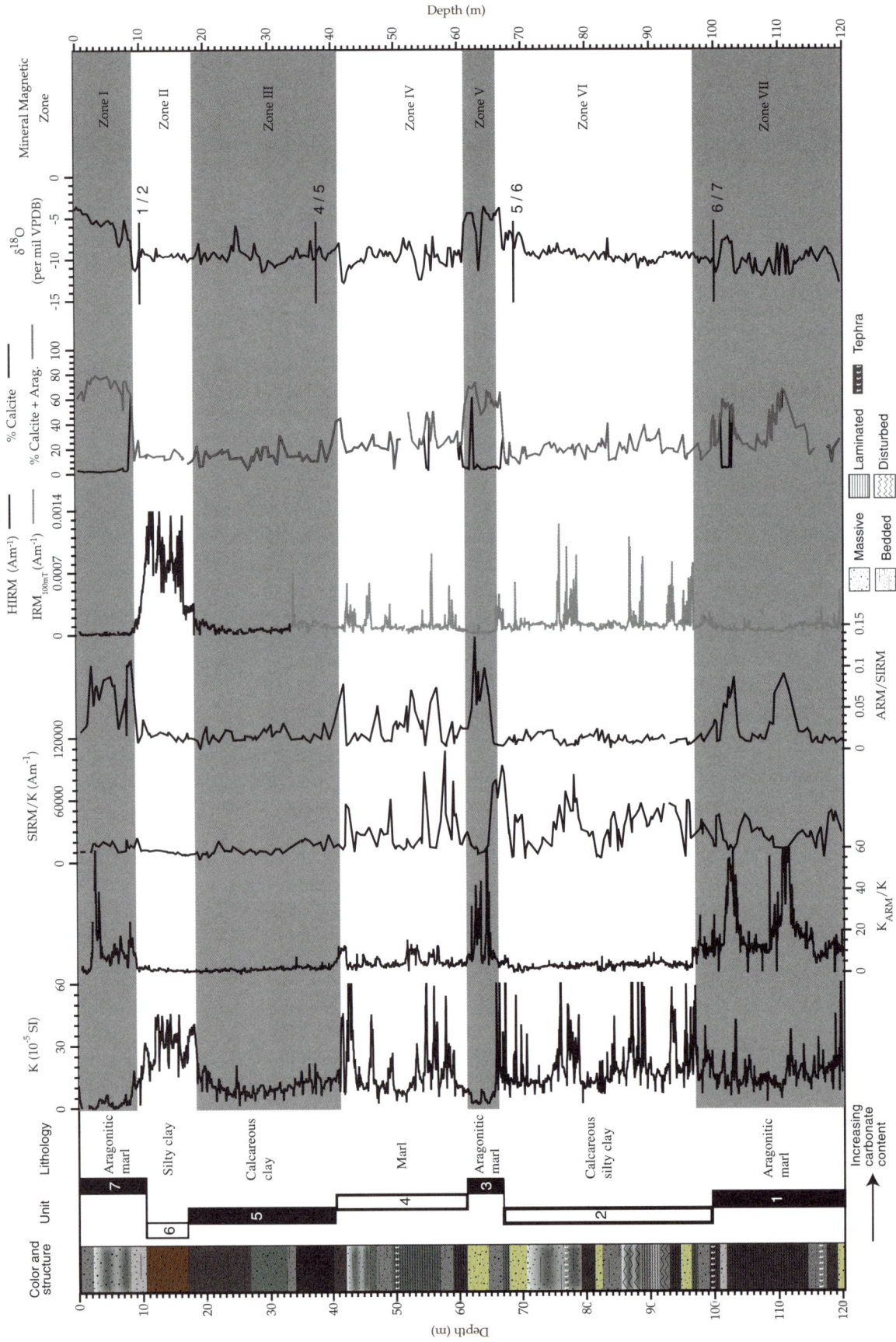

Figure 2. Magnetic properties, calcite and calcite + aragonite content from Dean (this volume), and δ^{18}O data with interpreted stage boundaries from Bright et al. (2006). The SIRM/K and ARM/SIRM data are from discrete samples and are lower resolution than the u-channel and core data (K, K$_{ARM}$/K, HIRM, and IRM$_{100mT}$). The sedimentologic description and summary (sediment color, structure, unit, and lithology) are taken from Kaufman et al. (this volume) and are based on the initial core descriptions conducted at the National Lacustrine Core Repository (LaCore) at the University of Minnesota, Minneapolis.

TABLE 1. GENERALIZED CHARACTERISTICS OF MAGNETIC MINERAL ZONES FROM BEAR LAKE

Magnetic zones	Magnetic concentration (K, M_{rs}, M_s)	Magnetic grain size (FORC, M_{rs}/M_s, H_{cr}/H_c)	Magnetic mineralogy (H_{cr}, petrography, IRM_{100mT})	Lake conditions
I: (0–9 m, 0–17 ka)	Low	SD and MD (0–2 m MD only)	High proportion of low-coercivity Mt and Ti-ht as well as Py	High salinity and reducing conditions during present interglacial
II: (9–18 m, 17–33 ka)	High	SD and MD	Mixture of low-coercivity, MD Mt and Ti-ht and high-coercivity, SD Ht	Lake transgression captures Bear River, which carries Fe-oxide minerals from Uinta Mts. during the LGM
III: (18–41 m, 33–76 ka)	Moderate	MD (occasional SD)	High proportion of low-coercivity, MD Ti-ht, as well as Py with very sparse higher-coercivity,* SD Gt	Saline and reducing conditions, but perhaps less than zone I because greigite is still sparsely present
IV: (41–61 m, 76–113 ka)	Variable	SD and MD	Mixture of low-coercivity MD Ti-ht and higher-coercivity, SD Gt	Fluctuating salinity; greigite formation during fresher intervals
V: (61–66 m, 113–122 ka)	Low	MD (and much less SD)	Mostly low-coercivity, MD Ti-ht and pyrite with sparse higher-coercivity, SD Gt	Similar to present interglacial conditions; high salinity and reducing conditions
VI: (66–97 m, 122–180 ka)	Variable	SD (and MD to a much lesser extent)	High proportion of higher-coercivity, SD Gt and sparse low-coercivity, MD Ti-ht and sparse Py	Salinity fluctuations similar to zone IV, with greigite more prominent in fresher intervals
VII: (97–120 m, 180–222 ka)	Variable	SD and MD	Alternates between higher-coercivity, SD Gt and low-coercivity, MD Ti-ht	High salinity and reducing conditions, perhaps slightly less than zones I and V but more than zones III, IV, and VI

Note: Magnetic grain size: SD—single domain; MD—multi-domain. Magnetic mineralogy: Mt—magnetite; Ht—hematite; Ti-ht—titanohematite; Py—pyrite; Gt—greigite. LGM—last glacial maximum. See text for explanation of other abbreviations.

*Although greigite is a low-coercivity mineral, single-domain greigite tends to have a higher coercivity than magnetite and titanohematite (Roberts, 1995). Here (and in the text) the phrase "higher coercivity" is used in regard to greigite, but is not to be confused with "high coercivity" as used in regard to hematite.

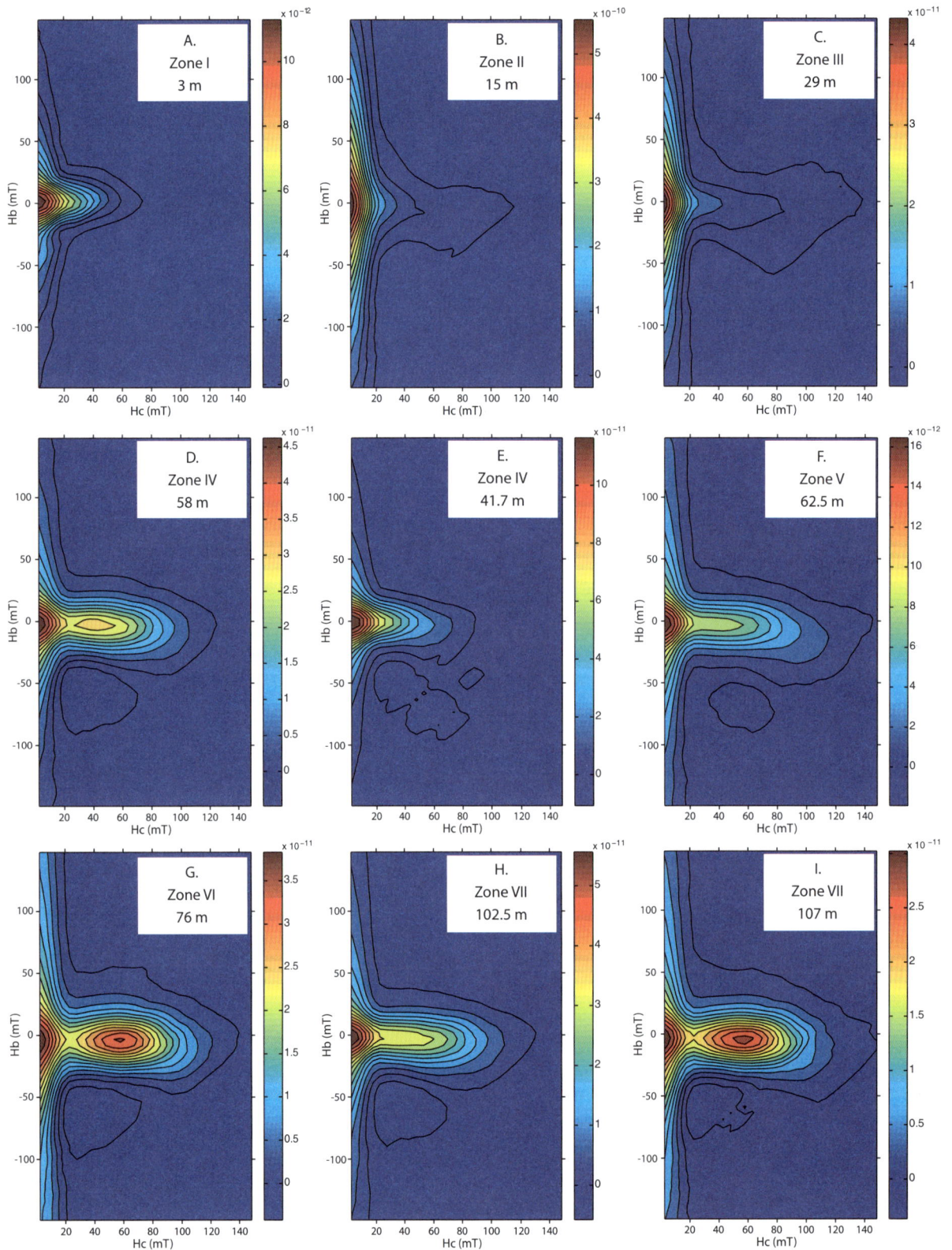

Figure 3. Representative first-order reversal curve (FORC) diagrams from each of the magnetic zones. The H_b axis shows the degree of grain-to-grain interaction by the amount of vertical spread in the contours. For instance, there will be little or no vertical spread in the contours for a sample consisting of non-interacting single-domain magnetic minerals. Also, the presence of a mean stabilizing field (such as that resulting in the alignment of the magnetic minerals in the sample) would be represented by a displacement of the distribution below the H_b axis. The H_c axis indicates the range of coercive fields (in mT) identified for the different minerals present in the sample, which is a function of magnetic grain size and mineralogy. The scale bar next to each FORC diagram indicates the proportion (or concentration) of the different coercive fields identified in a particular sample, with warmer colors indicating higher concentrations.

higher-coercivity minerals and MD, low-coercivity minerals is the cause of the intermediate values of ARM/SIRM.

Zone IV

This zone extends from 41 to 61 m. The zone is characterized by highly variable values of all magnetic parameters, indicating large variations in the types and quantities of magnetic minerals. The zone is a diatomaceous marl with centimeter-scale gray banding, variable bioturbation, and variations in $CaCO_3$ that often exceeds 20% (Dean, this volume; Kaufman et al., this volume). Magnetic separates were examined from two samples within this interval. Titanohematite appears to be the most abundant phase in a magnetic separate from a sample (41.59 m) (Reynolds and Rosenbaum, 2005) that has a low value of IRM_{100mT}. A small quantity of magnetite in this sample is preserved inside silicate grains. The other sample (43.00 m) has a high value of IRM_{100mT}, and petrographic observations show that greigite is the most abundant mineral in the magnetic separate. High values of IRM_{100mT} within this zone are therefore interpreted as an indicator of greigite and not hematite. Titanohematite is a minor component of the magnetic separate from this sample, and there is good evidence of magnetite dissolution. The FORC data indicate variations in the proportions of SD, higher-coercivity magnetic minerals and MD, low-coercivity magnetic minerals (Fig. 3D and E). Most samples with high K also have high IRM_{100mT} and a higher proportion of SD, higher-coercivity minerals (Fig. 3D); conversely, samples with lower K have low IRM_{100mT} values and a lower proportion of SD, higher-coercivity minerals (Fig. 3E). Based on similar magnetic properties and petrographic observations of the two samples, the samples in this zone with lower values of K are thought to have magnetic mineralogy similar to sediment in zone III. The variable magnetic properties in zone IV appear to reflect variable amounts of greigite.

Zone V

This thin zone, extending from 61 to 66 m, is characterized by magnetic properties like those of zone I including extremely low values of K. Petrography indicates that detrital magnetite was destroyed and that the most abundant magnetic mineral is titanohematite. Like zone I, the low content of magnetic minerals is due to the combination of destruction of Fe-oxide grains and dilution of the detrital component by a high carbonate content (>50%) composed largely of aragonite. The FORC diagram from this zone (Fig. 3F) indicates a slightly higher contribution of the SD, higher-coercivity minerals than in zone I (Fig. 3A).

Zone VI

This zone extends from 66 to 97 m. Magnetic properties are highly variable with similar characteristics to those in zone IV. Sedimentologically, this interval is made up of calcareous silty clay, which are gray to greenish-gray and massive with "whispy" black staining suggesting the presence of sulfides (Dean, this volume; Kaufman et al., this volume). Greigite is the most abundant mineral in magnetic separates from three samples with high val-

ues of IRM_{100mT} (67.10, 78.33, and 96.2 m), whereas titanohematite is the most abundant mineral in separates from intervals with low values of IRM_{100mT} (83.2 and 97.5 m). As in zone IV, the FORC data indicate variations in the proportions of SD, higher-coercivity minerals (high K and IRM_{100mT} values) and MD, low-coercivity minerals (low K and IRM_{100mT} values; Figs. 2 and 3G) that are apparently due to variations in the amount of greigite.

Zone VII

This zone spans the lowermost portion of the section from 97 m to the bottom of the core and is delineated by the transition to lower, less variable IRM_{100mT}. Although there is a thin interval of aragonitic marl (101.6–103 m), most of the unit is a calcitic marl that is diatomaceous and variably laminated and massive (Dean, this volume; Kaufman et al., this volume). Relative to zones IV and VI, the K values are intermediate and less variable, whereas the K_{ARM}/K values alternate between ~10 and 60 (Fig. 2). Reynolds and Rosenbaum (2005) determined from magnetic separates that greigite is the most abundant mineral in the higher-SIRM/K intervals and titanohematite is the dominant mineral in the lower-SIRM/K intervals. Of the five FORC samples taken within this zone, three were taken from high-SIRM/K intervals (99, 107, and 115.5 m) and two were taken from low-SIRM/K intervals (102.5 and 111 m). The FORC data for the greigite-bearing intervals show a higher proportion of SD and minerals having a higher coercivity, as well as a smaller fraction of MD, low-coercivity minerals (Fig. 3I). The FORC data from the low SIRM/K intervals also show SD, higher-coercivity minerals mixed with MD, low-coercivity minerals, but for these intervals the coarse fraction constitutes a higher proportion (contours are not closed in the higher coercivity range) (Fig. 3H). Despite having somewhat higher K values, the titanohematite intervals of this zone have FORC diagrams and mineral magnetic properties similar to zone V, in which titanohematite is the dominant magnetic mineral (Fig. 3F).

Paleomagnetism

Variable magnetic mineralogies influence the validity of the paleomagnetic directional data (Fig. 4). For instance, the sediments from intervals with a relative abundance of Fe-oxide minerals (zones II and III) record changes in field direction of ±30° around the geocentric axial dipole value, which is typical of secular variation behavior of Earth's magnetic field (Fig. 4; King, 1983; King and Peck, 2001; Lund, 1996). We interpret this behavior to reflect primary magnetization (a depositional remanence). On the other hand, changes in directional properties from sediments with abundant greigite (zones IV, VI, and VII) vary by ±15° (Fig. 4), casting doubt on the paleomagnetic integrity of these zones. The demagnetization behavior of Bear Lake sediments indicates that the magnetic minerals record a stable remanence (Fig. 5). Following the removal of a weak overprint at the 5 mT demagnetization step, the sediments generally demagnetized linearly toward the origin of the vector endpoint diagrams, with <40% of their initial values remaining after the 40 mT level.

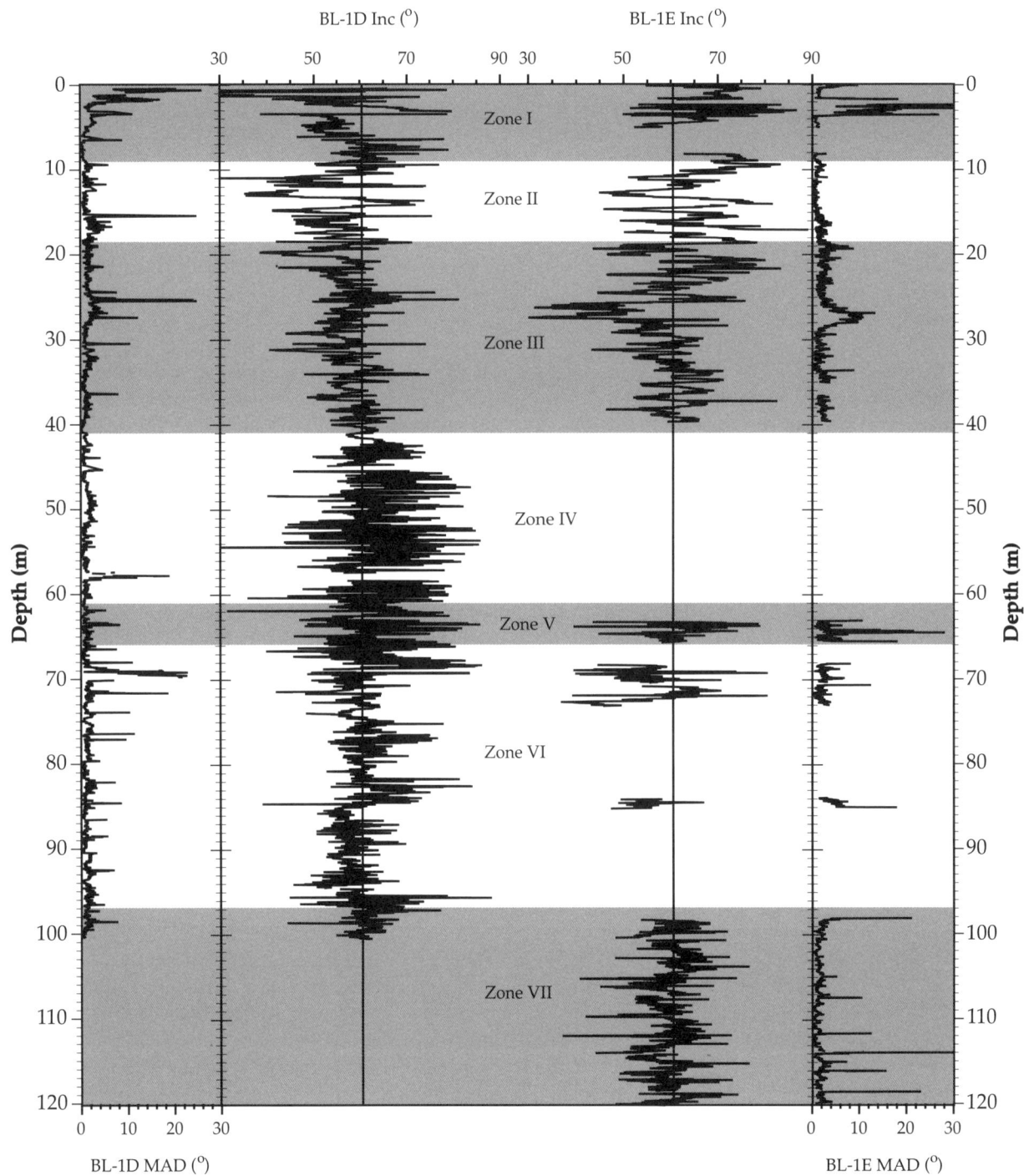

Figure 4. Paleomagnetic inclination data from cores BL00-1D and -1E and mineral magnetic zones from Figure 1. Hole 1E was measured to complete the lowermost 20 m of the Bear Lake section that was not recovered in hole 1D. In addition, the upper ~40 m were measured to illustrate the reproducibility and fidelity of the paleomagnetic record from that interval. The brief intervals of data shown for hole 1E between 40 and 100 m were generated in an attempt to duplicate potential geomagnetic features from hole 1D, which were later determined to reflect sediment disturbance. The maximum angular deviation (MAD) values are shown for each record. MAD values were calculated using the software created by Mazaud (2005).

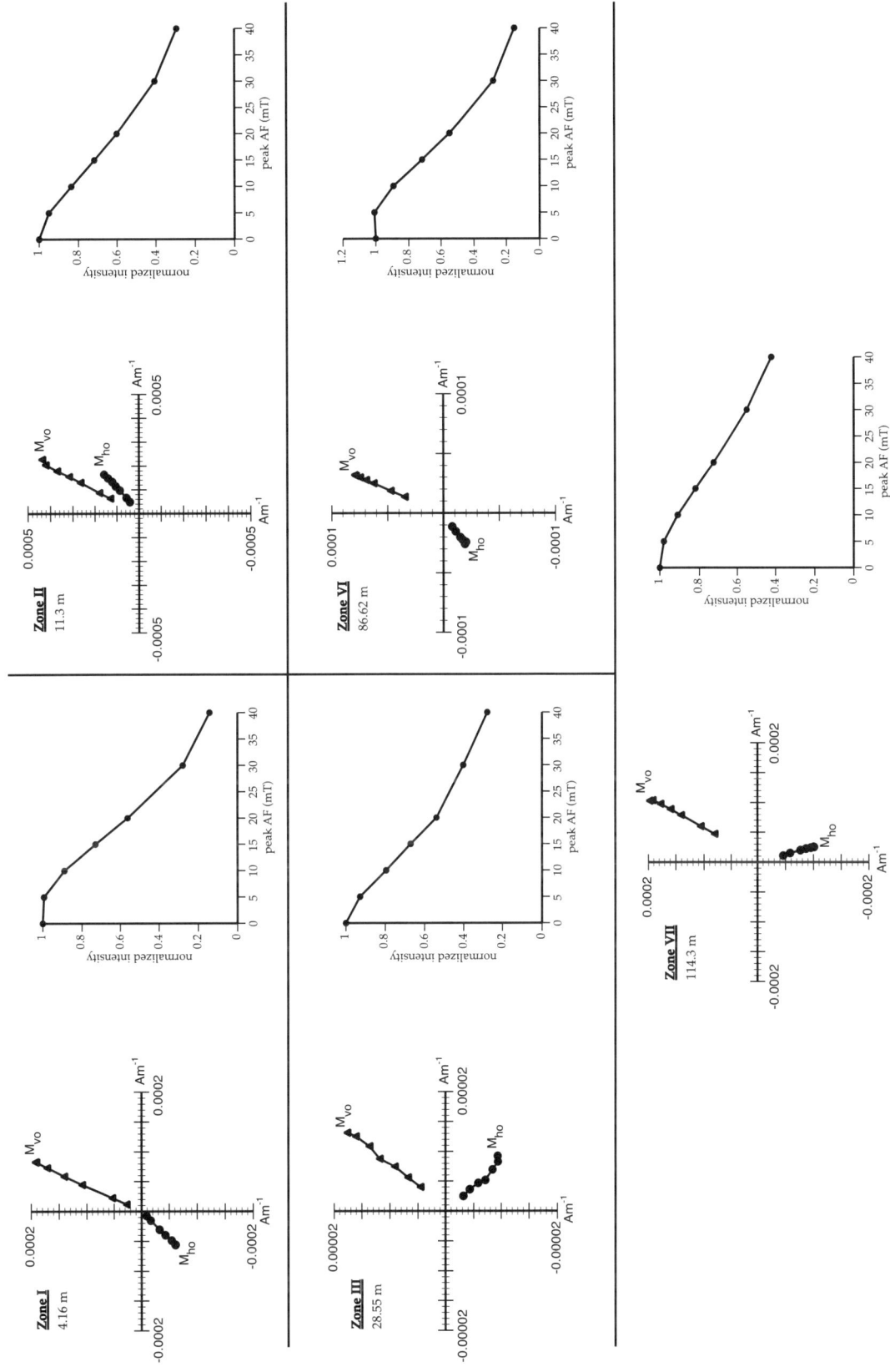

Figure 5. Typical Zijderveld plots from the zones I–III, VI, and VII. Zones IV and V are not shown because directional data from this interval are highly variable due to coring disturbance. M_{vo} —the initial magnetization in the vertical plane, M_{ho} —the initial magnetization in the horizontal plane, Am[-1]—amps per meter, and AF—alternating field.

A visual comparison of the inclination data from hole BL00-1D and -1E for the interval spanning 0–40 m indicates a reasonably close correspondence between the two data sets, particularly in the 8.5–20.5 m and 29–40 m intervals (Fig. 6). Maximum angular deviation values (MAD; Fig. 4) are generally low (~3°) except for portions of zone I and the interval between ~25–28 m. The high MAD values and the erratic inclination values in the top 4 m of zone I are the result of post-depositional destruction of the remanent carrying detrital Fe-oxides. Between 20.5 and 29 m, the records show poor correspondence. In particular, BL00-1D does not have the same high amplitude variations that are present in hole 1E (Fig. 6). Between 25.5 and 27 m in hole BL00-1E, there

is a distinct shallowing of the inclination (Figs. 4 and 6); however, the same feature is not present in the BL00-1D hole. The MAD values are generally higher in this interval for both holes; however, values in hole 1D are more variable, with maxima exceeding 25°, whereas hole 1E values are generally around 10° with one peak value of 13° (Fig. 4). Minimum susceptibility axes from the 20–40 m interval are mostly near vertical, but several samples between 24 and 27 m from hole 1D yielded shallower inclinations of susceptibility axes (Fig. 6). Although the sediments are not visually disturbed, the lack of sedimentologic evidence for rapid deposition (such as turbidites) suggests that these shallowed susceptibility axes result from deformation of the sedimentary

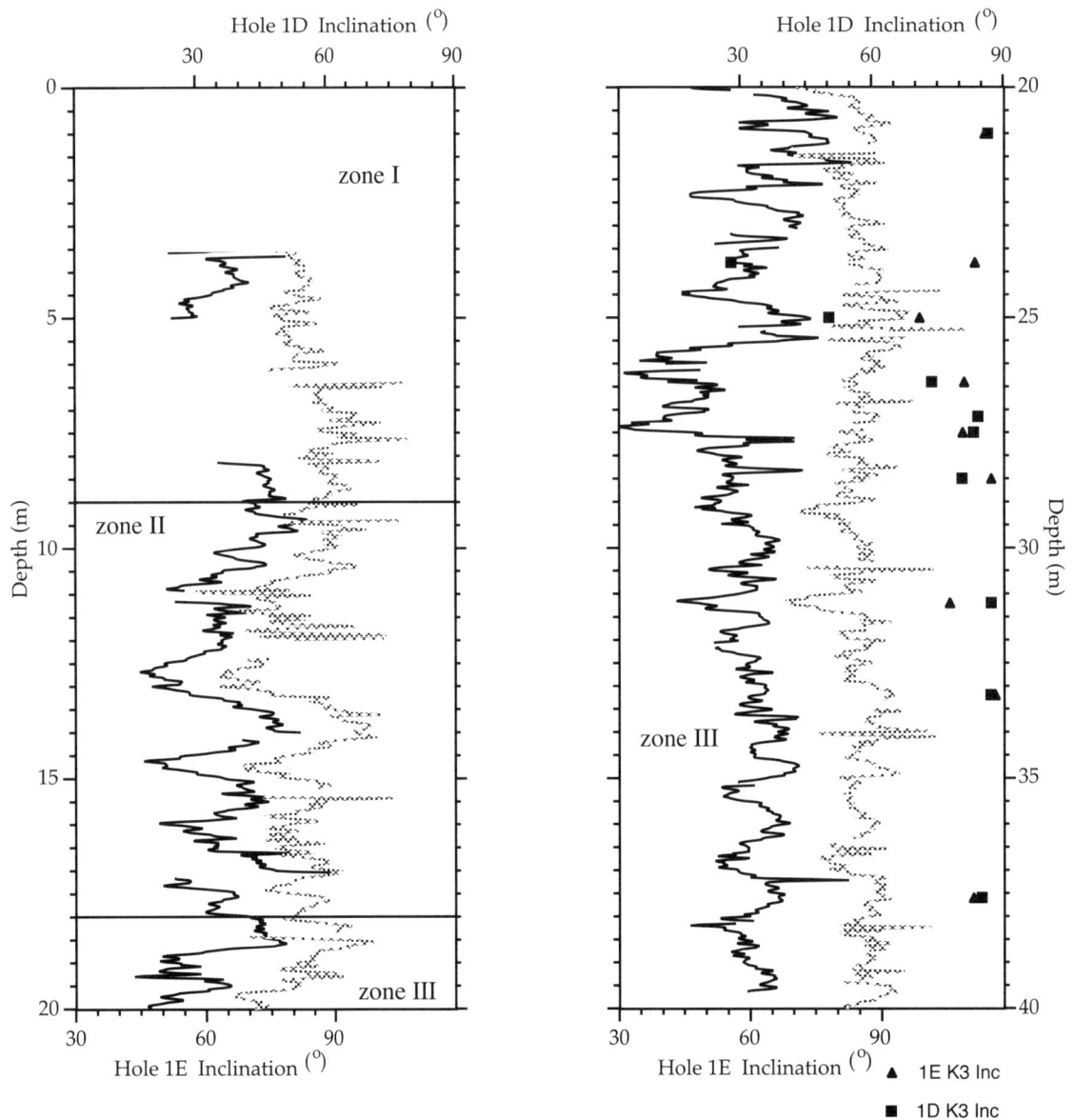

Figure 6. The uppermost 40 m of paleomagnetic inclination data and the inclinations of minimum magnetic susceptibility axes (K_3). Note: (1) There is generally good correlation of paleomagnetic data between the cores in zones II and III. (2) Low inclinations of K_3 axes occur between ~23.5 and 27 m in hole 1D.

fabric (Rosenbaum et al., 2000). From ~40–70 m, the inclination data are highly variable. This behavior is likely an artifact of the change from the hydraulic piston corer to the rotary corer that can cause the sediment to fracture, rotate, or tilt (Poag and Foss, 1985). The lowermost part of the sequence (70–120 m) was cored using the extended nose corer and, as a result, contains less variable paleomagnetic directions with lower amplitude peaks and troughs (Fig. 4). The MAD values below 40 m are generally low except for zone V, in which the environmental magnetic evidence suggests post-depositional destruction of Fe-oxides (Fig. 2) resulting in the loss of remanent-carrying minerals.

In addition to paleomagnetic inclination data, the ratios of NRM intensity to K and IRM and ARM intensities (Fig. 7B) were calculated as possible measures of relative paleointensity. Generally, the mineral magnetic properties in the Bear Lake section do not meet the criteria for geomagnetic paleointensity studies defined by King et al. (1982) that include (1) a uniform mechanism of detrital remanent magnetization, (2) uniform magnetic mineralogy (preferably magnetite), and (3) uniform magnetic grain size. The exception is zone III (18–33 m) in which the magnetic concentration (K), grain size (ARM/SIRM and FORC), and mineralogy (FORC, IRM_{100mT}, and petrography) are uniform, and in which low-coercivity, multi-domain minerals (probably ferrimagnetic titanohematite) are the dominant remanence carriers (FORC diagrams, Fig. 3). Within this zone, the NRM/ARM, NRM/IRM, and NRM/K data show distinctly low values between 25 and 27 m, coincident with low-inclination values (Fig. 7). Representative demagnetization curves for NRM/ARM from the interval indicate that the low values are present regardless of demagnetization level (Fig. 7C).

DISCUSSION

Magnetic Properties

Magnetic properties of the Bear Lake section reflect varying degrees of post-depositional alteration of detrital Fe-oxide minerals and varying quantities of the magnetic sulfide mineral, greigite. Zone II (9–18 m) is the only interval in which detrital Fe-oxide minerals are largely preserved. Sediments within this interval contain large amounts of glacial flour derived from the Uinta Mountains and carried to Bear Lake by the Bear River (Rosenbaum and Heil, this volume). Preservation of Fe-oxides reflects some combination of fresh water, rapid deposition (sulfidization may cease when sediment has been buried to the point that diffusion can no longer replenish sulfate in the pore waters, see Canfield and Berner, 1987), and perhaps large amounts of Fe-oxide minerals.

The relatively uniform magnetic properties through zone III (18–41 m) probably reflect post-depositional conditions. In this zone, detrital magnetite and hematite have been largely destroyed and the most important magnetic mineral is ferrimagnetic titanohematite. Titanohematite, which is relatively resistant to post-depositional alteration (Canfield et al., 1992; Reynolds et al.,

1994), probably occurs in small amounts throughout the entire Bear Lake section and becomes important phase when other detrital magnetic minerals have been destroyed.

Zones I and V (0–9 m and 61–66 m, respectively) have similar magnetic mineralogy to zone III but lower concentrations of detrital material due to high carbonate content. The upward coarsening of magnetic grain size in the uppermost 2 m of zone I (indicated by decreasing values of ARM/SIRM) coincides with increasing values of $\delta^{18}O$ (Fig. 2). Bright et al. (2006) suggest that this is an interval of increased salinity. Higher concentrations of sulfate in this interval may have caused more complete destruction of Fe-oxide grains including small magnetite grains enclosed in detrital rock particles. Zones IV (41–61 m) and VI (66–97 m) are characterized by intervals dominated by greigite and intervals with characteristics like those of zone III.

Zone VII (97–120 m) alternates between intervals with greigite (higher SIRM/K values) and intervals with titanohematite (lower SIRM/K values). Dean (this volume) identifies two intervals in this zone that are enriched in calcite + aragonite and interprets this enrichment (particularly the increased aragonite) as more saline lake conditions. The correspondence of the calcite + aragonite peaks with the relative dominance of titanohematite in zone VII suggests that the increased availability of sulfate contributed to sulfidization of the less-resistant Fe-oxides. The salinity of this zone is likely to have been similar, or just slightly less than that of zones I and V (based on the calcite + aragonite data), which are also dominated by the more-resistant titanohematite. The dominance of titanohematite in zone III is not as easily explained by higher salinity conditions since the calcite + aragonite content is not as high as zones I, V, and VII. In this case, the dominance of titanohematite may result from a combination of the reduction of the less-resistant Fe-oxides and a decrease in the amount of Fe-oxides deposited in the lake at that time, i.e., lower salinities (lower sulfate concentrations) are required for sulfidization of fewer Fe-oxides.

Variations in magnetic properties reflect differences in lake conditions during deposition. Zone II records freshwater conditions during a period when the Bear River delivered large volumes of glacial flour from the Uinta Mountains to Bear Lake (Rosenbaum and Heil, this volume). Zones I and V correspond closely to intervals previously correlated with marine isotope stages (MIS) 1 and 5e (Fig. 2; Bright et al., 2006). During these periods, the lake is thought to have been topographically closed, highly evaporative, and productive. Based on high aragonite content (Fig. 2), Dean (this volume) interprets these intervals to represent the most saline conditions in the cored section. Differences in magnetic properties in the other four zones largely reflect the presence or absence of greigite. The concentration of sulfate may be an important factor in determining the fate of Fe-oxide minerals and production of Fe sulfides (Snowball and Torii, 1999).

In a study of Owens Lake, Reynolds et al. (1998) suggest that the formation of greigite was controlled by sulfate availability. If reducing conditions exist and sulfate is not available, Fe may go into solution and be lost from the sediment. If sulfate is abundant,

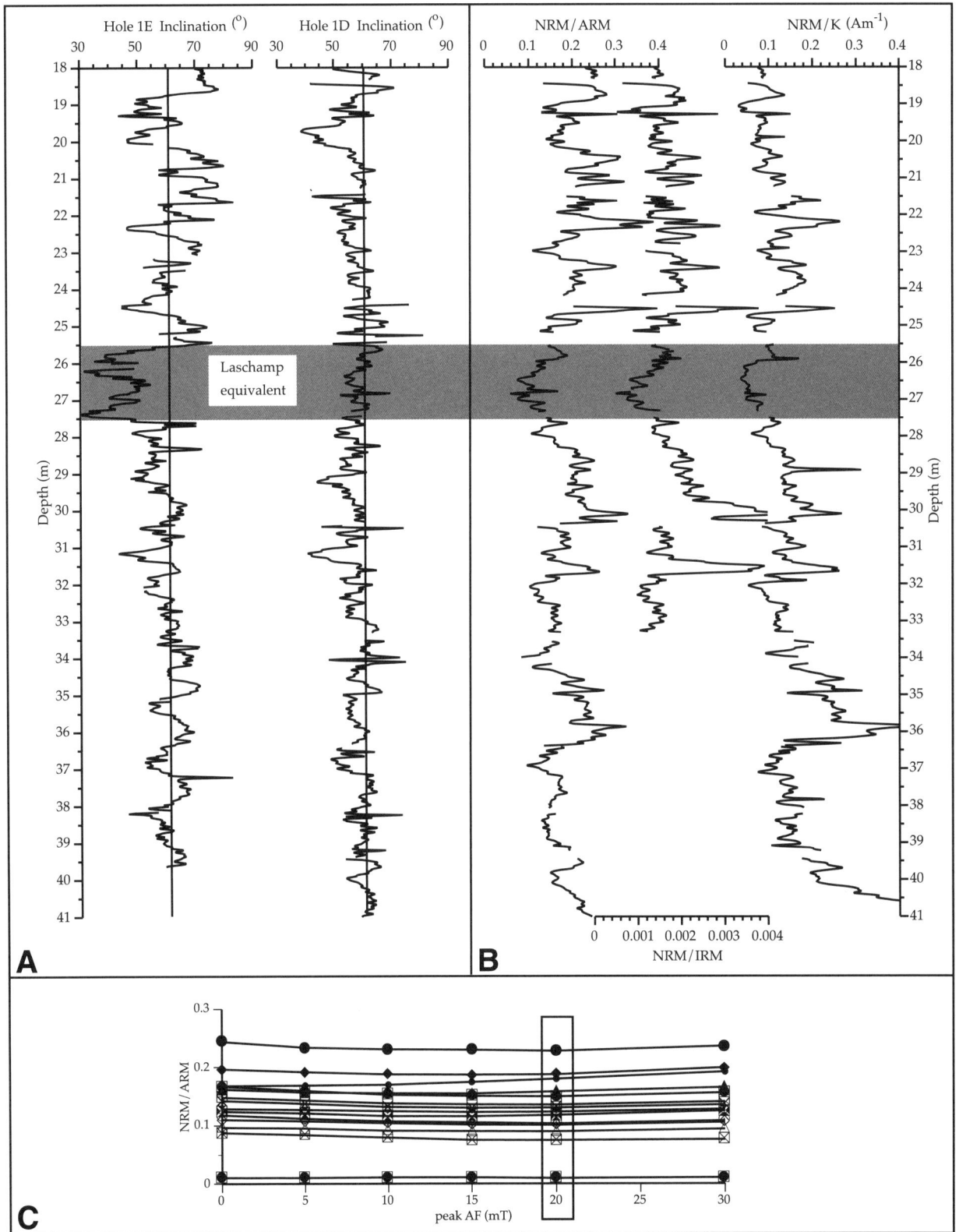

Figure 7. (A) Paleomagnetic inclination data from zone III for BL00-1E and -1D showing the geomagnetic feature interpreted as the time-equivalent of the Laschamp excursion found in higher-fidelity records. The expected geocentric axial dipole value (~61°) for Bear Lake at 42°57'06"N is indicated with solid vertical lines. (B) Relative paleointensity proxy data (NRM/ARM, NRM/IRM, and NRM/K) from BL00-1E showing low values at the time of the interpreted Laschamp equivalent. (C) Representative NRM/ARM data versus demagnetization level from the interval of low relative paleointensity. Constant ratio values regardless of demagnetization level illustrate the ability of the sediments within this interval to reliably record relative paleointensity variations (King et al., 1982). ARM, IRM, NRM—anhysteretic, isothermal, and natural remanent magnetization, respectively.

and therefore not a limiting factor, pyrite is likely to form at the expense of Fe-oxide minerals. Greigite may form under intermediate conditions, when sulfate is present in quantities less than needed to form pyrite. The absence of greigite in portions of the Bear Lake record could indicate more saline conditions (i.e., more sulfate leading to more pyrite upon reduction) or less saline (i.e., less sulfate and Fe-oxide preservation) than conditions for the greigite-bearing intervals. These possibilities cannot be distinguished on the basis of magnetic properties alone. We interpret greigite-free intervals (other than zone II) to indicate more saline conditions because (1) some pyrite occurs in these sediments (Reynolds and Rosenbaum, 2005), (2) intervals deposited under the most saline conditions, as indicated by high concentrations of aragonite, lack greigite, and (3) all other high-carbonate intervals, which reflect elevated salinities, also lack greigite.

Paleomagnetism

Although variable amounts of the detrital remanence-carrying minerals have been removed by post-depositional alteration, detrital Fe-oxide minerals are present in most of the sequence, even if only in small quantities as resistant titanohematite minerals (Reynolds and Rosenbaum, 2005). In addition, the inclination data are centered on the geocentric axial dipole value of 61° (Fig. 4). Excluding zone II, which has a relative abundance of magnetite and hematite, it is likely that a combination of titanohematite minerals and greigite contributes to the paleomagnetic record from Bear Lake. Greigite produces a strong and stable chemical remanence (CRM; Roberts, 1995) acquired post-depositionally. The difficulty associated with relying on greigite as a remanence carrier is the time it takes for the acquisition of the CRM. If greigite is formed either syndepositionally, or soon after deposition, then the CRM may be an accurate recorder of geomagnetic field behavior. If there is a significant lag time associated with the acquisition of the CRM, the paleomagnetic directional signal recorded in the sediments is not representative of the geomagnetic field behavior at the time of deposition.

It is evident that the fidelity of the paleomagnetic record is related to the magnetic mineralogy. The uppermost 4 m of the sequence (top of zone I) has highly variable inclination values that are not correlative between holes 1D and 1E, possibly reflecting a very low abundance of detrital Fe-oxide minerals. In contrast, zone II, which has abundant detrital Fe-oxide minerals, yields a record that resembles typical secular variation (King, 1983; Lund, 1996; King and Peck, 2001), and the inclination data are highly correlative between the two cores (Fig. 6). A close comparison of the 18–40 m interval (zone III) shows that, although the amplitude of the peaks and troughs is damped, the data can be correlated between the two holes (Fig. 6). From 40 to 120 m (zones IV through VII), the sediments have been variably affected by post-depositional alteration, and greigite is relatively abundant. The directional signal throughout this interval seems "muted" compared to zone II. This muted effect is likely the result of the CRM carried by greigite overprinting the DRM carried by the titanohematite.

Paleointensity interpretations from normalized NRM intensity data are problematic due to changes in mineralogy and grain size (Fig. 2). Because of post-depositional changes in magnetic minerals (King et al., 1982), Bear Lake is not a good location to obtain a paleointensity record. However, the sediments of zone III (18–41 m) meet the criteria of King et al. (1982) in that they have (1) a uniform mechanism of detrital magnetization, (2) uniform magnetic mineralogy, and (3) uniform magnetic grain size (Fig. 2). Although magnetite is the preferred magnetic mineral (King et al., 1982), the reproducibility of the inclination data (Fig. 6) and normalized NRM ratios (Fig. 7B) between the two holes suggests that titanohematite provides a sufficient detrital magnetization. In addition, the demagnetization behavior of the NRM/ARM data does not change with demagnetization level (Fig. 7C; see King et al., 1982). On the basis of this evidence, we interpret the NRM/ARM, NRM/IRM, and NRM/K ratios as relative paleointensity records for this interval in the Bear Lake sequence.

Despite post-depositional changes in magnetic mineralogy, the low-inclination values between 25.5 and 27.5 m in hole 1E appear to record a geomagnetic feature (Fig. 7). The absence of this low-inclination feature in hole 1D is probably due to core disturbance, as suggested by the AMS data (Fig. 6) and higher MAD values of hole 1D (Fig. 4). The stratigraphic position of this feature relative to the $\delta^{18}O$ data of Bright et al. (2006) suggests that it occurred during MIS 3. Guillou et al. (2004) identified a geomagnetic excursion in MIS 3, the Laschamp, with an $^{40}Ar/^{39}Ar$ age of 40.4 ± 2.0 ka. Although the low-inclination values from hole 1E do not deviate from the GAD value by 40°, as required for excursion classification (Barbetti and McElhinny, 1972), we interpret this feature to be synchronous with the Laschamp excursion of higher-fidelity paleomagnetic records (e.g., Lund et al., 2001; Guillou et al., 2004). The low paleointensity values from the same interval in Bear Lake (Fig. 7B) support our interpretation, because published paleointensity records (Guyodo and Valet, 1996; Channell et al., 1997; 2000; Stoner et al., 1998, 2000, 2002) indicate a paleointensity low ca. 40 ka.

The linear sedimentation rate (0.54 mm yr⁻¹) used by Kaufman et al. (this volume) indicates that the stratigraphic position of this geomagnetic feature corresponds to an age of 50 ka, ~7 k.y. older than the reported age of the Laschamp excursion (40.4 ± 2.0 ka; Fig. 8; Guillou et al., 2004). Alternatively, extrapolating the Colman et al. (this volume) ^{14}C age model through the U-series age at 67 m depth (127.7 ka; Colman et al., 2006) suggests an age for this geomagnetic feature that is somewhat closer to the reported age of the Laschamp excursion (Fig. 8). Although the geochronologic control is insufficient to assign an age to this geomagnetic feature, consideration of its stratigraphic position suggests that this feature is equivalent to the Laschamp excursion of other records.

Mineral Magnetic Zone Ages

In order to assign ages to our mineral magnetic zones (Table 1) we use the linear sedimentation rate age model

(0.54 mm yr⁻¹) used by Kaufman et al. (this volume). This sedimentation rate was constructed from a simple linear fit of four control points: 0 m = 0 ka, 11.3 m = 18.5 ka, 26.5 m = 41 ka, and 67 m = 127.7 ka. Although there are some discrepancies between this model and the ¹⁴C-based age model for the uppermost 20 m (Colman et al., this volume), a comparison of our magnetic data (K_{ARM}/K) and the $\delta^{18}O$ data of Bright et al. (2006) (both plotted using the linear sedimentation rate) with the $\delta^{18}O$ stack of Lisiecki and Raymo (2005) shows a relatively good correlation (Fig. 9). Rather than tuning the ages of our record to match the timing of global MIS stratigraphy, however, we use the linear-sedimentation-rate model to interpret our record in the context of regional and global climate without the circularity associated with a climatically tuned age model, as discussed by Kaufman et al. (this volume).

A linear sedimentation rate indicates that mineral magnetic zone I (0–9 m) represents the Holocene and latest Pleistocene (9 m = 17 ka). In zone II (9–18 m), high HIRM values (Fig. 10)

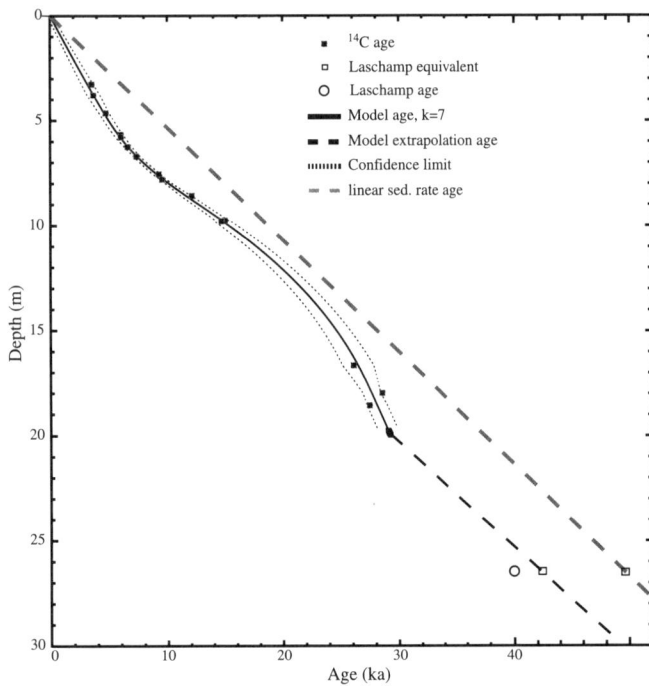

Figure 8. Age models for the uppermost 30 m of Bear Lake. The solid dark line is the ¹⁴C-based spline function of Colman et al. (this volume). The "k" value associated with the confidence limit refers to the number of spline functions used and controls the degree of smoothness. The dark dashed line is the extrapolation of the ¹⁴C age model through a U-series date at 67 m (127.7 ka from Colman et al., 2006). The light dashed line is the linear-sedimentation-rate model (0.54 mm yr⁻¹) used by Kaufman et al. (this volume). The open squares indicate the inferred age of our Laschamp equivalent feature for each age model. The open circle indicates the age of the Laschamp excursion (40.4 ± 2.0 ka) from Guillou et al. (2004). Considering the stratigraphic position of our Laschamp equivalent feature and the uncertainties associated with the different age models, it is likely that our feature formed around the time of the Laschamp excursion of other records.

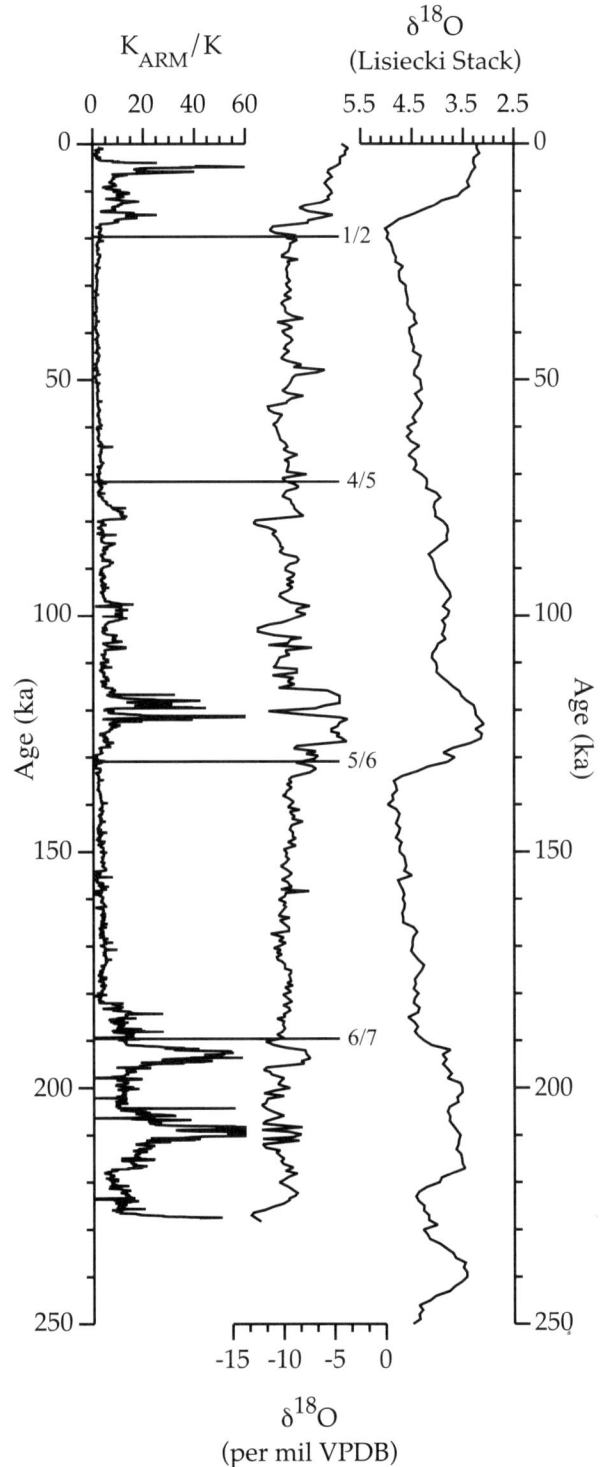

Figure 9. Comparison of K_{ARM}/K and $\delta^{18}O$ (Bright et al., 2006) from the Bear Lake record (using the linear sedimentation rate age model of Kaufman et al. [this volume]) to the $\delta^{18}O$ global stack of Lisiecki and Raymo (2005). The interpreted marine isotope stage (MIS) boundaries for the Bear Lake record (from Bright et al., 2006) are indicated. The comparison shows a relatively good correlation between the Bear Lake record and the global stack and suggests that the linear-sedimentation-rate age model is an appropriate first-order approximation for the age of Bear Lake sediments. VPDB—Vienna Pee Dee Belemnite.

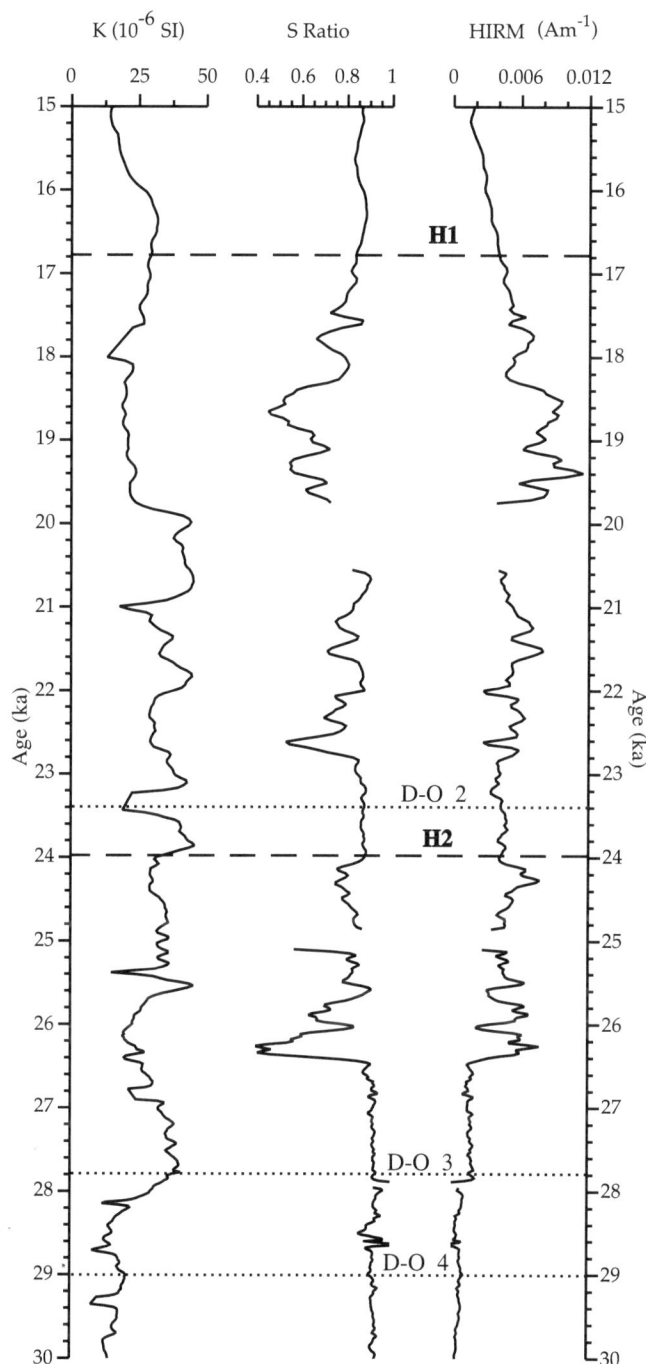

Figure 10. Magnetic properties (K, HIRM, and S ratio) versus age (age model from Colman et al., this volume), showing changes in the abundance of Fe-oxide minerals (Rosenbaum and Heil, this volume). The ages of Heinrich events 1–2 (Hemming, 2004, and references therein) and Dansgaard-Oeschger (D-O) oscillations 2–4 (Rahmstorf, 2003) are indicated with dashed lines and dotted lines, respectively. HIRM—hard isothermal remanent magnetization.

are interpreted by Rosenbaum and Heil (this volume) as glacial flour derived from the Uinta Mountains. The highest values occur around 10.8 m and are assigned an age of 20 ka, which is consistent with an age of 21 ± 3 ka assigned to the Pinedale glacial maximum in the Wind River Range (Pierce, 2004). The upper and lower boundaries for zone III (18–41 m) correspond to 33 ka and 76 ka, respectively. As discussed previously in this paper, the presence of a geomagnetic feature synchronous with the Laschamp excursion (ca. 40 ka) between 25.5 and 27.5 m in this zone suggests that the linear-sedimentation-rate age model does not provide the best fit for this interval (Fig. 8). Zone IV (41–61 m) represents most of MIS 5 (Bright et al., 2006) spanning the time interval of 76–113 ka. MIS 5e (ca. 117–131 ka; Lisiecki and Raymo, 2005) is represented by zone V (61–66 m) and spans the interval 113–122 ka. Zones VI and VII span MIS 6 and most of MIS 7, respectively (Bright et al., 2006). The upper boundary of zone VI (66 m) corresponds to 122 ka and the upper boundary of zone VII (at 97 m) corresponds to 180 ka. The extrapolated basal age of the GLAD800 core from Bear Lake is ca. 222 ka.

Millennial-Scale Variability during the Last Glacial Maximum

Hematite content (as indicated by HIRM and the S parameter) shows well-defined millennial-scale variability between 15 and 30 ka (Fig. 10). Here we use the ^{14}C-based age model of Colman et al. (this volume) because this model is more accurate in this interval than the linear-sedimentation-rate model used for the entire Bear Lake GLAD800 core. These variations during the last glacial maximum (LGM) are interpreted to reflect changes in the content of hematite-rich glacial flour derived from the Uinta Mountains (Rosenbaum and Heil, this volume). Changes in the amount of glacial flour are interpreted to reflect changes in the extent of glacier size in the Uinta Mountains. Using this interpretation, we compared our data from Bear Lake to the timing of North Atlantic Heinrich events 1–2 and Dansgaard-Oeschger (D-O) cycles, well-known millennial-scale climate phenomena observed during the LGM (Fig. 10). A link between alpine glaciation in the western United States and Heinrich events and D-O cycles has been made on the basis of evidence of episodic growth and retreat of alpine glaciers in the Rocky Mountains, the Cascade Range, and the Sierra Nevada (Clark and Bartlein, 1995; Benson et al., 2003). Given existing age constraints, these episodes are in phase with the growth and collapse of the Laurentide ice sheet. In addition, studies from the Great Basin show evidence of Heinrich events and D-O oscillations in the sediment records from several lakes (Benson et al., 1997, 1998, 2003; Benson, 1999; Zic et al., 2002).

North Atlantic Heinrich events are associated with periods of increased iceberg discharge (Heinrich, 1988) initiated by ice sheet instabilities following short-term growth of the Laurentide ice sheet (MacAyeal, 1993). Dansgaard-Oeschger oscillations are the result of either internal oscillations in Earth's climate system (i.e., the formation of North Atlantic Deep Water) or external forcings (i.e., changes in solar irradiance) (Broecker et al., 1990). Although the mechanism is not entirely understood, Benson et al. (2003) showed that the mean position of the polar jet stream (PJS) shifted between 35°N during D-O stadials (cold phase) and ≥43°N during interstadials (warm phase). Thompson et al. (1993) indicated a similar displacement of the PJS during the last glacial maximum. The position of the polar jet stream affects temperature and moisture, with cold, dry air north of the polar jet and cool, moist air south of the polar jet (Thompson et al., 1993). The central location of Bear Lake within the path of the polar jet stream may have made the lake sensitive to changes in the size of the Laurentide ice sheet. We speculate that changes in the position of the PJS, like those associated with D-O oscillations and Heinrich events, affected glaciation in the Uinta Mountains by altering both air temperature and the amount of moisture available for growth of the alpine glaciers. Although uncertainties in the age model of Colman et al. (this volume) are too large to make firm correlations, we note that, if the model is taken at face value, the major peaks in hematite content do not coincide with the timing of any Heinrich events or D-O cycles (Fig. 9). However, it appears that changes do occur on the order of 1–2 k.y., which is similar in duration to D-O cycles (Grootes and Stuiver, 1997). We emphasize that millennial-scale oscillations in the magnetic mineralogy of Bear Lake sediment during the LGM may reflect only regional climate changes within the Great Basin and not hemisphere-wide climatic events.

According to Bright et al. (2006), zone VI corresponds to MIS 6, the penultimate glacial period. Climatic correlations are not suggested for this zone because sediments from this interval do not have magnetite and hematite concentrations like those of zone II. In large part, this condition is related to preservation of Fe-oxide minerals in zone II and destruction of these minerals elsewhere. The high degree of preservation in zone II may reflect the absence of strongly reducing conditions due to fresh water (i.e., low sulfate content) and low productivity, or limited reduction related to rapid sedimentation linked to the influx of glacial flour. Perhaps the Bear River did not flow into the lake during MIS 6, thus preventing glacial flour from entering the lake, or other conditions that would have allowed the preservation of Fe-oxide minerals did not exist.

CONCLUSIONS

The sediments of Bear Lake show a complex record of environmental change. Variations in lake salinity caused by changing freshwater inputs and evaporative conditions associated with glacial and interglacial periods have affected the preservation and deposition of magnetic minerals. Preservation of Fe-oxide minerals during the LGM was likely related to rapid burial of glacial flour input from the Bear River. At other times, changes in the hydrologic balance of Bear Lake near marine isotope stage and substage boundaries led to significant changes in the sulfate concentrations of the lake waters, forming greigite during the fresher, colder stages, and pyrite in the more saline, warmer stages.

Preservation of Fe-oxide minerals in some intervals of the Bear Lake deposits, and varying degrees of post-depositional alteration in others, makes it impossible to establish a continuously reliable paleomagnetic record. Despite these complications, directional variations within an interval of relatively constant magnetic mineralogy (zone III) can be interpreted as a geomagnetic feature equivalent to the Laschamp excursion, thereby providing an important age constraint. Although the sediments of Bear Lake do not generally lend themselves to paleointensity studies because of the complex variations in magnetic minerals, normalized NRM intensities within zone III provide paleointensity information. Lows in NRM/K, NRM/ARM, and NRM/IRM values coincident with the directional variations support the interpretation of the Laschamp equivalent geomagnetic feature.

The sediment record from Bear Lake for the past 220,000 years is complex in both a paleomagnetic and environmental magnetic sense, but some intervals provide important information about climate change in the Great Basin and its relation to both regional and global temporal and spatial scales. Variations in the amount of glacially derived hematite (Fig. 10) delivered to Bear Lake from the Uinta Mountains (Rosenbaum and Heil, this volume) during the LGM occurred on millennial time scales, suggesting a potential link to millennial-scale internal oscillations of the Laurentide ice sheet and the resulting displacement of the polar jet stream. However, uncertainties in the age model prevent us from unequivocally correlating these magnetic features to Heinrich events or Dansgaard-Oeschger oscillations. Although the exact mechanisms of D-O oscillations are not entirely understood, they have a periodicity of ~1500 years (Grootes and Stuiver, 1997), which is strikingly similar to the duration of the hematite peaks in zone II from Bear Lake. At this point these links are speculative and are suggested only as feasible possibilities based on evidence from other lake records and glacial deposits within and around the Great Basin.

ACKNOWLEDGMENTS

We acknowledge the Limnological Research Center at the University of Minnesota for providing samples and the Institute for Rock Magnetism at the University of Minnesota for use of their lab and their help with the anisotropy of magnetic susceptibility analyses. We also acknowledge the thoughtful reviews of Jon Hagstrum and John Barron of the U.S. Geological Survey as well as the reviews and comments of Joseph Stoner, Stefanie Brachfeld, and one anonymous reviewer, all of which have made this paper a stronger contribution to this volume.

REFERENCES CITED

Barbetti, M.F., and McElhinny, M.W., 1972, Evidence for a geomagnetic excursion 30,000 yr B.P: Nature, v. 239, p. 327–330, doi: 10.1038/239327a0.
Benson, L., 1999, Records of millennial-scale climate change from the Great Basin of the Western United States, *in* Clark, P.U., Webb, R.S., and Keigwin, L.D., eds., Mechanisms of global climate change at millennial time scales: Washington, D.C., American Geophysical Union, Geophysical Monograph 112, p. 203–225.

Benson, L.V., Currey, D.R., Dorn, R.I., Lajoie, K.R., Oviatt, C.G., Robinson, S.W., Smith, G.I., and Scott, S., 1990, Chronology of expansion and contraction of four Great Basin lake systems during the past 35,000 years: Palaeogeography, Palaeoclimatology, Palaeoecology, v. 78, p. 241–286, doi: 10.1016/0031-0182(90)90217-U.

Benson, L., Burdett, J.W., Lund, S.P., Kashgarian, M., and Mensing, S., 1997, Nearly synchronous climate change in the Northern Hemisphere during the last glacial termination: Nature, v. 388, p. 263–265, doi: 10.1038/40838.

Benson, L.V., Lund, S.P., Burdett, J.W., Kashgarian, M., Rose, T.P., Smoot, J.P., and Schwartz, M., 1998, Correlation of late-Pleistocene lake-level oscillations in Mono Lake, California, with North Atlantic climate events: Quaternary Research, v. 49, p. 1–10, doi: 10.1006/qres.1997.1940.

Benson, L., Lund, S., Negrini, R., Linsley, B., and Zic, M., 2003, Response of North American Great Basin lakes to Dansgaard-Oeschger oscillations: Quaternary Science Reviews, v. 22, p. 2239–2251, doi: 10.1016/S0277-3791(03)00210-5.

Birdsey, P.W., 1989, The limnology of Bear Lake, Utah-Idaho, 1912–1988: A literature review: Salt Lake City, Utah Department of Natural Resources, Publication 89-5, 113 p.

Brachfeld, S.A., and Banerjee, S.K., 2000, A new high-resolution geomagnetic relative paleointensity record for the North American Holocene: A comparison of sedimentary and absolute intensity data: Journal of Geophysical Research, v. 105, n. B1, p. 821–834, doi: 10.1029/1999JB900365.

Bright, J., Kaufman, D.S., Forester, R.M., and Dean, W.E., 2006, A continuous 250,000 yr record of oxygen and carbon isotopes in ostracode and bulk-sediment carbonate from Bear Lake, Utah-Idaho: Quaternary Science Reviews, v. 25, p. 2258–2270, doi: 10.1016/j.quascirev.2005.12.011.

Broecker, W.S., Bond, G., Klas, M., Bonani, G., and Wolfi, W., 1990, A salt oscillator in the glacial northern Atlantic?: 1. The concept: Paleoceanography, v. 5, p. 469–477, doi: 10.1029/PA005i004p00469.

Canfield, D.E., and Berner, R.A., 1987, Dissolution and pyritization of magnetite in anoxic marine sediments: Geochimica et Cosmochimica Acta, v. 51, p. 645–659, doi: 10.1016/0016-7037(87)90076-7.

Canfield, D.E., Raiswell, R., and Bottrell, S., 1992, The reactivity of sedimentary iron minerals toward sulfide: American Journal of Science, v. 292, p. 659–683.

Channell, J.E.T., Hodell, D.A., and Lehman, B., 1997, Relative geomagnetic paleointensity and $\delta^{18}O$ at ODP Site 983 (Gardar Drift, North Atlantic) since 350 ka: Earth and Planetary Science Letters, v. 153, p. 103–118, doi: 10.1016/S0012-821X(97)00164-7.

Channell, J.E.T., Stoner, J.S., Hodell, D.A., and Charles, C.D., 2000, Geomagnetic paleointensity for the last 100 kyr from the sub-Antarctic South Atlantic: A tool for inter-hemispheric correlation: Earth and Planetary Science Letters, v. 175, p. 145–160, doi: 10.1016/S0012-821X(99)00285-X.

Clark, P.U., and Bartlein, P.J., 1995, Correlation of late Pleistocene glaciation in the western United States with North Atlantic Heinrich events: Geology, v. 23, p. 483–486, doi: 10.1130/0091-7613(1995)023<0483:COLPGI>2.3.CO;2.

Colman, S.M., Kaufman, D.S., Bright, J., Heil, C., King, J.W., Dean, W.E., Rosenbaum, J.G., Forester, R.M., Bischoff, J.L., Perkins, M., and McGeehin, J.P., 2006, Age models for a continuous 250-kyr Quaternary lacustrine record from Bear Lake, Utah-Idaho: Quaternary Science Reviews, v. 25, p. 2271–2282, doi: 10.1016/j.quascirev.2005.10.015.

Colman, S.M., Rosenbaum, J.G., Kaufman, D.S., Dean, W.E., and McGeehin, J.P., 2009, this volume, Radiocarbon ages and age models for the last 30,000 years in Bear Lake, Utah and Idaho, *in* Rosenbaum, J.G., and Kaufman, D.S., eds., Paleoenvironments of Bear Lake, Utah and Idaho, and its catchment: Geological Society of America Special Paper 450, doi: 10.1130/2009.2450(05).

Dean, W.E., 2009, this volume, Endogenic carbonate sedimentation in Bear Lake, Utah and Idaho over the last two glacial-interglacial cycles, *in* Rosenbaum, J.G., and Kaufman, D.S., eds., Paleoenvironments of Bear Lake, Utah and Idaho, and its catchment: Geological Society of America Special Paper 450, doi: 10.1130/2009.2450(07).

Dunlop, D., and Özdemir, Ö., 1997, Rock magnetism—Fundamentals and frontiers: New York, Cambridge University Press, 573 p.

Grayson, D.K., 1993, The desert's past: A natural pre-history of the Great Basin: Washington, D.C., Smithsonian Institution Press, 356 p.

Grootes, P.M., and Stuiver, M., 1997, Oxygen 18/16 variability in Greenland snow and ice with 10^{-3}- to 10^5-year time resolution: Journal of Geophysical Research, v. 102, n. C12, p. 26,455–26,470, doi: 10.1029/97JC00880.

Guillou, H., Singer, B.S., Laj, C., Kissel, C., Scaillet, S., and Jicha, B.R., 2004, On the age of the Laschamp geomagnetic excursion: Earth and Planetary Science Letters, v. 227, p. 331–343, doi: 10.1016/j.epsl.2004.09.018.

Guyodo, Y., and Valet, J.-P., 1996, Relative variations in geomagnetic intensity from sedimentary records: The past 200,000 years: Earth and Planetary Science Letters, v. 143, p. 23–36, doi: 10.1016/0012-821X(96)00121-5.

Heinrich, H., 1988, Origin and consequences of cyclic ice rafting in the northeast North Atlantic Ocean during the past 130,000 years: Quaternary Research, v. 29, p. 142–152, doi: 10.1016/0033-5894(88)90057-9.

Hemming, S.R., 2004, Heinrich events: Massive late Pleistocene detritus layers of the North Atlantic and their global climate imprint: Reviews of Geophysics, v. 42, RG1005, doi: 10.1029/2003RG000128.

Kaufman, D.S., Bright, J., Dean, W.E., Moser, K., Rosenbaum, J.G., Anderson, R.S., Colman, S.M., Heil, C.W., Jr., Jiménez-Moreno, G., Reheis, M.C., and Simmons, K.R., 2009, this volume, A quarter-million years of paleoenvironmental change at Bear Lake, Utah and Idaho, *in* Rosenbaum, J.G., and Kaufman D.S., eds., Paleoenvironments of Bear Lake, Utah and Idaho, and its catchment: Geological Society of America Special Paper 450, doi: 10.1130/2009.2450(14).

King, J.W., 1983, Geomagnetic secular variation curves for northeastern North America for the last 9,000 years B.P. [Ph.D. thesis]: Minneapolis, University of Minnesota, 195 p.

King, J.W., and Channell, J.E.T., 1991, Sedimentary magnetism, environmental magnetism, and magnetostratigraphy: Review of Geophysics, Supplement, p. 116–128.

King, J.W., and Peck, J., 2001, Use of paleomagnetism in studies of lake sediments, *in* Last, W.M., and Smol, J.P., eds., Basin analysis, coring, and chronological techniques, v. 1 of Tracking environmental change using lake sediments: Dordrecht, Kluwer Academic Publishers, p. 371–389.

King, J.W., Banerjee, S.K., Marvin, J., and Özdemir, Ö., 1982, A new rock-magnetic approach to selecting sediments for geomagnetic paleointensity studies: Application to paleointensity for the last 4000 years: Earth and Planetary Science Letters, v. 59, p. 404–419, doi: 10.1016/0012-821X(82)90142-X.

King, J.W., Banerjee, S.K., Marvin, J., and Lund, S., 1983, Use of small-amplitude paleomagnetic fluctuations for correlation and dating of continental climatic changes: Palaeogeography, Palaeoclimatology, Palaeoecology, v. 42, p. 167–183, doi: 10.1016/0031-0182(83)90043-3.

Laabs, B.J., and Kaufman, D.S., 2003, Quaternary highstands in Bear Lake Valley, Utah and Idaho: Geological Society of America Bulletin, v. 115, p. 463–478, doi: 10.1130/0016-7606(2003)115<0463:QHIBLV>2.0.CO;2.

Lehman, B., Laj, C., Kissel, C., Mazaud, A., Paterne, M., and Labeyrie, L., 1996, Relative changes of the geomagnetic field intensity during the last 280 kyr obtained from piston cores in the Açores area: Physics of the Earth and Planetary Interiors, v. 93, p. 269–284, doi: 10.1016/0031-9201(95)03070-0.

Lewis, C.F.M., Heil, C.W., Jr., Hubeny, J.B., King, J.W., Moore, T.C., Jr., and Rea, D.K., 2007, The Stanley unconformity in Lake Huron basin: Evidence for a climate-driven closed lowstand about 7500 ^{14}C BP, with similar implications for the Chippewa lowstand in Lake Michigan basin: Journal of Paleolimnology, v. 37, p. 435–452, doi: 10.1007/s10933-006-9049-y.

Liddicoat, J.C., 1992, Mono Lake Excursion in Mono Basin, California, and at Carson Sink and Pyramid Lake, Nevada: Geophysical Journal International, v. 108, p. 442–452, doi: 10.1111/j.1365-246X.1992.tb04627.x.

Liddicoat, J.C., and Coe, R.S., 1979, Mono Lake Excursion: Journal of Geophysical Research, v. 84, p. 261–271, doi: 10.1029/JB084iB01p00261.

Lisiecki, L.E., and Raymo, M.E., 2005, A Pliocene-Pleistocene stack of 57 globally distributed $\delta^{18}O$ records: Paleoceanography, v. 20, p. PA1003, doi: 10.1029/2004PA001071.

Lund, S.P., 1996, A comparison of Holocene paleomagnetic secular variation records from North America: Journal of Geophysical Research, v. 101, p. 8007–8024, doi: 10.1029/95JB00039.

Lund, S.P., Liddicoat, J.C., Lajoie, K.R., Henyey, T.L., and Robinson, S.W., 1988, Paleomagnetic evidence for long-term (10^4 year) memory and periodic behavior in the Earth's core dynamo process: Geophysical Research Letters, v. 15, p. 1101–1104, doi: 10.1029/GL015i010p01101.

Lund, S.P., Acton, G.D., Clement, B., Okada, M., and Williams, T., 2001, Brunhes Chron magnetic field excursions recovered from Leg 172 sediments, *in* Keigwin, L.D., Rio, D., Acton, G.D., and Arnold, E., eds., Proceedings of the Ocean Drilling Program, Scientific Results, v. 172, p. 1–18.

MacAyeal, D.R., 1993, Growth/purge oscillations of the Laurentide ice sheet as a cause of the North Atlantic's Heinrich events: Paleoceanography, v. 8, p. 775–784, doi: 10.1029/93PA02200.

Mazaud, A., 2005, User-friendly software for vector analysis of the magnetization of long sediment cores: Geochemistry Geophysics Geosystems, v. 6, doi: 10.1029/2005GC001036.

McCalpin, J.P., 1993, Neotectonics of the northeastern Basin and Range margin, western USA: Zeitschrift für Geomorphologie, N.F., Suppl. Bd., v. 94, p. 137–157.

Meynadier, L., Valet, J.-P., Weeks, R., Shackleton, N.J., and Lee Hagee, V., 1992, Relative geomagnetic intensity of the field during the last 140 ka: Earth and Planetary Science Letters, v. 114, p. 39–57, doi: 10.1016/0012-821X(92)90150-T.

Morrison, R.B., 1991, Quaternary stratigraphic, hydrologic, and climate history of the Great Basin, with emphasis on Lake Lahontan, Bonneville, and Tecopa, *in* Morrison, R.B., ed., Quaternary nonglacial geology: Conterminous U.S.: Boulder, Colorado, Geological Society of America, Geology of North America, v. K-2, p. 283–320.

Negrini, R.M., Davis, J.O., and Verosub, K.L., 1984, Mono Lake geomagnetic excursion found at Summer Lake, Oregon: Geology, v. 12, p. 643–646, doi: 10.1130/0091-7613(1984)12<643:MLGEFA>2.0.CO;2.

Oviatt, C.G., Currey, D.R., and Sack, D., 1992, Radiocarbon chronology of Lake Bonneville, eastern Great Basin, USA: Palaeogeography, Palaeoclimatology, Palaeoecology, v. 99, p. 225–241, doi: 10.1016/0031-0182(92)90017-Y.

Petersen, N., Von Dobonek, T., and Vali, H., 1986, Fossil bacterial magnetite in deep-sea sediments from the South Atlantic Ocean: Nature, v. 320, p. 611–615, doi: 10.1038/320611a0.

Pierce, K.L., 2004, Pleistocene glaciations of the Rocky Mountains, *in* Gillespie, A.R., Porter, S.C. and Atwater, B.F., eds., The Quaternary Period in the United States: Developments in Quaternary Science, v. 1: Elsevier, Amsterdam, p. 63–76.

Pike, C., Roberts, A., Dekkers, M., and Verosub, K., 2001, An investigation of multi-domain hysteresis mechanisms using FORC diagrams: Physics of the Earth and Planetary Interiors, v. 126, p. 11–25, doi: 10.1016/S0031-9201(01)00241-2.

Poag, C.W., and Foss, G., 1985, Explanatory notes, *in* Graciansky, P.C. de, Poag, C.W., et al., 1985, Initial Reports of the Deep Sea Drilling Project, v. 80: Washington, D.C., U.S. Government Printing Office, p. 15–31.

Rahmstorf, S., 2003, Timing of abrupt climate change: A precise clock: Geophysical Research Letters, v. 30, 1510, doi: 10.1029/2003GL017115.

Reynolds, R.L., and King, J.W., 1995, Magnetic records of climate change: Reviews of Geophysics, U.S. National Report to International Union of Geodesy and Geophysics 1991–1994, p. 101–110.

Reynolds, R.L., and Rosenbaum, J.G., 2005, Magnetic mineralogy of sediments in Bear Lake and its watershed, Utah, Idaho, and Wyoming: Support for paleoenvironmental and paleomagnetic interpretations: U.S. Geological Survey Open-File Report 2005-1406, 14 p.

Reynolds, R.L., Tuttle, M.L., Rice, C., Fishman, N.S., Karachewski, J.A., and Sherman, D., 1994, Magnetization and geochemistry of greigite-bearing Cretaceous strata, North Slope basin, Alaska: American Journal of Science, v. 294, p. 485–528.

Reynolds, R.L., Rosenbaum, J.G., Mazza, N., Rivers, W., and Luiszer, F., 1998, Sediment magnetic data (83 to 18 m depth) and XRF geochemical data (83 to 32 m depth) from lacustrine sediment in core OL-92 from Owens lake, California, *in* Bischoff, J.L., ed., A high-resolution study of climate proxies in sediments from the last interglaciation at Owens Lake, California: Core OL-92: U.S. Geological Survey Open-File Report 98-132, 20 p.

Reynolds, R.L., Sweetkind, D.S., and Axford, Y., 2001, An inexpensive magnetic mineral separator for fine-grained sediment: U.S. Geological Survey Open-File Report 01-281, 7 p.

Roberts, A.P., 1995, Magnetic properties of sedimentary greigite (Fe$_3$S$_4$): Earth and Planetary Science Letters, v. 134, p. 227–236, doi: 10.1016/0012-821X(95)00131-U.

Roberts, A.P., 2006, High-resolution magnetic analysis of sediment cores: Strengths, limitations and strategies for maximizing the value of long-core magnetic data: Physics of the Earth and Planetary Interiors, v. 156, p. 162–178, doi: 10.1016/j.pepi.2005.03.021.

Roberts, A., Pike, C., and Verosub, K., 2000, First-order reversal curve diagrams: A new tool for characterizing the magnetic properties of natural samples: Journal of Geophysical Research, v. 105, B12, p. 28,461–28,475, doi: 10.1029/2000JB900326.

Roberts, A.P., Jiang, W.T., Florindo, F., Horng, C.S., and Laj, C., 2005, Assessing the timing of greigite formation and the reliability of the Upper Olduvai polarity transition record from the Crostolo River, Italy: Geophysical Research Letters, v. 32, p. 1–4.

Rosenbaum, J.G., and Heil, C.W., Jr., 2009, this volume, The glacial/deglacial history of sedimentation in Bear Lake, Utah and Idaho, *in* Rosenbaum, J.G., and Kaufman D.S., eds., Paleoenvironments of Bear Lake, Utah and Idaho, and its catchment: Geological Society of America Special Paper 450, doi: 10.1130/2009.2450(11).

Rosenbaum, J., Reynolds, R., Smoot, J., and Meyer, R., 2000, Anisotropy of magnetic susceptibility as a tool for recognizing core deformation: Reevaluation of the paleomagnetic record of Pleistocene sediments from the drill hole OL-92, Owens Lake, California: Earth and Planetary Science Letters, v. 178, p. 415–424, doi: 10.1016/S0012-821X(00)00077-7.

Sagnotti, L., Roberts, A.P., Weaver, R., Verosub, K.L., Florindo, F., Pike, C.R., Clayton, T., and Wilson, G.S., 2005, Apparent magnetic polarity reversals due to remagnetization resulting from the late diagenetic growth of greigite from siderite: Geophysical Journal International, v. 160, p. 89–100, doi: 10.1111/j.1365-246X.2005.02485.x.

Snowball, I.F., 1997, The detection of single-domain greigite (Fe$_3$S$_4$) using rotational remanent magnetization (RRM) and the effective gyro field (*Bg*): Mineral magnetic and palaeomagnetic applications: Geophysical Journal International, v. 130, p. 704–716, doi: 10.1111/j.1365-246X.1997.tb01865.x.

Snowball, I., and Torii, M., 1999, Incidence and significance of magnetic iron sulphides in Quaternary sediments and soils, *in* Maher, B.A., and Thompson, R., eds., Quaternary climates, environments, and magnetism: Cambridge, UK, Cambridge University Press, p. 199–231.

Stoner, J.S., Channell, J.E.T., and Hillaire-Marcel, C., 1998, A 200 ka geomagnetic chronostratigraphy for the Labrador Sea: Indirect correlation of the sediment record to SPECMAP: Earth and Planetary Science Letters, v. 159, p. 165–181, doi: 10.1016/S0012-821X(98)00069-7.

Stoner, J.S., Channell, J.E.T., Hillaire-Marcel, C., and Kissel, C., 2000, Geomagnetic paleointensity and environmental record from Labrador Sea core MD95-2024: Global marine sediment and ice core chronostratigraphy for the last 110 kyr: Earth and Planetary Science Letters, v. 183, p. 161–177, doi: 10.1016/S0012-821X(00)00272-7.

Stoner, J.S., Laj, C., Channell, J.E.T., and Kissel, C., 2002, South Atlantic and North Atlantic geomagnetic paleointensity stacks (0–80 ka): Implications for interhemispheric correlation: Quaternary Science Reviews, v. 21, p. 1141–1151, doi: 10.1016/S0277-3791(01)00136-6.

Thompson, F., and Oldfield, F., 1986, Environmental magnetism: London, Allen and Unwin, 227 p.

Thompson, R., 1984, A global review of paleomagnetic results for wet lake sediments, *in* Haworth, E.Y., and Lund, J.W.G., eds., Lake sediments and environmental history: Minneapolis, University of Minnesota Press, p. 145–164.

Thompson, R., Whitlock, C., Bartlein, P.J., Harrison, S.P., and Spaulding, W.G., 1993, Climate changes in the Western United States since 18,000 yr B.P., *in* Wright, H.E., Jr., Kutzbach, J.E., Webb, T., III, Ruddiman, W.F., Street-Perrott, F.A., and Bartlein, P.J., eds., Global climates since the Last Glacial Maximum: Minneapolis, University of Minnesota Press, p. 468–513.

Verosub, K.L., and Roberts, A.P., 1995, Environmental magnetism: Past, present, and future: Journal of Geophysical Research, v. 100, n. B2, p. 2175–2192, doi: 10.1029/94JB02713.

Winklhofer, M., and Zimanyi, G.T., 2005, Extracting the intrinsic switching field distribution in perpendicular media: A comparative analysis, http://arxiv.org/abs/cond-mat/0509074 (accessed September 2005).

Yamazaki, T., and Ioka, N., 1994, Long-term secular variation of the geomagnetic field during the last 200 kyr recorded in sediment cores from the equatorial Pacific: Earth and Planetary Science Letters, v. 128, p. 527–544, doi: 10.1016/0012-821X(94)90168-6.

Zic, M., Negrini, R.M., and Wigand, P.E., 2002, Evidence of synchronous climate change across the Northern Hemisphere between the North Atlantic and the northwestern Great Basin, United States: Geology, v. 30, p. 635–638, doi: 10.1130/0091-7613(2002)030<0635:EOSCCA>2.0.CO;2.

MANUSCRIPT ACCEPTED BY THE SOCIETY 15 SEPTEMBER 2008

The Geological Society of America
Special Paper 450
2009

A *quarter-million years of paleoenvironmental change at Bear Lake, Utah and Idaho*

Darrell S. Kaufman
Jordon Bright
Department of Geology, Box 4099, Northern Arizona University, Flagstaff, Arizona 86011, USA

Walter E. Dean
Joseph G. Rosenbaum
U.S. Geological Survey, Box 25046, MS 980 Federal Center, Denver, Colorado 80225, USA

Katrina Moser
University of Western Ontario, Department of Geography, London, Ontario N5Y 2S9, Canada

R. Scott Anderson
Center for Environmental Sciences and Education, Northern Arizona University, Flagstaff, Arizona 86011, USA

Steven M. Colman
Large Lakes Observatory and Department of Geological Sciences, University of Minnesota, Duluth, Minnesota 02543, USA

Clifford W. Heil Jr.
School of Oceanography, University of Rhode Island, South Ferry Road, Narragansett, Rhode Island 02882, USA

Gonzalo Jiménez-Moreno
Center for Environmental Sciences and Education, Northern Arizona University, Flagstaff, Arizona 86011, USA

Marith C. Reheis
Kathleen R. Simmons
U.S. Geological Survey, Box 25046, MS 980 Federal Center, Denver, Colorado 80225, USA

ABSTRACT

A continuous, 120-m-long core (BL00-1) from Bear Lake, Utah and Idaho, contains evidence of hydrologic and environmental change over the last two glacial-interglacial cycles. The core was taken at 41.95°N, 111.31°W, near the depocenter of the 60-m-deep, spring-fed, alkaline lake, where carbonate-bearing sediment has accumulated continuously. Chronological control is poor but indicates an average sedimentation rate of 0.54 mm yr^{-1}. Analyses have been completed at multi-centennial to millennial scales, including (in order of decreasing temporal resolution) sediment magnetic properties,

Kaufman, D.S., Bright, J., Dean, W.E., Rosenbaum, J.G., Moser, K., Anderson, R.S., Colman, S.M., Heil, C.W., Jr., Jiménez-Moreno, G., Reheis, M.C., and Simmons, K.R., 2009, A quarter-million years of paleoenvironmental change at Bear Lake, Utah and Idaho, *in* Rosenbaum, J.G., and Kaufman, D.S., eds., Paleoenvironments of Bear Lake, Utah and Idaho, and its catchment, Geological Society of America Special Paper 450, p. 311–351, doi: 10.1130/2009.2450(14).

oxygen and carbon isotopes on bulk-sediment carbonate, organic- and inorganic-carbon contents, palynology; mineralogy (X-ray diffraction), strontium isotopes on bulk carbonate, ostracode taxonomy, oxygen and carbon isotopes on ostracodes, and diatom assemblages. Massive silty clay and marl constitute most of the core, with variable carbonate content (average = 31 ± 19%) and oxygen-isotopic values ($\delta^{18}O$ ranging from −18‰ to −5‰ in bulk carbonate). These variations, as well as fluctuations of biological indicators, reflect changes in the water and sediment discharged from the glaciated headwaters of the dominant tributary, Bear River, and the processes that influenced sediment delivery to the core site, including lake-level changes. Although its influence has varied, Bear River has remained a tributary to Bear Lake during most of the last quarter-million years. The lake disconnected from the river and, except for a few brief excursions, retracted into a topographically closed basin during global interglaciations (during parts of marine isotope stages 7, 5, and 1). These intervals contain up to 80% endogenic aragonite with high $\delta^{18}O$ values (average = −5.8 ± 1.7‰), indicative of strongly evaporitic conditions. Interglacial intervals also are dominated by small, benthic/tychoplanktic fragilarioid species indicative of reduced habitat availability associated with low lake levels, and they contain increased high-desert shrub and *Juniperus* pollen and decreased forest and forest-woodland pollen. The $^{87}Sr/^{86}Sr$ values (>0.7100) also increase, and the ratio of quartz to dolomite decreases, as expected in the absence of Bear River inflow. The changing paleoenvironments inferred from BL00-1 generally are consistent with other regional and global records of glacial-interglacial fluctuations; the diversity of paleoenvironmental conditions inferred from BL00-1 also reflects the influence of catchment-scale processes.

INTRODUCTION

Bear Lake, Utah and Idaho, is one of the longest-lived extant lakes on the North American continent. The half-graben that contains the lake is filled with an estimated 3 km of sediment (Evans et al., 2003). High-resolution seismic-reflection profiles of the upper 250 m reveal well-layered sediment with few unconformities (Colman, 2006). The upper part of the sedimentary sequence was recovered in a 120-m-long composite core drilled in 2000 (BL00-1; Dean et al., 2002). Sediment deposited in Bear Lake is strongly influenced by variable production of endogenic carbonate within the lake, by fluctuating input of fluvial and glacial-fluvial products from its headwaters in the Uinta Mountains and local catchment, and by periodic retraction of the lake into a topographically closed basin. These, in turn, are influenced by climatic and non-climatic processes. The extent to which changes in the lacustrine deposits can accurately be ascribed to particular causes depends on a multi-parameter investigation of its physical, chemical, and biological composition. The purpose of this chapter is to synthesize the available data on BL00-1, and to summarize our current understanding of paleoenvironmental change over the last two glacial-interglacial cycles at Bear Lake.

To date, most analyses of BL00-1 are based on sampling on the scale of decimeters (centuries) to meters (millennia). More detailed, centimeter-scale analyses have been completed on shorter cores from the lake (see overview of the lake-coring campaign by Rosenbaum and Kaufman, this volume). Our interpretations of BL00-1 rely heavily on analyses of these shorter cores, which in some cases benefited from parameters that were not measured in BL00-1 (e.g., elemental geochemistry; Rosenbaum et al., this volume). They also rely on an understanding of the modern composition and fluxes of sediment and water to the lake (e.g., Dean et al., 2007; Bright, this volume, Chapter 4). The rerouting of Bear River water into Bear Lake ca. 1912 had a pervasive effect on the lake. This event affords a whole-lake experiment on the response of lake sediment to a major hydrological change. The reader is referred to the other chapters and journal articles for a more detailed account of the methods and interpretations that are summarized in this chapter. The other works place the BL00-1 core site into a lake- and drainage-basin-wide context.

Setting

Bear Lake is located in an intermontane basin straddling the northeastern Great Basin and the Rocky Mountain physiogeographic provinces along the Utah-Idaho state border (Fig. 1). The primary surface-water inflow to Bear Lake Valley is the Bear River, which drains the northwest sector of the Uinta Mountains southeast of the lake. Within historical times, the river was not connected directly to the lake, but bypassed the lake en route to Great Salt Lake, located 100 km southwest of Bear Lake. When separated from Bear River, the primary surface inflows are streams that drain eastward from the Bear River Range, which are fed by springs that emerge from cavernous Paleozoic carbonate rocks. This "local" surface catchment area is relatively small, encompassing ~1300 km², or about five times the surface area of the lake (280 km²). When Bear River is a tributary to the lake, the drainage basin area expands by a factor of six. Reheis et al. (this volume) discuss mechanisms that might have caused Bear River to swing into and out of Bear Lake. Bear Lake Valley currently contains an alkaline lake (Bear Lake) that is confined to the southern end of the valley, with an overflow threshold at an elevation of 1805 m above sea level (asl). The lake is ~30 km long and 10 km wide, with a maximum depth of 63 m. The physical, chemical, and biological limnology of the lake is reviewed by Dean et al. (this volume).

At present, Bear Lake is nearly in hydrologic balance, and only a small increase in effective moisture would be required to cause the lake to overflow (Bright et al., 2006; Bright, this volume, Chapter 4). Intermittent overflow is also indicated by Lamarra et al.'s (1986) hydrological balance modeling, which showed that, without the diversion of Bear River, Bear Lake would have exceeded its threshold ~24% of the time during the 60 years prior to 1984. Furthermore, McConnell et al. (1957) provided historical reports

of northward drainage, and Reheis et al. (this volume) described north-trending channels that were active during the Holocene. With a transgression of just a few meters, the northern shoreline of Bear Lake would capture the channel of Bear River. Once the shoreline intersected the river, surface-water inflow to the lake would more than double, and the water residence time would be reduced by an order of magnitude. To disconnect the river from the lake requires the lake shoreline to regress south of where the river enters the valley. Climate-induced desiccation could cause a regression, as could erosion of the lake outlet where it previously fed now-abandoned, north-flowing channels. If the elevation of this outlet lowered, then lake level would lower despite inflow of the Bear River. In summary, the lake can alternate between topographically open (overflowing) and closed states. The lake may overflow without input from Bear River, but when the lake expands, it necessarily is

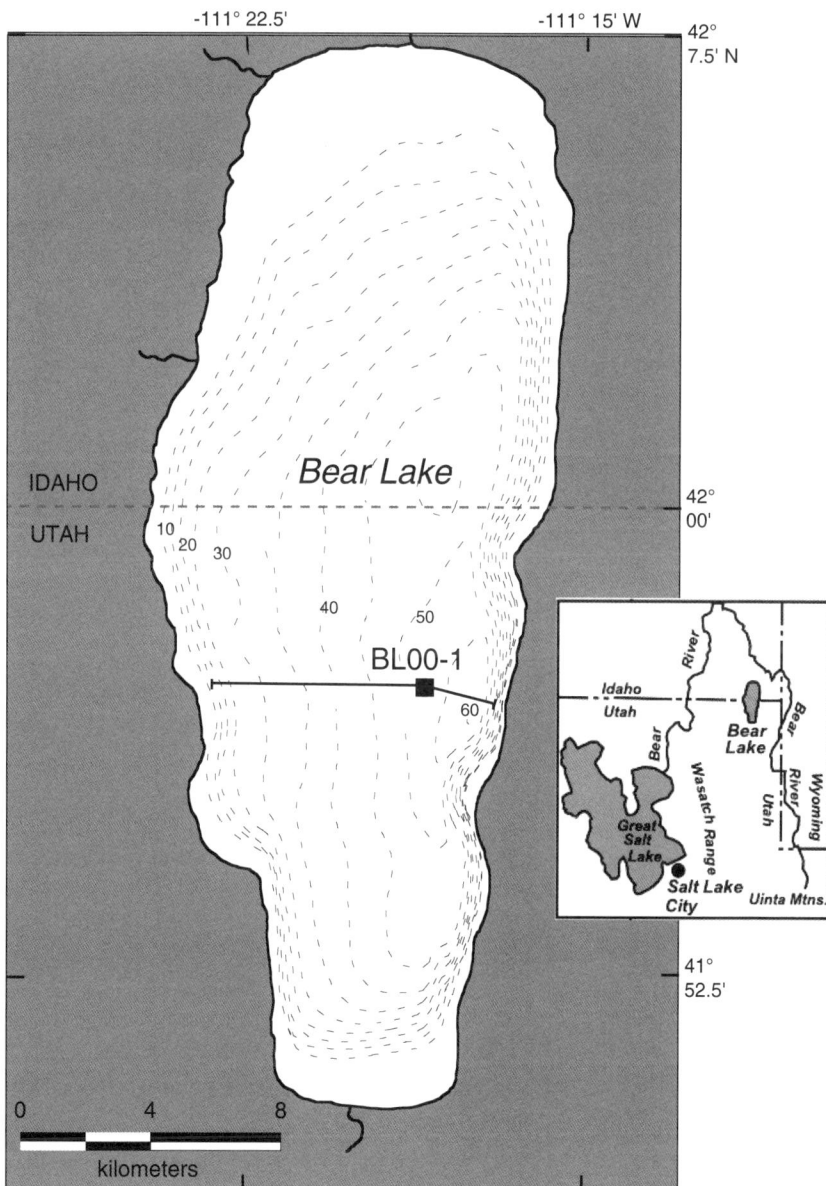

Figure 1. Location map showing Bear Lake and BL00-1 core site. Solid line is track of acoustic profile shown in Figure 2. Bathymetry (dotted lines) in meters from Denny and Colman (2003).

connected with the river. Lake-level changes in Bear Lake Valley are controlled by climatic changes as well as tectonic effects and the dynamics of erosion and deposition of the outlet stream that control the elevation of the basin threshold.

The core site (BL00-1; 41.9517°N, 111.3083°W) is in 54.8 m water depth (relative to full lake [1805 m asl]) near the lake depocenter, where sedimentation rates are highest. Acoustic subbottom profiles show that the stratigraphic units are laterally continuous and thin westward from the core site (Fig. 2; Colman, 2006). The site was located ~1.5 km west of the deepest part of the lake with the intent of reducing the impact of subaqueous mass wasting associated with the steep fault-bounded eastern margin of the lake. Two holes offset by a few meters were drilled by the GLAD800 (Global Lake Drilling to 800 m) platform during its trial engineering tests in September 2000 (http://dosecc.org/html/utah_lakes.html; Dean et al., 2002). Continuous coring in 2- or 3-m-long segments produced ~100 m of sediment from one hole (BL00-1D) and 121.07 m from the second (BL00-1E), both with essentially 100% recovery. Core breaks were offset vertically to provide a continuous record. Fluctuations in magnetic susceptibility were used to correlate between the holes, resulting in a composite record (hereafter, "BL00-1") based on the depth scale of BL00-1E.

SUMMARY OF PRIMARY FINDINGS

Geochronology

The currently accepted age model for BL00-1 was developed by Colman et al. (2006) (Fig. 3). The model is based on ^{14}C ages from shorter cores (Colman et al., this volume) projected onto the upper part of BL00-1 on the basis of correlations by magnetic susceptibility and other stratigraphic markers. Numerical age control beyond the range of ^{14}C dating includes one magnetic excursion correlated with the Laschamp excursion (Heil et al., this volume), and a single uranium-series age on aragonite from the last interglaciation (Colman et al., 2006). Amino acid racemization and tephrochronology support the overall sedimentation rates determined by the other three techniques (Colman et al., 2006). In addition, peaks and troughs in the carbonate content and stable isotope composition of BL00-1 can be correlated with those of the well-dated oxygen isotope record of vein calcite from Devils Hole (Winograd et al., 1992), located ~700 km southwest of Bear Lake. The correlations provide the only available age control for the lower half of the core, and they impose an age for Termination II that is ~12,000 years older than that inferred for the global marine oxygen isotope record. As emphasized by Colman et al. (2006), tuning the BL00-1 chronology to Devils Hole limits our ability to infer the timing of events from BL00-1 independently of the assumption that climate change is registered simultaneously at Bear Lake and Devils Hole.

To avoid the limitations of an age model predicated on climate correlations, for this synthesis we adopt an age model based on four secure ages only, and the assumption of a linear sedimentation rate (Fig. 3). The four points used to determine the sedimentation rate are (1) the surface (0 m below lake floor [blf] = 0 yr = A.D. 2000); (2) the prominent peak in sediment magnetic properties, which has been ascribed to glacier flour production (Rosenbaum and Heil, this volume) and dated by [14]C (Colman et

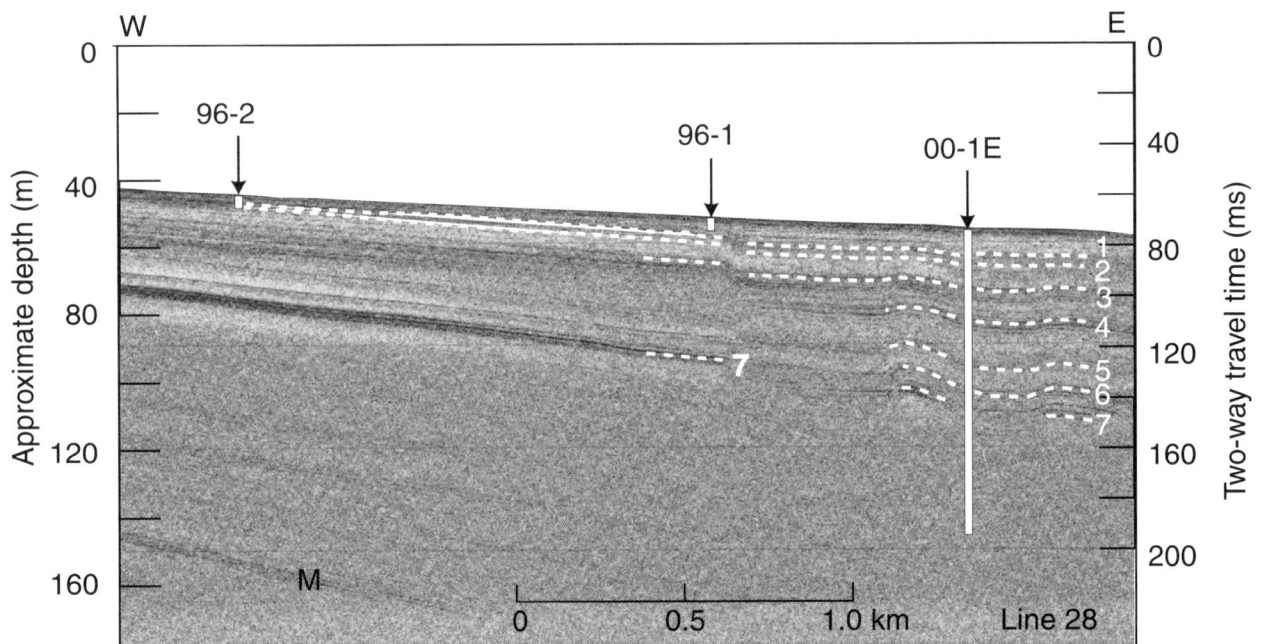

Figure 2. Acoustic profile along section line through core site BL00-1E (from Colman, 2006). Numbered dashed lines mark prominent reflectors. Approximate depths are calculated from two-way travel time, assuming a sound velocity of 1500 m s[-1]. 96-2 and 96-1 are locations of gravity cores collected in 1996 and discussed elsewhere in this volume. M—multiple reflection.

al., this volume) (11.38 m blf = 18.52 cal ka); (3) the correlated Laschamp excursion (26.5 m blf = 41 ka; Heil et al., this volume); and (4) the uranium-series age (67 m blf = 127.7 ka; Colman et al., 2006). A least-squares linear regression with y-intercept forced through zero results in an average sedimentation rate of 0.54 mm yr^{-1} over the upper half of the core (r^2 = 0.990), which we extrapolate over the lower half. This rate is identical to the average rate over the entire core based on the original age model of Colman et al. (2006) and, like the original model, is supported by the available tephrochronology.

The ages based on this simple linear model are approximate. The average absolute difference between the Devils-Hole-tuned and the linear age models is 4900 years as evaluated at 1 m intervals, with the largest differences (up to 12,000 yr) centered on 80 m blf. This difference is less than the average age uncertainty of ± 11,000 years for the original age model based on the 95% confidence intervals of the spline-fitting procedure used by Colman et al. (2006). We emphasize the importance of the broad age uncertainties for BL00-1, even though errors are not stated along with the approximate ages cited in this chapter. This chapter deals with trends over the entire 120-m-long core based on an approximate time scale with uncertainties of several millennia. References

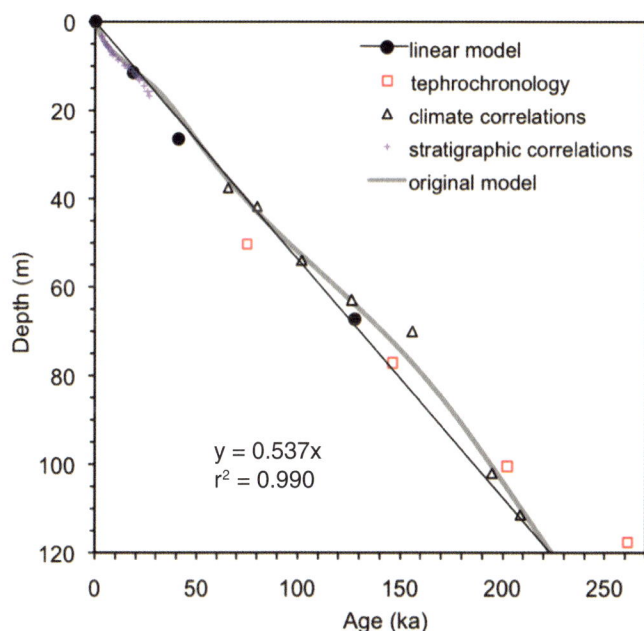

Figure 3. Geochronology of core BL00-1. The age model used in this chapter is based on the least-squares linear regression through the four points from the upper half of the core (solid circles) and is independent of any assumptions about regional or global climate correlations (see text for discussion). The original age model of Colman et al. (2006; thick gray curve) was tuned to correlate with the chronology from Devils Hole (triangles). Tephra ages (squares) are discussed by Colman et al. (2006); stratigraphic correlations based on ^{14}C ages from piston cores (small plus signs) over the upper 20 m are described by Colman et al. (this volume), and are not used in this chapter.

to marine oxygen isotope stages (MIS) are for convenience and do not imply strict correlations to the record of global ice volume.

Colman et al. (this volume) present a detailed age model for the top 20 m of BL00-1 (last 30,000 years), which is well constrained by ^{14}C in other piston cores and transferred to BL00-1 on the basis of correlations of stratigraphic horizons. The ^{14}C chronology shows that the average linear sedimentation rate is too low over the Holocene (Fig. 3); ages generated by the linear model are as much as 5000 years too old during the early Holocene (5–10 m blf). The linear model coincides with the ^{14}C chronology during the last local glacial maximum (11 m blf), but is 7000 years too old at the depth of the presumed Laschamp excursion (26.5 m blf).

Lithostratigraphy

The cores were curated at the Limnological Research Center (LRC), University of Minnesota, where they were processed according to standard initial core description (ICD) procedure (http://lrc.geo.umn.edu/corefac-icd.htm). The results of whole-core logging of porosity, bulk density, magnetic susceptibility, and P-wave velocity are available digitally, along with visual descriptions of the lithostratigraphy and photographs of each split-core segment (http://www.ngdc.noaa.gov/mgg/curator/curator.html). Smear slides were made at levels where lithology changed, and the observations were incorporated into the descriptions of the cores.

Briefly, the sediment consists mainly of massive gray to greenish-gray silty clay with carbonate contents ranging from calcareous silty clay (<30% CaCO$_3$) to marl (>30% CaCO$_3$) (Fig. 4). Most of the CaCO$_3$ is low-Mg calcite, but two of the marl units consist almost entirely of aragonite, and a third is aragonite-bearing. Most units contain centimeter-scale bands distinguished by degree of bioturbation or color, which probably reflect differences in the content of sulfide minerals, organic matter, and CaCO$_3$ abundance. The sedimentary sequence is subdivided into seven primary units (Fig. 4), which roughly coincide with the mineral magnetic zonation of Heil et al. (this volume): Unit 1 (from the base of the core to 101 m blf) is marl, composed mostly of aragonite near the top of the unit (103.1–101.6 m blf), with calcite below. Unit 2 (101–67 m blf) is calcareous silty clay. Unit 3 (67–61 m blf) is aragonitic marl. Unit 4 (61–40 m blf) is marl. Unit 5 (40–17 m blf) is calcareous clay. Unit 6 (17–10 m blf) is reddish-gray silty clay. Unit 7 (10–0 m blf) is light-gray aragonitic marl.

Organic and Inorganic Carbon

A total of 334 samples, at an average spacing of ~36 cm through core BL00-1, were analyzed by coulometry (e.g., Engleman et al., 1985) to measure the total carbon and inorganic carbon (IC) contents. Organic carbon (OC) was calculated as the difference between the two values, and weight percent CaCO$_3$ was calculated by dividing IC by the fraction of carbon in CaCO$_3$ (0.12). The same technique was used by Dean (this volume) and Dean et al. (2006, 2007) to analyze sediment from shorter piston cores. The results are tabulated in Appendix A and are shown in

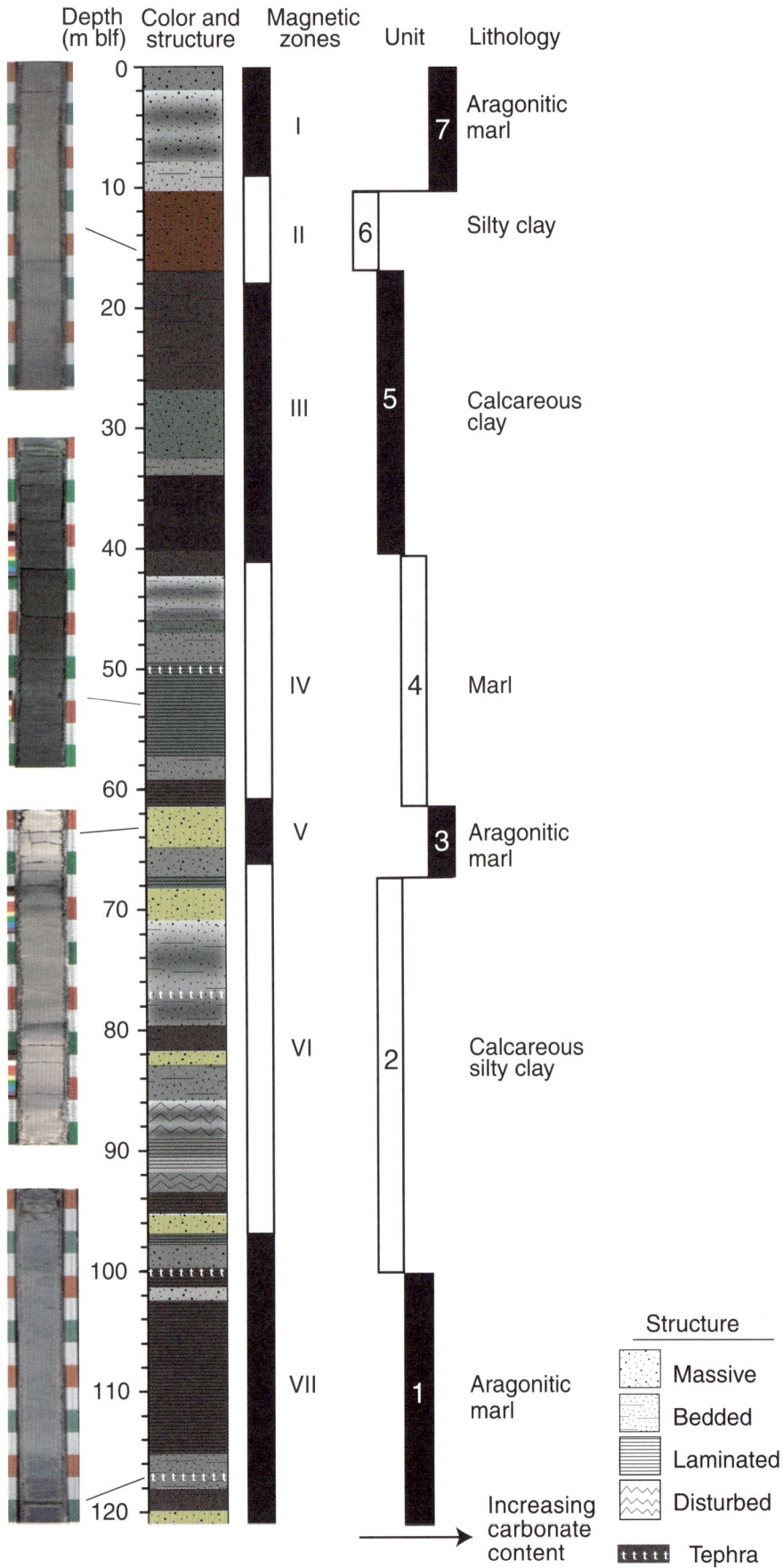

Figure 4. Lithostratigraphy of core BL00-1 dividing the core into seven lithologic units as summarized from the initial core descriptions (http://www.ngdc.noaa.gov/mgg/curator/curator.html). Depths are in meters below lake floor (m blf). Core photographs of four selected colors and features are shown on the left. Colors on lithologic log, including shades of gray, denote mud colors, but are not accurate representations. Magnetic zones (I–VII) are based on mineral magnetic properties discussed by Heil et al. (this volume).

Figure 5, together with the results of other analyses, where the sample depth has been transferred to age based on a linear sedimentation rate of 0.54 mm yr^{-1} (Fig. 3).

The abundance of OC in Bear Lake sediment reflects the flux of organic matter delivered to the lake and produced within the water column, relative to the rate at which it is decomposed. The CaCO$_3$ content is controlled primarily by the production rate of endogenic carbonate, and secondarily by the flux of allogenic carbonate delivered to the lake, reworking of previously deposited carbonate, carbonate dissolution, and dilution by the non-carbonate lithic input. The type and rate of endogenic carbonate produced by Bear Lake water are dictated by a variety of hydrogeochemical factors reviewed by Dean (this volume).

The CaCO$_3$ content in BL00-1 averages 31 ± 19% over the core (n = 334). Six intervals (111.7–109.0, 103.5–100.1, 67.4–61.4, 56.6–52.5, 41.9–41.2, and 8.9–0 m blf) constituting 29% of the samples, contain distinctly higher CaCO$_3$ contents that exceed 40% (average = 57 ± 11%; n = 97). Trends in CaCO$_3$ content mimic the global marine oxygen isotope record. The substages of MIS 5 are expressed clearly by the total CaCO$_3$ content (Fig. 5). Sediment deposited during MIS 7 exhibits two prominent peaks that are correlated with substages 7c and 7a, and the later part of substage 7e is indicated by increasing CaCO$_3$ content of the lowest four samples. The X-ray diffraction analyses (discussed below) indicate that five CaCO$_3$-rich intervals contain aragonite. They are correlated with the peak interglacial intervals, MIS 7c, 7a, 5e, 5c, and 1.

The OC content of BL00-1 sediment averages 1.3 ± 0.7%. The OC and CaCO$_3$ contents are not inversely related, as might be expected through a dilution effect, but show a weak positive correlation (r = 0.16; p = 0.07). The OC content tends to be slightly higher in the aragonite-rich intervals (median = 1.2%) compared with the calcite-dominated intervals (median = 1.0%).

Mineralogy

The relative abundance of the common minerals that make up Bear Lake sediment was analyzed by X-ray diffraction (XRD). A total of 329 samples (nearly all of the levels analyzed by coulometry) were analyzed according to standard techniques (e.g., Moore and Reynolds, 1989), which are described by Dean et al. (2006) for the 1996 cores. In addition, the silt-plus-clay fractions (<53 μm; hereafter "mud") of 53 samples of fluvial sediment collected from the bed (27 samples) and flood-plain banks (26 samples) of modern streams were also analyzed by XRD; the results of these and other analyses of physical properties of fluvial sediment are presented by Rosenbaum et al. (this volume). Results for quartz, calcite, aragonite, dolomite, and feldspar from BL00-1 samples are reported as a percentage of the sum of the main XRD peak intensities (Appendix A). These percentage values are not accurate representations of the concentration of mineral abundances and should be interpreted cautiously because they do not account for different X-ray mass absorption characteristics of different minerals.

Quartz is the dominant mineral in Bear Lake sediment (average = 50 ± 16% over the entire core), except in the six CaCO$_3$-rich

intervals. Quartz abundance estimated from XRD peak intensities is inversely related to the abundance of CaCO$_3$ as measured independently by coulometry, indicating the effect of dilution (r = –0.87, p < 0.001 for the correlation between log-transformed quartz versus CaCO$_3$ for the entire core) (Fig. 6). The XRD results also show that, of the six CaCO$_3$-rich (CaCO$_3$ > 40%) intervals, four are dominated by aragonite: 103.1–101.6, 67.4–61.0, 55.8–55.3, and 8.4–0 m blf, and a fifth interval contains some aragonite (111.0–110.5 m blf). Of these, aragonite is most abundant in the uppermost interval (Holocene) and least abundant in the lowermost (MIS 7c), possibly due in part to diagenetic conversion of aragonite to low-Mg calcite. One of the carbonate-rich intervals (MIS 5a) does not contain aragonite.

Most of the stream-sediment samples that we analyzed are also dominated by quartz. Mud carried by the Bear River in proximity to Bear Lake tends to contain a higher proportion of quartz (average = 80%) than that in the local tributaries (average = 65%) (Fig. 7). Perhaps the clearest mineralogical distinction of sediment carried by local streams is the dolomite content. Local stream mud contains an average of ~17% dolomite, with a higher proportion (25%) in the west-side streams, compared to 5% for Bear River mud. Dolomite is a significant constituent of Bear Lake sediment, averaging ~7% (when excluding the CaCO$_3$-rich intervals), and is thought to be entirely allogenic (Dean et al., 2006; Rosenbaum et al., this volume). The relative percentages of quartz and dolomite track each other closely (Fig. 5) because they both derive from allogenic sources, and because they are related inversely to the relative percentage of calcite, the second most abundant mineral.

We examined downcore trends in two mineral constituents (Fig. 5). First, the relative abundance of quartz might serve as an indicator of the input of the allogenic component relative to endogenic carbonate, although the carbonate has both allogenic and endogenic sources that vary in proportion through the core. Second, the ratio of quartz to dolomite might provide some information on the relative importance of input from the Bear River, which is depleted in dolomite, versus input from local streams, which carry a higher proportion of dolomite (Fig. 7).

In summary, the mineralogy of Bear Lake sediment is controlled by the production rate of endogenic carbonate and the extent to which it is diluted by allogenic input. Aragonite formation depends on whether the lake water contains sufficient Mg, which appears to be the case only when Bear Lake is isolated from Bear River (Dean et al., 2007; Dean, this volume). Aragonite is preserved in five intervals of BL00-1, and these have been correlated with peak interglaciations of MIS 7c, 7a, 5e, 5c, and 1. Quartz is the dominant mineral in the carbonate-poor lake sediment and stream mud. An increase in the discharge of the Bear River relative to local streams might have the dual effect of increasing the flux of quartz and decreasing the production of endogenic carbonate, because water carried by the Bear River tends to have a lower solute content than the local streams and springs (Dean et al., 2007; Bright, this volume, Chapter 4). An increase of Bear River discharge relative to local discharge should also be indicated by

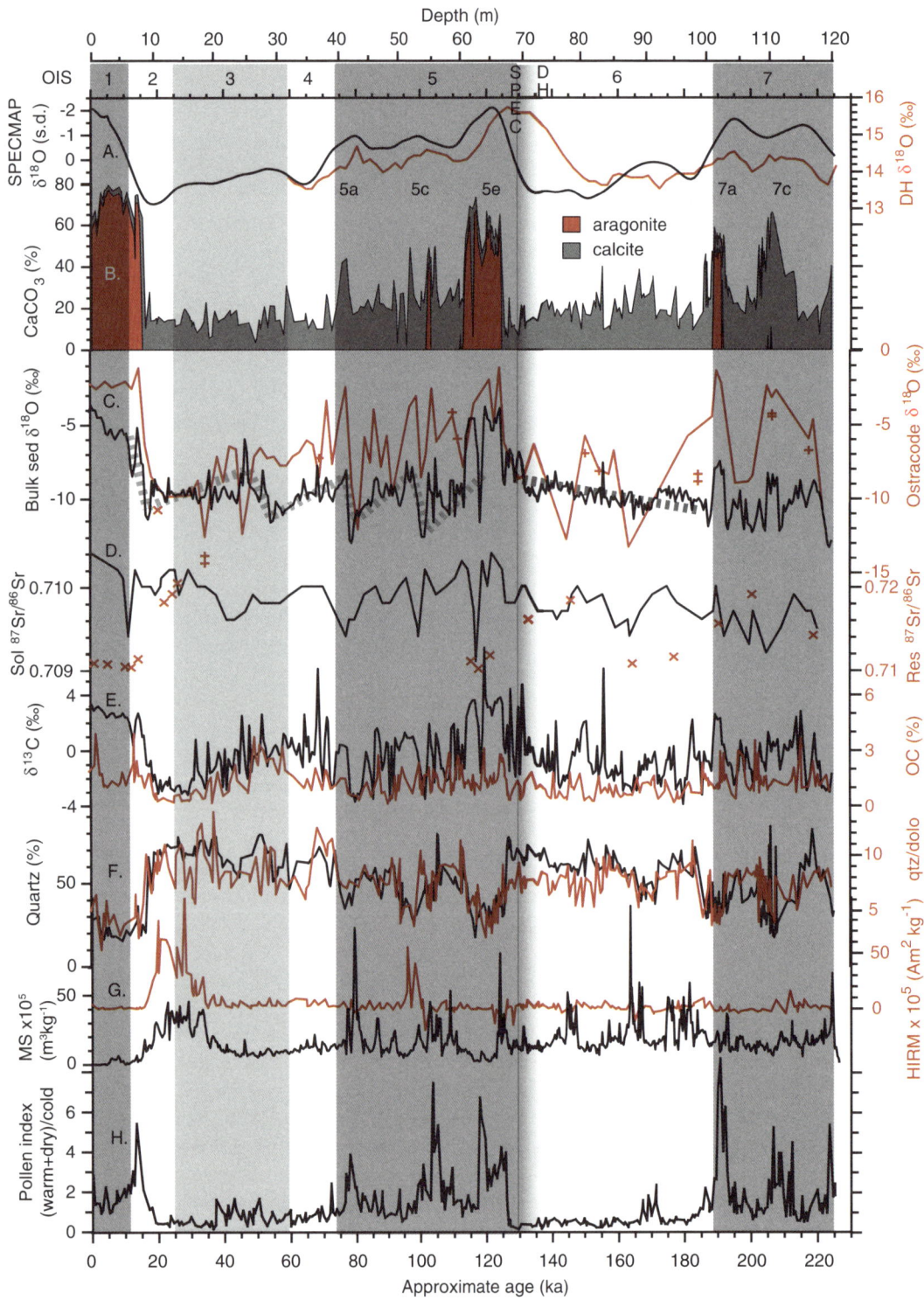

Figure 5. Downcore plots of mineralogic, isotopic, and magnetic properties of core BL00-1. Gray and white vertical intervals with numbers delimit marine isotope stages (MIS) according to SPECMAP chronology (SPEC = onset of MIS 5e in SPECMAP chronology; DH = onset of MIS 5e in Devils Hole chronology). Age model for BL00-1 is based on a linear sedimentation rate of 0.54 mm yr⁻¹ (Fig. 3). (A) Oxygen isotope values (δ¹⁸O) based on SPECMAP (black, standard deviation [s.d.] units) and Devils Hole calcite (red). (B) Calcite (gray) and aragonite (red) content, where the total CaCO₃ weight % is based on total inorganic carbon analysis and the relative proportion of the two mineral phases is based on X-ray diffraction (XRD) peak heights. (C) δ¹⁸O values in bulk-sediment carbonate (black), and ostracode calcite (red line—*Candona* sp. 1; red crosses—*Cytherissa lacustris*); gray dashed lines highlight trends discussed in text. (D) Strontium-isotope ratios for 5M-acetic-acid-soluble fraction (sol—black), and insoluble residue fraction (res—red Xs). (E) Carbon isotope ratios (δ¹³C) in bulk-sediment carbonate (black), and organic carbon (OC) content (red). (F) Approximate quartz percentage (black), and ratio of quartz to dolomite (qtz/dolo) (red) based on XRD peak intensities. (G) Magnetic susceptibility (MS) (black; from the initial core descriptions), and hard isothermal remanent magnetization (HIRM) (red). (H) Pollen index calculated as the ratio of "dry" (*Ambrosia*, Chenopodiaceae-*Amaranthus*, and *Sarcobatus*) plus "warm" (*Quercus* and *Juniperus*) to cold (*Picea*, other Asteraceae, and *Eriogonum*) pollen indicators. All data are listed in Appendices, except magnetic properties and pollen abundances.

the ratio of quartz to dolomite in Bear Lake sediment, which is independent of endogenic carbonate production.

Sediment Magnetic Properties

The magnetic properties of Bear Lake sediment have been studied in samples from the shorter piston cores (Rosenbaum and Heil, this volume) and in continuous channel samples from BL00-1 (Heil et al., this volume). The magnetic properties of stream sediment in the Bear Lake drainage basin also have been characterized (Dean et al., 2006; Rosenbaum et al., this volume). Here we focus on two key magnetic properties from BL00-1 (Fig. 5): magnetic susceptibility (MS) and hard isothermal remnant magnetization (HIRM). MS commonly reflects the content of ferrimagnetic minerals (e.g., magnetite, titanomagnetite, and greigite), whereas HIRM is a measure of high-coercivity minerals (e.g., hematite and goethite). Variations in these properties in Bear Lake sediment can be explained largely by (1) post-depositional destruction of detrital Fe-oxide minerals, (2) dilution of magnetic minerals by nonmagnetic endogenic $CaCO_3$, and (3) post-depositional formation of greigite (a ferrimagnetic Fe-sulfide mineral).

Magnetic properties of BL00-1 are highly variable. The most important magnetic minerals in much of the cored section are detrital titanohematite and authigenic greigite. Petrographic observations (Reynolds and Rosenbaum, 2005) indicate that most detrital magnetite has been destroyed (either dissolved or converted to pyrite), although a small quantity is present in some samples, mostly protected inside rock fragments. Variations in MS arise largely from differences in the quantity of authigenic greigite, which varies greatly, probably reflecting changes in salinity (i.e., availability of sulfate).

Two intervals exhibit exceptionally low values of MS and HIRM: from 66 to 61 m blf (MIS 5e) and from 9 to 0 m blf (younger than 15 ka). These reflect extensive alteration of detrital Fe-oxide minerals and dilution by high $CaCO_3$ contents. In contrast, the sediment from 18 to 9 m blf (MIS 2) exhibits high values of both MS and HIRM due to low $CaCO_3$ content and high magnetic mineral content. This is the only interval in which detrital Fe-oxide minerals are well preserved, and corresponds to high content of hematite-rich glacial flour from the headwaters of the Bear River in the Uinta Mountains (Rosenbaum and Heil, this volume). The preservation of the Fe-oxides in this interval reflects some combination of fresh water, rapid deposition, and perhaps greater quantities of detrital Fe-oxides.

Oxygen and Carbon Isotopes

Bright et al. (2006) reported the results of oxygen and carbon isotopic analyses from BL00-1. They analyzed bulk-sediment carbonate of 375 samples spaced at an average interval of 32 cm across the core (~700-year resolution). To compare the isotopic composition of purely endogenic carbonate with the bulk sediment, they also analyzed calcite in ostracode valves (*Candona* sp. 1; Bright et al., 2005) from 78 levels, mainly from the upper half of the core, where valves are better preserved. In addition, in this study we present new results on *Cytherissa* valves from 13 core levels to compare with the results on *Candona*. Values for the ratios of $^{18}O/^{16}O$ and $^{13}C/^{12}C$ are reported using standard permil (‰) δ notation, where the standards Vienna Standard Mean Ocean Water (VSMOW) and Vienna Pee Dee Belemnite (VPDB) were used for water and carbonate, respectively. Additional context for interpreting the downcore changes in δ^{18}O and δ^{13}C from BL00-1 is provided by (1) the isotopic composition of water that

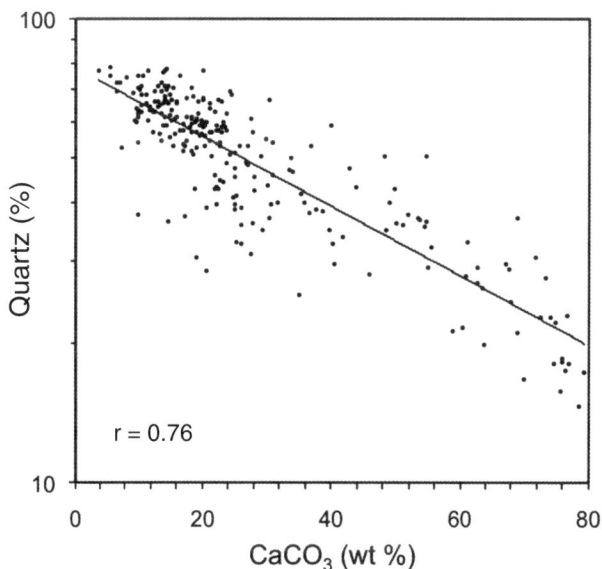

Figure 6. Percent $CaCO_3$ measured by coulometry versus quartz abundance estimated from X-ray diffraction peak intensities in samples from core BL00-1.

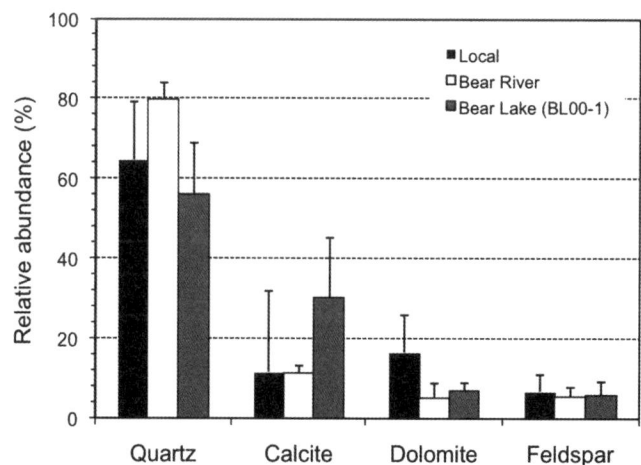

Figure 7. Average mineral assemblages for Bear Lake, Bear River, and local stream sources based on X-ray diffraction peak intensities. Bulk sediment was analyzed from BL00-1; the silt+clay fraction was analyzed from samples from Bear River and local streams. Errors bars are 1σ. Data for stream sediments are summarized from Rosenbaum et al. (this volume).

presently feeds the lake (Bright, this volume, Chapter 4), (2) the isotopic shift in shallow surface cores that span the ca. 1912 diversion of Bear River water into Bear Lake (Dean et al., 2007), and (3) the results of more detailed analyses of shorter cores taken in 1996 (Dean et al., 2006). The oxygen isotope composition of carbonate in Bear Lake sediment provides insights into the sources of water and the hydrologic budget of the lake, whereas the carbon isotopes reflect changes in the biologic activity as well as the hydrologic budget. We focus here on the $\delta^{18}O$ results because our understanding of the factors that control isotopic composition is better for oxygen than for carbon.

The $\delta^{18}O$ values range broadly, from about $-13‰$ to $-5‰$ in the bulk-sediment carbonate (Fig. 5; Appendix B). Enrichment in ^{18}O is significantly greater in the five intervals that contain aragonite (average $= -5.9 \pm 1.6‰$, $n = 52$) than in the rest of the core (average $= -9.8 \pm 1.1‰$, $n = 322$), an increase that cannot be attributed solely to the offset in mineralogic fractionation. The $\delta^{13}C$ values exhibit a similarly broad range, from about $-4‰$ to $7‰$, with higher values in the aragonite-bearing intervals ($2.4 \pm 0.9‰$) compared with the rest of the core ($-0.7 \pm 1.6‰$). The two isotopes correlate more strongly in the aragonite-bearing intervals ($r = 0.89$; $p < 0.001$) than over the rest of the core ($r = 0.42$; $p < 0.001$) (Fig. 8A), although the extent of the correlation varies among the oxygen isotope stages (Bright et al., 2006). The $\delta^{13}C$ values also correlate positively with OC content ($r = 0.40$; $p < 0.001$) (Fig. 5).

The $\delta^{18}O$ value of calcite in *Candona* is nearly always higher than that in the enclosing bulk-sediment carbonate (average difference $= 2.8 \pm 2.4‰$, $n = 78$) (Fig. 5), as expected from the vital effect (e.g., $+2.2‰$ for *Candona*; von Grafenstien et al., 1999). The new results on *Cytherissa* generally agree with those of *Candona*. Excluding the ten results from the aragonite-bearing intervals, the $\delta^{18}O$ values of ostracodes ($-7.1 \pm 2.6‰$, $n = 66$) vary more than the $\delta^{18}O$ of the enclosing bulk sediment ($-9.8 \pm 1.0‰$, $n = 66$), with the ostracodes typically within the range of $-12‰$ to $-2‰$, and the bulk carbonate within the range of $-9‰$ to $-11‰$ (Fig. 8B). The lower variability in the isotopic composition of bulk carbonate probably reflects the influence of allogenic carbonate. When the aragonitic intervals are excluded, the correlation between $\delta^{18}O$ of ostracode and bulk-sediment carbonate is relatively weak ($r = 0.33$; $p = 0.01$). The lower variability in the $\delta^{18}O$ values of bulk sediment might also result from greater time-averaging than for the ostracode samples, which comprised several valves.

Values of $\delta^{18}O$ and $\delta^{13}C$ in rocks, stream sediment, and water from the Bear Lake tributaries have been presented previously, and are summarized here (Table 1). The Paleozoic limestone and dolomite that underlie the Bear River Range, and the Mesozoic carbonate rock within the Bear Lake Plateau to the east, have $\delta^{18}O$ values ranging from -5.1 to $-13.3‰$, and averaging $-8.2 \pm 2.7‰$ ($n = 9$) (Bright et al., 2006). In contrast, mud carried by local streams contains carbonate with much lower $\delta^{18}O$ values. Bright et al. (2006) reported an average value of $-13.6 \pm 1.3‰$ ($n = 7$), whereas Dean et al. (2006; their Table 2) reported values based on a slightly more extensive data set: The average $\delta^{18}O$ value for west-side streams is $-14.3 \pm 1.1‰$ ($n = 5$; after excluding

Hobble Creek, which is not on the west side, and the one outlier from Paris Creek); the average $\delta^{18}O$ value for east-side streams is $-12.5 \pm 1.0‰$ ($n = 5$; this value is different than reported by Dean et al. [2006, p. 107] but is correct). $CaCO_3$ in mud carried by the Bear River north of the Uinta Mountains has slightly higher $\delta^{18}O$ values than $CaCO_3$ in mud from local streams ($-11.6 \pm 1.3‰$, $n = 11$). The results on the alluvial mud probably more closely reflect the allogenic input than do spot analyses on bedrock units. They are lower than the values in the rocks ($-8.2‰$) because the streams carry detrital tufa grains derived with $\delta^{18}O$ values of around $-16‰$ (Bright et al., 2006), closer to that of present-day water, which averages $-46.8‰$ VPDB ($= -17.3 \pm 0.7‰$ VSMOW, $n = 40$) for

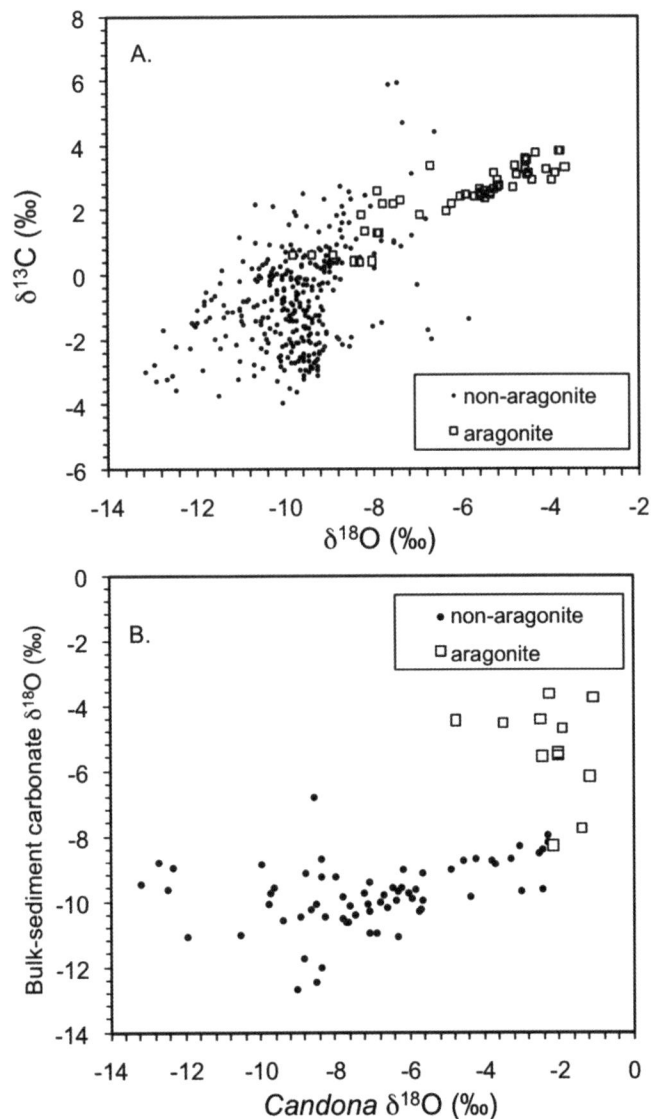

Figure 8. Stable isotope ratios from core BL00-1. (A) Cross plot of oxygen ($\delta^{18}O$) and carbon ($\delta^{13}C$) isotope ratios in bulk-sediment carbonate. (B) $\delta^{18}O$ in ostracode (*Candona*) calcite versus those in enclosing bulk sediment (includes allogenic and endogenic carbonates). Data from Bright et al. (2006), and listed in Appendix B.

TABLE 1. AVERAGE VALUES FOR OXYGEN, CARBON, AND STRONTIUM ISOTOPE RATIOS IN WATER, SEDIMENT, AND ROCKS FROM BEAR LAKE AREA

Province	Material	$\delta^{18}O$†	±1σ	n	$\delta^{13}C$†	±1σ	n	Soluble fraction or water $^{87}Sr/^{86}Sr$#	±1σ	n	Residue fraction $^{87}Sr/^{86}Sr$#	±1σ	n
West side	Spring and stream water	-17.4	0.7	30	-9.7	0.2	6	0.71084	0.00074	26	–	–	–
East side	Spring and stream water	-17.0	0.8	10	-9.2	1.2	2	0.70975	0.00113	20	–	–	–
Bear River	Stream water	-16.3	0.4	4	-8.9	–	1	0.70823	0.00050	6	–	–	–
Bear Lake	Lake water	-8.1	0.1	4	-2.2	–	1	0.70919	0.00002	11	–	–	–
East & west sides	Carbonate rocks	-8.2	2.7	9	0.4	1.4	9	0.70881	0.00102	16	0.71042	0.00346	16
West side	Stream sediment (<53 µm bulk)	-12.5	1.0	5	-5.3	1.1	5	0.71015	0.00015	4	0.71999	0.00262	4
East side	Stream sediment (<53 µm bulk)	-12.5	1.0	5	-5.3	1.1	5	0.70866	0.00088	3	0.71325	0.00143	3
East & west sides	Stream sediment (<53 µm bulk)	-13.6	1.3	7	-5.8	1.1	7	0.70951	0.00095	7	0.71710	0.00413	7
Bear River	Stream sediment (<53 µm bulk)	-11.6	1.3	11	-5.1	0.8	11	0.71027	0.00085	12	0.72451	0.00796	12
East & west sides	Stream sediment (picked tufa grains)	-15.7	0.4	3	-6.5	0.8	3	0.71054	0.00040	2	–	–	–

Note: Oxygen and carbon isotopes values are ‰ relative to Vienna standard mean ocean water for water and Vienna Peedee belemnite for carbonate.
n—number of samples used to calculate mean and standard deviation.
†Data used to calculate mean values for O and C isotope ratios are listed in Bright et al. (2006).
#Data used to calculate mean values for Sr isotope ratios are from: stream water (Bright, this volume, Chapter 4); carbonate rocks—soluble (Bright, this volume, Chapter 4); carbonate rocks—residue (this study); Bear Lake water (Dean et al., 2007); stream sediment soluble fraction (Dean, this volume); stream sediment residue fraction and tufa grains (this paper).

local streams and −45.9‰ VPDB (= −16.3 ± 0.4‰ VSMOW, $n = 4$) for the Bear River (Bright et al., 2006). Values of $\delta^{18}O$ in bulk carbonate in BL00-1 are almost always higher than those in bulk carbonate in modern stream mud, suggesting that the allogenic component of the bulk sediment moderates the fluctuations in the isotopic composition of endogenic carbonate, but that allogenic carbonate is not sufficiently abundant to entirely overwhelm the endogenic component.

In summary, the isotopic composition of Bear Lake water and sediment has fluctuated considerably over the last two glacial-interglacial cycles. Values of $\delta^{18}O$ and $\delta^{13}C$ were higher during interglacial intervals, when fractionation was clearly controlled by evaporative effects. Such effects are consistent with the presence, if not dominance, of aragonite, among other indicators of a topographically closed-basin lake which are discussed elsewhere in this volume. Outside of the aragonitic intervals, ostracode calcite exhibits larger fluctuations in $\delta^{18}O$ values than does the bulk-sediment carbonate, and can be interpreted without the uncertainty that results from fluctuating carbonate detritus. Ostracodes are more abundant in the upper half of the core, however, and the sampling resolution is presently too coarse throughout to allow detailed interpretation.

Strontium Isotopes

Sr isotopes ($^{87}Sr/^{86}Sr$) have been analyzed from the 5M-acetic-acid-soluble (hereafter "soluble") fraction of bulk sediment at an average spacing of 1.3 m in BL00-1. In addition, $^{87}Sr/^{86}Sr$ was analyzed in the acetic-acid-insoluble (hereafter "residue") fraction of 20 samples scattered across the core. The methods and precision of Sr isotope analyses of Bear Lake sediment by thermal ionization mass spectrometry are given by Dean et al. (2006, 2007). The interpretation of Sr isotope results, like those for the O and C isotopes, is guided by an understanding of the modern water and sediment input to the lake, and by more detailed analyses of other cores (e.g., Dean et al., 2006, 2007). Like O isotopes, Sr isotopes can provide insights into the source area of water; unlike O isotopes, Sr isotopes are not complicated by the fractionation effects that are influenced by temperature and evaporation.

Values of $^{87}Sr/^{86}Sr$ of the soluble fraction of sediment samples from BL00-1 range from 0.7091 to 0.7104, and average 0.7099 ± 0.0003 ($n = 93$) (Fig. 5; Appendix B). These values are bracketed by the two primary sources of water to the lake: the Bear River with present-day $^{87}Sr/^{86}Sr$ values averaging 0.7082 ± 0.0005 ($n = 6$), and local stream water, which averages 0.7104 ± 0.0011 ($n = 46$) (Bright, this volume, Chapter 4; note: Dean et al. (2007) listed a subset of these values, focusing on sites where additional solute data were also available). The $^{87}Sr/^{86}Sr$ value of endogenic carbonate that precipitates in Bear Lake depends directly on the $^{87}Sr/^{86}Sr$ value of the lake water. In turn, this depends on the flux of Sr, which is controlled by its concentration and the discharge from the various inflow sources. On the basis of the data that are presently available, west- and south-side surface-water sources account for greater Sr flux than east-side sources, and they are more radiogenic (west-side average = 0.7108 ± 0.0007, $n = 26$ compared with east-side average = 0.7098 ± 0.0011, $n = 20$) (Table 1). Bear River water tends to have a higher Sr concentration than the local streams (Bright, this volume, Chapter 4), and is probably the dominant solute source of Sr when it discharges into the lake.

Although the $^{87}Sr/^{86}Sr$ value of soluble Bear Lake sediment can be accounted for by various mixtures of water from Bear River and local stream sources, the analyses of bulk sediment are also influenced by the composition and abundance of soluble allogenic components. The $^{87}Sr/^{86}Sr$ value of the Paleozoic and Jurassic carbonate bedrock units in the local drainage basin averages 0.7088 ± 0.0010 ($n = 16$) (Bright, this volume, Chapter 4; Dean et al. (2006) calculated an average value of 0.7092, $n = 20$, but included data from units that do not crop out in the local drainage), slightly lower than the average value of the soluble fraction of sediment samples in BL00-1 (0.7099 ± 0.0003). The $^{87}Sr/^{86}Sr$ value of the soluble fraction of mud (<0.53 μm) collected from the bed of local streams reflects the composition of the carbonate rocks in the catchments. The value is higher for west-side streams (0.7101 ± 0.0001, $n = 4$) than for east-side streams (0.7087 ± 0.0009, $n = 3$) (Dean, this volume). The Bear River carries mud with the highest $^{87}Sr/^{86}Sr$ values in the Bear Lake drainage (0.7103 ± 0.0009; $n = 12$) (Dean, this volume). The $^{87}Sr/^{86}Sr$ values are higher for the residue fraction of the fluvial mud than for the soluble fraction, averaging 0.7272 ± 0.0076 ($n = 19$), and the values from the two fractions correlate (r = 0.60, $n = 12$) (Dean, this volume). Data on Sr concentrations in the two fractions are not available; presumably Sr is enriched in the soluble fraction.

The $^{87}Sr/^{86}Sr$ values of soluble Bear Lake sediment generally track the $\delta^{18}O$ values of the bulk-sediment carbonate (Fig. 5). The enrichment of both ^{18}O and ^{87}Sr in the aragonitic zones probably reflects the absence of Bear River input, which increases the water residence time and the consequent evaporative influence on $\delta^{18}O$ and eliminates the dominant source of solutes with low $^{87}Sr/^{86}Sr$ values. Without the input of solutes and clastic sediment from the Bear River, the soluble components of Bear Lake sediment acquire the $^{87}Sr/^{86}Sr$ value of west-side inflow. With the input of the Bear River, the $^{87}Sr/^{86}Sr$ value of the endogenic carbonate in Bear Lake decreases, whereas the $^{87}Sr/^{86}Sr$ value of the allogenic component increases. The net effect therefore depends on the balance of the two components, and we do not know the relative difference in the concentration of Sr in the endogenic and allogenic components.

The $^{87}Sr/^{86}Sr$ values for the residue fraction of lake sediment tend to track the relative proportion of quartz in BL00-1 (Fig. 5). This supports the interpretation that the quartz-rich units are dominated by sediment delivered by the Bear River, which carries mud with high $^{87}Sr/^{86}Sr$ values. Half of the samples from BL00-1 analyzed for the residue fraction yielded $^{87}Sr/^{86}Sr$ values less than 0.7119, which is lower than any value measured in stream sediment from the drainage basin (Dean, this volume). The source of the low $^{87}Sr/^{86}Sr$ values in the residue fraction is not clear, although the carbonate bedrock units that crop out in the Bear Lake drainage, especially the Jurassic units on Bear Lake Plateau, tend to have lower values than indicated by the analyses of stream sediment data (average = 0.7109 ± 0.0037, $n = 18$; Table 1).

Ostracodes

A discussion of the ostracodes from Bear Lake sediment—most of them endemic—along with photomicrographs of the valves, is presented by Bright (this volume, Chapter 8; Bright et al., 2005). We processed 138 sediment samples, with an average spacing of 88 cm and an average mass of 54 g, according to methods outlined by Bright (this volume, Chapter 8). Whole (>80% complete) adult ostracode valves >150 μm were counted and identified to genus, and where possible, to species level. In addition, the presence of juvenile valves was noted for *Cytherissa lacustris*, the only taxon for which paleoenvironmental inferences can be made, but the valves were not counted. The presence of only juveniles is assumed to indicate that the species was living in the lake, but in low numbers, or possibly that the environment at the core site was poor for ostracode longevity.

Adult ostracode valves were recovered in 85 (62%) of the samples, and an additional 16 samples contained fragmented or juvenile valves of *Cytherissa*. In all, 2355 valves were counted, of which 93% are *Candona* spp. Valves are more abundant in the upper half of the core; 94% are from the top 65 m. Only 19 samples contained >1 valve per gram of sediment (vpg), and about half of these were within the Holocene section (Fig. 9; Appendix C). The ostracode fauna consists of one unidentified genus, several unnamed species of *Candona*, two unnamed species of *Limnocythere*, and *Cytherissa lacustris*. With the exception of *Cytherissa lacustris*, all of the ostracode species are endemic (Bright, this volume, Chapter 8). The greater abundance of valves in the upper half of the core, and especially in the Holocene aragonite-rich mud, is probably a result of post-depositional preservation. Little can be inferred about paleohydrologic conditions from the abundances of the endemic forms; many taxa cross lithostratigraphic boundaries and seem to have inhabited the lake whether it was generating aragonite in a topographically closed basin, or whether it was receiving input from the Bear River.

Cytherissa lacustris is an extant species with known environmental tolerances (Bradbury and Forester, 2002). This species inhabits dilute (TDS < 365 mg L^{-1}) boreal lakes typically under the influence of polar-high and subpolar-low air masses. Lakes that support *C. lacustris* are stable, well oxygenated, and display little if any seasonal variation. The species' modern distribution is confined to northern Canada, Alaska, and a few isolated localities in the northern conterminous United States. Its presence in Pleistocene lake sediment from the western United States coincides with glacial periods, when the polar and subpolar air masses were forced southward by large ice sheets over northern North America (e.g., Bradbury and Forester, 2002). *C. lacustris* is absent from four of the five aragonite-bearing intervals, consistent with its preference for dilute lakes formed during glacial intervals. It is present, however, in sediment that correlates with the interglacial

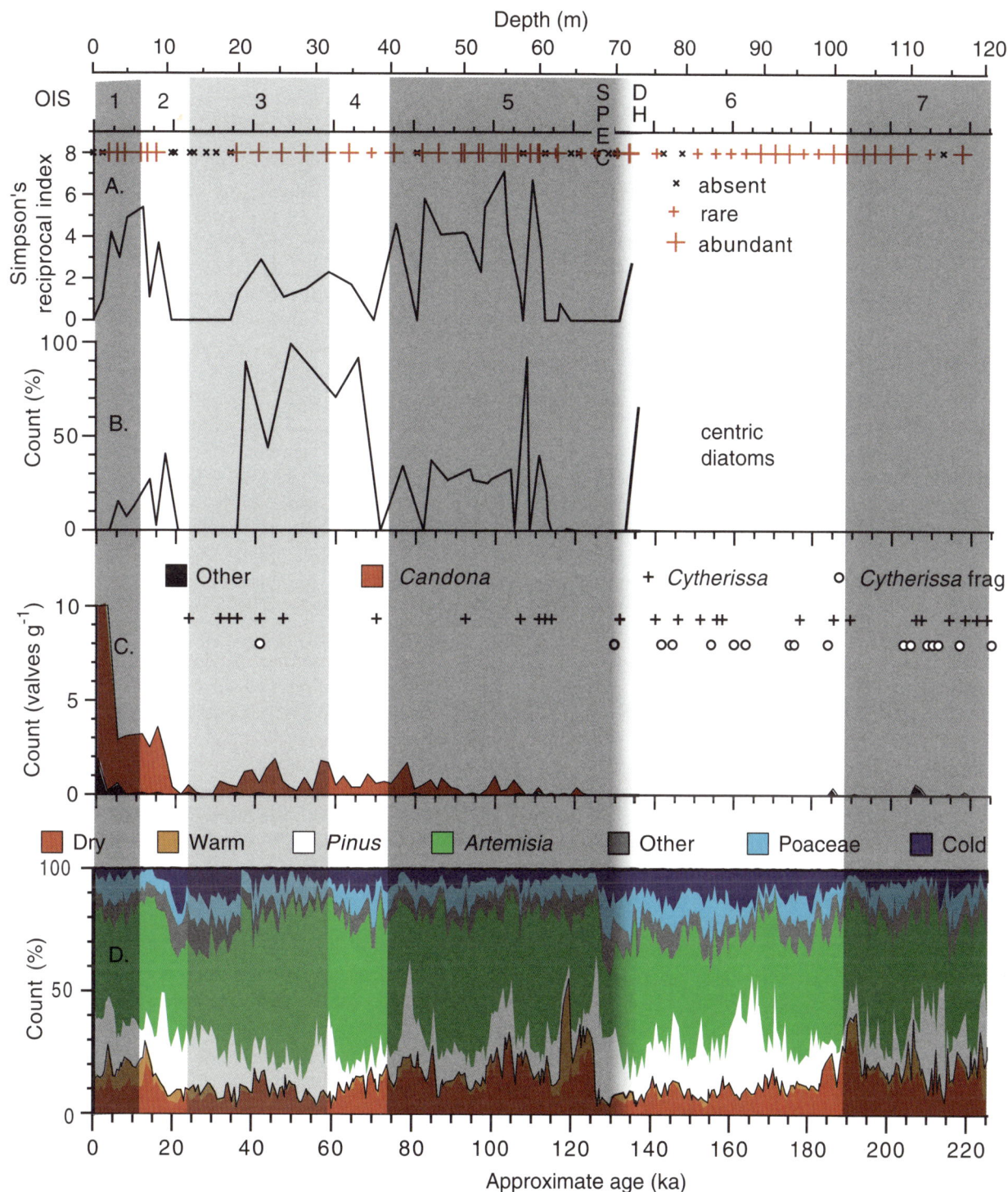

Figure 9. Paleobiological data from core BL00-1 with time scale and isotope stages as described for Figure 5. (A) Diatom diversity based on the Simpson's reciprocal index; plot symbols at top indicate relative abundance of diatoms (rare—dissolved or broken diatoms that were difficult or impossible to count). (B) Percent centric diatoms (approximately equivalent to % planktic diatoms, except that many of the non-centric planktic diatoms [<1% in any given sample] were counted as either tychoplanktic or planktic). (C) Abundance of all ostracode *Candona* spp. and other whole, adult ostracode taxa, with the presence of *Cytherissa lacustris* shown as whole (plus signs), and as fragmented or juvenile (open circles). (D) Relative percentages of selected pollen taxa where "dry" = *Ambrosia*, Chenopodiaceae-*Amaranthus,* and *Sarcobatus,* "warm" = *Quercus* and *Juniperus,* and "cold" = *Picea,* other Asteraceae, and *Eriogonum*. Abbreviations are the same as for Figure 5. Diatom and ostracode data are listed in Appendices C and D; pollen data are from Jiménez-Moreno et al. (2007) and are available from the North American Pollen Database (http://www.ncdc.noaa.gov/paleo/napd.html).

MIS 7c. To our knowledge, this is the first documented case of *C. lacustris* in the western contiguous United States during peak global interglacial conditions.

Pollen

The recent pollen study of BL00-1 by Jiménez-Moreno et al. (2007) provides one of the most detailed and continuous records of Quaternary vegetation change in North America, and Doner (this volume) analyzed pollen in a short piston core from the lake (BL96-2) extending back 19,000 yr. Pollen from BL00-1 was analyzed in 359 samples (intervals of 30 cm = ~600-yr resolution), each comprising 2 cm³. A minimum of 300 grains of terrestrial pollen were identified to the lowest taxonomic level possible, usually genus but sometimes family. The pollen data are available through the North American Pollen Database (http://www.ncdc.noaa.gov/paleo/napd.html).

Preservation of pollen in BL00-1 is excellent, with generally high concentrations averaging >40,000 grains cm⁻³. The assemblage is dominated by *Artemisia* (sagebrush), which averages 46% of the grains throughout the core. *Pinus* (pine) is the second most abundant type, averaging 17%. Other pollen types generally constitute <10% each, and their interpretation is discussed by Jiménez-Moreno et al. (2007). Here (Fig. 9), we summarize the pollen data by combining taxa that indicate similar climatic conditions. "Dry" indicators include *Ambrosia* (ragweed), Chenopodiaceae-*Amaranthus* (saltbush), and *Sarcobatus* (greasewood); "warm" indicators include *Quercus* (oak) and *Juniperus* (juniper); and "cold" indicators are *Picea* (spruce), other Asteraceae (sunflower family), and *Eriogonum* (buckwheat).

The pollen assemblages in BL00-1 can be subdivided into two general types: glacial and interglacial. The pollen spectra from interglacial periods contain higher percentages of "warm" and "dry" indicators, with higher *Juniperus* percentages during the early part of each interglacial interval. Vegetation interpretations suggest that valley bottoms were occupied by salt-tolerant, high-desert shrubs, and that *Juniperus* woodlands expanded locally during interglaciations. Pollen spectra of glacial intervals generally have higher percentages of "cold" indicators, suggesting that forest or forest-woodland conditions prevailed.

The pollen data can be further summarized by using a ratio of the "warm" plus "dry" indicators ("warm+dry") to "cold" indicators (Fig. 5). This index is nearly identical to the relative abundance of "warm-arid" taxa emphasized by Jiménez-Moreno et al. (2007), but the ratio avoids issues related to closed-array percentage data while integrating information on the abundance of cold indicators. Sediment deposited during glacial intervals generally contains fewer pollen grains of "warm+dry" indicator taxa than "cold" taxa (i.e., index values [(warm+dry)/cold] < 1), whereas interglacial periods are typified by more "warm+dry" than "cold" pollen grains (i.e., index values >1). Overall, the penultimate glacial period (MIS 6) shows lower index values, reflecting a higher proportion of *Picea* pollen than during the last glacial cycle (MIS 4–2), and suggesting that MIS 6 was generally colder. The high-

est index values are associated with peak interglacial substages, especially MIS 7a, 5e, and 5c. Index values for the early Holocene are intermediate among the peak interglacial intervals.

Diatoms

Diatoms were analyzed from 49 samples (25 core catcher samples and 24 core samples) from the upper 73 m of BL00-1 (Fig. 9; Appendix D). Of the samples analyzed, 28 (57%) contained sufficient diatoms for meaningful counts. Core catcher samples were also analyzed from below 78 m blf to assess qualitatively the relative abundance of diatoms. Samples were processed following standard methods, as described by Moser and Kimball (this volume). Diatom taxonomy was based primarily on Krammer and Lange-Bertalot (1986–1991). Because the main goal of this study was to infer climatic and hydrologic change, the diatom data are summarized using two indices: (1) percentage planktic diatoms, with higher percentages typically associated with higher lake levels (Wolin and Duthie, 1999), and (2) species diversity, which has been suggested to increase with greater habitat variability and nutrient availability (Douglas et al., 1994). Simpson's reciprocal index (SRI; Simpson, 1949) was calculated as: $SRI = 1/\Sigma p_i^2$, where p_i is the proportional numerical abundance of each species i.

The lowermost sample analyzed (72.7 m blf; MIS 6) contains abundant planktic diatoms (65%), mainly *Stephanodiscus medius*, suggestive of relatively high lake level and nutrient-rich conditions (Bradbury, 1997). However, samples from below this level will need to be analyzed to determine whether this is a long-term trend. Samples from 71.1 to 63.8 m blf (late MIS 6) contain few to no diatoms. The near absence of diatoms throughout this interval indicates poor preservation or a lack of diatoms living in Bear Lake. An absence or near absence of diatoms in lake sediments deposited during glacial periods has been observed at other sites, particularly deep lakes, and has been explained by increased dissolution (MacKay, 2007). Because there is no evidence of dissolution in the few diatoms observed in this interval, an absence of diatoms in the lake is favored. Conditions were probably cold, with an associated increase in ice coverage and turbidity, which would have lowered light conditions and prevented diatoms from inhabiting the lake (Bradbury et al., 1994; Bradshaw et al., 2000; Karabanov et al., 2004).

Samples from 63.8 to 60.5 m blf (MIS 5e) contain a mixture of well-preserved diatoms and broken diatoms, some with signs of dissolution. The broken pieces are typically from large pennate diatoms that live in the littoral zone of lakes, including *Epithemia* and *Cymbella*. The sample from 63.1 m blf contains well-preserved small, benthic/tychoplanktic fragilarioid species (>90%), mainly *Pseudostaurosira brevistriata*, indicative of a shallow lake (Bradbury, 1988). Unlike the previous interval, these data suggest that the lack of diatom valves is due to poor preservation. This interval contains aragonite-rich sediment, and the poor preservation of diatoms is probably caused by alkaline water, especially Mg-rich waters (Flower, 1993). The broken diatoms are consistent with low lake level, and may have been eroded and re-deposited as lake level fell. Alternatively, increased bioturbation due to shallower water and increased oxygen could result in broken diatoms.

From 60.5 to 39.6 m blf (MIS 5d–5a) the diatom assemblages include relatively low percentages of planktic taxa (average = 20%) and are dominated (typically >50%) by small, benthic/tychoplanktic fragilaroid species—mainly *Staurosirella pinnata* and *Pseudostaurosira brevistriata,* which are suggestive of shallow water (Bradbury, 1988). The main planktic diatoms, *Stephanodiscus medius* and *Cyclotella meneghiniana,* indicate high nutrient availability, and *Cyclotella meneghiniana* suggests rapidly fluctuating salinities. This interval is characterized by high diversity (average = 4.1) with assemblages composed of numerous planktic and benthic species. These results suggest an open-water lake with a significant littoral zone and a large macrophyte population, and warmer water than during MIS 6.

In contrast, samples from 36.6 to 20.0 m blf (MIS 4 and MIS 3) are dominated by planktic taxa (average = 78%), mainly *Stephanodiscus medius,* and low diversity (average = 1.8). *Stephanodiscus medius* requires enhanced phosphorus and is typically found in mesotrophic to eutrophic waters (Bradbury, 1997). The decrease in diversity occurred in both planktic and benthic species. These changes suggest an increase in open water and nutrients, indicative of a deeper lake. A deeper lake would increase stratification and increase nutrients during spring and fall overturn (Bradbury, 1997). A loss of macrophytes, perhaps due to lake transgression, could have also led to an increase in nutrient delivery to open water (e.g., Karst and Smol, 2000). More Bear River input during this period would also have increased nutrient delivery to the lake.

Samples from 20 to 10.1 m blf (late MIS 3 and early MIS 2) are characterized by an absence of diatoms. As in to MIS 6 and the lower part of zone 1 in BL96-2, the absence of diatoms is likely due to cold, turbid water (Moser and Kimball, this volume).

The interval from 10.1 to 1.1 m blf (late MIS 2 and MIS 1) is typified by assemblages with low percentages (average = 17.5%) of planktic diatoms and high species diversity (average = 3.3). Samples are dominated (average = 68%) by small, benthic *Staurosirella pinnata* and *Pseudostaurosira brevistriata,* suggestive of shallow water (Bradbury, 1988). The only planktic diatom with an abundance >1% is *Cyclotella meneghiniana,* which suggests more variable salinity than during MIS 3 or 4. These results indicate a warm and dry period. The uppermost sample, 0.5 m blf, is similar to the upper part of cores BL96-1 and 96-2, and is characterized by numerous, large broken pennate diatoms (Moser and Kimball, this volume). The cause of the broken diatoms is unclear, but is suggestive of drying carbonate sediments (Flower, 1993) or a high-energy environment.

In summary, the absence of diatoms in sediment deposited during glacial periods (MIS 6 and MIS 3/2) indicates low-light conditions associated with increased ice cover and increased turbidity, which limited the growing season and diatom population. Diatom assemblages of interglacial periods (MIS 5 and 1) indicate low lake levels and more saline conditions. The MIS 3 assemblage is distinct from MIS 5 and 1, and suggests that lake level was higher during this period than during the full interglacials.

DISCUSSION

Synthesis of Evidence for Paleoenvironmental Change

The data from BL00-1 document orbital- to millennial-scale fluctuations in the hydrology of Bear Lake and climate-related environmental changes in the catchment during the last quarter-million years. The broad-scale fluctuations (10^4 yr) in mineralogy, isotopes, and pollen generally coincide with summer insolation in the Northern Hemisphere driven by orbital variations, and with global ice volume. Suborbital variability (10^3 yr) in these indicators is also evident. Jiménez-Moreno et al. (2007) suggested that changes in the pollen spectra from BL00-1 coincide with ice-rafting (Heinrich) events in the North Atlantic, but Heil et al. (this volume) note that the timing of major peaks in hematite content (a proxy for glacial rock flour from the Uinta Mountains) does not. The uncertainty in the age model for BL00-1 is too high to determine whether the timing of any millennial-scale fluctuation coincides with a particular shift reported from other paleoclimate records. The chronology is sufficient to conclude only that the frequency of the variability in BL00-1 indicators is similar to stadial-interstadial cycles recognized in well-known records from other lacustrine (e.g., Benson et al., 2003) and marine (e.g., Hendy and Kennett, 1999) basins, and to Dansgaard-Oeschger oscillations exhibited in the isotope record from Greenland ice (Stuiver and Grootes, 2000). Unlike most terrestrial paleoclimate records, BL00-1 extends continuously through the penultimate glacial-interglacial cycle and reveals millennial-scale fluctuations throughout.

Not all of the variability can be attributed to climatic changes, however. Rearrangements of the confluence of Bear River and Bear Lake also affected the paleoenvironmental indicators (Fig. 10). The two end-member paleohydrogeographic configurations are topographically closed, and fluvially dominated; intermediate configurations involve various degrees of hydrologic exchange between the lake and the distributaries of Bear River where the river debouches onto the flat floor of the northern Bear Lake Valley. Presently, the wetland area separating the lake and the main channel moderates the influence of Bear River on the lake. In addition, some of the variability in BL00-1 reflects processes within the lake (as well as tectonic processes) that control the delivery of allogenic sediment to the core site, and the biogeochemical processes that influence the endogenic components.

The rearrangements of the Bear Lake drainage area are reflected in shifts in mineral assemblages, magnetic properties, and isotope ratios of O and Sr. Most of our isotope analyses are on bulk-sediment carbonate, which is confounded by a variable and unknown proportion of allogenic carbonate. Because the bedrock of the region is dominated by carbonate rocks, allogenic carbonate was probably always present in the lake sediment. The extent to which allogenic carbonate dominated the bulk sediment

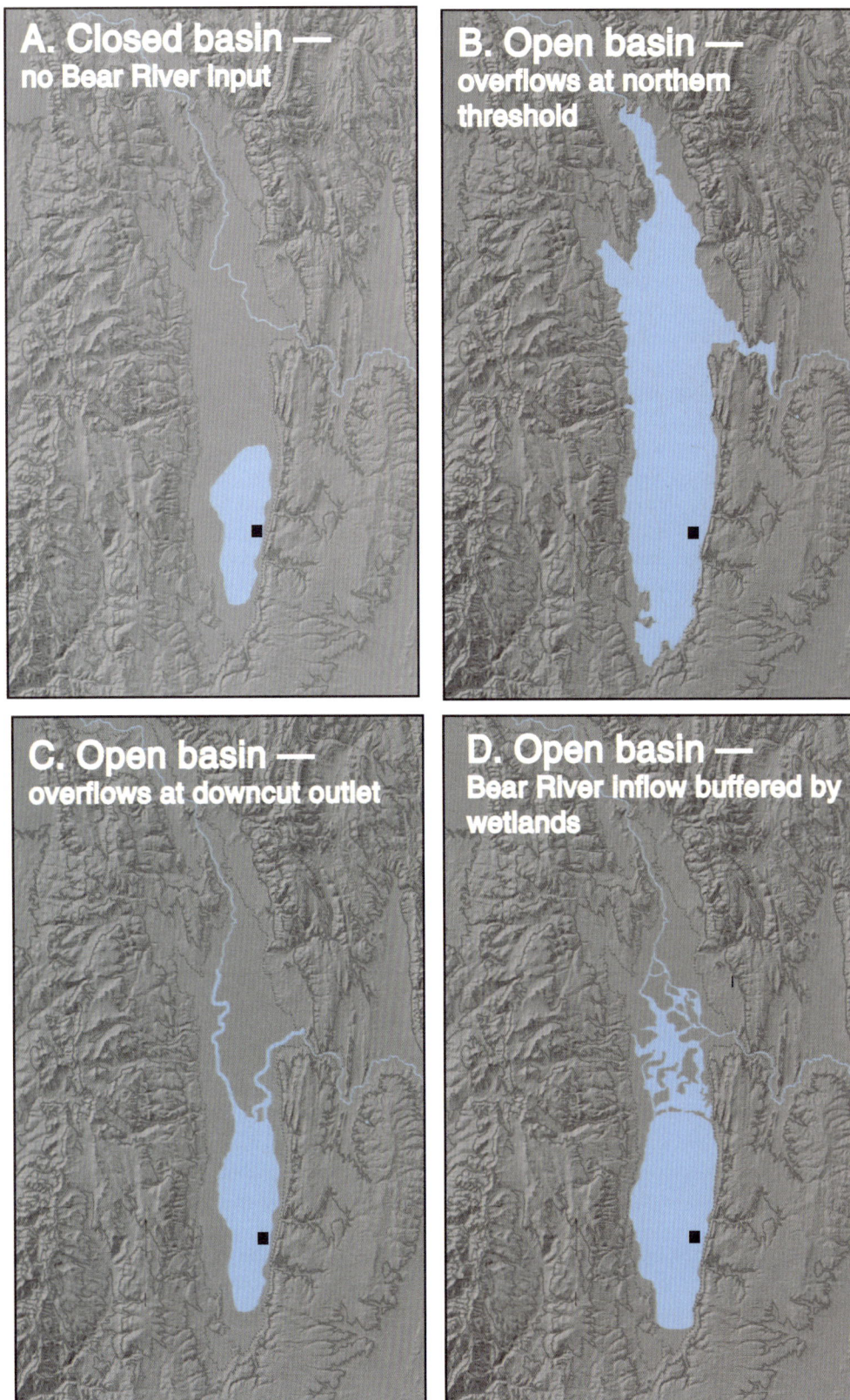

A. Closed basin —
no Bear River input

B. Open basin —
overflows at northern threshold

C. Open basin —
overflows at downcut outlet

D. Open basin —
Bear River inflow buffered by wetlands

Figure 10. Alternative conceptual hydrogeographies for the confluence of Bear River and Bear Lake. Black square—core site BL00-1. (A) Topographically closed-basin configuration with Bear Lake isolated from surface water flow of Bear River. Separation of Bear Lake from Bear River is caused by climatically controlled reduction in effective moisture that occurred during peak interglaciations. (B) Topographically open basin with the shoreline expanded to the basin threshold at ~1845 m asl. Capture of Bear River by Bear Lake is caused by the transgression of the northern shoreline to intercept the river channel, or by avulsion of the channel southward into the lake. (C) Topographically open basin with regressed shoreline. The direct connection with the outflow of Bear River increases delivery of fluvial sediment to the core site; increased outflow from the lake causes down-cutting at the outlet, which lowers lake level. (D) Intermediate configuration with Bear River confluent with Bear Lake via wetlands at the north end. The effect of the river is moderated by wetlands that impend sediment transport and solute exchange with the lake; this hydrogeography might have been the most common during the past quarter-million years.

depends on its proportion relative to the endogenic component, rather than on its absolute flux. At times when the production of endogenic carbonate was low, the isotope composition was more strongly influenced by the allogenic component. Changes in the proportion of allogenic versus endogenic material, and unknown differences in the concentration of Sr in the two components, complicate the interpretation for Sr isotopes measured in the acid-soluble bulk sediment.

Interglacial Intervals

At the broadest scale, the paleoenvironmental data document at least three distinct intervals when the lake retracted into a topographically closed basin, shut off from Bear River inflow. On the basis of the available age control, these coincide with peak global interglacial intervals of MIS 7, 5, and 1. Diatom assemblages indicate low lake levels during MIS 5e and 1, and pollen spectra from these intervals indicate an expansion of salt-tolerant plants in the valley bottoms and a reduction of coniferous forest in the uplands, consistent with the warm and dry conditions that were required for lake level to regress below the inflow of Bear River.

Once the lake regressed, Bear Lake and Bear River became disconnected. Evidence for the absence of Bear River inflow includes (1) production of aragonite, favored by evaporative conditions and Mg to Ca ratio like that of the lake prior to the artificial diversion of Bear River into the lake; (2) $\delta^{18}O$ values in bulk-sediment and ostracode calcite that are enriched in ^{18}O, with strongly covarying values of $\delta^{13}C$, indicating evaporation-dominated fractionation; (3) generally increased Sr isotope ratios in the soluble fraction, reflecting endogenic carbonate precipitated from water derived from the local drainage; and (4) decreased ratio of quartz to dolomite, indicating the influence of dolomite-rich, locally derived stream-sediment input. Furthermore, these parameters all reversed their trends in sediment deposited after the ca. 1912 diversion of Bear River into the lake (Dean et al., 2007).

In deep-marine sediment, MIS 7 includes three global ice-volume minima (MIS 7e, 7c, and 7a), of which the younger two and the termination of the oldest are represented in BL00-1. Sediment apparently deposited in Bear Lake during MIS 7c and 7a exhibits maxima in $\delta^{18}O$ and carbonate content, and both intervals contain aragonite, suggesting that salinity was relatively high. Sediment of the younger substage (MIS 7a) exhibits an increase in $^{87}Sr/^{86}Sr$ values, whereas the next older substage (MIS 7c) does not. The aragonite content is higher during MIS 7a than 7c, and the ratio of "warm+dry" to "cold" pollen indicators reaches its maximum value during MIS 7a. Bear River apparently did not discharge into Bear Lake during MIS 7a. During the older substage, however, the evidence for Bear River input is equivocal, although the presence of aragonite indicates that the salinity and Mg to Ca ratio of the lake were relatively high for at least brief intervals.

Sediment deposited during MIS 5 includes a prominent aragonite-rich zone deposited early during the interval, which we correlate with MIS 5e. Like other interglacial intervals, the pollen spectra of MIS 5e are dominated by high-desert shrub taxa, with abundant juniper, indicating warm and dry conditions. The diatom

data are all suggestive of a shallow lake. Dramatic shifts in mineralogy and isotopes occur in the middle of MIS 5e, and almost certainly record the input of the Bear River. We cannot discern whether the re-entry of the river resulted from a non-climatically influenced avulsion, or from a climate-induced lake-level transgression that captured the river. The return of $\delta^{18}O$ values to their pre-MIS 5 values suggests that effective moisture was high (similar to conditions during glacial intervals) during the excursion. Similar climate reversals during MIS 5e have been noted elsewhere (e.g., Thouveny et al., 1994; Seidenkrantz et al., 1995).

Later during MIS 5, two peaks in $CaCO_3$ abundance, including one with aragonite, correlate with peaks in the proportion of "warm" and "dry" pollen indicators, and apparently coincide with the interstadials MIS 5c and 5a. During MIS 5c, the ratio of quartz to dolomite and the ratio of "warm+dry" to "cold" pollen indicators both attain values similar to those of MIS 5e, suggesting the return of peak interglacial conditions, with an increase in locally derived, dolomite-rich fluvial sediment input relative to dolomite-poor Bear River sediment input. During the intervening stadial intervals, $\delta^{18}O$ and $^{87}Sr/^{86}Sr$ in bulk sediment attain values that are among the lowest in the core. We interpret the isotopic data from these stadial deposits as indicating high effective moisture, probably resulting from a combination of higher precipitation and lower temperature.

During the Holocene (MIS 1), $\delta^{18}O$ and $^{87}Sr/^{86}Sr$ attain maximum values similar to those during MIS 5e. $CaCO_3$ abundance (almost all aragonite) is the highest of the entire quarter-million-year sequence. In contrast, the ratio of "warm+dry" to "cold" pollen indicators is of an intermediate value for interglacial periods. Our sample interval is too coarse to capture centennial-scale climate changes during the last glacial-interglacial transition. A prominent reversal to calcite deposition during the early Holocene is well studied in shorter cores and has been interpreted as the re-entry of Bear River into Bear Lake (Dean et al., 2006; Dean, this volume), perhaps coinciding with the transgression to the Willis Ranch shoreline (Laabs and Kaufman, 2003; Reheis et al., this volume). The 1000-year-long freshening event is succeeded in the shorter cores by a coarsening of grain size, indicating lake-level lowering just prior to the return to the aragonite precipitation that dominated the Holocene (Smoot and Rosenbaum, this volume).

Glacial Intervals

With the exception of the three interglacial periods represented by the prominent aragonitic zones, we infer that the Bear River flowed into Bear Lake, either directly via a channel or indirectly via a wetland, throughout most of the last quarter-million years. The extent to which water and sediment input from Bear River influenced the sedimentation at the core site depended on both climatic and non-climatic factors. Consequently, the manifestations of the glacial intervals were markedly different from one another.

During the penultimate glacial period (MIS 6) the ratio of "warm+dry" to "cold" pollen indicators is lower overall than during the last glacial cycle (MIS 4–2). Values of $\delta^{18}O$ in bulk-sediment carbonate became progressively higher (gray dashed line in Fig. 5C), and the proportion of quartz increased. These trends are interpreted as an increase in allogenic input. The correlation between quartz abundance and $^{87}Sr/^{86}Sr$ values is consistent with the influence of allogenic carbonate, particularly the proportion of stream sediment from the Bear River, relative to endogenic carbonate. The lowest $\delta^{18}O$ values in ostracodes were registered during this interval, also indicative of increased meltwater input. We interpret these trends as the progressive buildup of glacial sediment and meltwater from the Uinta Mountains. Although the analyses of $^{87}Sr/^{86}Sr$ in the acid-insoluble (residue) fraction are sparse, the high $^{87}Sr/^{86}Sr$ values in MIS 6 indicate a strong meltwater signal from the Uinta Mountains. Magnetic properties are equivocal as to the source of the allogenic material, probably because Fe-oxide minerals have been altered in sediments older than MIS 3 (Heil et al., this volume). Quartz content reached its maximum value immediately prior to the rapid regression that led to the sudden precipitation of endogenic carbonate, mostly aragonite. The low $\delta^{18}O$ values and high quartz values might represent an increase in Bear River sediment over endogenic carbonate and might coincide with the high lake level that occurred late during the Bear Hollow phase of Bear Lake (Laabs and Kaufman, 2003). The age constraints on the shoreline deposits are broad (200–100 ka) and permissive of a range of possible correlations to events in BL00-1.

During the last glacial cycle, the 110,000-year period extending from MIS 5d through MIS 2, $\delta^{18}O$ values in bulk sediment exhibit four relatively evenly spaced minima ca. 100, 80, 55, and 15 ka (gray dashed lines in Fig. 5C). These minima terminate periods of decreasing $\delta^{18}O$ values in bulk sediment and are followed by rapid increases in $\delta^{18}O$, resulting in saw-toothed patterns reminiscent of those exhibited by $\delta^{18}O$ in Greenland ice (e.g., Stuiver and Grootes, 2000). These abrupt increases in $\delta^{18}O$ might relate to rapid warming following the peak periods of cold/wet climate and attendant mountain glaciation in the Bear Lake catchment. The penultimate minimum in bulk-sediment $\delta^{18}O$ values at 29 m blf (ca. 55 ka), unlike the preceding two minima, does not show a parallel decrease in ostracode $\delta^{18}O$ values (Fig. 5). Instead, $\delta^{18}O$ values in ostracode valves exhibit two prominent minima at 24 and 19 m blf (ca. 45 and 35 ka). These minima occur following the Laschamp excursion in BL00-1, and their approximated ages are probably somewhat too old. Nonetheless, they coincide broadly with the Jensen Spring highstand of Bear Lake, which is dated to 47–39 ka by ^{14}C and amino acid geochronology (Laabs and Kaufman, 2003). During the Jensen Spring phase, lake level rose by 11 m, producing well-developed spits and other shoreline features in northern Bear Lake Valley. The proportion of quartz is also high during this interval, consistent with enhanced discharge from the Bear River.

Between late MIS 3 and early MIS 2, the absence of diatoms suggests that cold conditions persisted. The proportion of pollen from conifers and other cold-tolerant taxa is high, further suggesting cold conditions. Magnetic properties, trace-element geochemistry, mineralogy, and grain size in core BL96-3 indicate that the amount of sediment derived from the Uinta Mountains varied during this interval (Rosenbaum et al., this volume). On the basis of hematite abundance (HIRM) in BL00-1, and following the inter-

pretation of HIRM in shorter piston cores (Rosenbaum and Heil, this volume), the delivery of glacial flour from the Uinta Mountains to Bear Lake increased dramatically late during MIS 3. In addition, this interval contains the only distinctively red sediment in BL00-1 (unit 6, Fig. 4), reflecting an increase in the proportion of Uinta-derived sediment transported to the core site or the preservation of Fe oxides, or most likely both. Lake level was stable near the present level when the red sediment was deposited (Smoot and Rosenbaum, this volume). We surmise that the overflow threshold of the lake was eroded to near its present elevation, and that the Bear River entered and exited the lake without the shoreline transgressing the wetlands north of the lake (Fig. 10).

MIS 2 was an unremarkable interval in the sedimentary record of Bear Lake, other than the preservation of a prominent red, siliciclastic unit. The absence of a peak in the proportion of quartz at the close of the glacial period is difficult to reconcile with the major transgression in the terminal Bonneville Basin. In contrast to Lake Bonneville, no highstand deposits dating to the last glacial maximum have yet been discovered in the Bear Lake Valley, although this might reflect the geomorphic controls on the threshold elevation rather than the hydrologic balance of the lake. As in the shorter piston cores (Rosenbaum and Heil, this volume), the timing of the most recent peak in HIRM in BL00-1 agrees with the cosmogenic exposure ages on moraines formed during the local last glacial maximum in the headwaters of the Bear River in the Uinta Mountains, which have been dated to ca. 19–17 ka (Laabs et al., 2007). This interval also contains high $^{87}Sr/^{86}Sr$ values in the residue fraction. The last $\delta^{18}O$ minimum took place ca. 15 ka, when the level of Bear Lake rose by 8 m and formed shoreline deposits ascribed to the Raspberry Square phase (Laabs and Kaufman, 2003). Sedimentologic evidence from a shallower core also indicates a sharp lake-level rise at this time (Smoot and Rosenbaum, this volume).

Comparison with the Great Salt Lake Subbasin

Site BL00-1 was drilled as part of the same Global Lakes Drilling (GLAD) operation that recovered cores from Great Salt Lake (Dean et al., 2002). Great Salt Lake is the terminus of the Bear River drainage, and it occupies the largest subbasin of the Bonneville drainage basin. At times during the Pleistocene, the Great Salt Lake was isolated from the upper part of the Bear River drainage (Bouchard et al., 1998). The magnitude of lake-level fluctuations in the Great Salt Lake (i.e., Lake Bonneville and its predecessors) was much greater than in Bear Lake because of differences in their topographic confinement. Bear Lake never entirely dried out during the last 220,000 years or longer. During arid intervals, it generated abundant carbonate, but it never evaporated enough to precipitate evaporite minerals (e.g., gypsum and halite). In contrast, salts precipitated and wetland deposits accumulated in the Great Salt Lake during arid intervals (e.g., Oviatt et al., 1999). Although their sedimentary records are distinct, the first-order changes at both lakes are influenced primarily by lake level, driven by periodic changes in the delivery of Pacific moisture. At Bear Lake, lake level dictated whether the Bear River was in or out of the lake; at Great Salt Lake, lake level determined whether a core site was submerged or subaerial. The two lakes both responded to climate changes, but differed in their response times. The drainage area of Great Salt Lake is much larger than that of Bear Lake, and its hydrologic budget is influenced more strongly by lake-effect precipitation, implying some inherent stability (Hostetler et al., 1994; Laabs et al., 2006). When comparing the sedimentary sequences at Bear Lake and Great Salt Lake, we therefore expect broad-scale similarities in sedimentary features driven by climate, along with secondary differences related to differences in local hydrology.

Recently, Balch et al. (2005) described the sedimentary sequence and analyzed the ostracode fauna in a 120-m-long, ~280,000-year core from Great Salt Lake (GSL00-4). Their data indicate that the environment at the core site, located 9.3 m below present-day lake level, mainly alternated between a saline/hypersaline lake and a saline marsh. In addition, on the basis of their coarse sampling interval (1 m spacing = 2000 yr), they identified four intervals of freshwater marsh deposits and paleosols formed when the site was above lake level, and two intervals when the site was below a deep, freshwater lake. We focus our comparison of the GLAD cores from Bear Lake and Great Salt Lake on these more extreme events, starting with the two deep-water lakes.

The penultimate deep-lake cycle in the Bonneville Basin is represented by the Little Valley shoreline, which has been dated to ca. 150 ka by amino acid geochronology and other evidence (Scott et al., 1983). Using the currently accepted age model for GSL00-4 in the Great Salt Lake, however, Balch et al. (2005) inferred an older age of 170 ± 20 ka for the penultimate deep-water lake. At Bear Lake, the quartz maximum and $CaCO_3$ minimum located just a few meters below the MIS 5e aragonite might correlate with the Little Valley shoreline in the Bonneville Basin. If the age of the Little Valley shoreline is closer to 170 ka, as suggested by Balch et al. (2005), then we find no extraordinary features around this time in BL00-1 to mark the event, other than the low $\delta^{18}O$ value in ostracodes at 88 m blf (ca. 165 ka), although an equally low value occurs again at 78 m blf (ca. 145 ka). Nor do we find evidence for extreme aridity in BL00-1 that might coincide with the prominent lower salt layer in GSL00-4 that formed sometime between 170 and 140 ka, immediately following that deep-lake event (Balch et al., 2005).

The Lake Bonneville deep-lake phase occurred between ca. 25 and 15 ka (Oviatt et al., 1992). The last 5000 years of this interval corresponds to the $\delta^{18}O$ minimum of the most recent 25,000-year "saw tooth" in BL00-1. The decreasing quartz content after 25 ka during MIS 2 is difficult to reconcile with the evidence of extremely high lake levels downstream in Lake Bonneville. It points to intra-regional heterogeneity in climate change and its hydrologic response.

In addition to deep lakes during MIS 6 and MIS 2, lake level in the Great Salt Lake subbasin was high early during MIS 3, ca. 60 ka, as indicated by high flood-plain deposits of Bear River near Great Salt Lake (Kaufman et al., 2001), and by freshening of lake water at site GSL00-4 at about the same time

(Balch et al., 2005). A correlative of this event might be represented by the minimum in $\delta^{18}O$ values in bulk-sediment carbonate and "warm+dry" to "cold" pollen indicators around 30 m blf (ca. 55 ka) in BL00-1. The $\delta^{18}O$ values in ostracode calcite do not exhibit minima until ~25 and 19 m blf (45 and 35 ka), however, and these ages are probably too old considering their superposition relative to the assumed Laschamp excursion.

The intervals of exceptionally low lake level identified in GSL00-4 (Balch et al., 2005) appear to correlate with evidence of aridity in BL00-1, at least in three of the four cases. According to the current age model for GSL00-4, these occurred at 215, 130, 105, and 45 ka. The oldest three correspond roughly to prominent $\delta^{18}O$ maxima and aragonite formation in BL00-1 during MIS 7c, 5e, and 5c (Fig. 5). The lower-than-present lake level in the Great Salt Lake basin ca. 45 ka might correspond to the maximum in bulk-sediment $\delta^{18}O$ value that occurred immediately following the presumed Laschamp excursion in BL00-1. On the other hand, ostracodes from the same level have $\delta^{18}O$ values that are among the lowest in the core, and low $^{87}Sr/^{86}Sr$ values indicate input from the Bear River. Furthermore, this interval coincides with the Jensen Spring highstand deposits of Bear Lake (47–39 ka; Laabs and Kaufman, 2003). Reconciling these apparently conflicting observations will require further analyses and an improved geochronology.

SUMMARY AND CONCLUSIONS

Core BL00-1 provides a continuous sedimentary record of hydrologic and paleoenvironmental changes in the Bear Lake catchment during the last quarter-million years. The sedimentary sequence from near the depocenter of the lake exhibits major changes in all of the mineralogical, geochemical, isotopic, and paleontological indicators that we analyzed. These variations reflect changes in the water and sediment discharged from the glaciated headwaters of the dominant tributary, Bear River, and the processes that influenced sediment delivery to the core site, including lake-level changes. The first-order fluctuations coincide with orbital cycles and global ice volume. Millennial-scale fluctuations are pervasive throughout the last two glacial cycles and might correspond to stadial-interstadial cycles recognized in other well-known paleoclimate records, but the age control for BL00-1 is presently too uncertain.

In addition to climatic controls, the hydrogeography of the Bear River also influenced the paleoenvironmental indicators in BL00-1, and the two probably worked in concert. Isolation of the lake from the Bear River likely could have been maintained only when effective moisture was low. During most of the last quarter-million years, the Bear River discharged into Bear Lake, although the connection might have been attenuated by wetlands that separated the lake and the main channel. Our data indicate that Bear Lake retracted into a topographically closed basin during portions of global interglaciations (MIS 7c, 7a, 5e, 5c, and 1). During these intervals, the lake generated abundant endogenic carbonate with aragonite and high values of $\delta^{18}O$ and $^{87}Sr/^{86}Sr$. The ratio of "warm+dry" to "cold" pollen indicators was highest during MIS 7a, 5e, and 5a; the present interglaciation (MIS 1) exhibits intermediate values. During these interglacial intervals, low ratios of quartz to dolomite (allogenic component) indicate the dominance of locally derived stream sediment over Bear River sediment. Both MIS 5e and MIS 1 intervals include excursions of the Bear River into the lake. The excursion during MIS 1 might correspond to global climate changes during the last glacial-interglacial transition; whether the excursion during MIS 5e was induced climatically remains to be determined, although the low $\delta^{18}O$ values suggests that it was.

Sediment deposited during the penultimate glacial period (MIS 6) contrasts with sediment deposited during the last glacial cycle (MIS 4–2), although diatom data suggest that the termination of both events was characterized by particularly harsh conditions. The ratio of "warm+dry" to "cold" pollen indicators is lower overall during MIS 6, and the sediment exhibits a progressive increase in values of $\delta^{18}O$ and $^{87}Sr/^{86}Sr$, with parallel increases in quartz content. We interpret these changes as the progressive increase in the proportion of clastic sediment derived from the Uinta Mountains, reflecting the increased influence of glacial erosional products. The 100,000-year period following the last interglaciation (i.e., MIS 5d through MIS 2) comprises four relatively evenly spaced intervals of slowly decreasing $\delta^{18}O$ values, followed by an abrupt increase. The cause of these saw-toothed cycles is not clear, but the last two terminate about the time of high lake stands in Bear Lake Valley, and at least the youngest one coincides with the end of a deep-lake cycle downstream in the Bonneville Basin. The youngest one is characterized by the only red mud with preserved Fe-oxide minerals in the core, indicative of a fluvially dominated, through-flowing lake connected to Bear River.

ACKNOWLEDGMENTS

We thank Dennis Nielson and the DOSECC drilling crew for their essential role in obtaining the Bear Lake GLAD800 cores, and D. Schnurrenberger, B. Haskell, B. Valero-Garcés, and others who worked on the drilling barge and performed much of the initial core description. G. Skipp and D. Thornbury performed the X-ray diffraction and carbon analyses. A. Cohen, R. Reynolds, S. Starratt, and J. Stone provided valuable reviews of this chapter.

ARCHIVED DATA

Archived data for this chapter can be obtained from the NOAA World Data Center for Paleoclimatology at http://www.ncdc.noaa.gov/paleo/pubs/gsa2009bearlake/.

REFERENCES CITED

Balch, D.P., Cohen, A.S., Schnurrenberger, D.W., Haskell, B.J., Valero-Garcés, B.L., Beck, J.W., Cheng, H., and Edwards, R.L., 2005, Ecosystem and paleohydrological response to Quaternary climate change in the Bonneville Basin, Utah: Palaeogeography, Palaeoclimatology, Palaeoecology, v. 221, p. 99–122, doi: 10.1016/j.palaeo.2005.01.013.

Benson, L., Lund, S., Negrini, R., Linsley, B., and Zic, M., 2003, Response of North American Great Basin lakes to Dansgaard-Oeschger oscillations: Quaternary Science Reviews, v. 22, p. 2239–2251, doi: 10.1016/S0277-3791(03)00210-5.

Bouchard, D.P., Kaufman, D.S., Hochberg, A., and Quade, J., 1998, Quaternary history of the Thatcher Basin, Idaho, reconstructed from the $^{87}Sr/^{86}Sr$ and amino acid composition of lacustrine fossils—Implications for the diversion of the Bear River into the Bonneville basin: Palaeogeography, Palaeoclimatology, Palaeoecology, v. 141, p. 95–114, doi: 10.1016/S0031-0182(98)00005-4.

Bradbury, J.P., 1988, Diatom biostratigraphy and the paleolimnology of Clear Lake County, California, *in* Sims, J.D., ed., Late Quaternary climate, tectonism and sedimentation in Clear Lake, northern California Coast Ranges: Geological Society of America Special Paper 214, p. 97–129.

Bradbury, J.P., 1997, A diatom record of climate and hydrology for the past 200 ka from Owens Lake, California with comparison to other great basin records: Quaternary Science Reviews, v. 16, p. 203–219, doi: 10.1016/S0277-3791(96)00054-6.

Bradbury, J.P., and Forester, R.M., 2002, Environment and paleolimnology of Owens Lake, California: A record of climate and hydrology for the last 50,000 years, *in* Hershler, R., Madsen, D.B., and Currey, D.R., eds., Great Basin aquatic systems history: Washington D.C., Smithsonian Institution Press, p. 145–173.

Bradbury, J.P., Bezrukova, Ye. E., Chernyaeva, G.P., Colman, S.M., Khursevich, G., King, J.W., and Likoshway, Ye. V., 1994, A synthesis of post-glacial diatom records from Lake Baikal: Journal of Paleolimnology, v. 10, p. 213–252, doi: 10.1007/BF00684034.

Bradshaw, E.G., Jones, V.J., Birks, H.J.B., and Birks, H.H., 2000, Diatom responses to late-glacial and early-Holocene environmental changes at Krakenes, western Norway: Journal of Paleolimnology, v. 23, p. 21–34, doi: 10.1023/A:1008021016027.

Bright, J., 2009, this volume, Chapter 4, Isotope and major-ion chemistry of groundwater in Bear Lake Valley, Utah and Idaho, with emphasis on the Bear River Range, *in* Rosenbaum, J.G., and Kaufman, D.S., eds., Paleoenvironments of Bear Lake, Utah and Idaho, and its catchment: Geological Society of America Special Paper 450, doi: 10.1130/2009.2450(04).

Bright, J., 2009, this volume, Chapter 8, Ostracode endemism in Bear Lake, Utah and Idaho, *in* Rosenbaum, J.G., and Kaufman, D.S., eds., Paleoenvironments of Bear Lake, Utah and Idaho, and its catchment: Geological Society of America Special Paper 450, doi: 10.1130/2009.2450(08).

Bright, J., Forester, R., and Kaufman, D., 2005, Ostracode analysis for cores BL96-1 and BL96-2 from Bear Lake, Utah and Idaho: U.S. Geological Survey Open-File Report 2005-1227, http://pubs.usgs.gov/of/2005/1227/.

Bright, J., Kaufman, D.S., Forester, R.M., and Dean, W.E., 2006, A continuous 250,000 yr record of oxygen and carbon isotopes in ostracode and bulk-sediment carbonate from Bear Lake, Utah-Idaho: Quaternary Science Reviews, v. 25, no. 17–18, p. 2258–2270, doi: 10.1016/j.quascirev.2005.12.011.

Colman, S.M., 2006, Acoustic stratigraphy of Bear Lake, Utah-Idaho—Late Quaternary sedimentation in a simple half-graben: Sedimentary Geology, v. 185, p. 113–125, doi: 10.1016/j.sedgeo.2005.11.022.

Colman, S.M., Kaufman, D.S., Bright, J., Heil, C., King, J.W., Dean, W.E., Rosenbaum, J.G., Forester, R.M., Bischoff, J.L., Perkins, M., and McGeehin, J.P., 2006, Age models for a continuous 250-kyr Quaternary lacustrine record from Bear Lake, Utah-Idaho: Quaternary Science Reviews, v. 25, p. 2271–2282, doi: 10.1016/j.quascirev.2005.10.015.

Colman, S.M., Rosenbaum, J.G., Kaufman, D.S., Dean, W.E., and McGeehin, J.P., 2009, this volume, Radiocarbon ages and age models for the last 30,000 years in Bear Lake, Utah and Idaho, *in* Rosenbaum, J.G., and Kaufman, D.S., eds., Paleoenvironments of Bear Lake, Utah and Idaho, and its catchment: Geological Society of America Special Paper 450, doi: 10.1130/2009.2450(05).

Dean, W.E., 2009, this volume, Endogenic carbonate sedimentation in Bear Lake, Utah and Idaho, over the last two glacial-interglacial cycles, *in* Rosenbaum, J.G., and Kaufman, D.S., eds., Paleoenvironments of Bear Lake, Utah and Idaho, and its catchment: Geological Society of America Special Paper 450, doi: 10.1130/2009.2450(07).

Dean, W., Rosenbaum, J., Haskell, B., Kelts, K., Schnurrenberger, D., Valero-Garcés, B., Cohen, A., Davis, O., Dinter, D., and Nielson, D., 2002, Progress in global lake drilling holds potential for global change research: Eos (Transactions, American Geophysical Union), v. 83, p. 85, 90–91.

Dean, W., Rosenbaum, J., Skipp, G., Colman, S., Forester, R., Simmons, K., Liu, A., and Bishoff, J., 2006, Unusual Holocene and late Pleistocene carbonate sedimentation in Bear Lake, Utah and Idaho, USA: Sedimentary Geology, v. 185, p. 93–112, doi: 10.1016/j.sedgeo.2005.11.016.

Dean, W.E., Forester, R.M., Bright, J., and Anderson, R.Y., 2007, Influence of the diversion of the Bear River into Bear Lake (Utah and Idaho) on the environment of deposition of carbonate minerals: Limnology and Oceanography, v. 53, p. 1094–1111.

Dean, W.E., Wurtsbaugh, W.A., and Lamarra, V.A., 2009, this volume, Climatic and limnologic setting of Bear Lake, Utah and Idaho, *in* Rosenbaum, J.G., and Kaufman, D.S., eds., Paleoenvironments of Bear Lake, Utah and Idaho, and its catchment: Geological Society of America Special Paper 450, doi: 10.1130/2009.2450(01).

Denny, J.F., and Colman, S.M., 2003, Geophysical surveys of Bear Lake, Utah-Idaho, September 2002: U.S. Geological Survey Open-File Report 03-150.

Doner, L.A., 2009, this volume, A 19,000-year vegetation and climate record for Bear Lake, Utah and Idaho, *in* Rosenbaum, J.G., and Kaufman, D.S., eds., Paleoenvironments of Bear Lake, Utah and Idaho, and its catchment: Geological Society of America Special Paper 450, doi: 10.1130/2009.2450(09).

Douglas, M.S.V., Smol, J.P., and Blake, W., Jr., 1994, Marked post-18th century environmental change in high-Arctic ecosystems: Science, v. 266, p. 416–419, doi: 10.1126/science.266.5184.416.

Engleman, E.E., Jackson, L.L., Norton, D.R., and Fischer, A.G., 1985, Determination of carbonate carbon in geological materials by coulometric titration: Chemical Geology, v. 53, p. 125–128, doi: 10.1016/0009-2541(85)90025-7.

Evans, J.P., Martindale, D.C., and Kendrick, R.D., Jr., 2003, Geologic setting of the 1884 Bear Lake, Idaho, earthquake: Rupture in the hanging wall of a basin and range normal fault revealed by historical and geological analyses: Bulletin of the Seismological Society of America, v. 93, p. 1621–1632, doi: 10.1785/0120020159.

Flower, R.J., 1993, Diatom preservation—Experiments and observations on dissolution and breakage in modern and fossil material: Hydrobiologia, v. 269–270, p. 473–484, doi: 10.1007/BF00028045.

Heil, C.W., Jr., King, J.W., Rosenbaum, J.G., Reynolds, R.L., and Colman, S.M., 2009, this volume, Paleomagnetism and environmental magnetism of GLAD800 sediment cores from Bear Lake, Utah and Idaho, *in* Rosenbaum, J.G., and Kaufman, D.S., eds., Paleoenvironments of Bear Lake, Utah and Idaho, and its catchment: Geological Society of America Special Paper 450, doi: 10.1130/2009.2450(13).

Hendy, I.L., and Kennett, J.P., 1999, Latest Quaternary North Pacific surface water responses imply atmosphere-driven climate instability: Geology, v. 27, p. 291–294, doi: 10.1130/0091-7613(1999)027<0291:LQNPSW>2.3.CO;2.

Hostetler, S.W., Giorgi, F., Bates, G.T., and Bartlein, P.J., 1994, Lake-atmosphere feedbacks associated with paleolakes Bonneville and Lahontan: Science, v. 263, p. 665–668, doi: 10.1126/science.263.5147.665.

Jiménez-Moreno, G., Anderson, R.S., and Fawcett, P.J., 2007, Orbital- and millennial-scale vegetation and climate changes of the past 225 ka from Bear Lake, Utah-Idaho (USA): Quaternary Science Reviews, v. 26, p. 1713–1724, doi: 10.1016/j.quascirev.2007.05.001.

Karabanov, E., Williams, D., Kuzmin, M., Sideleva, V., Khursevich, G., and Prokopenko, A.A., 2004, Ecological collapse of Lake Baikal and Lake Hovsgal, ecosystems during the last glacial and consequences for aquatic species diversity: Palaeogeography, Palaeoclimatology, Palaeoecology, v. 209, p. 227–243, doi: 10.1016/j.palaeo.2004.02.017.

Karst, T.L., and Smol, J.P., 2000, Paleolimnological evidence of limnetic nutrient concentration equilibrium in a shallow, macrophyte-dominated lake: Aquatic Sciences, v. 62, p. 20–38, doi: 10.1007/s000270050073.

Kaufman, D.S., Forman, S.L., and Bright, J., 2001, Age of the Cutler Dam Alloformation (late Pleistocene), Bonneville basin, Utah: Quaternary Research, v. 56, p. 322–334, doi: 10.1006/qres.2001.2275.

Krammer, K., and Lange-Bertalot, H., 1986–1991, Süßwasserflora von MitteEuropa, Bacillariophyceae Band 2/1-4: Gustav Fischer Verlag, Stuttgart, Germany.

Laabs, B.J., and Kaufman, D.S., 2003, Quaternary highstands in Bear Lake Valley, Utah and Idaho: Geological Society of America Bulletin, v. 115, p. 463–478, doi: 10.1130/0016-7606(2003)115<0463:QHIBLV>2.0.CO;2.

Laabs, B.J.C., Plummer, M.A., and Mickelson, D.M., 2006, Climate during the last glacial maximum in the Wasatch and southern Uinta Mountains inferred from glacier modeling: Geomorphology, v. 75, p. 300–317, doi: 10.1016/j.geomorph.2005.07.026.

Laabs, B.J.C., Munroe, J.S., Rosenbaum, J.G., Refsnider, K.A., Mickelson, D.M., Singer, B.S., and Chafee, M.W., 2007, Chronology of the last glacial maximum in the upper Bear River Basin, Utah: Arctic, Antarctic, and Alpine Research, v. 39, p. 537–548, doi: 10.1657/1523-0430(06-089)[LAABS]2.0.CO;2.

Lamarra, V., Liff, C., and Carter, J., 1986, Hydrology of Bear Lake basin and its impact on the trophic state of Bear Lake, Utah-Idaho: The Great Basin Naturalist, v. 46, p. 690–705.

MacKay, A.W., 2007, The paleoclimatology of Lake Baikal: A diatom synthesis and prospectus: Earth-Science Reviews, v. 82, p. 181–215, doi: 10.1016/j.earscirev.2007.03.002.

McConnell, W.J., Clark, W.J., and Sigler, W.F., 1957, Bear Lake: Its fish and fishing: Utah State Department of Fish and Game, Idaho Department of Fish and Game, Wildlife Management Department of Utah State Agricultural College, 76 p.

Moore, D.M., and Reynolds, R.C., Jr., 1989, X-ray diffraction and identification and analysis of clay minerals: Oxford, UK, Oxford University Press, 332 p.

Moser, K.A., and Kimball, J.P., 2009, this volume, A 19,000-year record of hydrologic and climatic change inferred from diatoms from Bear Lake, Utah and Idaho, *in* Rosenbaum, J.G., and Kaufman, D.S., eds., Paleoenvironments of Bear Lake, Utah and Idaho, and its catchment: Geological Society of America Special Paper 450, doi: 10.1130/2009.2450(10).

Oviatt, C.G., Currey, D.R., and Sack, D., 1992, Radiocarbon chronology of Lake Bonneville, eastern Great Basin, USA: Palaeogeography, Palaeoclimatology, Palaeoecology, v. 99, p. 225–241, doi: 10.1016/0031-0182(92)90017-Y.

Oviatt, C.G., Thompson, R.S., Kaufman, D.S., Bright, J., and Forester, R.M., 1999, Reinterpretation of the Burmester core, Bonneville basin, Utah: Quaternary Research, v. 52, p. 180–184, doi: 10.1006/qres.1999.2058.

Reheis, M.C., Laabs, B.J.C., and Kaufman, D.S., 2009, this volume, Geology and geomorphology of Bear Lake Valley and upper Bear River, Utah and Idaho, *in* Rosenbaum, J.G., and Kaufman, D.S., eds., Paleoenvironments of Bear Lake, Utah and Idaho, and its catchment: Geological Society of America Special Paper 450, doi: 10.1130/2009.2450(02).

Reynolds, R.L., and Rosenbaum, J.G., 2005, Magnetic mineralogy of sediments in Bear Lake and its watershed: Support for paleomagnetic and paleoenvironmental interpretations: U.S. Geological Survey Open-File Report 2005-1406, http://pubs.usgs.gov/of/2005/1406/.

Rosenbaum, J.G., and Heil, C.W., Jr., 2009, this volume, The glacial/deglacial history of sedimentation in Bear Lake, Utah and Idaho, *in* Rosenbaum, J.G., and Kaufman, D.S., eds., Paleoenvironments of Bear Lake, Utah and Idaho, and its catchment: Geological Society of America Special Paper 450, doi: 10.1130/2009.2450(11).

Rosenbaum, J.G., and Kaufman, D.S., 2009, this volume, Introduction to *Paleoenvironments of Bear Lake, Utah and Idaho, and its catchment, in* Rosenbaum, J.G., and Kaufman, D.S., eds., Paleoenvironments of Bear Lake, Utah and Idaho, and its catchment: Geological Society of America Special Paper 450, doi: 10.1130/2009.2450(00).

Rosenbaum, J.G., Dean, W.E., Reynolds, R.L., and Reheis, M.C., 2009, this volume, Allogenic sedimentary components of Bear Lake, Utah and Idaho, *in* Rosenbaum, J.G., and Kaufman, D.S., eds., Paleoenvironments of Bear Lake, Utah and Idaho, and its catchment: Geological Society of America Special Paper 450, doi: 10.1130/2009.2450(06).

Scott, W.E., McCoy, W.D., Shroba, R.R., and Rubin, M., 1983, Reinterpretation of the exposed record of the last two cycles of Lake Bonneville, western United States: Quaternary Research, v. 20, p. 261–285, doi: 10.1016/0033-5894(83)90013-3.

Seidenkrantz, P., Kristensen, P., and Knudsen, K.L., 1995, Marine evidence for climatic instability during the last interglacial in shelf records from northwest Europe: Journal of Quaternary Science, v. 10, p. 77–82, doi: 10.1002/jqs.3390100108.

Simpson, E.H., 1949, Measurement of diversity: Nature, v. 163, p. 688, doi: 10.1038/163688a0.

Smoot, J.E., and Rosenbaum, J.G., 2009, this volume, Sedimentary constraints on late Quaternary lake-level fluctuations at Bear Lake, Utah and Idaho, *in* Rosenbaum, J.G., and Kaufman, D.S., eds., Paleoenvironments of Bear Lake, Utah and Idaho, and its catchment: Geological Society of America Special Paper 450, doi: 10.1130/2009.2450(12).

Stuiver, M., and Grootes, P.M., 2000, GISP2 oxygen isotope ratios: Quaternary Research, v. 53, p. 277–284, doi: 10.1006/qres.2000.2127.

Thouveny, N., de Beaulieu, J.-L., Bonifay, E., Creer, K.M., Guiot, J., Icole, M., Johnsen, S., Jouzel, J., Reille, M., Williams, T., and Williamson, D., 1994, Climate variations in Europe over the last 140 kyr deduced from rock magnetism: Nature, v. 371, p. 503–506, doi: 10.1038/371503a0.

von Grafenstien, U., Erlernkeuser, H., and Trimborn, P., 1999, Oxygen and carbon isotopes in modern fresh-water ostracode valves: Assessing the vital offsets and autecological effects of interest for palaeoclimatic studies: Palaeogeography, Palaeoclimatology, Palaeoecology, v. 148, p. 133–152, doi: 10.1016/S0031-0182(98)00180-1.

Winograd, I.J., Coplen, T.B., Landwehr, J.M., Riggs, A.C., Ludwig, K.R., Szabo, B.J., Kolesar, P.T., and Revesz, K.M., 1992, Continuous 500,000-year climate record from vein calcite in Devils Hole, Nevada: Science, v. 258, p. 255–260, doi: 10.1126/science.258.5080.255.

Wolin, J.A., and Duthie, H.C., 1999, Diatoms as indicators of water level change in freshwater lakes, *in* Stoermer E.F., and Smol, J.P., eds., The diatoms: Applications for the environmental and earth sciences: Cambridge, UK, Cambridge University Press, p. 183–226.

MANUSCRIPT ACCEPTED BY THE SOCIETY 15 SEPTEMBER 2008

Printed in the USA

APPENDIX A. MINERALOGY, INORGANIC, AND ORGANIC CARBON CONTENT, BL00-1

Sample #	Depth (m blf)	Quartz (%)	Plag. (%)	Arag. (%)	Dolo. (%)	Calcite (%)	CaCO$_3$ (%)	Organic carbon (%)
1h1 5	0.45	27.9	4.9	48.8	5.3	13.1	60.9	1.9
1h1 89.9	1.29	29.4	5.3	50.4	4.4	10.4	66.9	2.2
1h1 110.2	1.50	21.3	0.0	63.5	4.8	10.4	58.8	3.9
1E-1h	2.15	15.8	0.0	63.8	9.5	10.9	75.6	1.7
1h2 60	2.51	22.6	0.1	62.9	4.6	9.9	72.3	1.8
1h2 90	2.81	18.1	0.1	66.8	4.8	10.3	76.8	1.1
1h2 106	2.97	22.9	0.1	63.9	3.8	9.4	76.7	1.2
1D-1h	3.40	17.4	6.2	64.6	4.4	7.4	79.3	1.0
2h1 70	4.10	17.5	3.6	68.5	3.6	6.9	76.3	1.4
1E-2h	5.15	14.8	0.0	69.0	5.0	11.2	78.4	1.1
2h2 60	5.51	18.5	4.1	61.4	5.3	10.8	75.8	1.3
2h2 85	5.76	21.1	4.1	58.4	5.4	11.1	68.9	1.8
1D-2h	6.40	18.2	3.5	63.1	4.4	10.8	75.8	1.1
3h2 109.6	7.57	27.0	0.2	45.1	5.9	21.8	62.6	2.3
3h2 117	7.64	35.8	7.1	35.1	5.4	16.7	51.0	3.9
3h2 125.4	7.72	22.6	4.1	58.0	4.5	10.8	74.1	1.3
3h2 148.5	7.96	22.2	0.2	61.3	4.5	11.7	74.7	1.8
1E-3h	8.15	18.1	4.0	60.5	6.7	10.8	74.5	2.4
3h3 40	8.39	24.6	5.7	47.0	6.2	16.5	67.8	1.6
3h3 90	8.89	19.8	2.7	0.0	3.4	74.1	63.6	1.6
1D-3h	9.40	65.3	7.4	0.0	6.3	21.0	10.4	1.0
4h1 63	10.03	45.6	4.0	0.0	6.2	44.2	27.9	1.8
4h1 84	10.24	68.2	6.5	0.0	10.1	15.2	13.9	1.7
4h1 140	10.80	66.5	4.5	0.0	7.0	22.0	14.2	0.6
1E-4h	11.15	67.2	5.0	0.0	7.9	20.0	15.2	0.4
4h2 30	11.20	69.5	4.3	0.0	6.6	19.6	12.5	0.7
4h2 57.4	11.47	65.7	7.7	0.0	7.5	19.2	15.9	0.2
1D-4h	12.40	72.8	6.7	0.0	6.5	14.0	13.4	0.4
5h1 54	12.94	70.7	5.3	0.0	7.8	16.3	14.3	0.3
5h2 20	14.10	71.5	5.0	0.0	9.9	13.6	14.5	0.2
1E-5h	14.15	77.1	3.9	0.0	7.6	11.4	11.3	0.5
5h2 110	15.00	63.3	11.9	0.0	5.7	19.1	12.3	0.6
1D-5h	15.40	67.5	6.7	0.0	6.3	19.5	17.3	0.3
6h1 30	15.70	66.7	5.1	0.0	9.4	18.9	19.8	
6h1 110	16.50	66.5	4.9	0.0	7.3	21.3	15.9	0.3
6h1 123	16.63	66.2	6.2	0.0	7.5	20.0	14.2	0.3
6h1 146	16.86	71.5	9.5	0.0	9.0	10.0	14.0	0.4
6h1 147.5	16.87	68.8	8.2	0.0	10.8	12.3	11.3	0.7
1E-6h	17.15	65.4	12.3	0.0	10.5	11.8	13.3	0.7
6h2 42	17.33	68.3	7.3	0.0	11.7	12.7		
6h2 80	17.71	74.9	8.4	0.0	6.1	10.7	8.2	0.9
1D-6h	18.40	75.1	6.4	0.0	5.8	12.8	10.6	0.7
7h1 40	18.80	70.0	6.0	0.0	6.9	17.1	13.4	0.4
7h1 80	19.20	66.2	6.1	0.0	6.8	21.0	22.8	0.4
7h1 126	19.66	75.5	7.2	0.0	8.2	9.1	5.6	1.3
1E-7h	20.15	73.0	10.6	0.0	9.4	7.0	6.4	1.5
7h2 42	20.33	77.3	6.6	0.0	5.5	10.6	14.0	0.2
7h2 67	20.58	68.0	6.2	0.0	5.9	19.9	18.5	0.6
7h2 130	21.21	64.8	7.5	0.0	9.0	18.7	13.1	1.8
1D-7h	21.40	59.1	13.9	0.0	7.4	19.7	14.0	1.5
8h1 74	22.14	63.6	4.8	0.0	7.3	24.3	15.2	0.5
8h2 20.5	23.10	60.4	6.0	0.0	6.9	26.7	19.0	0.8
1E-8h	23.15	57.6	4.6	0.0	7.9	29.8	16.4	1.2
8h2 120.3	24.10	52.9	5.3	0.0	6.5	35.2	19.0	2.2
1D-8h	24.40	64.4	6.0	0.0	7.4	22.2	11.8	1.4
9h2 26	26.15	66.6	6.0	0.0	7.9	19.5	13.0	0.8
1E-9h	26.15	66.3	6.5	0.0	8.5	18.7	12.6	0.7
9h2 128.2	27.16	77.5	6.3	0.0	7.7	8.5	3.6	3.7
1D-9h	27.40	78.4	7.1	0.0	8.7	5.8	5.5	2.6
10h1 36	27.76	78.3	10.1	0.0	11.6	0.0	14.3	2.2
10h1 57	27.97	75.6	9.7	0.0	11.4	3.3	9.8	3.3
10h1 120	28.60	67.7	9.8	0.0	6.1	16.4	20.8	
1E-10h	29.15	68.7	5.4	0.0	9.1	16.7	10.5	2.3
10h2 60	29.51	50.4	5.3	0.0	4.6	39.7	21.2	2.5
10h2 129.7	30.20	54.5	6.0	0.0	5.2	34.3	21.1	2.6
1D-10h	30.40	59.0	6.9	0.0	8.9	25.3	13.8	2.3
11h1 69.9	31.09	58.9	7.6	0.0	7.3	26.2	14.2	1.7
11h1 131.7	31.71	73.0	8.6	0.0	9.5	9.0	7.1	3.2
1E-11h	32.15	45.8	5.1	0.0	6.2	42.8	30.8	1.9
1D-11h	33.40	62.5	8.1	0.0	7.6	21.8	13.4	1.6
1E-12h	35.15	61.7	7.9	0.0	9.9	20.6	14.1	1.0

(continued)

APPENDIX A. MINERALOGY, INORGANIC, AND ORGANIC CARBON CONTENT, BL00-1 (*continued*)

Sample #	Depth (m blf)	Quartz (%)	Plag. (%)	Arag. (%)	Dolo. (%)	Calcite (%)	CaCO$_3$ (%)	Organic carbon (%)
12h2 33.1	35.24	61.4	6.4	0.0	8.6	23.7	17.5	1.0
12h2 79.2	35.70	63.8	7.1	0.0	8.5	20.6	10.0	1.5
1D-12h	36.40	66.0	6.8	0.0	6.3	20.9	14.6	0.8
13h1 66	37.02	71.1	7.9	0.0	5.6	15.4	9.9	2.0
1E-13h	38.15	61.6	16.3	0.0	5.5	16.6	10.0	0.9
13h2 83.8	38.63	48.9	6.1	0.0	4.8	40.2	26.6	1.9
13h2 101.7	38.80	47.5	7.1	0.0	4.2	41.1	24.8	1.9
14h1 9.8	39.49	69.0	7.2	0.0	5.9	17.9	9.2	0.9
14h1 99.6	40.39	44.4	5.8	0.0	5.5	44.3	23.2	1.2
1E-14h	41.15	34.0	5.6	0.0	4.6	55.9	41.8	1.5
14h2 100	41.91	43.3	5.1	0.0	6.0	45.7	43.8	1.5
14h2 134	42.25	36.3	4.4	0.0	4.6	54.7	27.8	0.8
1D-14h	42.40	39.1	4.8	0.0	5.0	51.1	25.8	0.7
15h1 50	42.90	51.6	6.2	0.0	6.7	35.4	16.8	0.6
15h1 139.5	43.79	51.8	7.2	0.0	7.1	33.9	18.3	0.5
1E-15h	44.15	56.7	4.3	0.0	6.9	32.1	20.9	0.3
15h2 87.5	44.77	53.1	5.3	0.0	5.7	35.9	20.3	0.9
1D-15h	45.40	50.8	5.5	0.0	6.2	37.5	24.0	1.3
16a1 75	46.11	58.5	5.6	0.0	7.0	28.9	18.0	0.5
1E-16h	47.15	40.2	3.9	0.0	6.1	49.8	31.5	1.5
16a2 50	47.30	42.9	4.8	0.0	5.1	47.2	22.2	2.7
16a2 107.5	47.87	57.4	6.1	0.0	7.4	29.2	22.1	0.7
1D-16a	48.36	55.9	6.2	0.0	7.4	30.5	20.2	0.8
17a1 59.3	48.94	59.5	5.6	0.0	7.8	27.1	19.7	0.1
17a1 109.3	49.42	59.8	5.0	0.0	6.0	29.1	18.8	0.5
17a1 139.2	49.71	47.2	8.1	0.0	5.3	39.4	33.4	0.4
1E-17e	50.15	42.8	7.1	0.0	6.7	43.4	21.7	1.6
17a2 52	50.33	39.3	32.3	0.0	0.0	28.4	3.2	0.3
17a2 99.2	50.78	49.3	5.6	0.0	5.0	40.1	20.4	1.6
18a2 11.5	50.82	31.0	8.6	0.0	7.7	52.7	26.8	0.8
1D-17a	51.36	40.1	5.6	0.0	7.3	46.9	27.3	2.1
18a2 21.5	51.91	36.0	33.1	0.0	0.0	30.9	1.6	
18a2 21.5	51.91						1.3	
18a2 90	52.52	27.8	4.4	0.0	4.9	62.9	49.0	1.1
1E-18e	53.15	30.5	7.2	0.0	8.1	54.2	25.9	0.7
18a3 133	54.22	42.6	4.7	0.0	4.0	48.8	19.0	2.3
1D-18a	54.36	44.7	4.5	0.0	5.3	45.5	28.0	0.9
1D-19A-2-5	54.48	43.7	4.0	0.0	5.1	47.2	23.5	1.3
1D-19A-2-26	54.68	53.4	0.0	0.0	6.1	40.6	23.8	0.6
19a2 35.5	54.76	50.5	4.7	0.0	5.2	39.6	22.5	0.7
1D-19A-2-45	54.86	47.9	4.6	0.0	5.7	41.9	24.9	1.0
1D-19A-2-65	55.06	41.6	3.9	0.0	5.0	49.5	30.4	1.6
1D-19A-2-85	55.25	38.5	4.6	9.5	6.8	40.6	36.6	1.7
1D-19A-2-105	55.44	42.7	0.0	31.6	7.9	17.9	45.6	1.7
1D-19A-2-125	55.63	41.0	5.0	31.8	6.0	16.3	49.0	1.7
1E-19e	55.65	42.9	5.1	29.2	5.7	17.2	48.3	1.6
1D-19A-2-145	55.82	50.2	6.0	27.7	9.7	6.4	41.8	2.1
1D-19A-3-5	55.92	39.0	5.9	0.0	6.1	49.1	44.3	1.5
1D-19A-3-25	56.11	49.2	0.0	0.0	5.6	45.2	25.1	0.7
1D-19A-3-45	56.30	40.5	0.0	0.0	4.8	54.7	32.1	0.8
1D-19A-3-65	56.50	78.8	11.4	0.0	9.8	0.0	48.8	2.0
19a3 79.3	56.62	77.6	9.6	0.0	12.8	0.0	49.6	1.5
1D-19A-3-105	56.88	48.6	4.5	0.0	5.8	41.0	28.5	1.8
1D-19A-3-125	57.07	48.6	5.3	0.0	6.2	40.0	25.3	1.7
1D-19A-3-145	57.26	57.6	5.8	0.0	6.3	30.4	20.5	1.4
1D-19a	57.36	59.0	4.9	0.0	6.4	29.7	19.9	1.0
1E-20e	58.15	55.2	3.4	0.0	5.9	35.5	23.1	0.7
20a2 24.3	59.04	48.9	5.6	0.0	5.4	40.2	29.7	2.9
20a2 55.7	59.35	53.0	5.6	0.0	5.7	35.7	23.5	0.6
20a2 62.5	59.42	51.2	6.7	0.0	5.6	36.5	25.6	1.2
20a2 92.6	59.71	46.9	12.7	0.0	5.2	35.2	24.8	0.8
1D-20a	60.36	43.5	5.8	0.0	5.2	45.5	33.9	2.7
1D-21A-1-5	60.41	58.0	0.0	0.0	7.0	35.1	20.2	2.7
1D-21A-1-25	60.60	62.8	0.0	5.9	5.5	25.9	22.9	2.4
1E-21e	60.65	53.9	5.2	0.0	8.0	32.9	21.9	1.4
1D-21A-1-45	60.80	52.9	4.3	0.0	4.9	38.0	24.5	1.7
21a1 67.5	61.00	50.6	5.8	4.0	4.8	34.8	22.7	2.2
1D-21A-1-85	61.19	48.9	6.3	21.5	5.5	17.8	38.5	1.6
21a1 113	61.44	37.3	6.3	34.7	4.8	16.9	54.6	1.0

(*continued*)

APPENDIX A. MINERALOGY, INORGANIC, AND ORGANIC CARBON CONTENT, BL00-1 (*continued*)

Sample #	Depth (m blf)	Quartz (%)	Plag. (%)	Arag. (%)	Dolo. (%)	Calcite (%)	CaCO₃ (%)	Organic carbon (%)
1D-21A-1-125	61.57	38.1	0.0	39.3	6.2	16.3	56.4	1.3
1D-21A-1-145	61.77	31.7	8.5	37.7	6.4	15.7	60.8	1.2
21a2 24.8	62.08	28.8	1.8	50.7	5.1	13.7	68.9	1.4
21a2 79.8	62.60	16.7	1.1	3.2	2.6	76.4	67.6	0.8
21a2 104.7	62.84	27.5	5.1	50.6	4.9	12.0	70.0	1.7
21a2 133.8	63.12	30.6	1.7	46.5	4.5	16.6	73.3	0.6
1E-22e	63.15	29.0	3.3	45.8	7.3	14.5	71.6	1.2
1D-21a	63.36	33.0	5.9	34.0	4.4	22.7	62.8	1.2
22a1 70.9	64.06	37.3	5.4	35.0	9.0	13.3	46.2	0.9
22a1 143.5	64.78	26.3	8.4	43.4	9.6	12.3	53.3	3.1
1D-22A-2-5	64.90	35.3	4.7	42.8	6.9	10.4	66.4	1.3
1D-22A-2-25	65.10	32.1	5.3	38.8	7.7	16.1	60.3	1.4
1D-22A-2-45	65.30	29.1	4.5	42.9	7.8	15.8	64.2	1.0
22a2 68.5	65.53	32.2	1.7	42.3	6.9	17.0	63.4	1.1
1E-23e	65.65	35.6	5.0	34.9	9.0	15.5	55.4	1.3
1D-22A-2-95	65.79	32.8	0.0	38.5	10.5	18.2	59.3	1.6
22a2 126.5	66.10	41.7	5.7	29.6	6.8	16.2	53.2	1.2
22a2 126.5	66.10						51.7	
1D-22A-2-145	66.29	30.6	11.6	32.2	7.6	18.0	54.2	1.2
1D-22a	66.36	37.8	5.5	30.7	8.4	17.5	54.3	1.8
1D-23A-1-5	66.40	34.4	4.7	37.2	7.1	16.7	58.0	1.2
1D-23A-1-25	66.58	35.9	5.4	33.2	6.3	19.2	54.4	1.3
1D-23A-1-45	66.75	35.5	5.3	34.7	8.8	15.7	51.7	1.3
1D-23A-1-65	66.93	43.6	0.0	37.9	6.1	12.4	58.1	1.4
23a1 87	67.12	29.2	6.9	43.1	5.1	15.7	65.0	1.7
23a1 117	67.38	30.7	5.2	10.6	5.3	48.2	46.4	
23a1 117	67.38						45.0	2.8
1D-23A-1-145	67.62	43.7	4.2	0.0	4.7	47.4	28.3	2.8
1D-23A-2-5	67.71	54.6	4.6	0.0	7.7	33.1	21.8	2.4
1D-23A-2-25	67.89	69.6	5.5	0.0	10.7	14.2	10.6	2.6
1D-23A-2-45	68.06	76.5	5.0	0.0	13.0	5.5	6.9	2.2
1E-24e	68.15	70.1	11.9	0.0	10.1	7.9	9.8	2.0
23a2 57.5	68.17	76.8	6.1	0.0	10.5	6.6	10.2	1.6
23a2 57.5	68.17						9.1	
1D-23A-2-85	68.41	74.7	9.1	0.0	10.6	5.6	5.8	1.4
23a2 110	68.63	65.0	7.9	0.0	8.2	18.8	24.8	0.5
23a2 110	68.63						23.0	0.8
1D-23a	68.86	63.1	9.2	0.0	8.2	19.5	13.8	0.3
24e2 14.3	69.08	63.6	8.2	0.0	8.0	20.2	12.3	0.3
24e2 14.3	69.08						10.8	0.5
24e2 62.7	69.47	70.1	6.7	0.0	9.8	13.4	11.7	
24e2 62.7	69.47						10.3	0.9
24e3 2.5	70.16	65.8	8.5	0.0	7.0	18.8	6.5	1.3
24e3 58.7	70.61	60.3	6.9	0.0	8.9	23.9	21.1	1.0
1E-25e	70.65	64.4	4.5	0.0	8.4	22.8	15.1	0.3
24e3 112.8	71.03	63.2	6.5	0.0	7.7	22.5	21.0	0.5
24e3 133.4	71.19	71.9	8.2	0.0	10.0	10.0	10.4	0.7
1D-24e	71.37	71.2	7.1	0.0	9.0	12.7	12.9	1.2
25e2 132.5	72.50	65.6	8.6	0.0	8.2	17.6	15.3	1.8
1E-26e	73.15	66.8	6.0	0.0	8.9	18.4	13.8	1.1
25e3 77	73.24	63.2	6.4	0.0	8.4	22.0	13.9	0.9
1D-25e	73.87	70.1	8.8	0.0	8.1	13.0	21.8	1.1
26e2 95.3	74.75	59.0	5.5	0.0	7.0	28.5	24.0	0.9
26e2 132.7	75.04	59.7	5.4	0.0	9.1	25.8	19.4	0.5
26e3 39	75.47	58.9	6.7	0.0	8.6	25.8	23.2	0.4
1E-27e	75.65	61.2	3.7	0.0	7.8	27.2	22.0	0.4
26e3 109	76.01	55.9	6.7	0.0	7.9	29.5	15.8	1.7
1D-26e	76.37	59.2	5.4	0.0	7.8	27.5	19.4	0.4
27e1 38	76.66	60.8	5.1	0.0	7.6	26.5	18.2	0.4
1D-27e	76.97	60.3	5.6	0.0	7.2	26.9	15.7	0.8
28e2 77.8	77.77	60.7	5.5	0.0	7.5	26.3	19.7	0.4
1E-28e	78.15	60.1	4.3	0.0	6.5	29.1	19.0	0.5
28e2 147	78.29	56.9	5.8	0.0	7.3	30.0	20.5	0.3
28e3 16.4	78.43	56.2	5.4	0.0	8.7	29.7	22.8	0.7
28e3 60.2	78.75	59.1	7.1	0.0	8.0	25.8	13.3	1.8
28e3 93.4	79.00	57.2	6.2	0.0	7.5	29.1	20.2	1.0
28e3 140.3	79.35	57.5	5.8	0.0	6.6	30.1	22.3	1.0
1D-28e	79.47	54.0	7.0	0.0	9.6	29.4	23.5	0.9
29e3 58.3	80.41	67.0	5.7	0.0	8.3	19.1	9.9	1.7
1E-29e	80.65	53.5	6.4	0.0	9.7	30.4	22.2	1.4

(*continued*)

APPENDIX A. MINERALOGY, INORGANIC, AND ORGANIC CARBON CONTENT, BL00-1 (*continued*)

Sample #	Depth (m blf)	Quartz (%)	Plag. (%)	Arag. (%)	Dolo. (%)	Calcite (%)	CaCO$_3$ (%)	Organic carbon (%)
29e2 110.3	81.18	75.4	6.8	0.0	9.3	8.6	16.8	0.3
1D-29e	81.97	62.7	5.5	0.0	8.2	23.7	16.4	0.5
30e2 38	82.49	58.2	5.9	0.0	10.6	25.3	23.6	0.2
29e3 136	82.80	60.9	6.6	0.0	7.9	24.6	20.1	0.2
1E-30e	82.90	57.0	4.2	0.0	5.8	33.0	27.5	0.4
30e3 26.4	83.51	59.1	5.3	0.0	8.8	26.9	18.9	1.6
30e3 73.7	83.86	59.9	8.4	0.0	8.4	23.3	40.0	0.7
30e3 96.2	84.02	71.8	7.7	0.0	7.2	13.3	9.5	0.7
1D-30e	84.47	65.6	7.9	0.0	8.0	18.5	10.3	1.0
31e1 60.5	84.94	64.3	11.7	0.0	6.3	17.6	11.3	0.7
31e1 81.8	85.10	57.0	9.2	0.0	6.2	27.6	9.7	0.9
1E-31e	85.15	68.6	6.2	0.0	7.8	17.5	12.3	1.2
31e2 16.8	85.75	61.6	6.4	0.0	7.0	25.0	24.4	0.5
31e2 66.6	86.14	60.1	6.0	0.0	6.5	27.4	18.2	1.1
1D-31e	86.47	55.4	6.2	0.0	6.4	32.0	20.2	0.7
32e1 59.3	86.90	58.2	6.8	0.0	7.0	28.0	14.7	0.4
32e1 123	87.35	50.3	7.0	0.0	6.1	36.5	30.9	0.6
1E-32e	87.40	54.1	3.8	0.0	6.3	35.8	29.0	0.7
32e2 59.9	87.99	54.6	6.2	0.0	7.5	31.6	17.9	0.6
1D-32e	88.47	58.6	5.7	0.0	6.7	29.1	17.3	0.6
33e1 35.9	88.75	53.3	7.8	0.0	6.5	32.5	23.4	0.6
33e1 141.8	89.60	38.4	5.4	0.0	7.1	49.2	36.8	1.5
1E-33e	89.65	38.7	3.6	0.0	5.7	52.0	38.6	1.3
33e2 43.6	90.02	53.0	8.4	0.0	7.0	31.6	24.8	0.7
33e2 91.2	90.40	39.7	4.8	0.0	6.0	49.6	26.9	1.2
1D-33e	90.47	48.2	5.9	0.0	6.7	39.2	25.0	1.2
34e1 119.8	91.33	49.9	6.0	0.0	7.5	36.6	27.0	1.3
34e2 47.1	91.90	41.7	4.9	0.0	6.9	46.5	33.8	1.1
1E-34e	91.90	36.7	5.5	0.0	5.5	52.4	35.3	1.3
34e2 111	92.36	53.0	15.2	0.0	7.4	24.4	14.6	1.6
1D-34e	92.47	62.8	6.2	0.0	7.3	23.8	15.4	1.0
35e1 82.2	93.07	60.2	5.9	0.0	7.6	26.4	19.1	0.3
1E-35e	94.15	57.4	4.4	0.0	7.0	31.3	18.8	0.8
35e2 86.6	94.19	56.2	10.9	0.0	6.5	26.3	16.8	1.4
36e1 51.5	94.83	66.8	6.7	0.0	7.7	18.8	15.7	0.8
36e2 19.1	95.65	60.6	9.3	0.0	7.0	23.1	30.3	0.2
1E-36e	95.90	62.4	5.6	0.0	10.1	21.9	11.4	0.3
36e2 92.6	96.16	68.7	8.2	0.0	7.9	15.2	9.8	0.8
36e2 109.5	96.28	61.8	7.4	0.0	6.6	24.3	14.7	0.4
1D-36e	96.47	58.4	11.7	0.0	6.8	23.2	14.8	0.3
37e2 45.4	97.98	53.0	7.7	0.0	5.6	33.7	22.8	1.3
1E-37e	98.15	59.5	5.1	0.0	7.1	28.3	17.2	1.1
37e2 92	98.34	60.5	7.9	0.0	5.3	26.3	16.8	0.9
37e2 94.1	98.35	70.7	9.6	0.0	6.9	12.8	9.3	0.7
38e2 52.8	99.95	32.9	10.8	0.0	7.2	49.1	18.1	0.7
38e2 75.8	100.11	28.1	1.6	0.0	5.4	65.0	40.3	2.3
1E-38e	100.40	28.6	3.4	0.0	5.4	62.6	45.9	1.6
1D-38e	100.47	45.8	6.3	0.0	7.6	40.3	20.5	1.4
1E-39E-2-9	100.59	31.6	3.2	0.0	4.9	60.3	42.1	1.7
1E-39E-2-25	100.71	49.1	4.8	0.0	7.2	39.0	20.3	1.5
1E-39E-2-45	100.85	48.7	0.0	0.0	6.9	44.5	24.0	1.4
39E-2-53.6-54.9	100.91	47.5	4.6	0.0	5.7	42.2	24.9	1.4
1E-39E-2-65	101.05	45.9	5.2	0.0	6.0	42.9	23.8	1.5
1E-39E-2-85	101.13	49.5	4.4	0.0	5.6	40.5	24.6	1.1
1E-39E-2-105	101.27	37.1	4.2	0.0	5.1	53.6	36.5	1.3
1E-39E-2-125	101.42	27.9	4.0	0.0	7.3	60.8	47.2	1.1
39E-2-144.9-145.7	101.56	37.8	8.3	9.9	7.1	36.9	42.7	1.2
1E-39E-3-5	101.64	40.4	6.5	23.5	7.8	21.9	45.8	1.1
1E-39E-3-25	101.78	29.9	3.8	45.8	7.1	13.4	58.7	1.1
1E-39E-3-45	101.92	37.2	4.3	38.0	6.5	14.0	57.4	1.0
39E-3-59.4-60	102.02	36.4	5.2	34.6	7.7	16.1	51.7	0.9
1E-39E-3-85	102.20	44.1	0.0	31.6	10.9	13.4	55.8	1.0
39E-3-96.6	102.29	36.6	4.1	35.4	9.1	14.8	50.1	0.9
39E-3-119.5	102.45	36.9	4.1	37.1	7.6	14.4	54.7	1.1
1E-39E-3-140	102.59	36.5	0.0	39.6	7.5	16.4	53.2	1.1
39E	102.65	35.2	4.7	38.4	6.4	15.3	53.6	1.2
1E-40E-1-5	102.69	32.0	5.0	39.1	8.2	15.7	55.8	0.9
1E-40E-1-25	102.84	38.8	0.0	34.1	7.7	19.5	51.2	1.1
1E-40E-1-45	102.99	54.4	0.0	0.0	10.5	35.2	53.0	1.0

(*continued*)

APPENDIX A. MINERALOGY, INORGANIC, AND ORGANIC CARBON CONTENT, BL00-1 (*continued*)

Sample #	Depth (m blf)	Quartz (%)	Plag. (%)	Arag. (%)	Dolo. (%)	Calcite (%)	CaCO3 (%)	Organic carbon (%)
40E-1-59.8-61	103.10	35.1	0.0	40.0	7.5	17.4	39.5	2.9
1E-40E-1-85	103.29	52.2	6.7	0.0	10.4	30.7	54.9	1.3
1E-40E-1-105	103.45	31.1	0.0	0.0	4.1	64.8	49.1	1.5
40E-1-130-130.8	103.63	33.1	3.8	0.0	3.8	59.3	29.2	2.8
1E-40E-1-145	103.75	36.3	3.9	0.0	5.4	54.4	35.4	1.4
40E-2-32	104.03	43.6	4.0	0.0	5.5	46.9	25.3	1.7
40E-2-45.1	104.13	39.7	3.5	0.0	4.7	52.1	30.1	1.5
40E	104.90	53.3	4.8	0.0	6.5	35.4	22.1	0.6
41E-2-79.6	105.62	56.6	4.5	0.0	6.4	32.5	22.4	0.4
41E-2-129	105.95	37.4	4.9	0.0	5.0	52.7	19.6	3.0
41E-3-111.9	106.84	39.7	4.1	0.0	4.7	51.4	17.2	2.6
41E	107.15	41.4	3.5	0.0	4.6	50.5	30.5	1.3
42E-1-33-34	107.40	46.1	4.3	0.0	5.4	44.1	25.2	1.0
42E-2-99.4-100.5	107.90	41.1	3.7	0.0	4.9	50.3	21.7	0.8
1E-42E-2-5	108.32	45.1	0.0	0.0	5.2	49.6	24.7	1.2
1E-42E-2-25	108.47	51.5	4.8	0.0	6.3	37.4	19.1	0.8
42E-2-52.8	108.68	32.6	3.6	0.0	3.5	60.3	24.6	3.5
1E-42E-2-70	108.81	33.7	4.0	0.0	3.7	58.6	35.6	1.4
1E-42E-2-100	109.04	23.1	3.4	0.0	5.6	67.9	50.0	1.4
42E-2-122.9	109.21	35.2	3.4	0.0	3.5	57.9	26.0	2.8
1E-42E-2-141	109.35	31.2	10.0	0.0	3.4	55.3	40.3	0.7
42E	109.40	29.1	3.9	0.0	4.7	62.3	48.4	1.2
1E-43E-1-4	109.43	47.2	0.0	0.0	6.9	45.9	24.5	1.2
1E-43E-1-24	109.58	28.0	3.2	0.0	3.7	65.2	53.8	1.6
1E-43E-1-45	109.73	31.5	0.0	0.0	3.6	64.9	39.5	1.2
1E-43E-1-65	109.87	24.5	0.0	0.0	4.0	71.6	47.1	1.1
1E-43E-1-85	110.02	30.5	3.4	0.0	3.8	62.2	37.4	1.1
1E-43E-1-104	110.16	31.6	0.0	0.0	3.9	64.6	38.2	1.5
1E-43E-1-124	110.31	72.2		15.3	12.5	0.0	57.0	1.4
1E-43E-1-145	110.46	19.0	0.0	3.6	3.2	74.2	62.0	1.6
1E-43E-2-4	110.53	83.4	0.0	0.0	16.7	0.0	63.3	1.2
1E-43E-2-24	110.68	25.7	0.0	0.0	3.4	70.9	47.9	0.8
1E-43E-2-44	110.82	25.8	0.0	0.0	3.2	71.0	50.8	0.9
43E-2-49.2-49.9	110.86	21.5	2.5	0.0	3.2	72.8	54.9	1.0
1E-43E-2-64	110.97	17.4	0.0	5.0	3.4	74.2	65.1	1.1
1E-43E-2-84	111.11	16.6	3.3	0.0	3.5	76.6	65.9	1.1
1E-43E-2-104	111.26	18.6	0.0	0.0	3.9	77.5	65.3	1.1
1E-43E-2-124	111.41	71.1	13.7	0.0	15.3	0.0	64.0	1.3
1E-43E-2-144	111.55	19.5		0.0	4.8	75.7	61.8	1.3
43E	111.65	25.3	2.3	0.0	3.1	69.3	60.3	1.4
42E-2-59.9	113.20	38.1	5.5	0.0	6.2	50.3	35.0	1.1
44E	113.90	38.9	3.8	0.0	6.4	50.9	36.6	1.1
45E-2-69.5-70.4	114.54	37.2	3.6	0.0	4.5	54.8	37.6	1.4
45E-2-129-130.2	114.94	43.1	4.7	0.0	6.3	46.0	30.3	2.0
45E-3-3.1	115.11	54.7	6.2	0.0	10.5	28.6	18.7	1.5
45E-3-32.8	115.31	61.2	7.1	0.0	11.2	20.6	13.5	0.7
45E-3-132.8	115.99	57.1	5.3	0.0	7.2	30.4	18.1	3.6
45E	116.15	66.4	10.5	0.0	7.8	15.3	18.3	1.0
46E-1-81.2-82.1	116.76	59.5	9.1	0.0	8.3	23.1	13.8	
46E-1-140.8-141.3	117.21	81.5	18.5	0.0	0.0	0.0		
46E-2-91.1	117.97	56.7	4.8	0.0	6.5	32.0	19.3	2.4
46E	118.40	52.8	4.8	0.0	6.2	36.3	21.3	1.0
47E-2-29.5-30.6	118.74	51.0	4.6	0.0	5.6	38.9	7.3	1.7
47E-2-64.8-65.8	118.98	57.1	4.0	0.0	7.3	31.5	18.3	1.1
47E-3-80	120.12	40.3	3.9	0.0	4.5	51.3	27.7	1.6
47E-3-144.9	120.57	29.5	3.9	0.0	3.7	62.9	35.8	0.8
47E	120.65	32.1	3.1	0.0	4.4	60.4	40.3	0.9

Note: Mineralogic composition based on X-ray diffraction peak intensities and are approximations only. Organic and inorganic carbon contents are based on coulometry.

APPENDIX B. OXYGEN, CARBON, AND STRONTIUM ISOTOPIC RATIOS OF BULK SEDIMENT AND OSTRACODES, BL00-1

Depth (m blf)	δ¹⁸O (‰VPDB)	δ¹³C (‰VPDB)	Depth (m blf)	δ¹⁸O (‰VPDB)	δ¹³C (‰VPDB)	Depth (m blf)	δ¹⁸O (‰VPDB)	δ¹³C (‰VPDB)	Depth (m blf)	δ¹⁸O (‰VPDB)	δ¹³C (‰VPDB)	Depth (m blf)	δ¹⁸O (‰VPDB)	δ¹³C (‰VPDB)
						Bulk-sediment carbonate								
0.20	-4.0	3.2	15.00	-9.6	-3.1	29.10	-9.7	-0.5	43.76	-11.1	-3.2	58.20	-10.6	-2.2
0.50	-3.6	3.3	15.40	-9.3	-2.9	29.50	-11.5	-0.2	44.20	-10.7	-2.7	58.55	-10.8	-2.0
0.80	-3.9	2.9	15.70	-9.7	-2.9	29.90	-11.4	0.2	44.90	-10.3	-0.1	58.80	-11.0	-1.2
1.20	-3.9	3.1	16.00	-9.6	-3.2	30.20	-10.9	0.4	45.20	-10.3	-0.7	59.00	-8.7	1.6
1.47	-4.4	3.2	16.40	-9.2	-2.4	30.40	-10.1	0.3	45.40	-10.0	-1.0	59.36	-8.7	1.9
2.20	-4.4	2.9	16.70	-9.6	-2.0	30.80	-10.4	-0.3	45.80	-10.4	-2.1	59.70	-11.1	-2.1
2.50	-4.8	2.7	17.00	-9.9	-2.1	31.20	-10.6	-0.2	46.10	-9.9	-3.5	60.00	-9.0	1.5
2.85	-5.5	2.4	17.30	-9.2	0.8	31.50	-11.0	1.2	46.35	-8.7	1.4	60.19	-9.4	-0.4
3.20	-5.6	2.5	17.60	-9.9	-1.4	31.80	-10.7	1.0	46.70	-8.5	1.6	60.40	-9.6	-1.6
3.40	-5.4	2.6	17.90	-10.2	-1.7	32.20	-10.6	0.2	47.00	-8.7	1.2	60.70	-8.9	0.4
3.70	-5.2	2.7	18.20	-9.9	-1.9	32.50	-10.5	0.3	47.20	-9.4	0.6	61.08	-10.5	-1.3
3.97	-5.6	2.7	18.40	-9.8	-2.1	32.80	-10.4	0.9	47.62	-9.8	-1.4	61.40	-5.7	2.4
4.40	-5.9	2.4	18.80	-9.8	-2.7	33.20	-10.3	0.1	47.90	-9.8	-0.9	61.70	-5.1	2.9
5.20	-5.3	2.5	19.00	-9.6	-2.4	33.60	-9.8	-0.3	48.30	-8.9	0.6	61.95	-4.5	3.3
5.50	-5.2	2.7	19.30	-9.1	0.6	34.00	-9.6	0.5	48.55	-9.1	0.5	62.40	-4.5	3.6
5.90	-5.6	2.4	19.60	-8.0	-1.6	34.30	-10.0	-0.6	48.90	-9.7	-3.6	62.84	-4.5	3.6
6.15	-5.5	2.4	20.00	-10.6	-1.5	34.60	-9.5	-1.2	49.20	-9.0	0.3	63.10	-6.0	2.4
6.60	-6.3	2.0	20.30	-9.6	-1.3	34.90	-8.6	2.6	49.47	-9.7	-2.7	63.40	-6.9	1.8
6.90	-7.8	1.3	20.60	-9.1	-1.1	35.20	-9.3	0.3	49.80	-9.1	0.9	63.72	-11.5	-1.6
7.15	-8.4	0.4	21.00	-10.1	-0.1	35.50	-9.9	1.0	50.20	-9.6	0.3	64.10	-7.9	1.3
7.60	-7.5	2.2	21.40	-9.8	-0.7	35.70	-9.5	-0.8	50.50	-10.3	0.0	64.40	-4.5	7.4
8.00	-5.1	2.7	21.70	-9.3	-2.6	36.10	-9.5	-0.1	50.80	-10.4	0.5	64.55	-3.7	3.8
8.25	-6.2	2.2	22.00	-9.1	0.1	36.45	-9.6	-0.3	51.20	-9.8	0.8	65.00	-4.5	3.1
8.60	-7.5	1.0	22.30	-9.3	-2.8	36.80	-9.7	1.8	51.45	-9.3	1.3	65.37	-4.7	3.1
8.90	-7.6	1.1	22.60	-9.7	-0.6	37.10	-7.7	5.9	51.90	-7.4	0.9	65.70	-5.3	2.5
9.25	-11.0	-1.0	23.00	-9.9	-1.1	37.54	-10.1	-0.8	52.20	-8.6	0.9	66.00	-5.3	3.1
9.70	-11.3	-1.1	23.30	-9.4	-1.4	37.90	-9.8	-2.2	52.45	-8.8	0.9	66.21	-4.5	3.5
10.00	-10.8	-0.7	23.70	-9.9	0.1	38.20	-9.0	0.8	52.80	-7.8	1.0	66.50	-4.8	3.4
10.26	-8.7	-2.2	24.10	-10.2	0.0	38.50	-8.5	2.4	53.20	-8.8	0.5	66.70	-4.3	3.7
10.60	-9.7	-1.7	24.40	-9.6	-1.6	38.65	-8.7	2.1	53.55	-9.7	-0.1	67.07	-3.8	3.8
10.80	-8.6	-2.2	24.80	-8.2	2.5	39.00	-9.7	-2.1	53.80	-10.3	-0.7	67.40	-6.7	3.4
11.20	-9.3	-3.1	25.10	-9.0	0.5	39.30	-9.9	-2.7	54.00	-11.4	-1.3	67.60	-8.7	0.4
11.45	-9.3	-2.1	25.40	-5.9	-1.3	39.60	-10.1	-2.5	54.36	-12.5	-2.2	67.92	-9.0	-0.2
11.80	-9.4	-2.9	25.70	-6.7	-2.0	40.00	-9.3	-0.1	54.80	-12.5	-3.5	68.20	-6.8	1.8
12.10	-9.5	-2.5	26.00	-8.8	1.8	40.40	-9.2	-0.2	55.20	-10.9	-0.8	68.50	-7.8	-1.5
12.45	-10.1	-2.2	26.30	-9.1	-1.5	40.60	-9.0	0.4	55.50	-8.2	1.9	68.74	-6.6	4.4
12.70	-10.1	-2.5	26.60	-9.2	-0.5	41.00	-8.9	0.4	55.70	-9.3	0.6	69.10	-7.0	-0.3
13.00	-8.5	-1.9	27.00	-9.4	-0.3	41.50	-8.4	0.3	56.10	-9.6	-0.5	69.48	-8.8	-1.9
13.20	-9.6	-2.7	27.40	-9.7	-0.2	41.75	-8.0	0.2	56.30	-11.5	-2.3	69.90	-7.2	3.1
13.50	-9.4	-2.8	27.70	-10.1	0.0	42.30	-12.7	-3.2	56.60	-7.2	1.2	70.22	-6.8	-1.7
13.80	-9.6	-2.9	28.00	-9.6	2.6	42.60	-12.9	-3.3	56.86	-9.8	0.6	70.60	-7.4	4.7
14.10	-9.7	-2.6	28.30	-8.0	0.7	42.76	-12.6	-3.1	57.20	-9.7	-0.5	71.06	-9.2	-0.3
14.40	-9.6	-2.5	28.60	-9.2	0.2	43.10	-11.5	-3.7	57.40	-9.4	-1.3	71.40	-9.5	-0.3
14.70	-9.6	-2.5	28.90	-9.6	-1.1	43.50	-10.7	-3.1	57.72	-10.4	-1.3	71.70	-9.2	0.9

(continued)

338

APPENDIX B. OXYGEN, CARBON, AND STRONTIUM ISOTOPIC RATIOS OF BULK SEDIMENT AND OSTRACODES, BL00-1 *(continued)*

Bulk-sediment carbonate												Cytherissa		
Depth (m blf)	δ¹⁸O (‰VPDB)	δ¹³C (‰VPDB)	Depth (m blf)	δ¹⁸O (‰VPDB)	δ¹³C (‰VPDB)	Depth (m blf)	δ¹⁸O (‰VPDB)	δ¹³C (‰VPDB)	Depth (m blf)	δ¹⁸O (‰VPDB)	δ¹³C (‰VPDB)	Depth (m blf)	δ¹⁸O (‰VPDB)	δ¹³C (‰VPDB)
71.87	−9.7	−0.1	85.68	−9.8	−1.1	99.70	−10.2	−3.2	113.42	−10.6	0.0	16.7	−9.4	−4.1
72.30	−9.2	0.2	86.10	−10.1	−1.8	100.04	−10.7	−0.1	113.70	−10.3	0.0	19.0	−14.3	−6.5
72.72	−9.7	−0.9	86.45	−10.4	−1.6	100.40	−10.0	−0.8	113.90	−10.0	0.7	19.0	−13.8	−6.9
73.00	−9.8	−1.1	86.70	−10.5	−0.1	100.80	−11.8	−0.8	114.21	−10.3	0.2	37.5	−7.2	−1.1
73.20	−10.1	−1.1	86.90	−9.4	−2.9	101.20	−11.0	−1.5	114.50	−10.1	0.9	59.4	−4.1	0.4
73.48	−9.5	−1.1	87.22	−10.4	−2.4	101.52	−9.8	0.6	114.89	−9.1	2.4	60.2	−5.9	0.0
73.90	−9.0	0.1	87.99	−9.5	−2.6	101.80	−7.9	2.6	115.10	−9.5	0.4	81.1	−6.9	−0.5
74.24	−9.3	−1.2	88.40	−9.5	−3.0	102.23	−7.8	2.2	115.30	−9.9	−0.8	83.4	−8.1	−2.4
74.50	−9.6	−1.6	88.73	−11.3	−1.8	102.70	−7.4	2.3	115.56	−8.7	2.8	99.3	−8.8	−2.0
74.80	−8.4	1.1	89.10	−10.0	−2.9	103.03	−8.3	0.4	116.00	−9.6	0.6	99.3	−8.3	−1.4
74.99	−9.4	−1.5	89.47	−10.5	−0.8	103.30	−8.0	0.4	116.30	−10.0	−1.4	111.2	−4.2	−2.7
75.30	−9.2	−1.4	89.70	−9.8	−1.8	103.74	−11.7	−0.8	116.60	−10.5	−2.9	111.2	−4.4	−1.9
75.50	−8.9	−0.1	90.00	−10.3	−1.1	104.00	−12.0	−1.1	117.05	−9.6	−1.9	117.1	−6.7	−2.8
75.80	−9.2	−2.3	90.27	−11.0	−1.4	104.20	−11.4	−0.9	117.81	−8.7	1.2			
76.10	−8.1	2.1	90.50	−10.3	−0.3	104.54	−11.2	−1.1	118.10	−8.9	−0.5			
76.54	−9.2	−2.4	90.80	−10.7	−1.4	104.90	−10.0	−2.5	118.40	−9.5	−1.3			
76.70	−9.3	−2.7	91.03	−10.2	−0.6	105.29	−10.4	−1.9	118.61	−9.5	−0.8			
77.00	−9.4	−2.4	91.30	−10.7	−0.9	105.60	−10.6	−1.4	119.00	−10.1	−2.7			
77.29	−9.1	−1.8	91.60	−10.6	−0.9	105.96	−9.4	0.7	119.30	−11.0	−2.6			
77.50	−9.1	−2.2	91.84	−10.2	−0.5	106.30	−11.0	−0.2	119.70	−11.9	−2.9			
77.80	−9.1	−1.8	92.10	−9.9	−0.1	106.63	−12.0	−1.6	119.98	−13.0	−2.7			
78.04	−8.8	−2.0	92.40	−9.9	−1.2	106.90	−10.0	0.2	120.30	−13.2	−3.0			
78.40	−9.6	−1.1	92.57	−9.9	−2.4	107.20	−11.6	−0.6	120.59	−12.8	−1.7			
78.83	−8.8	−0.1	92.80	−9.3	−1.8	107.36	−11.7	−1.3						
79.20	−10.0	−0.4	93.10	−9.3	−2.6	107.70	−11.8	−1.4						
79.57	−9.7	0.8	93.27	−9.5	−2.0	107.90	−12.2	−2.3						
80.00	−9.8	−0.1	93.60	−9.4	−1.8	108.12	−12.0	−1.5						
80.31	−9.5	1.5	93.96	−9.3	−0.7	108.40	−11.6	−2.0						
80.70	−9.8	2.0	94.20	−9.7	−0.7	108.74	−11.0	−1.1						
81.10	−10.3	2.1	94.50	−9.1	0.2	109.10	−10.0	0.5						
81.50	−9.7	−0.8	94.74	−9.9	−1.4	109.40	−10.9	−0.8						
81.87	−9.2	−1.0	95.00	−9.5	−1.0	109.69	−11.8	−0.9						
82.20	−9.5	−1.7	95.20	−9.3	0.3	110.00	−12.1	−1.5						
82.62	−10.2	−1.5	95.49	−10.1	−2.8	110.42	−8.2	1.3						
82.90	−9.6	0.3	96.27	−9.2	−2.1	110.70	−8.9	0.6						
83.10	−9.5	−1.9	96.70	−10.1	−3.9	110.90	−9.6	−0.1						
83.41	−9.2	−1.2	96.99	−9.5	−2.0	111.16	−8.3	0.6						
83.70	−10.3	−0.1	97.30	−9.4	−2.1	111.50	−9.1	0.3						
83.90	−7.5	5.9	97.73	−10.0	0.0	111.70	−8.8	0.4						
84.14	−9.8	−1.2	98.00	−9.9	1.6	111.94	−8.5	0.8						
84.50	−9.9	−2.0	98.20	−10.7	2.2	112.30	−12.1	−1.4						
84.87	−10.5	−1.8	98.54	−10.0	−0.5	112.67	−11.8	−0.5						
85.20	−9.8	−1.4	98.90	−10.1	−0.7	113.00	−10.4	0.4						
85.50	−10.0	−2.2	99.31	−9.9	−0.1	113.20	−10.8	0.2						

(continued)

APPENDIX B. OXYGEN, CARBON, AND STRONTIUM ISOTOPIC RATIOS OF BULK SEDIMENT AND OSTRACODES, BL00-1 (continued)

Candona

Depth (m blf)	δ18O (‰VPDB)	δ13C (‰VPDB)
0.5	-2.2	-5.2
1.5	-2.5	-5.2
2.9	-2.0	-3.1
4.0	-2.4	-3.8
4.0	-1.7	-4.7
6.2	-2.0	-5.4
7.2	-2.5	-4.3
8.3	-1.1	-4.9
9.3	-6.3	-5.4
10.3	-8.4	-5.2
12.5	-9.8	-4.8
16.7	-9.7	-4.7
17.9	-8.7	-3.3
19.0	-12.5	-6.2
20.0	-9.4	-3.4
21.0	-7.1	-3.1
22.0	-8.8	-6.3
23.0	-6.4	-2.5
24.1	-6.6	-3.5
25.1	-12.3	-6.5
26.0	-10.0	-5.0
27.0	-7.1	-1.7
28.0	-6.5	-4.0
29.1	-7.2	-4.7
30.2	-7.1	-4.9
31.2	-7.7	-4.4
32.2	-7.7	-4.7
33.2	-7.1	-2.7
34.3	-6.8	-3.8
35.5	-6.0	-0.8
36.5	-6.3	-2.0
37.5	-8.5	-3.7
38.7	-3.3	-1.7
39.6	-7.6	-3.7
40.6	-4.9	-2.6
41.8	-2.4	-3.8
42.8	-9.1	-6.8
43.8	-12.0	-8.3
44.9	-5.8	-2.5
45.8	-7.5	-3.7
46.4	-7.8	-4.0
47.6	-3.9	-2.1
48.6	-5.7	-3.3
49.5	-9.7	-5.3
52.5	-3.7	-1.7
53.6	-3.0	-2.5
54.4	-8.5	-5.9
55.2	-6.9	-5.0
56.1	-2.5	-3.5
56.9	-6.1	-2.7
59.4	-4.3	-1.4
61.1	-7.8	-3.4
62.8	-3.5	-9.2
65.4	-1.9	-5.5
66.2	-4.8	-0.4
67.1	-1.1	-5.4
67.9	-6.2	-4.3
70.2	-8.6	-2.2
71.1	-8.4	-2.6
72.7	-6.3	-1.6
78.0	-12.7	-7.4
81.1	-5.7	-0.2
83.4	-8.0	-1.2
84.9	-8.3	-2.1
85.7	-6.7	-2.2
88.0	-13.2	-6.8
88.0	-13.8	-5.6
97.7	-5.7	-1.5
101.5	-4.4	-2.1
102.2	-1.3	-4.5
103.0	-2.1	-2.2
105.3	-8.9	-3.7
107.4	-8.8	-4.5
108.1	-8.4	-3.6
110.4	-2.3	-1.4
111.2	-3.1	-1.7
111.9	-2.6	-1.8
117.1	-5.9	-1.4
117.8	-4.6	-0.7
119.3	-10.5	-5.5

Soluble Sr isotopes

Depth (m blf)	87Sr/86Sr
1.20	0.71039
3.40	0.71030
5.51	0.71024
6.40	0.71010
7.20	0.70943
7.96	0.70995
8.30	0.71017
9.40	0.71000
10.80	0.71004
11.50	0.70994
12.40	0.71000
12.50	0.71013
13.80	0.71016
14.10	0.71019
14.70	0.71018
15.40	0.70990
16.86	0.71016
18.40	0.71000
19.66	0.71003
21.40	0.70990
23.10	0.70962
24.40	0.70960
26.15	0.70965
27.40	0.70990
28.60	0.70984
30.40	0.70980
31.71	0.70976
33.40	0.70990
35.24	0.71000
36.40	0.71000
38.63	0.70996
42.40	0.70940
42.90	0.70956
43.79	0.70958
45.40	0.70990
47.30	0.71000
48.36	0.70977
49.71	0.70984
51.36	0.71000
52.52	0.71005
54.36	0.70950
54.50	0.70943
55.60	0.71018
57.36	0.70990
59.04	0.71000
60.36	0.71010
61.44	0.71026
62.40	0.71022
63.36	0.70970
63.70	0.70913
64.78	0.71034
64.90	0.71025
65.10	0.71024
65.70	0.70996
66.36	0.71040
68.17	0.71020
68.70	0.70978
68.86	0.70980
70.16	0.70975
71.37	0.71000
71.90	0.71003
72.50	0.70985
73.87	0.70970
74.75	0.70974
76.37	0.70970
76.97	0.70960
77.77	0.70969
78.80	0.70972
79.47	0.70990
80.41	0.71002
81.97	0.70980
84.47	0.70990
86.47	0.70950
88.47	0.70960
88.70	0.70941
90.47	0.70970
92.47	0.70990
94.83	0.70996
95.50	0.70988
96.47	0.70980
100.47	0.70960
102.29	0.70970
102.70	0.70995
104.03	0.70940
105.62	0.70970
108.10	0.70932
108.68	0.70970
110.86	0.70920
113.20	0.70950
115.31	0.70990
117.21	0.70970
118.10	0.70965
118.98	0.70950

Residue Sr isotopes

Depth (m blf)	87Sr/86Sr
63.7	0.71028
7.2	0.71054
6.2	0.71062
88.7	0.71090
3.4	0.71092
1.2	0.71095
62.4	0.71124
8.3	0.71149
95.5	0.71169
65.7	0.71189
118.0	0.71431
102.7	0.71566
71.9	0.71621
12.5	0.71829
78.8	0.71847
108.1	0.71920
13.8	0.71932
14.7	0.72055
11.5	0.72944
68.7	0.73443

Note: m blf—meters below lake floor; VPDB—Vienna Peedee Belemnite.

APPENDIX C. OSTRACODE FAUNA FROM BL00-1

Depth (m blf)	Wt (g)	C sp. 1 (vpg)	C sp. 2 (vpg)	C sp. 3 (vpg)	C sp. 4 (vpg)	C sp. 5 (vpg)	C sp. 7 (vpg)	Cyth lac (vpg)	Unk sp (vpg)	Limno 1 (vpg)	Total (vpg)	Juvenile Limno	Juvenile Cyth
0.5	9.1	2.1	6.4	3.0	0.2	0.7	0.2	0.0	1.4	0.4	14.4	x	
1.5	9.7	2.5	9.3	0.8	0.4	0.0	0.3	0.0	0.0	0.2	13.5		
2.9	8.2	1.1	0.0	0.0	1.1	0.0	0.1	0.0	0.1	0.5	2.9		
4.0	12.0	2.1	0.1	0.1	0.8	0.0	0.0	0.0	0.0	0.0	3.1	x	
6.2	13.4	1.3	0.8	0.1	0.6	0.0	0.2	0.0	0.0	0.1	3.2		
7.2	14.6	2.1	0.3	0.1	0.0	0.0	0.1	0.0	0.0	0.0	2.5		
8.3	18.9	1.7	1.6	0.1	0.1	0.0	0.1	0.0	0.0	0.1	3.6	x	
9.3	26.7	2.2	0.0	0.0	0.0	0.0	0.0	0.0	0.0	0.0	2.2		
10.3	38.8	0.4	0.0	0.0	0.0	0.0	0.0	0.0	0.0	0.0	0.4		
11.5	62.8	0.0	0.0	0.0	0.0	0.0	0.0	0.0	0.0	0.0	0.0		
12.5	62.3	0.3	0.0	0.0	0.0	0.0	0.0	0.0	0.0	0.0	0.4		
13.5	69.5	0.1	0.0	0.0	0.0	0.0	0.0	0.0	0.0	0.0	0.1		
14.7	61.0	0.0	0.0	0.0	0.0	0.0	0.0	0.0	0.0	0.0	0.0		
15.7	57.4	0.0	0.0	0.0	0.0	0.0	0.0	0.0	0.0	0.0	0.0		
16.7	66.8	0.4	0.1	0.0	0.0	0.2	0.0	0.0	0.0	0.0	0.8		
17.9	60.6	0.5	0.0	0.0	0.0	0.0	0.0	0.0	0.0	0.0	0.5		
19.0	64.9	0.2	0.1	0.0	0.0	0.0	0.0	0.1	0.0	0.0	0.4		
20.0	50.5	1.0	0.0	0.2	0.0	0.0	0.0	0.0	0.0	0.0	1.2		
21.0	54.4	1.0	0.0	0.3	0.0	0.0	0.0	0.0	0.0	0.0	1.3		
22.0	65.3	0.5	0.0	0.0	0.0	0.0	0.0	0.1	0.0	0.0	0.6		x
23.0	49.7	1.3	0.0	0.2	0.0	0.0	0.0	0.0	0.0	0.0	1.4		
24.1	49.2	1.7	0.0	0.2	0.0	0.0	0.0	0.0	0.0	0.0	1.9		
25.1	71.5	0.6	0.0	0.0	0.0	0.0	0.0	0.0	0.0	0.0	0.7		
26.0	59.9	0.4	0.0	0.0	0.0	0.0	0.0	0.0	0.0	0.0	0.4		
27.0	50.1	0.2	0.0	0.0	0.0	0.0	0.0	0.0	0.0	0.0	0.2		
28.0	48.7	0.9	0.0	0.0	0.0	0.0	0.0	0.0	0.0	0.0	0.9		
29.1	45.2	0.2	0.0	0.0	0.0	0.0	0.0	0.0	0.0	0.0	0.2		
30.2	56.4	1.8	0.0	0.1	0.0	0.0	0.0	0.0	0.0	0.0	1.8		
31.2	58.1	1.7	0.0	0.1	0.0	0.0	0.0	0.0	0.0	0.0	1.7		
32.2	50.8	0.4	0.0	0.1	0.0	0.0	0.0	0.0	0.0	0.0	0.5		
33.2	51.2	0.8	0.0	0.2	0.0	0.0	0.0	0.0	0.0	0.0	1.0		
34.3	56.0	0.4	0.0	0.0	0.0	0.0	0.0	0.0	0.0	0.0	0.4		
35.5	51.5	0.4	0.0	0.1	0.0	0.0	0.0	0.0	0.0	0.0	0.4		
36.5	51.3	0.9	0.0	0.1	0.0	0.0	0.0	0.0	0.0	0.0	1.1		
37.5	65.2	0.4	0.0	0.2	0.0	0.0	0.0	0.0	0.0	0.0	0.6		
38.7	47.1	0.7	0.0	0.0	0.0	0.0	0.0	0.0	0.0	0.0	0.7	x	
39.6	48.4	0.5	0.0	0.1	0.0	0.0	0.0	0.0	0.0	0.0	0.6		
40.6	48.9	0.9	0.0	0.1	0.0	0.0	0.0	0.0	0.0	0.0	1.1		
41.8	60.1	1.3	0.0	0.4	0.0	0.0	0.0	0.0	0.0	0.0	1.7	x	
42.8	49.0	0.1	0.0	0.1	0.0	0.0	0.0	0.0	0.1	0.0	0.3		
43.8	64.2	0.4	0.0	0.0	0.0	0.0	0.0	0.0	0.0	0.0	0.5		
44.9	57.8	0.7	0.0	0.0	0.0	0.0	0.0	0.0	0.0	0.0	0.8	x	
45.8	55.5	0.2	0.0	0.1	0.0	0.0	0.0	0.0	0.0	0.0	0.3	x	
46.4	52.4	0.9	0.0	0.0	0.0	0.0	0.0	0.0	0.0	0.0	0.9		
47.6	61.0	0.4	0.0	0.0	0.0	0.0	0.0	0.0	0.0	0.0	0.4		
48.6	54.6	0.3	0.0	0.0	0.0	0.0	0.0	0.0	0.0	0.0	0.3		
49.5	62.4	0.0	0.0	0.0	0.0	0.0	0.0	0.0	0.0	0.0	0.0	x	
50.5	50.9	0.1	0.0	0.0	0.0	0.0	0.0	Cyth	0.0	Limno	0.1		
51.5	43.7	0.0	0.0	0.0	0.0	0.0	0.0	0.0	0.0	0.0	0.0		
52.5	57.7	0.3	0.0	0.0	0.0	0.0	0.0	0.0	0.0	0.0	0.3	x	
53.6	57.4	1.0	0.0	0.0	0.0	0.0	0.0	0.0	0.0	0.0	1.0	x	
54.4	48.5	0.1	0.0	0.0	0.0	0.0	0.0	0.0	0.0	0.0	0.2		
55.2	55.8	0.3	0.0	0.0	0.0	0.0	0.0	0.0	0.0	0.0	0.3	x	
56.1	64.0	0.8	0.0	0.0	0.0	0.0	0.0	0.0	0.0	0.0	0.8	x	
56.9	57.2	0.3	0.0	0.0	0.0	0.0	0.0	0.1	0.0	0.0	0.4	x	
57.7	53.5	0.0	0.0	0.0	0.0	0.0	0.0	0.0	0.0	0.0	0.0	x	
58.6	48.3	0.0	0.0	0.0	0.0	0.0	0.0	0.0	0.0	0.0	0.0		
59.4	45.0	0.2	0.0	0.0	0.0	0.0	0.0	0.2	0.0	0.0	0.4	x	
60.2	48.3	0.0	0.0	0.0	0.0	0.0	0.0	0.0	0.0	0.0	0.0	x	
61.1	55.2	0.1	0.0	0.0	0.0	0.0	0.0	0.0	0.0	0.0	0.1		
62.0	53.8	0.0	0.0	0.0	0.0	0.0	0.0	0.0	0.0	0.0	0.0	x	
62.8	49.6	0.1	0.0	0.0	0.0	0.0	0.0	0.0	0.0	0.0	0.1	x	
63.7	61.8	0.0	0.0	0.0	0.0	0.0	0.0	0.0	0.0	0.0	0.0	x	
64.6	57.0	0.0	0.4	0.0	0.0	0.0	0.0	0.0	0.0	0.0	0.4	x	
65.4	52.8	0.1	0.0	0.0	0.0	0.0	0.0	0.0	0.0	0.0	0.1	x	
66.2	47.6	0.0	0.0	0.0	0.0	0.0	0.0	0.0	0.0	0.0	0.0	x	
67.1	49.6	0.0	0.0	0.0	0.0	0.0	0.0	0.0	0.0	0.0	0.1	x	
67.9	49.3	0.0	0.0	0.0	0.0	0.0	0.0	0.0	0.0	0.0	0.0	x	
68.7	71.2	0.0	0.0	0.0	0.0	0.0	0.0	0.0	0.0	0.0	0.0	x	
69.5	53.9	0.0	0.0	0.0	0.0	0.0	0.0	0.0	0.0	0.0	0.0		x
70.2	62.4	0.0	0.0	0.0	0.0	0.0	0.0	0.0	0.0	0.0	0.1		

(continued)

APPENDIX C. OSTRACODE FAUNA FROM BL00-1 (*continued*)

Depth (m blf)	Wt (g)	C sp. 1 (vpg)	C sp. 2 (vpg)	C sp. 3 (vpg)	C sp. 4 (vpg)	C sp. 5 (vpg)	C sp. 7 (vpg)	Cyth lac (vpg)	Unk sp (vpg)	Limno 1 (vpg)	Total (vpg)	Juvenile Limno	Juvenile Cyth
71.1	61.9	0.0	0.0	0.0	0.0	0.0	0.0	0.0	0.0	0.0	0.0		
71.9	38.1	0.0	0.0	0.0	0.0	0.0	0.0	0.0	0.0	0.0	0.0		
72.7	60.5	0.0	0.0	0.0	0.0	0.0	0.0	0.0	0.0	0.0	0.0		
73.5	39.8	0.0	0.0	0.0	0.0	0.0	0.0	0.0	0.0	0.0	0.0		
74.2	59.5	0.0	0.0	0.0	0.0	0.0	0.0	0.0	0.0	0.0	0.0		
75.0	79.4	0.0	0.0	0.0	0.0	0.0	0.0	0.0	0.0	0.0	0.0		
75.8	62.0	0.0	0.0	0.0	0.0	0.0	0.0	0.0	0.0	0.0	0.0	x	x
76.5	67.3	0.0	0.0	0.0	0.0	0.0	0.0	0.0	0.0	0.0	0.0	x	
77.3	64.8	0.0	0.0	0.0	0.0	0.0	0.0	0.0	0.0	0.0	0.0		x
78.0	72.5	0.0	0.0	0.0	0.0	0.0	0.0	0.0	0.0	0.0	0.0		
78.8	69.5	0.0	0.0	0.0	0.0	0.0	0.0	0.0	0.0	0.0	0.0		
79.6	64.1	0.0	0.0	0.0	0.0	0.0	0.0	0.0	0.0	0.0	0.0		
80.3	64.6	0.0	0.0	0.0	0.0	0.0	0.0	0.0	0.0	0.0	0.0		
81.1	63.7	0.0	0.0	0.0	0.0	0.0	0.0	0.0	0.0	0.0	0.0	x	
81.9	71.8	0.0	0.0	0.0	0.0	0.0	0.0	0.0	0.0	0.0	0.0		
82.6	75.3	0.0	0.0	0.0	0.0	0.0	0.0	0.0	0.0	0.0	0.0		x
83.4	61.5	0.0	0.0	0.0	0.0	0.0	0.0	0.0	0.0	0.0	0.1		
84.1	59.5	0.0	0.0	0.0	0.0	0.0	0.0	0.0	0.0	0.0	0.0		
84.9	64.0	0.0	0.0	0.0	0.0	0.0	0.0	0.0	0.0	0.0	0.0		
85.7	55.3	0.0	0.0	0.0	0.0	0.0	0.0	0.0	0.0	0.0	0.0	x	x
86.5	49.9	0.0	0.0	0.0	0.0	0.0	0.0	0.0	0.0	0.0	0.0	x	
87.2	61.3	0.0	0.0	0.0	0.0	0.0	0.0	0.0	0.0	0.0	0.0		x
88.0	65.1	0.0	0.0	0.0	0.0	0.0	0.0	0.0	0.0	0.0	0.0	x	
88.7	55.2	0.0	0.0	0.0	0.0	0.0	0.0	0.0	0.0	0.0	0.0		
89.5	56.4	0.0	0.0	0.0	0.0	0.0	0.0	0.0	0.0	0.0	0.0		
90.3	63.5	0.0	0.0	0.0	0.0	0.0	0.0	0.0	0.0	0.0	0.0		
91.0	58.6	0.0	0.0	0.0	0.0	0.0	0.0	0.0	0.0	0.0	0.0		
91.8	57.4	0.0	0.0	0.0	0.0	0.0	0.0	0.0	0.0	0.0	0.0		
92.6	61.6	0.0	0.0	0.0	0.0	0.0	0.0	0.0	0.0	0.0	0.0		
93.3	60.5	0.0	0.0	0.0	0.0	0.0	0.0	0.0	0.0	0.0	0.0		x
94.0	69.8	0.0	0.0	0.0	0.0	0.0	0.0	0.0	0.0	0.0	0.0		x
94.7	68.7	0.0	0.0	0.0	0.0	0.0	0.0	0.0	0.0	0.0	0.0		
95.5	58.8	0.0	0.0	0.0	0.0	0.0	0.0	0.0	0.0	0.0	0.0		
96.3	66.7	0.0	0.0	0.0	0.0	0.0	0.0	0.0	0.0	0.0	0.0		
97.0	53.3	0.0	0.0	0.0	0.0	0.0	0.0	0.0	0.0	0.0	0.0		
97.7	55.3	0.0	0.0	0.0	0.0	0.0	0.0	0.0	0.0	0.0	0.0	x	
98.5	74.6	0.0	0.0	0.0	0.0	0.0	0.0	0.0	0.0	0.0	0.0		x
99.3	69.2	0.0	0.0	0.0	0.0	0.0	0.0	0.4	0.0	0.0	0.4		
100.0	47.3	0.0	0.0	0.0	0.0	0.0	0.0	0.0	0.0	0.0	0.0		
100.8	55.2	0.0	0.0	0.0	0.0	0.0	0.0	0.0	0.0	0.0	0.0		
101.5	64.9	0.0	0.0	0.0	0.0	0.0	0.0	0.0	0.0	0.0	0.0		
102.2	66.2	0.1	0.0	0.0	0.0	0.0	0.0	0.0	0.0	0.0	0.1	x	
103.0	56.4	0.0	0.0	0.0	0.0	0.0	0.0	0.0	0.0	0.0	0.0	x	
103.7	66.4	0.0	0.0	0.0	0.0	0.0	0.0	0.0	0.0	0.0	0.0		
104.5	49.5	0.0	0.0	0.0	0.0	0.0	0.0	0.0	0.0	0.0	0.0		
105.3	63.6	0.0	0.0	0.0	0.0	0.0	0.0	0.0	0.0	0.0	0.0	x	
106.0	43.5	0.0	0.0	0.0	0.0	0.0	0.0	0.0	0.0	0.0	0.0		
106.6	60.8	0.0	0.0	0.0	0.0	0.0	0.0	0.0	0.0	0.0	0.0		
107.4	55.6	0.0	0.0	0.0	0.0	0.0	0.0	0.0	0.0	0.0	0.0		
108.1	56.1	0.0	0.0	0.0	0.0	0.0	0.0	0.0	0.0	0.0	0.0		
108.7	64.1	0.0	0.0	0.0	0.0	0.0	0.0	0.0	0.0	0.0	0.0		x
109.7	68.5	0.0	0.0	0.0	0.0	0.0	0.0	0.0	0.0	0.0	0.0		x
110.4	40.0	0.0	0.0	0.0	0.0	0.0	0.0	0.6	0.0	0.0	0.7	x	
111.2	57.4	0.0	0.0	0.0	0.0	0.0	0.0	0.4	0.0	0.0	0.5	x	
111.9	55.5	0.0	0.0	0.0	0.0	0.0	0.0	0.0	0.0	0.0	0.0	x	x
112.7	38.7	0.0	0.0	0.0	0.0	0.0	0.0	0.0	0.0	0.0	0.0		x
113.4	49.7	0.0	0.0	0.0	0.0	0.0	0.0	0.0	0.0	0.0	0.0		x
114.2	46.4	0.0	0.0	0.0	0.0	0.0	0.0	0.0	0.0	0.0	0.0		
114.9	37.2	0.0	0.0	0.0	0.0	0.0	0.0	0.1	0.0	0.0	0.1		
115.6	37.6	0.0	0.0	0.0	0.0	0.0	0.0	0.0	0.0	0.0	0.0		
116.3	46.3	0.0	0.0	0.0	0.0	0.0	0.0	0.0	0.0	0.0	0.0		x
117.1	45.0	0.0	0.0	0.0	0.0	0.0	0.0	0.2	0.0	0.0	0.2		
117.8	31.9	0.0	0.0	0.0	0.0	0.0	0.0	0.0	0.0	0.0	0.0		
118.6	48.2	0.0	0.0	0.0	0.0	0.0	0.0	0.0	0.0	0.0	0.0		
119.3	43.3	0.0	0.0	0.0	0.0	0.0	0.0	0.0	0.0	0.0	0.0	x	
120.0	56.0	0.0	0.0	0.0	0.0	0.0	0.0	0.0	0.0	0.0	0.0		
120.6	45.6	0.0	0.0	0.0	0.0	0.0	0.0	0.0	0.0	0.0	0.0		x

Note: C—*Candona*; Cyth lac—*Cytherissa lacustris*; Unk—unknown; vpg—valves per gram of sediment; x—present. Photo documentation of endemic species in Bright et al. (2005).

APPENDIX D. DIATOM FLORA FROM BL00-1

			Benthic Taxa			
Sample ID	Depth	Comments	No. of diatoms counted	No. of spp. not including unid	Simpson reciprocal index	*Achnanthes curtissima*
Core catcher samples						
1E-26E	1.72		498	22	1.0	0.0
1E-27E	4.97		634	45	4.9	0.5
1E-28E	8.02		541	25	1.1	0.0
1E-29E	11.01	no visible diatoms or pieces	1	0	0.0	0.0
1E-30E	14.03	no visible diatoms or pieces	1	0	0.0	0.0
1E-31E	17.04	no visible diatoms or pieces	1	0	0.0	0.0
1E-32E	20.09		372	12	1.3	0.0
1E-33E	23.07		563	19	2.9	0.0
1E-34E	26.10		680	6	1.1	0.0
1E-35E	29.09		619	19	1.5	0.0
1E-36E	32.09		622	25	2.3	0.0
1E-37E	35.09		570	18	1.7	0.5
1E-38E	38.08	few diatoms	1	0	0.0	0.0
1E-39E	41.09		610	33	4.6	0.3
1E-40E	43.90	no visible diatoms or pieces	1	0	0.0	0.0
1E-41E	47.10		558	25	4.2	0.4
1E-42E	50.09		571	29	4.2	0.0
1E-43E	52.99		595	41	5.4	1.0
1E-44E	55.60		536	38	7.1	0.4
1E-45E	58.10	no visible diatoms or pieces	1	0	0.0	0.0
1E-46E	60.60		527	37	3.4	0.4
1E-22E	63.10	dissolution of some cells present, mostly on broken diatoms	417	15	0.8	0.0
1E-23E	65.60	no visible diatoms or pieces	1	0	0.0	0.0
1E-24E	68.10	very small pieces, some dissolution evident	1	0	0.0	0.0
1E-25E	70.61	no visible diatoms or pieces	1	0	0.0	0.0
1E-26E	76.32	very few diatoms, mostly *Stephanodiscus* and small fragilaroid, some dissolution	NA			
1E-27E	76.90	no diatoms, no pieces	NA			
1E-28E	79.43	no diatoms, no pieces	NA			
1E-29E	81.91	some diatoms, mostly small fragilaroid and *Stephanodiscus*, lots of cysts	NA			
1E-30E	84.42	very sparse, mostly small fragilaroid	NA			
1E-31E	86.42	very few diatoms, mostly *Stephanodiscus* and small fragilaroid, some dissolution	NA			
1E-32E	88.43	very sparse, centrics and pennates, mostly broken, some dissolution	NA			
1E-33E	90.42	abundant diatoms, both centrics and pennates, very diverse	NA			
1E-34E	92.43	abundant diatoms, both centrics and pennates, very diverse	NA			
1E-35E	94.43	abundant diatoms, mostly planktonic, dissolution and breakage present	NA			
1E-36E	96.43	sparse diatoms	NA			
1E-37E	98.42	sparse diatoms, mostly small fragilaroid, dissolution present	NA			
1E-38E	100.43	abundant diatoms and cysts, many *Stephanodiscus* and *Aulacoseira* (*granulata*?)	NA			
1E-39E	102.20	abundant *Stephanodiscus*	NA			
1E-40E	104.50	abundant *Stephanodiscus*	NA			
1E-41E	106.00	abundant *Stephanodiscus*	NA			
1E-42E	108.10	abundant diatoms, mostly small fragilaroid, few centrics	NA			
1E-43E	110.40	abundant diatoms, both pennates and centrics, very diverse	NA			
1E-44E	113.40	abundant diatoms, both *Fragilaria* and a few centrics	NA			
1E-45E	114.90	no diatoms, no pieces	NA			
1E-46E	117.80	abundant diatoms, many big centrics and *Aulacoseira*	NA			
Non-core catcher						
1H-1 50-54	0.50	many broken pieces, mostly *Navicula oblonga*, *Pinnularia viridis*, and *Cymbella*, very few others	0	0	0.0	0.0
2H-1 80-84	2.85	1/2 transect	723	44	4.2	0.3
2H-2 28-30	3.97	1/2 transect	622	38	3.0	0.6
3H-2 50-52	7.15	1/2 transect	683	35	5.4	0.0
4H-1 110-112	9.25	2 transects	645	28	3.7	0.2
5H-1 30-32	11.45	no diatoms, no pieces, no cysts, plenty of sediment	0	0	0.0	0.0
5H-2 80-83	13.50	no diatoms, no pieces, no cysts, plenty of sediment	0	0	0.0	0.0
6H-2 0-3	15.70	no diatoms, no pieces, no cysts, plenty of sediment	0	0	0.0	0.0
7H-1 70-73	19.00	mostly broken *Stephanodiscus*, very little dissolution visible	0	0	0.0	0.0
16E-2 60-63	44.90	15 transects	577	28	5.8	0.3
18E-1 30-33	50.50	1 transect	754	27	4.1	0.0
18E-2 80-83	52.45	1/2 transect	603	31	2.3	0.0
20E-1 50-53	56.10	1 transect	629	30	4.2	0.0
20E-2 100-103	57.72	4 transects	539	16	1.3	0.0
21E-1 150-153	59.36	3 transects, lots of pieces	615	41	6.7	0.0
21E-2 120-123	60.39	2 transects	614	26	3.9	0.0
22E-1 50-53	61.08	many well-preserved periphytic diatoms, but dissolution of some planktic diatoms	0	0	0.0	0.0
22E-2 100-103	62.84	many whole and well-preserved diatoms, including *Amphora*, *Aulacoseira* and small benthic/tychoplanktic fragilarioid species, but also many broken and dissolved large *Cymbella* and *Epithemia* diatoms	0	0	0.0	0.0
23E-2 20-23	64.55	no diatoms, no pieces, no cysts, plenty of sediment	0	0	0.0	0.0
24-1 67-70	66.21	diatoms are rare, no evidence of dissolution	0	0	0.0	0.0
24E-2 120-123	67.92	no diatoms, no pieces, no cysts, plenty of sediment	0	0	0.0	0.0
25E-2 30-33	69.48	no diatoms, no pieces, no cysts, plenty of sediment	0	0	0.0	0.0
26E-2 50-53	71.06	few diatoms observed, some breakage, but no evidence of dissolution	0	0	0.0	0.0
26E-2 100-103	72.72	no dissolution evidenced	802	25	2.7	0.0

		Benthic Taxa								
Sample ID	Depth	A. lanceolata	A. saccula	Amphora inariensis	A. ovalis	A. pediculus	Caloneis schumanniana	Caloneis sp. BL1	Cocconeis placentula	C. placentula var. lineata
Core catcher samples										
1E-26E	1.72	0.2	0.4	1.4	0.4	0.0	0.0	1.4	0.0	0.0
1E-27E	4.97	0.6	1.9	1.4	0.3	5.8	0.0	0.9	0.3	0.3
1E-28E	8.02	0.0	0.0	0.7	6.8	2.0	2.0	0.9	0.0	2.6
1E-29E	11.01	0.0	0.0	0.0	0.0	0.0	0.0	0.0	0.0	0.0
1E-30E	14.03	0.0	0.0	0.0	0.0	0.0	0.0	0.0	0.0	0.0
1E-31E	17.04	0.0	0.0	0.0	0.0	0.0	0.0	0.0	0.0	0.0
1E-32E	20.09	0.0	0.0	0.3	0.0	0.0	0.0	0.0	0.0	2.2
1E-33E	23.07	0.2	1.2	1.8	0.0	0.7	0.0	0.0	0.0	0.5
1E-34E	26.10	0.0	0.1	0.0	0.0	0.0	0.0	0.0	0.0	0.0
1E-35E	29.09	0.0	0.3	0.6	0.2	0.0	0.0	0.0	0.0	0.3
1E-36E	32.09	1.3	0.6	1.3	0.0	0.8	0.0	0.0	0.0	1.1
1E-37E	35.09	0.5	0.5	0.5	0.0	0.5	0.0	0.0	0.0	0.7
1E-38E	38.08	0.0	0.0	0.0	0.0	0.0	0.0	0.0	0.0	0.0
1E-39E	41.09	0.5	0.5	2.1	0.0	0.8	0.2	0.3	0.2	0.0
1E-40E	43.90	0.0	0.0	0.0	0.0	0.0	0.0	0.0	0.0	0.0
1E-41E	47.10	1.1	0.0	0.7	0.0	1.8	0.0	0.0	0.0	0.4
1E-42E	50.09	1.1	0.4	0.4	0.0	1.1	0.0	0.0	0.0	0.0
1E-43E	52.99	0.7	0.8	0.8	0.7	1.2	0.0	0.0	0.2	0.2
1E-44E	55.60	1.1	0.7	2.6	0.9	0.9	0.2	0.0	1.1	0.7
1E-45E	58.10	0.0	0.0	0.0	0.0	0.0	0.0	0.0	0.0	0.0
1E-46E	60.60	0.4	0.0	1.5	0.0	6.6	0.0	0.4	1.9	0.6
1E-22E	63.10	0.0	0.0	0.0	0.2	0.0	0.0	0.0	0.0	0.0
1E-23E	65.60	0.0	0.0	0.0	0.0	0.0	0.0	0.0	0.0	0.0
1E-24E	68.10	0.0	0.0	0.0	0.0	0.0	0.0	0.0	0.0	0.0
1E-25E	70.61	0.0	0.0	0.0	0.0	0.0	0.0	0.0	0.0	0.0
1E-26E	76.32									
1E-27E	76.90									
1E-28E	79.43									
1E-29E	81.91									
1E-30E	84.42									
1E-31E	86.42									
1E-32E	88.43									
1E-33E	90.42									
1E-34E	92.43									
1E-35E	94.43									
1E-36E	96.43									
1E-37E	98.42									
1E-38E	100.43									
1E-39E	102.20									
1E-40E	104.50									
1E-41E	106.00									
1E-42E	108.10									
1E-43E	110.40									
1E-44E	113.40									
1E-45E	114.90									
1E-46E	117.80									
Non-core catcher										
1H-1 50-54	0.50	0.0	0.0	0.0	0.0	0.0	0.0	0.0	0.0	0.0
2H-1 80-84	2.85	0.1	0.4	1.0	1.8	4.1	0.4	1.7	0.0	0.1
2H-2 28-30	3.97	0.2	0.8	0.2	1.0	2.6	0.5	0.3	0.0	0.2
3H-2 50-52	7.15	0.3	0.0	0.1	0.4	1.0	0.3	0.0	0.0	0.0
4H-1 110-112	9.25	0.0	0.2	0.9	0.3	1.2	0.0	0.0	0.0	0.2
5H-1 30-32	11.45	0.0	0.0	0.0	0.0	0.0	0.0	0.0	0.0	0.0
5H-2 80-83	13.50	0.0	0.0	0.0	0.0	0.0	0.0	0.0	0.0	0.0
6H-2 0-3	15.70	0.0	0.0	0.0	0.0	0.0	0.0	0.0	0.0	0.0
7H-1 70-73	19.00	0.0	0.0	0.0	0.0	0.0	0.0	0.0	0.0	0.0
16E-2 60-63	44.90	0.0	0.0	0.3	0.2	0.9	0.0	0.0	0.2	0.3
18E-1 30-33	50.50	0.1	0.0	0.0	0.1	0.5	0.0	0.0	0.0	0.0
18E-2 80-83	52.45	0.5	0.3	0.0	0.5	0.3	0.0	0.0	0.0	0.2
20E-1 50-53	56.10	0.0	0.0	0.0	0.0	0.0	0.0	0.0	0.0	0.0
20E-2 100-103	57.72	0.0	0.0	0.2	0.0	0.6	0.0	0.0	0.0	0.0
21E-1 150-153	59.36	0.2	0.0	1.0	0.0	2.6	0.0	0.0	0.0	2.9
21E-2 120-123	60.39	0.3	0.3	0.7	0.0	0.3	0.0	0.0	0.0	0.3
22E-1 50-53	61.08	0.0	0.0	0.0	0.0	0.0	0.0	0.0	0.0	0.0
22E-2 100-103	62.84	0.0	0.0	0.0	0.0	0.0	0.0	0.0	0.0	0.0
23E-2 20-23	64.55	0.0	0.0	0.0	0.0	0.0	0.0	0.0	0.0	0.0
24E-1 67-70	66.21	0.0	0.0	0.0	0.0	0.0	0.0	0.0	0.0	0.0
24E-2 120-123	67.92	0.0	0.0	0.0	0.0	0.0	0.0	0.0	0.0	0.0
25E-2 30-33	69.48	0.0	0.0	0.0	0.0	0.0	0.0	0.0	0.0	0.0
26E-2 50-53	71.06	0.0	0.0	0.0	0.0	0.0	0.0	0.0	0.0	0.0
26E-2 100-103	72.72	0.0	0.1	0.2	0.0	0.0	0.1	0.0	0.2	0.0

(*continued*)

APPENDIX D. DIATOM FLORA FROM BL00-1 (*continued*)

			Cymbella mesiana	Diatoma tenuis var. elongatum	Diploneis elliptica	Epithemia adnata
Sample ID	Depth	Comments				
Core catcher samples						
1E-26E	1.72		0.0	0.0	0.6	0.0
1E-27E	4.97		1.1	1.1	0.6	0.0
1E-28E	8.02		0.0	0.0	2.4	0.0
1E-29E	11.01	no visible diatoms or pieces	0.0	0.0	0.0	0.0
1E-30E	14.03	no visible diatoms or pieces	0.0	0.0	0.0	0.0
1E-31E	17.04	no visible diatoms or pieces	0.0	0.0	0.0	0.0
1E-32E	20.09		0.0	0.0	0.3	0.3
1E-33E	23.07		0.0	0.0	0.0	0.0
1E-34E	26.10		0.0	0.0	0.0	0.0
1E-35E	29.09		0.0	0.0	0.0	0.0
1E-36E	32.09		0.0	0.0	0.0	0.0
1E-37E	35.09		0.0	0.0	0.0	0.0
1E-38E	38.08	few diatoms	0.0	0.0	0.0	0.0
1E-39E	41.09		0.0	0.0	0.0	0.0
1E-40E	43.90	no visible diatoms or pieces	0.0	0.0	0.0	0.0
1E-41E	47.10		0.0	0.0	0.0	0.5
1E-42E	50.09		0.0	0.0	0.0	0.7
1E-43E	52.99		0.0	0.0	0.3	0.3
1E-44E	55.60		0.0	0.0	0.0	0.0
1E-45E	58.10	no visible diatoms or pieces	0.0	0.0	0.0	0.0
1E-46E	60.60		0.0	0.0	0.0	3.4
1E-22E	63.10	dissolution of some cells present, mostly on broken diatoms	0.0	0.0	0.0	0.7
1E-23E	65.60	no visible diatoms or pieces	0.0	0.0	0.0	0.0
1E-24E	68.10	very small pieces, some dissolution evident	0.0	0.0	0.0	0.0
1E-25E	70.61	no visible diatoms or pieces	0.0	0.0	0.0	0.0
1E-26E	76.32	very few diatoms, mostly *Stephanodiscus* and small fragilaroid, some dissolution				
1E-27E	76.90	no diatoms, no pieces				
1E-28E	79.43	no diatoms, no pieces				
1E-29E	81.91	some diatoms, mostly small fragilaroid and *Stephanodiscus*, lots of cysts				
1E-30E	84.42	very sparse, mostly small fragilaroid				
1E-31E	86.42	very few diatoms, mostly *Stephanodiscus* and small fragilaroid, some dissolution				
1E-32E	88.43	very sparse, centrics and pennates, mostly broken, some dissolution				
1E-33E	90.42	abundant diatoms, both centrics and pennates, very diverse				
1E-34E	92.43	abundant diatoms, both centrics and pennates, very diverse				
1E-35E	94.43	abundant diatoms, mostly planktonic, dissolution and breakage present				
1E-36E	96.43	sparse diatoms				
1E-37E	98.42	sparse diatoms, mostly small fragilaroid, dissolution present				
1E-38E	100.43	abundant diatoms and cysts, many *Stephanodiscus* and *Aulacoseira* (*granulata*?)				
1E-39E	102.20	abundant *Stephanodiscus*				
1E-40E	104.50	abundant *Stephanodiscus*				
1E-41E	106.00	abundant *Stephanodiscus*				
1E-42E	108.10	abundant diatoms, mostly small fragilaroid, few centrics				
1E-43E	110.40	abundant diatoms, both pennates and centrics, very diverse				
1E-44E	113.40	abundant diatoms, both *Fragilaria* and a few centrics				
1E-45E	114.90	no diatoms, no pieces				
1E-46E	117.80	abundant diatoms, many big centrics and *Aulacoseira*				
Non-core catcher						
1H-1 50-54	0.50	many broken pieces, mostly *Navicula oblonga*, *Pinnularia viridis*, and *Cymbella*, very few others	0.0	0.0	0.0	0.0
2H-1 80-84	2.85	1/2 transect	0.1	0.1	2.8	0.6
2H-2 28-30	3.97	1/2 transect	0.0	0.0	1.8	0.0
3H-2 50-52	7.15	1/2 transect	0.1	0.1	0.0	0.0
4H-1 110-112	9.25	2 transects	0.0	0.0	0.0	0.0
5H-1 30-32	11.45	no diatoms, no pieces, no cysts, plenty of sediment	0.0	0.0	0.0	0.0
5H-2 80-83	13.50	no diatoms, no pieces, no cysts, plenty of sediment	0.0	0.0	0.0	0.0
6H-2 0-3	15.70	no diatoms, no pieces, no cysts, plenty of sediment	0.0	0.0	0.0	0.0
7H-1 70-73	19.00	mostly broken *Stephanodiscus*, very little dissolution visible	0.0	0.0	0.0	0.0
16E-2 60-63	44.90	15 transects	0.0	0.0	0.0	0.2
18E-1 30-33	50.50	1 transect	0.0	0.0	0.1	0.0
18E-2 80-83	52.45	1/2 transect	0.0	0.0	0.0	0.2
20E-1 50-53	56.10	1 transect	0.0	0.0	0.0	0.0
20E-2 100-103	57.72	4 transects	0.0	0.0	0.0	0.0
21E-1 150-153	59.36	3 transects, lots of pieces	0.0	0.0	0.2	0.7
21E-2 120-123	60.39	2 transects	0.0	0.0	0.0	0.0
22E-1 50-53	61.08	many well-preserved periphytic diatoms, but dissolution of some planktic diatoms	0.0	0.0	0.0	0.0
22E-2 100-103	62.84	many whole and well-preserved diatoms, including *Amphora*, *Aulacoseira* and small benthic/tychoplanktic fragilarioid species, but also many broken and dissolved large *Cymbella* and *Epithemia* diatoms	0.0	0.0	0.0	0.0
23E-2 20-23	64.55	no diatoms, no pieces, no cysts, plenty of sediment	0.0	0.0	0.0	0.0
24E-1 67-70	66.21	diatoms are rare, no evidence of dissolution	0.0	0.0	0.0	0.0
24E-2 120-123	67.92	no diatoms, no pieces, no cysts, plenty of sediment	0.0	0.0	0.0	0.0
25E-2 30-33	69.48	no diatoms, no pieces, no cysts, plenty of sediment	0.0	0.0	0.0	0.0
26E-2 50-53	71.06	few diatoms observed, some breakage, but no evidence of dissolution	0.0	0.0	0.0	0.0
26E-2 100-103	72.72	no dissolution evidenced	0.0	0.0	0.2	0.0

			Pseudostaurosira		P.	P.				S.
		E.	brevistriata	P.	brevistriata	brevistriata	Ulnaria	Fragilariforma	Staurosira	construens
Sample ID	Depth	frickeii	total	brevistriata	(rhombic)	(oval)	biceps	constricta	construens	var. venter

Benthic Taxa

Core catcher samples

Sample ID	Depth	E. frickeii	Pseudostaurosira brevistriata total	P. brevistriata	P. brevistriata (rhombic)	P. brevistriata (oval)	Ulnaria biceps	Fragilariforma constricta	Staurosira construens	S. construens var. venter
1E-26E	1.72	0.2	83.1	45.0	33.5	4.6	0.0	0.0	0.0	0.0
1E-27E	4.97	0.0	33.1	12.1	14.4	6.6	0.0	0.0	0.5	0.0
1E-28E	8.02	2.8	63.4	54.5	7.9	0.9	0.0	0.6	0.0	0.0
1E-29E	11.01	0.0	0.0	0.0	0.0	0.0	0.0	0.0	0.0	0.0
1E-30E	14.03	0.0	0.0	0.0	0.0	0.0	0.0	0.0	0.0	0.0
1E-31E	17.04	0.0	0.0	0.0	0.0	0.0	0.0	0.0	0.0	0.0
1E-32E	20.09	0.5	2.4	2.4	0.0	0.0	0.0	0.0	0.0	0.8
1E-33E	23.07	0.0	17.4	16.0	0.5	0.9	0.0	0.5	0.7	0.4
1E-34E	26.10	0.0	0.0	0.0	0.0	0.0	0.0	0.0	0.0	0.0
1E-35E	29.09	0.0	6.6	2.3	3.7	0.6	0.0	0.0	0.6	0.2
1E-36E	32.09	0.0	13.2	8.4	4.0	0.8	0.0	0.2	1.8	0.6
1E-37E	35.09	0.0	0.9	0.9	0.0	0.0	0.0	0.0	0.2	0.0
1E-38E	38.08	0.0	0.0	0.0	0.0	0.0	0.0	0.0	0.0	0.0
1E-39E	41.09	0.0	27.7	12.8	12.3	2.6	0.0	0.5	4.6	3.8
1E-40E	43.90	0.0	0.0	0.0	0.0	0.0	0.0	0.0	0.0	0.0
1E-41E	47.10	0.0	26.9	15.1	9.5	2.3	0.7	0.4	0.0	2.9
1E-42E	50.09	0.0	32.6	17.2	14.5	0.9	5.6	0.4	1.8	8.1
1E-43E	52.99	0.0	18.3	5.9	7.9	4.5	0.5	0.5	2.0	1.0
1E-44E	55.60	0.0	22.6	7.1	12.9	2.6	9.1	0.0	1.5	1.9
1E-45E	58.10	0.0	0.0	0.0	0.0	0.0	0.0	0.0	0.0	0.0
1E-46E	60.60	0.0	43.8	17.5	20.9	5.5	0.9	0.9	2.7	1.9
1E-22E	63.10	1.7	91.6	54.7	31.4	5.5	0.0	0.0	0.0	0.0
1E-23E	65.60	0.0	0.0	0.0	0.0	0.0	0.0	0.0	0.0	0.0
1E-24E	68.10	0.0	0.0	0.0	0.0	0.0	0.0	0.0	0.0	0.0
1E-25E	70.61	0.0	0.0	0.0	0.0	0.0	0.0	0.0	0.0	0.0
1E-26E	76.32									
1E-27E	76.90									
1E-28E	79.43									
1E-29E	81.91									
1E-30E	84.42									
1E-31E	86.42									
1E-32E	88.43									
1E-33E	90.42									
1E-34E	92.43									
1E-35E	94.43									
1E-36E	96.43									
1E-37E	98.42									
1E-38E	100.43									
1E-39E	102.20									
1E-40E	104.50									
1E-41E	106.00									
1E-42E	108.10									
1E-43E	110.40									
1E-44E	113.40									
1E-45E	114.90									
1E-46E	117.80									

Non-core catcher

Sample ID	Depth	E. frickeii	Pseudostaurosira brevistriata total	P. brevistriata	P. brevistriata (rhombic)	P. brevistriata (oval)	Ulnaria biceps	Fragilariforma constricta	Staurosira construens	S. construens var. venter
1H-1 50-54	0.50	0.0	0.0	0.0	0.0	0.0	0.0	0.0	0.0	0.0
2H-1 80-84	2.85	0.0	36.4	21.0	8.7	6.6	0.0	0.1	0.0	0.0
2H-2 28-30	3.97	0.0	44.7	22.2	12.5	10.0	0.0	0.0	0.3	0.0
3H-2 50-52	7.15	0.0	20.8	3.7	8.8	8.3	0.0	0.0	1.0	1.8
4H-1 110-112	9.25	0.0	28.2	14.9	10.9	2.5	0.0	0.0	1.7	2.2
5H-1 30-32	11.45	0.0	0.0	0.0	0.0	0.0	0.0	0.0	0.0	0.0
5H-2 80-83	13.50	0.0	0.0	0.0	0.0	0.0	0.0	0.0	0.0	0.0
6H-2 0-3	15.70	0.0	0.0	0.0	0.0	0.0	0.0	0.0	0.0	0.0
7H-1 70-73	19.00	0.0	0.0	0.0	0.0	0.0	0.0	0.0	0.0	0.0
16E-2 60-63	44.90	0.0	18.9	5.9	10.4	2.6	0.2	0.7	0.9	3.1
18E-1 30-33	50.50	0.0	34.1	16.8	11.7	5.6	0.5	2.1	3.2	2.0
18E-2 80-83	52.45	0.0	51.9	30.7	16.3	5.0	0.7	0.0	3.2	1.5
20E-1 50-53	56.10	0.0	0.0	0.0	0.0	0.0	0.0	0.0	0.0	0.0
20E-2 100-103	57.72	0.0	2.8	0.2	2.0	0.6	0.0	0.0	0.0	0.0
21E-1 150-153	59.36	0.2	26.8	8.6	13.7	4.6	3.1	0.7	0.2	0.7
21E-2 120-123	60.39	0.0	35.5	14.0	10.9	10.6	0.0	1.8	4.1	6.7
22E-1 50-53	61.08	0.0	0.0	0.0	0.0	0.0	0.0	0.0	0.0	0.0
22E-2 100-103	62.84	0.0	0.0	0.0	0.0	0.0	0.0	0.0	0.0	0.0
23E-2 20-23	64.55	0.0	0.0	0.0	0.0	0.0	0.0	0.0	0.0	0.0
24E-1 67-70	66.21	0.0	0.0	0.0	0.0	0.0	0.0	0.0	0.0	0.0
24E-2 120-123	67.92	0.0	0.0	0.0	0.0	0.0	0.0	0.0	0.0	0.0
25E-2 30-33	69.48	0.0	0.0	0.0	0.0	0.0	0.0	0.0	0.0	0.0
26E-2 50-53	71.06	0.0	0.0	0.0	0.0	0.0	0.0	0.0	0.0	0.0
26E-2 100-103	72.72	0.4	15.0	7.2	5.4	2.4	0.0	0.0	0.7	2.2

(continued)

APPENDIX D. DIATOM FLORA FROM BL00-1 (*continued*)

			Benthic Taxa			
Sample ID	Depth	Comments	*Staurosirella leptostauron*	*S. pinnata*	*S. pinnata* var. *accuminata*	*Fragilaria tenera*
Core catcher samples						
1E-26E	1.72		0.0	4.0	0.6	0.0
1E-27E	4.97		0.0	19.2	1.9	0.9
1E-28E	8.02		0.0	0.6	0.0	0.0
1E-29E	11.01	no visible diatoms or pieces	0.0	0.0	0.0	0.0
1E-30E	14.03	no visible diatoms or pieces	0.0	0.0	0.0	0.0
1E-31E	17.04	no visible diatoms or pieces	0.0	0.0	0.0	0.0
1E-32E	20.09		0.0	1.3	1.1	0.0
1E-33E	23.07		0.0	28.8	0.9	0.0
1E-34E	26.10		0.0	0.0	0.0	0.0
1E-35E	29.09		0.0	4.0	0.6	0.0
1E-36E	32.09		0.0	5.5	0.3	0.0
1E-37E	35.09		0.0	0.4	1.1	0.0
1E-38E	38.08	few diatoms	0.0	0.0	0.0	0.0
1E-39E	41.09		0.0	18.0	1.6	0.0
1E-40E	43.90	no visible diatoms or pieces	0.0	0.0	0.0	0.0
1E-41E	47.10		0.0	31.7	3.0	0.0
1E-42E	50.09		0.0	10.9	2.1	0.0
1E-43E	52.99		0.2	33.9	1.5	0.0
1E-44E	55.60		0.0	12.1	0.7	0.0
1E-45E	58.10	no visible diatoms or pieces	0.0	0.0	0.0	0.0
1E-46E	60.60		0.0	12.1	1.1	1.3
1E-22E	63.10	dissolution of some cells present, mostly on broken diatoms	0.0	2.9	0.2	0.0
1E-23E	65.60	no visible diatoms or pieces	0.0	0.0	0.0	0.0
1E-24E	68.10	very small pieces, some dissolution evident	0.0	0.0	0.0	0.0
1E-25E	70.61	no visible diatoms or pieces	0.0	0.0	0.0	0.0
1E-26E	76.32	very few diatoms, mostly *Stephanodiscus* and small fragilaroid, some dissolution				
1E-27E	76.90	no diatoms, no pieces				
1E-28E	79.43	no diatoms, no pieces				
1E-29E	81.91	some diatoms, mostly small fragilaroid and *Stephanodiscus*, lots of cysts				
1E-30E	84.42	very sparse, mostly small fragilaroid				
1E-31E	86.42	very few diatoms, mostly *Stephanodiscus* and small fragilaroid, some dissolution				
1E-32E	88.43	very sparse, centrics and pennates, mostly broken, some dissolution				
1E-33E	90.42	abundant diatoms, both centrics and pennates, very diverse				
1E-34E	92.43	abundant diatoms, both centrics and pennates, very diverse				
1E-35E	94.43	abundant diatoms, mostly planktonic, dissolution and breakage present				
1E-36E	96.43	sparse diatoms				
1E-37E	98.42	sparse diatoms, mostly small fragilaroid, dissolution present				
1E-38E	100.43	abundant diatoms and cysts, many *Stephanodiscus* and *Aulacoseira* (*granulata*?)				
1E-39E	102.20	abundant *Stephanodiscus*				
1E-40E	104.50	abundant *Stephanodiscus*				
1E-41E	106.00	abundant *Stephanodiscus*				
1E-42E	108.10	abundant diatoms, mostly small fragilaroid, few centrics				
1E-43E	110.40	abundant diatoms, both pennates and centrics, very diverse				
1E-44E	113.40	abundant diatoms, both *Fragilaria* and a few centrics				
1E-45E	114.90	no diatoms, no pieces				
1E-46E	117.80	abundant diatoms, many big centrics and *Aulacoseira*				
Non-core catcher						
1H-1 50-54	0.50	many broken pieces, mostly *Navicula oblonga*, *Pinnularia viridis*, and *Cymbella*, very few others	0.0	0.0	0.0	0.0
2H-1 80-84	2.85	1/2 transect	1.2	18.1	1.2	0.0
2H-2 28-30	3.97	1/2 transect	1.4	22.5	1.3	0.0
3H-2 50-52	7.15	1/2 transect	0.0	31.9	1.0	0.1
4H-1 110-112	9.25	2 transects	0.3	15.8	0.3	0.0
5H-1 30-32	11.45	no diatoms, no pieces, no cysts, plenty of sediment	0.0	0.0	0.0	0.0
5H-2 80-83	13.50	no diatoms, no pieces, no cysts, plenty of sediment	0.0	0.0	0.0	0.0
6H-2 0-3	15.70	no diatoms, no pieces, no cysts, plenty of sediment	0.0	0.0	0.0	0.0
7H-1 70-73	19.00	mostly broken *Stephanodiscus*, very little dissolution visible	0.0	0.0	0.0	0.0
16E-2 60-63	44.90	15 transects	0.2	28.1	1.0	0.0
18E-1 30-33	50.50	1 transect	0.0	25.2	1.2	0.0
18E-2 80-83	52.45	1/2 transect	0.3	8.8	2.2	0.0
20E-1 50-53	56.10	1 transect	0.0	0.0	0.0	0.0
20E-2 100-103	57.72	4 transects	0.0	1.1	0.9	0.4
21E-1 150-153	59.36	3 transects, lots of pieces	0.0	6.5	1.0	0.0
21E-2 120-123	60.39	2 transects	0.0	20.8	3.9	0.0
22E-1 50-53	61.08	many well-preserved periphytic diatoms, but dissolution of some planktic diatoms	0.0	0.0	0.0	0.0
22E-2 100-103	62.84	many whole and well-preserved diatoms, including *Amphora*, *Aulacoseira* and small benthic/tychoplanktic fragilarioid species, but also many broken and dissolved large *Cymbella* and *Epithemia* diatoms	0.0	0.0	0.0	0.0
23E-2 20-23	64.55	no diatoms, no pieces, no cysts, plenty of sediment	0.0	0.0	0.0	0.0
24E-1 67-70	66.21	diatoms are rare, no evidence of dissolution	0.0	0.0	0.0	0.0
24E-2 120-123	67.92	no diatoms, no pieces, no cysts, plenty of sediment	0.0	0.0	0.0	0.0
25E-2 30-33	69.48	no diatoms, no pieces, no cysts, plenty of sediment	0.0	0.0	0.0	0.0
26E-2 50-53	71.06	few diatoms observed, some breakage, but no evidence of dissolution	0.0	0.0	0.0	0.0
26E-2 100-103	72.72	no dissolution evidenced	0.0	5.5	1.2	0.0

				Benthic Taxa						
Sample ID	Depth	F. ulna	F. virescens	Navicula capitata var. lueneburgensis	N. cryptocephela	N. cryptotenella	N. oblonga	N. pupula	N. rhyncocephela	N. tuscula
Core catcher samples										
1E-26E	1.72	0.0	0.0	0.2	0.0	0.0	0.2	0.0	1.6	0.0
1E-27E	4.97	0.0	0.0	2.8	0.0	1.1	0.0	0.0	4.1	0.6
1E-28E	8.02	0.0	0.0	0.0	0.0	0.0	4.1	0.0	0.0	0.0
1E-29E	11.01	0.0	0.0	0.0	0.0	0.0	0.0	0.0	0.0	0.0
1E-30E	14.03	0.0	0.0	0.0	0.0	0.0	0.0	0.0	0.0	0.0
1E-31E	17.04	0.0	0.0	0.0	0.0	0.0	0.0	0.0	0.0	0.0
1E-32E	20.09	0.0	0.0	0.0	0.0	0.0	0.0	0.0	0.0	0.0
1E-33E	23.07	0.0	0.0	0.0	0.0	0.0	0.0	0.2	0.0	0.2
1E-34E	26.10	0.0	0.0	0.0	0.0	0.0	0.0	0.0	0.0	0.0
1E-35E	29.09	0.0	0.0	0.0	0.0	0.0	0.0	0.0	0.2	0.2
1E-36E	32.09	0.0	0.0	0.0	0.0	0.0	0.0	0.2	0.0	0.0
1E-37E	35.09	0.0	0.0	0.0	0.0	0.0	0.0	0.2	0.0	0.0
1E-38E	38.08	0.0	0.0	0.0	0.0	0.0	0.0	0.0	0.0	0.0
1E-39E	41.09	0.0	0.0	0.3	0.0	0.2	0.0	0.2	0.7	0.3
1E-40E	43.90	0.0	0.0	0.0	0.0	0.0	0.0	0.0	0.0	0.0
1E-41E	47.10	0.0	0.0	0.0	0.0	0.0	0.0	0.0	0.0	0.4
1E-42E	50.09	0.0	0.0	0.4	0.0	0.0	0.0	0.0	0.0	0.5
1E-43E	52.99	0.0	0.0	0.8	0.0	0.0	0.0	0.2	0.7	0.7
1E-44E	55.60	1.9	0.0	1.1	0.0	0.0	0.0	0.2	0.7	0.7
1E-45E	58.10	0.0	0.0	0.0	0.0	0.0	0.0	0.0	0.0	0.0
1E-46E	60.60	0.2	0.0	0.0	0.0	0.6	0.0	0.4	0.2	0.0
1E-22E	63.10	0.0	0.0	0.0	0.0	0.0	0.0	0.0	0.5	0.0
1E-23E	65.60	0.0	0.0	0.0	0.0	0.0	0.0	0.0	0.0	0.0
1E-24E	68.10	0.0	0.0	0.0	0.0	0.0	0.0	0.0	0.0	0.0
1E-25E	70.61	0.0	0.0	0.0	0.0	0.0	0.0	0.0	0.0	0.0
1E-26E	76.32									
1E-27E	76.90									
1E-28E	79.43									
1E-29E	81.91									
1E-30E	84.42									
1E-31E	86.42									
1E-32E	88.43									
1E-33E	90.42									
1E-34E	92.43									
1E-35E	94.43									
1E-36E	96.43									
1E-37E	98.42									
1E-38E	100.43									
1E-39E	102.20									
1E-40E	104.50									
1E-41E	106.00									
1E-42E	108.10									
1E-43E	110.40									
1E-44E	113.40									
1E-45E	114.90									
1E-46E	117.80									
Non-core catcher										
1H-1 50-54	0.50	0.0	0.0	0.0	0.0	0.0	0.0	0.0	0.0	0.0
2H-1 80-84	2.85	0.0	0.1	2.1	0.4	0.0	0.0	1.2	1.1	1.1
2H-2 28-30	3.97	0.0	1.6	1.8	2.1	0.0	0.2	0.0	2.4	0.8
3H-2 50-52	7.15	0.0	1.2	1.9	0.0	0.0	0.0	0.0	0.1	0.3
4H-1 110-112	9.25	0.0	0.0	0.3	0.0	0.0	0.0	0.3	0.0	0.0
5H-1 30-32	11.45	0.0	0.0	0.0	0.0	0.0	0.0	0.0	0.0	0.0
5H-2 80-83	13.50	0.0	0.0	0.0	0.0	0.0	0.0	0.0	0.0	0.0
6H-2 0-3	15.70	0.0	0.0	0.0	0.0	0.0	0.0	0.0	0.0	0.0
7H-1 70-73	19.00	0.0	0.0	0.0	0.0	0.0	0.0	0.0	0.0	0.0
16E-2 60-63	44.90	3.3	0.0	0.0	0.0	0.0	0.0	0.2	0.0	0.0
18E-1 30-33	50.50	0.0	0.3	0.0	0.0	0.0	0.0	0.1	0.0	0.0
18E-2 80-83	52.45	0.0	0.2	0.3	0.0	0.0	0.0	0.2	0.0	0.0
20E-1 50-53	56.10	0.0	0.0	0.0	0.0	0.0	0.0	0.0	0.0	0.0
20E-2 100-103	57.72	0.0	0.0	0.0	0.0	0.0	0.0	0.2	0.0	0.0
21E-1 150-153	59.36	0.0	0.0	0.0	0.0	0.0	0.2	0.2	0.3	0.2
21E-2 120-123	60.39	0.0	1.1	0.2	0.0	0.0	0.0	0.2	0.0	0.0
22E-1 50-53	61.08	0.0	0.0	0.0	0.0	0.0	0.0	0.0	0.0	0.0
22E-2 100-103	62.84	0.0	0.0	0.0	0.0	0.0	0.0	0.0	0.0	0.0
23E-2 20-23	64.55	0.0	0.0	0.0	0.0	0.0	0.0	0.0	0.0	0.0
24E-1 67-70	66.21	0.0	0.0	0.0	0.0	0.0	0.0	0.0	0.0	0.0
24E-2 120-123	67.92	0.0	0.0	0.0	0.0	0.0	0.0	0.0	0.0	0.0
25E-2 30-33	69.48	0.0	0.0	0.0	0.0	0.0	0.0	0.0	0.0	0.0
26E-2 50-53	71.06	0.0	0.0	0.0	0.0	0.0	0.0	0.0	0.0	0.0
26E-2 100-103	72.72	0.0	0.7	0.0	0.0	0.0	0.0	0.2	0.0	0.0

(continued)

APPENDIX D. DIATOM FLORA FROM BL00-1 (*continued*)

			Nitzschia fonticola	*N. unid*	*Pinnularia viridis*	*Rhopalodia gibba*	*Surirella angusta*
Sample ID	Depth	Comments					
Core catcher samples							
1E-26E	1.72		0.8	0.2	0.4	0.0	0.0
1E-27E	4.97		0.5	1.1	0.3	0.0	0.0
1E-28E	8.02		0.0	0.7	1.7	0.0	0.0
1E-29E	11.01	no visible diatoms or pieces	0.0	0.0	0.0	0.0	0.0
1E-30E	14.03	no visible diatoms or pieces	0.0	0.0	0.0	0.0	0.0
1E-31E	17.04	no visible diatoms or pieces	0.0	0.0	0.0	0.0	0.0
1E-32E	20.09		0.0	0.0	0.0	0.0	0.0
1E-33E	23.07		0.0	0.2	0.0	0.0	0.0
1E-34E	26.10		0.0	0.0	0.0	0.0	0.0
1E-35E	29.09		0.0	0.0	0.0	0.0	0.2
1E-36E	32.09		0.0	0.0	0.0	0.0	0.2
1E-37E	35.09		0.0	0.7	0.0	0.0	0.0
1E-38E	38.08	few diatoms	0.0	0.0	0.0	0.0	0.0
1E-39E	41.09		0.3	0.0	0.0	0.0	0.5
1E-40E	43.90	no visible diatoms or pieces	0.0	0.0	0.0	0.0	0.0
1E-41E	47.10		1.1	0.4	0.0	0.0	0.2
1E-42E	50.09		0.0	0.0	0.0	0.0	0.7
1E-43E	52.99		0.5	0.3	0.0	0.0	0.0
1E-44E	55.60		1.1	0.0	0.0	0.0	0.0
1E-45E	58.10	no visible diatoms or pieces	0.0	0.0	0.0	0.0	0.0
1E-46E	60.60		0.4	0.6	0.0	2.5	0.0
1E-22E	63.10	dissolution of some cells present, mostly on broken diatoms	0.2	0.0	0.0	0.0	0.0
1E-23E	65.60	no visible diatoms or pieces	0.0	0.0	0.0	0.0	0.0
1E-24E	68.10	very small pieces, some dissolution evident	0.0	0.0	0.0	0.0	0.0
1E-25E	70.61	no visible diatoms or pieces	0.0	0.0	0.0	0.0	0.0
1E-26E	76.32	very few diatoms, mostly *Stephanodiscus* and small fragilaroid, some dissolution					
1E-27E	76.90	no diatoms, no pieces					
1E-28E	79.43	no diatoms, no pieces					
1E-29E	81.91	some diatoms, mostly small fragilaroid and *Stephanodiscus*, lots of cysts					
1E-30E	84.42	very sparse, mostly small fragilaroid					
1E-31E	86.42	very few diatoms, mostly *Stephanodiscus* and small fragilaroid, some dissolution					
1E-32E	88.43	very sparse, centrics and pennates, mostly broken, some dissolution					
1E-33E	90.42	abundant diatoms, both centrics and pennates, very diverse					
1E-34E	92.43	abundant diatoms, both centrics and pennates, very diverse					
1E-35E	94.43	abundant diatoms, mostly planktonic, dissolution and breakage present					
1E-36E	96.43	sparse diatoms					
1E-37E	98.42	sparse diatoms, mostly small fragilaroid, dissolution present					
1E-38E	100.43	abundant diatoms and cysts, many *Stephanodiscus* and *Aulacoseira* (*granulata*?)					
1E-39E	102.20	abundant *Stephanodiscus*					
1E-40E	104.50	abundant *Stephanodiscus*					
1E-41E	106.00	abundant *Stephanodiscus*					
1E-42E	108.10	abundant diatoms, mostly small fragilaroid, few centrics					
1E-43E	110.40	abundant diatoms, both pennates and centrics, very diverse					
1E-44E	113.40	abundant diatoms, both *Fragilaria* and a few centrics					
1E-45E	114.90	no diatoms, no pieces					
1E-46E	117.80	abundant diatoms, many big centrics and *Aulacoseira*					
Non-core catcher							
1H-1 50-54	0.50	many broken pieces, mostly *Navicula oblonga*, *Pinnularia viridis*, and *Cymbella*, very few others	0.0	0.0	0.0	0.0	0.0
2H-1 80-84	2.85	1/2 transect	2.1	0.4	0.0	0.0	0.0
2H-2 28-30	3.97	1/2 transect	1.0	0.3	0.0	0.0	0.0
3H-2 50-52	7.15	1/2 transect	0.0	0.0	0.0	0.0	0.0
4H-1 110-112	9.25	2 transects	0.3	0.0	0.0	0.3	0.0
5H-1 30-32	11.45	no diatoms, no pieces, no cysts, plenty of sediment	0.0	0.0	0.0	0.0	0.0
5H-2 80-83	13.50	no diatoms, no pieces, no cysts, plenty of sediment	0.0	0.0	0.0	0.0	0.0
6H-2 0-3	15.70	no diatoms, no pieces, no cysts, plenty of sediment	0.0	0.0	0.0	0.0	0.0
7H-1 70-73	19.00	mostly broken *Stephanodiscus*, very little dissolution visible	0.0	0.0	0.0	0.0	0.0
16E-2 60-63	44.90	15 transects	0.0	0.0	0.0	0.0	0.0
18E-1 30-33	50.50	1 transect	0.1	0.0	0.0	0.0	0.0
18E-2 80-83	52.45	1/2 transect	0.3	0.0	0.0	0.0	0.0
20E-1 50-53	56.10	1 transect	0.0	0.0	0.0	0.0	0.0
20E-2 100-103	57.72	4 transects	0.0	0.0	0.0	0.0	0.2
21E-1 150-153	59.36	3 transects, lots of pieces	0.2	0.0	0.0	0.0	1.0
21E-2 120-123	60.39	2 transects	0.0	0.0	0.0	0.0	0.0
22E-1 50-53	61.08	many well-preserved periphytic diatoms, but dissolution of some planktic diatoms	0.0	0.0	0.0	0.0	0.0
22E-2 100-103	62.84	many whole and well-preserved diatoms, including *Amphora*, *Aulacoseira* and small benthic/tychoplanktic fragilarioid species, but also many broken and dissolved large *Cymbella* and *Epithemia* diatoms	0.0	0.0	0.0	0.0	0.0
23E-2 20-23	64.55	no diatoms, no pieces, no cysts, plenty of sediment	0.0	0.0	0.0	0.0	0.0
24E-1 67-70	66.21	diatoms are rare, no evidence of dissolution	0.0	0.0	0.0	0.0	0.0
24E-2 120-123	67.92	no diatoms, no pieces, no cysts, plenty of sediment	0.0	0.0	0.0	0.0	0.0
25E-2 30-33	69.48	no diatoms, no pieces, no cysts, plenty of sediment	0.0	0.0	0.0	0.0	0.0
26E-2 50-53	71.06	few diatoms observed, some breakage, but no evidence of dissolution	0.0	0.0	0.0	0.0	0.0
26E-2 100-103	72.72	no dissolution evidenced	0.0	0.0	0.0	0.0	0.0

		Planktic Taxa							
Sample ID	Depth	Aulacoseira granulata	A. islandica	Cyclotella bodanica	C. glabriuscula	C. meneghiniana	C. michiganiana	C. ocellata	C. planktonica
Core catcher samples									
1E-26E	1.72	0.0	0.0	0.0	0.0	0.0	0.0	0.0	0.0
1E-27E	4.97	0.0	0.0	0.0	0.0	9.9	1.6	0.0	0.0
1E-28E	8.02	0.0	0.0	0.0	0.0	0.4	0.0	0.0	0.0
1E-29E	11.01	0.0	0.0	0.0	0.0	0.0	0.0	0.0	0.0
1E-30E	14.03	0.0	0.0	0.0	0.0	0.0	0.0	0.0	0.0
1E-31E	17.04	0.0	0.0	0.0	0.0	0.0	0.0	0.0	0.0
1E-32E	20.09	0.0	0.0	0.0	0.0	0.0	0.0	0.0	0.0
1E-33E	23.07	0.0	0.0	0.0	0.0	0.0	0.0	0.0	0.0
1E-34E	26.10	0.0	0.0	0.0	0.0	0.0	0.0	0.0	0.0
1E-35E	29.09	0.0	0.0	0.2	0.0	0.0	0.0	0.0	0.0
1E-36E	32.09	0.0	0.0	0.2	0.0	0.2	0.0	2.4	0.0
1E-37E	35.09	0.0	0.0	0.4	0.0	1.1	0.0	0.0	0.0
1E-38E	38.08	0.0	0.0	0.0	0.0	0.0	0.0	0.0	0.0
1E-39E	41.09	0.0	0.0	0.8	0.0	2.3	0.0	3.9	0.0
1E-40E	43.90	0.0	0.0	0.0	0.0	0.0	0.0	0.0	0.0
1E-41E	47.10	0.0	0.0	0.9	0.0	3.0	0.0	2.5	0.0
1E-42E	50.09	24.3	1.9	0.0	0.0	0.7	0.0	0.2	0.0
1E-43E	52.99	4.7	0.0	0.0	2.4	1.5	0.0	0.0	0.0
1E-44E	55.60	0.0	0.0	0.0	0.0	5.2	0.0	2.2	0.0
1E-45E	58.10	0.0	0.0	0.0	0.0	0.0	0.0	0.0	0.0
1E-46E	60.60	0.0	0.0	4.9	0.0	0.2	0.0	0.0	0.0
1E-22E	63.10	0.0	0.0	0.2	0.0	0.0	0.0	0.0	0.0
1E-23E	65.60	0.0	0.0	0.0	0.0	0.0	0.0	0.0	0.0
1E-24E	68.10	0.0	0.0	0.0	0.0	0.0	0.0	0.0	0.0
1E-25E	70.61	0.0	0.0	0.0	0.0	0.0	0.0	0.0	0.0
1E-26E	76.32								
1E-27E	76.90								
1E-28E	79.43								
1E-29E	81.91								
1E-30E	84.42								
1E-31E	86.42								
1E-32E	88.43								
1E-33E	90.42								
1E-34E	92.43								
1E-35E	94.43								
1E-36E	96.43								
1E-37E	98.42								
1E-38E	100.43								
1E-39E	102.20								
1E-40E	104.50								
1E-41E	106.00								
1E-42E	108.10								
1E-43E	110.40								
1E-44E	113.40								
1E-45E	114.90								
1E-46E	117.80								
Non-core catcher									
1H-1 50-54	0.50	0.0	0.0	0.0	0.0	0.0	0.0	0.0	0.0
2H-1 80-84	2.85	0.0	0.0	0.0	0.0	9.4	0.7	0.0	0.0
2H-2 28-30	3.97	0.0	0.0	0.0	0.0	6.6	0.6	0.0	0.0
3H-2 50-52	7.15	0.0	0.0	0.0	0.0	8.1	0.0	1.6	0.3
4H-1 110-112	9.25	0.0	0.0	0.0	0.0	0.5	0.0	0.8	0.0
5H-1 30-32	11.45	0.0	0.0	0.0	0.0	0.0	0.0	0.0	0.0
5H-2 80-83	13.50	0.0	0.0	0.0	0.0	0.0	0.0	0.0	0.0
6H-2 0-3	15.70	0.0	0.0	0.0	0.0	0.0	0.0	0.0	0.0
7H-1 70-73	19.00	0.0	0.0	0.0	0.0	0.0	0.0	0.0	0.0
16E-2 60-63	44.90	0.0	0.0	1.9	0.0	2.4	0.0	0.0	0.0
18E-1 30-33	50.50	8.2	0.7	0.0	0.0	0.3	0.1	0.0	0.0
18E-2 80-83	52.45	19.1	1.0	0.2	0.0	0.3	0.0	0.0	0.0
20E-1 50-53	56.10	0.0	0.0	0.0	0.0	0.0	0.0	0.0	0.0
20E-2 100-103	57.72	0.0	87.8	0.0	0.0	0.4	0.0	0.0	2.0
21E-1 150-153	59.36	0.0	15.6	1.8	0.0	1.1	0.0	0.0	5.4
21E-2 120-123	60.39	0.0	0.0	0.7	0.0	0.5	0.2	0.0	0.0
22E-1 50-53	61.08	0.0	0.0	0.0	0.0	0.0	0.0	0.0	0.0
22E-2 100-103	62.84	0.0	0.0	0.0	0.0	0.0	0.0	0.0	0.0
23E-2 20-23	64.55	0.0	0.0	0.0	0.0	0.0	0.0	0.0	0.0
24E-1 67-70	66.21	0.0	0.0	0.0	0.0	0.0	0.0	0.0	0.0
24E-2 120-123	67.92	0.0	0.0	0.0	0.0	0.0	0.0	0.0	0.0
25E-2 30-33	69.48	0.0	0.0	0.0	0.0	0.0	0.0	0.0	0.0
26E-2 50-53	71.06	0.0	0.0	0.0	0.0	0.0	0.0	0.0	0.0
26E-2 100-103	72.72	0.0	0.0	1.9	0.0	0.7	0.2	0.1	0.0

(continued)

APPENDIX D. DIATOM FLORA FROM BL00-1 (*continued*)

			Planktic Taxa		
Sample ID	Depth	Comments	*Cyclotella rossii*	*C. unid*	*Stephanodiscus hantzschii*
Core catcher samples					
1E-26E	1.72		0.0	0.0	0.0
1E-27E	4.97		0.0	0.5	0.0
1E-28E	8.02		0.0	0.0	0.0
1E-29E	11.01	no visible diatoms or pieces	0.0	0.0	0.0
1E-30E	14.03	no visible diatoms or pieces	0.0	0.0	0.0
1E-31E	17.04	no visible diatoms or pieces	0.0	0.0	0.0
1E-32E	20.09		0.0	0.0	0.0
1E-33E	23.07		0.0	0.0	0.0
1E-34E	26.10		0.0	0.0	0.0
1E-35E	29.09		2.3	0.0	0.0
1E-36E	32.09		4.8	0.0	0.0
1E-37E	35.09		0.0	0.2	5.1
1E-38E	38.08	few diatoms	0.0	0.0	0.0
1E-39E	41.09		0.3	0.0	0.0
1E-40E	43.90	no visible diatoms or pieces	0.0	0.0	0.0
1E-41E	47.10		0.0	0.0	0.0
1E-42E	50.09		0.0	0.0	3.5
1E-43E	52.99		0.0	0.0	2.5
1E-44E	55.60		0.6	4.1	0.0
1E-45E	58.10	no visible diatoms or pieces	0.0	0.0	0.0
1E-46E	60.60		0.0	0.0	0.0
1E-22E	63.10	dissolution of some cells present, mostly on broken diatoms	0.0	0.5	0.0
1E-23E	65.60	no visible diatoms or pieces	0.0	0.0	0.0
1E-24E	68.10	very small pieces, some dissolution evident	0.0	0.0	0.0
1E-25E	70.61	no visible diatoms or pieces	0.0	0.0	0.0
1E-26E	76.32	very few diatoms, mostly *Stephanodiscus* and small fragilaroid, some dissolution			
1E-27E	76.90	no diatoms, no pieces			
1E-28E	79.43	no diatoms, no pieces			
1E-29E	81.91	some diatoms, mostly small fragilaroid and *Stephanodiscus*, lots of cysts			
1E-30E	84.42	very sparse, mostly small fragilaroid			
1E-31E	86.42	very few diatoms, mostly *Stephanodiscus* and small fragilaroid, some dissolution			
1E-32E	88.43	very sparse, centrics and pennates, mostly broken, some dissolution			
1E-33E	90.42	abundant diatoms, both centrics and pennates, very diverse			
1E-34E	92.43	abundant diatoms, both centrics and pennates, very diverse			
1E-35E	94.43	abundant diatoms, mostly planktonic, dissolution and breakage present			
1E-36E	96.43	sparse diatoms			
1E-37E	98.42	sparse diatoms, mostly small fragilaroid, dissolution present			
1E-38E	100.43	abundant diatoms and cysts, many *Stephanodiscus* and *Aulacoseira* (*granulata*?)			
1E-39E	102.20	abundant *Stephanodiscus*			
1E-40E	104.50	abundant *Stephanodiscus*			
1E-41E	106.00	abundant *Stephanodiscus*			
1E-42E	108.10	abundant diatoms, mostly small fragilaroid, few centrics			
1E-43E	110.40	abundant diatoms, both pennates and centrics, very diverse			
1E-44E	113.40	abundant diatoms, both *Fragilaria* and a few centrics			
1E-45E	114.90	no diatoms, no pieces			
1E-46E	117.80	abundant diatoms, many big centrics and *Aulacoseira*			
Non-core catcher					
1H-1 50-54	0.50	many broken pieces, mostly *Navicula oblonga*, *Pinnularia viridis*, and *Cymbella*, very few others	0.0		0.0
2H-1 80-84	2.85	1/2 transect	0.0		0.1
2H-2 28-30	3.97	1/2 transect	0.0		0.0
3H-2 50-52	7.15	1/2 transect	0.0		0.1
4H-1 110-112	9.25	2 transects	0.0		2.0
5H-1 30-32	11.45	no diatoms, no pieces, no cysts, plenty of sediment	0.0		0.0
5H-2 80-83	13.50	no diatoms, no pieces, no cysts, plenty of sediment	0.0		0.0
6H-2 0-3	15.70	no diatoms, no pieces, no cysts, plenty of sediment	0.0		0.0
7H-1 70-73	19.00	mostly broken *Stephanodiscus*, very little dissolution visible	0.0		0.0
16E-2 60-63	44.90	15 transects	0.0		14.7
18E-1 30-33	50.50	1 transect	0.0		9.3
18E-2 80-83	52.45	1/2 transect	0.0		2.2
20E-1 50-53	56.10	1 transect	0.0		0.0
20E-2 100-103	57.72	4 transects	0.0		0.4
21E-1 150-153	59.36	3 transects, lots of pieces	0.0		3.4
21E-2 120-123	60.39	2 transects	0.0		0.0
22E-1 50-53	61.08	many well-preserved periphytic diatoms, but dissolution of some planktic diatoms	0.0		0.0
22E-2 100-103	62.84	many whole and well-preserved diatoms, including *Amphora*, *Aulacoseira* and small benthic/tychoplanktic fragilarioid species, but also many broken and dissolved large *Cymbella* and *Epithemia* diatoms	0.0		0.0
23E-2 20-23	64.55	no diatoms, no pieces, no cysts, plenty of sediment	0.0		0.0
24E-1 67-70	66.21	diatoms are rare, no evidence of dissolution	0.0		0.0
24E-2 120-123	67.92	no diatoms, no pieces, no cysts, plenty of sediment	0.0		0.0
25E-2 30-33	69.48	no diatoms, no pieces, no cysts, plenty of sediment	0.0		0.0
26E-2 50-53	71.06	few diatoms observed, some breakage, but no evidence of dissolution	0.0		0.0
26E-2 100-103	72.72	no dissolution evidenced	0.0		0.4

Note: Species are divided into benthic and planktic species. Only taxa that occurred in at least 1 sample in amounts ≥1% are included.

		Planktic Taxa								
Sample ID	Depth	*S. medius*	*S. minutulus*	*S. niagarae*	*S. neoastraea*	*S.* sp. BL1	*S. unid*	Unid pennates	Unid centrics	% Planktonics
Core catcher samples										
1E-26E	1.72	0.0	0.0	0.0	0.0	0.0	0.0	2.2	0.0	0.0
1E-27E	4.97	0.0	0.6	0.0	0.0	0.0	0.0	0.9	0.0	12.6
1E-28E	8.02	1.8	0.0	0.0	0.0	0.0	0.0	2.8	0.4	2.6
1E-29E	11.01	0.0	0.0	0.0	0.0	0.0	0.0	0.0	0.0	0.0
1E-30E	14.03	0.0	0.0	0.0	0.0	0.0	0.0	0.0	0.0	0.0
1E-31E	17.04	0.0	0.0	0.0	0.0	0.0	0.0	0.0	0.0	0.0
1E-32E	20.09	89.0	0.5	0.0	0.0	0.0	0.0	1.1	0.0	89.5
1E-33E	23.07	40.1	0.0	3.6	0.0	0.0	0.0	2.1	0.4	44.0
1E-34E	26.10	95.7	2.8	0.1	0.0	0.0	0.0	0.4	0.4	99.1
1E-35E	29.09	81.6	1.0	0.2	0.0	0.0	0.0	0.5	0.0	85.1
1E-36E	32.09	62.7	0.0	0.3	0.0	0.0	0.0	1.6	0.0	70.6
1E-37E	35.09	75.6	8.2	0.0	0.0	0.0	1.2	1.1	0.0	91.8
1E-38E	38.08	0.0	0.0	0.0	0.0	0.0	0.0	0.0	0.0	0.0
1E-39E	41.09	26.2	0.7	0.0	0.0	0.0	0.0	1.3	0.0	34.3
1E-40E	43.90	0.0	0.0	0.0	0.0	0.0	0.0	0.0	0.0	0.0
1E-41E	47.10	16.7	2.9	0.0	0.0	0.0	0.0	1.3	0.0	26.0
1E-42E	50.09	1.4	0.5	0.0	0.0	0.0	0.0	0.0	0.0	32.6
1E-43E	52.99	13.8	2.5	0.0	0.0	0.0	0.0	2.0	0.0	27.4
1E-44E	55.60	19.2	1.3	0.0	0.0	0.0	0.0	1.7	0.0	32.6
1E-45E	58.10	0.0	0.0	0.0	0.0	0.0	0.0	0.0	0.0	0.0
1E-46E	60.60	0.0	0.0	0.0	0.0	0.0	0.0	4.0	0.4	6.1
1E-22E	63.10	0.2	0.0	0.0	0.0	0.0	0.0	0.5	0.0	1.0
1E-23E	65.60	0.0	0.0	0.0	0.0	0.0	0.0	0.0	0.0	0.0
1E-24E	68.10	0.0	0.0	0.0	0.0	0.0	0.0	0.0	0.0	0.0
1E-25E	70.61	0.0	0.0	0.0	0.0	0.0	0.0	0.0	0.0	0.0
1E-26E	76.32									
1E-27E	76.90									
1E-28E	79.43									
1E-29E	81.91									
1E-30E	84.42									
1E-31E	86.42									
1E-32E	88.43									
1E-33E	90.42									
1E-34E	92.43									
1E-35E	94.43									
1E-36E	96.43									
1E-37E	98.42									
1E-38E	100.43									
1E-39E	102.20									
1E-40E	104.50									
1E-41E	106.00									
1E-42E	108.10									
1E-43E	110.40									
1E-44E	113.40									
1E-45E	114.90									
1E-46E	117.80									
Non-core catcher										
1H-1 50-54	0.50	0.0	0.0	0.0	0.0	0.0	0.0	0.0	0.0	0.0
2H-1 80-84	2.85	4.1	0.1	0.0	0.0	0.0	0.0	1.2	0.8	15.4
2H-2 28-30	3.97	0.0	0.0	0.0	0.0	0.0	0.0	1.3	0.0	7.2
3H-2 50-52	7.15	12.7	2.9	0.0	0.1	0.0	0.0	0.9	1.0	27.1
4H-1 110-112	9.25	35.8	0.0	0.0	1.4	0.0	0.0	1.4	0.2	40.6
5H-1 30-32	11.45	0.0	0.0	0.0	0.0	0.0	0.0	0.0	0.0	0.0
5H-2 80-83	13.50	0.0	0.0	0.0	0.0	0.0	0.0	0.0	0.0	0.0
6H-2 0-3	15.70	0.0	0.0	0.0	0.0	0.0	0.0	0.0	0.0	0.0
7H-1 70-73	19.00	0.0	0.0	0.0	0.0	0.0	0.0	0.0	0.0	0.0
16E-2 60-63	44.90	13.3	2.4	0.2	1.7	0.0	0.0	1.6	0.5	37.3
18E-1 30-33	50.50	6.0	1.6	0.0	0.1	0.1	0.0	0.4	0.4	26.9
18E-2 80-83	52.45	2.3	0.0	0.0	0.0	0.0	0.0	1.8	0.2	25.2
20E-1 50-53	56.10	0.0	0.0	0.0	0.0	0.0	0.0	0.0	0.0	1.3
20E-2 100-103	57.72	1.3	0.2	0.0	0.0	0.0	0.0	0.4	0.2	92.2
21E-1 150-153	59.36	8.8	0.0	0.3	0.0	1.6	0.0	2.9	1.8	39.8
21E-2 120-123	60.39	19.1	0.7	0.0	0.0	0.0	0.0	1.1	0.3	21.3
22E-1 50-53	61.08	0.0	0.0	0.0	0.0	0.0	0.0	0.0	0.0	0.0
22E-2 100-103	62.84	0.0	0.0	0.0	0.0	0.0	0.0	0.0	0.0	0.0
23E-2 20-23	64.55	0.0	0.0	0.0	0.0	0.0	0.0	0.0	0.0	0.0
24E-1 67-70	66.21	0.0	0.0	0.0	0.0	0.0	0.0	0.0	0.0	0.0
24E-2 120-123	67.92	0.0	0.0	0.0	0.0	0.0	0.0	0.0	0.0	0.0
25E-2 30-33	69.48	0.0	0.0	0.0	0.0	0.0	0.0	0.0	0.0	0.0
26E-2 50-53	71.06	0.0	0.0	0.0	0.0	0.0	0.0	0.0	0.0	0.0
26E-2 100-103	72.72	58.0	3.1	0.4	0.0	0.0	0.0	2.5	0.4	65.2